Macromolecular Symposia

190

WILEY-VCH

pp. 1–200

November 2002

Macromolecular Symposia publishes lectures given at international symposia and is issued irregularly, with normally 14 volumes published per year. For each symposium volume, an Editor is appointed. The articles are peer-reviewed. The journal is produced by photo-offset lithography directly from the authors' typescripts.
Further information for authors can be obtained from:
Editorial office "Macromolecular Symposia"
Wiley-VCH
Boschstrasse 12, 69469 Weinheim,
Germany
Tel. +49 (0) 62 01/6 06-2 38 or -5 81; Fax +49 (0) 62 01/6 06-3 09 or 5 10;
E-mail: macro-symp@wiley-vch.de
http://www.ms-journal.de
Suggestions or proposals for conferences or symposia to be covered in this series should also be sent to the Editorial office at the address above.

Macromolecular Symposia:
Annual subscription rates 2003 (print only or online only)*
Germany, Austria € 1318; Switzerland SFr 2168; other Europe € 1318; outside Europe US $ 1568.
Macromolecular Package, including Macromolecular Chemistry & Physics (18 issues), Macromolecular Bioscience (12 issues), Macromolecular Rapid Communications (18 issues), Macromolecular Theory & Simulations (9 issues) is also available. Details on request.
* For a 5 % premium in addition to **Print Only** or **Online Only**, Institutions can also choose both print and online access.
Packages including Macromolecular Symposia and Macromolecular Materials & Engineering are also available. Details on request.
Single issues and back copies are available. Please, inquire for prices.

Orders may be placed through your bookseller or directly at the publishers:
WILEY-VCH Verlag GmbH & Co. KGaA, P. O. Box 10 11 61, 69451 Weinheim, Germany, Tel. +49 (0) 62 01/6 06-4 00, Telefax +49 (0) 62 01/6 06-1 84, E-mail: service@wiley-vch.de

Macromolecular Symposia (ISSN 1022-1360) is published with 14 volumes per year by WILEY-VCH Verlag GmbH & Co. KGaA, P. O. Box 10 11 61, 69451 Weinheim, Germany. Air freight and mailing in the USA by Publications Expediting Inc., 200 Meacham Ave., Elmont, NY 11003. Periodicals postage pending at Jamaica, NY 11431. POSTMASTER: send address changes to Macromolecular Symposia, Publications Expediting Inc., 200 Meacham Ave., Elmont, NY 11003.

Printing: Strauss Offsetdruck, Mörlenbach. Binding: J. Schäffer, Grünstadt

International Union of Pure and Applied Chemistry

Macromolecular Division

Invited lectures presented at the

20th Discussion Conference of the Prague Meetings on Macromolecules
Scattering Methods for the Investigation of Polymers

held in
Prague, Czech Republic
July 9–12, 2001

Symposium Editor

Jaroslav Kahovec
Institute of Macromolecular Chemistry
Academy of Sciences of the Czech Republic
Heyrovského nám. 2
CZ-162 06 Praha 6
Czech Republic

ISBN 3-527-30480-0

The **tables of contents** of the published issues are displayed on the WWW.

This service as well as further information on our journals can be found at the following WWW address:

http://www.ms-journal.de

The service is free for everybody.

Contents of Macromol. Symp. 190

20th Discussion Conference of the Prague Meetings on Macromolecules
Scattering Methods for the Investigation of Polymers
Prague, 9–12 July 2001

Preface
J. Pleštil, J. Kahovec

1. Neutron Reflectivity of Polymer-Plasticiser Diffusion 1
*J. S. Higgins, S. A. Butler, D. G. Bucknall**

2. "Sponge-Like" Structures in Polymer Blends: Visualization, Physico-Mathematical Analysis, and Universality 9
Takeji Hashimoto, Hiroshi Jinnai, Yukihiro Nishikawa, Tsuyoshi Koga*

3. Structure and Dynamics of Silica-Filled Polymers by SANS and Coherent SAXS 23
Erik Geissler, Anne-Marie Hecht, Cyrille Rochas, Ferenc Horkay, Françoise Bley, Frédéric Livet*

4. SANS from Adsorbed Polymer Layers 33
Stephen King, Peter Griffiths, John Hone, Terence Cosgrove*

5. A Few Remarks on Classic Problems of Scattering by Polymer Solutions and Mixtures 43
Henri Benoît, Gérard Jannink*

6. Linear, Branched and Hyperbranched Macromolecules in Dilute Solution 55
Fabio Ganazzoli

7. Light Scattering Investigation of Association Phenomena in Saline Carrageenan Solutions 65
*Fabio Cuppo, Harry Reynaers**

8. The Crystallization and Melting of Linear Polyethylene Studied by Temperature-Modulated SAXS and WAXD 75
Bart Goderis, Harry Reynaers, Michel H. J. Koch*

9. Some Properties of Aqueous Xanthan Gels Formed under the Action of Aluminium(III) Ions 79
*Andrew B. Rodd, Dave E. Dustan, David V. Boger, Juergen Schmidt, Walther Burchard**

10. Phase Separation Mechanism in Gelling Aqueous Biopolymer Mixture Probed by Light Scattering 93
*D. Leisner, M. C. Blanco, M. A. López Quintela**

11. The Use of Monte-Carlo Simulations to Calculate Small-Angle Scattering Patterns 117
*Bryan C. McAlister, Brian P. Grady**

12. Monitoring Kinetic Processes in Polymer Solutions with Time Dependent Static Light Scattering (TDSLS) 131
Wayne F. Reed

13. The Effect of Water on Cellulose Solutions in DMAc/LiCl 151
Thomas Röder, Antje Potthast, Thomas Rosenau, Paul Kosma, Thomas Baldinger, Bernd Morgenstern, Otto Glatter*

14. Structure and Ordering Kinetics of Micelles in Triblock Copolymer Solutions in Selective Solvents 161
R. Bansil, H. Nie, Y. Li, G. Liao, K. Ludwig, M. Steinhart, Č. Koňák, J. Lal*

15. Time-Resolved SAXS Studies on Crystallization Behavior of Poly(ethylene isophthalate-*co*-ethylene terephthalate)s 173
*B. Lee, T. J. Shin, S. W. Lee, J. W. Lee, M. Ree**

16. Progress in SANS Studies of Polymer Systems (Panel Discussion) . . . 185
G. D. Wignall, H. Benoît, T. Hashimoto, J. S. Higgins, S. King, T. P. Lodge, K. Mortensen, A. J. Ryan

* The asterisk indicates the name of the author to whom inquiries should be addressed

Author Index

Baldinger, Thomas 151
Bansil, R. 161
Benoît, Henri 43, 185
Blanco, M. C. 93
Bley, Françoise 23
Boger, David V. 79
Bucknall, D. G. 1
Burchard, Walther 79
Butler, S. A. 1
Cosgrove, Terence 33
Cuppo, Fabio 65
Dustan, Dave E. 79
Ganazzoli, Fabio 55
Geissler, Erik 23
Glatter, Otto 151
Goderis, Bart 75
Grady, Brian P. 117
Griffiths, Peter 33
Hashimoto, Takeji 9, 185
Hecht, Anne-Marie 23
Higgins, J. S. 1, 185
Hone, John 33
Horkay, Ferenc 23
Jannink, Gérard 43
Jinnai, Hiroshi 9
King, Stephen 33, 185
Koch, Michel H. J. 75
Koga, Tsuyoshi 9
Koňák, Č. 161
Kosma, Paul 151
Lal, J. 161
Lee, B. 173
Lee, J. W. 173
Lee, S. W. 173
Leisner, D. 93
Li, Y. 161
Liao, G. 161
Livet, Frédéric 23
Lodge, T. P. 185
López Quintela, M. A. 93
Ludwig, K. 161
McAlister, Bryan C. 117
Morgenstern, Bernd 151
Mortensen, K. 185
Nie, H. 161
Nishikawa, Yukihiro 9
Potthast, Antje 151
Ree, M. 173
Reed, Wayne F. 131
Reynaers, Harry 65, 75
Rochas, Cyrille 23
Rodd, Andrew B. 79
Röder, Thomas 151
Rosenau, Thomas 151
Ryan, A. J. 185
Schmidt, Juergen 79
Shin, T. J. 173
Steinhart, M. 161
Wignall, G. D. 185

Preface

The 20th Discussion Conference "Scattering Methods for the Investigation of Polymers" was organized as the 60th meeting in the series of the Prague Meetings on Macromolecules. The meeting was held under the auspices of the International Union of Pure and Applied Chemistry in the Institute of Macromolecular Chemistry, Academy of Sciences of the Czech Republic, on July 9–12, 2001.

The meeting aimed to provide a forum for scientists specializing in scattering methods. These methods are important tools for characterization of the structure and dynamics of polymers. Not only the knowledge of the final structure, but also the structure evolution during the preparation and processing of polymeric materials are a prerequisite for understanding, predicting and controlling their properties and functions. Rapid development of instrumentation for time-resolved scattering experiments has extended significantly the accessibility of such information.

The topics of the conference included various aspects of light, X-ray and neutron scattering techniques, as well as applications of these methods to a variety of polymeric materials and systems. At the conference, 7 main lectures, 23 special lectures and 41 posters were presented. In addition, two interesting panel discussions organized and led by G. D. Wignall, USA (*Progress in SANS Studies of Polymer Systems*) and W. Burchard, Germany (*Weighing up New Possibilities of Scattering and Simulation Techniques for Determination of Structures and Chain Dynamics*) were included.

In the present volume, the papers based on main lectures and special lectures are collected together with the article written by G. D. Wignall and S. King, UK, summarizing the main results of the panel discussion devoted to neutron scattering. In this paper, not only the present state of the art and prospects for future developments in the field are discussed but also the predictions made at the "Neutron Scattering in the Nineties" symposium held in the mid-eighties are reviewed.

One of the events contributing to a friendly atmosphere at the conference was an unpretentious celebration of the 80th birthday of Professor Henri Benoît, one of the leading people in scattering methods.

The organizers of the conference wish to express their gratitude to the sponsoring companies and institutions: Sigma-Aldrich Ltd, Czech Republic; Bayer A. G., Germany; ExxonMobil Chemical, Belgium; Léčiva a. s., Czech Republic; Molecular Metrology Inc., USA; Polymer Institute, Brno, Czech Republic.

Josef Pleštil
Conference Chairman

Jaroslav Kahovec
Conference Editor

Neutron Reflectivity of Polymer-Plasticiser Diffusion

J.S. Higgins,[1] *S.A. Butler*[1] *and D.G. Bucknall**[2]

[1] Department of Chemical Engineering, Imperial College, Prince Consort Road, London, SW7 2BY, U.K.

[2] Department of Materials, Oxford University, Parks Road, Oxford, OX1 3PH, U.K.

SUMMARY: Neutron reflectivity (NR) has been widely exploited to look at polymer thin films and in many ways is an ideal technique for studying polymer interfaces and surfaces, providing high-resolution concentration - depth profiles across the film thickness. Most NR studies to date have concentrated on thin films of amorphous polymers which possess T_g values well above room temperature. These polymers are ideally suited to NR measurements, firstly because they form homogeneously flat films and, secondly, heat-quench cycles can be used to study time-dependent processes. This has been used to great effect in NR studies of the initial stages of polymer - polymer interdiffusion or the kinetics of surface segregated layers for instance. One of the biggest drawbacks to this approach is that in polymer systems where one or more of the components has a T_g close to or less than room temperature, the polymers can still move during the measurement time of an NR profile, which typically takes 1-2 h for a full profile. Therefore, in order to study such systems, we have developed an approach to NR measurements that allows us to investigate diffusion processes *in situ*. Our new approach allows us to take NR profiles in only 20 s. This paper describes the method of real-time NR measurements in detail and illustrates the capabilities of the technique with highlights from some of our recent work on the early stages of polymer-plasticiser interdiffusion.

Introduction

Neutron reflectivity has been widely used to study polymer - polymer interfaces,[1,2] including detailed investigation of interdiffusion.[3-7] To date most of these studies have utilised 'anneal-quench' procedures where the sample is heated for a given time above the T_g's of the polymers, rapidly quenched to room temperature and then a full reflectivity profile collected. This process is repeated for each successive annealing time, allowing a detailed picture of the change in interfacial composition as a function of time to be built up. Although this

§ Current address: Department of Chemical Engineering, University of Cambridge, Pembroke St., Cambridge CB2 3RA, UK

 CCC 1022-1360/00/$ 17,50+.50/0

measurement procedure has proved to be highly successful for amorphous polymer systems, where the T_g's of both polymers are well above room temperature,[3-7] it is not applicable for the study of systems where the T_g of one or both of the polymers is close to or less than room temperature, or where small molecule ingress into a polymer layer is to be studied. The problem lies in the mobility of the interface at room temperature, which would then require a fast quench to a temperature below which the diffusing species is immobilised. Experimentally this creates a problem if either the quench is not rapid enough or the interface is disturbed by this thermal cycling processes, which undoubtedly occurs when the sample is heated for the next annealing stage.

An alternative approach and one we have been developing recently is *in-situ* real-time reflectivity measurements. We have shown that by utilising the white beam of a pulsed neutron source, it is possible to collect a limited reflectivity profile at a fixed angle within a few minutes or less. This very rapid measurement time compares to the 1-2 h which is typical for a full reflectivity profile over an extended momentum transfer range. To make these very rapid measurements, only one angle is used which gives a fixed range of momentum transfer values, Q (= $(4\pi/\lambda)\sin\theta$, where λ is the neutron wavelength and θ is the angle of incidence). The length of the measurement time per reflectivity profile is simply limited by the time required to gain adequate statistics. To measure diffusion processes, it is important to remember that for simple systems diffusion will initially appear as an apparent increase in interfacial width with increasing time. In a reflectivity profile this manifests itself as an increase in dampening of the high Q fringes which themselves are associated with one of the polymer layer thicknesses. An assumption which must therefore be made is that the rest of the low Q data is unaffected by this increase in interfacial width. This is true for simple systems but may become increasingly less reliable as time increases or for more complex systems. This therefore remains a limitation for the current technology and can only be alleviated by increased neutron flux and, at the same time, utilising a larger wavelength range which would allow a much larger Q window to be measured.

Nevertheless, we have shown the effectiveness of this technique for a number systems. Our approach was first demonstrated by studying the interdiffusion of normal polystyrene (PS) and deuterated PS (dPS) at 115 °C, *i.e.* above the T_g of both polymers, using the SURF reflectometer (ISIS Facility, Rutherford Appleton Laboratory).[8] This allowed us to determine a diffusion coefficient of 2×10^{-17} cm^2s^{-1}, entirely consistent with expectations compared to other techniques, validating the capability of these real-time NR measurements. We have also extended this method to study ingress of low-molecular-weight oligostyrene into high-molecular-weight dPS,[8] where asymmetric interfacial composition profiles were observed. More recently, we have also studied ingress of oligo(methyl methacrylate) and oligo(ethylene glycol) into high-molecular-weight deuterated poly(methyl methacrylate) (dPMMA).[9]

This paper describes the early results of studies of a series of phthalate plasticisers into high-molecular-weight dPMMA. Whereas our initial measurements for the PS-dPS system were about 6 min per reflectivity profile, we have reduced this to 20 s, thus opening up tremendous possibilities in many areas of diffusion studies.

Experimental

The characteristics of the dialkyl phthalate plasticisers used in this study are listed in Table 1.

Table 1: Dialkyl phthalate plasticisers characteristics, including bulk scattering length density value (ρ_s) used to fit the NR data.

Alkyl	Name	Molar Mass (g/mol)	Density (g/cm^3)	ρ_s x 10^6 (Å^{-2})
C_9H_{19}	diisononyl phthalate (DINP)	418.6	0.97	0.54
$CH_2CH(C_2H_5)(CH_2)_3CH_3$	dioctyl phthalate (DOP)	390.5	0.98	0.62
$(CH_2)_3CH_3$	dibutyl phthalate (DBP)	278.4	1.04	1.06

Deuterated PMMAs of two molecular weights were used: M_w = 188 000 g/mol and 100 000 g/mol. Thin films of these PMMA films were spun-cast onto 70×50×5 mm polished silicon substrates from dilute toluene solutions with a concentration of 20 mg/ml. After spin casting the samples were dried at 55 °C for several days before use in the reflectivity experiments. Full NR profiles of the as-made dPMMA films were measured using three-angle measurements at room temperature. Real-time NR measurements at temperatures above room temperature were made using a specially constructed cell,[10] where the dPMMA-coated silicon is held inverted and facing a small trough containing the plasticiser. The silicon and plasticiser are held apart until the moment the reflectivity measurements are ready to start. Then the two are brought together remotely and the data collection started together with the clock at the moment the two touch each other.

Reflectivity measurements were initially taken at a fixed incident angle of 0.8° giving a fixed window in Q of approximately 0.027-0.08 Å^{-1}. This angle was chosen so as to place the

Q window in the region of most interest along an imagery full reflectivity profile, where the interference fringes are expected to be seen to dampen. The incident angle was incrementally decreased in steps of 0.2° as the most significant changes in the reflectivity profile progressed to lower Q values as the diffusion process occurs. The lowest incident angle used was 0.4° giving a Q range of approximately 0.013-0.05 $Å^{-1}$. At this lowest angle and a higher flux, measurement times of just 20 s were possible.

The preliminary analysis of these data has been performed using standard single-layer models assuming a simple Debye-Waller Gaussian interfacial profile[1] between the dPMMA and the plasticiser. As discussed below, use of a single-layer model is only adequate for an initial period of time, which depends on the plasticiser used. Although the analysis of the later-time NR profiles is still ongoing, it is clear that a much more complex composition profile through the dPMMA layer is evident than could be approximated by using a single-layer model.

Results and Discussion

The reflectivity profiles of the plasticiser diffusion into the dPMMA layers were measured at T = 70 °C for DINP and DOP studies and 32 °C for DBP. Clearly, for direct comparison, the same temperature for all the samples would have been ideal, but at the higher temperature the diffusional processes in the dPMMA/DBP system were so fast that all reflectivity interference fringes were lost within the first few minutes. Therefore, even at the fastest measurement times of 20 s per reflectivity profile, there are significant changes in the composition profile of the system, invalidating any data we can collect over this period. We therefore had to choose a lower temperature where the motion was slow enough so that there is not a noticeable change in reflectivity over the time scale of each individual measurement.

A series of selected reflectivity profiles for DINP and DOP systems are shown in Figs 1 and 2. The limited Q range windows for each incident angle are evident from these plots. As can be seen, the interference fringe spacing decreases with increasing time indicating that the layer thickness of the dPMMA, d, increases. The rate at which this happens for the two plasticisers is clearly different. A quantitative analysis of the initial stages of the ingress of the plasticiser into the dPMMA was achieved as described above by fitting each reflectivity curve using a single-layer model. As can be seen from Figs 1 and 2, even within the limits imposed by the statistics of the data associated with these very short collection times, this simple model adequately describes the data at very early stages, but becomes increasingly less accurate for later stages. The results of the fitting analysis for the DOP and DINP are shown in Fig. 3, where the time dependence of the change in dPMMA thickness, Δd, is plotted. Both plasticisers show a linear dependence of Δd as a function of time, suggesting that this early-stage diffusion process is Case II. Although Fickian behaviour showing a $t^{0.5}$ dependence is typical for solvent and gas diffusion into polymers,[11,12] this is normally only true for polymers above their T_g and is common for rubbers. For glassy amorphous polymers, Case II behaviour is often observed. In Case II sorption of fluids into polymers, there is expected to be an induction period where the fluid concentration at the film surface increases forming an asymmetric interfacial profile or front. Once this fluid front has developed, it then propagates into the film at constant speed with an equilibrium concentration of fluid.[12] We do not observe the development of the fluid front in our data which would manifest itself as an asymmetric interface with a long concentration tail into the bulk plasticiser. It is entirely possible that this does in fact happen but, without extended Q range data, we cannot say with any certainty whether this happens.

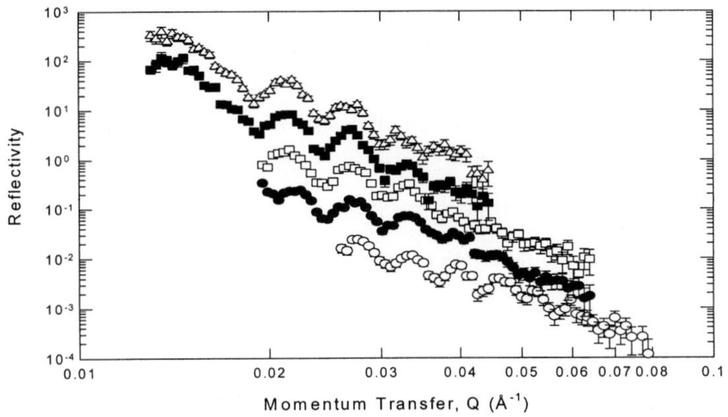

Figure 1: Reflectivity profiles of dPMMA/DOP measured in real time at 70 °C after 2.6 (○), 8.3 (●), 13.6 (□), 18.9 (■) and 22.8 (△) min. The plots have been scaled for clarity.

Circumstantial evidence to support the development of a fluid front during the initial stages of the sorption process can be seen from the DINP system where an induction period is observed. During this induction period over the initial 4 min, the value of the dPMMA thickness does not appear to change. Only after this period does the film thickness begin to increase linearly with time, as expected for Case II behaviour.

The differences in the rates of film thickness increase for DINP and DOP are evident from the linear fits to the data shown in Fig. 3. The DOP system shows a marked deviation from this linear behaviour after about 6 min, and this reflects the breakdown in the adequacy of the single-layer model to describe the data, which after about 10 min is not adequate to describe the DOP system data.

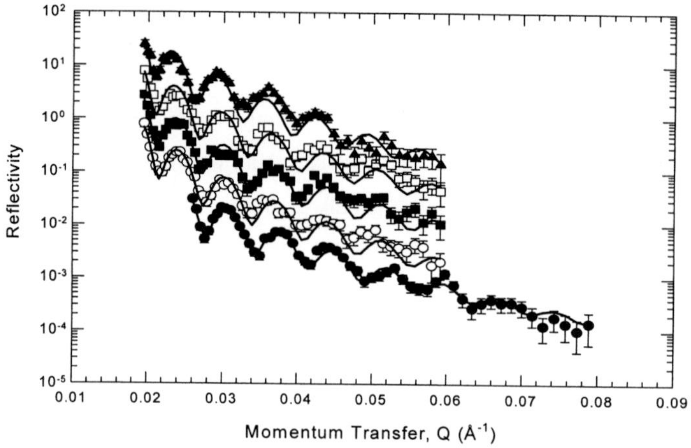

Figure 2: Reflectivity profiles of dPMMA/DINP measured in real time at 70 °C after 0.5 (●), 5.6 (○), 9.4 (■), 12.8 (□) and 26.0 (▲) min. The plots have been scaled for clarity.

On the basis of the gradient of the initial portion of the DOP system plot, *i.e.* the first 6 min and over the entire DINP range, rates of change of thickness, $\dot{s}$, ($=\Delta d/\Delta t$) equal to $+(3.5\pm0.1)\times10^{-12}$ and $+(2.33\pm0.05)\times10^{-11}$ m/s for the DINP and DOP systems were obtained, respectively. The positive sign of $\dot{s}$ indicates that the dPMMA/plasticiser interface is moving towards the plasticiser and is indicative of the polymer swelling. The values of $\dot{s}$ are consistent with the change in molar masses of the plasticisers, where the smaller plasticiser is expected to penetrate into the layer fastest.

Single-layer fits to the DBP data do not adequately model the data, except for the single reflectivity profile taken in the first 30 s. In this case the initial dPMMA thickness of 87.4 nm has increased to 98.6 nm over that time period. As a first approximation, we can estimate the rate of change of the dPMMA layer, $\dot{s}$, using these values, to give a value of $+(7.5\pm0.5)\times10^{-9}$ m/s at 32 °C. There is still a significant difference in the measured values of $\dot{s}$, between the DINP and DOP and the very fast process observed for DBP, even with the latter measured at a reduced temperature compared to the other two systems. Unlike the DINP and DOP systems after the very early times, the dPMMA layer when exposed to DBP begins to decrease in thickness, as observed by an increase in the reflectivity interference fringe spacing. This suggests that this is no longer a swelling process caused by sorption of the plasticiser. Without further detailed analysis, it is difficult to determine whether this apparent decrease in the dPMMA layer thickness is a dissolution process or not.

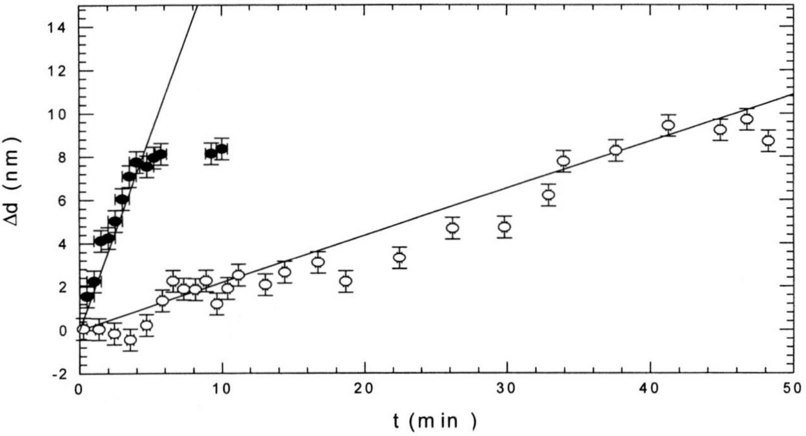

Figure 3: Plot of change in dPMMA layer thickness as a function of time for DINP (○) and DOP (●) systems.

As the plasticiser front moves into the dPMMA during the sorption process, an equilibrium concentration is expected to result. This can be calculated from the change in scattering length density of the dPMMA layer, which is directly related to the volume fraction of plasticiser sorbed into the polymer through the relationship, $\rho_s^T = (1-\phi_s)\rho_P + \phi_s \rho_s$, where ρ_S^T is the scattering length density obtained from the model fit, ϕ_s is the volume fraction of the plasticiser and ρ_i are the scattering length densities of the plasticiser (i = s) and dPMMA (i = P). Based on the fitting parameters for the DINP (after 50 min) and DOP systems (after 10 min), the values of ϕ_s are 0.02 ± 0.007 and 0.05 ± 0.01, respectively. The apparent large error on each of these values derives from the lack of extended Q range data, and again demonstrates the limitations that are incurred when using only restricted data sets. However, given that these data are only for very early times, it is unlikely that either of these values of ϕ_s represents the true value anticipated from the Case II model.

Similar analysis of the DBP system is not possible using the assumptions made for DINP and DOP. Indeed, the fits to the data do not support a simplistic single-layer model except for the initial reflectivity profile. Further work is required on this system and for the later stages for the other plasticisers, but initial work shows that a front does appear to move into the polymer, as expected for the Case II sorption process.

Conclusions

We have shown that real-time neutron reflectivity can be used to study plasticisier ingress into deuterated poly(methyl methacrylate). Three plasticisers have been studied, diisononyl phthalate, dioctyl phthalate and dibutyl phthalate, with measurements made in situ at 70 °C, except for the latter which was measured at 32 °C.

Single-layer fits to the restricted momentum transfer reflectivity data were only applicable for the early times of the sorption process for all the systems. Where a single-layer model could be used to fit the data, an increase in dPMMA layer thickness was observed for all three plasticiser systems. The DINP and DOP systems demonstrate a linear growth rate of the dPMMA layer thickness which is characteristic of Case II diffusion behaviour. As may be expected intuitively, the rate of plasticiser ingress into the polymer film, $\dot{s}$, is largest for DBP which is the smallest and more mobile plasticiser of the three studied. Similarly, $\dot{s}$ is smallest for DINP which is the largest and therefore least mobile plasticiser.

References

1. D.G. Bucknall and J.S. Higgins, in *Polymers and Surfaces - A Versatile Combination*, H. Hommell (Ed.), Research Signpost, Trivandrum, India, 1998.
2. T.P. Russell: *Physica B* **221**, 267 (1996).
3. A. Karim, A. Mansour, G.P. Felcher and T.P. Russell: *Physica B* **156**, 430 (1989).
4. M.L. Fernandez, J.S. Higgins, J. Penfold and C. Shackleton: *J. Chem. Soc., Faraday Trans.* **87**, 2055 (1991).
5. Y. Liu, G. Reiter, K. Kunz and M. Stamm: *Macromolecules* **26**, 2134 (1993).
6. A. Karim, G.P. Felcher and T.P. Russell: *Macromolecules* **27**, 6973 (1994).
7. K. Kunz and M. Stamm: *Macromolecules* **29**, 2548 (1996).
8. D.G. Bucknall, S.A. Butler and J.S. Higgins: *Macromolecules* **32**, 5453 (1999).
9. D.G. Bucknall, J.S. Higgins and B.A. Butler: *Chem. Eng. Sci.* **56**, 1 (2001).
10. D.G. Bucknall, S.A. Butler and J.S. Higgins: *J. Phys. Chem. Solids* **60**, 1273 (1999).
11. G.S. Park, in *Diffusion in Polymers*, J. Crank, G.S. Park (Eds.), Academic Press, London, 1968.
12. C.J. Durning, M.M. Hassan, H.M. Tong and K.W. Lee: *Macromolecules* **28**, 4234 (1995).

"Sponge-Like" Structures in Polymer Blends: Visualization, Physico-Mathematical Analysis, and Universality

Takeji Hashimoto[*1,2], *Hiroshi Jinnai*[1,a], *Yukihiro Nishikawa*[1,b] *Tsuyoshi Koga*[1,c]

[1]Hashimoto Polymer Phasing Project, ERATO, JST, Japan
[2]Department of Polymer Chemistry, Graduate School of Engineering, Kyoto University, Kyoto 606-8501, Japan

SUMMARY: Mesoscopic structures formed during an ordering process in thermodynamically unstable, isometric, binary molecular mixtures were explored by time-resolved scattering (TRS) and laser scanning confocal microscopy (LSCM). Three-dimensional (3D) bicontinuous structures, which were constructed for the first time by time-resolved LSCM, were found to have a "sponge-like" structure composed of two phases. The structure factor obtained by 3D Fourier transformation of the sponge was found to be identical to that obtained by TRS, confirming that the sponge truly reflects the structural entities evolving in the system. Furthermore, the sponge was shown for the first time to be theoretically predictable by using 3D computer simulations based on the time-dependent Ginzburg-Landau theory. The sponge was subjected to differential geometrical analysis: its Gaussian curvature K, mean curvature H, and their distributions were successfully determined for the first time. The result revealed that the sponge has hyperbolic interfaces with area-averaged curvatures satisfying $\langle K \rangle < 0$ and $\langle H \rangle \cong 0$ and that its interface has some deviations from a minimal surface. The sponge was found to be strikingly similar to that occurring in oil/water/surfactant systems at the hydrophile-lipophile-balance, though their characteristic length scales are diversely different (μm vs nm), implying universality of the sponge.

Introduction

Binary mixtures of molecules (or atoms) A and B become thermodynamically unstable, depending on the balance between the cooperative forces associated with intermolecular forces and the forces associated with the entropy of the systems. From a thermodynamic point of view, this instability occurs inside the spinodal curve. The unstable systems undergo phase separation via spinodal decomposition (SD; Ref.1) into two macroscopic phases rich in A and B molecules at thermal equilibrium. The SD process involves the time evolution of phase-separating structures, starting from thermally activated dynamic composition fluctuations of A and B molecules in space (defined as thermal concentration fluctuations) to coexisting domains rich in A and B on a mesoscopic scale (between microscopic and macroscopic scales) and eventually to the two macroscopic phases. This self-organization process of

[a] Present addresses: Department of Polymer Science and Engineering, Kyoto Institute of Technology, Matsugasaki, Kyoto 606-8585, Japan
[b] Structural Biochemistry Laboratory, The Institute of Physical and Chemical Research (RIKEN), 1-1-1 Kouto, Mikaduki, Sayo, Hyogo 679-5148, Japan
[c] Department of Polymer Chemistry, Graduate School of Engineering, Kyoto University, Kyoto 606-8501, Japan

 CCC 1022-1360/00/$ 17,50+.50/0

molecules is one of the themes of non-linear and non-equilibrium phenomena which are left to be answered by modern science.

Many works have been devoted to this non-equilibrium, non-linear phenomenon (Refs. 2-4). One of fundamental questions to be answered is what kind of self-organized structures evolves in mixtures giving rise to two phases which have nearly equal phase-volume (i.e., isometric systems). Experimental studies, in particular time-resolved light scattering (TRLS) studies (one of the method involving the structure analysis in Fourier space), have revealed that the structure evolved is periodic and bicontinuous. The bicontinuity of the structure has been elucidated with the aid of theories (Ref. 5) and computer simulations (Refs. 6,7). The structures evolved at the late stage of SD have been found (Refs. 3,4,8) to obey the dynamic scaling hypothesis (Ref. 9): the evolving structures show a dynamic self-similarity (dynamic fractality) and are primarily scaled with time- and temperature-dependent length scales.

Questions to be answered are: (i) what kind of bicontinuous structures can be expected ? (ii) what is the physical basis of the structures? (iii) how can the structures be characterized mathematically ? (iv) how universal could the structures be ?

To answer question (i), we need to investigate the structures *in situ* and *at real time* using real-space analysis (microscopy methods). Unfortunately, little work along this line has been done up to the present time owing to experimental difficulties. Although many works have been reported on time evolution of scattering structure factors (Fourier-space analysis by scattering methods) (see for example Ref. 3), one cannot visualize real structures directly from the structure factors. To answer question (ii), we should conduct three-dimensional (3D) computer simulations based on a non-linear time evolution equation and find statistical identity between the 3D structures simulated and experimentally determined in real space. Again no work has been reported along this line. To answer question (iii), we should analyze mean principal curvatures and curvature distributions from the mean values for interfaces which separate A-rich and B-rich domains on the basis of differential geometry (see Fig. 1). Again, no work has been reported so far. At the late stage of the SD process the interfaces are well defined and the excess Gibbs energy of our system is localized at the interfaces. Under this situation, *the curvatures of the interfaces and their distributions are the most important information* which characterizes our system. Nevertheless, little information has been available so far because of experimental difficulties. The information should be hidden in the scattering structure factors. However, relationships between the curvature information and the structure factor have not been fully explored. Thus *the curvatures and the structure factors are considered to be essentially independent physical quantities* at the moment. To answer question (iv), we attempted to compare the structures of phase-separating mixtures of simple liquids and polymers with a bicontinuous structure in oil/water/surfactant systems having diversely different physical origin and length scales. No such attempts have been made so far.

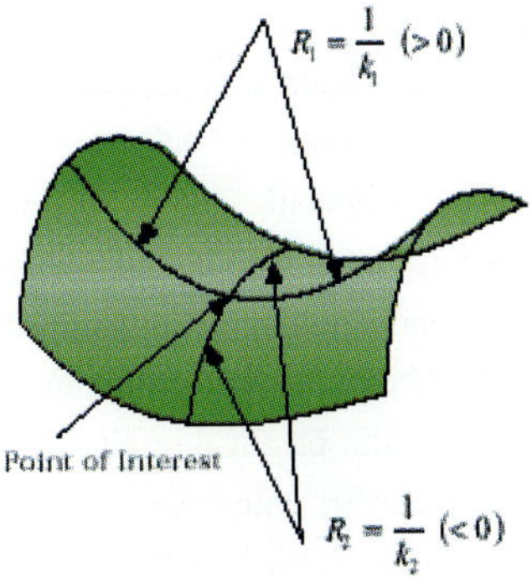

Fig. 1. Definition of two principal curvatures $k_1(\mathbf{r})$ and $k_2(\mathbf{r})$ or mean curvature $H = (k_1 + k_2)/2$ and Gaussian curvature $K = k_1k_2$ at a point of interest located at **r** on the interface.

The present work is aimed to explore the four unsolved problems described above and to pioneer a new interdisciplinary problem of chemistry, physics and mathematics with respect to the time evolution of form in soft-condensed matter. We will answer each question below.

What Kind of Bicontinuous Structures Do We Expect to Have ?

In this study we have overcome the experimental difficulties described in the previous section by using polymer mixtures as a model system. We took advantage of polymers having large characteristic lengths and long characteristic times: *large structures grow very slowly in polymer mixtures* as reviewed in Ref. 3. For this purpose we made a lot of effort in chemistry: we synthesized a particular pair of polymers which have a well-defined phase diagram to facilitate the dynamic experiments as described in the previous section. A pair of polymers such as polybutadiene (PB) and polyisoprene (PI) or poly(styrene-*ran*-butadiene) and PB or deuterated PB (dPB) and nondeuterated PB with a narrow molecular weight distribution, were selected as simple hydrocarbon polymers which are amorphous liquids far above glass transition temperature and which exhibit only van der Waals interactions.

We had difficulties investigating 3D periodic, bicontinuous structures with transmission optical microscopy, simply because an overlap of the structures along the depth direction of the sample (i.e., along the incident beam direction) smears real structural entities.
This is one of the primary experimental difficulties in the studies of 3D bicontinuous structures and, in fact, almost no good microscopy studies have been reported so far. We circumvented this difficulty by using laser scanning confocal microscopy (LSCM) (Ref. 10), a recently developed technique which has been applied especially in the field of biology, to our mixtures. The technique enables a 3D scanning of objects and a reconstruction of a 3D image of them. The fact that the time required to capture the 3D image is much shorter than the time scale of the structure growth enables *in situ* observation of 3D optical images.

Figure 2 shows a typical result obtained with LSCM for a polymer mixture at a given time at a late stage of SD (Ref. 11). The figure shows a series of optically sliced images obtained with a focal depth of 0.5 μm and a distance between the successive images of 0.5 μm. The evolving structure is composed of two isometric domains having a spacing about 10 μm: bright domains (domains rich in one polymer) and dark domains (domain rich in the other). The structure grows in size in the time scale of hours, days and years, depending on temperature but the growing structure has dynamic self-similarity (Refs. 3,4,8). In the image at the top-left corner, the dark domain marked **X** is surrounded by the bright domain, thus appearing to be isolated from the neighboring dark domains. However, when we scan the images along the depth direction (corresponding to a shift of the image from a to f), we find that these two dark domains are interconnected in the fourth slice of image d (i.e., 1.5 μm below the first slice).

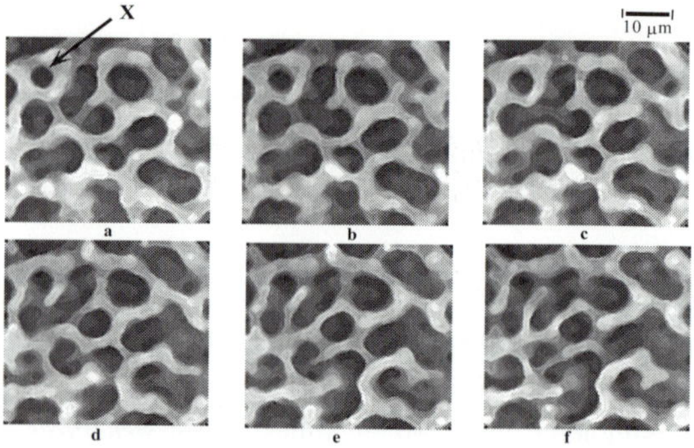

Fig. 2. Optically sliced images taken by laser scanning confocal microscopy (LSCM) of an isometric polymer mixture at a late stage of the SD process. A series of the images were taken at 0.5-μm intervals along the sample thickness direction with focal depth 0.5 μm. The depth where the image is taken increases from a to f.

Thus we find that the two domains are really interconnected and periodically arranged, forming a 3D "maze". We define this characteristic structure as *sponge-like structure* according to the term "sponge" in the field of topology and differential geometry (see, e.g., Ref.12).

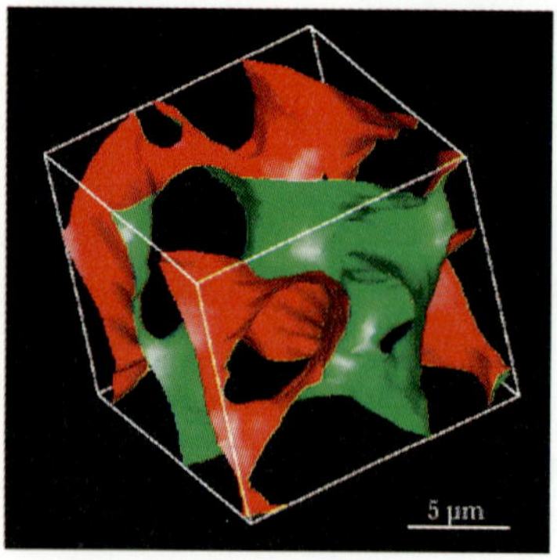

Fig. 3. 3D LSCM image of the interface for the isometric polymer mixture at a late stage of SD having average spacing $\Lambda_m = 12.1$ μm (Ref. 13). The bar corresponds to 5 μm.

Figure 3 shows a typical 3D image of the interface constructed by LSCM which separates the two phases, one side of the interface being red and the other green. We can clearly see (i) the bicontinuous feature of the red and green phases, i.e., a 3D continuity of the two phases, and (ii) that most of the interface is saddle-shaped or hyperbolic with $K < 0$. The image in the figure shows the size of one average spacing. The image representing a much larger length scale is shown in Fig. 4.

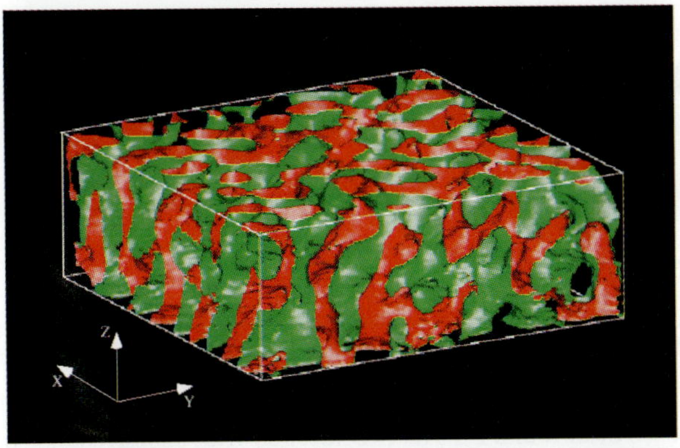

Fig. 4. 3D LSCM images of the interface shown over a larger scale than that in Fig. 3. The sample observed is the same as that in Fig. 3. The image has a volume of 80 × 80 × 30 μm³.

In order to confirm whether or not the 3D phase structure constructed with LSCM truly reflects a real structural entity, we calculated the scattering structure factor from the 3D digital image and compared it with the structure factor obtained from a TRLS experiment. For this purpose, we calculated scattering structure factor $S(q) \sim |\Psi(q)|^2$ where $\Psi(q)$ is the 3D Fourier transformation of the binarized 3D digital image where q is wavenumber. The results are shown in Fig. 5 where the scaled structure factor $S(q/q_m)$ obtained at various times at the late stage of SD of a 50/50 mixture of dPB (weight-average molecular weight M_w=1.43x10^5) and PB labelled with anthracene (PB-AN, M_w =9.5x10^5) by using fluorescent LSCM (black symbols) is compared with that obtained from TRLS (red symbols) (Ref. 14). Naturally, the structure factor obtained from TRLS is universal with time and, similarly, the structure factor obtained from LSCM is also universal with time. Moreover, it is quite impressive that two universal structure factors agree quite well over a large intensity scale, as large as five orders of magnitude, and a large spatial scale, as large as two orders of magnitude. Thus we conclude that the 3D LSCM image truly reflects the real structure entity. For the definition of the scaled structure factor, the reader is referred to Refs. 2-4, for example.

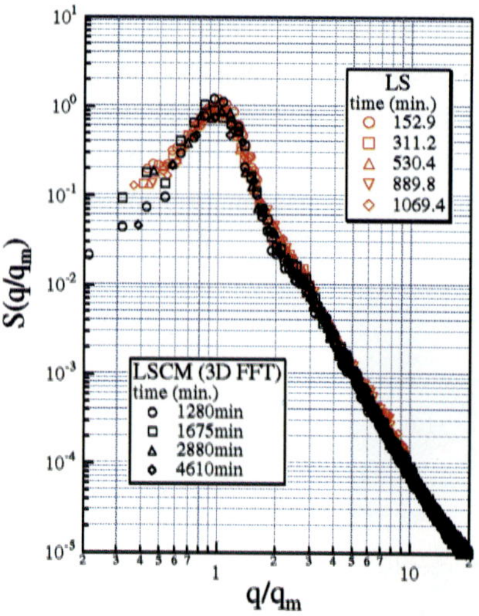

Fig. 5. Comparisons of scaled structure factors $S(q/q_m)$ obtained from the 3D LSCM image (black symbols, ○, □, △, ◇) and the TRLS experiments (red symbols, ○, □, △, ▽, ◇) for an isometric polymer mixture composed of dPB and PB-AN (PB labelled with anthracene). The variable q is the wavenumber of a Fourier mode of the structure or magnitude of scattering vector and q_m is q at which the structure factor $S(q)$ becomes maximum, i.e., the wave number for the dominant Fourier modes. ($q_m = 2\pi/\Lambda_m$, where Λ_m is average spacing or characteristic length of the structure.).

Figure 6 demonstrates time evolution of 3D structures observed for the 50/50 v/v mixture of dPB and PB-AN. It is shown that the sponge-like structure grows self-similarly: The scaled structure factors obtained from the 3D images constructed at different times at the late stage of SD are universal with time (see Fig. 5). Therefore, the shape of the sponge-like structure is universal with time but only the characteristic length scale $\Lambda_m = 2\pi/q_m$ increases with time according to the scaling law of $\Lambda_m \sim t$.

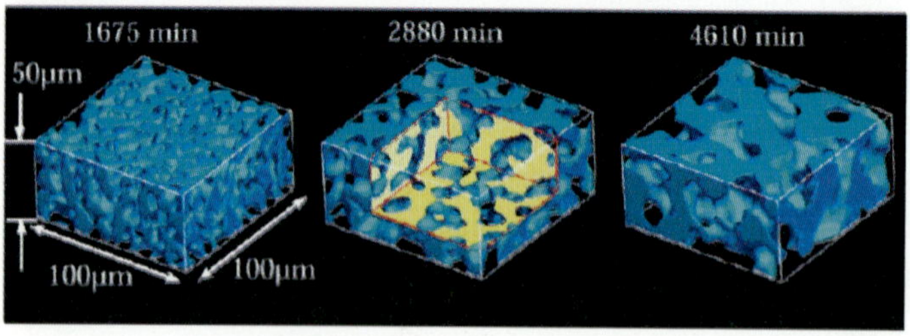

Fig. 6. Time evolution of sponge-like structure observed for dPB/PB labelled with anthracene (PB-AN): the 50/50 v/v mixture undergoing the late-stage spinodal decom-position. The PB-AN phase is blue, while the dPB phase is the empty space (by fluorescent laser scanning confocal microscopy).

What is the Physical Basis of Sponge-like Structure?

Having firmly established the 3D optical image of the sponge-like structure, we now address the question concerning the physical basis of the structure. For this purpose, we studied whether or not the structure can be predicted by a field theory called the time-dependent Ginzburg-Landau theory (TDGL) with hydrodynamic interactions (Ref. 7). The TDGL theory is based on a non-linear time evolution equation of the order of parameter field, i.e., spatial composition fluctuations of molecules A and B. Since the equation is non-linear and non-local, we cannot expect analytical solutions. Thus we solved the problem by 3D computer simulations. The detailed methodology and detailed comparison of the results obtained by simulations and scattering experiments were reported elsewhere (Refs. 7,15).

Figure 7 compares $S(q/q_m)$ obtained from the computer simulation (solid line) with that obtained from LSCM (shown by data points), showing again an excellent agreement with $S(q/q_m)$ obtained from the LSCM and TRLS experiments.

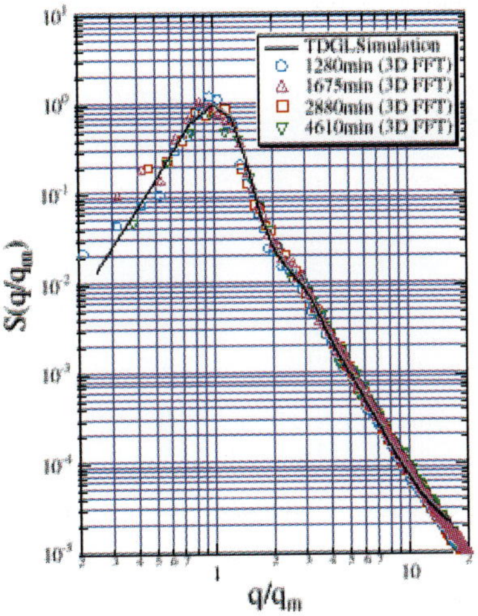

Fig. 7. Comparison of scaled structure factors $S(q/q_m)$ obtained from the 3D computer simulation with system size of 128^3 (solid line) with that obtained from the 3D LSCM image (data points ○, △, □, ▽).

The result thus demonstrates that the sponge-like structure is theoretically predictable: the sponge formation discussed here is driven by thermodynamic instability of the isometric systems. The simulated result was obtained for a reduced shear viscosity of $\acute{\eta}$ where $\acute{\eta}$ is the only relevant parameter in the TDGL model associated with hydrodynamic interactions and measures the inverse of the strength of the hydrodynamic interactions (Ref. 7). It was also elucidated that the shape of evolving structure, $S(q/q_m)$, is rather insensitive to the value of $\acute{\eta}$, although the time evolution of the size of the structure $\Lambda_m(t) = 2\pi/q_m(t)$ is very sensitive to the value of $\acute{\eta}$ (Refs. 7,15). Molecular parameters and thermodynamic parameters such as temperature and pressure also affect the time evolution of the characteristic length Λ_m but the sponge structure scaled with Λ_m, i.e., the shape of the structure, is universal with these parameters. Thus we can conclude that the form of the evolving structures bears universality in the scaling regime.

Figure 8a shows a 3D image of the interface constructed by simulations for an isometric binary mixture at the late stage of SD, while Fig. 8b shows corresponding 2D sliced images demonstrating the time evolution of morphology. Note that features of the computer-simulated interface are quite similar to those constructed by LSCM, as is naturally expected from the identity in $S(q/q_m)$ shown in Fig. 7.

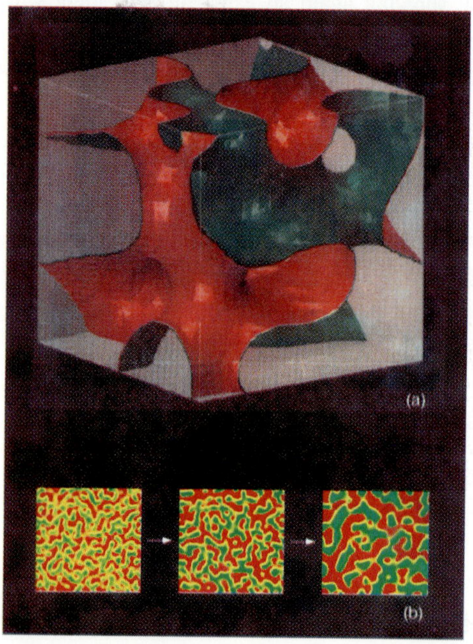

Fig. 8. 3D image of the interface constructed by computer simulations using (a) the TDGL model for an isometric binary mixture in the late-stage SD, and (b) 2D sliced images showing time evolution of the morphology (time elapsing from left to right). The edge length of the box in (a) corresponds to about 1.5 Λ_m.

The TDGL model is quite general, applicable to all mixtures of small molecules as well as polymer molecules. When the model is applied to polymer mixtures, the two polymers are assumed to have equal static and dynamic properties, i.e., polymer pairs being "symmetric", having nearly equal molecular weights and viscosities. The real polymer mixtures used in this study are also symmetric in this sense. The scaled structure factors obtained for such symmetric polymer mixtures were shown to be essentially identical to those obtained for simple liquid mixtures (Refs. 4,16) and metalic alloys (Ref. 2). Thus the sponge-like structure shown in Figs. 3, 4 and 6 is universal for any phase-separating isometric mixtures via the SD process. Thus we succeeded for the first time in visualizing the real-space images of spinodally decomposed bicontinuous two-phase structures for simple liquid mixtures, metalic alloys and inorganic glasses, etc. by using polymer mixtures as model systems.

How Can the Sponge-like Structures Be Characterized and Analyzed Mathematically?

Now that we have determined the 3D structure, which accurately reflects real structures from the global to local scale, and that we have obtained a sound physical basis for this structure, we proceed to analyze the structure in terms of interface curvatures, based on differential geometry. We first used the "parallel surface method" to determine the area-averaged mean curvature $<H>$ and Gaussian curvature $<K>$,

$$A(d) = A_0\,(1 + 2 <H> d + <K> d^2) \qquad (1)$$

where A_0 is interface area, and $A(d)$ is the area of the surface which is parallel to and located at a distance d from the interface. H and K are local mean curvature and Gaussian curvature defined earlier in the caption to Fig. 1.

We used the "marching-cube algorithm" (Ref. 17) to construct the 3D interface structure in that the interface is composed of triangles, as schematically shown in Fig. 9a. To construct the parallel surface at a displacement vector $\boldsymbol{d}$ ($|\boldsymbol{d}| = d$) from the interface, we first calculate normal vectors $\boldsymbol{n}$ ($|\boldsymbol{n}| = 1$) at the vertices of the triangles (red arrows), as will be detailed elsewhere (Refs. 18,19). Then $\boldsymbol{d}$ is given by $\boldsymbol{d} = d\boldsymbol{n}$, and the parallel surface, composed of yellow triangles as shown in Fig.9b, is constructed by connecting the ends of the displacement vectors. $A(d)$ is then calculated by summing the areas of the triangles.

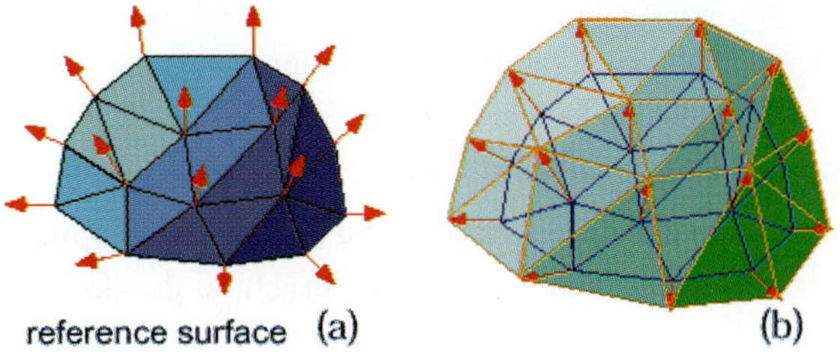

Fig. 9. Schematic representation of (a) the interface (blue reference surface) composed of triangles and normal vectors (red arrows) at their vertices and (b) the parallel surface constructed by connecting the normal vectors.

The best fit of the results of $A(d)$ obtained from the 3D images shown in Figs. 3 and 4 with Eq. 1 yielded (Ref. 19)

$$A_0/V = 2.2 \times 10^{-1}\ \mu m^{-1},$$

$<H> = 7.0 \times 10^{-3}\ \mu m^{-1}$,

$<K> = -6.2 \times 10^{-2}\ \mu m^{-2}$ (or $[-<K>]^{1/2} = 2.5 \times 10^{-1}\ \mu m^{-1}$),

where A_0/V is the interface area per unit volume. These parameters depend on time, and the absolute values definitely decrease with time (Ref. 19). We can also conclude that the sponge-like structure has a hyperbolic interface having a negative area-averaged Gaussian curvature $<K>$ and almost zero area-averaged mean curvature $<H>$, i.e., $r \equiv <H>/[-<K>]^{1/2} \cong 0$. The small positive value of r (ca. 0.03) may reflect portions of bicontinuous domains having cylinder-like structures which tend to burst during the domain growth.

The time evolution of these characteristic parameters has been analyzed as well (Ref. 19). It is intriguing to note that $r \cong 0$, irrespective of time during the self-organization process. This may be the most rational path taken by the evolving structures in the isometric systems. The fact that $<H>$ is close to zero does not necessarily mean that the interface of the evolving non-equilibrium domains is close to minimal surface. To understand this point, one must determine distribution functions of the local curvatures H and K. We have attempted to determine the distributions both from the 3D optical images (Ref. 20), also based on a principle of differential geometry, and from the 3D computer simulations (Ref. 21). The results revealed that the distribution of H for our interface is broader than that for the periodic minimal surface of gyroid, for example, implying distortions of our interface from the ideal mathematical minimal surface.

The area-averaged curvatures $<H>$ and $<K>$ and the distributions of the curvatures from the mean values determined from the 3D optical images are found to be consistent with the so-called "scattering-mean-curvature" obtained from scattering experiments (Refs. 13,22), based on the Kirste-Porod-Tomita theory (Refs. 23,24). This evidence further reinforced the merit of our approach presented here in two senses: (i) it supports the scattering theory and proves that the scattering method provides a particular moment of the curvature distributions which we designate "scattering-mean-curvature" $<K_S{}^2> \equiv (3 <H^2> - <K>) / 2$; (ii) it further supports the accuracy of our methods and analyses because the scattering method is essentially independent of the present method in terms of the curvature determination.

How Universal Could The Sponge-like Structures Be?

We have already noted that the sponge-like structure found for the spinodally decomposed symmetric polymer mixtures is universally applied to bicontinuous structures developed for critical mixtures of small molecules or atoms via the SD processes. In order to further address this question, we compare the bicontinuous structures in microemulsion systems, e.g., a ternary mixture of heavy water (D_2O), *n*-octane, and tetraethylene glycol monodecyl ether (C_{10}-E_4, a non-ionic surfactant) at the hydrophile-lipophile balance (HLB) temperature, with those in our simple molecular mixtures composed of hydrocarbon polymers. The microemulsion system also forms structures similar to those in the polymer mixtures, i.e., isometric bicontinuous domains of water and oil with most of the surfactant molecules at the interface, despite the fact that intermolecular interactions in the system are more complex than those in polymer mixtures. It should be noted here that the bicontinuous structures in the microemulsion systems are equilibrium structures, while those formed via the SD processes are nonequilibrium ones.

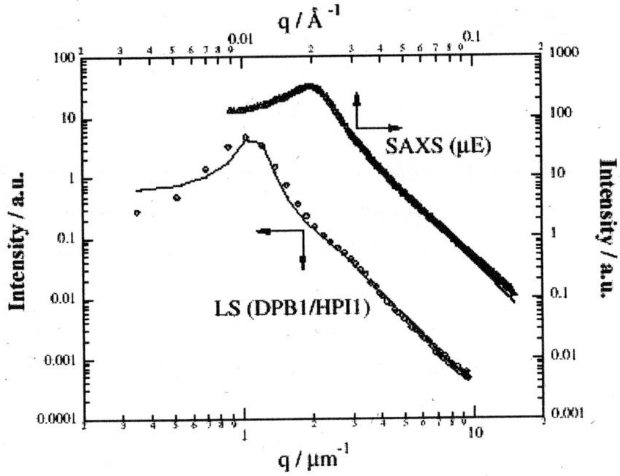

Fig. 10. Comparison of the SAXS function from the equilibrium isometric microemulsion system at the HLB point (μE) with the TRLS function from the isometric polymer mixture at a particular time in the late-stage SD (both shown by data points (Ref. 25)). The solid lines are the best fit with the RGW theory.

Figure 10 shows a small-angle X-ray scattering (SAXS) function from the microemulsion system (triangles) together with TRLS functions (circles) from the isometric polymer mixture at the late stage of SD (Ref. 25). The solid lines show the best fits of the random Gaussian wave (RGW) theory (Refs. 26,27) for the experimental scattering functions. The theory assumes a suitable form of the spectral function *f(k)* which is an inverse sixth-order polynomial in wavenumber *k* and which is an extension of the well-known structure factor proposed by Teubner and Strey (Ref. 28). The scattering function as well as the real-space structure can be calculated from *f(k)*.

We choose the following approach in order to extract useful information about the structures occurring in the systems in the context of the RGW theory. The best fit of the calculated scattering function for the experimental scattering function determines *f(k)*, and the *f(k)* in turn determines important physical parameters such as an average spacing of the structure, Λ_m, the coherence length of the local order, ξ (Ref. 29), and $\langle K \rangle$ (Ref. 25) as well as the real-space structures. The real-space structures thus determined are shown in Fig. 11.

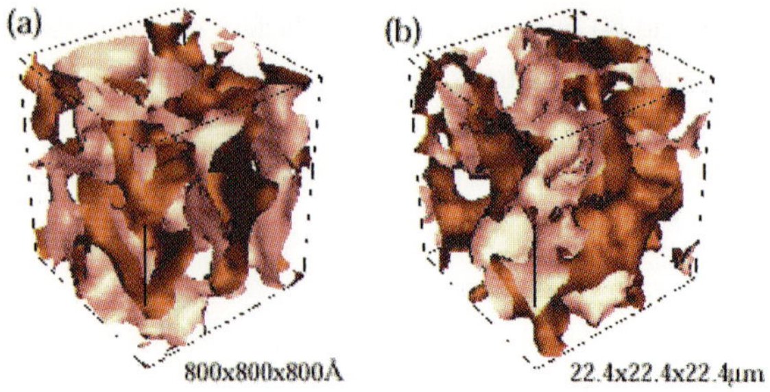

Fig. 11. 3D image of the interfaces constructed on the basis of Gaussian random wave theory: (a) for an isometric microemulsion system at the HLB point and (b) for an isometric mixture at the late stage of SD (Ref. 25).

At first glance, the two profiles shown in Fig. 11 are very similar, except for the absolute length scales (about 10 nm vs. 1 μm). Similarly, the two real-space structures constructed on the basis of the RGW theory show a striking resemblance, having sponge-like characteristics (see Fig. 11). The real-space structures constructed using the RGW theory are expected to reflect real structural entities because of good agreement with the theoretical and experimental scattering functions.

Quantitative comparisons of the two systems are summarized in Table 1. Note the similarity of the two sponge-like structures in the reduced Gaussian curvature $[-\langle K\rangle]^{1/2}/q_m$ on a quantitative basis as well. Again, we should note that the values of $\langle K\rangle$ thus estimated are negative for the two structures. An important difference in the two structures is discernible in the reduced coherence length, ξ/Λ_m. The polymer system has a larger coherence length (a higher degree of the long-range order) than the microemulsion system. This aspect is manifested by the fact that the low-q part of $I(q)$ in the polymer system cannot be fitted well by the theoretical scattering function as shown in Fig. 10. This point raises an important question in theoretical physics of the RGW theory when it is applied to polymer systems with a seemingly smaller random thermal noise effect compared with the low-molecular-weight microemulsion system. In this regard, the TDGL theory gives a better prediction than the RGW theory.

Table 1. Comparison of sponge-like structures in the microemulsion system and the polymer mixture (polymer).

Physical quantity	Polymer	Microemulsion	Polymer/ microemulsion[a]
q_m[b]	1 μm^{-1}	2×10^{-2} Å^{-1}	5×10^{-3}
Λ_m[b]	6.28 μm	3.14×10^{2} Å	2×10^{2}
$[-\langle K \rangle]^{1/2}$	1.44 μm^{-1}	1.07×10^{-2} Å^{-1}	4×10^{-3}
$[-\langle K \rangle]^{1/2}/q_m$	0.44 (0.48[c])	0.54	0.8 (0.9[c])
ξ/Λ_m[d]	1	0.6	1.7

[a] The ratio of a physical quantity for polymer and that for microemulsion; [b] $q_m = 2\pi/\Lambda_m$ where q_m and Λ_m are the characteristic wavenumber and length, respectively; [c] the value for the polymer mixture as determined by direct real-space analysis; [d] reduced coherence length.

Concluding Remarks

We have constructed three-dimensional (3D) real-space images of bicontinuous domain structures in phase-separating polymer mixtures via spinodal decomposition (SD). The 3D images exhibit "sponge-like" structures. They were found to provide a visualized model for the universal scaled structure factors established for critical mixtures of symmetric polymer molecules, small molecules or atoms undergoing the late-stage process of SD. The sponge-like structures well predictable by the nonlinear time-evolution equation based on the TDGL theory are universal for systems with thermodynamic instability which phase-separate into bicontinuous domains with well-defined interface. They compose of saddle-shape interfaces with negative area-averaged Gaussian curvature $\langle K \rangle$ and almost zero area-averaged mean curvature $\langle H \rangle$.

Acknowledgements

The authors thank Dr. S-H. Chen and Dr. S.T. Hyde for their collaborations and enlightening comments.

References

1. J.W. Cahn, *J. Chem. Phys.* **1965**, 42, 93.
2. See, e.g., J.D. Gunton, M. San Miguel, and P.S. Sahni, in *Phase Transitions and Critical Phenomena*, (C. Domb and J.L. Lebowitz, Eds.), Academic Press, New York, **1983**, pp. 269-482; K. Binder, in *Materials Science and Technology*, (R.W. Cahn, P. Haasen, and E.J. Kramer, Eds.), Vol. 5 *Phase Transformations in Materials*, VCH, Weinheim, **1991**, pp. 143-212; S. Komura and H. Furukawa, *Dynamics of Ordering Processes in Condensed Matter*, Plenum, New York, **1988**.
3. T. Hashimoto, in *Materials Science and Technology*, (R.W. Cahn, P. Haasen, and E.J. Kramer, Eds.), Vol. 12, *Structure and Properties of Polymers*, VCH, Weinheim, **1993**, pp.251-300.
4. T. Hashimoto, *Phase Transitions* **1988**, 12, 47.
5. See, e.g., H. Furukawa, *J. Phys. Soc. Jpn.* **1989**, 58, 216.
6. A. Shinozaki and Y. Oono, *Phys. Rev. E* **1993**, 48, 2622.
7. T. Koga and K. Kawasaki, *Physica A* **1993**, 196, 389.

8. T. Hashimoto, H. Jinnai, H. Hasegawa, and C.C. Han, *Physica A* **1994**, 204, 261.
9. K. Binder and D. Stauffer, *Phys. Rev. Lett.* **1974**, 33, 1006.
10. See, e.g., T. Wilson, *Confocal Microscopy*, Academic Press, London, **1990**, pp. 1- 64.
11. A. Ribbe and T. Hashimoto, *Macromolecules* **1997**, 30, 3999.
12. S. Hyde, S. Anderson, K. Larsson, Z. Blum, T. Landh, S. Lidin, and B.W. Ninham, *The Language of Shape*, Elsevier, Amsterdam, **1997**.
13. T. Hashimoto, H. Jinnai, Y. Nishikawa, T. Koga, and M. Takenaka, *Prog. Colloid Polym. Sci.* **1997**, 106, 118.
14. T. Hashimoto, T. Koga, H. Jinnai, and Y. Nishikawa, *Nuovo Cimento Soc. Ital. Fis., D* **1998**, 20, 1947.
15. T. Koga, K. Kawasaki, M. Takenaka, and T. Hashimoto, *Physica A* **1993**, 198, 473.
16. M. Takenaka and T. Hashimoto, *J. Chem. Phys.* **1992**, 96, 6177.
17. W. E. Lorensen and H. E. Cline, *Computer Graphics SIGGRAPH'87* **1987**, 21, 163.
18. Y. Nishikawa, H. Jinnai, T. Koga, T. Hashimoto, and S.T. Hyde, *Langmuir* **1998**, 14, 1242.
19. H. Jinnai, Y. Nishikawa, T. Koga, T. Hashimoto, and S.T. Hyde, *Phys. Rev. Lett.* **1997**, 78, 2248.
20. Y. Nishikawa, T. Koga, T. Hashimoto, and H. Jinnai, *Langmuir* **2001**, 17, 3254.
21. T. Koga, H. Jinnai, and T. Hashimoto, *Physica A* **1999**, 263, 369.
22. M. Takenaka and T. Hashimoto, *Physica A*, **2000**, 276, 22.
23. B. Kirste and G. Porod, *Kolloid-Z. Z. Polym.* **1962**, 184, 1.
24. H. Tomita, *Prog. Theor. Phys.* **1984**, 72, 656.
25. H. Jinnai, T. Hashimoto, D.D. Lee, and S.H. Chen, *Macromolecules* **1997**, 30, 130.
26. N.F. Berk, *Phys. Rev. Lett.* **1987**, 58, 2718.
27. S.H. Chen, D.D. Lee, and S.L.Chang, *J. Mol. Struct.* **1993**, 296, 259.
28. M. Teubner and R. Strey, *J. Chem. Phys.* **1987**, 87, 3195.
29. S.H. Chen, S.L. Chang, and R. Strey, *J. Appl. Crystallogr.* **1991**, 24, 721.

Structure and Dynamics of Silica-Filled Polymers by SANS and Coherent SAXS

Erik Geissler, Anne-Marie Hecht, Cyrille Rochas, Ferenc Horkay, Françoise Bley and Frédéric Livet*

Laboratoire de Spectrométrie Physique UMR CNRS 5588, Université J. Fourier de Grenoble, B.P.87, 38402 St Martin d'Hères Cedex, France

Laboratory of Integrative and Medical Biophysics, National Institutes of Health, 13 South Drive, Bethesda MD 20892, USA

Laboratoire de Thermodynamique et Physico-chimie Métallurgiques, CNRS UMR 4777, INPG, 38402 St Martin d'Hères Cedex, France

SUMMARY: Random crosslinking in elastomers gives birth to local variations in the crosslink density. When the network is swollen in a low-molecular-weight solvent, competition between the osmotic pressure and the local elastic constraints transforms these variations into differences in polymer concentration, the range and amplitude of which can be measured by small-angle X-ray or neutron scattering (SAXS or SANS). In filled systems, the distribution both of the polymer and of the elastic constraints is modified. By varying the proportion of deuterated solvent in the network, the scattering function of the polymer can be distinguished from that of the filler using SANS. Such measurements yield not only the internal surface area of the filler particles but also the fraction of that surface in contact with the polymer. The recently developed technique of quasi-elastic SAXS detects slow dynamic processes at wave vectors larger than those accessible with visible light lasers. This technique is used to investigate the dynamics of filler particles in uncrosslinked polymer melts. It is directly shown that the structural reorganization process of the filler following an external mechanical perturbation is diffusion-controlled.

Introduction

According to the scaling theory [1], thermodynamic properties of polymer solutions depend on a single characteristic length. Under the semidilute conditions, where different coils overlap, this distance is the thermal correlation length, ξ. When crosslinks are introduced into the polymer in a sufficient number to generate a network, the osmotic properties of the solution persist, but a new length scale Ξ is required to describe the elastic perturbations imposed by the crosslinks. In a dry rubber without solvent, neither of these length scales is directly visible. If, however, a low-molecular-weight solvent is introduced into the network, the resulting osmotic pressure is balanced by the local elastic pressure. As crosslinks are not uniformly distributed, random permanent local variations in concentration develop as the

 CCC 1022-1360/00/$ 17,50+.50/0

network swells, which scatter radiation. If filler particles are also present, the distribution of the polymer is modified. The characteristic lengths, which are generally significant in rubbers, typically range from 1 to 100 nm. This size range coincides with the optimum sensitivity of small-angle neutron or X-ray scattering (SANS or SAXS) [2]. Furthermore, by using mixtures of normal and deuterated polymers in the solvent, it is possible to discriminate between the signal from the polymer and that of the filler. The slow dynamics of the filler particles in the polymer matrix can be explored by means of X-ray correlation spectroscopy.

Experimental Part

Sample preparation

Linear vinyl-terminated poly(dimethylsiloxane) (viscosity 20 Pas) was blended with 9 % w/w of fumed silica filler. Two different fillers were used; the first was an untreated hydrophilic silica (AE300, Degussa, specific surface area 300 m^2g^{-1}, sample 1), and the other with octamethyltetrasiloxane grafted to its surface (AE60, Degussa, specific surface area 60 m^2g^{-1}, sample 2).

Both samples were prepared by mixing the filler and the polymer melt in a ball mill under dry nitrogen at 70 °C for 3 h. The resulting mixtures form thixotropic pastes that can be swollen in toluene. The measurements reported here were performed at polymer volume fraction $\varphi = 0.2$. The samples were swollen in mixtures of toluene (Prolabo, Analytical Grade) and deuterated toluene (Acros) with deuterium content > 99 %. The specimens for SANS were contained in a cell consisting of 1-mm thick quartz windows separated by a 1-mm-spacer sealed with a Viton O-ring.

Small-angle neutron scattering measurements

SANS measurements were made on the NG3 instrument at NIST, Gaithersburg, using an incident wavelength of $\lambda = 8$ Å and selector bandwidth $\Delta\lambda/\lambda = 10$ %. The detector was placed at three distances, 1.8 m, 6 m and 13.1 m from the sample. The range of wave vector q [= $(4\pi/\lambda)\sin\theta/2$, λ being the incident wavelength and θ the scattering angle] explored was 0.005 $Å^{-1} \leq q \leq 0.23$ $Å^{-1}$, with counting times of between 10 min and 1 h. Measurements were also made at the Institut Laue Langevin, Grenoble, on the D22 instrument with incident wavelengths 8 Å and 14 Å with three sample-detector distances, 2 m, 6 m and 17 m, corresponding to an explored wave vector range 0.001 $Å^{-1} \leq q \leq 0.15$ $Å^{-1}$. The sample temperature during the experiment was maintained at 25° ± 0.1 °C.

The sample cell consisted of 1-mm thick quartz windows separated by a 1-mm thick spacer, sealed with a Viton O-ring. After radial averaging, corrections for incoherent background, detector response and cell window scattering were applied [3]. Calibration of the scattered neutron intensity was performed using the signal either from standard samples of silica aerogel or water [4,5].

Small-angle coherent X-ray scattering measurements

SAXS observations were made on the BM2 beamline at the ESRF, Grenoble, France (http://www.esrf.fr/exp_facilities/BM2/BM2.html) with an incident wavelength $\lambda = 1.5$ Å and resolution $\Delta\lambda/\lambda = 2 \times 10^{-4}$. The primary beam diameter was reduced by slits to 100 μm immediately after the focusing mirror and a 10-μm diaphragm was placed 10 m downstream just before the sample. A CCD detector was located 2.3 m beyond the sample. The spatially coherent beam thus obtained had a coherence area approximately equal to the detector pixel size, 22 μm. The beamstop was a gold cylinder of diameter 0.5 mm, glued to a thin Kapton foil. The evacuated flightpath between the mirror and the beamstop was free of windows.

Small-angle coherent light scattering measurements

Small-angle dynamic light scattering measurements were made with a 2-mW HeNe laser source. Light scattered from the sample was focused by a Fresnel lens on a CCD camera which collected successive images at a adjustable rate (ca. 1 s^{-1}) and then transferred them to a computer to calculate the intensity correlation function for each of the 512 x 512 pixels of the CCD. These correlation functions were grouped into subsets corresponding to different values of q, thus yielding simultaneous ensemble-averaged correlation functions for a wide range of q values [6].

Background

In a binary neutral gel system consisting of a polymer network and diluent, the scattering intensity can be described to a good approximation by the expression

$$I(q) = \Delta\rho^2 \left[\frac{kT\varphi^2}{M_{os}} \frac{1}{1+q^2\xi^2} + \frac{8\pi\Xi^3\langle\delta\varphi^2\rangle}{\left(1+q^2\Xi^2\right)^2} \right] \quad (1)$$

where $\Delta\rho^2$ is the contrast factor between polymer and solvent. The first and second terms in the square brackets come respectively from the osmotic fluctuations and the static variations in concentration due to random elastic constraints. As the gel swells, the mean square static concentration fluctuation $\langle\delta\varphi^2\rangle$ in the second term of Eq. 1 increases. This behavior is illustrated in Fig. 1, where the scattering intensity $I(q)$ of a solution of uncrosslinked silicone sample is compared with that of the corresponding gel swollen to the same degree in the same solvent. Although the liquid-like feature near $q = 0.1$ Å^{-1} is clearly distinguishable in the gel, a strong increase in intensity dominates the low-q region of the gel spectrum, corresponding to the second term in Eq. 1.

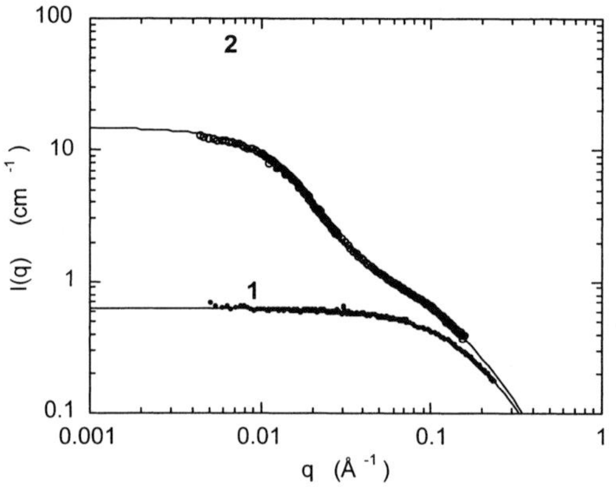

Figure 1. SANS spectra from a silicone solution in deuterated acetone (**1**), compared with that of a silicone gel swollen in the same solvent at the same volume fraction $\varphi = 0.33$ (**2**). The lines are fits of the solution data to the first term in Eq. 1 and of the gel data to the full Eq. 1.

The thermal correlation lengths ξ found from the fits to the data in Fig. 1 are 7 Å for the solution and 9 Å for the gel. When a filler is introduced into the polymer, the system becomes ternary (i.e. polymer, filler and solvent). The scattering response is then

$$I(q) = kT\,[(\rho_p - \rho_s)^2 S_{pp}(q) + (\rho_p - \rho_s)(\rho_f - \rho_s)\, S_{pf}(q) + (\rho_f - \rho_s)^2 S_{ff}(q)] \qquad (2)$$

where, ρ_p, ρ_f and ρ_s are the scattering length densities of the polymer, filler and solvent respectively. To solve for the three partial structure factors S_{pp}, S_{ff} and S_{pf}, which define the scattering from the polymer component, that from the filler and that of the interface between polymer and filler, respectively, $I(q)$ must be measured for at least three different values of the solvent scattering length ρ_s. This is achieved by swelling the sample to the same degree in a solvent containing different proportions of deuterated solvent in order to modify the scattering contrast of different components. Typical results of this procedure [7] are shown in Fig. 2, where the polymer partial structure factor $S_{pp}(q)$ is plotted for a filled crosslinked and a filled uncrosslinked PDMS system (sample 2) at approximately identical concentrations ($\varphi \approx 0.2$). Both spectra exhibit a characteristic curvature at the lowest values of q, revealing the presence of objects having a radius close to 500 Å. This feature is associated with a polymer shell that forms around the filler aggregates. This scattering region is followed at higher q by a steep drop in intensity with a slope of approximately –4, indicative of scattering from smooth polymer surfaces (Porod scattering) [8]. The area of these polymer surfaces may be estimated from the relation

$$\Sigma_p = \frac{kT}{2\pi} \lim_{q\to\infty} S_{pp}(q) q^4 \qquad (3)$$

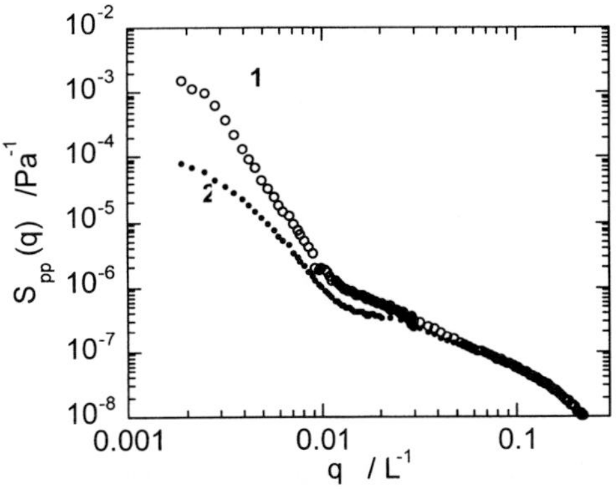

Figure 2. Polymer-polymer partial structure factors $S_{pp}(q)$ of a filled PDMS gel (1) and the corresponding filled uncrosslinked solution, sample 2 (2).

At intermediate values of q (0.01 Å^{-1} $\leq q \leq$ 0.03 Å^{-1}), a shoulder is visible in the spectrum from the gel, caused by the elastic constraints imposed by the crosslinks, corresponding to the second term in Eq. 1. As expected, this feature is absent, or at least less pronounced than in the spectrum of the uncrosslinked solution. At the high q-end of the curves, a further drop in intensity can be observed, which is defined by the thermal correlation length of the osmotic fluctuations [1]. This behavior is described by the first term in Eq. 1, with $\xi \approx 15$ Å.

For the silica structure factor $S_{ff}(q)$ (not shown here), a Porod scattering regime with slope -4 is also observed, but only at the highest values of q. The specific surface area of these filler particles is then given by [8]

$$\frac{\Sigma_f}{Vd} = \frac{\pi}{d} \frac{\lim_{q\to\infty} S_{ff}(q) q^4}{\int S_{ff}(q) q^2 \,\mathrm{d}q} \qquad (4)$$

where V is the scattering volume of the silica in the sample and d is its density.

Discussion

Figure 3 shows the partial structure factors found on applying the contrast variation technique to an uncrosslinked PDMS sample in which the silica filler particles have undergone a second, additional treatment with hexamethyldisilazane to improve the interaction with the surrounding polymer.

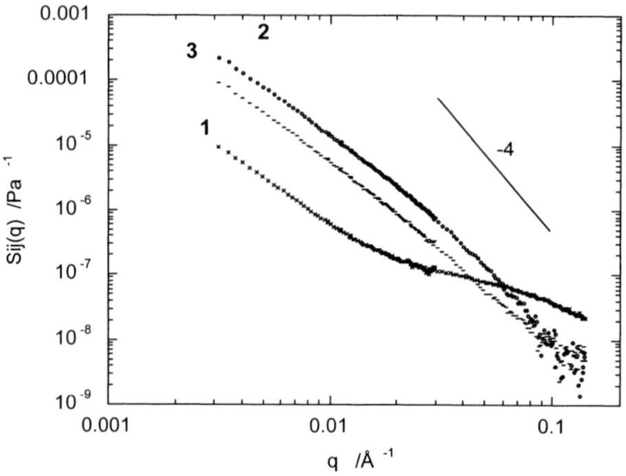

Figure 3. Partial structure factors $S_{pp}(q)$ (1), $S_{ff}(q)$ (2) and $S_{pf}(q)$ (3) in PDMS sample with hexamethyldisilazane-treated filler.

Comparison with the straight line of slope -4 in this figure allows to identify regions where Porod scattering prevails. This behavior shows up only for the silica structure factor at high values of q. The specific surface area determined from Eq. 4 for this sample is 300±30 m^2/g, in agreement with the results of independent BET surface area measurements.

For sample 2, the specific surface area of the silica is found to be 440 m^2/g. However, the ratio of the surface area of the polymer to that of the filler, Σ_p/Σ_f, (Eq. 3), is less than 0.02 in this sample. This surprising result shows that, when this filled system is swollen in toluene, polymer chains separate from the aggregate structures and form a smooth barrier at the outer edge. But in the sample containing the hexamethyldisilazane-treated filler, no equivalent smooth surface is visible in the polymer structure factor, indicating that the coupling agent prevents separation of the polymer from the filler. As shown in Fig. 4, $S_{pp}(q)$ can be fitted to an expression similar to Eq. 1, but where the exponent of the second term is smaller than 2:

$$I(q) = (\rho_p - \rho_s)^2 \left[\frac{kT\varphi^2}{M_{os}} \frac{1}{1+q^2\xi^2} + \frac{A}{(1+q^2\Xi^2)^{3/2}} \right] \qquad (5)$$

Equation 5 is the result expected for star polymers [9] in a semidilute polymer solution. This finding implies that polymer chains do not form a distinct outer sheath around silica aggregates but penetrate inside and adhere to the surface. In this case, since the Porod scattering region for the polymer is not clearly distinguishable from the intensity scattered by thermal fluctuations, it is not easy to evaluate quantitatively the area of polymer in direct contact with the solid.

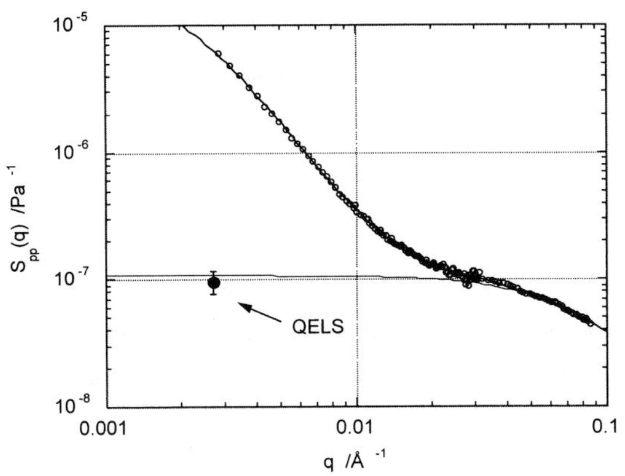

Figure 4. Partial structure factor $S_{pp}(q)$ from Fig. 3 showing fit of Eq. 5 (line) together with the osmotic component (first term in Eq. 5) and the osmotic susceptibility measured from dynamic light scattering (filled circle).

Figure 4 also shows the term arising from osmotic concentration fluctuations (first term in Eq. 5) resulting from the fit of Eq. 5 to the data. The amplitude of this term can also be determined independently by dynamic light scattering [10]. The result of this measurement, shown as a filled circle in Fig. 4, lies on the horizontal plateau of the dynamic component. This agreement lends credibility to the physical model on which the data analysis is based, and illustrates the consistency between the neutron and light scattering results.

Coherent X-ray scattering spectroscopy was used to probe the dynamics of structural relaxation processes in the filled PDMS samples, in the absence of solvent [11,12]. This technique, which is identical in concept to visible dynamic light scattering spectroscopy, relies on a synchrotron source for the X-rays. The intensity correlation that is measured is given by [13]:

$$G(t) = \langle I(q,0)I(q,t)\rangle / \langle I(q,t)\rangle^2 = 1 + \beta \exp(-2\Gamma t) \qquad (6)$$

where $I(q,t)$ is the intensity of a speckle at wave vector q and time t. The factor β is an instrumental constant and

$$\Gamma = Dq^2 \qquad (7)$$

is the characteristic relaxation rate of light scattered by particles having a diffusion coefficient D. The use of X-rays permits the study of relaxation processes at values of q beyond the range accessible to visible light scattering and provides a unique means of investigating the dynamics of structural rearrangements at high resolution.

In the present system, structural relaxation occurs when the samples are mechanically stirred. After the initial blending process, the silica aggregates in the polymer melt cluster into loose larger units called agglomerates. These associations can be disrupted by stirring, but immediately afterwards they start to reform. It is therefore expected that the observed diffusion rate of the particle in suspension will be higher immediately after stirring but the relaxation rate will subsequently decrease as the agglomerates reform. To account for this process, we set

$$\Gamma = Dq^2\,[1+a\,\exp(-t_0/\tau_s)] \qquad (8)$$

where D is the diffusion coefficient of the agglomerates and the factor $(1+a)$ describes the increase in diffusion rate due to the fragmentation into aggregates. In Eq. 8, t_0 is the time lapse between the mechanical perturbation and the measurement and τ_s is the structural relaxation time.

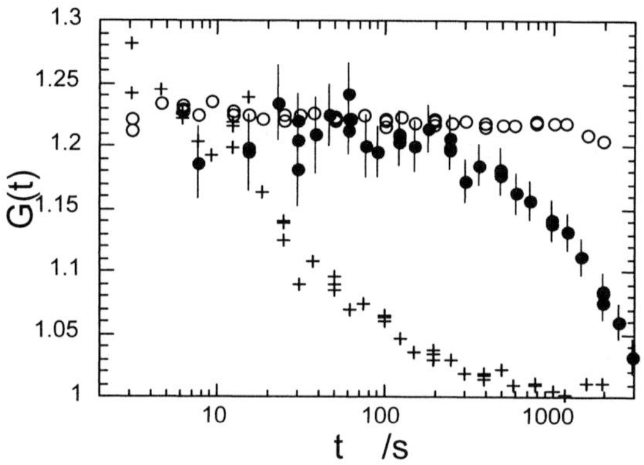

Figure 5. Coherent X-ray correlation functions $G(t)$ measured for dry aerosol powder (open circles), untreated silica dispersed in neat PDMS (+) and neat sample 2 (filled circles).

Figure 5 shows the intensity correlation functions obtained with coherent X-rays for $q = 2\text{x}10^{-3}$ Å^{-1} from three different samples. For correlator delay times t greater than one hour, the correlation function of a dry aerosol powder displays no relaxation, as is expected from a system that is completely frozen. For the two silica suspensions in PDMS, however, relaxation is clearly visible: the suspension of the untreated fumed silica relaxes in about 100 s, while the uncrosslinked sample 2 displays a much slower relaxation.

Figure 6 compares the relaxation behavior of sample 1 (as a function of q) for the data obtained both from coherent X-ray scattering and from small-angle light scattering.

In this representation, each data set approaches the same limiting curve with slope 2, corresponding to Eq. 7 in which $D = 4.6 \times 10^{-14}$ cm^{2}s^{-1} is the diffusion coefficient of large agglomerates.

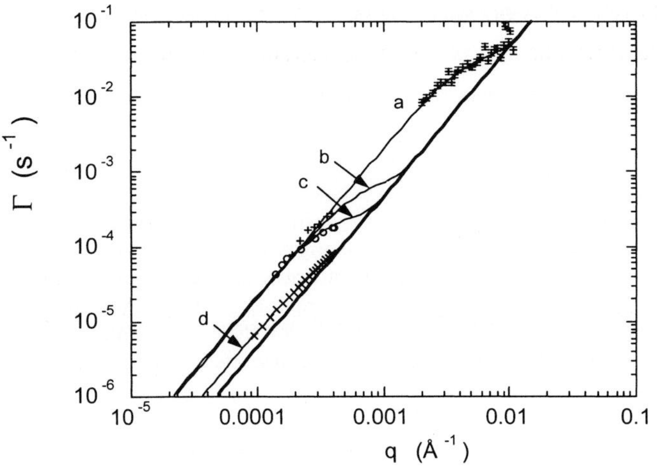

Figure 6. Relaxation rates $\Gamma(q)$ measured for sample 1. The thick straight line is defined by Eq. 7 where $D = 4.6 \times 10^{-14}$ cm^{2}s^{-1}. The upper set of data is from coherent X-ray spectroscopy with time lapse t_0 = 3000 s after stirring, the lower data sets are from small-angle dynamic light scattering with t_0 = 10^5 s , 3×10^5 s and 10^6 s, respectively. Thin lines are calculated from Eqs. 8 and 9 with D_s = 1×10^{-15} cm^{2}s^{-1}.

The thin lines drawn through the data sets are obtained from Eq. 8 for the measured time lapses t_0 = 3000 s, 10^5 s, $3{\times}10^5$ s and 10^6 s. To yield a consistent description of the experimental observations, the structural relaxation rate must obey a relation of the form

$$1/\tau_s = D_s q^2. \qquad (9)$$

where D_s = 10^{-15} cm^{2}s^{-1}. This result shows that over the whole investigated q range, the structural relaxation process is diffusion-controlled.

Conclusions

It is demonstrated that small-angle neutron scattering and contrast variation allow to make structural characterization of filled polymer systems. In this study, differences are highlighted between the structure of polymers containing fumed silica with different surface properties. It is found that the distribution of the polymer is perturbed in the vicinity of solid surfaces. Swelling these rubbers and estimation of the surface area of the filler and the surface coverage of the polymer reveals further structural details. In the case of weak interactions, the solvent causes the polymer to desorb almost completely from the filler particles. However, in fillers treated with coupling agents, thorough penetration of the polymer into the filler structure is observed.

It is shown that coherent X-ray scattering is a powerful method for investigating the relaxation processes in filled polymers. It is found that structural relaxation following an external mechanical perturbation proceeds by a diffusion-controlled mechanism.

Acknowledgements

The authors acknowledge the support of the National Institute of Standards and Technology, U.S. Department of Commerce, the Institut Laue-Langevin, Grenoble and the European Synchrotron Radiation Facility, Grenoble, France, in providing access to the small-angle neutron scattering instruments NG3 and D11 and to the small-angle X-ray instrument at BM2. We are grateful to Drs B. Hammouda and J.F. Bérar for their invaluable help in these measurements. This work is partly based upon activities supported by the National Science Foundation under agreement No. DMR-9423101.

1. P.G. de Gennes, Scaling Concepts in Polymer Physics, Cornell, Ithaca 1979.
2. J.S. Higgins and H.C. Benoît, Polymers and Neutron Scattering, Oxford Science Publications, Oxford, 1994.
3. A Computing Guide for Small Angle Neutron Scattering Experiments, R.E. Ghosh, S.U. Egelhaaf, A.R. Rennie, Institut Laue Langevin, June 2000.
4. NIST Cold Neutron Research Facility, NG3 and NG7 30 meter SANS Instruments Data Acquisition Manual, January 1999.
5. P. Lindner, F. Leclercq and P. Damay, Physica B, **291**, 152 (2000).
6. L. Cipelletti and D.A. Weitz, Rev. Sci. Instrum. **70**, 3214 (1999).
7. A.M. Hecht, E. Geissler and F. Horkay, Phys. Rev. E **59**, 1976 (1999).
8. G. Porod in Small Angle X-ray Scattering (O. Glatter and O. Kratky, Eds), Academic Press 1985.
9. G.S. Grest, K. Kremer and T.A. Witten, Macromolecules **20**, 1376 (1987).
10. F. Horkay, W. Burchard, E. Geissler and A.M. Hecht, Macromolecules **26**, 1296 (1993).
11. M. Sutton, S.E. Nagler, S.G.J. Mochrie, T. Greytak, L.E. Bermann, G. Held and G.B. Stephenson, Nature **352**, 608 (1991).
12. E. Geissler, A.-M. Hecht, C. Rochas, F. Bley, F. Livet and M. Sutton, Phys. Rev. E **62**, 8308 (2000).
13. B.J. Berne and R. Pecora, Dynamic Light Scattering, Wiley, New York 1976.

SANS from Adsorbed Polymer Layers

*Stephen King** [a], *Peter Griffiths* [b], *John Hone* [c] *and Terence Cosgrove* [c]

[a] ISIS Facility, Rutherford Appleton Laboratory, Didcot, OX11 0QX, UK
[b] Chemistry Department, Cardiff University, Cardiff, CF10 3TB, Wales, UK
[c] School of Chemistry, University of Bristol, Bristol, BS8 1TS, UK

SUMMARY: One of the areas in which small-angle neutron scattering (SANS) has had a profound impact over the last twenty years is the investigation of polymers and surfactants adsorbed at the solid/liquid and liquid/liquid interfaces. This is of course also an area of considerable commercial importance. Control of the structure of an adsorbed layer is thus a key priority of the colloid scientist. Equally important, however, is the ability to probe the arrangement of the adsorbed polymer molecules and to extract quantitative information from what might well be a complex, multicomponent system. SANS in combination with the principle of deuterium contrast variation is uniquely placed to provide this information.

Introduction

The first objective of any SANS experiment is to determine the differential scattering cross-section as a function of the scattering vector, $(d\Sigma/d\Omega)$ (Q), for a sample, since it is this quantity which contains all the information about the shape, size, and interactions between the scattering centres (here given the subscript "p" for particles) [1,2]. It may be expressed thus

$$\frac{\partial\Sigma}{\partial\Omega}(Q)=N_{\mathrm{p}}V_{\mathrm{p}}^{2}(\Delta\rho)^{2}P(Q)S(Q)+B \tag{1}$$

where N_{p} is the number concentration of particles, V_{p} is the volume of one particle, $(\Delta\rho)^2$ is the contrast ($= (\rho_{\mathrm{p}} - \rho_{\mathrm{m}})^2$ where ρ_{p} is the neutron scattering length density of the particles and ρ_{m} that of the dispersion medium), $P(Q)$ is the form (or shape) factor, $S(Q)$ is the structure factor (normally engineered to have a minimal contribution to the scattering in the experiments under discussion in this paper), and B is the background signal.

From a properly designed and executed SANS experiment, it may be possible to obtain values for the surface area per unit volume of the substrate, estimates of the thickness of the adsorbed layer, a measure of how much polymer or surfactant is present within the layer, and a description of how it is arranged - the *volume fraction profile,* $\Phi(z)$ [3]. It is $\Phi(z)$ which will be the focus of this paper [4-6].

CCC 1022-1360/00/$ 17,50+.50/0

SANS From Adsorbed Polymer Layers

Before considering methods of analysis involving $\Phi(z)$, it is worth remembering that, just as in any conventional light scattering experiment, SANS can be used to elucidate the shape of the particles in a dispersion. If that shape is not too complex - and emulsion droplets, polymer latices and pigment particles can usually be well approximated as spheres, for example - then a measure of the size of those particles can be deduced by the application of, say, the Guinier approximation [7]. If the measurement is repeated in the presence of a layer of adsorbed polymer or surfactant, then the difference between the two sizes may be used to estimate the thickness of the layer. The drawback with this approach is that SANS is relatively insensitive to the few, well-solvated, but highly-extended, polymer "tails" at the periphery of the adsorbed layer.

The "Core-Shell" Model

This approach assumes that a particle and its adsorbed layer (given the subscript "l") can be modelled as two homogeneous, concentric spheres; see Fig. 1a.

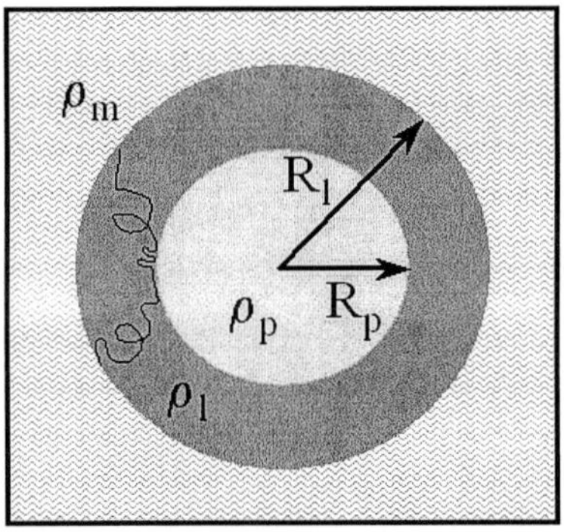

Figure 1a. A schematic representation of the "core-shell" model of a particle with an adsorbed layer. Notice how the model under-describes the actual structure in the layer. See the main text for an explanation of the symbols.

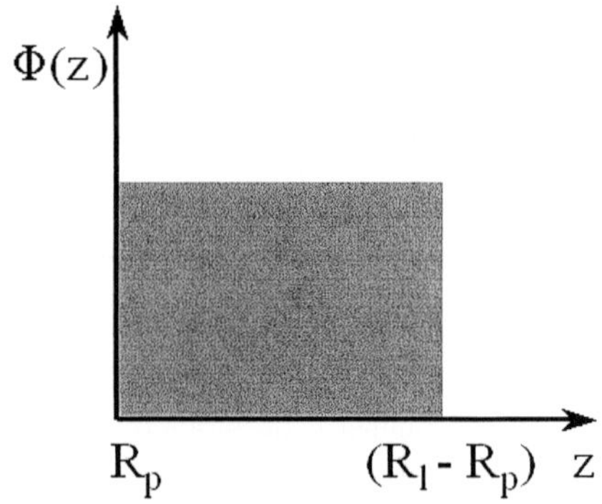

Figure 1b. The volume fraction profile corresponding to the "core-shell" model depicted in Fig. 1a.

If $P(Q,R_l)$ and $P(Q,R_p)$ are the form factors for spheres of radii R_l and R_p, respectively, then the overall scattering from such a composite particle is

$$\frac{\partial\Sigma}{\partial\Omega}(Q)=\frac{16\pi^2}{9}N_p P(Q)S(Q)+B \tag{2}$$

where

$$P(Q)=\left[(\rho_l-\rho_m)^2F(Q,R_l)^2\right]+\left[2(\rho_l-\rho_m)(\rho_p-\rho_l)F(Q,R_p)F(Q,R_l)\right]+\left[(\rho_p-\rho_l)^2F(Q,R_p)^2\right] \qquad (3)$$

and

$$F(Q,R_l)^2=R_l^6P(Q,R_l)$$

$$F(Q,R_p)^2=R_p^6P(Q,R_p)$$

Clearly, when $\rho_l = \rho_m$, that is, when the adsorbed layer is contrast-matched to the dispersion medium, only the scattering from the core of the composite particle is observed. A more useful scenario, though, occurs when $\rho_p = \rho_m$, i.e., when the core is contrast-matched to the dispersion medium. Then, the form factor describes the scattering from a spherical shell of thickness $(R_l - R_p)$ and scattering length density ρ_l:

$$\begin{aligned} P(Q) &= \left[(\rho_l-\rho_m)F(Q,R_l)-(\rho_l-\rho_m)F(Q,R_p)\right]^2 \\ &=\left[(\rho_l-\rho_m)F(Q,R_l)+(\rho_p-\rho_l)F(Q,R_p)\right]^2 \end{aligned} \qquad (4)$$

The drawback with this approach is that it assumes the scattering length density within the adsorbed layer is uniform (i.e., that $\Phi(z)$ is a step function; more commonly referred to as a "block" profile); see Fig. 1b.

As will become apparent, this assumption is not very satisfactory in the case of polymers, although the "core-shell" model does work reasonably well for adsorbed layers of short-chain surfactants. Two excellent examples of the application of this approach to such systems are the work of Ottewill and co-workers [8-12] and Cummins *et al.* [13].

The "Inversion" Technique

The great advantage of this approach is that it does not presuppose a form for $\Phi(z)$. In fact, this technique is actually capable of determining it. In addition to showing how the average density of polymer segments varies with distance, z, in the direction perpendicular to the interface, $\Phi(z)$ also provides several measures of the layer thickness, the adsorbed amount (from the integral under the curve) and the average fraction of bound segments (from the proportion of segment density within a finite distance of $z = 0$).

The inversion technique has essentially evolved in two forms; one due to Crowley [14] and co-workers from studies of both grafted [15,16] and, in particular, physically adsorbed polymers [3-5,17-19], and the other due to Auvray [20] and co-workers, principally from studies of grafted polymer layers [21-29]. The principle difference between the two forms is that Auvray and co-workers specifically consider the contribution to the scattering that arises from segment density fluctuations in the adsorbed layer, whereas Crowley and coworkers generally ignore this contribution. The latter argue that such fluctuations are generally unimportant except in a few special cases, such as in systems with dense grafted layers [21,30,31] and indeed the scattering from the fluctuation term is very weak for polymer layers that are neither very dense nor comprise high-molecular-weight polymers.

The starting point for both derivations is a variation of Eq. 1, *viz.*

$$\frac{\partial \Sigma}{\partial \Omega}(Q) = N_{\mathrm{p}} P(Q) S(Q) + B \tag{5}$$

where (cf. Eqs 2-4)

$$P(Q) = \left[(\rho_{\mathrm{p}} - \rho_{\mathrm{m}}) F_{\mathrm{p}}(Q) + (\rho_{\mathrm{l}} - \rho_{\mathrm{m}}) F_{\mathrm{l}}(Q)\right]^2 \tag{6}$$

$F_{\mathrm{p}}(Q)$ represents the *intraparticle* form factor for the core particle and $F_{\mathrm{l}}(Q)$ the *intralayer* form factor for the adsorbed polymer. The inversion technique therefore properly considers the separate contributions to the scattering from both the particle and the adsorbed layer.

If the curvature of the surface is small compared to Q, that is, if $QR_{\mathrm{p}} >> 1$, then Eq. 6 may be expanded to give

$$P(Q) = \left[(\rho_{\mathrm{p}} - \rho_{\mathrm{m}})^2 F_{\mathrm{p}}(Q)^2\right] + \left[2(\rho_{\mathrm{p}} - \rho_{\mathrm{m}})(\rho_{\mathrm{l}} - \rho_{\mathrm{m}}) F_{\mathrm{p}}(Q) F_{\mathrm{l}}(Q)\right] + \left[(\rho_{\mathrm{l}} - \rho_{\mathrm{m}})^2 F_{\mathrm{l}}(Q)^2\right] \tag{7}$$

from which it can be seen that the first term (called $I_2(Q)$ by Crowley *et al.* or $I_{\mathrm{gg}}(Q)$ by Auvray *et al.*; the subscript "g" is for grain, i.e. particle) describes the contribution to $P(Q)$ arising from the core particle, the third term (called $I_0(Q)$ by Crowley *et al.* or $I_{\mathrm{pp}}(Q)$ by Auvray *et al.*; the subscript "p" in their case is for polymer) describes the contribution from the adsorbed layer, and the second term (called $I_1(Q)$ by Crowley *et al.* or $I_{\mathrm{pg}}(Q)$ by Auvray *et al.*) is a particle-surface interference term.

Using an independent notation, where the subscript "p" signifies the particle and the subscript "l" signifies the (adsorbed) layer, these three terms may be written explicitly thus:

$$I_{\mathrm{pp}}(Q) = (\delta_{\mathrm{p}} - \delta_{\mathrm{m}})^2 \frac{2\pi A_{\mathrm{p}}}{Q^4}\left[1 + \frac{1}{Q^2 R_{\mathrm{p}}^2}\right] \tag{8}$$

$$I_{\mathrm{pl}}(Q) = (\delta_{\mathrm{p}} - \delta_{\mathrm{m}})(\delta_{\mathrm{l}} - \delta_{\mathrm{m}}) \frac{4\pi A_{\mathrm{p}}}{Q^4}\left[\int_0^t \Phi(z)\cos(Qz)\,\mathrm{d}z - QR_{\mathrm{p}} \int_0^t \Phi(z)\sin(Qz)\,\mathrm{d}z\right] \tag{9}$$

$$I_{\mathrm{ll}}(Q) = (\delta_{\mathrm{l}} - \delta_{\mathrm{m}})^2\, 2\pi A_{\mathrm{p}}\left[\frac{1}{Q^2}\left|\int_0^t \Phi(z)\exp(\mathrm{i}Qz)\,\mathrm{d}z\right|^2 + \tilde{I}_{\mathrm{ll}}\right] \tag{10}$$

where A_{p} is the surface area per unit volume of a particle (usually expressed as $S_{\mathrm{p}}/V_{\mathrm{p}}$ in the Auvray derivation, where $S_{\mathrm{p}} = 4\pi R_{\mathrm{p}}^2$ for spherical particles), t is the maximum extent of the adsorbed layer and $\tilde{I}_{ll}$ is related to the density-density correlation function that describes fluctuations in the adsorbed layer. It is the $\tilde{I}_{ll}$ term in Eq. 10 that the Crowley derivation does not treat implicitly.

The $I_{pp}(Q)$ term

Equation 8 is simply a statement of the well-known Porod law [32], the Q^{-4} dependence of the particle scattering being general result for a surface fractal.

The $I_{pl}(Q)$ term

If $QR_p >> 1$, Eq. 9 reduces to the sine Fourier transform of $\Phi(z)$, that is

$$I_{pl}(Q) \approx -(\rho_p - \rho_m)(\rho_l - \rho_m)\frac{4\pi A_p}{Q^3}\int_0 \Phi(z)\sin(Qz)\,\mathrm{d}z \tag{11}$$

Thus, by subtracting the scattering obtained at contrast match for the particles ($\rho_p = \rho_m$) from the scattering obtained "off-contrast" ($\rho_l \neq \rho_p \neq \rho_m$), multiplying the result by Q^3 and taking the Fourier transform, it is possible to obtain $\Phi(z)$. In fact, this was the method used to derive the very first volume fraction profile from SANS data [15,16]. This route to $\Phi(z)$ has both advantages and disadvantages. In its favour are that inversion only requires a Fourier transform, the procedure is not affected by the form of $\Phi(z)$, and (importantly) it is independent of the fluctuation term $\tilde{I}_{ll}$. The downside, however, is that when the particles are not at contrast match, the $I_{pp}(Q)$ term is the dominant contribution to the observed scattering. This much stronger signal can be subtracted (either by using the scattering obtained when the adsorbed layer is at contrast match or by using Porod's law), but the resulting $I_{pl}(Q)$ scattering can then suffer from a poor signal-to-noise ratio. One way to overcome this drawback is to maximize A_p and Auvray and co-workers have demonstrated that $I_{pl}(Q)$ can be obtained with good statistics when the polymers are adsorbed in dispersions of macroporous particles [25,26].

The $I_{ll}(Q)$ term

The more direct route for obtaining $\Phi(z)$ is from the scattering when the particles are at contrast match ($\rho_p = \rho_m$). Under this condition, $P(Q)$ in Eq. 7 reduces to Eq. 10. Unfortunately, because it is the *modulus* of the integral that is measured experimentally, the transformation of the first term in the brackets is complicated by the need to introduce a phase factor, $\exp(i\varphi Q)$, where φ is unknown. Fortunately the particle surface can be used as a phase reference point (since $\Phi(z) = 0$ for $z < 0$) and this allows φ to be determined using a dispersion integral relationship [14,16]. $\Phi(z)$ can then be obtained by Hilbert transformation of the $I_{\parallel}(Q)$ scattering data. This procedure always generates a possible $\Phi(z)$ though some rapidly decaying functional forms for $\Phi(z)$ can be problematical. In practice, very few functions are limited by this constraint. A physically realistic $\Phi(z)$ can also be back-transformed into reciprocal space and compared with the actual experimental scattering data, for example using a maximum entropy minimization [33].

Figure 2a shows the volume fraction profile for a block copolymer adsorbed at the liquid-liquid interface in an emulsion [34] and was obtained by inversion of the $I_{\parallel}(Q)$ scattering. The shape of this profile, whilst typical of a physically adsorbing block copolymer, is clearly very different to the block profile assumed by the "core-shell" model. When the polymer chains in the adsorbed layer are "grafted" (terminally-attached) to the interface, the volume fraction profile often displays a maximum, see Fig. 2b.

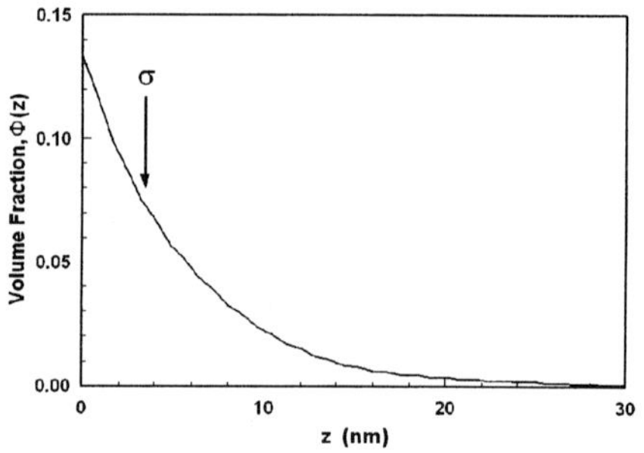

Figure 2a. The volume fraction profile for the PEO-PPO-PEO block copolymer Pluronic F127 (Poloxamer 407) adsorbed at the interface in a perfluorodecalin-in-water mesoscopic emulsion system (from Ref. 34). Also shown is the position of the second moment of the profile.

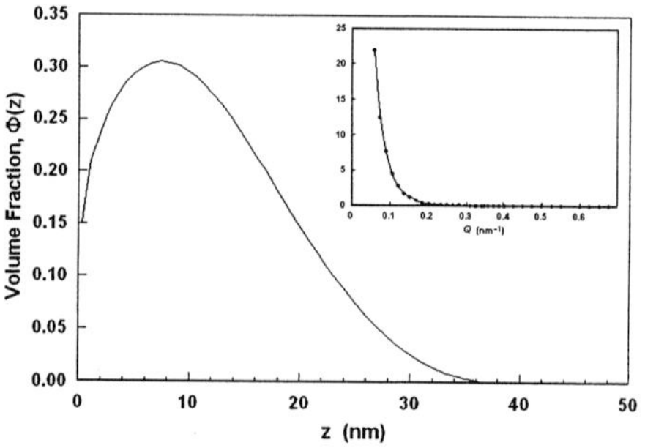

Figure 2b. The volume fraction profile that best describes the structure of the adsorbed layer formed by perdeuterated PS (M_w~24000) grafted to colloidal silica particles dispersed in DMF (from Ref. 31). Γ~6·5 mg m^{-2}. The inset shows the corresponding experimental (•) and calculated (—) scattering data.

Neglecting fluctuations, and provided $R_p >> t$ and $Q\sigma < 1$, the exponential term in Eq. 10 can be expanded to yield

$$I_{\parallel}(Q) \approx (\rho_l - \rho_m)^2 \frac{2\pi A_p M^2}{Q^2} \exp\left(-Q^2\sigma^2\right) \tag{12}$$

where the normalisation constant, M (Auvray *et al.* use the symbol γ), is given by

$$M = \int_0^t \Phi(z)\mathrm{d}z = \frac{\Gamma}{\delta} \tag{13}$$

Γ is the mass of polymer adsorbed per unit area and δ is the bulk density. σ, *the second moment about the mean,* of Φ(*z*) is defined as

$$\sigma^2 = \langle z^2 \rangle - \langle z \rangle^2 \qquad \text{where } \langle z^n \rangle = M^{-1} \int_0^t \Phi(z) z^n \mathrm{d}z \tag{14}$$

and the *root-mean-square thickness*, t_{rms}, as

$$t_{\mathrm{rms}} = \langle z^2 \rangle^{1/2} \tag{15}$$

Physically, σ provides an estimate of the average distance of the centre-of-mass of an adsorbed layer from the interface. For a "block" profile of thickness *t*, $\sigma^2 = t^2/12$.

Substituting expressions for A_p and *M* (Eq. 13) into Eq. 12, and noting that the latter expression is per unit volume, then leads to the very useful result

$$\frac{\partial \Sigma}{\partial \Omega}(Q) \approx (\rho_\mathrm{l} - \rho_\mathrm{m})^2 \left[\frac{6\pi\phi_\mathrm{p}}{Q^2} \frac{\Gamma^2}{\delta^2 R_\mathrm{p}} \exp\left(-Q^2\sigma^2\right) \right] + B \tag{16}$$

Equation 16 describes the scattering from the adsorbed polymer layer in a dilute, non-interacting system where the substrate is at contrast match. This expression can be model-fitted to the observed scattering to obtain values for Γ and σ, and in the case of the data responsible for Fig. 2a yielded values of $\Gamma = 1.9$ mg m^{-2} and $\sigma = 2.3$ nm.

An additional parameter that can be determined with a knowledge of the volume fraction profile is the average bound fraction, <*p*>, given by the expression

$$< p > = M^{-1} \int_0^l \Phi(z)\mathrm{d}z \qquad \text{where } l \approx 1 \text{ nm [4]} \tag{17}$$

This is the average fraction of segments in adsorbed polymer molecules that are bound to the interface or, in other words, the fraction of segments in "trains". As an example, for the Φ(*z*) depicted in Fig. 2a, it is found that <*p*> ≈ 14 %.

Equations 12 and 16 work well if the adsorbed layer has a reasonable extent but the limiting behaviour of Eq. 10 causes problems if the adsorbed layer is relatively thin, as is often the case with short-chain surfactants. Acknowledging this, Crowley has recently formulated an extension to the method outlined above [35]. It relies on the fact that there is a narrow range of *Q* over which the "sheet-like" scattering from the adsorbed layer dominates other contributions (specifically, the particle and particle-layer interference scattering when $Q \leq \sigma^{-1}$, and the specular Porod's law scattering when $Q \geq \pi/R_\mathrm{p}$). However, this new formalism cannot provide the volume fraction profile for the adsorbed layer.

The $\tilde{I}_{II}$ term

Auvray *et al.* have shown that it is possible to draw analogies between de Gennes scaling description of semidilute polymer solutions [36] and the structure of adsorbed layers [25,26]. They show that any concentration fluctuations in the adsorbed layer manifest themselves as a contribution to the scattering (cf. Eqs 10 and 12) of the form

$$\tilde{I}_{II} \propto \frac{1}{(aQ)^n} \tag{18}$$

where a is the size of a polymer segment. A slightly modified functional form is necessary with dense adsorbed layers, as is discussed below. Three different categories of fluctuation contribution have been identified, each characterized by its own scaling exponent n. Only two have been verified experimentally. In all cases, n is positive, and so $\tilde{I}_{II}$ decays as Q increases.

(a) Physically adsorbed layers where $R_p >> t$

When the thickness of the adsorbed layer is relatively small compared to the radius of the substrate particle, $\Phi(z)$ is predicted to decay as $z^{-4/3}$ over the range $a < z < Q^{-1}$. Under such conditions, $n = {}^4/_3$. However, experimental evidence suggests that this behaviour is not universal [3].

(b) Physically adsorbed layers where $R_p \leq t$

When the thickness of the adsorbed layer is of the same order as the radius of the substrate particle, $n = {}^5/_3$. However, the scaling of $\Phi(z)$ has not yet been established.

(c) Grafted layers

When the polymer is grafted (terminally-attached) to the surface of the substrate, or possibly where a block copolymer is adsorbed at high coverage, a "brush-like" adsorbed layer develops. This type of adsorbed layer structure can be characterized by parabolic or Gaussian volume fraction profiles (i.e., $\Phi(z) \propto z^{-2}$) and a length scale, ξ, related to the distance between anchor points. In this category of adsorbed layer the fluctuation contribution varies as

$$\tilde{I}_{II} \propto \frac{1}{1+(\xi Q)^2} \tag{19}$$

The "Scaling" Methodologies

By combining Eqs 10 and 18, a very generic description of the scattering from an adsorbed layer can be formulated, *viz.*

$$\frac{\partial \Sigma}{\partial \Omega}(Q) = \frac{U}{Q^2} + \frac{V}{Q^n} + B \tag{20}$$

where U and V are constants and $^4/_3 \leq n \leq 2$. Multiplying Eq. 20 through by Q^n, and neglecting the term in Q^{n-2} (except when $n = 2$), then gives

$$Q^n \frac{\partial \Sigma}{\partial \Omega}(Q) \approx W + B\,Q^n \tag{21}$$

A graph of Q^n $(\mathrm{d}\Sigma/\mathrm{d}\Omega)(Q)$ versus Q^n can thus be used to estimate the magnitude of the fluctuation contribution.

SANS from Adsorbed Layers in Ultra-Concentrated Dispersions

Crowley and co-workers have also developed a description of the scattering from film-forming polymer latex dispersions [37]. As the dispersion medium evaporates, capillary forces drive the polymer latex particles together. Locally-flat bilayers (Plateau borders [38]) develop where the particle surfaces meet and, in some instances, the initially spherical particles may deform into rounded polyhedra. This obviously has implications for any adsorbed layer.

This treatment relies on the fact that when $QR_p > \pi$, only specular scattering from the bilayer region is observed. This scattering is the sum of two contributions. The first of these is simply the self-scattering from the individual particle surfaces forming the bilayer (i.e., $I_{pp}(Q)$, Eq. 8). The second contribution, $\Delta I_{pp}(Q)$, which is the important one when $\phi_p > 0.74$ is due to interference effects between adjacent particle surfaces in the bilayer and is given by

$$\Delta I_{pp}(Q) = -(\rho_p - \rho_m)^2 \frac{4\pi A_B}{Q^4} \cos(Q\Delta_0) \tag{22}$$

where A_B is the surface area of a bilayer of thickness Δ_0.

Review

Small-angle neutron scattering (SANS) is one of the most successful experimental techniques for probing the number and arrangement of polymer or surfactant molecules in an adsorbed layer and can yield quantitative information that can be directly compared with other measurements.

References

1. S.M. King, in *Modern Techniques for Polymer Characterisation*, R.A. Pethrick, J.V. Dawkins, Eds., John Wiley, **1999**. Also see the references and bibliography therein.
2. R.H. Ottewill, *J. Appl. Crystallogr.* **1991**, *24*, 436.
3. G.J. Fleer, M.A. Cohen-Stuart, J.M.H.M. Scheutjens, T. Cosgrove, B. Vincent, *Polymers at Interfaces*, Chapman & Hall, London, **1993**.
4. T. Cosgrove, *J. Chem. Soc., Faraday Trans.* **1990**, *86*, 1323.
5. T. Cosgrove, S.M. King, P.C. Griffiths, in *Colloid-Polymer Interactions: From Fundamentals to Practice*, R.S. Farinato, P.L. Dubin, Eds., John Wiley, **1999**.
6. S.M. King, P.C. Griffiths, T. Cosgrove, in *Applications of Neutron Scattering to Soft Condensed Matter*, B.J. Gabrys, Ed., Chap. 4, Gordon & Breach, **2000**.
7. A. Guinier, G. Fournet, *Small Angle Scattering of X-rays*, John Wiley, **1955**.
8. I. Markovic, R.H. Ottewill, D.J. Cebula, I. Field, J.F. Marsh, *Colloid Polym. Sci.* **1984**, *262,* 648.
9. I. Markovic, R.H. Ottewill, *Colloid Polym. Sci.* **1986**, *264*, 65.
10. I. Markovic, R.H. Ottewill, *Colloid Polym. Sci.* **1986**, *264*, 454.
11. I. Markovic, R.H. Ottewill, *Colloids Surf.* **1987**, *24*, 69.
12. R.H. Ottewill, E. Sinagra, I.P. MacDonald, J.F. Marsh, R.K. Heenan, *Colloid Polym. Sci.* **1992**, *270*, 602.
13. P.G. Cummins, E. Staples, J. Penfold, *J. Phys. Chem.* **1990**, *94*, 3740.
14. T.L. Crowley, *Ph.D. Thesis*, University of Oxford, U.K. 1984.
15. T. Cosgrove, T.L. Crowley, B. Vincent, K.G. Barnett, T.F. Tadros, *Faraday Symposia of the Chemical Society*, No. 16, Royal Society of Chemistry, London 1981.
16. K.G. Barnett, *Ph.D. Thesis*, University of Bristol, U.K. 1982.
17. K.G. Barnett, T. Cosgrove, B. Vincent, A.N. Burgess, T.L. Crowley, T. King, J.D. Turner, T.F. Tadros, *Polymer* **1981**, *22*, 283.
18. T. Cosgrove, K. Ryan, *Langmuir* **1990**, *6*, 136.
19. T. Cosgrove, T.L. Crowley, K.Ryan, J.R.P. Webster, *Colloids Surf.* **1990**, *51*, 255.
20. L. Auvray, *C. R. Acad. Sci.,Vie Sci.* **1986**, *302*, 859.
21. L. Auvray, P.G. de Gennes, *Europhys. Lett.* **1986**, *2*, 647.
22. L. Auvray, J.P. Cotton, *Macromolecules* **1987**, *20*, 202.
23. L. Auvray, P. Auroy, M. Cruz, *J. Phys. I (France)* **1992**, *2*, 943.
24. P. Auroy, L. Auvray, L. Léger, *J. Phys.: Condens. Matter* **1990**, *2*, SA317.
25. P. Auroy, L. Auvray, L. Léger, *Macromolecules* **1991**, *24*, 2523.
26. P. Auroy, L. Auvray, L. Léger, *Physica A* **1991**, *172*, 269.
27. P. Auroy, L. Auvray, *Macromolecules* **1992**, *25*, 4134.
28. P. Auroy, Y. Mir, L. Auvray, *Phys. Rev. Lett.* **1992**, *69*, 93.
29. I. Caucheteux, H. Hervet, F. Rondelez, L. Auvray, J.P. Cotton, *Proc. Int. Symp. Trends Phys. Phys. Chem. Polym.* **1989**, 63.
30. T. Cosgrove, T.G. Heath, K. Ryan, T.L. Crowley, *Macromolecules* **1987**, *20*, 2879.
31. T. Cosgrove, T.G. Heath, K. Ryan, *Langmuir* **1994**, *10*, 3500.
32. G. Porod, *Kolloid-Z. Z. Polym.* **1951**, *124*, 83.
33. N.A. Finch, *Ph.D. Thesis*, University of Bristol, U.K. 1988.
34. C. Washington, S.M. King, R.K. Heenan, *J. Phys. Chem.* **1996**, *100*, 7603.
35. J. M. Gladman, T.L. Crowley, J.D. Schofield, A. Eaglesham, *Phys. Scr.* **1995**, *T57*, 146.
36. P.G. de Gennes, *Scaling Concepts in Polymer Physics*, Cornell University Press, Ithaca, New York, 1979.
37. A.R. Sanderson, T.L. Crowley, J.D. Morrison, M.D. Barry, A.J. Morton-Jones, A.R. Rennie, *Langmuir* **1992**, *8*, 2110.
38. J. Plateau, *Mem. Acad. R. Soc. Belg., 6th Ser.* **1861**, 33; G.L. Brown, *J. Polym. Sci., Part A* **1956**, *22*, 423.

A Few Remarks on Classic Problems of Scattering by Polymer Solutions and Mixtures

Henri Benoît[*1] *and Gérard Jannink*[2]

[1] Institut Charles Sadron, 6, rue Boussingault, 67083 Strasbourg, France
[2] Léon Brillouin Laboratory, C.E. Saclay, 91191 Gif-sur-Yvette, Cedex, France

SUMMARY: The problem of scattering by multicomponent systems has been the object of many papers. Here it will be shown that the scattering can be split into two terms, one corresponding to the compressibility of the system and the other depending only on fluctuations of the number of molecules of different constituents. In neutron scattering on samples far from critical conditions, the compressibility term is negligible. In the second part, a simple demonstration of the random phase approximation theorem is proposed, which makes the theorem more understandable for nonspecialists.

Introduction

Being at the end of our career, it is no longer time to discuss new theories and new fields of activity. We will try in this paper to be more modest and to present problems on which we have been working for years hoping to have made some progress. We will present a semiquantitative treatment using a minimum of mathematics to explain classic results used every day in interpretation of scattering data. Two problems will be discussed: the interpretation of scattering by multicomponent systems and the other which is usually called the random phase approximation (RPA) introduced by de Gennes[(1)]. It is used by all experimentalists mainly for the interpretation of neutron scattering data as a function of angle. These remarks do not have the pretension to replace what has been written in the classic papers. Since real understanding of a theory includes the ability to explain it in relatively simple terms, we hope that this approach is not useless.

Multicomponent systems

When I (H. B.) began to work in the field of polymers, Professor C. Sadron asked me to study the potentials of light scattering as a tool in macromolecular physical chemistry. I began to learn the subject and to teach it in the frame of initiation of chemistry students to the light scattering. Bravely I decided to demonstrate all the rules used for the interpretation of light scattering data beginning with the perfect gas where the scattering is the sum of the scattering by individual molecules and obtaining thus a very simple equation. This technique was used by

 CCC 1022-1360/00/$ 17,50+.50/0

a French professor J. Cabannes[2,3] in 1920 who used it for the evaluation of Avogadro's number and showed that in a pure liquid, there should be no scattering if its density is constant. This is difficult to prove experimentally since the liquid has to be perfectly clean, without any impurity. Even under these circumstances, as it has been demonstrated by Einstein[4], there is some scattering produced by the density fluctuations due to thermal agitation. This has been verified in benzene by Cabannes[5] only approximately since this scattering is very low. Under the best experimental conditions that could be realized at that time, the ratio between the intensities of incident light and scattered light is of the order of 10^4. This was very difficult to measure with the techniques available (the light source was the sun and the detector a photographic plate which is far from being a linear detector). It is well known that a dilute solution is equivalent to the perfect gas. Therefore, since Avogadro's number can be calculated for a gas with known molar mass, one can also, knowing Avogadro's number, measure the molar mass of an unknown polymer. This is the general idea which was the basis of the work of Puttees and Brosteaux[6] followed by Debye[7]. When you teach students on this problem, you want to make the subject complete and to work with both a pure liquid and a dilute solution. This leads to the question: what happens when you have both phenomena together. In an old textbook, you could find the following statement: it is evident that there are no correlations between density fluctuations (responsible for the scattering by pure liquid and depending on the compressibility) and concentration fluctuations making possible the measurements of molar masses. The author was optimistic since, so far, it has been impossible to find how he did reach this conclusion by simple arguments. This was an open question which has been solved only in the 1950's. Since that time many papers have been devoted to the problem and its generalization to scattering by multicomponent systems. The best-known of these papers have been written by Stockmayer[8] and by Kirkwood et al.[9], followed by many others[10-12]. The interest in this problem was renewed during the last decade for the following reason. As it is well known, the scattered intensity makes it possible to determine not only the molar mass but also the second virial coefficient or the covolume. This leads first to an estimation of the famous χ parameter of Flory[13] (which is often called the black box where one puts all that is not clearly understood). Different methods leading to different values of χ are still a puzzling problem. It was suggested that the discrepancies between the values obtained by different methods were due to incorrect treatment of the scattering data and to the influence of density fluctuations on composition fluctuations. Now, we are going back to the general problem of multicomponent compressible systems to show again, without complex calculations, that there is complete independence between composition and density fluctuations, justifying a posteriori the work of Stockmayer, which shows that one can completely forget about the effect of density fluctuations in the classic studies on light and neutron scattering by polymers.

The equation

Let us assume that the scattering system consists of n scattering objects (which will be called molecules), having b as scattering length per unit volume.

$$I(q) = \sum_{i=1}^{n} \sum_{j=1}^{n} b_i b_j \exp(-\mathrm{i} q \mathbf{r}_{ij}) \qquad (1)$$

calling $\mathbf{q}$ the scattering vector ($q = \frac{4\pi}{\lambda} \sin(\theta/2)$) and $\mathbf{r}_{ij}$ the vector joining the scattering points i and j having neutron scattering length b_i and b_j (or a quantity related to the refractive index increment in light scattering or to the electron density in X-ray scattering).

Since we are interested in the scattering at zero angle ($\theta = 0$), this relation is simplified (the exponential term being equal to unity) and one can collect the terms corresponding to i and j, either on the same species of molecule or on different species of molecules[14,15]:

$$I(0) = \sum_{i,j} \sum b_i b_j \overline{\int \mathrm{d}^3\mathbf{r}\, \delta(\mathbf{r} - \mathbf{r}_i - \mathbf{r}_j)} \qquad (2)$$

where the double sum is extended to the p^2 pairs of molecules of different nature and the integral is extended to all the pairs of monomeric units of the molecules of type i and j in the scattering volume. We have written Δn_i and Δn_j since it is well known that it is only the fluctuations in n, the number of molecules, that scatter light. Let us now call S_{ij} the term

$$S_{ij} = \overline{\Delta n_i \Delta n_j} \qquad (3)$$

Equation (2) becomes

$$I(0) = \sum_{r,s}^{p} b_r b_s S_{rs} \qquad (3')$$

This is the starting equation which will allow us to explicitly express the result.

The first remark is that, due to symmetry of the system, $S_{ij} = S_{ji}$. The S_{ij}'s can be collected in a p-by-p matrix and the problem is to evaluate this matrix by the classic techniques of statistical mechanics in order to find a relation between the properties of the system and the scattered intensity.

This is a classic exercise which can be found in practically all textbooks and we will just give here the result as a function of chemical potential μ_i

$$\overline{\Delta n_i^2} = k_B T \left(\frac{\partial \overline{n_i}}{\partial \mu_i} \right)_{V,T,k \neq 1} \tag{4}$$

$$\overline{\Delta n_i \Delta n_j} = k_B T \left(\frac{\partial \overline{n_i}}{\partial \mu_j} \right)_{V,T,k \neq i,j}$$

where k_B is the Boltzmann constant and $\frac{\partial \overline{n_i}}{\partial \mu_j}$ the derivative of the average number of particles of species i ($\overline{n_i}$) per unit volume. The second equation is of the same type but corresponds to a cross-term. In order to solve this problem, one has first to transform the quantities $\left(\frac{\partial \overline{n_i}}{\partial \mu_i} \right)$ and $\left(\frac{\partial \overline{n_i}}{\partial \mu_j} \right)$ into the reciprocals, i.e., $\frac{\partial \mu_i}{\partial \overline{n_i}}$ and $\frac{\partial \mu_j}{\partial \overline{n_i}}$. The main reason for the transformation is that the latter expressions are more familiar and can be handled more easily. But even when put into this form, the quantities are not very useful since they require to work at constant V and T. Changing the number of molecules without changing the volume is practically impossible if the system is incompressible. One has therefore to make a second step and replace $\left(\frac{\partial \mu_i}{\partial \overline{n_i}} \right)_{V,T}$ with $\left(\frac{\partial \mu_i}{\partial \overline{n_i}} \right)_{P,T}$. The first step is a classic procedure for functions with more than one variable and is symbolized by the equation which allows to evaluate the $\frac{\partial \overline{n_i}}{\partial \mu_i}$ matrix as a function of the matrix $\frac{\partial \mu_i}{\partial \overline{n_i}}$. One has just to write

$$\mathbf{S} = \left[\frac{\partial \overline{n_i}}{\partial \mu_i} \right] = \left[\left(\frac{\partial \mu_i}{\partial \overline{n_i}} \right) \right]^{-1} \tag{5}$$

In other words, the matrix **S** of the inverse is equal to the inverse of the matrix $\mathbf{S}^{-1}$.

The second step is to go from constant volume and temperature to constant pressure and temperature. This is also a classic result of thermodynamics[16]:

$$\left(\frac{\partial \mu_i}{\partial \overline{n_j}} \right)_{V,n,T} = \left(\frac{\partial \mu_i}{\partial \overline{n_j}} \right)_{P,n,T} + \frac{v_i v_j}{\chi_T V} \tag{6}$$

where v_i and v_j are the molecular volumes of constituents i and j, V is the total volume of the scattering system and χ_T the isothermal compressibility of the system,

$$\chi_T = \frac{-1}{V} \left(\frac{dV}{dP} \right)_T$$

We have now all the ingredients to write the general formula allowing to evaluate $I(0)$ If one writes for simplification

$$a_{ij} = \left(\frac{\partial \mu_i}{\partial n_j}\right)_{P,n,T} \tag{7}$$

one obtains for a three-component system

$$\mathbf{S} = \begin{pmatrix} a_{11} + \frac{v_1^2}{V\chi_T} & a_{12} + \frac{v_1 v_2}{V\chi_T} & a_{13} + \frac{v_1 v_3}{V\chi_T} \\ a_{21} + \frac{v_2 v_1}{V\chi_T} & a_{22} + \frac{v_2^2}{V\chi_T} & a_{23} + \frac{v_2 v_3}{V\chi_T} \\ a_{31} + \frac{v_3 v_1}{V\chi_T} & a_{32} + \frac{v_3 v_2}{V\chi_T} & a_{33} + \frac{v_3^2}{V\chi_T} \end{pmatrix}^{-1} \tag{8}$$

The calculation of the S_{ij} seems easy but, even for this simple three-component system, it is rather tedious. The classic method to evaluate the inverse matrix of a given matrix is to write the adjunct matrix and divide it by the determinant of the matrix.

Calculation of the denominator

Since each term is the sum of two terms, one can write the determinant of the matrix **S** as the sum of p^2 determinants and to evaluate them separately. Let us follow this method and write the first determinant of the denominator:

$$\text{Det} = \begin{pmatrix} a_{11} & a_{12} & a_{13} \\ a_{21} & a_{22} & a_{23} \\ a_{31} & a_{32} & a_{33} \end{pmatrix} \tag{9}$$

This determinant is zero since, at constant pressure and temperature, the Gibbs-Duhem equation can be written as $\Sigma n_i \mathrm{d}\mu_i = 0$ or, after derivation with respect to n_j and using our notations,

$$\sum_{j-1}^{p} n_i \left(\frac{\partial \mu_i}{\partial n_j}\right)_{P,n,T} = \sum_{j=1}^{p} n_i a_{ij} = 0 \tag{10}$$

One can easily see that multiplying the first column by N_1, the second by N_2 and the third by N_3 and adding the three columns gives a line of zero.

In order to study systematically these sums, let us consider the three determinants having one column of v and the other of a_{ij}. They have no reason to be zero and one obtains, after factorization of $V\chi_T$

$$\chi_T V\,\mathrm{Det} = v_1\begin{vmatrix} v_1 & a_{12} & a_{13} \\ v_2 & a_{22} & a_{23} \\ v_3 & a_{23} & a_{33} \end{vmatrix} + v_2\begin{vmatrix} a_{11} & v_1 & a_{13} \\ a_{21} & v_2 & a_{23} \\ a_{31} & v_3 & a_{33} \end{vmatrix} + v_3\begin{vmatrix} a_{11} & a_{12} & v_1 \\ a_{21} & a_{22} & v_2 \\ a_{31} & a_{32} & v_3 \end{vmatrix} \qquad (11)$$

Finally, the determinants having two or more columns in v have, after factorization, two identical columns; they are equal to zero. We are therefore left with Eq. (11) to express the denominator. This shows that the inverse of the compressibility (χ_T^{-1}) factorizes the denominator of the S_{ij} and we can write that the denominator Det of all S_{ij} is of the form

$$\mathrm{Det} = \frac{1}{V\chi_T}\alpha \qquad (12)$$

where α is given in the case of $p = 3$ by Eq. (11).

It is evident that this result can be generalized to any value of p.

Calculation of the numerator

The determination of **S** requires the evaluation of the adjunct matrix of the matrix of Eq. (8). The method is well known: one deletes in the determinant of the matrix of Eq. (8) column i and row j obtaining a new determinant C_{ij} which is proportional to the coefficient S_{ij} and has the form

$$C_{ij} = \frac{1}{V\chi_T}C_{ij}^1 + C_{ij}^2 \qquad (13)$$

It is similar to the denominator of Eq. (11). The main difference results from the fact that the component of the determinant with no v columns is not zero since one cannot use the Gibbs-Duhem equation. Therefore we are left with two terms, one independent of χ_T and the other proportional to $(\chi_T V)^{-1}$. This allows to write for the numerator C_{11} of $\mathbf{S}_1$ for instance

$$C_{11} = \begin{vmatrix} a_{22} + \dfrac{v_2^2}{\chi_T} & a_{12} + \dfrac{v_2 v_1}{\chi_T} \\ a_{12} + \dfrac{v_2 v_1}{\chi_T} & a_{33} + \dfrac{v_3^2}{\chi_T} \end{vmatrix} \qquad (14)$$

$$C_{11} = a_{22}a_{33} - a_{23}^2 + \frac{1}{V\chi_T}\left[v_2^2 a_{33} - 2v_2 v_3 a_{23} + v_3^2 a_{22}\right] \qquad (14')$$

confirming the form of Eq. (13), where the C_{ij}^1 and C_{ij}^2 have a clear meaning. One can write, regardless of the value of p,

$$S_{ij} = \frac{V\chi_T C_{ij}^2 + C_{ij}}{\alpha} = S_{ij}^1 + S_{ij}^2 \qquad (15)$$

calling S_{ij}^2 the contribution of compressibility to scattering and S_{ij}^1 the contribution of composition fluctuations.

Density fluctuations

This result, which was obtained by different authors(8,10,11), is important since it is in contradiction with recent calculations. It seems that the method that has been used is correct and cannot be criticized since it does not use any new variable and its derivation is straightforward. The result proves unambiguously that, when dealing with compressible systems, one has just to add to the term corresponding to composition fluctuations a term taking into account the compressibility, which obeys the classic relation

$$I(0) = \left(\sum n_i v_i b_i\right)^2 \frac{\chi_T k_B T}{V} = V\bar{b}^2 \chi_T k_B T \tag{16}$$

after introducing the average scattering length $\bar{b}$

$$\bar{b} = \frac{1}{V} \sum n_i v_i b_i \tag{16'}$$

Note that the vector **v**, defined by its components v_i, is a null space of $\mathbf{S_1}$ for compressible systems and a null space of **S** for incompressible systems.

A recent work(17) has shown that the contribution of Eq. (16) is practically negligible in all actual experiments on neutron scattering by polymers. It can be remarked that if one subtracts from the scattering of the sample the scattering by a system without structure in order to eliminate the incoherent scattering, one subtracts also the effect of compressibility. This means that this correction is in some cases already made.

Composition fluctuations

Now we are left with the second part of the S_{ij} which corresponds to the classic fluctuations of composition. If, starting from Eq. (8), all the terms containing χ_T are eliminated, one is left with a matrix which, because of the Gibbs-Duhem equation, is singular and cannot be used to evaluate S_{ij}^1. One could take Eq. (13) to evaluate these terms but this is very cumbersome and not easy to read. The problem is that we have too many S_{ij} which are not independent.

If, for instance, $p = 2$, we have three different S_{ij} but need only two parameters for the evaluation of $I(0)$ - the bulk compressibility and the scattering by the solution. The extra parameters must therefore be eliminated. The choice of variables is a difficult problem. Two decades ago, des Cloizeaux and Jannink(11), using linear algebra methods, have shown that there are two ways to perform this elimination. The first is to use the contrast between the scattering species and the average value of the scattering length. Having been used by Williams et al.(18), this is especially useful in the studies on copolymers. The other, the most frequently used method consists in treating one of the constituents as a solvent and to express the result as a function of the contrast between the scatterer and the solvent. More generally, unless there is a large compressibility contribution, the simplest way to eliminate the corresponding term is to utilize the classic methods of neglecting the contribution of compressibility except when one is interested in the effect of pressure on the scattering (in the latter case, the influence of pressure on $(\frac{\partial \mu_i}{\partial n_j})_{P,T}$ must be taken into account). Summarizing the results of the preceding calculations, we write for the scattering matrix **S**

$$\mathbf{S}^{-1} = \mathbf{A} + \frac{\mathbf{v}\mathbf{v}^{\mathrm{Tr}}}{\chi_T V} \qquad \text{with the property } \mathbf{An} = 0,$$

or

$$\mathbf{S} = \mathbf{B} + \mathrm{nn}^{\mathrm{Tr}} \frac{\chi_T}{V} \qquad \text{with the property } \mathbf{Bv} = 0$$

where **B** stands for $\mathbf{S}^{-1}$ in Eq. (15). The system displays symmetry, the condition **Bv** = 0 being conjugated to the Gibbs-Duhem relation **An** = 0.

The method we have used is in some respect pedestrian. Our colleague Prof. Akcasu(19) reminded us of a formula established by Sherman and Morrison(20) which allows to study this problem with more efficient tools. If a matrix **M** can take the form

$$\mathbf{M}^{-1} = [\mathbf{A} + \frac{1}{\Gamma}\,\mathbf{v}\,\mathbf{v}^{\mathrm{Tr}}] \tag{17}$$

where **v** is a vector and $\mathbf{v}^{\mathrm{Tr}}$ its transform, and if the matrix **A** is not singular, its inverse can be written as

$$\mathbf{M} = \frac{\mathbf{A}^{-1} + \mathbf{v}\mathbf{v}^{\mathbf{Tr}}\mathbf{A}^{-1}}{\Gamma + \mathbf{v}^{\mathbf{Tr}}\mathbf{A}^{-1}\mathbf{v}} \tag{18}$$

Unfortunately, our matrix **A** (which is the matrix a_{ij}) is singular and this formula cannot be used without modifications.

The random phase approximation

So far we have only discussed the behavior at zero angle but, even though it is an interesting problem, a majority of the experiments require the interpretation of the scattering in a large domain of the scattering vector **q**.

The exploration of this domain was limited to dilute solutions because there was no theory (besides the second-order in concentration) to explain the results. The situation changed dramatically when P G. de Gennes published[21] in 1970 the method he called the random phase approximation (RPA). This method gives correct results for concentrated solutions and for a large domain of q allowing to demonstrate at low q the results which we have just discussed. Unfortunately, its demonstration is rather complex and there are not many experimentalists which are ready to read the de Gennes paper. The idea is that, since the result is simple[15,21], a simple demonstration to prove it should exist and it is worth spending some time to try to find it. In this last part, it will be shown that this demonstration does exist and one can hope that it is easy to understand.

Let us consider a system made by an incompressible mixture of two polymers. There are N_A and N_B molecules of each species. We cut them into units of the same volume, their number being n_a for polymer A and n_b for B. Their scattering length per unit volume are $\tilde{a}$ and $\tilde{b}$. Following the preceding discussion, we can rewrite the equation for the scattering as

$$I(0) = \tilde{a}^2 \sum_i^{n_a} \sum_j^{n_a} \overline{\exp(-iqr_{ij})} + \tilde{b}^2 \sum_i^{n_b} \sum_j^{n_b} \overline{\exp(-iqr_{ij})} + 2\tilde{a}\tilde{b} \sum_i^{n_a} \sum_j^{n_b} \overline{\exp(-iqr_{ij})}$$

$$= \tilde{a}^2 S_{aa} + 2\tilde{a}\tilde{b}\, S_{ab} + \tilde{b}^2\, S_{bb} \tag{19}$$

where the average has to be taken for all pairs of monomers a-a, b-b, a-b. Consider the term where the average is taken on the monomers a. If the points i_a and j_b are on the same molecule, there are n_a^2 couples and the average value of the interference term will be called $P_a(q)$. Therefore the part corresponding to the couple a-a in the same molecule can be written as $N_A n_a^2 P_a(q)$. But one has also to consider the interferences between different molecules of the same type and this leads to $N_A^2 n_a^2 Q_{aa}(q)$ where $Q_{aa}(q)$ is the average value taken on one pair of molecules. This leads to

$$\begin{aligned} S_{aa}(q) &= N_A n_a^2 P_a(q) + N_A^2 n_a^2 Q_{aa}(q) \\ S_{bb}(q) &= N_B n_b^2 P_b(q) + N_B^2 n_b^2 Q_{bb}(q) \\ S_{ab}(q) &= N_A N_B n_a n_b Q_{ab}(q) \end{aligned} \tag{20}$$

since there are no molecules having simultaneously a and b monomers. As we have seen, the S_{ij} are not independent. It is easy to show that we have the two relations:

$$\begin{aligned} S_{aa} + S_{bb} &= 0 \\ S_{bb} + S_{ab} &= 0 \end{aligned} \tag{21}$$

which are a generalization of the Gibbs-Duhem equation (Eq. 10) for $q \neq 0$ and lead to

$$\frac{I(q)}{\left(\tilde{a}-\tilde{b}\right)^2} = N_{\mathrm{A}} n_{\mathrm{a}}^2 P_{\mathrm{a}}(q) + N_{\mathrm{A}}^2 n_{\mathrm{a}}^2 Q_{\mathrm{aa}}(q) + N_{\mathrm{B}} n_{\mathrm{b}}^2 P_{\mathrm{b}}(q) + N_{\mathrm{B}}^2 n_{\mathrm{b}}^2 Q_{\mathrm{bb}}(q) + N_{\mathrm{A}} N_{\mathrm{B}} n_{\mathrm{a}} n_{\mathrm{b}} Q_{\mathrm{ab}}(q) \qquad (22)$$

We still need another equation to be able to write the result as a function of the terms P_{b}, P_{ab}, Q_{aa}, Q_{bb}, and Q_{ab}. For this purpose, let us consider the diagram of Fig. 1.

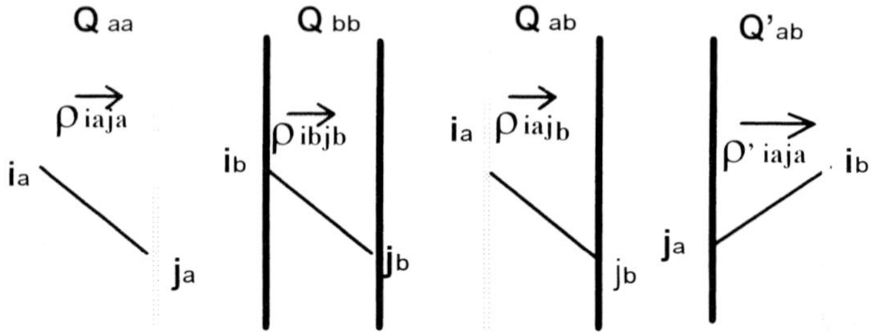

Figure 1 Schematic representation of the interactions between the two species of molecules. The vector joining the two interacting points is called ρ_{ij}. Type A molecules are represented by a gray line, type B by a black line.

This figure is a schematic representation of the terms Q_{aa}, Q_{bb} and Q_{ab} ($= Q'_{\mathrm{ab}}$). The vertical lines represent the molecules of type A and B. From this diagram one can easily write:

$$\begin{aligned} Q_{\mathrm{aa}} &= \overline{\exp(-iq\rho_{i_{\mathrm{a}}j_{\mathrm{a}}})} \\ Q_{\mathrm{bb}} &= \overline{\exp(-iq\rho_{i_{\mathrm{b}}j_{\mathrm{b}}})} \\ Q_{\mathrm{ab}} &= Q'_{\mathrm{ab}} = \overline{\exp(-iq\rho_{i_{\mathrm{a}}j_{\mathrm{b}}})} = \overline{\exp(-iq\rho_{i_{\mathrm{b}}j_{\mathrm{a}}})} \end{aligned} \qquad (22')$$

One writes now the quantities $Q_{\mathrm{aa}}Q_{\mathrm{bb}}$ and $Q_{\mathrm{ab}}Q'_{\mathrm{ab}}$, using a common arbitrary origin 0 and this allows to write $\rho_{ij} = r_j - r_i$.

$$\begin{aligned} Q_{\mathrm{aa}}Q_{\mathrm{bb}} &= \overline{\exp(-iqr_{j_{\mathrm{a}}})\exp(-iqr_{i_{\mathrm{a}}})\exp(-iqr_{j_{\mathrm{b}}})\exp(-iqr_{i_{\mathrm{b}}})} \\ Q_{\mathrm{ab}}Q'_{\mathrm{ab}} &= \overline{\exp(-iqr_{j_{\mathrm{a}}})\exp(-iqr_{i_{\mathrm{a}}})} \end{aligned} \qquad (23)$$

One sees clearly that they are identical if one neglects the average signs; however, this cannot be done without a hypothesis. The random phase approximation introduced by de Gennes[21] in 1970 is just this hypothesis. It assumes that one can write

$$Q_{aa}Q_{bb} = Q_{ab}^2 \tag{24}$$

Assuming this relation to be correct, it is easy to eliminate the Q functions and one obtains

$$\frac{(\tilde{a}-\tilde{b})^2}{I(q)} = \frac{1}{N_A n_a^2 P_a(q)} + \frac{1}{N_B n_b^2 P_b(q)} \tag{25}$$

which is the classic formula for a system with two components. This result has been obtained assuming no interaction between molecules A and B. It can be shown that the χ parameter of Flory can be easily introduced leading to

$$\frac{(\tilde{a}-\tilde{b})^2}{I(q)} = \frac{1}{N_A n_a^2 P_a(q)} + \frac{1}{N_B n_b^2 P_b(q)} - 2\chi \tag{26}$$

The procedure we have used can be extended to any systems containing copolymers and a large number of components. One has only to assume that each pair of constituents satisfies Eq. (24) but the calculations become rapidly tedious.

Assuming that component B is a small molecule ($n_B = P_B = 1$) and that the volume of the A is equal to the volume of solvent molecule, one obtains, using the volume fraction $\varphi = \frac{N_A n_a}{N_A n_a + N_B}$ and introducing the scattering per monomer unit $i(q)$, $i(q) = \frac{I(q)}{N_A n_a + N_B}$ and thus recovers the classic result:

$$\frac{(\tilde{a}-\tilde{b})^2}{i(q)} = \frac{1}{N_A P(q)} + \frac{1}{1-\varphi} - 2\chi \tag{27}$$

This gives a simple explanation to the random phase approximation. It can be reminded that it is a very good approximation, giving at zero angle the rigorous formula obtained by thermodynamics, which is also rigorous when the two polymers are identical. It can be extrapolated to an arbitrary number of constituents. Efforts have been made to improve it but, so far, the results have kept the same form, the improvements consisting mainly in a more rigorous evaluation of the excluded volume parameter $v_s = \frac{1}{\varphi_s} - 2\chi$.

References

(1) P.G. de Gennes, *Scaling Concepts in Polymer Physics*, Cornell University Press, Ithaca (NY) 1979.
(2) J. Cabannes, C. R. Acad. Sci. (Paris) **171**, 852 (1920).
(3) J. Cabannes, *La diffusion moléculaire de la lumière*, Les Presses universities de France, Paris 1939.
(4) A. Einstein, Ann. Phys. **33**, 1275 (1910).
(5) J. Cabanas and P. Dare, J. Phys. **8**, 321 (1927).
(6) P. Puttees and J. Brosteaux, Trans. Faraday Soc. **31**, 1334 (1945).
(7) P. Debye, J. Appl. Phys. **15**, 338 (1944).

(8) W.H. Stockmayer, J. Chem. Phys. **18**, 58 (1950).
(9) J. G. Kirkwood and R.J. Goldberg, J. Chem. Phys. **18**, 54 (1950).
(10) T.L. Hill, *Statistical Mechanics*, McGraw Hill and Dover Publications, Mineola, New.York 1987, pp 113-121.
(11) J. des Cloizeaux and G. Jannink, Physica A **102**, 120 (1980).
(12) H. Benoît, Polymer **32**, 579 (1991).
(13) P.J. Flory, *Principles of Polymer Chemistry*, Cornell University Press, Ithaca (NY) 1953.
(14) J.S. Higgins and H. Benoît, *Polymers and Neutron Scattering*, Oxford University Press, London 1994.
(15) H. Benoît and G. Jannink, Eur. Phys. J. E **3**, 281 (2000).
(16) H. Yamakawa, *Modern Theory of Polymer Solutions*, Harper and Row, New York 1971.
(17) J.K. Taylor, P.G. Debenetti, W.W.Grasseley and S. Kumar, Macromolecules **29**, 764 (1996).
(18) C.E. Williams, M. Nierlich, J.P. Cotton, G. Jannink, F. Boué, M. Daoud et al., J. Polym. Sci., Polym. Lett. **17**, 379 (1979).
(19) Z. Akcasu, private communication.
(20) G.H. Golub and C.F. Van Loan, *Matrix Computations,* John Hopkins Press, 1983.
(21) P.G. de Gennes, J. Phys. **31**, 235 (1970).

Macromol. Symp. ***190***, *55–64 (2002)*

Linear, Branched and Hyperbranched Macromolecules in Dilute Solution

Fabio Ganazzoli

Dipartimento di Chimica, Politecnico di Milano
via L. Mancinelli 7, 20131 Milano, Italy

SUMMARY: We report some theoretical results obtained for star polymers and dendrimers under different solvent conditions using the Gaussian self-consistent approach. The focus is mainly on the model assumptions and approximations, and on the results. We discuss the Θ state in terms of both the Θ temperature and the molecular size, and then we consider the good-solvent conditions also through the intramolecular dynamics.

Introduction

In recent years branched and hyperbranched polymers such as stars and dendrimers (see Scheme 1) were synthesized with an essentially complete control over the molecular architecture, reviving the theoretical interest in such systems.[1-3] As a result of both theory and experiment, some simple models and general results had to be revised.

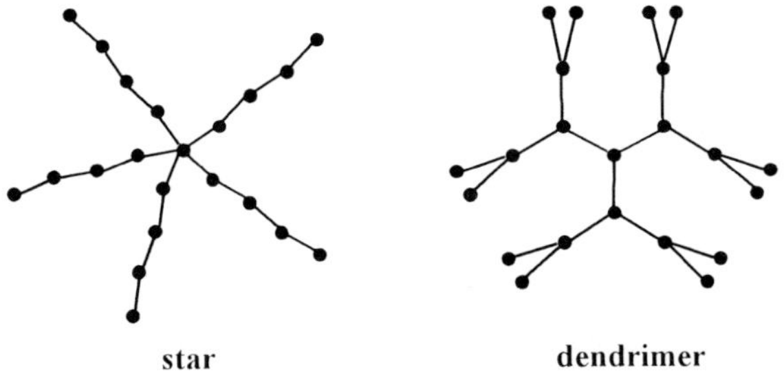

Scheme 1

We report here theoretical results obtained in our group for branched polymers using a coarse-grained model to study equilibrium and dynamics in a unified approach. A basic assumption is the Gaussian distribution of the distances among the beads. The molecular conformation is obtained from self-consistent free-energy minimization, which provides the name of Gaussian self-consistent (GSC) approach.[4] The method correctly accounts for the relevant degrees of freedom within the configurational entropy, and includes two- and three-body interactions. In dynamics, we adopt the Langevin equation taking into account the equilibrium results.

 CCC 1022-1360/00/$ 17,50+.50/0

Stars and dendrimers. Mathematical background

We consider regular stars formed by f equal arms, each one comprising N/f beads, while dendrimers are formed by f arms of order m (the dendrons) for a total of N bonds in both cases. The dendrimer topology can be described in terms of a sequence of concentric layers, the outmost one defining the generation g, numbered from 0 onward, with the number of bonds that increases exponentially with generation.

In linear and star polymers, the mean-square radius of gyration $\langle R_S^2 \rangle$ increases with N through a simple power law[5,6]

$$\langle R_S^2 \rangle \propto N^{2\nu} \tag{1}$$

where the Flory exponent ν is 0.5 for the random-walk (RW) model and in the Θ state, while in a good solvent $\nu \cong 0.6$ (with f-dependent prefactors). Therefore, we introduce the topological ratio[1] $g_S = \langle R_S^2 \rangle_{\text{star}} / \langle R_S^2 \rangle_{\text{lin}}$; for the RW model, $g_S^{\text{RW}} = (3f-2)/f^2$, showing the increasing compactness of the star compared to linear chains.

Cascade molecules[7] do follow Eq. (1) with a ν exponent that can be as low as 0.25, whereas dendrimers belong to a different universality class. In fact, in the RW model they follow a different behavior[8,9]

$$\langle R_S^2 \rangle \propto (g+1) \propto \ln N \tag{2}$$

Therefore, they form an extreme case of regularly branched molecules whose size increases logarithmically with molar mass, more slowly than the weakest power law.

Using Eq. (1), the average density within a molecule is $d \propto N / \langle R_S^2 \rangle^{3/2} \propto N^{1-3\nu}$. The exponent $\nu = 1/3$ corresponds to a constant density, while for $\nu > 1/3$ the average density goes to zero with increasing molar mass (as in linear and star polymers), and for $\nu < 1/3$ it diverges (as in cascade molecules and *a fortiori* in dendrimers). Because of the nonphysical nature of a diverging density, we conclude that the RW model can never apply to such molecules, and that a limiting generation must eventually be achieved.

We obtain the molecular conformation by self-consistent free-energy minimization[4,9,10] under the assumption of a Gaussian distribution of the distances among the beads. With reference to the RW model, the free energy of the single molecule is written as $\mathcal{A} = \mathcal{A}_{\text{el}} + \mathcal{A}_{\text{intra}}$. $\mathcal{A}_{\text{el}}$ accounts for the configurational entropy, which determines the molecular elasticity (hence its name), while the intramolecular contribution includes the long-range two- and three-body interactions, $\mathcal{A}_{intra} = \mathcal{A}_2 + \mathcal{A}_3$. $\mathcal{A}_2$ and $\mathcal{A}_3$ are given by the probability densities of contact among all the beads pairs and triplets, the coefficients being the irreducible binary and ternary cluster integrals of the beads β_2 and β_3. The former parameter is temperature-dependent and vanishes at $T = \Theta_{bead}$. In a good solvent, $\beta_2 > 0$ and the two-body interactions are repulsive, whereas in a poor solvent they are attractive and $\beta_2 < 0$. The Θ state, corresponding to vanishing interactions *between two molecules*, is more complex[4,6,11] and does not simply reduce to $\beta_2 = 0$. As to β_3, it is always positive and temperature-independent: these interactions are repulsive, since they correct the extra contribution of simultaneous two-body interactions among three beads, and account for the lesser room available for the third bead when the other two are in contact.

The variational unknowns describing the molecular conformation can be the average scalar products among the bond vectors. Their number, however, is too large to study the asymptotic behavior, even exploiting the molecular symmetry. Therefore, in stars we optimize the eigenvalues of the matrix of the average scalar products, assuming the eigenvectors to be known by symmetry.[4,10,11] The bond vectors are thus transformed into the configurational normal modes, and the eigenvalues are their strain ratios with respect to the RW conformation. This procedure can be followed provided the molecule is conformationally homogeneous and there are no long-range correlations. These requirements[4] are not met: (i) with polymers in a good solvent and (ii) with copolymers, because of the unlike monomers. In the latter case we must adopt the full procedure, while in the former one, the normal modes can be used for linear and star polymers in the crossover regime, but not for dendrimers due to the large correlations imposed by steric constraints.[9,10]

The dynamics is studied following Zimm's approach with a bead-and-spring model, each bond being replaced with a harmonic spring, using the Langevin equation.[8,9,12] The equilibrium results previously obtained affect both the intramolecular elasticity and the hydrodynamic interaction, accounted for in the preaveraging approximation. The dynamic problem reduces to solving an eigenvalue equation, which yields a zero eigenvalue related to the overall diffusion and the relaxation times.

In the present paper, we only report the dynamic structure factor $S(q,t)$, measured in quasi-elastic scattering experiments. Here, q is the modulus of the scattering vector: $q = |\mathbf{q}| = 4\pi \sin(\vartheta/2)/\lambda$, ϑ being the scattering angle and λ the radiation wavelength. In the Gaussian approximation, $S(q,t)$ can be written as[4,12]

$$S(q,t) = \exp\left(-q^2Dt\right)\frac{1}{(N+1)^2}\sum_{i\neq j}\exp\left[-\frac{q^2}{6}\left\langle r_{ij}^2(t)\right\rangle\right] \tag{3}$$

Here we separate the motion of the whole molecule, with diffusion coefficient D, from the intramolecular dynamics, with the time-dependent mean-square distances $\left\langle r_{ij}^2(t)\right\rangle = \left\langle\left[\mathbf{r}_j(t)-\mathbf{r}_i(0)\right]^2\right\rangle - 6Dt$. For $t\to0$, $S(q,t)$ reduces to the static structure factor, or form factor. The dynamic structure factor is characterized through its first cumulant $\Omega(q)$[12,13]

$$\Omega(q) = -\frac{\partial}{\partial t}\ln\left(\frac{S(q,t)}{S(q,0)}\right)\Bigg|_{t\to0} \tag{4}$$

Since $\Omega(q)$ can also be calculated without preaveraging the hydrodynamic interaction,[13] this enables us to quantify the preaveraging error.

The unperturbed Θ state of stars and dendrimers

The unperturbed state is achieved when the second virial coefficient A_2 vanishes, and the solution shows a (pseudo)ideal behavior.[5] The temperature producing $A_2 = 0$ is the Θ temperature, generally taken as independent of molar mass, consistent with a vanishing β_2: attractive and repulsive interactions exactly cancel, the molecules do not interact and have a RW conformation, leading to the notion of universality of the Θ state. Recent experiments contradict these conclusions:[1,14] stars exhibit an f-dependent Θ depression compared with linear chains, although the same value is asymptotically reached. Also, in this limit, stars do show a larger size than expected from the RW model and, accordingly, display residual interactions, unlike linear chains.

To explain this behavior, we calculate the Θ temperature that makes $A_2 = 0$ through a vanishing interaction free energy between two molecules.[4,5] We find that Θ is lower than Θ_{bead}, because temperature-dependent two-body attractions must balance the three-body repulsions between two molecules. In branched molecules, the larger multiplicity of interactions between two molecules compared with linear chains brings about larger three-body repulsions (an effect of molecular connectivity), and the two-body attractions must correspondingly be larger, with a lower Θ. Numerical calculation of Θ is coupled with the issue of the molecular conformation, since it requires the knowledge of the intramolecular probability density of contact. In turn, this is obtained by minimization of the intramolecular free energy at the true Θ temperature.[11]

We find a striking difference between stars and dendrimers. In stars, Θ changes with the arm length N/f according to a square-root law,

$$\Theta = \Theta_\infty - \varphi\,\chi_f\,(N/f)^{-1/2} \qquad (5)$$

where χ_f depends only on the arm number, and φ is a parameter characteristic of the monomer/solvent pair. For $f > 4$, Θ increases with molar mass to the same Θ_∞ as in linear chains, because the star topology becomes irrelevant for very long arms. Moreover, at a fixed arm length, the Θ depression increases with the arm number as an effect of the larger multiplicity of intermolecular interactions.

On the other hand, Θ in dendrimers decreases without limit with increasing generation g. (Actually, only finite g values can be realized because of the increasingly larger crowding of the monomers.) The reason for this different behavior is apparent: increasing g means also increasing the interaction multiplicity, so that dendrimers can be compared with stars with an increasingly large number of arms.[11] This opposite behavior of the Θ temperature is also expected from the RW model if we assume the probability density of contact between two beads being simply proportional to V^{-1} (V is the molecular volume). We get $\Theta = \Theta_{\text{bead}}/\left[1 + C \cdot d\right]$, C being a constant. Thus, in linear and star polymers, Θ should tend to the limit $\Theta_\infty \equiv \Theta_{\text{bead}}$ from below with increasing molar mass since the average density scales as $d \propto N^{-1/2}$. Conversely, in cascade polymers, where $d \propto N^{1/4}$, and in dendrimers, where d diverges, Θ decreases without limit.

As for the molecular dimensions at $T = \Theta$, the compensation between two-body attractions and three-body repulsions in linear chains yields a RW conformation. On the other hand, residual three-body repulsions among three beads on three different arms are non-compensated in stars and dendrimers, lacking any *inter*molecular counterpart. This produces a finite star swelling[1,11] with expansion factor $\alpha_S^2 = \langle R_S^2 \rangle / \langle R_S^2 \rangle^{\text{RW}}$ and in stars a topological ratio $g_S = g_S^{\text{RW}} \cdot \alpha_{S,\text{star}}^2 / \alpha_{S,\text{lin}}^2$. Thus, we get $g_S^\Theta > g_S^{\text{RW}}$, since $\alpha_{S,\text{lin}} = 1$ to a very good approximation. The calculated g_S^Θ agrees well with experimental results, which show the non-universality of the Θ state. Indeed, we argue that in general the RW size is only achieved in a melt because of an essentially complete screening of all interactions, unlike what happens in the Θ state. In dendrimers, the larger multiplicity of three-body interactions leads to a sharp increase in molecular size over the RW value, although eventually a maximum size is reached, dictated by molecular connectivity.[11]

A potential problem in the study of the Θ state, in particular for computer simulations, is the determination of Θ, often established in single molecules through the relationship $\langle R_S^2 \rangle \propto N$, whereas its thermodynamic definition requires considering a two-chain system, which is more difficult to simulate. Also, the nonuniversality of the Θ state is related with the three-body interaction parameter, which strongly depends on the ratio between local thickness to "bond" length. Therefore, slight differences among different monomers may significantly affect the Θ behavior.

Equilibrium and dynamics of stars and dendrimers in a good solvent

In a good solvent, the two-body repulsions swell the molecules, the expansion being qualitatively different from that in the Θ state. Within the GSC approach, the molecular conformation is obtained neglecting the three-body repulsions absorbed in the renormalized two-body interactions. The result is a non-affine expansion concentrated at the molecular center where the bead density is the largest.[4] Thus, the arms are stretched outward in stars with little correlation among their initial directions, as nicely confirmed by computer simulations, so that both intra- and interarm repulsions are minimized.[10] In dendrimers, the repulsive interactions among the dendrons, the subdendrons, etc., strongly stretch them outward with an incipient dendron segregation, again in agreement with computer simulations. The large density of beads in branched polymers gives rise to a peak in the Kratky plot of the static structure factor, reported in Fig. 1. A larger peak sharpness for branched molecules indicates a larger bead density; therefore, the peak broadens in a good solvent because the molecular expansion relieves the steric congestion near the core.

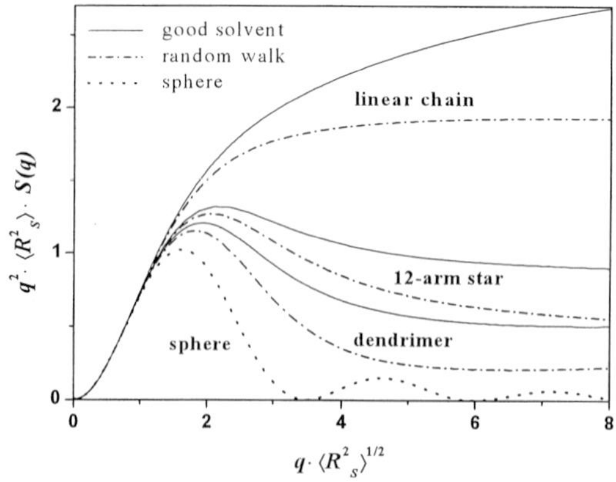

Fig. 1: Kratky plot of the static structure factor $S(q)$ for linear chains, a twelve-arm star, a sixth-generation dendrimer and a compact sphere for comparison.

In spite of the non-affine expansion, linear and star polymers in a good solvent show a universal behavior if the expansion factor α_S^2 is reported as a function of the reduced variable $z \propto \beta_2\sqrt{N}$. The asymptotic dependence is simply $\alpha_S \propto z^{2\nu-1}$, yielding Eq. (1). Our results in the crossover region show that α_S has roughly the same z dependence for linear chains and stars due to the compensating effect of arm expansion and loss of correlation across the branch point; accordingly, the topological ratio $g_S^{GS} = g_S^{RW} \times \alpha_{S,\mathrm{star}}^2/\alpha_{S,\mathrm{lin}}^2$ is close to the RW value, in keeping with previous theoretical results.[1]

Dendrimers do follow a universal behavior, too[9], but the appropriate reduced variable is $\beta_2 N^{\varphi}$, with an empirical exponent $\varphi = 0.923$. This value is much larger than $\varphi = 0.5$, but close to $\varphi = 1$ for polymers in two dimensions,[6] consistent with the fact that a large fraction of the beads lies near the outer surface. Moreover, the dendrimer expansion reaches an asymptotic maximum value, in contrast to stars, where the arm length can increase without limit. This finite swelling is consistent with the finite extension of the real dendrons, whose local stereochemistry is not affected by the solvent quality.

Turning now to the dynamics, a characteristic feature of branched polymers consists in the degeneracy of their relaxation times, strictly related with molecular symmetry. In stars, there are non-degenerate modes related to the concerted motion of the arms, and (f–1)-degenerate modes describing the independent motion of the arms, while the complex structure of dendrimers produces a whole hierarchy of relaxation times.[9]

The relaxation times become larger with molecular expansion, at least for collective modes, due to both the change in intramolecular elasticity, and the decrease in hydrodynamic interaction. Therefore, in a good solvent, the intramolecular dynamics approaches in part the free-draining regime. Incidentally in our procedure, the main error due to preaveraging[9] is largest in dendrimers in view of their large bead density, although being less dramatic under good-solvent expansion.

The intrinsic viscosity is obtained from the sum of the relaxation times. We define its expansion factor with respect to the RW value as $\alpha_\eta^3 = [\eta]/[\eta]^{RW}$, which is a universal function of z. Asymptotically, we have Zimm's result[4,12] $[\eta] \propto N^{0.8}$. As to the topological ratio $g_\eta = [\eta]_{\mathrm{star}}/[\eta]_{\mathrm{lin}}$, we find that at first g_η^{GS} increases with z, in keeping with the perturbative analysis, but eventually it decreases to a constant value,[10] consistent with experimental results.[1]

In dendrimers, α_η increases to an asymptotic maximum value with increasing molecular expansion, just as α_S. A universal curve is obtained using the universal variable $\beta_2 N^\xi$, with ξ = 0.735. This exponent is lower than the φ exponent required for α_S because of the small length of the dendrons *and* of the complex hydrodynamic behavior of these crowded molecules. As for the dendron length, we get an apparent exponent ν_{app} of the radius of gyration, which decreases from 0.43 for g = 1 to 0.23 for g = 6 at intermediate solvent quality,[9] in agreement with simulations. If we used the scaling expression $[\eta] \propto R^3/N$ in terms of some radius R proportional to the radius of gyration, we would get $[\eta] \propto N^{3\nu_{app}-1}$. Thus, $[\eta]$ should increase as $N^{0.29}$ at low g, but decrease as $N^{-0.31}$ at large g, passing through a maximum, as observed in some cases.[2] Though affected by the preaveraging approximation, our calculations disagree with this result, predicting that $[\eta]$ should monotonically increase with g, showing that at small dendron length the scaling relationships are not valid.

Finally, we discuss the first cumulant $\Omega(q)$ for dendrimers (reported in Fig. 2) to estimate also the preaveraging error. Both at very small and at very large q, we have $\Omega(q) \propto q^2$ because of the diffusion of the whole molecule (low q) or of the single beads (high q) before they experience any connectivity constraint.[12] In the intermediate q range, the dependence switches to $\Omega(q) \propto q^3$ for linear chains.

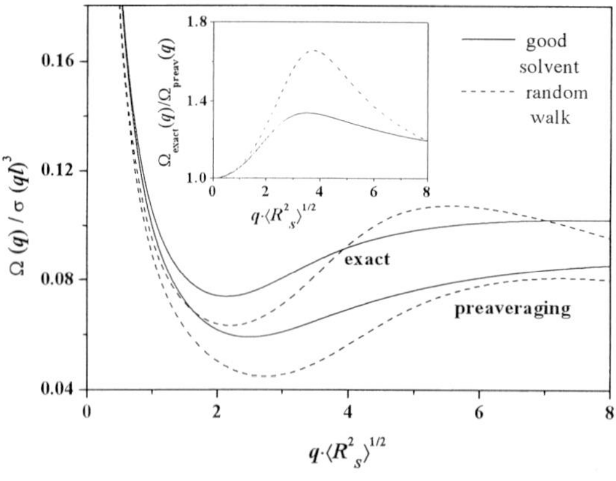

Fig. 2: The first cumulant of the dynamic structure factor for the sixth-generation dendrimer. The exact (in the GSC approach) and the preaveraged results are shown, together with the preaveraging error in the inset. Here, σ^{-1} is the time unit.

The characteristic feature of branched polymers consists in a shallow minimum in the plot of $\Omega(q)/q^3$ at $q\cdot\left\langle R_S^2\right\rangle^{1/2} \cong 2\div3$, probing distances of the order of the radius of gyration, as predicted and observed in stars.[7,15] Such minimum, which matches the position of the peak in the Kratky plot of Fig. 1, deepens with an increasing bead density, hence with dendrimer generation, and becomes shallower with good-solvent expansion. The region of this minimum is also roughly the q range where the preaveraging error is largest, as shown in the inset; however, as anticipated, the error becomes significantly smaller in a good solvent.

Concluding remarks

The main weakness of the GSC approach consists in using the Gaussian approximation and, additionally, in dynamics on preaveraging the hydrodynamic interaction. On the other hand, the method allows to investigate both equilibrium and dynamics within a unified scheme. The main advantages showing the rich potentialities of the method are: it accounts for all the configurational degrees of freedom and for molecular connectivity, it permits to study molecules of any topology with an arbitrary sequence of unlike beads, and it is consistent with the linear Langevin equation in dynamics. Improvements beyond the Gaussian approximation are not trivial, since it is unclear whether the correct distribution function of linear chains applies also to branched molecules. Also, it does not yield a linear elastic force in the Langevin equation, making the consistent study of equilibrium and dynamics a difficult task. On the other hand, computer simulations do not suffer from this limitation, but the ergodicity problem becomes relevant for branched molecules, while simulating the dynamics is by no means trivial. On the experimental side, we note that the unperturbed state of dendrimers was hardly investigated, while such data can provide an even more stringent test of theoretical predictions.

Acknowledgments We gratefully thank Prof. G. Allegra, Dr. G. Raos, Dr. G. Raffaini, Dr. R. La Ferla and Dr. E.G. Timoshenko for useful discussions. This work was financially supported by MURST (Ministry of University and of Scientific and Technological Research, Italy) and by CNR (Consiglio Nazionale delle Ricerche, Italy) in the frame of the project "Materiali Speciali per Tecnologie Avanzate II".

References

[1] J.F. Douglas, J. Roovers, K.F. Freed, *Macromolecules* **1990**, *23*, 4168.
[2] D.A. Tomalia, A.M. Naylor, W.A. Goddard III, *Angew. Chem. Int. Ed. Engl.* **1990**, *29*, 138.
[3] G.S. Grest, L.J. Fetters, J.S. Huang, D. Richter, *Adv. Chem. Phys.* **1996**, *94*, 67; J.J. Freire, *Adv. Polym. Sci.* **1999**, *143*, 35.
[4] G. Allegra, F. Ganazzoli, *Adv. Chem. Phys.* **1989**, *75*, 265; G. Allegra, F. Ganazzoli, *Prog. Polym. Sci.* **1991**, *16*, 463; F. Ganazzoli, G. Raos, G. Allegra *Macromol. Theory Simul.* **1999**, *8*, 65.
[5] H. Yamakawa, *Modern Theory of Polymer Solutions*, Harper & Row, New York, 1971.
[6] P.-G. de Gennes, *Scaling Concepts in Polymer Physics*, Cornell University Press, Ithaca (NY), 1979.

[7] W. Burchard, *Adv. Polym. Sci.* **1983**, *48*, 1.
[8] R. La Ferla, *J. Chem. Phys.* **1997**, *106*, 688.
[9] F. Ganazzoli, R. La Ferla, G. Terragni, *Macromolecules* **2000**, *33*, 6611; F. Ganazzoli, R. La Ferla, G. Raffaini, *Macromolecules* **2001**, *34*, 4222.
[10] G. Allegra, E. Colombo, F. Ganazzoli, *Macromolecules* **1993**, *26*, 330; F. Ganazzoli, G. Allegra, E. Colombo, M. De Vitis, *Macromolecules* **1995**, *28*, 1076.
[11] F. Ganazzoli, *Macromolecules* **1992**, *25*, 7357; F. Ganazzoli, R. La Ferla, *J. Chem. Phys.* **2000**, *113*, 9288.
[12] M. Doi, S.F. Edwards, *The Theory of Polymer Dynamics*, Oxford Scientific Publications, Oxford, 1986.
[13] Z. Akcasu, H.J. Gurol *J. Polym. Sci., Polym. Phys. Ed.* **1976**, *14*, 1.
[14] B.J. Bauer, N. Hadjichristidis, L.J. Fetters, J.E.L. Roovers, *J. Am. Chem. Soc.* **1980**, *102*, 2410.
[15] D. Richter, B. Stühn, B. Ewen, D. Nerger, *Phys. Rev. Lett.* **1987**, *58*, 2462.

Macromol. Symp. ***190***, *65–74 (2002)*

Light Scattering Investigation of Association Phenomena in Saline Carrageenan Solutions

*Fabio Cuppo and Harry Reynaers**

Laboratory of Macromolecular Structural Chemistry, Department of Chemistry, Catholic University of Leuven, Celestijnenlaan 200F, B-3001 Heverlee, Belgium

SUMMARY: Intermolecular association of ι- and κ-carrageenan in aqueous solution has been studied by wide-angle laser light scattering as a function of salt composition and temperature. For ι-carragenan, the effect of three different salts (LiCl, NaCl, NaI) on the associative properties was investigated. As to κ-carrageenan, the association-inducing effect of Cs^+ ions was studied, with particular interest in the time dependence of the aggregation process.

Introduction

Carrageenans are a family of sulfated polysaccharides extracted from marine red algae. Three main species can be identified, namely ι-, κ- and λ-carrageenan. This study is restricted to ι- and κ-carrageenan, the primary structure of which is shown in Fig. 1. Despite their widespread application, especially in food industry as thickening and stabilizing agents, the mechanism of gelation has been a matter of discussion until very recently, and further investigation is still needed for a full comprehension of the process[1-4]. The controversial data so far reported in literature must be related to the extreme sensitivity of these polymers to preparative and experimental conditions, as already pointed out in previous work[1-4].

Compositional (polymer concentration, nature and concentration of added salts, etc.) as well as physical parameters (temperature) play a key role in the gelling behavior of κ- and ι-carrageenan[1-5]. The mechanism of gelation has currently been shown to consist of a two-step process[1-4]: first, a conformational transition from the disordered to an ordered (single-helix) state takes place; then, interchain association between helical stretches gives rise to the three-dimensional gel network. However, it is worth noting that interchain association may occur on the microscopic scale without the development of a macroscopic gel behavior, *e.g.*, at high dilution and/or in the presence of suitable co-solutes.

 CCC 1022-1360/00/$ 17,50+.50/0

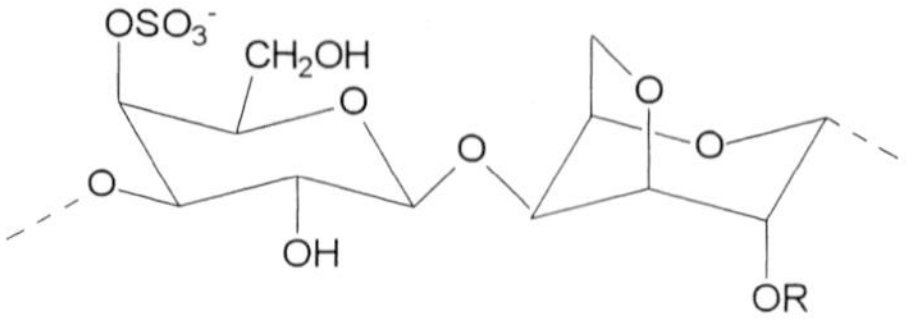

Figure 1. Repeating unit of κ-carrageenan (R = H) and ι-carrageenan (R = SO_3^-).

Carrageenans are known to modulate their conformational-associative behavior depending on the specific nature (and concentration) of ionic species in solution[1-5]. Whereas older studies suggested a different behavior of ι- and κ-carrageenan in this respect[6], more recent work has led to opposite conclusions, i.e., only a small difference is observed between the two forms[7,8]. In general, it can be stated that the alkali metal ions with higher atomic number (K^+, Rb^+, Cs^+) are more effective in inducing gelation than the lighter ones (Na^+, Li^+). ι-Carrageenan, however, shows a higher tendency to interchain association, which is reflected by a lower ion specificity in comparison with κ-carrageenan.

The present research focuses on solution properties of ι-carrageenan and κ-carrageenan by exploring the influence of different physicochemical parameters, such as the nature of added salts, ionic strength, temperature and polysaccharide concentration.

A light scattering approach (wide-angle laser light scattering, WALLS) has been employed, inasmuch as this method offers the remarkable advantage of allowing the study of essentially unperturbed systems in comparison with other experimental techniques (osmometry, viscosimetry, chromatographic techniques, etc.).
Preliminary results are reported on the associative behavior of κ-carrageenan in aqueous NaI/CsI mixed systems. This aspect has recently found increasing interest due to the observation of a nematic phase of κ-carrageenan[9] under those salt conditions (although at higher polymer concentrations).

Materials and Methods

ι- and κ-carrageenan were supplied by Sigma-Aldrich (ι-carrageenan type V, from *Eucheuma spinosa*, lot no. 27F0373; κ-carrageenan type III, from *Eucheuma cottonii*, lot no. 73H0840). The commercial samples were purified following a general procedure which had already proved to be safe and reliable avoiding undesired aggregation phenomena[1-3,10]. The purified ι-carrageenan sample was subjected to combined NMR and IR analyses. It turned out that the contaminating κ-type fraction amounted to 8 % at most. The polydispersity of the sample was checked by gel permeation chromatography (GPC-LALLS) the found polydispersity index being 2.0. All chemicals used were of analytical grade; ultrapure water (Milli-Q Reagent Grade Water Systems, Millipore) was used throughout.

Light scattering measurements were performed as previously described[2,11]; the polymer solutions were prepared according to the procedure discussed in the following paragraph.

Sample Preparation and Handling

Polysaccharide solutions are particularly difficult to handle due to their high viscosity and their tendency to form entanglements, aggregates and other forms of interchain association. As a consequence, the method for preparation of polysaccharide solutions is very critical, and can possibly affect the reliability of the experimental results when proper procedures are not employed.

When ι- or κ-carrageenan solutions are prepared starting from an aqueous carrageenan solution of a known concentration, to which suitable volumes of salt solutions are carefully added (under stirring at sufficiently high temperature), reliable light scattering data are obtained. On the contrary, light scattering measurements show poor reproducibility and, in particular, a higher molar mass when the freeze-dried polymer is dissolved directly in an aqueous salt solution. Similar observations are reported by Lecoutier *et al.* for light scattering data on lyophilized xanthan[12].

In the present case, light scattering measurements were carried out first starting from aqueous salt solutions to which a known amount of lyophilized ι- or κ-carrageenan was added under vigorous stirring. A dilution series was then obtained by addition of increasing amounts of the 1:1 electrolyte solutions. All solutions underwent light-scattering measurements immediately after preparation, starting with the lowest polymer concentration.

From the initial light scattering measurements unusual Zimm plots were obtained. In particular, a general tendency to an increase in the resulting molar mass as a function of increasing salt concentration appeared. All the experimental observations, including time-dependent effects, pointed to the occurrence of intermolecular association. It was then concluded that the preparation procedure was inappropriate. It should be noted furthermore that only a comparison of repeated experiments could lead to this conclusion, as a single occasional experiment might give apparently reliable data.

After several trials with different samples and under different conditions (polymer and salt composition, curing times, etc.), the following procedure turned out to give the best results: an aqueous (ι- or κ-) carrageenan solution of a known concentration was diluted in order to obtain solutions of the desired final concentrations. A suitable volume of salt solutions was slowly added to the half-filled flasks (*i.e.* at half dilution), under stirring at 65 °C, so as to reach the proper final ionic composition. Using a peristaltic pump in a close circuit, the solutions were filtered several times through a Millipore filter (pore size 0.22 μm) directly in the light scattering cells to obtain dust-free solutions. In all cases, the polymer concentration was lower than 0.50 g L^{-1}. In the case of ι-carrageenan, the solutions were stored at room temperature for 24 h before the measurements since it was found that, under the investigated conditions, an equilibrium state could be reached within that period. Hence, the light scattering data on ι-carrageenan can be safely regarded as equilibrium data. On the other hand, it turned out from several experiments that the equilibration time for κ-carrageenan can often be much longer than 24 h under aggregating conditions (while no time dependence immediately after the preparation of the samples is observable under non-aggregating conditions). It was then chosen to start the measurements on the κ-carrageenan samples immediately after their preparation, studying the time dependence of association under different experimental conditions and performing consecutive experiments on the same samples at different times.

Analysis of Scattering Data: The Open Association Model of Elias

The processing of the light scattering data was performed in the framework of the Elias open association model (OAM), which has been demonstrated to give the best description of the experimental data for ι- and κ-carrageenan under associating conditions[2-4]. As a more detailed theoretical discussion of the model can be found in the original treatment by Elias[13] as well as in the quoted previous work, only the fundamental features and equations will be reported here.

The basic assumption of the open association model is that the association process occurs through a series of thermodynamically equivalent steps, leading from the fundamental non-associated form, the "unimer", to dimers, trimers, tetramers, etc. Each step is described by an equilibrium constant, which is assumed to be independent of the extent of association: $K_1 = K_2 = K_3 \ldots = K_{\text{assoc}}$.

The weight-average molar mass of the associating polymer system, $(\overline{M}_w)_{\text{assoc}}$, can thus be defined, according to the model, as

$$\left(\overline{M}_{\mathrm{w}}\right)_{\mathrm{assoc}} = \sqrt{\left(\overline{M}_{\mathrm{w}}\right)_0^2 + 4000 \cdot K_{\mathrm{assoc}} \cdot PI \cdot \left(\overline{M}_{\mathrm{w}}\right)_0 \cdot C_{\mathrm{p}}} \qquad (1)$$

where $\left(\overline{M}_{\mathrm{w}}\right)_0$ is the weight-average molar mass of the non-associated species, PI is the polydispersity index and C_{p} is the polymer concentration.

The reduced scattering intensity, extrapolated to zero angle, is given by the following equation:

$$\left.\frac{K \cdot C_{\mathrm{p}}}{R_\theta}\right|_{\theta=0°} = \frac{1}{\left\{\left(\overline{M}_{\mathrm{w}}\right)_0^2 + 4000 \cdot K_{\mathrm{assoc}} \cdot \mathrm{PI} \cdot \left(\overline{M}_{\mathrm{w}}\right)_0 \cdot C_{\mathrm{p}}\right\}^{1/2}} + 2 \cdot A_2 \cdot C_{\mathrm{p}} \qquad (2)$$

where A_2 is the second virial coefficient.

Equation (2) as a two-parameter equation (K_{assoc} and A_2) was used to fit the experimental data. $\left(\overline{M}_{\mathrm{w}}\right)_0$ was obtained under non-associating conditions. The value of PI for the ι-carrageenan sample was determined as described above (see Materials and Methods), while for κ-carrageenan the value 2.27 was taken from GPC-LALLS results previously obtained for a similar sample[14].

Results and discussion

ι-Carrageenan

The conformational-associative behavior of ι-carrageenan has been studied in a wide range of salt and temperature conditions. In particular, the effect of three different salts, namely NaI, NaCl and LiCl, has been investigated as a function of the ionic strength (and temperature), in order to span the region of conformational transition from the disordered to the ordered (single-)helical state. In fact, from optical activity measurements[15] it follows that at 25 °C, independently of the type of the 1:1 supporting electrolyte in solution, the biopolymer is in a fully disordered state at ionic strength equal or lower than 0.02 M, while, on the contrary, a fully ordered state is always reached at ionic strength equal or higher than 0.09 M. The experimental light scattering parameters under these two extreme conditions are given in Table 1.

Table 1. Light scattering parameters for ι-carrageenan under different salt conditions at 25 °C.

	0.02 M LiCl	0.02 M NaCl	0.02 M NaI	0.09 M LiCl	0.09 M NaCl	0.09 M NaI
$K_{\mathrm{assoc}} \cdot 10^{-5}$	$\le 0.01 \pm 0.00$	$\le 0.01 \pm 0.00$	$\le 0.01 \pm 0.00$	6.6 ± 0.9	5.34 ± 0.18	2.29 ± 0.27
$A_2 \cdot 10^3$ mol cm^3 g^{-2}	15.51 ± 0.04	9.15 ± 0.13	9.14 ± 0.16	4.8 ± 0.1	2.43 ± 0.70	2.36 ± 0.23

The associative behavior of ι-carrageenan, under the given experimental conditions, seems to vary between the limits of LiCl and NaI, where the highest and the lowest degree of association were respectively observed for all the investigated ionic strengths (0.02÷0.09 M); NaCl represents an intermediate state between the two limits. Hence, it can be concluded that Li^+ is more effective than Na^+ in promoting aggregation, and Cl^- more than I^-. The effect of the ionic strength reflects the progressive conformational ordering of macromolecules upon increasing salt concentration, since interchain association requires the transition to the helical state as a preliminary stage[1-3,5,11].

As to the second virial coefficient, a decrease in the order LiCl > NaCl > NaI was observed over the whole range of ionic strength. This result, at first glance, appears to be in contradiction with the association-inducing properties of the three salts. It must be noted, however, that the physicochemical meaning of the second virial coefficient cannot be regarded in an oversimplified way, especially in a rather complex system such as a polyelectrolyte in aqueous salt solution undergoing intermolecular association. Several effects should be taken into account in such cases, like chain conformation and stiffness, excluded-volume effects, specific interactions between the polymer and ions, etc. Proper interpretation of the experimental data in terms of molecular parameters would then require a deeper theoretical approach, which is beyond the purpose of this work.

Some examples of Zimm plots, showing typical effects of intermolecular association on the angular and concentration dependence of the light scattering data, are given in Figs 2, 3 and 4.

Figure 2 shows how the NaCl salt concentration affects the Zimm plots of ι-carrageenan (treated according to the Elias method, as described above) at constant temperature, 25 °C.

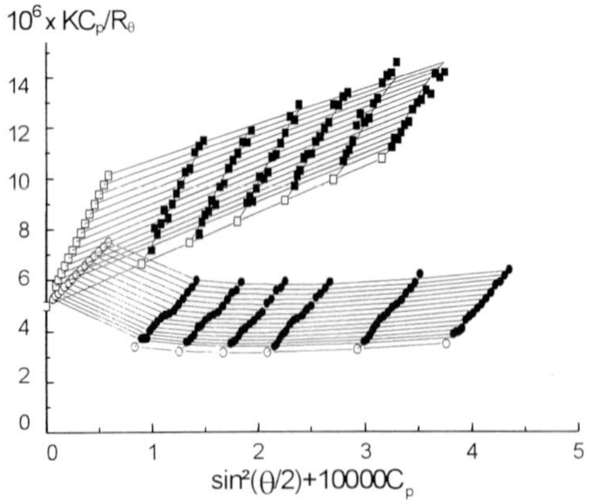

Figure 2. Zimm plots of ι-carrageenan in 0.02 M NaCl (squares) and in 0.09 M NaCl (circles). (The solid symbols represent the experimental data and the open symbols the extrapolated values.)

In 0.09 M NaCl, an upward curvature can be observed; the scattered intensities are higher than in 0.02 M NaCl, where the curvature is absent (*i.e.*, little or no association occurs). From the slope of the values extrapolated to zero angle, it is evident that the second virial coefficient is larger in 0.02 M than in 0.09 M NaCl (Table 1). It can also be seen, from the slope of the angular dependence extrapolated to $C_p = 0$, that the radius of gyration is larger in 0.02 M than in 0.09 M NaCl (115.3 ± 1.6 nm and 88.4 ± 5.7 nm, respectively).

Figure 3 shows a similar Zimm plot of ι-carrageenan in 0.07 M NaI at 25 °C. Here the association effect is also clearly visible.

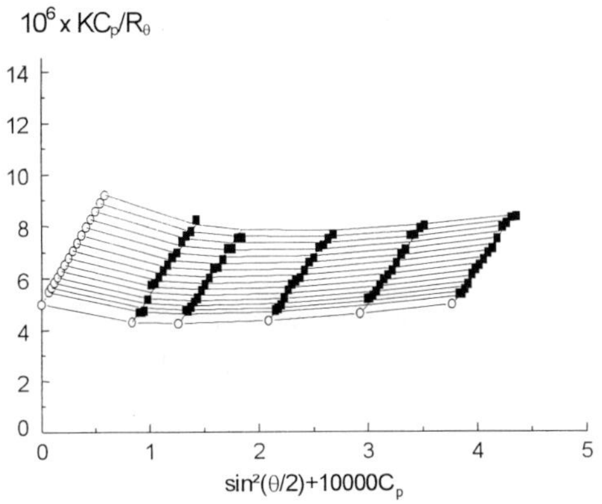

Figure 3. Zimm plot of ι-carrageenan in 0.07 M NaI. (The solid squares are the experimental data and the open circles the extrapolated values.)

In Fig. 4, the influence of temperature on the associative behavior of ι-carrageenan is clearly manifested: in 0.08 M NaCl, an upward curvature of the Zimm plots, especially in the low-concentration region, appears at 11 °C.

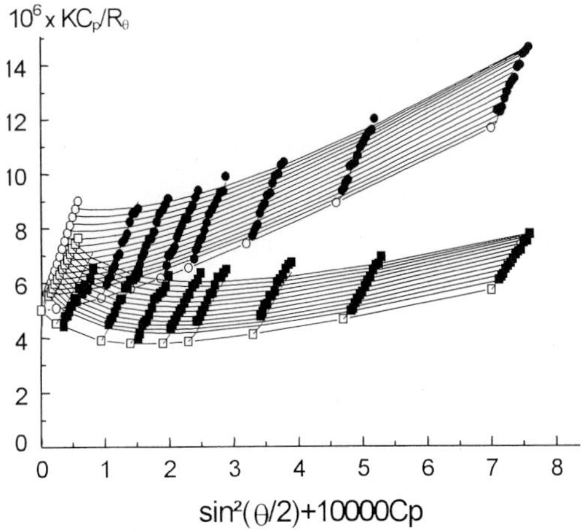

Figure 4. Zimm plots of ι-carrageenan in 0.08M NaCl at 11 °C (squares) and at 55 °C (circles). (The solid symbols represent the experimental data and the open symbols the extrapolated values.)

The scattered intensities clearly decrease upon raising temperature to 55 °C; from the slope at zero polymer concentration, it can be seen that the radius of gyration increases with increasing temperature. At 55 °C, a certain curvature is still present in the Zimm plot, pointing to a small but persistent association effect.

κ-Carrageenan

The experimental work currently carried out in our laboratory deals with the associative behavior of κ-carrageenan as a function of salt concentration and temperature. In particular, the effect of a gradual change from non-associating (NaI 0.1 M) to increasingly associating conditions (NaI/CsI mixtures, at constant 0.1 M ionic strength and increasing the CsI mole fraction, X_{Cs}) is being studied. Some preliminary results will be shown hereinafter.

The first characterization of the κ-carrageenan sample under non-associating conditions confirmed the fundamental features already reported in previous work performed in our laboratory[1-2,11] (Table 2).

Table 2. Light scattering parameters for κ-carrageenan under non-associating conditions at 25 °C.

	0.1 M NaCl	0.1 M NaI $\Delta t = 0$	0.1 M NaI $\Delta t = 5$h
$M_w \cdot 10^{-3}$ / g mol^{-1}	227 ± 4	211 ± 5	211 ± 3
$A_2 \cdot 10^3$ / mol cm^3 g^{-2}	3.72 ± 0.05	1.88 ± 0.07	1.85 ± 0.08

In particular, the conformational transition between a disordered state (NaCl 0.1 M) and a helical state (NaI 0.1 M), already demonstrated by several authors on the basis of optical rotation and molar mass measurements[1,5,11], is reflected by the experimental light scattering data. No change of the weight-average molar mass is observable between the two conformational states beyond the experimental error. This finding provides further evidence of the presence of a single-helical conformation as the fundamental ordered state in aqueous solution[1-4]. The thermodynamic stability of the system in NaI was tested over a period of 5 h, after which no significant change of the experimental parameters was detected. As to the second virial coefficient, its values in NaI are remarkably lower in comparison with those in NaCl. This result is very close to those previously reported under same or very similar conditions[1,11]. They indicate that iodide is a poorer solvent for the ordered polymer conformation than chloride is for the disordered one.

The occurrence of intermolecular association becomes evident when a part of the Na^+ ions is replaced by Cs^+ ions. In Table 3 and in Fig. 5, the results at $X_{Cs} = 0.28$ are shown.

Table 3. Light scattering parameters obtained for κ-carrageenan in mixed 0.1 M CsI/NaI solution at CsI mole fraction $X_{Cs} = 0.28$, at 25 °C.

	$\Delta t = 0$ h	$\Delta t = 24$ h	$\Delta t = 45$ h	$\Delta t = 70$ h	After heating
$K_{assoc} \cdot 10^{-3}$	115 ± 10	260 ± 28	514 ± 70	737 ± 159	39.7 ± 22.0
$A_2 \cdot 10^3$ / mol cm^3 g^{-2}	1.28 ± 0.04	-0.26 ± 0.04	-0.29 ± 0.05	-0.30 ± 0.04	2.03 ± 0.20

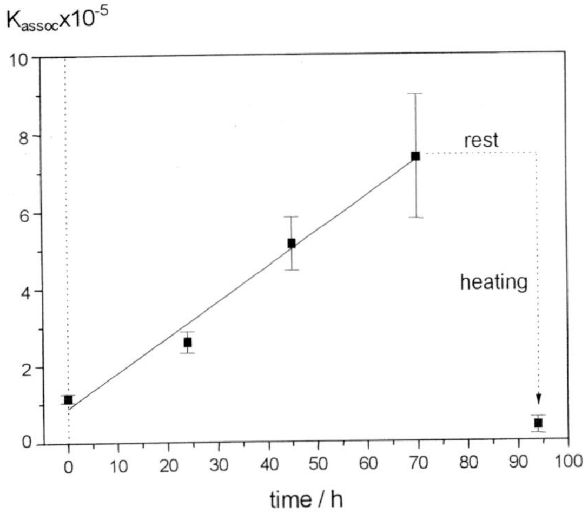

Figure 5. Time evolution of the "apparent" association constant for κ-carrageenan at X_{Cs} = 0.28, at 25 °C and after heating for 1 h at 60 °C. The solid line is a linear fit of the experimental data.

A continuous increase in the "apparent" association constant (which is actually a function of time, slowly approaching a true equilibrium value at $t \rightarrow \infty$) can be observed; this clearly indicates that the system is still far from reaching equilibrium state after 70 h. Meanwhile, the second virial coefficient falls to a negative value already at t = 24 h and remains constant thereafter.

At t = 94 h (*i.e.*, 24 h after the last experiment), the solutions at different C_p for X_{Cs} = 0.28 were heated for 1 h at 60 °C and then immediately subjected to light scattering measurement at 25 °C. The resulting association constant is even lower than that at t = 0 h, clearly pointing to a thermally induced dissolution of polymer aggregates.

Even though at a preliminary stage, the results reported above clearly demonstrate that (thermoreversible) intermolecular association of κ-carrageenan under the investigated experimental conditions occurs below the "threshold value" of X_{Cs} = 0.4 recently suggested by Viebke *et al.* from DSC measurements[16]. This finding, anyway, deserves further study and analysis, which will be fully reported in a forthcoming paper.

Conclusions

The associative behavior of ι-carrageenan, at 25 °C, is strongly enhanced by increasing the ionic strength of the supporting salt (LiCl, NaCl, NaI) in the range 0.02÷0.09 M. However, the ability to induce association follows the order LiCl > NaCl > NaI.

From preliminary experimental results, it turns out that κ-carrageenan undergoes remarkable intermolecular association in NaI/CsI mixed systems at 0.1 M ionic strength and the CsI mole fraction, X_{Cs}, equal to 0.28. The associative process, under the investigated conditions, is rather slow, as the system is still far from reaching equilibrium after 70 h. This result clearly indicates that association occurs at cesium concentration lower than the threshold value (X_{Cs} = 0.4) recently suggested by other authors.

Measurements performed under non-associating conditions (NaCl or NaI, 0.1 M) show no increase in the molar mass of κ-carrageenan upon conformational ordering. This finding provides further evidence of the monomolecular nature (single helix) of the conformational transition.

References

1. K. Bongaerts, H. Reynaers, F. Zanetti, S. Paoletti, *Macromolecules* **1999**, 32, 675.
2. K. Bongaerts, Ph.D. Thesis, Katholieke Universiteit Leuven, Heverlee, Belgium, 1999.
3. B. Denef, Ph.D. Thesis, Katholieke Universiteit Leuven, Heverlee, Belgium, 1997.
4. K. Bongaerts, S. Paoletti, B. Denef, K. Vanneste, F. Cuppo, H. Reynaers, *Macromolecules* **2000**, 33, 8709.
5. M. Ciancia, M. Milas, M. Rinaudo, *Int. J. Biol. Macromol.* **1997**, 20, 35.
6. D.A. Rees, E.R. Morris, D. Thom, J.K. Madden, in: *The Polysaccharides*, G. O. Aspinall, Ed., Academic Press, New York 1982.
7. L. Piculell, C. Håkansson, S. Nilsson, *Int. J. Biol. Macromol.* **1987**, 9, 297.
8. M.J. Ridout, S. Garza, G.J. Brownsey, V.J. Morris, *Int. J. Biol. Macromol.* **1996**, 18, 5.
9. J. Borgström, P.-O. Quist, L. Piculell, *Macromolecules* **1996**, 29, 5926.
10. K. Bongaerts, H. Reynaers, F. Zanetti, S. Paoletti, *Macromolecules* **1999**, 32, 683.
11. D. Slootmaekers, C. De Jonghe, H. Reynaers, F.A. Varkevisser, C.J. Bloys van Treslong, *Int. J. Biol. Macromol.* **1988**, 10, 160.
12. J. Lecoutier, G. Chauveteau, G. Muller, *Int. J. Biol. Macromol.* **1986**, 8, 306.
13. H.G. Elias, in: *Light Scattering from Polymer Solutions*, M. B. Huglin, Ed., Academic Press, London 1972, Chap. 9, p.397.
14. D. Slootmaekers, J.A.P.P. van Dijk, F.A. Varkevisser, C.J. Bloys van Treslong, H. Reynaers, *Biophys. Chem.* **1991**, 41, 51.
15. B. Denef, A. Gamini, F. Delben, S. Paoletti, H. Reynaers, K. Vanneste, *Biopolymers* **1998**, 45, 105.
16. C. Viebke, J. Borgström, I. Carlsson, L. Piculell, P. Williams, *Macromolecules* **1998**, 31, 1833.

The Crystallization and Melting of Linear Polyethylene Studied by Temperature-Modulated SAXS and WAXD

Bart Goderis[a], Harry Reynaers[a], Michel H.J. Koch[b]*

[a] Catholic University of Leuven, Laboratory of Macromolecular and Structural Chemistry, Celestijnenlaan 200F, B-3001 Heverlee-Leuven, Belgium
[b] European Molecular Biology Laboratory, EMBL c/o DESY, Notkestraβe 85, D-22603 Hamburg, Germany

SUMMARY: Temperature-reversible and -irreversible morphological events could be separated in the case of linear polyethylene during quasi-isothermal crystallization by using simultaneous temperature-modulated synchrotron SAXS and WAXD. Crystallization and subsequent annealing was followed at 126 °C for 90 min while applying a temperature modulation with an amplitude of 1 °C and a period of 2 min. The crystal growth rate associated with the irreversible part of the crystallization decreases with increasing temperature in a cycle. The crystalline lamellae irreversibly thicken with time. The actual crystallite thickness, however, exhibits a superimposed modulation out of phase with that of the temperature modulation. Melting was studied during heating at 1 °C/min after cooling at 10 °C/min. A temperature modulation was superimposed with an amplitude of 2 °C and a period of 3 min. Once again temperature-reversible crystal thickness changes and irreversible crystal thickening could be observed.

Introduction

There is growing evidence for temperature-reversible melting and crystallization phenomena in semicrystalline polymers, based on temperature-modulated differential scanning calorimetry[1] (TM-DSC) experiments. The complex heat capacity (also referred to as the reversing heat capacity) measured by TM-DSC under quasi-isothermal conditions often displays an excess heat capacity superimposed to the baseline heat capacity[2-4]. This has been interpreted as being due to reversible melting and crystallization, which implies crystallization without the need for supercooling. This concept falls out of the realm of classic polymer crystallization theories which rely on nucleation concepts[5,6]. The mechanism of reversible melting and crystallization is not settled. The literature on the subject is rather speculative since the reversing excess heat capacity signal in TM-DSC does not provide any direct information about the morphological or molecular aspects of interest. The present paper sheds a light on this matter in the case of linear polyethylene (LPE). The structural changes during quasi-isothermal crystallization and during melting under typical TM-DSC conditions are reported. Temperature-reversible

 CCC 1022-1360/00/$ 17,50+.50/0

morphological events are separated from irreversible ones. Simultaneous synchrotron wide-angle X-ray diffraction (WAXD) and small-angle X-ray scattering (SAXS) were used to meet this goal in conjunction with a high-brilliance synchrotron X-ray source (the X33 double focusing camera of the EMBL in HASYLAB, on the storage ring DORIS of the Deutsches Elektronen Synchrotron (DESY) at a wavelength of 1.5 Å[7]) to collect data at short sampling intervals during the rather fast-oscillating temperature programs.

Experimental

All measurements were performed on a linear polyethylene (DSM code JW1114), synthesized using a single-site vanadium-based catalyst, with a weight-average molar mass of 52 300 g/mol and a polydispersity (M_w/M_n) of 3.2.

The experimental conditions and the data processing procedures have been described earlier[8]. The WAXD data were analyzed to provide information on changes of the crystalline density, the crystalline fraction and the lateral dimensions of the crystals. SAXS data yielded values for the crystalline (T_c) and amorphous layer thickness (T_a), the long period (LP), the specific inner surface (O_S), the fraction of semicrystalline regions (α), the local (ϕ) and overall crystallinity. SAXS data processing involves the calculation and evaluation of the total scattering power or invariant (Q), of linear correlation functions (CF(x))[9] and interface distribution functions (IDF(x))[8], with x representing the distance in real space.

Crystallization and subsequent annealing was followed at 126 °C for 90 min while applying a temperature modulation with an amplitude of 1 °C and a period of 2 min. A Mettler FP82-HT controlled the temperature. Sets of successive frames of 20 s were recorded. One frame consisted of 14-s waiting time and 6-s acquisition time. A mini-shutter placed upstream from the sample was closed during the waiting periods. Melting was studied during heating at 1 °C/min between 80 °C and 140 °C after cooling at 10 °C/min from 200 °C down to 40 °C. A temperature modulation was superimposed with an amplitude of 2 °C and a period of 3 min. SAXS and WAXD patterns were collected consecutively every 6 s.

Results and Discussion

The findings associated with the quasi-isothermal crystallization experiment were discussed in detail in an earlier paper and can be summarized as follows[8]. Crystallization is mainly irreversible, i.e. the crystallinity progressively increases with time. The instantaneous crystallization rate, however, depends on the actual temperature: a decrease is observed with increasing temperature in a cycle. The lamellar crystals irreversibly thicken with time.

Superimposed to this overall thickening process, a small reversible change is observed out of phase with the temperature excursions, i.e. T_c slightly decreases with increasing temperature and vice versa. This activity at the fold surface of the lamellar crystals appears to be the key mechanism behind reversible melting and crystallization as observed by TM-DSC.

A similar observation can be reported for the TM-DSC-like melting experiment. Figure 1 displays the evolution of T_c and it also illustrates the used temperature program.

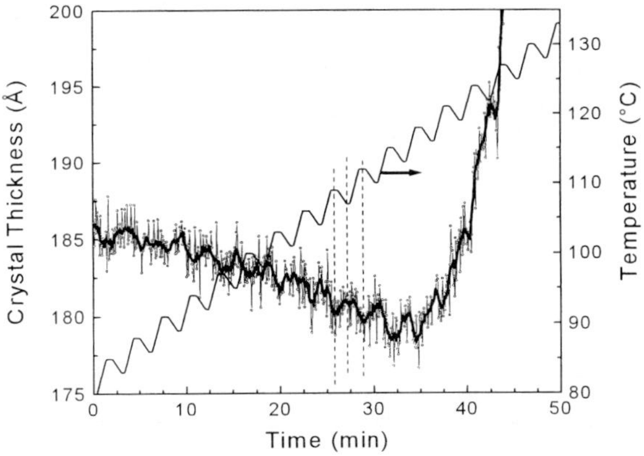

Figure 1: Evolution of T_c (small circles) during the melting of a linear polyethylene heated according to the temperature program illustrated by the oscillating full line. The thick full line represents a smoothing of T_c by a seven-point averaging. The dashed lines are guides to the eye.

An overall decrease in T_c of up to 35 min (± 115 °C) can be observed with an additional small oscillation out of phase with that of temperature. This clearly illustrates the reversible character of fold surface melting and crystallization. At longer times (higher temperatures), the crystals start to thicken. This irreversible thickening is an annealing effect associated with the relatively long residence time at each temperature. After all, the underlying heating rate is only 1 °C/min. This thickening can be avoided by increasing the heating rate. Figure 2 shows the evolution of T_c as a function of temperature together with the evolution of T_c when heated at 10 °C/min without the application of a temperature modulation. In the latter case surface melting continues even beyond 115 °C and crystal thickening is not observed. Note that the irreversible thickening essentially occurs during the heating step of a temperature cycle. During subsequent cooling, T_c slightly decreases as expected for reversible fold surface activity.

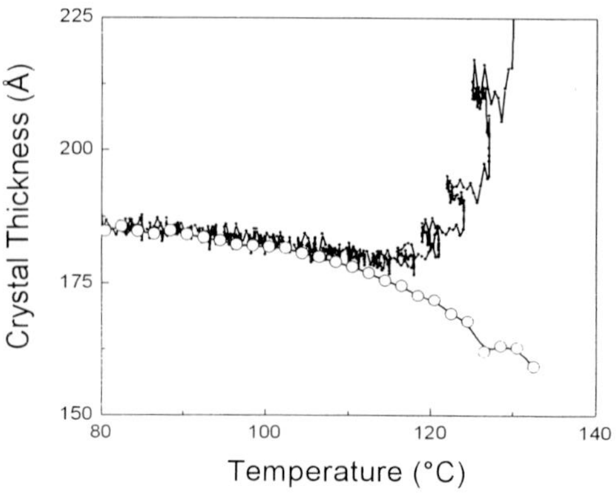

Figure 2: T_c data of Fig. 1 but plotted as a function of temperature (small connected closed circles). The large open circles represent the evolution of T_c when heated at a rate of 10 °C/min. Both samples where cooled to 10 °C prior to heating.

An analysis of the total scattering power reveals that when crystals start to thicken, there is some concomitant irreversible melting involved, which gains importance with increasing temperature.

Conclusions

The present data illustrate the metastable character of lamellar PE crystals. There is a tendency to gain stability by thickening during (quasi-)isothermal crystallization as well as during slow heating. This process takes time and competes with melting during heating. Because of this natural thickening tendency, complete (irreversible) melting cannot be the limit case of fold surface melting. In other words, a thinner crystal cannot be more stable than a thicker one. The limit temperature for stability of a lamellar crystal with a given thickness most likely follows the trend indicated by the well-known Gibbs-Thomson equation. Reversible fold surface activity (melting) in the case of linear polyethylene was recently associated with the temperature-dependent amount and distribution of gauche defects in the crystals[10].

References

[1] Reading, M. *Trends Polym. Sci.* **1993**, 8, 248.
[2] Okazaki, I.; Wunderlich, B. *Macromolecules* **1997**, 30, 1758.
[3] Ishikiriyama, K.; Wunderlich, B. *Macromolecules* **1997**, 30, 4126.
[4] Scherrenberg, R.; Mathot V.; Steeman P. *J. Therm. Analysis* **1998**, 54, 477 and Scherrenberg, R.; Mathot, V.; van Hemelrijck, A. *Thermochim. Acta* **1999**, *330*, 3.
[5] Hoffman, J.D. *Polymer* **1982**, 23, 656; Hoffman, J.D. *Polymer* **1983**, 24, 3.
[6] Wunderlich, B. *Macromolecular Physics*, Vol. 2, Academic Press, New York, 1976.
[7] Koch, M.; Bordas, *J. Nucl. Instrum. Methods* **1983**, *208*, 435.
[8] Goderis, B.; Reynaers, H.; Scherrenberg, R.; Mathot, V.B.F.; Koch, M.H.J. *Macromolecules* **2001**, *34*, 1779.
[9] Goderis, B.; Reynaers, H.; Koch, M.H.J., Mathot, V.B.F. *J. Polym. Sci., Part B: Polym. Phys.* **1999**, 37, 1715.
[10] Pak, B.; Wunderlich, B. *Macromolecules* **2001**, 34, 4492.

Some Properties of Aqueous Xanthan Gels Formed under the Action of Aluminium(III) Ions

Andrew B. Rodd[1], *Dave E. Dustan*[1], *David V. Boger*[1], *Juergen Schmidt*[2], *Walther Burchard**[2]

[1] Department of Chemical Engineering, University of Melbourne, Melbourne, Australia
[2] Institute of Macromolecular Chemistry, University of Freiburg, 79104 Freiburg, Germany

SUMMARY: Aqueous xanthan solutions form gels when aluminium salts are added and the solutions are heated above 45 °C. The gelation process was followed by dynamic light scattering. Characterization was based on the heterodyne and nonergodic approaches. Both techniques gave the same fast relaxation times, but for the slow motion much larger values were found in the heterodyne than in the nonergodic approach. The heterodyne fraction 1-X was found to correlate closely with the plateau height of the time correlation function (TCF) at large delay times in the nonergodic experiments. Three methods of gel point determination are demonstrated: (i) onset of heterodyne/nonergodic behavior, (ii) observation of a sharp maximum for the fast relaxation time at the gel point, (iii) observation of power-law behavior of the TCF. The statistics of nonergodic fluctuations were examined and evaluated. The potential of this procedure for detailed structure evaluation of inhomogeneities in the gel is emphasized.

Introduction

Gels are polymer networks, highly swollen in suitable good solvents. Most theories developed so far consider these systems as homogeneous in crosslinking density. However, in common crosslinking reactions such homogeneity can never be realized because once a crosslink has been formed, the local density of reactive monomer units becomes considerably larger than the overall concentration averaged over the whole system. As a consequence, the reaction around such crosslinks is favored and may also cause extensive intramolecular reaction around this domain.

Such heterogeneities can be detected by static light scattering[1,2] or by SANS and SAXS techniques, but far more efficiently by dynamic light scattering. The technique for detecting these phenomena has a long history starting probably in 1987 with Sellen[3] and reaching at present a state of two partially competing conceptions which are the *heterodyne* and the

 CCC 1022-1360/00/$ 17,50+.50/0

nonergodic approaches, associated with the names of Erik Geissler[4] and Peter Pusey[5] and his group. They may be considered as representatives of polymer and colloid sciences, respectively. The polymer scientist has to deal with flexibility and segmental mobility in a macromolecule in addition to the translational motion of the center of gravity and naturally is interested in the mobile part of a network. Pusey, on the other hand, studied glass transition and crystallization of hard colloidal particles which possess no internal mobility. He wondered how much of the material may be frozen-in so that no translational motion remained possible and was strongly interested in the immobilized fraction of the material.

Both scientists started with the same assumption that the inhomogeneous domains are drastically impeded in their mobility and can be, in a first approximation, considered as fully frozen-in. The scattering from such a system then can be treated as arising from mobile and immobilized scattering centers and can be written as

$$i(q,t) = i_F(q,t) + i_C(q) \tag{1}$$

where the subscripts F and C indicate the fluctuating and constant, i.e., frozen-in parts. Depending on whether time or ensemble averages are considered, this assumption gives rise to a derivation of the following field - time correlation functions (TCF)

$$g_H(q,t) = 1 + \frac{1}{X}[(g_2^T(q,t) - (g_2^T(q,0) - 1))^{1/2} - 1] \tag{2}$$

$$g_E(q,t) = 1 + \frac{1}{Y}[(g_2^E(q,t) - (g_2^E(q,0) - 1))^{1/2} - 1] \tag{3}$$

where $g_2(q,t)$ denotes the intensity TCF and the subscripts H and E refer to the heterodyne (H) and non-ergodic (E) approaches. These equations look mathematically equivalent. One has to look rather carefully at these equations to recognize the differences. They include on the one hand the superscripts T and E which denote time and ensemble averages and, on the other, the parameters X and Y which are given by

$$X = \frac{< i_F(q) >_T}{< i(q) >_T} \tag{4}$$

$$Y = \frac{< i(q) >_E}{< i(q) >_T} = \frac{< i_F(q) + i_C(q) >_E}{< i(q) >_T} \tag{5}$$

The fluctuating intensity is always ergodic, i.e., time and ensemble averages are the same. Hence the difference between X and Y consists in the fact that in Y the frozen-in part of the scattering intensity, which varies in different spatial positions, is incorporated in the ensemble average, whereas in *X* the time average at one special cell position is taken. Clearly, the two equations become identical when $i_C(q) = 0$, i.e., for systems without immobilized domains. Actually, Eqs (5) and (6) relate to two different types of experiments: In the *heterodyne* experiment the intensity TCF is measured at one position of the scattering cell over a fairly long delay time and finally it is normalized by the time-average scattering intensity at infinitely long delay times.

In the *nonergodic* experiment the intensity TCFs are measured for a large number of different positions and the ensemble average is formed which is then normalized by the ensemble average of the static light scattering. This procedure is time-consuming.

Pusey noticed that often it is sufficient to carry out the measurement at only one position and to perform the ensemble average of the total scattering intensity at the end of these measurements by collecting the scattering intensity from a rotating cell. Nonetheless, we decided to carry out the dynamic measurements at 100 and sometimes at 500 different positions. This was possible by a special step motor that was driven by a computer program to make a rotational step of 3° when the one measurement was finished and the next run was started. To the best of our knowledge, no direct comparison of the two techniques has yet been performed with the same gel. As will be shown later, our more laborious decision proved to be appropriate.

Demonstration of two types of experiment on the time correlation functions

We applied both techniques to an aqueous xanthan solution to which $Al_2(SO_4)_3$ was added[6]. Xanthan is a microbial polysaccharide with a cellulose backbone in which every other glucose unit carries a trisaccharide as a side chain. This side chain contains one glucuronic acid unit and at the chain end a pyruvate group. Thus xanthan is a polyelectrolyte of a fairly low charge density. The chemical formula of its repeating unit is shown below.

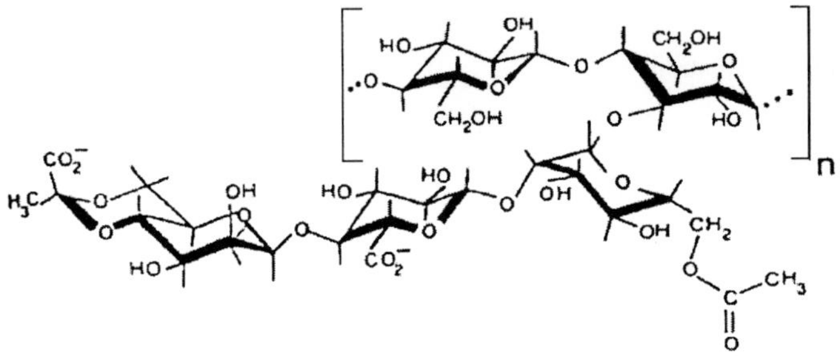

Xanthan easily forms a double-stranded helix and has a high tendency to aggregation, but it does not form gels by itself. However, when a trivalent metal salt is added, gelation can occur by electrolyte complex formation. In the present study aluminium sulfate was used as gel promoter. The actual structure of the complex is not yet known and remains as an issue to be solved by inorganic chemists using crystalline compounds of well defined oligosaccharides. The complex is not formed at low temperatures, and $Al_2(SO_4)_3$ could be added at 15 °C without causing a significant change in solution properties[6].

A technical xanthan was used. Because of remarkable chain stiffness due to the double-helix conformation, xanthan forms rather compact aggregates by side-by-side alignment. Static light scattering of untreated xanthan gave a molar mass of $M_w = 1.4\times10^9$ g/mol which corresponds to about 600 chains aggregated into one cluster. After a heat treatment at 130 °C for 30 min, a decrease in the molar mass to $M_w = 79\times10^6$ g/mol was obtained. The analysis of the scattering curve on the basis of a Casassa-Holtzer plot[7,8] permitted the determination of the linear mass density from which a lateral alignment of 16 double strands on average and a total aggregation of 34 double strands were obtained[6].

These two types of dynamic light scattering experiments led to strikingly different curves. In the *heterodyne* experiment the intercept at zero delay time decreased from the theoretical value of $g_2^{\mathrm{T}}(q,0) = 2$ to a lower value and decayed for sufficiently long delay times in most cases to the base line of unity. In the *nonergodic* experiment the TCF always started at $g_2^{\mathrm{T}}(q,0) = 2$ but apparently approached a plateau value of $g_2^{\mathrm{E}}(q,\infty) > 1$. The following two graphs demonstrate the observations.

Heterodyne registration In Fig. 1, a number of time correlation functions from xanthan solutions containing a certain amount of Al(III) salts is shown which demonstrates the mentioned decrease in the intercept when the temperature was increased beyond 44 °C. From the intercept the magnitude of X (homodyne fraction) or 1-X (heterodyne fraction) can be calculated from the equation

$$X = 1 - \sqrt{2 - g_2^{\mathrm{T}}(q,0)} \tag{6}$$

Nonergodic registration This result is contrasted in Fig. 2 with the TCFs from the same aqueous xanthan solution. The upper part shows the intensity TCF and the lower the field TCF. As already mentioned, all curves start at the value of 2, or 1, respectively, but only for temperatures up to about 45 °C a complete relaxation to the theoretical base line was obtained. At higher temperatures the curves no longer fully decay but seem to reach a finite plateau, whose height is easily determined. The magnitude of parameter *Y* requires a fit of the TCF.

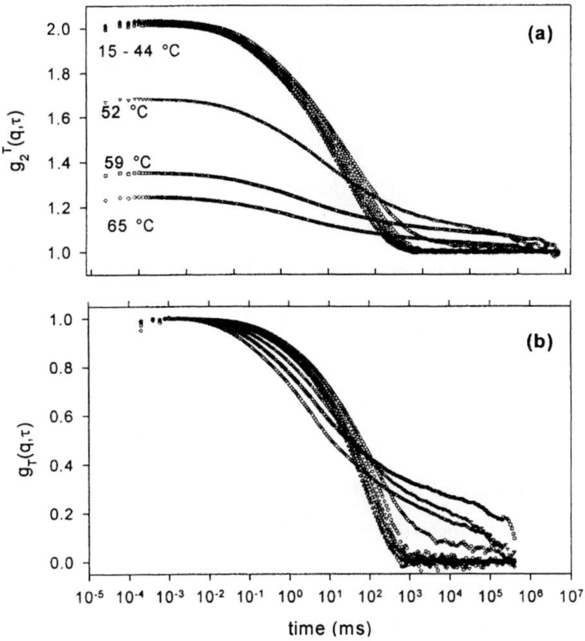

Fig. 1: *Heterodyne experiment:* (a) intensity $g_2^T(q,t)$ and (b) field $g_T(q,t)$ time correlation functions for 1.0 mg/mL xanthan and 2.8 mg/mL Al(III) ions in water for a range of temperatures. Conditions: argon ion laser ($\lambda_0 = 488$ nm) equipped with a pinhole, collection time 40 min before gelation, $T < 45$ °C and 2 h in the gel region, $T > 45$ °C, 30 min thermal equilibration between temperatures.

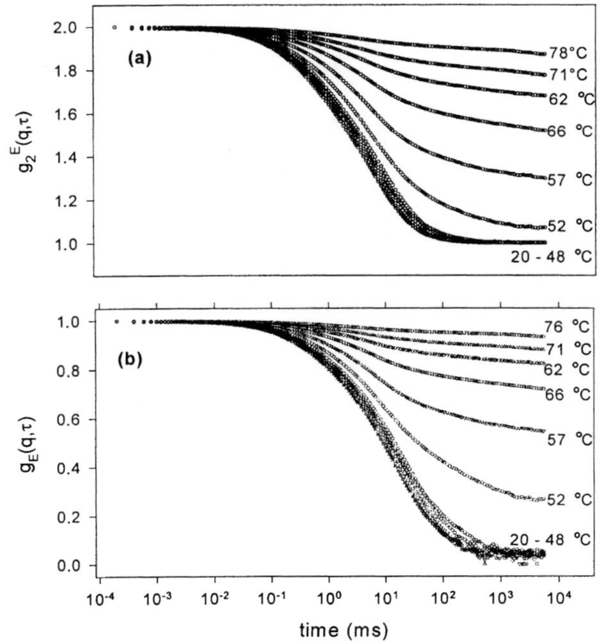

Fig. 2: *Nonergodic experiment:* (a) intensity $g_2^E(q,t)$ and (b) field $g_E(q,t)$ time correlation functions for 1.0 mg/mL xanthan and 2.8 mg/mL Al(III) ions in water for a temperature range. Conditions: krypton ion laser ($\lambda_0 = 647$ nm) with monomode fiber optics, 100 sample rotations with a collection time of 40 s per rotation and 30 min thermal equilibration between temperatures.

Gel point

The observation of nonergodic behavior, i.e., deviation of the ensemble average from the time average, coincided with typical gel behavior. No flow was observed when tilting the cell containing the sample. This behavior was obtained when heating the solution above 44-48 °C. The gel did not dissolve again when the sample was cooled back to room temperature. This qualitative observation could be made more precise in two ways. In one case the heterodyne fraction 1-X was plotted against the temperature and for the nonergodic experiment the plateau height was chosen. The value of X could be determined from the intercept in the intensity TCF of the heterodyne experiments as given by Eq. (6). For the other procedure, the fit of the time correlation functions with suitable model functions had to be performed.

Figure 3 shows a plot of the 1-X values against the temperature and the same for the plateau height in the nonergodic experiment. Let us take the midpoint of the two transitions as the point of gelation. Then, according to the heterodyne data, the gel point occurs close to 46 °C and, according to the nonergodic approach, at about 48 °C. Both data lie in the range of experimental errors. If the onset temperature of heterodyne or nonergodic behavior is taken as characteristics for the gel point, then the temperature would be between 43 and 44 °C.

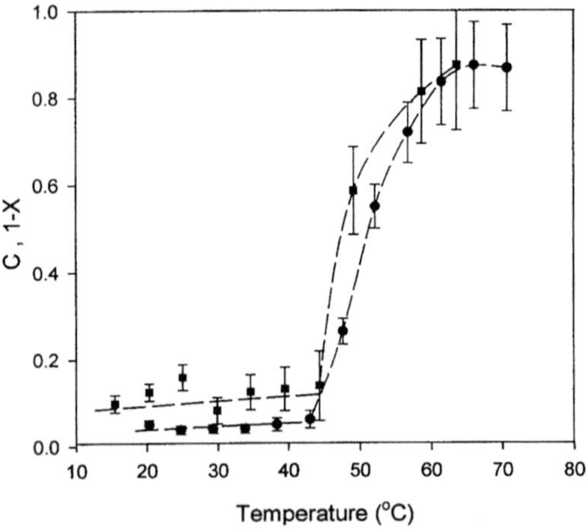

Fig. 3: Plot of the heterodyne fraction 1-X (squares) and the nonergodic plateau height (circles) as a function of temperature.

Concerning the second procedure for quantitative estimation of the gel point, we used the fit of the field TCFs with stretched exponentials. In the *nonergodic* experiment, the field TCF was fitted by two stretched exponentials[9,10,11] plus an additive constant C which described the plateau height $g_{\mathrm{E}}(q, t \to \infty)$

$$g_1^{\mathrm{E}}(t) = A \exp[-(t/\tau_1)^{\beta_1}] + B \exp[-(t/\tau_2)^{\beta_2}] + C \tag{7}$$

where subscripts 1 and 2 indicate the two components, τ is the relaxation time and the value of $\beta \leq 1$ is a measure for the width of the relaxation spectrum. Both relaxations were q^2-dependent and are therefore diffusive. The diffusion coefficients are given by Eq. 8[12]

$$D_{1,2} = \frac{1}{< \tau_{1,2} > q^2} \tag{8}$$

with the mean relaxation times[11]

$$< \tau_{1,2} > = \frac{\tau_{1,2}}{\beta_{1,2}} \Gamma(1/\beta_{1,2}) \tag{9}$$

and $\Gamma(x)$ being the gamma function.

In the *heterodyne* approach, first the value of X given by Eq. (6) has to be determined and then introduced into Eq. (2) to obtain the field TCF, $g_{\mathrm{T}}(q,t)$ This could be described by a relationship similar to Eq. (7) but without the constant term

$$g_1^{\mathrm{T}}(t) = A \exp[-(t/\tau_1)^{\beta_1}] + B \exp[-(t/\tau_2)^{\beta_2}] \tag{10}$$

The relaxation times and the stretched exponents may not be the same as in the nonergodic approach. The fast and slow relaxations will now be considered separately.

The time dependence of the relaxation times of the fast motion is shown in Fig. 4. Remarkably, almost identical behavior was found in both approaches. Strikingly a maximum occurred near the temperature $T = 42$ °C. These findings are important and make clear that in both experiments the same fast relaxation times are registered.

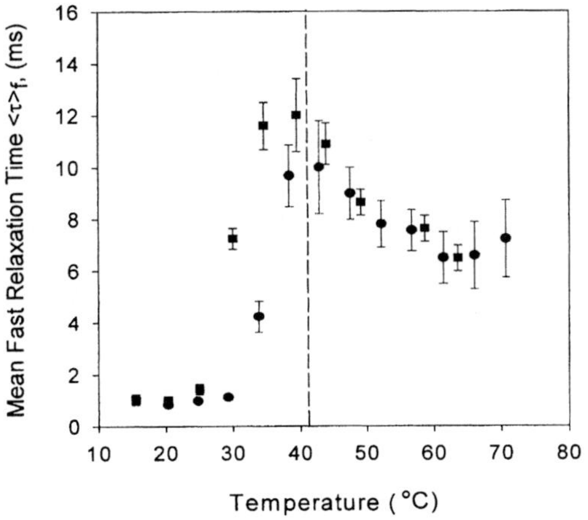

Fig. 4: Mean fast relaxation times (at 90°) from heterodyne (squares) and nonergodic (circles) experiments. The vertical line indicates the gel point.

The occurrence of the maximum can be interpreted as follows. There is much evidence that the fast motion in the pregel state arise from individual chains which become crosslinked at random via, say, point-like bonds. The clusters grow in size with progressing crosslinking as the temperature increases. Eventually, at the gel point, the sizes span the whole space of the light scattering cell, the translational motion of the center of gravity goes to zero and the relaxation time should go to infinity. Once the gel point is passed, dynamic light scattering now probes the mesh size[13] of the formed network, which becomes narrower with progressing crosslinking, and the correlation time decreases again. However, at the gel point the relaxation time actually never goes to infinity, because the network chains are at least semiflexible and the internal modes of motion still remain even when the translational motion of the center of gravity has completely stopped. Similar behavior to that shown in Fig. 4 was observed also with other gelling[14,15] samples. Thus the maximum in the relaxation time can be taken as the gel temperature, which narrows this temperature together with the data from Fig. 4 to $T_{\text{gel point}} = 42\text{-}44$ °C if the onset of heterodyne behavior is taken as the gel point. Another criterion for the gel point followed the suggestion of Martin and Wilcoxon[16], and a theoretical prediction by Muthukumar who claimed power-law behavior of the TCF at the gel point with an exponent n which is determined by the fractal dimension d_f of clusters near the gel point as[17]

$$d_f = \frac{7.5 - 3n}{3 - n} \tag{11}$$

We found such power-law behavior with an exponent $n = 0.63 \pm 0.01$ with a fractal dimension of $d_f = 2.37 \pm 0.06$ which is typical of branched clusters[18-20].

Next the slow motion may be considered. Figure 5 shows the result from the two types of experiments.

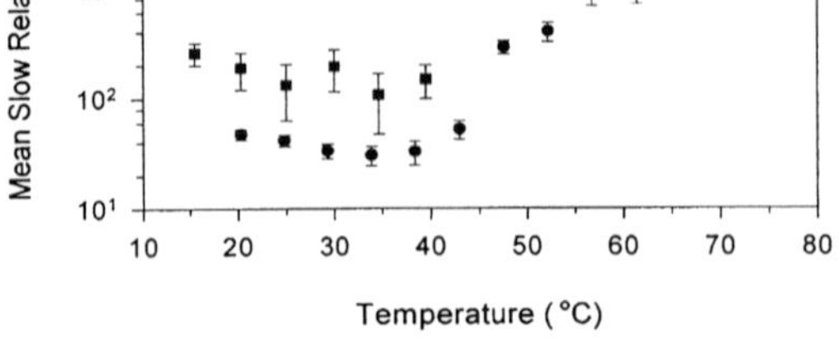

Fig. 5: Mean slow relaxation times as a function of temperature as registered in the heterodyne experiment (squares) and the nonergodic experiment (circles). Note the logarithmic scale in comparison with the linear scale in Fig. 4.

Now the relaxation times determined by the heterodyne approach increase much stronger with temperature and the values are up to 150 times larger than those observed in the nonergodic experiment. The explanation for this stunning discrepancy follows from the initial conditions of the two experiments. The nonergodic experiment was time-consuming because the intensity TCF was measured at 100 different positions and, in the end, the ensemble average was taken. Therefore, the range of delay times had to be truncated. In the present case to 40 s which added up to 1000 s for measurement at one scattering angle. In the heterodyne experiment the longest delay time was 45 min = 2700 s, i.e., about 70 times longer[6]. Thus everything that moves more slowly than 40 seconds is considered in the nonergodic experiment as an immobile rock. This limit was shifted in the heterodyne experiment to 2700 s and what appeared to be a rock was actually still slowly moving. Similar observations were recently made by Cipelletti et al.[21] in small-angle dynamic light scattering from a glassy colloidal material, and even here he found a complete decay to the baseline with a correlation time of about 24 h.

At this point the question arises: what may be the origin of the slow motion ? In xanthan a slow motion is present already at low temperatures without added aluminium salt and causes a huge molar mass of about 1.4×10^9 g/mol with a radius of gyration[6] R_g = 409 nm. This corresponds to an aggregate of about 600 xanthan chains, partly side-by-side-aligned (molar mass of a double-helix (DH) xanthan $M_{w,DH}$ = 3.5×10^6 g/mol [22-25]). Upon a 30-min heat treatment at 130 °C the molar mass was reduced to 79×10^6 g/mol and a radius of gyration of R_g = 315 nm, i.e. corresponding to aggregates (about 16) of about 34 now mainly laterally aligned chains. The aggregates remain present during the crosslinking reaction and apparently grow in size. These rather compact aggregates differ from speckles, which give rise to the nonergodic fluctuations.

Statistics of speckles and evaluation of the fluctuating and constant contributions

The next question to be answered is what the actual size and structure of these heterogeneities may be. This question is a rather involved problem which has not yet been fully solved. We followed a suggestion by Joosten and Pusey[26] and evaluated the spatial fluctuation by a statistical treatment. Like Joosten et al., who studied polyacrylamide gels, we registered the scattering intensity from the xanthan gel at 57 °C at 520 positions time-averaged over 40 s. The result is shown in Fig. 6.

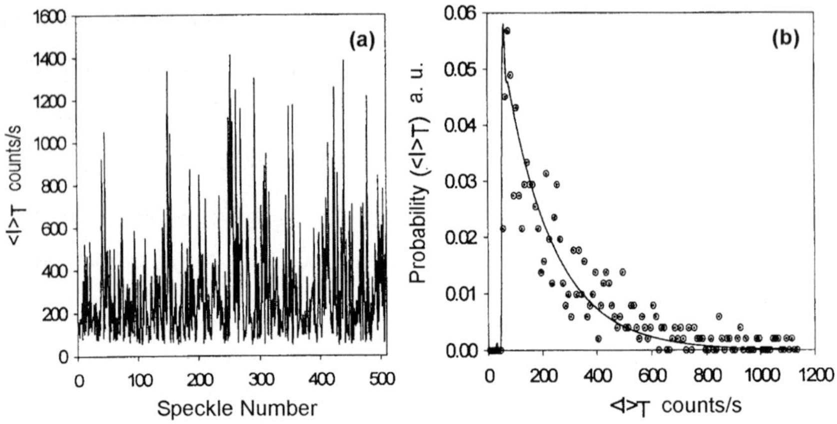

Fig. 6: (a) Scattering intensities at 520 different cell positions averaged over 40 s. Measurements were made with the xanthan gel at 57 °C. (b) Histogram of the peak distribution in (a). The full line corresponds to Eq. (12) derived by Pusey[5,27]

$$P(<I(q)>_\mathrm{T}) \propto H(<I(q)>_\mathrm{T}) - <(I(q)>_\mathrm{F} \exp\left[-\frac{<I(q)>_\mathrm{T} - <I(q)>_\mathrm{F}}{<I(q)>_\mathrm{E} - <I(q)>_\mathrm{F}}\right] \quad (12)$$

In Fig. 6a one notices a fairly constant minimum scattering intensity upon which a strong scatter in the intensity is superimposed. Evidently the constant part represents the fluid term for which the time average equals the ensemble average and the strongly fluctuating peaks relate to the speckles where the peak height is the product of the molar mass of the non-moving particle times the partial concentration of the non-moving heterogeneities. Hence the peak height may correspond to very large domains with a low concentration in the detection volume or it can arise from many but much smaller immobilized domains. Statistical techniques can be applied to get information on the size distribution of these heterogeneities which is shown in Fig. 6b. This distribution could be described[5,27] by Eq. (12) which allowed to extract the desired information on the average size times the amount of immobile domains, which is an ensemble average, and on the contribution of the ergodic scattering intensity. For the xanthan gel, the ratio $<i_\mathrm{F}(q)>_\mathrm{T}/<i_\mathrm{C}(q)>_\mathrm{E} = 0.2$ was obtained[6] at 90°, whereas Joosten[26] obtained for the acrylamide gel at 150° $<i_\mathrm{F}(q)>_\mathrm{T}/<i_\mathrm{C}(q)>_\mathrm{E} = 0.6$. Apparently the polyacrylamide gel contained less rock-like heterogeneities than the xanthan gel.

Angular dependence

Both the fluid part and the large heterogeneities will have a different angular dependence. Measurement of these two components as a function of the scattering angle will provide us with detailed information on the structure of the two components separately. Panyukov and Rabin[28] tried to set up a basic theory of this angular dependence. Their theory is basically an extension of the well-known Tanaka[29] theory of networks who neglected the existence of immobile heterogeneities. The theory is complex and resulted in an angular dependence which is difficult to interpret in physical terms. Their result does not appear to be in agreement with preliminary observations. Probably the theory could be considerably simplified when starting with reasonable assumptions on the structure of the two components. Although the study of the angular dependence will be considerably time-consuming, the invested effort will be worthwhile because, in addition to the information on size and structure, the distribution of sizes of the heterogeneities would be possibly accessible to determination.

Conclusions

Two main results can be obtained from this comparison of the heterodyne and nonergodic experiment: (i) The fast motion is correctly determined in both experiments. It represents the fluid part in a gel which is ergodic and the time and ensemble averages coincide. (ii) The value of the heterodyne fraction $1\text{-}X = 1 - \langle i_F(q)\rangle_T/\langle i(q)\rangle_T$ correlates strongly with the asymptotic plateau height in the nonergodic experiment. Furthermore, dynamic light scattering allows to determine the point of gelation in several ways, among others by the onset of heterodyne or nonergodic behavior, by the maximum of the correlation time corresponding to translational diffusion and by checking when the power-law behavior of the TCF is observed.

It does not seem to make any serious difference which type of experiment is chosen as long as the heterodyne experiment is conducted to very long delay times. The heterodyne experiment is easier carried out and is less time-consuming, but it can no longer be satisfactorily applied in detectors equipped with the monomode fiber optics. Only with pinhole optics we could register a sensibly smooth heterodyne signal. If monomode fiber optics is used, one obtains either a heterodyne signal or just a homodyne response. The reason for this becomes clear from Fig. 7 which shows the time dependence of the scattered light at one cell position. With monomode-fiber optics only a tiny scattering volume is probed. Thus in one case only the liquid ergodic fraction is registered, in another only a speckle. Thus sensible nonergodic observations at one position can be made when either a sufficient number of speckles is present in the detection volume or the detection volume has to be increased to the size of the scattering volume if possible. This was achieved with the pinhole setup. This statement holds true for *pseudo*-nonergodic systems which eventually become ergodic at sufficiently long delay times.

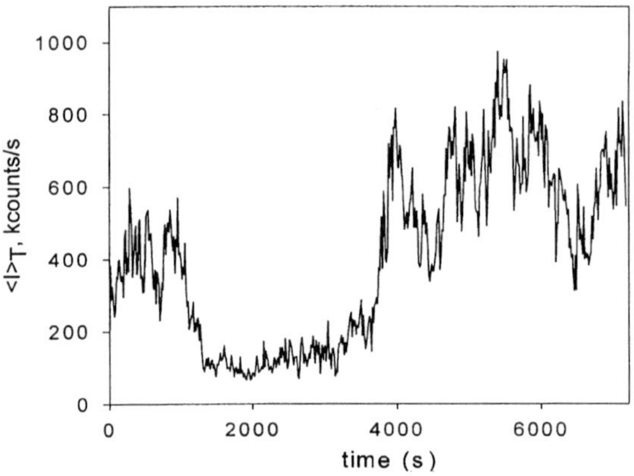

Fig. 7: Change of the scattering intensity from xanthan gel with time observed with monomode fiber optics (57 °C). Each value was time-averaged over 40 s. Note the time scale in the beginning when one speckle is registered. After 1500 s = 25 min, the speckle moved out of the volume of observation, and approximately after 30 min only the ergodic fluid part is registered, until another, bigger speckle entered the observation volume.

References

1. T. Coviello, W. Burchard, E. Geissler, D. Maier, *Macromolecules* **30**, 2008 (1997)
2. W. Burchard, *Biomacromolecules* **2**, 342 (2001)
3. D. B Sellen, *J. Polym. Sci. B: Polym. Phys.* **25**, 699 (1987)
4. E. Geissler, *Dynamic Light Scattering from Gels*, in: *Dynamic Light Scattering*, W. Brown (Ed), Clarendon Press, Oxford 1993, Chap. 13
5. P. N. Pusey, W. van Megen, *Physica A* **157**, 705 (1998)
6. A. B. Rodd, D. E. Dunstan, D. V. Boger, J. Schmidt, W. Burchard, *Macromolecules* **34**, 3339 (2001)
7. E. F. Casassa, *J. Chem.Phys.* **23**, 596 (1952)
8. A. Holtzer, *J. Polym. Sci.* **17**, 432 (1952)
9. R. Kohlrausch, *Poggendorff Anal. Chem.* **91**, 179 (1854); **119**, 337 (1863)
10. G. Williams, D. C. Watts, (a) *Trans. Faraday Soc.* **66**, 80 (1970); (b) G. Williams, D. C. Watts, A. M. Worth, *Trans. Faraday Soc.* **67**, 1323 (1971)
11. G. D. Patterson, *Adv. Polym. Sci.* **48**, 124 (1983)
12. See books on dynamic light scattering, e.g., B. J. Berne, R. Pecora, *Dynamic Light Scattering*, Wiley, New York 1976
13. P.-G. de Gennes, *Scaling Concepts in Polymer Physics*, Cornell University Press, Ithaca, NY, 1979
14. T. Fuchs, W. Richtering, W. Burchard, K. Kajiwara, *Polym. Gels Networks* **5**, 541 (1997);

15. (a) W. Burchard, P. Lang, L. Schulz, T. Coviello, *Macromol. Symp.* **58**, 21 (1992); (b) W. Burchard, T. Aberle, T. Fuchs, W. Richtering, T. Coviello, E. Geissler, L. Schulz, *The Wiley Polymer Networks Group Review Series* **1**, 13 (1998)
16. J. E. Martin, J. Wilcoxon, *Phys. Rev. Lett.* **62**, 373 (1988)
17. (a) M. Muthucumar, *J. Chem. Phys.* **83**, 367 (1985); (b) *Macromolecules* **22**, 4656 (1989)
18. D. Stauffer, *Introduction to Percolation Theory*, Taylor & Francis, London 1985
19. M. Daoud, J. E. Martin, in: *The Fractal Approach to Heterogeneous Chemistry*, D. Avnir (Ed), Wiley, New York 1989
20. W. Burchard, *Adv. Polym. Sci.* **143**, 113 (1998)
21. L. Cipelletti, S. Manley, R. C. Ball, D. A. Weitz, *Phys. Rev. Lett.* **84**, 2275 (2000)
22. G. Holzwarth, *Carbohydr. Res.* **66**, 173 (1978)
23. M. Dentini, T. Coviello, K. Kajiwara, V. Crescenzi, W. Burchard, *Macromolecules* **21**, 312 (1982)
24. G. Berth, H. Dautzenberg, B. E, Christensen, S. E. Harding, G. Rother, O. Smidsrod, *Macromolecules* **29**, 3491 (1996)
25. I. Capron, G. Brigand, G. Muller, *Polymer* **38**, 5289 (1997)
26. J. G. H. Joosten, J. L. McCarthy, R. N. Pusey, *Macromolecules* **24**, 6690 (1991)
27. P. N. Pusey, *Macromol. Symp.* **79**, 10 (1994)
28. S. Panyukov, Y. Rabin, *Macromolecules* **29**, 7960 (1996)
29. T. Tanaka, L. O. Hocker, G. B. Benedek, *J. Chem. Phys.* **59**, 5151 (1973)

Phase Separation Mechanism in Gelling Aqueous Biopolymer Mixture Probed by Light Scattering

*D. Leisner, M.C. Blanco, M.A. López Quintela**

Dept. Química Física, Facultad de Química, Universidad de Santiago, Av. das Ciencias s/n, E-15782 Santiago de Compostela, Spain

SUMMARY: Using time-resolved static and dynamic light scattering (DLS) we have studied the kinetics of phase separation in an aqueous gelatin/maltodextrin mixture upon fast cooling. The time evolution of the droplet radius is modelled for the monodisperse case under reaction-limited and diffusion-limited conditions and compared with the observed evolution of the mode associated with the droplet diffusion. For quenches to above the gelatin ordering temperature, nucleation and rather reaction-limited than diffusion-limited growth and late-stage coalescence of droplets with diameters up to 90 μm were concluded. Quenches to well below the gelatin ordering temperature seem to induce diffusion-limited growth or (delayed) spinodal decomposition (SD) to a phase-separated microstructure with slow late-stage coarsening. In deep quenches, a second slow SD or diffusion-limited cluster aggregation (DLCA) process becomes apparent from the evolution of the static structure factor; the process seems to be related to the maltodextrin gelation in the composite.

1. Introduction

The stabilization of mixtures containing partly incompatible biopolymers is of interest in the development of formulations with controlled textural properties for dairy products. Just a relatively high kinetic stability of a microphase-separated mixture is required: an intermediate morphology, evolved by a slow phase separation process, must be stabilized by a gelation process that proceeds in parallel and is controlled by certain processing variables (e.g. temperature). A gelation process modifies the thermodynamic compatibility of the polymers in the evolving phases and thus the driving force for the phase separation. The process must be designed to finally trap the dispersed phase in the meshes of the gel at the desired scale of microphase separation. Pinning of the intermediate structure results from inhibition of hydrodynamic flows[1] which otherwise would drive its fast coarsening by coalescence[2].

In this work, we want to take a glance how physical gelation influences the kinetics of the phase separation process in an aqueous gelatin (LH1e)/maltodextrin (SA2) system. This process has already been investigated by time-resolved confocal laser scanning microscopy (CLSM) and turbidity measurements[3-7]. The thermodynamic incompatibility of the partly hydrolysed potato starch (maltodextrin SA2) and gelatin in the coiled conformation (above the triple helix melting temperature, ≅ 30 °C) has been investigated by quantitative analysis of coexisting phases[8] and mixing enthalpic measurements[9]. Based on CLSM, turbidity[6] and

 CCC 1022-1360/00/$ 17,50+.50/0

optical rotation results, an experimental binodal curve and its displacement upon gelatin triple helical ordering are available[3].

The gelation of the microphase-separated systems has been monitored by rheology[10,11]; on this basis, a gelation temperature diagram $T_g(c_1, c_2)$ is available[4]. Most of this work has been done applying an identical temperature vs. time profile for processing: individual polymer/water sample preparation at above 60 °C, careful (low-shear) mixing at 60 °C, linear cooling to the final temperature at -1 K/min only few minutes after mixing. However, measurements have also been performed in slower (-0.3 K/min) and faster (< -3 K/min) cooling conditions for comparison[4,5]. The results show that the standard cooling rate (-1 K/min) is mostly to be considered as a fast cooling if the final temperature is only slightly below the gelation temperature or higher if phase separation occurs by a slower nucleation and growth (NG) mechanism.

In the absence of gelatin, the kinetics of the maltodextrin gelation (which is accompanied by the formation of crystallites and thus turbidity increase) has been investigated by Kasapis[12]. For the concentration (c) and temperature (T) dependence of isothermal gel times (t_g), he found the relation

$$\lg(t_g/\mathrm{s}) = 0.038(T/°\mathrm{C}) - 4.53\lg(c/\%\mathrm{SA2}) + 8.67 \tag{1}$$

The maltodextrin gelation in the absence of gelatin is thus very slow: even at 5 °C and concentration c = 12 %, it needs 3 h, and for c = 7 % already 24 h.

In the present work, we try to detect the phase separation mechanism as function of the depth of a fast quench into an immiscible region of the phase diagram, for a single near-critical composition. Cluster growth and phase separation will be detected by measuring the time evolution of the static and dynamic structure factors in light scattering, and by comparing with the turbidity evolution. Gel times will be estimated from time-resolved viscosity measurements. The observed changes in the scattering functions will be compared with the expected growth rates and time-dependent concentration distributions from two simple models for the growth of monodisperse spherical droplets in the reaction-controlled and diffusion-controlled limits. These models will therefore be presented in the following.

2. Growth simulation for monodisperse spherical droplets

Since the growth of nucleated colloidal particles in a matrix of the material (here biopolymer monomers) is a heterogeneous reaction, the most general mechanism involves diffusion of bulk material to the colloid surface, modification at its surface (active site) and finally aggregation to the colloid as a colloid constituent. Once aggregated to the colloid, the structure of the colloid may evolve by internal rearrangement in order to minimize the surface energy. We will compare here only two extreme cases: diffusion-limited growth with fast internal colloid rearrangement as a consequence of an almost zero activation energy for the monomer uptake into the colloid or a considerably low bulk mobility, and reaction-limited growth with a considerably high activation energy for the modification/uptake reaction at the colloid surface, which allows the bulk concentration to equilibrate all the time. We do not consider cluster-cluster aggregation: clusters may grow only by monomer uptake. This corresponds to a Smoluchowski kernel[13] with $k(i,j) \propto i^{\mu}j^{\nu}$, $\mu << 0$ so that $k(i,1) >> k(i,j>1)$ where $k(i,j)$ is the bimolecular rate constant for aggregation of clusters consisting of i and j monomers.

2.1. Diffusion-limited growth in three dimensions with fast colloid rearrangement

Irreversible diffusion-limited colloid aggregation (DLCA) is a mechanism which leads to branched fractal colloids[14]; it includes formation of a system-spanning cluster above a critical concentration[15,16]. In the reversible case and if colloidal rearrangement is at least as fast as diffusion of bulk material to the liquid-like colloid surface, spherical droplets may evolve in order to minimize the surface energy. In a simplified model, we calculate the evolution of the droplet radius upon spontaneous nucleation of monodisperse spheres fixed on a ccp lattice, at time $t = 0$. The droplets grow only by diffusion of monodisperse monomers from an initially homogeneous liquid matrix to their surface. Although the late-stage behavior of a real process is dominated by cluster-cluster aggregation and the early-stage behavior is usually influenced by continuous nucleation and the resulting polydispersity, the early-stage characteristics of a triggered nucleation event that converts dilute inert, repulsive interacting macromolecules to sinks attractive for another major fraction of the macromolecules could be described satisfactorily by such a model.

Considering only the osmotic force, the integral monomer flux (dN_c/dt) to a droplet with radius R_c is proportional to its surface area $A = 4\pi R_c^2$ and to the concentration gradient at this surface, $(\partial c/\partial r)_{Rc}$. The diffusion equation for the monomers diffusing in three dimensions with a concentration-independent diffusion coefficient D along the radial coordinate r reads:

$$\frac{\partial c}{\partial t} = D\nabla^2 c = D\frac{\partial}{\partial r}\left(\frac{\partial}{\partial r} + \frac{2}{r}\right)c \tag{2}$$

We consider a non-adiabatic boundary at $r = R_c$ and a semi-adiabatic boundary at $r = d_{cc}$. This second sink for the monomer concentration, taken also as radial symmetric, represents the first coordination sphere occupied with 12 neighbouring droplets. $n_c = 4\cdot(4\pi/3)/2^{3/2}$ droplets have to be alimented exclusively from the sphere around a central droplet: the integral monomer flux to the shell at d_{cc} is n_c-1 or 4.924 times higher than that to the central sink at $r = R_c$. Starting values of droplet radius and homogeneous bulk monomer concentration are $R_c(t = 0) = R_{c0}$ and $c(r>R_c,t=0) = c_0$, corresponding to a final droplet volume fraction ϕ_∞ at the bulk monomer concentration c_{eq}. The sphere d_{cc} was divided into 10^4 shells of thickness dr. In one of about 10^5 equidistant time steps dt, the excess content of any shell, $(c-c_{eq})4\pi r^2 dr$, was allowed to walk the average walking distance $2(Ddt/\pi)^{1/2}$ in positive and negative r directions weighted by the volumes of the final shells, and being reflected at the semi-adiabatic boundary. Since only walking distances of entire shell thickness dr appear as changes, the content is partitioned into neighbouring shells in that way that the average displacement is satisfied. If the jump ends at a shell with $r < R_c$, the corresponding amount of material enters the sink. R_c is updated after each time step and the (n_c-1)-fold quantity of what has entered to the central sink is withdrawn from the outer shells with $r > d_{cc}-R_c$. Figure 1 shows some results for the droplet radius evolution.

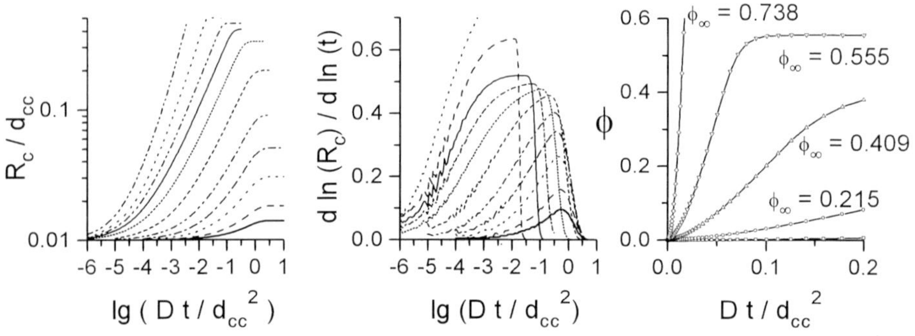

Fig. 1: Evolution of the droplet radius R_c and of the power of its change (dlnR_c/dln$t$) with time, for initial droplet radius R_{c0} = 0.01 d_{cc}. Maximum attainable ϕ from bottom to top: 1.65e-5, 3.55e-5, 1.67e-4, 7.79e-4, 4.45e-3, 0.0474, 0.215, 0.409, 0.555, 0.738, 0.888. The last graph shows some $\phi(t) \propto R_c^3(t)$ on linear scale.

We observe that the power exponent 0.5 is usually the upper limit for the growth of the radius with time for droplet volume fractions below the droplet-overlap threshold. For low final droplet volume fractions, the maximum slope could be as low as 0.1 or less, and the time range where it could be observed is much less than one decade. The early stage increase in the droplet volume $\phi \propto R_c^3$ is faster than linear, with the limit $\phi(t) \propto t^{3/2}$, but never exponential. An exponential increase in $\phi(t)$ might result if continuous nucleation would be considered.

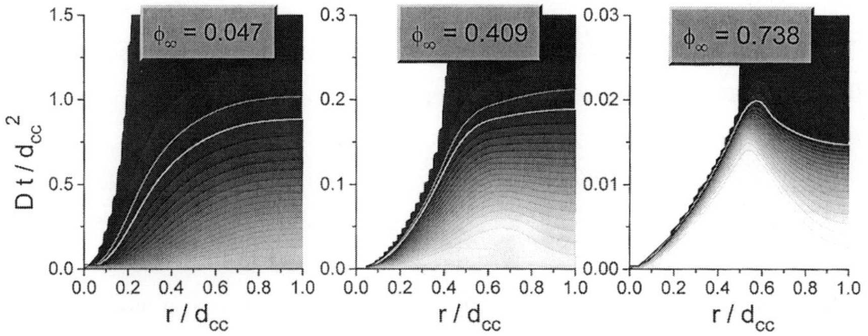

Fig. 2: Time dependence of the radial distribution of the bulk concentration of the monomer. From light grey to black: 1 to 0 times the initial excess concentration; white: central droplet.

Figure 2 shows three examples for the time dependence of the radial distribution of the bulk monomer concentration. The concentration at the colloid surface immediately drops to the equilibrium concentration c_{eq}, and a quasi-stationary monomer concentration gradient around the droplet establishes (depletion layer), which decreases proportionally to the average excess bulk concentration $\langle c(t)\rangle$ - c_{eq}. If the bulk concentration is not much lower than the concentration in the droplet (at high ϕ_∞), the rapidly expanding droplet always lets the gradient high; the depletion layer is a thin coating on the droplet. With decreasing ϕ_∞, the depletion layer becomes thicker and provides less contrast against the bulk. Generally, when ϕ has reached about 95 % of its final value ϕ_∞, depletion layers overlap and the growth slows down considerably.

The depletion layer provides a characteristic length in the radial distribution function and thus contributes a nonzero maximum to the static structure factor $S(q)$, at $q_m \geq 4\pi/d_{cc}$.

The time evolution of this maximum should be characterized by a continuous left-shift of q_m and an intermediate maximum of $S(q_{m},t)$. q_m should be highest for high ϕ_∞, and the peak disappears when depletion layers overlap. This breakdown is almost sudden at high ϕ_∞. In any case, there should be a second maximum of $S(q)$ due to the growth of the spheres, which is expected to evolve at lower q. We do not regard the periodic part from the ccp lattice model.

2.2. Reaction-limited growth in three dimensions

The growth of monodisperse spherical droplets with initial mass M_{c0} at time t=0 and mass density ρ, in a matrix with bulk monomer concentration c, is proportional to the actual excess bulk concentration c - c_{eq} and to the colloid surface, $(36\pi)^{1/3}(M_c/\rho)^{2/3}$:

$$\frac{d(M_c / M_{cmax})}{dt} = k'(1 - M_c / M_{cmax})(M_c / M_{cmax})^{2/3} \tag{3}$$

with the rate constant

$$k' = k_1(c_0 - c_{eq})\frac{(36\pi/\rho^2)^{1/3}}{1 - M_{c0} / M_{cmax}} \tag{4}$$

$c_0 - c_{eq}$ provides material to let the droplets grow to a maximum mass M_{cmax}; k_1 is the rate constant for the uptake of a bulk monomer to the droplet at its surface, and $k_{-1} = k_1 c_{eq}$ is the rate for the dissociation of a monomer from the droplet which is also proportional to the surface, neglecting the effects of curvature since we are interested in the equilibrium for the final droplets with radii of several micrometers.

When at $M_c/M_{cmax} << 1$, the bulk concentration remains still close to its initial value, integration leads to an expression for the asymptotic behavior

$$\lim_{M_c \to M_{c0}} \frac{M_c}{M_{cmax}} = \left[\frac{k't}{3} + \left(\frac{M_{c0}}{M_{cmax}}\right)^{1/3}\right]^3 \tag{5}$$

Numerical integration yields the results presented in Fig. 3: the maximum exponent for the cluster mass increase with time is 3 and may be observed for even more than one decade of times, until the bulk concentration has decreased by about 10 %. The droplet radius $R_c \propto M_c^{1/3}$ increases about linearly with time in the range up to $R_{cmax}/2$. The maximum growth rate for the mass is at $M_c/M_{cmax} = 0.4$ whereas for the radius it is at $t = 0$.

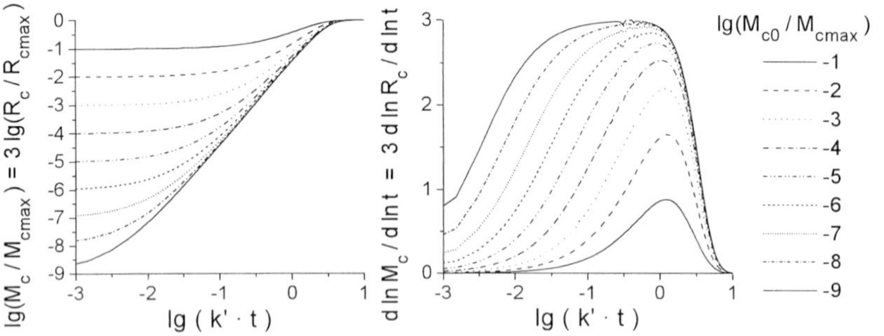

Figure 3: Evolution of the droplet mass M_c and of the power of its change ($d\ln M_c/d\ln t$) with time for different initial droplet radii or masses M_{c0}.

3. Experimental

3.1. Sample preparation

Limed hide gelatin (LH1: Bloom 240, pI 4.7, M_n = 83300, M_w = 146 000) was provided from SKW Biosystems. Potato starch maltodextrin (Paselli SA2: glucose equivalent = 2, $M_n \cong$ 9000, highly polydisperse) was provided from Avebe. Samples (pH 6-7, ionic strength 0.01 M by NaCl) were prepared by mixing biopolymer and a filtered salt solution in bidistilled water, then gently stirring for 30 min at 60 °C (gelatin) or 95 °C (maltodextrin). The solutions were carefully mixed at 60 °C (about 1:1) and let in a stove for some 3 min with gentle shaking. Cylindrical cuvettes (1 cm external diameter, filled to 1/3, for light scattering) and standard quartz cuvettes (d = 2 mm, for UV/VIS spectroscopy) were stoppered tightly in order to avoid solvent evaporation.

3.2. Thermal ramping

"Rapid" exponential cooling from $T(t = 0)$ = 60 °C to temperatures T_∞, starting at time t = 0, was performed at cooling rates of $dT/dt = -5\ \text{min}^{-1}\cdot(T-T_\infty)$ (light scattering) or $-0.8\ \text{min}^{-1}\cdot(T-T_\infty)$ (UV/VIS); standard prethermostatted cell holders were used. 99 % (light scattering, viscosity) or 45 % (UV/VIS) of the initial temperature difference were equilibrated within 1 min. All final temperatures were maintained with at least 0.25 K absolute precision.

3.3. Dynamic light scattering

An ALV SP-86 goniometer equipped with a PMT detector and a Spectra Physics 127 helium-neon laser, operating at wavelength λ_o = 632.8 nm, was employed. Intensity data from θ = 90° angle reflections (at a scattering vector q = 18.7 µm^{-1}) were recorded and their time correlation analyzed by an ALV-5000 multiple-tau correlator (ALV, Langen, Germany). Normalized time-averaged intensity correlation functions $g_2^T(t)$ were accumulated initially for 30 s, the accumulation time increasing geometrically by a factor 1.05. All data were analyzed within the ergodicity approximation $g_2^T = g_2^E$, where g_2^E is the ensemble-averaged intensity autocorrelation function. The latter is linked to the electric field autocorrelation function g_1 by the Siegert relation: $g_1 = (g_2^E - 1)^{0.5}$. g_1, proportional to the dynamic structure factor, was thus fitted to a spectrum of exponential decays: $g_1 = \int A(\tau)\cdot\exp(-t/\tau)\mathrm{d}\ln\tau$ by the CONTIN algorithm. We have obtained relaxation time spectra $A(\tau)$ in the τ range from 10^{-6} s up to 10^3 s on logarithmic scale. Assuming homodyne light beating and the absence of multiple scattering, the distribution of diffusivities D was obtained applying the relation $D = 1/(q^2\tau)$, where $q = (4\pi n/\lambda_o)\sin(\theta/2)$ is the scattering vector and n is the refractive index of the solution. The apparent correlation lengths $(\xi\cdot\eta/\eta_0)$ for cooperative displacements $\Delta r = q^{-1}$ of inhomogeneities in the optical density were obtained applying the Stokes-Einstein relation, $D = kT/6\pi\eta\xi$, where the effective (dynamic-process-dependent) viscosity η takes values in the range from the solvent viscosity η_0 to the macroscopic viscosity.

3.4. Static light scattering

Ensemble-averaged intensity data from a slowly rotating sample were collected at 10 scattering angles θ; a single scanning cycle was completed within 84 s. Data from seven identical quenching experiments with coherent results were collected to enhance the time resolution.

3.5. Turbidity measurement

The simultaneous time (t) evolution of the transmittance of light at several wavelengths λ (532, 582, 632, 800 nm, away from absorption bands of the polymers) through a thermostatted quartz cuvette (path length d = 2 mm) was measured using an HP 8452A spectrophotometer equipped with deuterium lamp and diode-array detector, with the transmittance of the water-filled cuvette set to 1. The turbidity coefficient τ is calculated as $\tau(\lambda,t) = -\ln(\text{transmittance}(\lambda,t))/d$.

3.6. Viscosity measurement

Seven ml of a hot sample is poured into the cup of an SSA21 measurement system of a Brookfield DV2+ viscometer and the data acquisition is started ($t = 0$). The cup is lifted to its fixed position in the thermostatted jacket and the prethermostatted rotating spindle dives in. Temperature equilibration in the 1.2 mm thick gap is at least as fast as in the light scattering experiments. Continuous shear of less than 2 Hz (usually 0.28 Hz) is applied. The gel time t_g is approximated (though underestimated) by the time when a first linear increase in the measured torque starts to develop.

4. Results

We present time-resolved data for an off-critical composition in a ternary system of water and the biopolymers gelatin LH1 (4 %) and maltodextrin SA2 (7 %). The turbidity evolution will be analysed as a measure of generated interface by phase separation. The viscosity evolution was observed as a sensor for gelation. Dynamic light scattering results will be presented for the binary subsystems in order to identify the modes for the single polymers under identical conditions. Then the modes in the decomposed DLS spectra of the ternary system will be identified and analysed, taking into account the turbidity and viscosity information. Finally static light scattering results are presented. The results are classified into two groups: (i) quenches above the gel melting temperature of gelatin which are generally quenches into a metastable region of the phase diagram, ii) quenches below this temperature where the evolution is influenced by gelation of the continuous gelatin-rich phase.

4.1. Turbidity evolution during phase separation

The optical density contrast on the length scale comparable to the wavelength of the incident light provides light scattering and thus turbidity. The relative refractive index difference of the phases is low enough to consider Rayleigh-Gans scattering, at least until SA2-rich droplets collapse and the carbohydrate starts to crystallize. The turbidity at a given wavelength then increases proportionally to the concentration of the interface and to the square of the refractive index difference between the separated domains. Figure 4 shows the evolution of the turbidity with time after fast quenches of the system to different final temperatures.

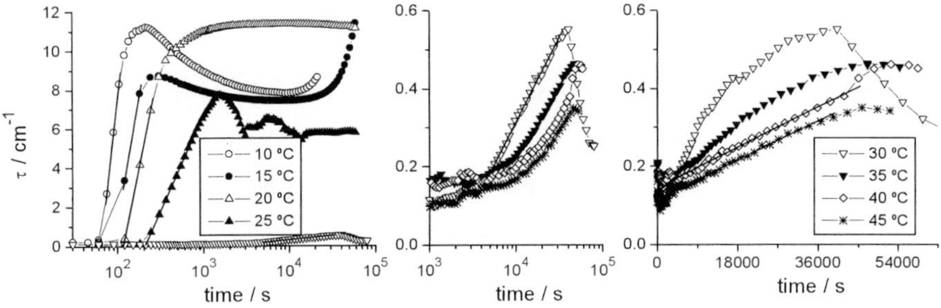

Fig. 4: Turbidity evolution at λ = 800 nm with time after a rapid quenching from 60 °C to different final temperatures.

Table 1: Delays for onsets of turbidity increase and rates of the turbidity increase.

T / °C	Onset t_1 / s	$(d\tau / dt)_{t>t1}$ / $cm^{-1}s^{-1}$	Onset t_2 / s	$(d\tau / d\ln t)_{t>t2}$ / cm^{-1}
45	2100 ± 500	$(5.6 \pm 0.1)\cdot 10^{-6}$	--	--
40	1500 ± 500	$(5.8 \pm 0.1)\cdot 10^{-6}$	(42000 ± 3000)	(0.17 ± 0.02)
35	1500 ± 1000	$(6.8 \pm 0.2)\cdot 10^{-6}$	9400 ± 500	0.17 ± 0.02
30	700 ± 500	$(26 \pm 1)\cdot 10^{-6}$	4400 ± 300	0.19 ± 0.01
25	--	--	220 ± 9	4.24 ± 0.03
20	--	--	117 ± 7	9.5 ± 0.2
15	--	--	96 ± 5	11.0 ± 2.0
10	--	--	66 ± 5	16.7 ± 0.4

After a delay of t_1 which is the longest for the shallowest quench (to 45 °C), the turbidity increases. The increase is linear in time for shallow quenches to final temperatures where

gelatin ordering does not yet occur (above 30 °C). In this regime, the rate of the linear turbidity increase, $d\tau/dt$, increases only slightly with the depth of the quench (see Table 1).
For a quench to 30 °C or below, another turbidity increase starts after a delay t_2. The rate of this turbidity increase *decreases* exponentially with time – the turbidity data (up to 70 % of the maximum) yield a straight line if displayed on logarithmic time scale. The data for 35 °C and 40 °C show both effects: first a linear turbidity increase with the rate similar to that at the higher temperature 45 °C and, after about 3 h (35 °C) or 12 h (40 °C), a faster but more time-dependent turbidity increase. The final turbidity value which corresponds to that of the upper phase in a bulk liquid-liquid phase separated system is identical for both temperatures but already higher than for the shallow quench to 45 °C.
For the quenches to 30 and 25 °C, a turbidity decrease is observed after a *late-stage* maximum was reached. The decrease was due to the beginning bulk phase separation. Whereas for the quench to 30 °C the final turbidity observed is very close to its equilibrium value in the upper phase, an equilibrium was not reached after the quench to 25 °C because, after about 1 h, the gelatin-enriched upper phase gels (see viscosity increase, Fig. 5), which precludes further sedimentation of droplets of the maltodextrin-enriched phase. Instead, the gelation boosts further incompatibility and a second wave of phase separation in the actual upper phase, as can be concluded from the observed new turbidity increase. This turbidity increase is not delayed because the concentration of small droplets (nuclei) is still high.
For the quenches to below 25 °C, the rapid viscosity increase (gelation) has the same delay as the first turbidity increase. This suggests that the large pregel clusters provide the necessary incompatibility for a subsequent fast phase separation parallel to the gelation process. The turbidity increases exponentially rather than linearly at an early-stage time and then turns to an intermediate-state maximum which is much higher than that observed at above 25 °C. For the quench to 20 °C, the turbidity maintains its high maximum value for a long time. Bulk phase separation is not observed after three days, not even for deeper quenches. Then, for quenches to 15 and 10 °C, turbidity decreases in a characteristic way, after the maximum of a similar height has been reached. This turbidity decrease will be discussed later (Section 5.2).

4.2. Viscosity evolution during phase separation and gelation

The gelation process is monitored by second-order changes in the viscosity evolution. The molecular weight of the gelling polymer is expected to critically increase to "infinite" in the vicinity of the sol-gel transition and formation of a percolation cluster. Thus an accelerated viscosity increase precedes gelation. A viscosity decrease would be expected to occur if the porosity of the interface between phases of different viscosity decreases. Upon bulk phase separation, the measured system viscosity would approach the viscosity of the less viscous phase, which in a shear field could organize itself as a continuous film normal to the gradient of the shear if the shear force is stronger than the sedimentation force.

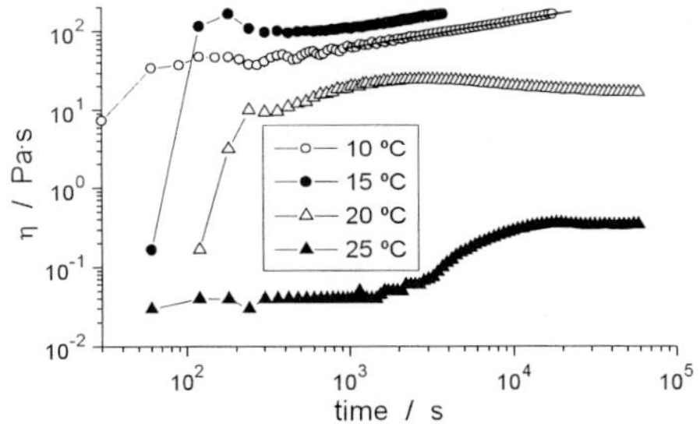

Figure 5: Time evolution of viscosity at a shear rate of 0.28 Hz or 2.8 Hz (▲) after rapid quench to various final temperatures

The viscosity increase due to gelation is observed after about 45 s (10 °C), 100 s (15 °C), 170 s (20 °C) or 4000 s (25 °C). For the lowest temperatures, the viscosity increases further with a power-law behaviour at a late stage. The fitted straight line in Fig. 5 yields the power exponent 0.332 ± 0.003. This might correspond to gelation of the dispersed phase. If the viscosity of the composite is proportional to the molecular weight of the filler, the dispersed carbohydrate clusters, at a constant monomer concentration (Staudinger relation), then the molecular weight of the clusters increases also with $t^{1/3}$, just as predicted for the Lifshitz-Slyozov diffusion-limited coarsening mechanism with a conserved order parameter[17]. In the case of the weaker gels formed at 20 and 25 °C, the system viscosity decreases slowly after a late-stage maximum was reached. The maximum itself is reached with a non-constant power of the curing time t. The power exponent is first larger than 1/3, much closer to 1 at $T = 25$ °C and, just before the maximum, it decreases to a value which is again close to 1/3. The exponent of 1 is indeed expected for a coarsening process driven by hydrodynamic flow[2,18], which at 25 °C is not immediately inhibited in the still large pores of the weak gel.

4.3. Dynamic changes upon cooling of *binary* aqueous biopolymer mixtures

Figure 6 shows dynamic light scattering data from quenching experiments with only one (though polydisperse) biopolymer present. Even at 60 °C, the temporary networks exhibit slow optical density fluctuations with relaxation times in the range from 5 to 50 ms, indicating inhomogeneities with Stokes radii of the order of 1 μm. In the maltodextrin samples, upon cooling, the Stokes radius shrinks to about half of its high-temperature value and then increases with time, whereas the scattering intensity already increases from the very beginning. One can almost certainly attribute the observed inhomogeneities (mode **d**) to large branched amylopectin molecules. Upon a temperature decrease, the solvent becomes poorer in this carbohydrate, the branches collapse and let the molecule assume a more compact and dense structure (**d₁**), which would also cause an increase in scattering intensity. The enhanced solvent-solvent interaction makes the molecules sensitive to hydrophobic intra- and intermolecular interactions and subsequent aggregation, as concluded from the further increasing scattering intensity and a slowing-down of mode $\mathbf{d_1}$.

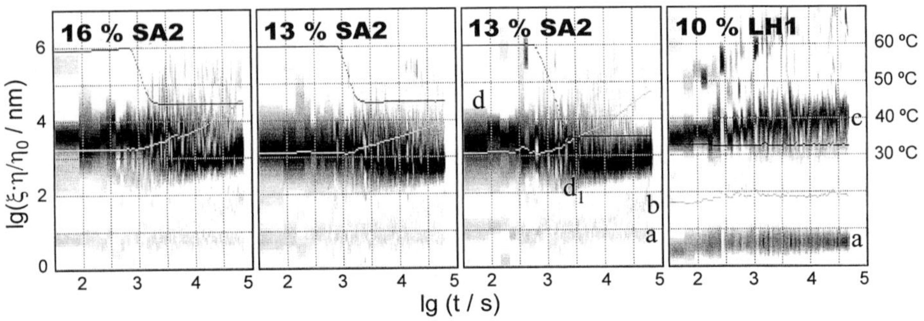

Figure 6: Time evolution of the dynamic structure factor for maltodextrin / water and gelatin / water binary mixtures upon slow (SA2) or fast (LH1) cooling from 60 °C. Grey scale: g_1 amplitude; dark grey line: temperature; light grey line: logarithmic static scattering intensity (two grids/decade).

In the gelatin solution, the slow mode (**c**) is due to the temporary gel which has to rearrange its crosslinks to fully relax fluctuation. The observed relaxation time of 30 - 100 ms in the 10 % LH1-containing aqueous solution then is the typical average lifetime of the temporary crosslinks[19)]. This average increases with time since the most stable configurations (which might already include hydrogen bonds) survive longest. Below the helix-coil transition temperature (30 °C), helical ordering of the collagen-like molecules is favoured by synergistic hydrogen bonding in long-living triple-helical segments, and by percolation of molecules connected by these junctions physical gelation may occur.

The fast mode (**a**) indicates the correlation length ξ, the average mesh size of the temporal gel. Its increase with time in the case of the gelatin sample is in accord with the decrease in the concentration of total chain length in the system, due to partial triple-helical alignment. In the maltodextrin samples, the correlation length ξ diverges with time (modes **a**, **b**). Although this would have to be investigated in more detail, it could be related to partial gelation upon double-helical ordering of amylose and amylopectin fragments.

4.4. Quench of the LH1/SA2 mixture to above the gelatin ordering temperature

From CLSM with stained polymers, we know that upon shallow quenches (to between 50 and 30 °C), disperse maltodextrin-enriched spherical droplets from 2 to > 20 μm in diameter develop in the gelatin-enriched continuous phase and finally sediment as a bulk liquid phase. Our dynamic light scattering results for fast quenches to such temperatures are shown in Fig. 7. The gelatin modes are denoted **a** and **c**. The maltodextrin fast modes are **a** and **b**. The most intense mode, denoted **d**, belongs to the maltodextrin subsystem (amylopectin). As in the binary system, the major part ($\mathbf{d_1}$) of this mode becomes more rapid upon cooling (collapse of branches). The mode then by accident superimposes the less intense gelatin slow mode **c**. In contrast to the binary system, part of mode **d** maintains the high-temperature value and starts to slow down directly (mode $\mathbf{d_2}$). At 45 °C, this is the major part, whereas at 30 °C the minor one.

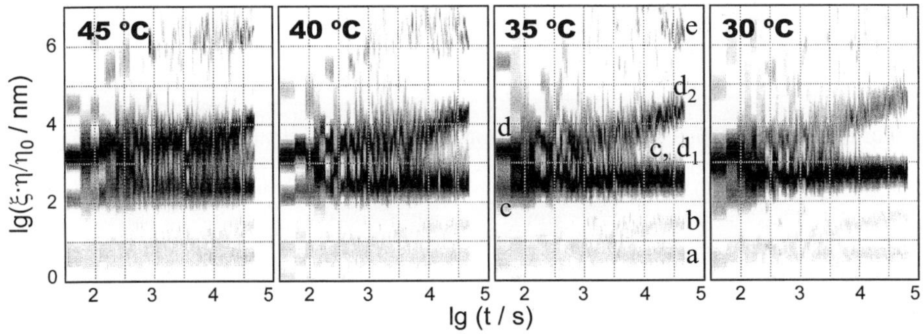

Fig. 7: Evolution of the relaxation spectrum with time after fast quench to various temperatures, at $q = 18.7\ \mu m^{-1}$. The grey scale is the almost linear g_1 amplitude, but the lowest amplitudes (> 0.002 A_{max}) appear enhanced for suitable contrast.

Obviously the presence of the gelatin inhibits the fast collapse of the amylopectin branches and makes them more susceptible to aggregation of faster diffusing linear amylose fragments, so that they grow to clusters and then droplets with Stokes radius $R_c = \xi(\mathbf{d_2})$. Since the size-dependent cluster viscosity is in between that of the solvent (water), η_0, and that of the macroscopic system viscosity, η, the figures indicate the upper size limit – the real radius could be about one order of magnitude smaller. The later observed mode **e** is due to sedimenting droplets formed by bulk liquid-liquid phase separation. We will focus our attention on the evolution of the cluster diffusion mode $\mathbf{d_2}$ presented in Fig. 8.

The apparent droplet radius is initially $\eta/\eta_0 \cdot (3.6 \pm 0.6)\ \mu m$, increasing linearly with time. When it reaches about half of its observed maximum value, the growth rate decreases. This behaviour is similar to that predicted by the reaction-limited growth model (Section 2.2.). The initial growth rate and the observed "final" radius (after 50 000 s) increase with decreasing temperature (see Table 2), whereas the intensity scattered from the droplets at $q = 18.7\ \mu m^{-1}$ remains almost unaffected – it even decreases with decreasing temperature and reaction time. Interestingly, the radial growth rate normalised to the droplet surface area remains virtually constant, indicating a curvature contribution to the kinetic rate constant.

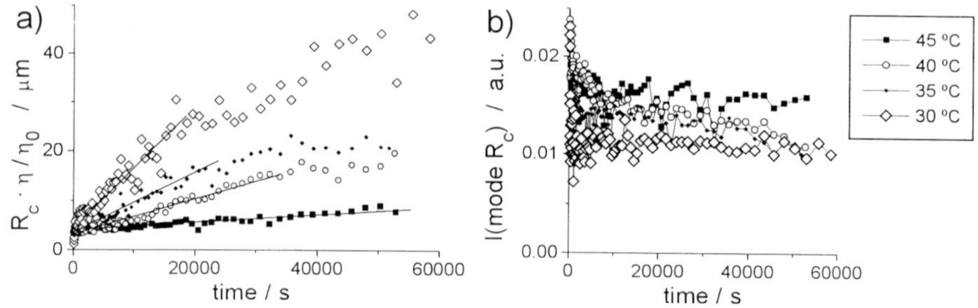

Fig. 8: Evolution of the droplet mode **d** – **d₂** with time after quench. a) apparent Stokes radius, b) the mode amplitude in g_1 multiplied by the static scattering intensity.

Table 2: Initial and final (50 000 s) droplet radii, growth rate and intensity from Fig. 8.

T / °C	η/η_0	$R_{c0}\cdot\eta/\eta_0$ / μm	$R_{c\infty}\cdot\eta/\eta_0$ / μm	$dR_c/dt\cdot\eta/\eta_0$ / $nm\cdot s^{-1}$	$dR_c/dt/(\eta/\eta_0)/R_{c\infty}^2$ / $m^{-1}\cdot s^{-1}$	I_∞ / I_0
45	12 ± 2	4.1 ± 0.4	9 ± 1	0.08 ± 0.01	0.99 ± 0.34	0.90
40	12 ± 2	3.2 ± 0.4	19 ± 3	0.36 ± 0.03	0.98 ± 0.39	0.56
35	13 ± 2	3.6 ± 0.6	23 ± 3	0.60 ± 0.05	1.13 ± 0.39	0.66
30	14 ± 2	4.2 ± 0.9	45 ± 5	1.24 ± 0.10	0.61 ± 0.18	0.91

Since the droplets are at least four times larger than the inverse of the scattering vector, the mode scattering intensity should be proportional to the integral droplet surface (Porod's law). The decrease in the intensity shows that droplets either sediment or coagulate at lower temperatures. The enhanced tendency to coagulation can be explained by less repulsive droplet-droplet interaction potentials, which should decrease in the same manner as the activation energy for the monomer uptake by the droplet. In fact, the late-stage evolution of the droplet **d2** radius produces a slope close to 0.7 on the double-logarithmic plot of R_c vs. time (Fig. 7). A slope between 0.5 and 1 is expected for a coagulation mechanism in liquids[20)], and a slope of 1/3 would be expected for the much slower Lifshitz-Slyozov ripening[17)]. Since the system above 30 °C is still a rapidly fluctuating temporal network, hydrodynamic flows are not inhibited and activate coagulation and sedimentation. Mode **e** appears together with macroscopic phase separation being indicative of macroscopic sedimenting droplets, since we observe the upper region of the sample.

4.5. Quench to below the gelatin helical ordering temperature

Figure 9 shows the same system quenched to below the helix-coil and gel-sol transition temperatures of gelatin. The gelation time coincides with the time when the fast slow-down of mode **c** ends; the mode then disappears because the infinite cluster becomes a homogeneous system and the pores become large-scale inhomogeneities. There is a crossover of modes **c** and **d** at the beginning of gelation. The stability of mode **d** indicates that amylopectin clustering is inhibited by the gelation of the metastable system. Large inhomogeneities, which could be attributed to dispersed clusters partially trapped in the polydisperse pore system, become apparent from mode $\mathbf{d_2}$ (phase microseparation).

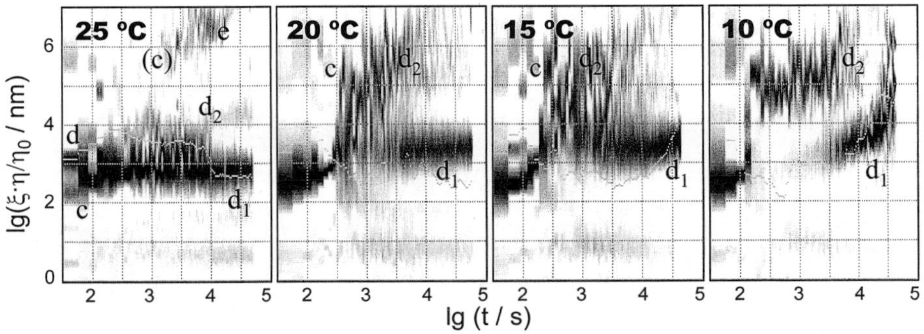

Fig. 9: Evolution of DLS spectra after temperature jump from 60 °C (cf. Fig. 5).

The evolution of the microdroplet radius R_c derived from the correlation length ξ of mode $\mathbf{d_1}$ ($R \cong \xi \cong$ 1-15 μm, using solvent viscosity) agrees with results from time-resolved confocal laser scanning microscopy (CLSM)[4]. This suggests that mode $\mathbf{d_1}$ corresponds to the fraction of dispersed clusters that can diffuse freely in gel pores on a length scale larger than $2\pi/q$; in this case larger than 336 nm. It is interesting to see that, even more extremely than at higher temperatures, in the late-stage evolution mode $\mathbf{d_1}$ is gaining intensity at the expense of mode $\mathbf{d_2}$. This indicates that smaller pores of the gelatin network, which trap the clusters of fraction $\mathbf{d_2}$, tend to disappear and the average pore size increases so that more and more carbohydrate clusters move "freely" on the observed length scale $2\pi/q$. The coarsening of the gelatin network microphase boundary is most likely due to a ripening mechanism with breaking and new formation of gelatin crosslinks.

Once the carbohydrate clusters are no longer isolated in separate pores of the early-stage gel, they become again subject to large-scale diffusion and cluster-cluster aggregation. Either diffusion-limited cluster aggregation (DLCA) or spinodal decomposition (SD) becomes evident from the new exponential increase in both the relaxation time and the intensity of mode $\mathbf{d_1}$ in the late-stage evolution at 15 and 10 °C. At 10 °C, even the tendency to form an entire-pore-filling "infinite" cluster is evident from the divergent slowing-down of mode $\mathbf{d_1}$; this would correspond to the gelation of the dispersed phase.

A late-stage turn to a delayed isothermal SD process is unlikely except if the spinodal of the non-equilibrium system is moved by a slow process like polymerization, supramolecular complexation or physical gelation.

Time- and angle-resolved static light scattering results for quenches of increasing depth into the region below the gelatin ordering temperature are presented in Fig. 10. Although the data are very noisy and the resolution is poor, it can be seen that the static structure factor $S(q)$ increases first at higher q (corresponding to correlation distances $2\pi/q$ of about 400 nm).

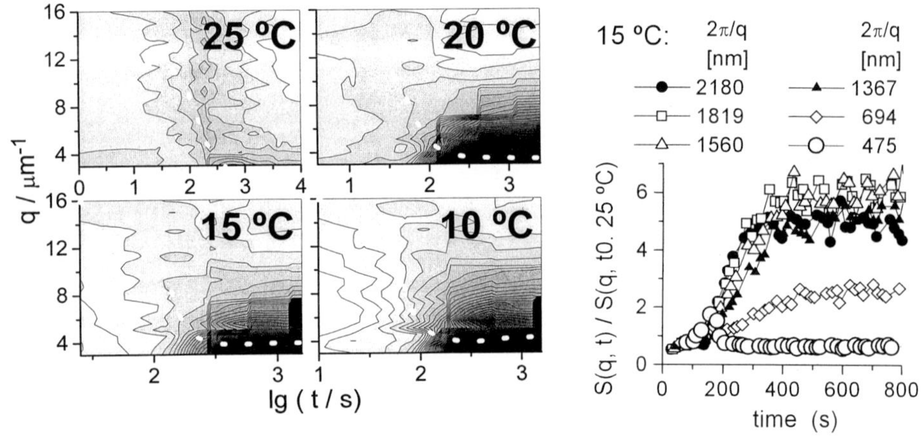

Fig. 10: Evolution of the static structure factor, $S(q,t) = I(q,T,t)/I(q,T_0{=}25\ °C,\ t_0{=}0)$, from 0.25 (white) to 6.25 (black), step size 0.25. The estimated $q_m(t)$ is indicated by a dotted line. Right-hand side: data obtained at 15 °C, on linear scale.

At $t \cong 150$ s, the intensity at $q > 10\ \mu m^{-1}$ passes through a sharp maximum and then decreases suddenly whereas, at the same time, the intensity at $q < 7\ \mu m^{-1}$ increases rapidly and from then on dominates the scattering function. Within the experimental uncertainty of the cooling rate, this is the time when the turbidity increases sharply. The position of the maximum of $S(q)$ at a given time t, $q_m(t)$, is roughly indicated by a white dashed line. It can be observed that q_m moves far out of the observed window at 25 °C, indicating a continued growth of the main correlation distance (e.g. of the droplet radius). Instead, at 20 °C and below, q_m apparently remains then close to 4 μm^{-1} and, at 10 °C, even turns to increase at later times.

There is a broad rather than a peaked maximum at the early stage, and possibly a second maximum at small angles just outside the experimental window.

The time evolution of the small correlation lengths was followed at enhanced time resolution. At fixed $q = 18.7\ \mu m^{-1}$ (see Fig. 11 and Table 3), rather an exponential than linear increase in the structure factor with time was observed – an exponential increase at q_m would be predicted for either the early stage of spinodal decomposition (SD) from the linear Cahn-Hilliard approach and, approximately, for DLCA. The PS mechanism will be discussed in Section 5.3.

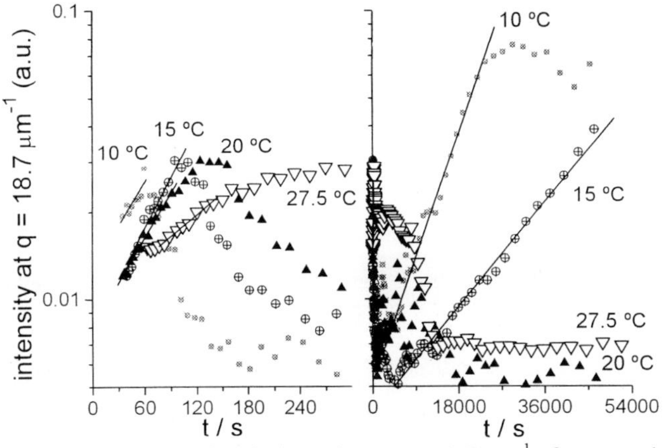

Fig. 11: Time evolution of static intensity at $q = 18.7\ \mu m^{-1}$ after quench.

Table 3: Fitting results from Fig. 11[a].

T / °C	1st SD or DLCA			2nd SD or DLCA		
	t_{start} / s	d ln$I$/dt /mHz	t_{max} / s	t_{start} / s	d ln$I$/dt /mHz	t_{max} / s
10	< 30	≥ 12	60	1500 ± 800	0.111 ± 0.003	30000
15	< 30	16 ± 2	100	4200 ± 500	0.047 ± 0.001	> 50000
20	≤ 30	12 ± 1	135	-- (> 50000)	--	--
27.5	70 ± 10	6 ± 1	280	-- (> 50000)	--	--

[a] t_{start} - beginning of the early-stage evolution during which the structure factor S increases exponentially; t_{max} - transition to late-stage behavior (maximum of scattering intensity I proportional to the product interface area x phase contrast product).

Both the first and the second exponential intensity increase become faster with decreasing temperature (increasing quench depth), ignoring the uncertain result for the first effect at 10 °C. The second process is about 300 times slower. The ratio of the rates can be taken as the ratio of the effective diffusion coefficients in the two PS processes. This means that in the second process clusters rather than monomers must move and that the viscosity has increased considerably: the process starts after gelation of the continuous phase.

5. Discussion

The turbidity data show three different kinds of time evolution, indicating different phase separation (PS) processes to take place, sometimes even two at the same time. They will be examined by comparison of the simultaneous evolution of the scattering function $I(q)$ and its integral (turbidity), the decomposed dynamic structure factor, and the macroscopic viscosity.

5.1. Evolution after shallow quench

Above 27.5 °C, where no specific gelatin self-association to triple-helical segments was observed from optical rotation measurements[3)], the turbidity increase starts linearly with time, indicating a linear increase in weight-average molecular weight M_w of the clusters and almost constant optical contrast. The evolution is quite similar to that observed by other authors for smaller concentrations of maltodextrin[3,6)]. This PS process can be followed by the evolution of mode **d2** in the decomposed DLS spectra (Fig. 7). A late-stage power exponent ($d\ln(\xi\eta)/d\ln t = 0.7 \pm 0.07$) of this mode indicates that ripening of the corresponding droplet fraction by a combination of diffusion and hydrodynamically activated coagulation takes place, since the viscosity η is not expected to change so much. The droplet radii evolve from 1 to 45 μm (Table 2) before the mode becomes again faster and intensity decreases due to sedimentation of large droplets.

The delayed faster increase in the turbidity with the exponentially decaying rate under isothermal conditions has not been reported before (in this system). It is related to the evolution of mode **d1** (Fig. 7) and occurs also in the absence of gelatin (**d1** in Fig. 6). The turbidity growth law has the character of a slow relaxation process triggered by a critical thermodynamical shift at time t_2. This shift is apparently related to a slow sterically controlled collapse of some fraction of maltodextrin, starting at the temperature quench and proceeding at least until maltodextrin gelation. If a critical (concentration-dependent) degree of dehydration (deswelling) is reached at t_2, the intermolecular interaction becomes attractive and a NG process becomes nucleated: an Avrami-type kinetic model of the form $\tau = \tau_\infty \cdot \left[1 - \exp(-Z(t - t_2)^n)\right]$ (with Avrami exponent $n = 1$, indicative of a short grain-boundary-nucleation period just at t_2)[21)] seems to fit the turbidity (τ) data well. Mode **d1** becomes the most intense one and slows down by no more than a factor of 1.5, corresponding to a maximum cluster mass increase by a factor of 3.4 (at 30 °C), whereas the scattering intensity (turbidity) increases also by a factor between 2 and 3.6. This indicates that (if they are dense spherical droplets) the number of the small clusters remains constant after their nucleation at t_2, in accord with the Avrami exponent 1. The apparent critical radius of the nuclei (initial $\xi\eta/\eta_0$ of mode **d1**) is between 200 nm (at 45 °C) and 400 nm (at 30 °C). In CLSM experiments (with 500 nm resolution) under slow cooling conditions (-1 K/min), Lorén and Hermansson have seen "spherical droplets coming up all at the same time"[22), which confirms the triggered nucleation to take place. The carbohydrate fraction that forms the core of the nuclei is most likely the highly branched amylopectin: these molecules have a high molecular weight like gelatin, which suggests that they are more compatible with this biopolymer and phase-separate later than the short-chain linear amylose segments. The low-molecular-weight amylose segments in turn may rapidly form the liquid droplets of mode **d2**, and later also build up the periphery of **d1** clusters. Before amylopectin collapses, its diffusion is very slow because of the swelled branches (mode **d** in Figs 6, 7), and its bimolecular interaction is repulsive because of the excluded volume. And finally, because of the long branches, it is the best candidate for a collapse delayed by the temporal network of the other polymer even after its equilibrium excluded volume would turn towards zero. A delayed collapse can be concluded from the delayed switch of mode **d** to mode **d1**.

5.2. Evolution after a deeper quench affected by gelation

After quenches to below 30 °C, gelatin is poised to transform to a physical gel with triple-helical junction zones. In the course of that transformation, first some coils are rearranged to helices. Then larger and larger clusters of this polymer are formed which get more and more incompatible with the low- and average-molecular weight fraction of the carbohydrate. Finally the clusters connect to a permanent gel with inclusions of dispersed maltodextrin-enriched droplets. However, bulk PS occurs instead of the formation of a system spanning strong cluster if the concentration of helical gelatin is not large enough for rapid gelation; this is the case at 27.5 and 25 °C.

In the decomposed DLS spectra (Fig. 8), mode **d1** behaves similarly to the **d1** at higher temperatures. However, the viscosity increase in the continuous gelatin phase shifts this mode to longer relaxation times and apparent hydrodynamic radii $\xi\eta/\eta_0$, except in the early stage. If the critical size of the nuclei is not altered by temperature, this observation can serve to estimate the change in mobility of these clusters due to gelation. The effective relative viscosity η/η_0 would be about 2.5 at 25 °C or 8 at 20 °C and below, considering a critical hydrodynamic radius of 400 nm just as at 30 °C. Such a "low" viscosity value in a gel of much higher system viscosity (see below) indicates that the environment of the cluster is large pores filled with a semidilute polymer solution.

The time evolution of the droplet mode **d2** is apparently different from the linear NG evolution at above 30 °C. Just together with the gelation (viscosity increase, slow-down and disappearance of mode **c**), this part of mode **d** slows down within a minute by more than two orders of magnitude. The time resolution of the data is insufficient to resolve the growth law. We must note that also the system viscosity increases by up to 3.5 orders of magnitude upon gelation. A comparison of onsets and maxima of turbidity, viscosity, and viscosity-corrected ξ(**d2**) is given in Table 4. The ξ(**d2**), however, is underestimated if $\eta < \eta_{max}$.

Table 4: Comparative evolution of turbidity, system viscosity, and apparent mode **d2** droplet radius ξ(**d2**) calculated for $\eta = \eta_{max}$ (estimates by visual analysis of presented figures).

T / °C	Turbidity			Viscosity				DLS
	t_{onset} / s	T_{max} / s	τ_{max} / cm^{-1}	t_{onset} / s	t_{max} / s	η_{max} / Pa·s	η_0 / mPa·s	ξ(d2) / nm
10	66	200	11.5	20	100	50	1.32	5.3
15	96	280	8.8	60	200	200	1.15	1.8
20	117	4000	11.5	110	300	10	1.01	20
25	220	1900	7.7	1500	2500	0.05	0.90	285

The estimated ξ(**d2**) indicate that the mode is due to clusters of the same size as those for mode **d1** (or even less) with the only difference that they are trapped in the gel instead of in large pores.

The turbidity evolution yields the Avrami-type NG behavior (with Avrami exponent 1), before a high and roughly temperature-independent maximum value is reached. The gelation (viscosity increase) and the NG process (turbidity increase) are apparently not highly coupled. The viscosity evolution might seem more rapid than the turbidity evolution at the lowest temperatures because the cooling rate is effectively higher in the viscometer than in the UV/VIS cuvette. But even at 20 and 25 °C, when 40 s is not much compared with the time scale of the PS and gelation processes, onsets and maxima of turbidity and viscosity increase occur at significantly different times. The gelation process is much more sensitive to temperature changes than the NG process in this temperature interval.

From the temperature dependence of the initial slope of the turbidity increase (see Fig. 4), it becomes apparent that there is a critical temperature between 25 and 20 °C where the dependence increases very fast with decreasing temperature. At this temperature one could expect that the long-range cluster interaction potential turns from repulsive to attractive. At 10 °C, the slope is very similar to that at 15 °C, indicating that the growth process has almost turned to its diffusion-controlled limit. The significant intermediate-stage decrease in the turbidity at 10 and 15 °C, where no bulk phase separation (or droplet sedimentation) occurs, can be satisfactorily explained. Aymard and Williams[6)] have suggested that this is a result of a droplet growth to radii of several micrometers, which were observed by CLSM. The form factors of these droplets would peak entirely at q below a critical q value corresponding to the acceptance angle of the spectrometer, so that their scattering contribution to the turbidity is no longer considered. They checked the phenomenon with latex spheres. We do not observe any intensity decrease in static light scattering. Hence, a phase-contrast-changing phenomenon can be excluded, as it also would change the static light scattering intensity. Thus the phenomenon can qualitatively be taken as indication of continued coarsening, generating droplets with diameters of several micrometers.

The late turbidity increase (after more than 10^4 s) seems to be due to spinodal decomposition of the gelling dispersed phase, resulting in the formation of carbohydrate crystallites. It was also observed during gelation of high-concentrated ($\geq$ 10 % SA2) gelatin-free systems quenched to these temperatures, after comparable times.

5.3. Identification of PS mechanism upon deeper quench: DLCA vs. SD

Upon the quenches to below 25 °C, the static structure factor starts its evolution with a broad maximum at $q_m \cong 15\ \mu m^{-1}$ (Fig. 10). At a wave vector slightly higher (18.7 μm^{-1}), an almost exponential (accelerated) increase in the scattering intensity is observed at the early stage. Although both features could be explained by a SD process at its early stage[1)], the high value of q_m would be exceptional for phase separation in not very viscous liquids. In the SD process, at the intermediate stage, q_m would be expected to shift to lower values with $S(q_m)$ still increasing exponentially. $S(q)$ should obey a dynamic scaling rule: $S(x = q/q_m) = S(q_m)\cdot 4x^2/(3+x^6)$, for an off-critical quench in binary mixtures[23)]. For $q > 2\ q_m$, an asymptotic behavior $S(q) \propto q^{-4}$ should be observed, the same as the Porod regime for objects with smooth surfaces. For a quench through the critical point, the corresponding asymptotic behavior would be $S(q) \propto q^{-6}$. We observe instead $S(q/q_m > 2)$ decaying fastest with $q^{-3.25}$ in the early intermediate and with $q^{-5.05}$ at later stages, with the asymptotic behavior not holding for more than half decade of q. This indicates that the early generated interface is not smooth but is probably the interface of fractal clusters of aggregated colloids, and the later-stage, slowly changing $S(q)$ is influenced by minima from a hard-sphere-type form factor.

A more plausible interpretation of the data is thus in terms of a DLCA process. The broad peak at about 15 μm^{-1} then corresponds to the mixed peak of fast-growing (diffusion-limited) small clusters (presumably around collapsing amylopectin) and of depletion layers around them (depleted in carbohydrate), of comparable thickness as the initial cluster radius. At some critical cluster size and when depletion layers overlap and thus disappear, long-range cluster-cluster interaction potentials become attractive and unscreened, and clusters aggregate to large clusters by DLCA. This is the moment when the peak, due to the depletion layers, breaks down and q_m shifts to low values, even to q's outside the observed window, and an accelerated intensity increase is observed at q near the new q_m, changing later to something like a $t^{1/2}$ behavior Note the similarity of the observed intensity evolution at $2\pi/q > 1$ µm (Fig. 10b) and the DLCA-like modeled time evolution of $\phi(t)$ or the cluster volume (Fig. 1c) for $\phi_\infty \cong 0.5$. Final cluster radii of up to 4 µm were observed by CLSM for systems with 6 instead of 7 % of SA2 under otherwise identical conditions[4]. The radii would correspond to a final q_m of 1.5 µm^{-1}; SALS measurements would be helpful to confirm details of the PS process. So far, the late-stage coarsening has been indicated only by a characteristic slow turbidity decrease.

Acknowledgement

This study has been carried out with financial support from the Commission of the European Communities, Agriculture and Fisheries (FAIR) specific RTD programme, CT 96 1015, "Mixed Biopolymers - Mechanism and Application of Phase Separation". The study does not necessarily reflect Commission views and in no way anticipates its future policy in the area.

References

1. R. H. Tromp, A. R. Rennie, R. A. L. Jones, *Macromolecules* **28**, 4129 (1995)
2. E. D. Siggia, *Phys. Rev. A* **20**, 595 (1979)
3. N. Lorén, A-M. Hermansson, M. A. K. Williams, L. Lundin, T. J. Foster, C. D. Hubbard, A. H. Clark, I.T. Norton, E. T. Bergström, D. M. Goodall, *Macromolecules* **34**, 289 (2001)
4. N. Lorén, A.-M. Hermansson, *Int. J. Biol. Macromol.* **27**, 249 (2000)
5. N. Lorén, M. Langton, A.-M. Hermansson, *Food Hydrocolloids* **13**, 185 (1999)
6. P. Aymard, M.A.K. Williams, A. H. Clark, I. T. Norton, *Langmuir* **16**, 7383 (2000)
7. M. A. K. Williams, D. Fabri, C. D. Hubbard, L. Lundin, T. J. Foster, A. H. Clark, I. T. Norton, N. Lorén, A.-M. Hermansson, *Langmuir* **17**, 3412 (2001)
8. S. Kasapis, E. R. Morris, I. T. Norton, M. J. Gidley, *Carbohydr. Polym.* **21**, 249 (1993)
9. A. Cesàro, F. Cuppo, D. Fabri, F. Sussich, *Thermochim. Acta* **328**, 143 (1999)
10. S. Kasapis, E. R. Morris, I. T. Norton, R. T. Brown, *Carbohydr. Polym.* **21**, 261 (1993)
11. V. Normand, P.D.A. Pudney, P.Aymard, I.T. Norton, *J. Appl. Polym. Sci.* **77**, 1465 (2000)
12. S. Kasapis, E. R. Morris, I. T. Norton, A. H. Clark, *Carbohydr. Polym.* **21**, 243 (1993)
13. M. Tirado-Miranda, A. Schmitt, J. Callejas-Fernández, A. Fernández-Barbero, *Langmuir* **15**, 3437 (1999)
14. J. L. Burns, Y. Yan, G. J. Jameson, S. Biggs, *Langmuir* **13**, 6413 (1997)
15. J.-M. Jin, K. Parbhakar, L. H. Dao, K. H. Lee, *Phys. Rev. E* **54**, 997 (1996)
16. R. Jullien, A. Hasmy, *Phys. Rev. Lett.* **74**, 4003 (1995)
17. R. H. Tromp, R. A. L. Jones, *Macromolecules* **29**, 8109 (1996)
18. A. J. Bray, *Phys. Rev. E* **47**, 3191 (1993)
19. I. Dierking, *J. Phys. Chem. B* **104**, 10642 (2000)
20. M. C. Blanco, D. Leisner, C Vázquez, M. A. López-Quintela, *Langmuir* **16**, 8585 (2000)

21. G. J. Hopkin, H. K. D. H. Bhadeshia, in: *Proc. 21st Risø Int. Symp. Mater. Sci.: Recrystallization – Fundamental Aspects and Relations to Deformed Microstructure*, N. Hansen, X. Huang, D. Juul Jensen, E. M. Lauridsen, T. Leffers, W. Pantleon, T. J. Sabin, J. A. Wert (Eds.), Risø National Laboratory, Roskilde (Denmark) 2000
22. A.-M. Hermansson, private communication
23. H. Furukawa, *Physica A* **123**, 497 (1984)

The Use of Monte-Carlo Simulations to Calculate Small-Angle Scattering Patterns

Bryan C. McAlister[a 1] *and Brian P. Grady**[,b]

[a] Rogers Corporation, Microwave Division, 100 S. Roosevelt Avenue, Chandler, AZ 85226, U.S.A.
[b] Max Planck Institute for Colloid and Interface Science, Potsdam D-14424, Germany

SUMMARY: Determining the morphology that corresponds to an experimentally observed scattering pattern is a non-trivial problem in small-angle scattering. The purpose of this paper is to describe methods to calculate small-angle scattering patterns from a given morphology via Monte-Carlo methods. In these methods, scattering intensities are calculated from pairs of randomly selected points. This paper reviews the different formulas and procedures that have been used. Multiple studies have shown that the Monte-Carlo method gives accurate results for various morphologies having no interparticle interference. In certain cases, these calculations have been used to fit experimental patterns according to a minimization criterion. For morphologies with interparticle interference, these methods have not been very successful for reasons that are not entirely clear.

Introduction

The number of approaches to the analysis of data from small-angle scattering experiments is substantial; primarily because scattering due to the shape of particles (i.e. form-factor scattering) as well as scattering due to the arrangement of particles (i.e. structure-factor scattering) are relevant in small-angle experiments. The most common approach used to completely describe the morphology (as opposed to determining only certain parameters such as the isothermal compressibility or the z-average radius of gyration) involves fitting one or more mathematical expressions to the observed scattering intensity, or alternatively, the Fourier transform of the observed intensity. This expression is almost always a closed-form mathematical description of scattering from some postulated morphology, i.e. a single sphere, a single cylinder, a collection of spheres, etc. Deriving these mathematical descriptions is extremely difficult, and the number of systems with such descriptions are quite limited, especially for systems relevant to polymers. An essentially complete list of morphologies with mathematical descriptions of the scattering pattern are given in the recent paper by Pederson.[1]

---

[1] Permanent address: School of Chemical Engineering and Materials Science, University of Oklahoma, Norman, OK 73019-0628, U.S.A.

 CCC 1022-1360/00/$ 17,50+.50/0

The focus of this paper is to describe an approach where explicit mathematical descriptions are not used to describe scattering from a given morphology; rather the scattering pattern is calculated by randomly choosing pairs of points and calculating the scattering from those points using fundamental expressions from scattering theory. Because the points are randomly selected, this type of procedure is properly categorized as a Monte-Carlo (MC) method. A review of these Monte-Carlo methods is the main focus of this paper.

In order to introduce Monte-Carlo methods, two more common cases where MC methods are used in conjunction with scattering from polymers will be briefly described. Both methods use Monte-Carlo approaches to determine *morphology*, and the scattering pattern is calculated by choosing pairs of points and using a fundamental expression of scattering intensity. These methods are *not* the same as the methods that comprise the main focus of this paper because the pairs of points used to calculate the scattering pattern are *not* selected randomly; one tacitly assumes that the morphologies are simple enough so that the scattering intensity can be accurately calculated within a reasonable time frame using all scattering points. These methods divide the object into identical discretized objects, and scattering is calculated using the mathematical description of scattering from these discretized objects, as well as a term involving the location of each object's center relative to one another. In general, the more objects used, the better accuracy obtained. This approach to calculating scattering patterns have been termed Debye methods for reasons given below.

The most common use of MC methods to determine morphology in polymer physics is in the field of polymer dilute solutions. The conformation of the polymer chain is determined through random movements of individual, discretized sections of the chain coupled with restrictions on those random steps, for example self-avoidance. Scattering is then calculated by application[2] of the Debye formula for the form factor $F^2(q)$:

$$F^2(q) = I_o(q)\left[n + 2\sum_{i=1}^{n-1}\sum_{j=i+1}^{n} f_i f_j \frac{\sin(qr_{ij})}{qr_{ij}}\right] \qquad (1)$$

In this formula, n is the number of discretized objects, $q(4\pi\sin\theta/\lambda)$ is the magnitude of the scattering vector $\underline{q}$, r_{ij} is the distance between the i-th and j-th objects, $I_o(q)$ is the scattering in the given experimental setup due to one discretized object (in this case one chain section), and finally f_i and f_j are the scattering powers of the i-th and j-th objects. The discretized objects must all be identical and scattering from an object ($I_o(q)$) must be describable by a closed-form mathematical expression. Further, the discretized object must have a uniform scattering power across its volume. For x-rays the scattering power is electron density, while for neutrons the scattering power is scattering length density. The scattering power of the solvent is assumed to be zero; in fact of course, the scattering powers in this formula are really the difference between the scattering power of the discretized object and that of the solvent. In the limit of no interparticle interference, the form factor is the same as the scattering pattern when multiplied by the number of scattering objects N (in this case, the number of polymer chains). Finally, in practice often only normalized curves are presented, i.e. the q-dependent shape is retained but factors such as the intensity of the incident x-ray beam and the number of scattering objects are ignored. This latter simplification is used for almost all simulations presented in the literature, unless a direct comparison to experimental data is performed.

The double summation is quite time-consuming to calculate and does not need to be carried out because Eq. (1) can be rewritten as:

$$F^2(q) = I_o(q)\left[N_{\mathrm{bin}\,s} + 2\sum_{i=1}^{N_{\mathrm{bin}\,s}} g(r_i)\frac{\sin(qr_i)}{qr_i}\right] \tag{2}$$

where $g(r_i)$ is the pair-distance histogram and $N_{\mathrm{bin}\,s}$ is the number of bins in the histogram. The pair-distance distribution function $g(r)$ has as its x-axis the distance (r) between two scattering objects and as its y-axis the normalized numbers of pairs of objects that have the distance r. For polymer chains, the discretized objects are assumed to be infinitely thin rods; however, at high angles, the finite width of the chain must be accounted for. Vanishingly small concentrations (to the point where measuring the signal can be extremely difficult) must be used in order to avoid a contribution to the scattering pattern due to pairs of points on different chains; otherwise the single-contact approximation should be used. A closed-form analytical expression has been mapped out from the results of Monte-Carlo simulations to aid in data analysis. [3,4]

The second case where Monte-Carlo methods are used to determine morphology is the so-called reverse Monte-Carlo (RMC) method initially described by McGreevy and Pusztai.[5] Instead of allowing the MC simulation to go to completion and then calculating the scattering pattern as in the example above, after each step (or series of steps) another restriction is introduced: the decision to allow a step is based on whether the resulting scattering pattern is closer to the experimentally observed scattering pattern. Again, the scattering pattern is calculated using an exact approach; most often the Debye method where spheres are the discretized objects. An example of the implementation of a RMC method using a large number of discretized spheres is given in papers by Pantos et al.[6,7]

Two subtleties are important in RMC methods: first, similar to more conventional MC simulations, a proportion of moves that lead to worse agreement are accepted according to some exponential weighting factor, and second, the function to be minimized is not the sum of the squares of the difference between the experimental and model pattern, but rather this sum divided by some measure of the experimental error. As the co-originator of the technique points out in his excellent review paper,[8] dividing by the experimental error is used because it works. The popularity of this technique is partially explained by the fact that this technique can be used for any experiment that yields a radial distribution function or structure factor, i.e. EXAFS, anomalous scattering, etc. The RMC approach has been applied primarily to wide-angle neutron or x-ray scattering patterns from amorphous small-molecule liquids or solids; however, this method has also been used to analyze small-angle light scattering patterns from colloidal polymer systems.[9,10,11] Some weighted variation of either the form factor or structure factor is almost always used as the objective function rather than the scattered intensity itself. In a paper by Doniach et al.,[12] protein structures in solution are modelled and, in this case, the objective function is the scattered intensity weighted by user-defined functions. This paper, however, does not term the method an RMC method, but the methodology is similar.

Work on the RMC method is relevant to the main focus described in this paper because an RMC approach will be necessary in order to fit experimental scattering patterns with scattering determined via Monte-Carlo methods for most of the morphologies of interest. The general procedure would be to: (1) move the objects in the morphology according to an MC method (2) calculate the scattering pattern using a MC method, (3) compare the calculated scattering pattern with the experimental scattering pattern to decide whether the movement should be accepted, and (4) go to step 1. Computationally, this effort would be enormous and the reader is correct if he guessed that this effort is computationally beyond the capabilities of current computers for the systems of interest. However, within the professional lifetime of the authors, we expect that computers will become fast enough so that these methods can and will be used.

The fundamental systems of interest for a method that calculates the small-angle scattering pattern (i.e. not the morphology!) via Monte-Carlo methods are those morphologies for which a simple mathematical expression cannot be written for the scattering pattern. The MC methods described herein are so computationally intensive that a closed-form mathematical expression should be used if possible. However, the number of systems without closed-form mathematical expressions is extremely large. For example, any densely concentrated system of objects other than spheres cannot be treated in a conventional manner and the scattering pattern can only be calculated via MC methods. This type of morphology is very common in solid polymers. Even in systems with no interparticle interference, if the distribution is anything other than perfect, or perfectly random, the scattering pattern cannot be calculated exactly except with Monte-Carlo methods. Conventional methods cannot be used to calculate scattering for many simple one-particle systems with off-axis orientation, i.e., a cylinder where the long axis is not normal to the incident x-ray beam. Finally, something very common in biological systems, dilute systems of single particles with unique shapes in solution are treated effectively with MC methods, although other methods, including the Debye methods represented by Eq. (2), can be used.

As clearly delineated earlier, one substantial drawback to using Monte-Carlo methods to calculate scattering patterns is that these methods are computationally intensive. However, the accuracy by which MC methods calculate the patterns should not be seen as a drawback, since inherently the methods can be as accurate as any "exact" method that uses a computer to calculate irrational numbers. The only question (assuming the code is free from errors) is how much computational time is required to achieve an acceptable level of accuracy. This use of Monte-Carlo methods to calculate scattering patterns is much different than for moving objects around in a polymer chain for example, because one can always question whether the right restrictions are being used on placements and/or movements. In this case, restrictions are not a relevant concept and Monte-Carlo methods can correctly be regarded as "exact" in theory.

In practice, however, eliminating errors in code and determining the minimum simulation time required for an acceptable level of accuracy is typically the first task of anyone trying to calculate scattering patterns from a given morphology using Monte-Carlo methods. Completing both of these tasks is straightforward; since exact analytical expressions exist for the scattering patterns from many objects. We have found that a single sphere provides an excellent check, since correctly describing the shape of the sharp drop in intensity near the extinction points severely tests any code. These tests demonstrate that the linear congruential generator used by most computer languages to generate random numbers does not produce random enough random numbers, some sort of shuffling as described in the excellent monograph by Newman and Barkema[13] is required. Our group's solution to this problem was perhaps a bit too severe, since instead of shuffling entries in an array of random numbers we shuffled individual digits determined from these numbers. In our defense, the resulting scattering patterns are extremely sensitive to the randomness of the numbers and extreme care must be exercised in this seemingly trivial matter.

The final consideration is how to describe the goodness of fit, whether the fit is to experimental data or to an analytical expression.[14] In either case, a mathematical expression should be developed to quantify the goodness of fit; this step assumes extreme importance in an RMC algorithm because this expression also serves to drive the routine towards convergence. Even if such an expression is not needed for a convergence routine, a one-parameter measure of the discrepancy is useful for determining minimal simulation times. In experiments performed in our laboratory, we used a variation on the coefficient of determination to quantitatively measure the error:

$$R = 1 - \frac{\sum \left(\log(y_i) - \log(\bar{y}_i)\right)^2}{\sum \log(y_i^2) - \frac{\left(\sum \log(y_i)\right)^2}{n}} \tag{3}$$

For the MC simulations described in this paper, this curve rapidly approaches one, usually monotonically, then plateaus at some value typically slightly less than one. The plateau is not monotonic; random (to the naked eye) jumps occur in R around the plateau value. The minimum time for simulation is set as the first time where the R value is at this plateau.

Monte-Carlo Simulations Used to Calculate Small-Angle Scattering Patterns

Morphologies without interparticle interference

The simplest morphology to consider is a single object that is spherically symmetric, or a collection of objects that are spherically symmetric in the sense that orientations are rotationally averaged and the concentration range is dilute enough so that interparticle interference can be neglected. Hansen[15] published the first effort in this area, while Henderson [16] generalized this approach to morphologies with more than one electron density. As in the Debye method, the object is divided into simpler shapes, but in this case the shapes are simply used to build the arbitrary morphology. In other words, one uses simple objects (spheres, cylinders, ellipses, etc.) to build the morphology much as one might do with wooden blocks and glue. The objects do not have to be identical in size or type. A difference between the two approaches is that the Hansen approach considers only two electron densities and hence overlap of the shapes is allowed in building the model but is accounted for when calculating the scattering pattern, while in the Henderson approach the objects must be small enough so that the morphology is accurately described without using overlapping objects.

In both approaches, the centers of each object are translated to the origin and random points are generated to fill the objects. The range of the allowed coordinates for the points must obviously be large enough to encompass the largest object. The same array of random numbers is used for each object, although not all members of the array are used for each object. In the Hansen paper, the number of random coordinates in each object is proportional to the volume of the object. To subtract points in different objects that belong to overlapping regions in space, each object is checked for overlap prior to translation to the origin and random points for a particular object are removed if another object already contains that particular set of random points. In the Henderson paper, the model forces the number of random points in a particular region to be proportional to the product $|\rho_i - \rho_o| * V_i$, where ρ_i is the electron density of the region, ρ_o is the electron density of the medium and V_i is volume of the region that has electron density ρ_i. To account for electron density differences of opposite sign, in the binning process $g(r)$ is reduced by one instead of increased by one if the signs of $(\rho_i - \rho_o)$ are different for the two points. Binning into a pair-distance distribution function is identical to the procedure described earlier; note that all pairs of points, i.e. both those within a particular subvolume as well as between subvolumes, are used to calculate $g(r)$. With n pts the number of simulations points, the scattering pattern is calculated by the following:

$$\frac{I(q)}{I_o(q)} = 4\pi V \sum_{i=1}^{n\,\mathrm{pts}} g(r_i) \left[\frac{\sin(qr_i)}{qr_i} \right] \tag{4}$$

The time required to simulate a pattern using Monte-Carlo methods and the formulations above is on the order of seconds with today's personal computers. Henderson[16] shows quantitatively the effect of changing the number of simulation points on simulation accuracy. Recently published papers by Spinnozi et al. used a similar method to the one described by Hansen to simulate scattering from inhomogeneous rotationally averaged particles[17] and globular proteins.[18]

Clearly, there are many similarities between the Debye method exemplified by Eq. (2) and the Monte-Carlo method exemplified by Eq. (4). The most substantial difference is that the result directly depends on the characteristics of the constituent objects in the Debye method, while this statement is not true for the MC methods. For isotropic scattering patterns without interparticle interference, one relevant question is whether a method that artificially divides the scattering volume into small spheres or some other arbitrary object shapes and then calculates the pattern using all objects is better than Monte-Carlo methods. The term "better" is really only a question about which method is faster for a given level of accuracy, since both are exactly correct at infinite computer time. Such a comparison has not been done to our knowledge. We suspect that a comparison would show that a Monte-Carlo method is a more efficient approach to this problem.

The Debye method cannot be used with anisotropic objects that have some preferential degree of orientation. Hence, Monte-Carlo methods have the potential to be much more useful in systems with preferential orientation. Most industrial processes produce a polymer with some level of residual orientation. The MC methods described above cannot be easily used for anisotropic scattering, a different approach is required. The approach described here describes scattering from oriented single-particle systems and was developed by the authors of this monograph. The method is explained more fully in a recent publication[19] and another one which will appear shortly.[20] The following equation is true for any centrosymmetric particle and was used to simulate scattering:

$$\frac{I(\underline{q})}{I_e(\underline{q})} = \left(\sum_{k=1}^{n\text{pts}} f_k (\cos \underline{q} \bullet \underline{r}_k) \right)^2 \tag{5}$$

$I_o(\underline{q})$ is the scattering from one electron or neutron that would be found in the same experimental setup, and $\underline{r}_k$ is a vector from the center of symmetry to the k-th simulation point. Once again the scattering power in this formula is the difference between the scattering power of the object and that of the solvent, the k-th simulation point must be contained within the object of interest. Note that pairs of points are not selected, in essence the second point is the center of symmetry. For synthetic polymers, the restriction of centrosymmetry is not a large one since nearly all anisotropic objects are centrosymmetric. For biological polymers, this restriction is much more significant since a great number of biological structures are not centrosymmetric.

Experiments in our lab show that simulating scattering from a sphere requires 10-100 times more computational time using the centrosymmetric method vs. the methods developed by Hansen and Henderson. The longer amount of time is not surprising, since in essence Eq. (4) is the rotational average of Eq. (5). Equation (5) allows any scattering object with any orientation to be simulated as long as interparticle interference is not present and the scattering particles are centrosymmetric with respect to scattering power. Simulating scattering from morphologies with any sort of orientation distribution is straightforward,[20] although again interparticle interference must be negligible. Incorporating a MC method based on this equation into a RMC algorithm is feasible and run-times would be reasonable, most likely on the order of a few days with a high-end PC and considerably shorter with a supercomputer. If this procedure were implemented, this effort would represent the first time to our knowledge that a least-squares algorithm was used to fit the entire 2-d anisotropic scattering pattern from a non-crystalline system.

In theory, objects that are not centrosymmetric and oriented could be simulated using a Monte-Carlo approach, although to our knowledge nothing has been published. In this case, the relevant formula would be the following:[21]

$$\frac{I(\underline{q})}{I_e(\underline{q})} = \left(\sum_{i=1}^{n\,\mathrm{pts}_i} \sum_{j=1}^{n\,\mathrm{pts}_j} f_i f_j (\cos \underline{q} \bullet \underline{r}_{ij}) \right)^2 \tag{6}$$

The double sum makes this expression quite tedious to calculate. When using this expression, however, there is a real question concerning the strategy for random number generation that can have a large impact on simulation time. Specifically, is the most inefficient approach of choosing two new random numbers for each simulated scattering point used the best approach, or should random numbers be stored in an array and reused. The problem with the latter approach is that if not all random numbers are paired with all other random numbers, then the possibility of artifactual correlations affecting the result are increased if the random numbers are truly not random. In unpublished numerical experiments which should be regarded as preliminary, we used Eq. (6) to calculate scattering patterns from a sphere. This simulation times were approximately 10 times longer than those using Eq. (5).

Patterns with interparticle interference

From the previous section, it is obvious that simulating objects without interparticle interference has met with great success. Scattering patterns for all three cases - rotationally averaged particles, oriented centrosymmetric single particles, and oriented non-centrosymmetric particles - can be determined successfully. Although extremely computationally intensive, a distribution of sizes and/or orientation could easily be handled simply through the appropriate averaging of scattering patterns from the constituent particles. At this stage, however, MC methods have not been generally successful at simulating scattering for morphologies with interparticle interference. More specifically, we believe it is not interparticle interference per se that is causing the problem, rather it is the fact that the relevant scattering area is in theory infinite on morphological length scales. Again, evaluation of the quality of the simulation is easily done, since analytical expressions have been calculated for scattering from an ensemble of randomly-placed hard spheres. The statement that success has not been reached is based on the observation that scattering patterns from simulations do not agree either qualitatively or quantitatively with those from analytical expressions in the angular region where interparticle interference becomes significant. Although the analytical expressions such as that from Ashcroft and Lekner[22] are probably not exact, the deviations are much too large to believe that they are the result of anything other than a problem with the simulation technique. The agreement is much worse at higher number densities, which for polymers is typically the region of significantly more interest.

Much of this next section would better be termed speculation, because until something is done correctly, one cannot truly know what is wrong. However, based on our extensive experience in attempting to simulate scattering from these systems, we feel that what we are about to present is likely to be correct.

The procedure is straightforward, first a box needs to be filled with objects to a certain number density and according to some orientation in a random fashion. If the number density is above some value, which we found to be approximately 0.4 for spheres, then placement must not occur randomly, rather one must place the objects in some crystalline arrangement, typically so that the distance between objects is uniform. This ensemble of particles must then be allowed to thermally equilibrate by moving particles randomly for an appropriate time. To calculate scattering from this morphology, the equivalent of Eq. (6) for discrete objects (rather than points) is used:

$$\frac{I(\underline{q})}{I_e(\underline{q})} = \sum_{i=1,j=1} F_i(\underline{q}) F_j(\underline{q}) \cos\left[\underline{q} \bullet (\underline{R}_i - \underline{R}_j)\right] \tag{7}$$

Sjöberg[23] et al. used an alternative mathematical expression for Eq. (7) because this alternative was faster to calculate:

$$\frac{I(\underline{q})}{I_e(\underline{q})} = \left[\sum_{k=1} F_k(\underline{q}) \cos(\underline{q} \bullet \underline{R}_k)\right]^2 + \left[\sum_{i=1} F_i(\underline{q}) \sin(\underline{q} \bullet \underline{R}_i)\right]^2 \tag{8}$$

The meanings of the symbols inside these expressions are identical with those given earlier, with R describing a vector from the center of the simulation box to the discrete object in question. The key questions that need to be asked concerning this type of simulation are: (1) How big a box should be used ? (2) Is it more computationally efficient to use a smaller box, and then place the objects multiple times to the given number density, or use a bigger box, and then only place the objects once ? and finally (3) What adjustments need to be made to account for the finite size of the box ? Both Sjöberg and our group independently came to the conclusion that computationally it is more efficient to use a smaller box and place the objects many times; we believe the reason is that in any random configuration, low-amplitude oscillations in the radial distribution function go out to very long distances. Hence, in essence the random placement of the first few objects sets the morphology, and resetting the morphology by starting over is more efficient than making the box very large. The size of the box is typically chosen to make run-times reasonable, while periodic boundary conditions are used to help alleviate the effect of the edge of the box. Finally, for the size boxes that are typically used, one must subtract the scattering caused by the box itself, i.e. volume scattering.

Sjöberg placed elliptical objects using a Monte-Carlo procedure to determine their position, then calculated the scattering directly according to Eq. (8). In some sense, one can envision this morphology as one large object made up of a large number of unconnected discrete objects. The important difference is that in this case the large object is of infinite size. Further, the result for systems with interparticle interference was very different; the Monte-Carlo simulations did not reproduce the known solutions. Using hard spheres, deviations starting at about $qR_g < 1.5$ are clearly apparent in Fig. 5 in the Sjöberg paper. Similar disagreements are found in a paper by Bertram;[24] however, the comparisons in this paper are meaningless since the objects were not allowed to relax and hence were not at thermodynamic equilibrium.[25] In unpublished work in our own lab, we have gone to much higher volume fractions using a Monte-Carlo method and the deviations are even greater, and start at higher qR_g. Specifically, we are not even close to reproducing the interparticle interference peak in these systems. The fundamental question is why these simulations are not successful.

We are not yet able to give a definitive answer to this question. At very low angles, this type of approach will not be correct, since particles are not fixed in space and move due to thermal fluctuations. Sjöberg23] presented a numerical approach to this problem based on statistical mechanics, which seemed to work quite well. Although we never tested this hypothesis, we are confident that performing a fast Fourier transformation directly as in the paper by Frenkel236] instead of calculating via Eq. (8) would eliminate the disagreement between simulations and experiment; however, this approach would be extremely unwieldy for anisotropic scatterers. Our best guess is that the problem is the simulation time available; however, admittedly, this could just be a problem in the accuracy of the code and/or the method used to calculate the scattering patterns.

Projections for the Future

In our opinion, the real power of Monte-Carlo methods as related to scattering is to simulate scattering from morphologies that do not have analytical solutions. One category of systems having this characteristic are rotationally averaged systems with complicated scattering objects and no interparticle interference. Because of the importance of this problem in biological systems, these types of systems have been studied more thoroughly than oriented systems. For isotropic scatterers, a non-Monte-Carlo approach is possible since one can subdivide the object into identically shaped discretized objects with different centers, and all the objects can be used in calculating the scattering pattern. As stated earlier, a direct comparison to show which method is computationally more efficient has not been performed. Another alternative for rotationally-averaged systems is to use spherical harmonics;[27,28,29] in our opinion this method is more useful for systems where little is known *a priori* about the morphology or the scattering power at every point in space cannot be easily determined. Clearly inherent in the Monte-Carlo simulations is that a starting point is known, and the scattering power at every point in space can easily be calculated. Fractal-type approaches have also not been discussed throughout this paper for a similar reason; to determine the electron density at every point in space for a given overall simulation volume is non-trivial and hence fractal methods do not really compete with the Monte-Carlo methods described herein.

The most important use of Monte-Carlo methods in our opinion will eventually come for systems with only one continuous phase having anisotropic scattering objects and hence anisotropic scattering patterns. Fundamentally, the information contained in an anisotropic scattering pattern is greater than that in a scattering pattern where the objects have been rotationally averaged; however, almost no one purposefully orients a rotationally averaged system to improve quantitative analysis of a scattering pattern because data analysis is much more difficult. Other methods exist to analyze scattering for oriented polymer morphologies, such as those described by Wilke,[30,31] the latter of which uses paracrystalline concepts introduced by Hosemann.[32] Both methods have the flaw that only perfectly oriented, or nearly perfectly oriented, systems can be studied. Also interparticle interference cannot be handled. In theory, Monte-Carlo methods have no such restrictions. However, for Monte-Carlo methods to be used successfully, a procedure must be developed so that modelling infinite structures accurately can be performed. Until this task is completed, the outlook for these methods, at least in the synthetic polymer area, is probably quite limited.

References

[1] J.S. Pederson, *Adv. Colloid Interface Sci.* **1997**, *70,* 171.
[2] Summations instead of integrals are used since this paper describes computer simulations, and hence summations are more appropriate in this context.
[3] J.S. Pederson, P. Shurtenberger, *Macromolecules* **1996**, *29,* 7602.
[4] D. Pötschke, P. Hickl, M. Baullauff, P.O. Åstrand, J.S. Pederson, *Macromol. Theory Simul.* **2000**, *9,* 345.
[5] R.L. McGreevy, L. Pusztai, *Mol. Simul.* **1988**, *1,* 359.
[6] E. Pantos, J. Bordas, *Pure Appl. Chem.* **1994**, *66,* 77.
[7] E. Pantos, H.F. von Garderen, P.A.J. Hilbers, T.P.M. Beelen and R.A. Santen, *J. Mol. Struct.* **1996**, *383,* 303.
[8] R.L. McGreevy, *Nucl. Inst. Methods Phys. Res. A,* **1995**, *354,* 1.
[9] G. Tóth, L. Pusztai, *J. Phys. Chem.* **1992**, *96,* 7150.
[10]G. Tóth, *Chem. Phys. Lett.* **1997**, *269*, 413.
[11] G. Tóth, *Langmuir* **1999**, *15*, 6718.
[12] D. Walther, F.E. Cohen, S. Doniach, *J. Appl. Crystallogr.* **2000**, *33*, 350.
[13] M.E.J. Newman and G.T. Barkema, *Monte Carlo Methods in Statistical Physics*, Clarendon Press, Oxford, 1999, Chap. 16.
[14]Of course, if the data is fit to experimental data, the MC-generated data will need to be smeared to account for finite beam size etc, and normalized intensities cannot be used.
[15] S. Hansen, *J. Appl. Crystallogr.* **1990**, *23*, 344.
[16] S.J. Henderson, *Biophys. J.* **1996**, *70*, 1618.
[17] F. Spinnozi, F. Carsughi, P. Mariani, C. Teixeira and L.Q. Amaral, *J. Appl. Crystallogr.* **2000**, *33*, 556.
[18] P. Mariani, F. Carsughi, F. Spinnozi, S. Romanzetti, G. Meier, R. Casadio, C.M Bergamini, *Biophys J.* **2000**, *78*, 3240.
[19] B. McAlister, B.P. Grady, *J. Appl. Crystallogr.* **1998**, *31,* 594.
[20] B. McAlister, B.P. Grady, submitted to *J. Polym. Sci., Polym. Phys.*
[21] This expression is written as a single sum, rather than a double sum because this expression as written assumes that two new random numbers are generated for each calculation and hence a single sum is actually calculated. For a further discussion of this point, see the text.
[22] N.W. Ashcroft, J. Lekner, *Phys. Rev.* **1966**, *145*, 83.
[23] B. Sjöberg, *J. Appl. Crystallogr.* **1999**, *32,* 917.
[24] W.K. Bertram, *J. Appl. Crystallogr.* **1996**, *29*, 682.
[25] J.S. Pederson, *J. Appl. Crystallogr.* **1998**, *31,* 488.
[26] D. Frenkel, R.J. Vos, C.G. de Kruif, A. Vrij, *J. Chem. Phys.* **1986,** *84*, 4625.
[27] H.B. Stuhrmann, *Acta Crystallogr., Sect. A* **1970**, *26,* 297.
[28] D.I. Svergun, H.B. Stuhrmann, *Acta Crystallogr., Sect. A* **1991**, *47,* 736.
[29] D.I. Svergun, V.V. Volkov, M.B. Kozin, H.B. Stuhrmann, *Acta Crystallogr., Sect. A* **1996**, *52,* 419.
[30] A. Karl, S. Cunis, R. Gehrke, G. V. Krosigk, U. Lode, I. Luzinov, S. Minko, T. Pomper, V. Senkovsky, A. Voronov, W. Wilke, *J. Macromol. Sci.-Phys.* **1999**, B38, 901.
[31] W. Fronk, W. Wilke, *Colloid Polym. Sci.* **263**, 97 (1985).
[32] R. Hosemann and S.N. Bagchi, *Direct Analysis of Diffraction by Matter*, North-Holland, Amsterdam, 1962.

Monitoring Kinetic Processes in Polymer Solutions with Time Dependent Static Light Scattering (TDSLS)

Wayne F. Reed
Physics Dept.
Tulane University
New Orleans, La. 70118, USA

SUMMARY: An overview of recent developments in TDSLS and auxiliary techniques is given. This includes background theory, and examples of results and applications for online monitoring of polymerization reactions, kinetics and structural determinations from degrading polymers, aggregation, and dissolution of dry polymer powders. Also, automatic dilution techniques allow for continuous characterization of the equilibrium states of multi-component solutions. Illustrations are given for the case where the change in polyelectrolyte dimensions, interactions and hydrodynamics can be probed.

Introduction

Light scattering techniques have traditionally been used for characterizing the equilibrium properties of polymer solutions. Since changes in light scattering intensity and angular distribution can easily be monitored on the time scale of many kinetic processes, TDSLS is a useful tool for characterizing such phenomena. These latter include the polymerization process itself, polymer degradation, microgel formation, aggregation, and dry polymer dissolution. Additionally, TDSLS lends itself naturally to automated forms of equilibrium characterization. There has been a recent increase in activity in TDSLS and related techniques, including instrumentation.[1,2,3] Formation of complexes and aggregation in both bio- and synthetic polymer solutions has been reported,[4,5,6,7,8,9,10,11] as well as phase separation phenomena,[12,13,14,15,16,17] crystallization,[18,19] and degradation.[20]

Use of the Zimm approximation

The most widely used light scattering approximation for polymers, due to Zimm,[21] is normally expressed as the excess Rayleigh scattering $I(q,c_p)$, which is total solution scattering minus solvent scattering background, and is a function of polymer concentration c_p and scattering vector q. In TDSLS, the scattering is also a function of time, so that it can be written in terms of q, c_p and t, $I(q,c_p,t)$. Zimm's equation becomes:

 CCC 1022-1360/00/$ 17,50+.50/0

$$\frac{Kc_p}{I(q,c_p,t)} = \frac{1}{MP(q)} + 2A_2c_p + \left[3A_3Q(q) - 4A_2{}^2MP(q)(1-P(q))\right]c_p{}^2 \qquad (1)$$

where it is understood that all parameters K, c_p, M, P(q), etc. are potentially functions of t. This equation forms the basis of the Zimm plot, which, at low concentrations and for $q^2<S^2><<1$ can be written, for a polydisperse polymer population as

$$\frac{Kc_p}{I(q,c_p)} = \frac{1}{M_w}\left(1 + \frac{q^2 <S^2>_z}{3}\right) + 2A_2c_p \qquad (2)$$

which directly permits determination of weight averaged molecular mass M_w, second virial coefficient A_2, and the z-averaged mean square radius of gyration $<S^2>_z$,. K is an optical constant, given for vertically polarized incident light by

$$K = \frac{4\pi^2n^2(dn/dc_p)^2}{N_A\lambda^4} \qquad (3)$$

where n is the solvent refractive index, λ is the laser vacuum wavelength, dn/dc_p is the differential refractive index for the polymer/solvent, and $q=(4\pi n/\lambda)\sin(\theta/2)$, where θ is the scattering angle.

This equation is based on the Rayleigh-Debye approximation, in which it is assumed that the phase shift accrued by a scattered wave traversing the scattering particle itself is due purely to the geometrical path length covered; i.e the difference in index of refraction between the scatterer and the solvent adds negligibly to the phase shift. For polymers, this approximation is usually quite good, since the polymer is a thread-like entity of scatterers immersed in a solvent. When the scatterers become dense, or solid, such as can occur in aggregation and crystallization phenomena, this approximation quickly breaks down, and the scattering problem requires a full solution of Maxwell's equations within and without the scatterer, and matched boundary conditions. This latter approach is often termed 'Mie scattering'.[22,23]

Instrumentation and methods

The detection portion of all the monitoring experiments involves some combination of the following instruments: a multi-angle TDSLS detector, a viscometer, a refractometer (RI), and an ultra-violet absorption detector (UV). TDSLS detectors used have included both home-built units, previously described in detail,[24] and a Wyatt Technologies Dawn-

F. In fact, this same detector train is also used for traditional multi-detector Gel Permeation Chromatography (GPC).[25,26]

The significant differences between the monitoring experiments involve sample conditioning and handling. The main requirement for monitoring is to produce a continuous stream of dilute sample, so that absolute light scattering, and viscosities approaching intrinsic viscosity, can be measured. In some instances the polymer solution in the reactor is so concentrated it must be continuously diluted. In others there are macroscopic particles in the polymer solution that need to be filtered (e.g. dissolution of dried polymers), whereas in other cases temperature conditions might be desired that are difficult to create and maintain in the detectors (e.g. high and low temperature degradation reactions). Means of sample conditioning have been developed for each scenario, although it is a continuing challenge to find means of handling the sampling aspect for increasingly complex systems.

Online monitoring of polymerization reactions

The ability to perform automatic, continuous online monitoring of polymerization reactions (ACOMP) is of interest in three main areas: At the fundamental level, polymer chemists can immediately observe the effects of different synthetic strategies, monomers, initiators, and other factors, and how they influence kinetics and polymer properties. This should prove to be a significant advance for new material discovery. Industrial scientists can use monitoring as a tool for pilot plant optimization of industrial processes. One of the ultimate goals of ACOMP is to apply it to full scale industrial reactors, integrated into a feedback loop that can control reactions. This promises to have a great economic and ecological impact on how large industrial polymer producers operate. Monitoring kinetics of polymerization reactions requires that measurements be made either periodically or continuously on a reacting solution. One of the first works is due to Flory[27], who manually withdrew aliquots from a glycol/dibasic acid polyesterification reaction every ten minutes and performed a KOH titration to estimate M_w and rate constants. Bresler et al.[28] made periodic ESR measurements to follow radical concentrations in homogeneous polymerization reactions of methyl methacrylate and vinyl acetate. Shen and Tian used ESR simultaneously to measure radical concentration and monomer conversion in methyl methacrylate polymerization at intervals a few minutes apart.[29] Ballard and van Lienden[30] used dilatometry to estimate initial polymerization rates of vinyl monomers by metal compounds.

Real-time monitoring of polymerization is a more recent development. Near-IR[31,32,33,34,35,36,37,38,39,40] (near IR), ultra-violet (uv)[41,42] and fluorescence[43] monitoring are proving particularly useful for measurements of monomer conversion and process completion. Fluorescence techniques are also being used for monitoring curing reactions of epoxies.[44,45] IR studies have also been used to estimate copolymer composition and rheological properties[46], whereas direct rheological techniques are being used to estimate molecular weight online.[47,48,49] Monitoring post-pulse processes in laser induced polymerization has also been reported.[50] The author was previously involved in surfactant vesicle laser-induced polymerization, in which laser flash photolysis was used to follow monomer conversion rates.[51,52]

ACOMP has recently been introduced.[53] It allows M_w, monomer conversion f, intrinsic viscosity [η], and certain measures of polydispersity to be followed in near real-time, without the use of chromatographic columns. The method requires that a continuous stream of dilute polymer/monomer solution pass through a train of detectors, normally comprising TDSLS, refractometer (RI), ultra-violet absorbance (UV), and viscosity detectors. For chain growth reactions, the UV detector has been used to follow the disappearance of double bonds in many types of polymerization reactions, which allows both the conversion to be monitored, and M_w to be computed, since only the polymer concentration figures into the computation of M_w. In contrast, for step-growth reactions both monomer and polymer concentrations figure into the computation of M_w, so that no concentration detectors are needed.

The dilute stream is produced by continuously withdrawing a small amount of reactor liquid (typically 50 μL/min) and mixing it with a much larger volume of pure solvent (typically 1.95 mL/min). This has been achieved both with low pressure mixing, involving a programmable gradient mixer in conjunction with a single isocratic pump, and a high pressure scheme using two isocratic pumps and a micro-mixing chamber.

Raw data from a typical polymerization reaction, that of acrylamide,[54] is shown in figure 1. The stabilization of detectors with pure solvent for the first 500 seconds is seen, followed by stabilization with a stream consisting of 4% reactor liquid and 96% water from a reservoir (total monomer concentration is 0.0014 g/cm^3). At 1600s a persulfate initiator was added and the polymerization can be seen by an increase in light scattering and viscosity signals, and a decrease in the UV, as the monomer double bond absorption is lost when the monomer is incorporated into a polymer chain.

Figure 1. Raw data for an acrylamide polymerization reaction at T=70C (adapted from Giz et al., ref. 54)

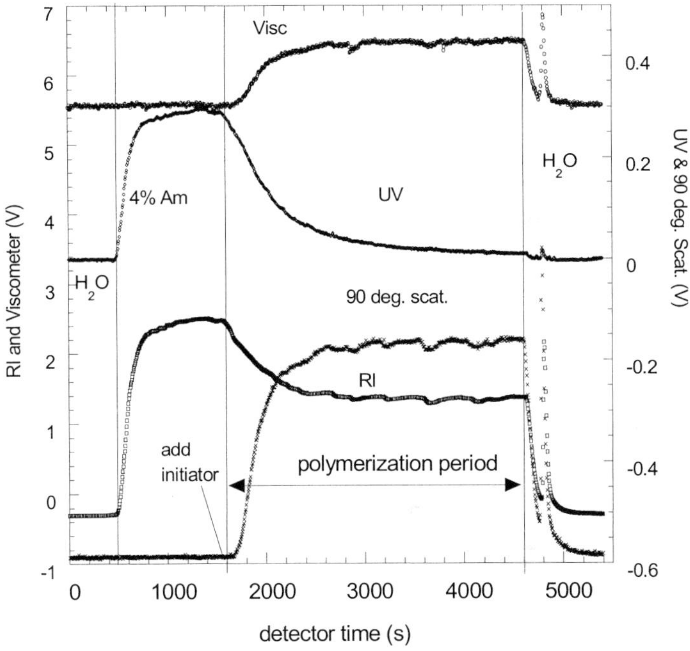

Figure 2 shows conversion curves for acrylamide polymerization reactions. The conversion is first order (exponential) when a high concentration of initiator is used, which is expected for ideal free radical polymerization.[55]

Figure 2. Monomer conversion vs. time for the data of figure 1. (Adapted from Giz et al., ref. 54)

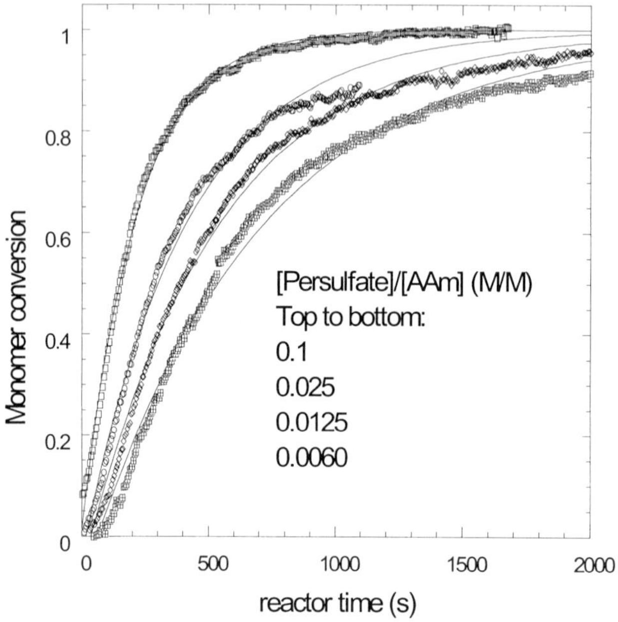

At lower initiator concentration, deviations from first order behavior are apparent at early and late conversion. This has been interpreted in terms of initial competition for radicals from impurities, and cage effects.[54,56,57] M_w vs. conversion for typical reactions is shown in figure 3. Over the majority of conversion the decrease of M_w closely follows the predictions of ideal free radical polymerization, namely[54]

$$M_w(f) = M_w(0)(1 - f/2) \qquad (4)$$

The deviation from this occurs at early conversion.

Figure 3. M_w **vs. conversion for various acrylamide polymerization reactions. The linear regime that sets in at early conversion reflects ideal free radical polymerization. (Adapted from Giz et al., ref. 54)**

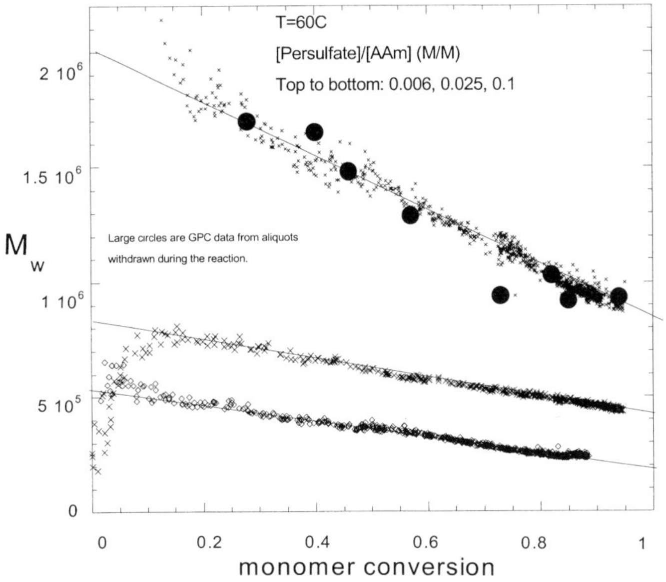

Because ACOMP uses no chromatographic columns or other separation techniques, to which it owes its near real-time advantages, it is interesting to see how much information it can provide on the evolution of polydispersity during polymerization reactions. This area was recently considered, and it was found there are at least three immediate approaches.[58] The first is applicable when the polymers are large enough to give a measurable angular dependence for scattered light, which usually requires $<S^2>^{1/2}$ to be ten nanometers or higher. The slope of Kc/I(q,c) is

$$\frac{d(Kc_p / I(q))}{dq^2} = \frac{< S^2 >_z}{3M_w} = A\frac{< M^\beta >_z}{M_w} \qquad (5)$$

where β is the exponent linking $<S^2>$ to M; $<S^2>=3AM^\beta$. For the case of ideal coils β=1, so that the slope is directly proportional to M_z/M_w. For coils with excluded volume β~1.2, and so the slope will be proportional to a quantity slightly higher than M_z/Mw.

This measure of polydispersity is automatically collected during ACOMP, as long as the polymers have a measurable $<S^2>^{1/2}$ (typically a minimum of 10nm).

A second means of monitoring the evolution of polydispersity is by combining the simultaneously measured viscosity and scattering data. It is then possible to obtain the quantity M_w/M_η, where M_η is the viscosity average mass, which often falls between M_n and M_w. In principle, the combination of viscosity and M_w data should also prove valuable for following branching during polymerization reactions, since M_w combined with $[\eta]$ yields the polymer's hydrodynamic volume. Light scattering and viscosity have frequently been used to determine branching.[59,60,61]

For chain growth reactions where dead polymer chains are rapidly produced compared to the time needed for total monomer conversion, it is possible to compute the instantaneous weight average molecular weight, $M_{w,inst}$, according to:

$$M_{w,inst}(f) = M_w(f) + f\left.\frac{dM_w}{df}\right|_f \qquad (6)$$

Figure 4 shows data for the polymerization of poly(vinyl pyrrolidone), or PVP,[53] and the computation of $M_{w,inst}$.[58] It was found that M_w stayed at a constant level throughout conversion, although no convincing explanation for this was found. In fig. 4 PVP polymerization occurred at a constant M_w, and then a boost of hydrogen peroxide initiator was given to the reactor at 20% conversion. The effect was to make the remaining monomer build to polymers of lower M_w for the remainder of the conversion. The cumulative M_w decreases smoothly, as shown, so it is not evident that a bimodal population is produced. $M_{w,inst}$, plotted on the same curve, however, clearly shows that $M_{w,inst}$ quickly plunges to a much lower value, as soon as initiator is injected.

The computation of $M_{w,inst}$ from the ACOMP data requires no model dependent assumptions. If one wishes to obtain absolute polydispersity indices, such as M_w/M_n or M_z/M_w, then model dependent assumptions must be made. For free radical polymerization, where termination is by disproportionation, the instantaneous values of $M_{z,inst}$, $M_{w,inst}$ and $M_{n,inst}$ stand in the ratio 3:2:1. For termination by recombination M_w/M_n= 3:2. Hence, knowing $M_{w,inst}$ allows computation of $M_{z,inst}$ and $M_{n,inst}$, as well as any other conceivable average resulting from ideal free radical polymerization. The insets to figure 4 show histogram distributions of the polymer population at 20% and 90% monomer conversion, derived from $M_{w,inst}$.

ACOMP has also been used to follow step growth reactions,[62] and for controlled radical polymerization (CRP).[63] CRP combines the advantages of narrow polydispersity and controllable chain architecture of living polymerization with the robust, economical nature of free radical polymerization. ACOMP has been adapted to continuous reactors,[64] and used to determine chain transfer constants.[65] Current and future applications will include copolymerization and inhomogeneous phase polymerization.

Figure 4. Cumulative M_w and $M_{w,inst}$ for a vinyl pyrrolidone polymerization reaction, and distribution histograms (Adapted from ref. 58)

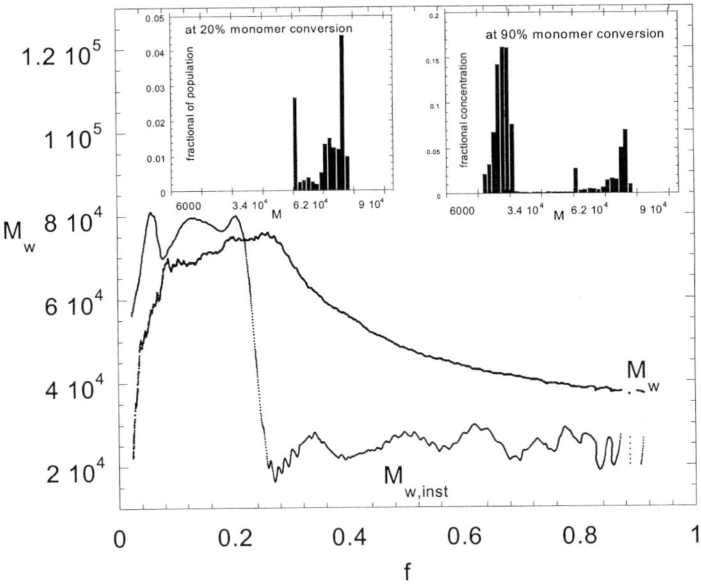

Degradation of polymers

When polymers are degraded under the influence of agents such as acids, bases, enzymes, heat, and radiation, the light scattering intensity will drop. The physical problem is to find a quantitative expression linking the time course of light scattering to the kinetics and mechanisms causing the degradation, and to the polymer structure; i.e. whether the polymer is singly or multiply stranded, branched, etc.

The first deductions about polymer structure made from periodic sampling during a degradation reaction were due to Thomas,[66] who used the quadratically increasing number of cleavages in time to deduce that DNA was double-stranded. Since then, a general formalism for computing the time and wave vector dependent scattering I(q,t) has been presented,[67] and signatures for a variety of degradation mechanisms and polymer structures have been predicted and found experimentally.[68,69,70,71,72]

Figure 5 (adapted from ref. 72) shows the UV degradation of polystyrene that has a small percentage of methyl vinyl ketone (MVK) links. The initial degradation rate increases proportionately with the percentage of MVK links. Degradation is by random cleavage of single polymer chains.

Figure 5. UV degradation of MVK substituted PS. **Figure 6. PG degradation**

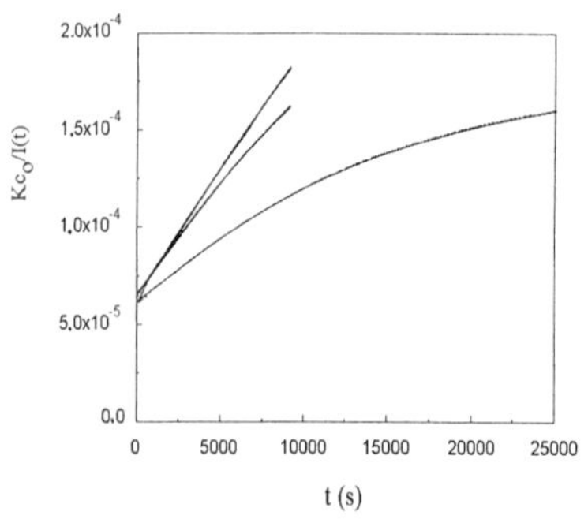

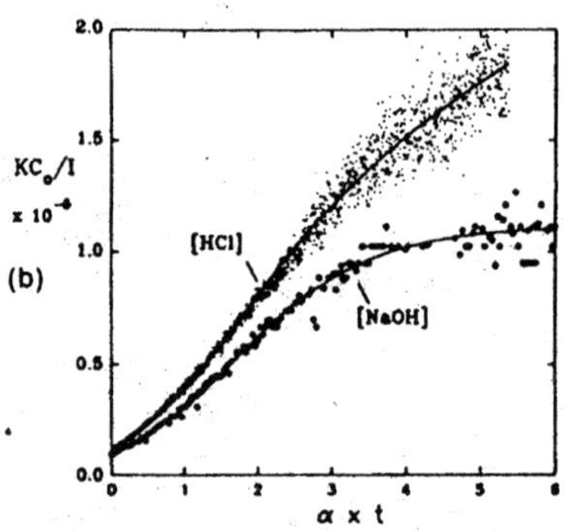

Figure 6 (adapted from ref. 71) shows the degradation of proteoglycan subunits (PG) under HCl and NaOH. The PG are composed of a protein backbone, to which short (2-20 kg/mole), sulfonated polysaccharide chains are covalently linked. The signature for NaOH indicates that degradation is via stripping of sidechains at the covalent link, whereas the signature for HCl shows that both sidechain stripping and random degradation of the protein backbone are occurring. The expressions for the signatures allow degradation rates and fraction of material contained in the backbone and sidechains to be determined, as well as the degradation mechanism. Distinct signatures are calculated and found both for different types of degradation (e.g. endocytic and exocydic enzymatic cleavage, combinations of enzymes,[73] center-weighted scission, etc.), but also for different architectures; single and multiple strand chains and different types of branching. Hence, degradation studies can also be useful for determination of structure, such as branching, in addition to kinetics.

Dissolution of dry polymeric powders

The rate at which dry polymers dissolve in a solvent is an important characteristic in pharmaceutical, food, and general polymer sectors, and has been studied theoretically and experimentally.[74,75,76] For possible uses of ACOMP where polymer is formed in the solid phase (e.g. fluidized bed), the dissolution time will be of critical importance in determining feasibility of ACOMP.

Michel and Reed[77] constructed a dissolution reactor by using a stirred beaker in a thermostated bath, and a peristaltic pump that filtered and circulated reactor liquid through the detector train. It was shown that the concentration of dissolved polymer $C_p(t)$, followed the expected third order dependence in time,[78]

$$C_p(t) = C_o\left[1-(1-kt/R_o)^3\right],\quad kt \le R_o \qquad (7)$$

and, was used to explore the effect of ionic strength on dissolution of polyelectrolytes. Figure 7 shows dissolution data for sodium poly(styrene sulfonate) (PSS). The RI data in the inset show that there is no appreciable difference in dissolution rates whether the PSS is dissolved in pure water or salt water (0.1M NaCl), and is also independent of the pore size of the inline filtration membrane. The TDSLS data, however, depend significantly on whether pure water or salt water is used, and also on the size of the filter. In pure water there are initial spikes in the light scattering, which are so large that a logarithmic scale is used for the TDSLS voltage. The larger the pore size, the larger the effect. The spike is due to a small population of transient aggregates present as the polyelectrolyte dissolves in pure water. The aggregates eventually disappear over a long enough period of time. These experiments strengthened earlier reports that the so-called 'slow mode' of diffusion[79,80] is due to removable aggregates,[81,82,83] rather than to any fundamental equilibrium nature of the polyelectrolyte solutions.

Figure 7. Dissolution of PSS in pure water and water with NaCl. Light scattering and refractometer monitoring results. (Adapted from ref. 77)

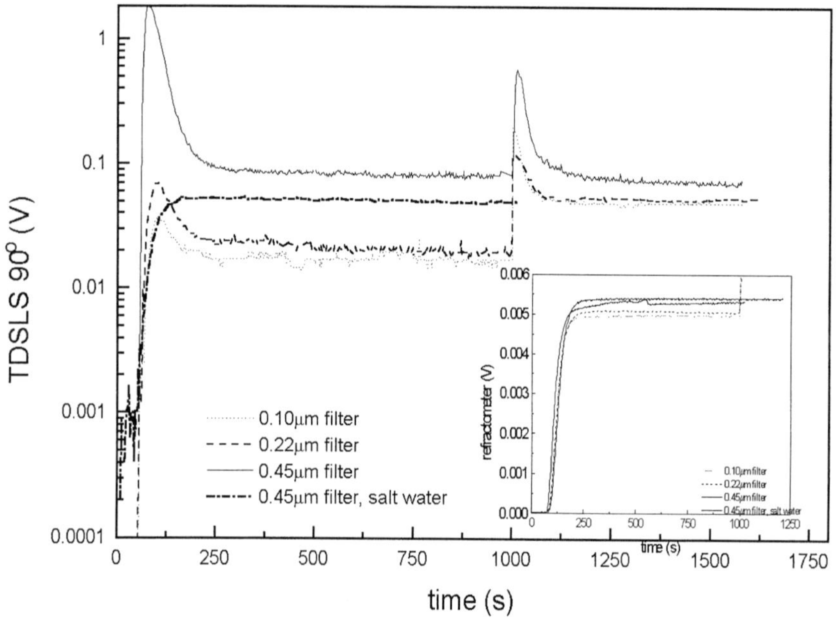

Aggregation

At low angles and high dilution, scattering within the Rayleigh-Debye approximation increases as the square of the scatterer's mass. Hence, TDSLS is exquisitely sensitive to aggregation, oligomerization, micro-crystallization, micro-gelation, and other processes where polymer mass increases in time.

An example of the aggregation of a fluoropolymer in DMSO is shown in figure 8, expressed as I/Kc vs. t. The signature suggests diffusion controlled aggregation of chains into larger and larger structures, which then attain a maximum value. These aggregates are themselves only metastable, as the solutions precipitates after several days.

Heteogeneous Time Dependent Light Scattering (HTDSLS)

Traditionally, the presence a population of large colloid scatterers amidst a population of weakly scattering polymers would render it impossible to make useful light scattering measurements. The idea of HTDSLS is to make the solution flow through a small scattering volume (around 10 nanoliters) so that the passage of large particles produces recognizable peaks amidst the polymer scattering background. With intensity sampling fast enough to recognize these peaks, both the scattering background due to pure polymer can be recovered, at the same time that the number density of large scatterers is determined. This has utility in situations where large scatterers are an integral part of a polymer solution; e.g. bacteria in a fermentation reactor that produce or hydrolyze biopolymers such as polysaccharides or proteins, or the formation of microgels or microcrystals in a polymerization reactor. Other applications include cases where solutions are so contaminated by large particles they would normally be unmeasurable.

Figure 8. Aggregation kinetics of a fluoropolymer (Moyses and Reed, unpublished)

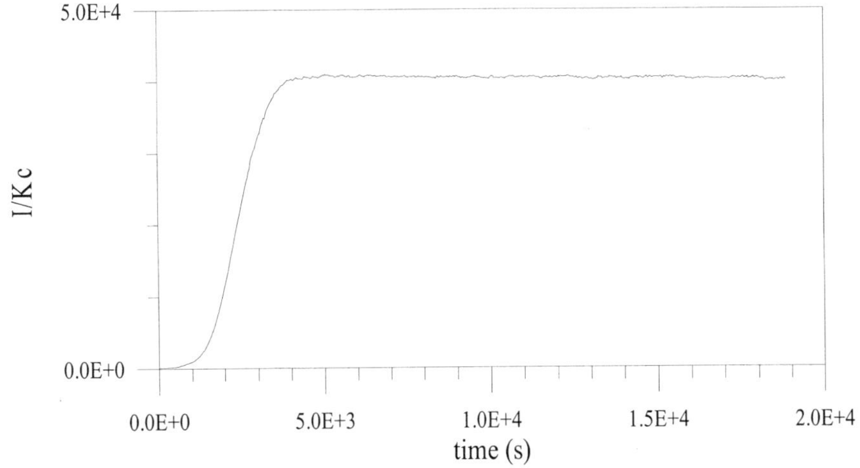

Fig. 9 shows an example of HTDSLS data, adapted from Schimanowski et al.,[84] in which a 0.1 mg/ml solution of PVP (650 Kg/mole) co-exists with a population of growing E.

Coli bacteria at 38^{0}C. The PVP and E. Coli do not interact. The main portion of the data show the spikes due to E. Coli, and how the density of the spikes increases in time as the number density of bacteria increases. Despite the increase in E. Coli, the baseline scattering, which is due to the PVP, is easily recovered. The insets to the figure show the simultaneously obtained determination of Kc/I (yielding $<S^2>$ and M_w if A_2 is known), and the density of bacteria.

Figure 9. Co-existing polymers (PVP) and growing bacterial population (E.Coli). Spike density increases with bacterial growth. Insets show number density of bacteria and Kc/I for the polymer. (Adapted from ref. 84)

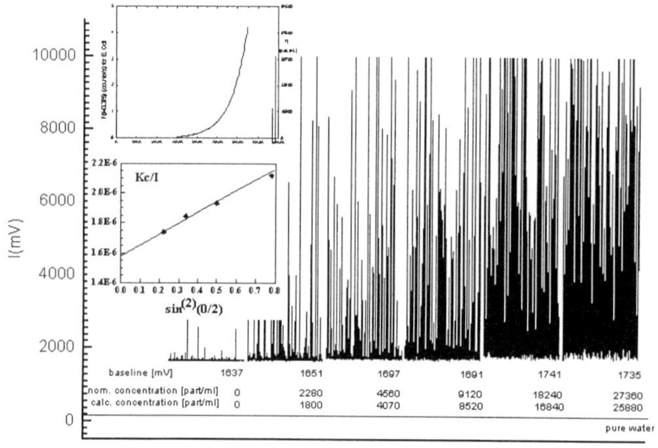

Automated methods for batch characterization

The automatic dilution used in ACOMP lends itself naturally to automated batch characterization when a sample is mixed in a continuous (or stepped) gradient with another sample solution, keeping the same detector train used in ACOMP. This yields a high resolution, continuous record of how light scattering and viscosity change as a function of a given solute; e.g. polymer, salt, or surfactant concentration. This significantly expands the scope of experimental investigation of simple and complex polymer solutions, for which, traditionally, individual concentration samples would be manually prepared and measured.

Automated Zimm plot/intrinsic viscosity determination

Ref. 24 detailed the use of continuous dilution to produce a complete light scattering and viscosity characterization of water soluble polymers. Figure 10 shows the Zimm plot data for PVP in water. The inset shows the raw scattering intensity. A maximum value is reached, followed by a decline. This is due to the third virial coefficient A_3.

Effect of changing ionic strength on polyelectrolyte conformations, interactions and hydrodynamics

It is well known that polyelectrolytes are sensitive to the ionic strength of the medium they are in. An extensive theoretical and experimental literature exists on how conformations, persistence lengths (PL), excluded volume, and hydrodynamics depend on electrostatic effects, whose magnitude, in turn, depends on ionic strength.[85,86,87,88,89] Recently, experiments were performed in which the concentration of polyelectrolyte was held fixed, and the ionic strength was increased continuously.[90] Figure 11 shows raw data on LS and viscosity vs. [NaCl] for sodium hyaluronate (HA). As expected, the light scattering intensity increases as salt increases, since the large, electrostatic contribution to excluded volume, and hence A_2, decreases. At the same time, the contraction of the polyelectrolyte dimensions with increased screening decreases the hydrodynamic volume, and hence the polyelectrolyte's reduced viscosity.

The inset to the figure shows the viscosity as a function of [NaCl] for HA, xanthan and PSS, which immediately reveals the relative stiffness of these polymers. The xanthan is stiffest, with an intrinsic PL over 100nm. It quickly looses its electrostatically enhanced PL at low [NaCl] and its intrinsic stiffness allows no further contraction as [NaCl] increases. HA has intermediate flexibility, an intrinsic PL of 10nm, and it is seen that it contracts over the whole range of [NaCl], beginning to flatten out at the highest values. PSS has an intrinsic PL of about 1nm, and the viscosity continues to drop unabatedly over the whole range of [NaCl].

Figure 10. Automated Zimm plot determination for PVP.
Inset shows scattering maximum due to A_3 (Adapted. from ref. 24)

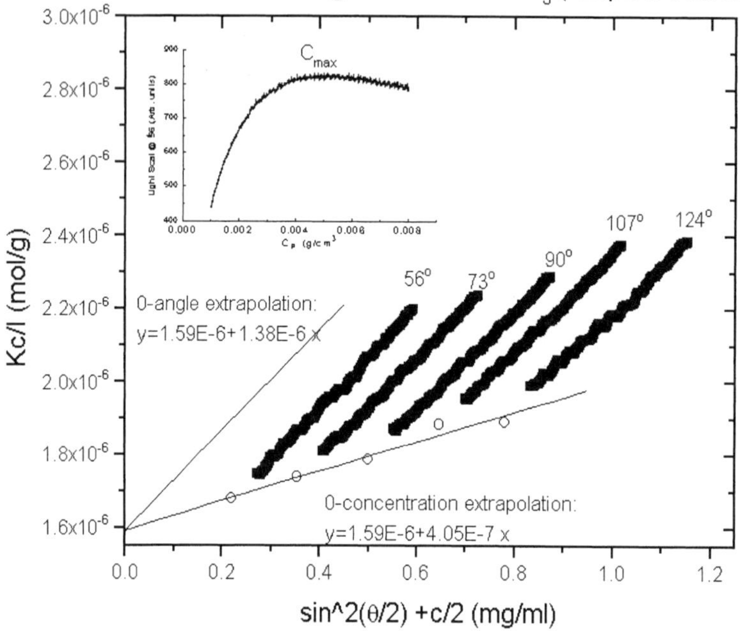

Fig. 12 shows data on the electroviscous effect for polyelectrolytes.[91,92,93] In this effect, as the polyelectrolyte solution is diluted with low ionic strength aqueous solvent there is a maximum value of η_r at a low polyelectrolyte concentration, after which further dilution leads to a decrease in η_r. The effect is usually attributed to the fact that because the counterions of the polyelectrolyte contribute to solution ionic strength, as dilution with a fixed ionic strength solvent proceeds, the net solution ionic strength decreases, leading to an expansion of the polyelectrolyte coils with a concomitant increase of both intrinsic viscosity and the interchain hydrodynamic interactions.[93,94] Other areas where automated batch characterization is useful include multi-component systems, where complex interactions are expected. Any parameterized curves in the phase diagram can be programmed into the method. Currently, we are using the method to determine the phase diagram, and the nature of aggregates formed between surfactant micelles and neutral and charged polymers.

Figure 11. Light scattering and viscometer response of a linear polyelectrolyte to increasing ionic strength. (Adapted from ref. 90)

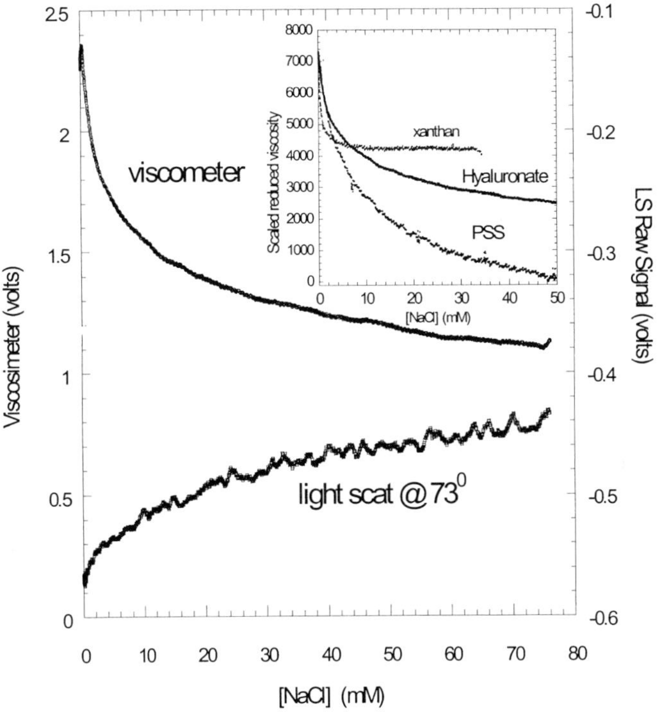

Figure 12. Electroviscous effect for hyaluronic acid (Strelitzki and Reed, unpublished)

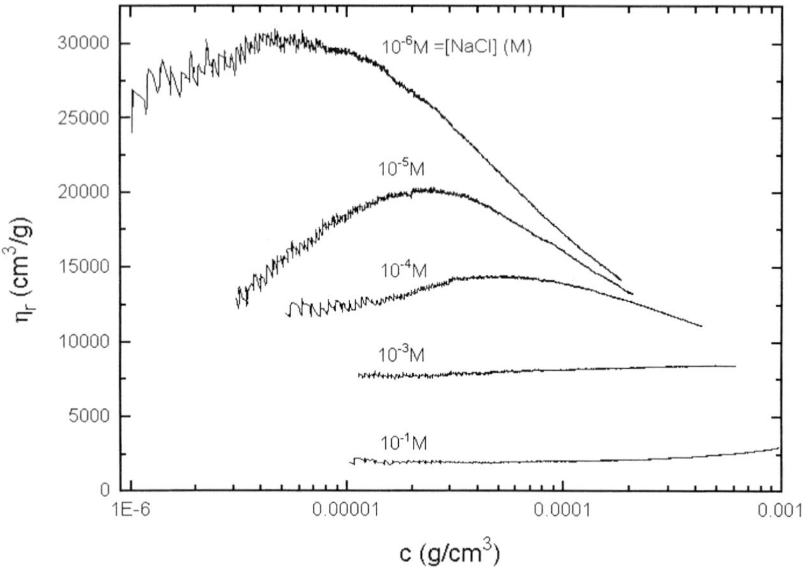

Acknowledgments

I am grateful to many colleagues over the past few years who have participated in this work. These include Alan Parker, David Norwood, Corinne Vinches, Roland Strelitzki, Ruth Schimanowski, Fabio Florenzano, Ricardo Michel, Jean-Luc Brousseau, Stephan Moyses, Bruno Grassl, Joana Ganter, Erica Bayly, Gina Sorci, Alina Alb, Huceste and Ahmet Giz, and Florence Chauvin. Support from the U.S. National Science Foundation grant 9877206, Louisiana Board of Regents, Elf Atochem, Totalfina, Brookhaven Instruments Corp., Firmenich, and Int'l. Specialty Products is acknowledged.

[1] H. Holthoff, S.U. Egelhaaf, M. Borkovec, P. Schurtenberger, H. Sticher, *Langmuir*, *12*, 5541-5549, 1996
[2] S.U. Egelhaal, P. Schurtenberger, *Rev. Sci. Instrum.*, *67*, 2, 540-545, 1996
[3] L.S. Wright, A. Chowdhury, P. Russo, *Rev. Sci. Instrum.*, *67*, 10, 3645-3648, 1996
[4] E. Lai, J.H. Van zanten, *Biophys. J.*, 80, 2, 864-873, 2001
[5] C. LeBon, T. Nicolai, D. Durand, *Macromolecules,* 32, 19, 6120-6127, 1999
[6] R.M. Murphy, M.M. Pallitto, *J. Struct. Biol.*, 130, 2, 109-122, 2000
[7] S. He, P.G. Arscott, V.A. Bloomfield, *Biopolymers*, 53, 4, 329-341, 2000
[8] B. Barham, K. Fosser, G. Voge, D. Waldow, A. Halasa, *Macromolecules*, 34, 3, 514-521, 2001

[9] A.J. O'Connor, T. A. Hatton, A. Bose, *Langmuir*, 13, 26, 6931-6940, 1997
[10] S. Bernocco, S. Ferri, A. Profumo, C. Cuniberti, M. Rocco, *Biophys. J.*, 79, 561-583, 2000
[11] T. Norisuye, M. Shibayama, S. Nomura, *Polymer*, *39*, 13, 2769-2775, 1998
[12] R. Bansil, L. Guangdon, P. Falus, *Physica A: Stat. and Theor. Phys.*, 231, 3, 346-358, 1996
[13] J. Kojima, Y. Nakayama, M. Takenaka, T. Hashimoto, *Rev. Sci. Inst.*, 66, 8, 4066-4072, 1995
[14] H. Takeshita, T. Kanaya, K. Nishida, K. Kaji, *Macromolecules*, 32, 23, 7815-1819, 1999
[15] R. Kita, T. Kaku, K. Kubota, T. Dobashi, *Phys. Lett. A,* 259, 4, 302-307, 1999
[16] G.D. Merfeld, D.R. Paul, *Polymer,* 41, 2, 649-661, 2000
[17] F. Mallamace, N. Micali, S.H. Chen, *Physica A: Stat. Theor. Phys.*, 235, 2, 170-185, 1997
[18] R.S. Stein, J. Cronauer, H.G. Zachmann, *J. Molec. Struct.*, 383, 3, 19-22, 1996
[19] M. Okamoto, T. Inoue, *Polymer*, 36, 14, 2739-2744, 1995
[20] A. Niu, C. Li, Y. Zhao, J. He, Y. Yang, C. Wu, *Macromolecules,* 34, 3, 460-464, 2001
[21] B.H. Zimm, *J. Chem. Phys.*, 16, 12, 1093-1114, 1948
[22] H.C. van de Hulst, *Light Scattering by Small Particles,* Wiley, N.Y., 1957
[23] M. Kerker, *The Scattering of Light and Other Electromagnetic Radiation,* Academic Press, N.Y. 1969
[24] R. Strelitzki, W.F. Reed, *J. App. Polym. Sci.*, *73*, 2359-2368, 1999
[25] W.F. Reed, *Strategies for Size Exclusion Chromatography,* ACS ser. 635; Potschka, M., Dubin, P., Eds.; ACS: Wash. D.C., 1996; pp 7-34
[26] D.P. Norwood, W.F. Reed, *Int. J. Polym. Ana. and Char.*, 4, 2, 99-132, 1997
[27] P.J. Flory, *J. Am. Chem. Soc.*, 61, 3334-3340, 1939
[28] S.E. Bresler, E.N. Kazbekov and V.N. Shadrin, *Makromole. Chem.,* 175, 2875-2880, 1974
[29] J. Shen, Y. Tian, *Makromole. Chem, Rapid Comm.,* 8, 615-620, 1987
[30] D.G.H. Ballard, P.W. van Lienden, *Makromole. Chem.*, 154, 177-190, 1972
[31] C. Jones, J.A. Brown, *Adv. in Instrumentation.* 38, Part 1. Proc. of ISA Int. Conf., p. 705, 1983
[32] R.L. Grob, D.J. Skahan, K. Dix, K. Nielsen, *Process Control & Quality*, 2, 3, 225-35, 1992
[33] D.J. Ehntholt, R.F. Tayor, E.V. Miseo, *ISA Trans.*, 32, 2, 183-8, 1993
[34] R.F. Storey, A.B. Donnalley, T.L. Maggio, *Macromolecules*, 31, 5, 1523-1526, 1998
[35] J. Mijovic, S. Andjelic, J.M. Kenny, *Polym. Adv. Technol.*, 7, 1, 1-16, 1996
[36] C. Gwosdz, M.S. Francis, A. Bickel, *Proc. Control Qual.*, 4, 1, 31-35, 1992
[37] M. Scott, *Control & Instrumentation*, 23, 3, 40-43, 1991
[38] R. Reshadat, S.T. Balke, *App. Spectroscopy*, 53, 10, 1309-1311, 1999
[39] P. Dallin, *Process Cont. and Qual.*, 9, 4, 167-172, 1997
[40] D.J.T. Hill, L.Y. Shao, P.J. Pomery, A.K. Whittaker, *Polymer*, 42, 11, 4791-4802, 2001
[41] A. Penlidis, J.F. MacGregor, A.E. Hamielec, *Proc. of Amer. Control Conf.,* 2, 878-880, 1985
[42] R.S. Saltzman, *I&CS,* 67, 2, 49-51, 1994
[43] Y.S. Kim, C. Sook, C. P. Sung, *J. Appl. Polym. Sci.* ,57,3, 363-70, 1995
[44] H.J. Paik, N.H. Sung, *Polymer Eng. Sci.*, 34, 1025-1032, 1994
[45] A. Fuchs, N.H. Sung, *Polym. Mater. Sci. Eng.,* 71, 439-440, 1994
[46] M.G. Hansen, S. Vedula, *Polym. Process Eng.* 97, 89-102, 1997
[47] J.M. Starita, C.L. Rohn, *Plast. Compd.*, 10, 2, 46-51, 1987
[48] O. Brand, J.M. English, S.A. Bidstrup, M.G. Allen, *Sensors and Actuators.,* 1, 121-124 , 1997
[49] S. Ponnuswamy, S.L. Shah, C. Kiparissides, *J. Appl. Polym. Sci.*, 32, 1, 3239-53, 1986
[50] I.N. Kotchetov, D.C. Neckers, *J. Imaging Sci. & Tech.*, 37, 2, 156-163, 1993
[51] W.F. Reed, L. Guterman, P. Tundo, J.H. Fendler , *J. Am. Chem. Soc.*, 106, 1897-1907, 1984
[52] W. F. Reed, *Macromolecules*, 18, 2402-2409, 1985
[53] F.H.Florenzano, R. Strelitzki, W.F. Reed, *Macromolecules*, 1998, 31, 7226-7238
[54] A. Giz, H.C. Giz H.C., A. Alb, J.L. Brousseau, W.F. Reed, *Macromolecules,* 34, 1180-1191, 2001
[55] N.A.Dotson, R.Galvan, R.L.Laurence, M.Tirrel, *Polymerization Process Modeling*, VCH Pub. N.Y. 1996
[56] T. Ishige, A.E. Hamielec, *J. App. Polym. Sci.* 17, 1479 -1506, 1973
[57] J.P. Riggs, F. Rodriguez, *J. Polym. Sci,* A-1, 5, 3151 -3181, 1967
[58] W.F. Reed, *Macromolecules*, 33, 7165-7172, 2000
[59] W. Burchard, *Static and Dynamic Light Scattering from Branched Polymers and Biopolymers*, in *Advances in Polymer Sci.,* 48, Springer Verlag, Berlin, 1983
[60] J.F. Douglas, J. Roovers, K.F. Freed, *Macromolecules*, 23, 4168-80, 1990
[61] C.E. Ioan, T. Aberle and W. Burchard, *Macromolecules*, 33, 5730-39, 2000

[62] A. Giz, H. Giz, J.L. Brousseau, A. Alb and W.F. Reed, *J. Appl. Polym. Sci.*, in press, 2001
[63] F. Chauvin, A. Alb, D. Bertin, P. Tordo, W.F. Reed, *Macromolecules*, submitted
[64] B. Grassl, W.F. Reed, *Am Inst. Chem. Eng. J.*, submitted
[65] B. Grassl, A. Alb, W.F. Reed, *Makromole. Chem. Phys.*, in press, 2001
[66] C.A. Thomas, *J. Amer. Chem. Soc.*, 78, 1861-1866, 1956
[67] W.F. Reed, *J. Chem. Phys.*, 103, 7576-7584, 1995
[68] C.E. Reed, W.F. Reed, *J. Chemical Physics*, 91, 7193-7199, 1989.
[69] W.F. Reed, C.E. Reed, L. Byers, *Biopolymers*, 30, 1073-1082, 1990
[70] S. Ghosh, I. Kobal, D. Zanette, W. F. Reed, *Macromolecules*., 26, 17, 4685-4693, 1993
[71] S. Ghosh, W.F. Reed, *Biopolymers*, 5, 435-450, 1995
[72] L.H. Catalani, A.M. Rabello, F.H. Florenzano, M.J. Politi, W.F. Reed, *Int. J. Polym. Char. Anal.*, 3, 2, 231-247, 1997
[73] J.L. Ganter, J. Sabi, J., W.F. Reed, *Biopolymers*, in press, 2001
[74] D.J. Buckley, M. Berger, *J. Polym. Sci.*, 56, 175-179, 1962
[75] F. Brochard, P.G. De Gennes, *Physicochem. Hydrodynam.*, 4, 313-319, 1983
[76] I. Devotta, V. Premnath, M.V. Badiger, P.R. Rajamohanan, S. Ganapathy, R.A. Mashelkar, *Macromolecules*, 27, 532-539, 1994
[77] R.C. Michel, W.F. Reed, *Biopolymers*, 53, 19-39, 2000
[78] A. Parker, F. Vigouroux, W.F. Reed, *Am. Inst. Chem. Eng. Journal*, 46, 7, 1290-1299, 2000
[79] S.C. Lin, W.I. Lee, J.M. Schurr,, *Biopolymers*, 17, 1041-1064, 1978
[80] M. Drifford, J.P. Dalbiez, *Biopolymers*, 24, 1501-1514, 1984
[81] S. Ghosh, R.M. Peitzsch, W.F. Reed, *Biopolymers*, 32, 1105-1122, 1992
[82] X. Li, W.F. Reed, *J. Chem. Phys.*,94, 4568-4580, 1991
[83] R.G. Smits, M.E. Kuil, M. Mandel, *Macromolecules*,27, 5599-5608, 1994
[84] R. Schimanowski, R. Strelitzki, D.A. Mullin, W. F. Reed, *Macromolecules*, 32, 21, 7055-7063, 1999
[85] T. Odijk, *J. Polym. Sci. Phys. Ed.*, 10, 9444-9451, 1977
[86] M. Fixman, J. Skolnick, *Macromolecules*, 11, 8638-8644, 1978
[87] W.F. Reed, S. Ghosh, G. Medjhadi, J. Francois, *Macromolecules*, 24, 6189-6202, 1991
[88] S. Förster, M. Schmidt, *Adv. Polym. Sci.*, 120, 51-133, 1995
[89] C.E. Reed, W.F. Reed, *J. Chem. Phys.*, 94, 8479-8486, 1991
[90] Bayly, E., Brousseau, J.L., Reed, W.F., *Int. J. Polym. Char. and Anal.*, in press, 2001
[91] S. Basu, *Nature*, 168, 341-343, 1951
[92] W.F. Reed, *J.Chem. Phys.*, 101,3, 2515-2521, 1994
[93] R.M. Fuoss, V.P.Strauss, *J. Polym. Sci.*, 3, 602-609, 1948
[94] M. Rinaudo, M. Milas, M. Jouon, R. Borsali, *Polymer*,34, 17, 3710-3718, 1993

Macromol. Symp. **190**, 151–159 (2002)

The Effect of Water on Cellulose Solutions in DMAc/LiCl

Thomas Röder[*1,3], *Antje Potthast*[1], *Thomas Rosenau*[1], *Paul Kosmsa*[1], *Thomas Baldinger*[3], *Bernd Morgenstern*[4], *Otto Glatter*[2]

[1] University of Agricultural Sciences Vienna, Christian Doppler Laboratory, Muthgasse 18, A-1190 Vienna, Austria
[2] University of Graz, Department of Chemistry, Heinrichstrasse 28, A-8010 Graz, Austria
[3] Lenzing AG, A-4860 Lenzing, Austria
[4] FILK GmbH, Postfach 1144, D-09581 Freiberg, Germany

SUMMARY: The solution state of cellulose in the system *N,N*-dimethylacetamide/lithium chloride (DMAc/LiCl) depends on various factors such as cellulose concentration, provenience (cotton, hardwood, softwood) and chemical history (pulping, pretreatment, bleaching) of cellulose, LiCl concentration, activation method, dissolution conditions (time, shaking), and water content. In particular the influencing of the latter has been intensively investigated in our present studies. Working in anhydrous conditions is not practicable for routine size exclusion chromatography (SEC) analysis. Especially in solutions diluted to SEC levels (0.9 wt% LiCl), an aggregation induced by water was observed. Depending on the time of dissolution and on the amount of water, changes in the solution state were observed. In some cases the amount of aggregates increases within a few minutes. This is reflected by a time-dependent increase in the scattering intensity and quantitatively proved by an increase in the aggregation peak in the calculated intensity distributions. With less soluble pulps, traces of water (lower than 0.01 M) can already suffice to induce and promote aggregation. To disturb a "good" stock solution, the concentration of water must be higher than 0.05 M. The aggregates formed correspond to the model of the fringed micelle.

Introduction

The solvent system *N,N*-dimethylacetamide/lithium chloride (DMAc/LiCl) is commonly used for the determination of the molecular weight distribution of cellulosic substrates by size exclusion chromatography (SEC)[1-4].

As the mixture is able to effect dissolution of cellulose within a certain concentration range of LiCl and cellulose, it is widely used for analytical purposes, *e.g.*, SEC measurements, and in organic synthesis, *e.g.*, for homogeneous derivatization reactions[5]. The published solubility limits of LiCl in "dry" DMAc range from 9 to even 13 wt%. The procedures proposed for drying DMAc and LiCl to obtain a water-free mixture are similarly inconsistent: different distillation protocols and various azeotropic distillation procedures with different water-trapping agents have been proposed to remove traces of water from DMAc. Drying at

 CCC 1022-1360/00/$ 17,50+.50/0

different temperatures, flame drying and even melting are suggested to remove residual water from LiCl. Therefore, also the literature values of the limiting solubility of defined celluloses disagree with each other. More serious, however, SEC data of celluloses dissolved in DMAc/LiCl, which were obtained by different groups under seemingly identical conditions, show strong deviations[6], which might, among other reasons[7], be also a result of differing water contents in the solvents used[8]. Therefore the SEC-relevant solutions were investigated in this study with respect to water content and time stability. Dynamic light scattering (DLS) is a very fast and sensitive method to gain information about the solution state.

Since SEC in DMAc/LiCl is the most frequently used method for characterization of pulps in research and industry, our results have a bearing on scientific research as well as industrial purposes.

Experimental

Materials and cellulose dissolution: Chemicals were obtained from commercial sources and of the highest purity available. For dynamic light scattering (DLS) experiments, LiCl was dried 6 h at 200 °C *in vacuo*. Absolute DMAc as used in the experiments was obtained by refluxing commercial DMAc over CaH_2 for 2 h, followed by distillation onto CaH_2. In some cases, refluxing and distillation was preceded by stirring for two days at room temperature. The dry DMAc was stored over CaH_2 and handled according to standard protocols in organometallic synthesis. For the solubility measurements, defined amounts of distilled water were added to the dry DMAc with syringe to obtain H_2O concentrations in the range 0.1 - 10 M. A beech sulfite pulp (Lenzing AG, Austria) with a degree of polymerization of 1500 was used as cellulose sample (KZO_3 - ozone-bleached). For a better dissolution in DMAc/LiCl, the pulp was activated by solvent exchange (H_2O, acetone, DMAc). Completely dry samples were produced by freeze-drying the activated samples.

Light scattering measurements: Viscosities (Table 1) were measured with a SR-5000-NF stress rheometer (Rheometric Scientific, Munich, Germany). Refractive indices were determined with an Atago RX-5000 digital refractometer (Japan).

Table 1: Viscosities and refractive indices for the used solvents at 20 °C

[LiCl] in DMAc (wt%)	Viscosity (Pa s)	Refractive index
2.6	0.0034	1.4443
0.9	0.0013	1.4360

The DLS experiments were performed on a laboratory-built goniometer equipped with a 5-W argon ion laser (λ_0 = 514.5 nm, Spectra Physics, USA). The scattering cells (10-mm cylindrical cuvettes, Hellma, Germany) were immersed in a thermostatted index-matching bath (decalin). The scattering angle was set to 90°. The q-dependence of the correlation time was proved for one example.

Detection was performed *via* a single-mode fiber (OZ from GMP, Zurich, Switzerland) with a grin lens coupled to a Thorn-Emi photomultiplier (type B2FBK/RFI), the output of which was analyzed by an ALV-5000 digital multiple-τ correlator (ALV, Langen, Germany) with 256 quasi-exponentially spaced channels[9)]. In the case of depolarized (VH) measurements, the laser power was set to 1 W. For depolarized dynamic light scattering (DDLS) measurements (VH), the primary beam and the scattered light passed through Glan-Thomson polarizers (Halle, Berlin, Germany) with an extinction coefficient better than 10^{-6}. The first polarizer guaranteed that only vertically polarized light meets the sample; the orientation of the second polarizer (analyzer) was carefully adjusted to crossed position with minimum scattering intensity.

Sample preparation: All measurements were performed at 20 °C. The activated pulp samples were dissolved and filtered through an 0.45-μm PTFE filter prior to measurement. For dissolution of the pulp, a stock solution with 8 wt% LiCl in carefully dried DMAc was prepared. Another stock solution, containing 9 wt% LiCl in commercially available DMAc with 0.03 wt% of water, was prepared by the SCAN-CM 15:99 method. The latter stock solution corresponded to traditionally used methods, and was used to obtain results comparable with those in common practice[6)]. For DLS measurements, the stock solutions were diluted with DMAc to a LiCl content of 2.6 wt% (injection concentration in SEC) and 0.9 wt% (detector concentration in SEC), respectively. The DMAc used for dilution had the same water concentration as that used for the stock solutions.

Theoretical background

Dynamic light scattering (DLS) was employed to examine the effect of water in the solvent system DMAc/LiCl on the solution state of cellulose. This topic is of special interest with regard to SEC measurements that use this solvent system. The possible occurrence of large particles or solvent aggregates in DMAc/LiCl/water was investigated, and the influence of trace amounts of water on the particle size distribution of dissolved cellulose was studied.

The quantity measured in DLS is the angular-dependent auto-correlation function $g_2(q,t)$ of the coherently scattered light; $g_2(q,t)$ is related to the autocorrelation function of the electric field $g_1(q,t)$ in a homodyne experiment by Eq. (1):

$$g_2(q,t)=A\left(1+B\left|g_1(q,t)\right|^2\right) \tag{1}$$

A is the base line and B is the coherence factor. For a polydisperse system and at a constant scattering angle, $g_1(t)$ may be expressed as superposition of the contributions of all particle sizes present to the field correlation function:

$$g_1(t)=\int G(\Gamma_i)\exp(-\Gamma_i t)\,\mathrm{d}\Gamma \tag{2}$$

Equation (2) represents the Laplace transformation of the probability density function $G(\Gamma)$ of the translational diffusion coefficients D_i. With $\Gamma_i = q^2D_i$, the diffusion coefficient D can be related to the hydrodynamic radius R_H with the Stokes-Einstein relation $D = k_BT / 6\pi\eta_o R_H$, where η_o is the dynamic viscosity of the solvent, T the temperature, and k_B Boltzmann's constant. By inverse Laplace transformation of Eq. (2), the intensity-weighted size distribution of the hydrodynamic radii of the particles present in the solution can be estimated[10]. This graphic representation accentuates larger particle populations by the intensity - radius relationship ($I \propto R_H^6$ for globular particles).

The DLS method is an appropriate tool to gain qualitative information about the distribution of the sizes of particles in dilute solutions on the basis of measurement of only one sample solution. Absolute analysis of concentrated solutions is more complicated; however, solutions of equal concentrations (i.e., of equal viscosities) can be compared with each other. As a result, apparent hydrodynamic radii of the particles are obtained.

Results and discussion

Dilution of the stock solutions to 2.6 wt% and 0.9 wt% LiCl and adjustment of the water content was performed as described in the Experimental Section. Two time frames are important for SEC measurements in the system cellulose/DMAc/LiCl: 1 h - the maximum elution time for the SEC solutions (0.9 wt% LiCl) and 5 days - the maximum storage time for the injected solutions (2.6 wt% LiCl).

The influence of water and aggregation time on the size distribution of cellulose particles in solutions of 2.6 wt% LiCl as a function of the apparent radius is shown in Fig. 1. One hour after diluting the stock solution to SEC injection concentration (2.6 wt% LiCl), no significant differences between completely dry solutions (filled triangles) and solutions 0.05 M in water (filled circle) were detected. However, while the dry solution (empty triangles) remained nearly unaltered over longer periods of time (5 days in this measurement), the moist solution (empty circles) showed pronounced aggregation to larger particles.

Also for cellulose solutions at the SEC detection concentration (0.9 wt% LiCl), the solution in dry DMAc (filled triangles) was stable over time, whereas the solution in moist DMAc showed a distinct tendency to aggregation (fresh: filled circle, 1 day: half-filled circle, 5 days: empty circle). The aggregation process was largely complete after one day. At longer times of standing, the aggregates grew even larger, as demonstrated by the peak broadening at high R_H in Fig. 2. The two maxima around 15 and 40 nm should be interpreted as one polydisperse particle population.

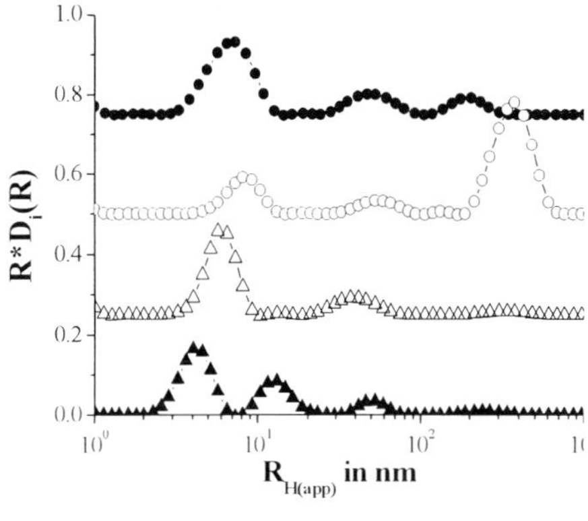

Fig. 1: DLS intensity-weighted size distribution of dissolved cellulose as a function of the apparent radius: the influence of water on the size and number of cellulose particles (2.6 wt% LiCl and 0.325 wt% pulp).

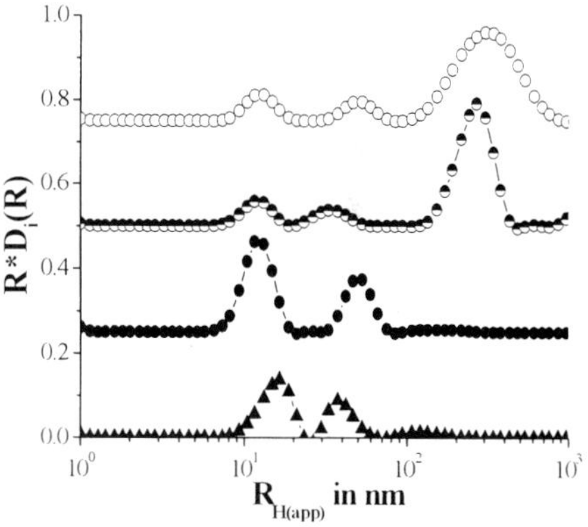

Fig. 2: DLS intensity-weighted size distribution of dissolved cellulose as a function of the apparent radius: the influence of water on the size and number of cellulose particles (0.9 wt% LiCl and 0.028 wt% pulp).

(8 wt% LiCl stock solution with 1 wt% of pulp in dry DMAc; scattering angle 90°.)

Figure 3 shows the results for solutions with 2.6 wt% LiCl, prepared from the 9 wt% LiCl solution in commercial DMAc; they are less unambiguous than those for solutions prepared from dried DMAc stock solutions. Nevertheless, the instability of the solutions can clearly be seen. Not only prolonged times of storage, but also additional water will decrease the quality of the solution state, and lead to the formation of larger aggregates. However, storage time, LiCl content, water content, and cellulose content are important factors for the stability of the solutions.

Solutions with 0.9 wt% LiCl were prepared by diluting the stock solutions with 9 wt% LiCl (commercial DMAc). While the solutions without additional water (regular) showed no signs of aggregation, already the freshly prepared solution with 0.05 M did so (cf. peaks around R_H of 100-200 nm; Fig. 4). The very fast aggregation process was already complete during the dilution of the stock solution, so that a time dependence of the aggregation was not observed. By filtration through 0.45-μm filters, the cellulose aggregates formed in solution are either permanently destroyed or completely retained. Re-aggregation after filtration was not observed, which favours the interpretation of a complete removal of larger particles over a temporary destruction. SEC measurements of the unfiltered solution showed an increased high-molecular-weight portion, while experiments with the unfiltered solution gave a decreased recovery rate.

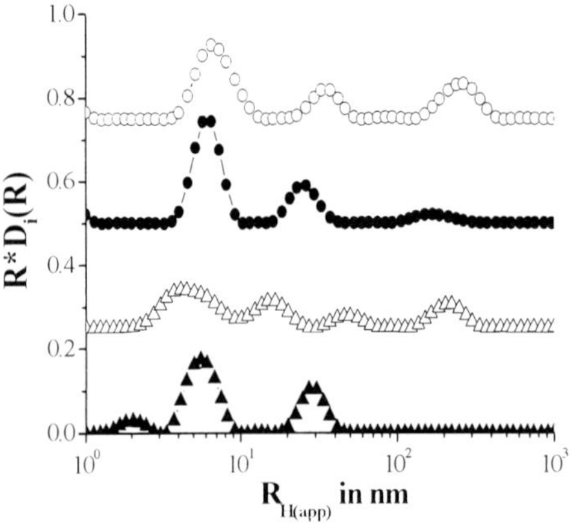

Fig. 3: DLS intensity-weighted size distribution of dissolved cellulose as a function of the apparent radius: the influence of water on the size and number of cellulose particles (2.6 wt% LiCl and 0.346 wt% pulp).

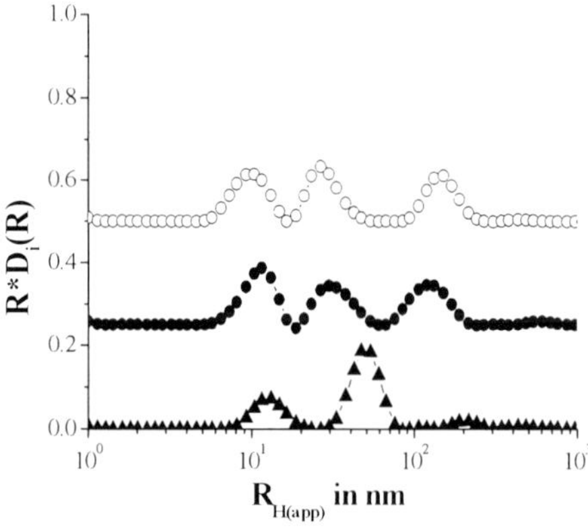

Fig. 4: DLS intensity-weighted size distribution of dissolved cellulose as a function of the apparent radius: the influence of water on the size and number of cellulose particles (0.9 wt% LiCl and 0.028 wt% pulp) .

• 0.05 M in water (fresh), ○ 0.05 M in water (5 days), Δ regular (5 days), ▲ regular (fresh). (9 wt% LiCl stock solution with 1 wt% of pulp in commercial DMAc; scattering angle 90°.)

Influence of dissolution time, an example of what can happen under unfavorable conditions: The stock solutions used before were produced using a laboratory shaker overnight. To demonstrate the influence of the stock solution quality on the SEC measurements, the dissolution time was reduced to 2 h. After this time, the solutions were optically clear. Figure 5 shows the scattering intensity at 90° as a function of time in the SEC solutions with 0.9 wt% LiCl. A few minutes after dissolution, a steady rise in intensity was already observed for solutions containing water. In contrast, the measurements of the dry solution do not show any changes in intensity with time.

The quality of the stock solution was profoundly disturbed by water. However, it was not possible to reproduce the aggregation times. In some cases, aggregation starts after 3 h or even later. The system is obviously rather complex, and the aggregation behavior is dependent on many factors.

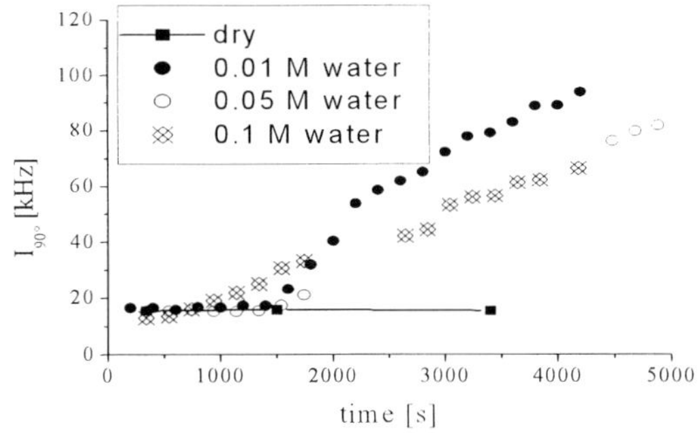

Fig. 5: Time dependence of the scattering intensity of SEC cellulose solutions in DMAc with 0.9 wt% LiCl, 0.028 wt% KZO_3 (freeze-dried) and with various water contents; scattering angle 90°.

Nevertheless, the results represent the solution conditions quite well. As an example, the solution with 0.05 M water content was chosen. Approximately 25 min after dissolution to SEC level, a displacement of the correlation function $g_2(t)$ to longer times occurred (Fig. 6). The scattering intensity of such highly dilute solutions is very low resulting in a low intercept. This intercept increases with increasing aggregation. This effect has no essential influence on the principles of the analysis, only a slight shift in the size distribution by factor 2 is possible. Twenty-five minutes after dilution, a second population of particles around 100 nm is formed (Fig. 7).

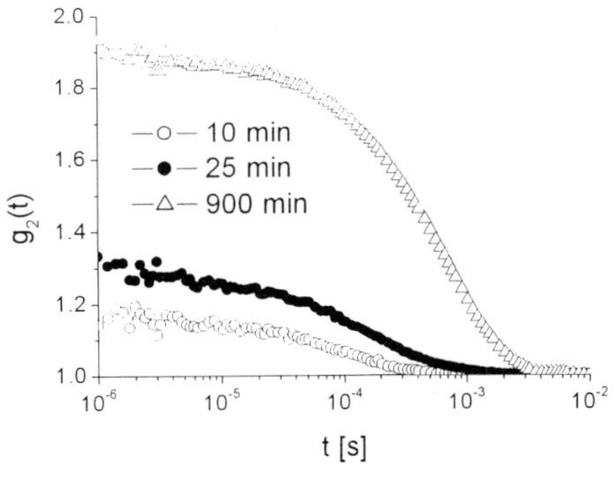

Fig. 6: Time dependence of the intensity autocorrelation function for cellulose (0.028 wt%) in 0.9 wt% DMAc/LiCl with 0.05 M water.

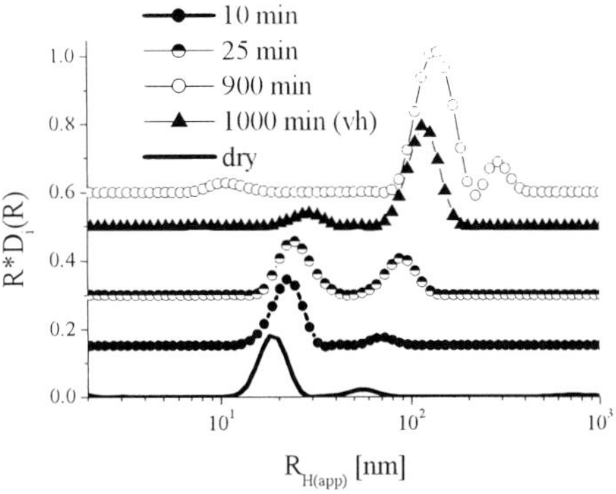

Fig. 7: Time dependence of the intensity-weighted size distribution of dissolved cellulose as a function of the apparent radius (for details, see Fig. 6).

Almost one day later, the radius of the second fraction showed a distribution from 80 to 200 nm. By depolarized dynamic light scattering (DDLS) measurements of the same solution with crossed polarizers, a correlation function was measured which indicated the existence of optically anisotropic particles. The apparent radii of particles range between 70 and 180 nm. Assuming the model of the fringed micelles (a relatively hard and well oriented swollen core with flexible fringes like a hedgehog with soft spines) for the cellulose particles, the DDLS measurements showed the core and the DLS measurements the whole particle. Quantitative estimation of the dimensions failed due to the instability of the system over long times, but the result is yet another indication of the formation and presence of fringed micelles in cellulose solutions[11,12].

Conclusions

An insufficiently long dissolution time and a water content of ≥ 0.01 M can result in cellulose aggregation in the SEC solutions (0.9 wt%) during the regular time window of SEC measurements. All measurements in this study were performed with a standard sample, for which normally no aggregation is observed. Nevertheless, aggregation phenomena were induced. To disturb a "good" stock solution, i.e., well soluble pulps, the concentration of water must be higher than 0.05 M. For pulps with less than optimum solubility, already traces of water (lower than 0.01 M) can suffice to induce and promote aggregation.

Long storage times of the injected solution sometimes result in lower retrieval rates in SEC. Aggregation in the DMAc/LiCl /cellulose system is a very complex process, the influence of water is one factor and should be regarded if SEC problems occur. The following general rules should be considered:

(1) the water content of the used DMAc should be below 0.9 wt%;
(2) sufficient time of dissolution (overnight) is necessary;
(3) freeze-drying of cellulose is not absolutely necessary, but it has no negative effect;
(4) the higher the LiCl amount, the longer the possible storage time before aggregation sets in.

If aggregation problems occur in SEC analysis, reduction of the storage time of the solutions with injection concentration (if possible), or reduction of the cellulose content (probably resulting in lower resolution) are worth considering.

Acknowledgements

The authors are grateful to the Austrian Christian-Doppler-Forschungsgesellschaft and the Lenzing AG, Austria, for financial support. We are grateful to Univ.-Doz. Dr. Herbert Sixta (Lenzing AG) for helpful discussions. We would also like to thank Walter Milacher (Lenzing AG) and Gerald Ebner (University of Agricultural Sciences Vienna, Austria) for preparation of solutions.

References

1. A. L. Kvernheim, E. Lystad, *Acta Chem. Scand.* **43**, 209 (1989)
2. J. D. Timpa, H. H. Jr. Ramey, *Textile Res. J.* **59**, 661 (1989)
3. A. A. Silva, M. L. Laver, *Tappi J.* **80**, 173 (1997)
4. J. L. Ekmanis, *Polym. Notes, Waters* **1** (1986)
5. T. R. Dawsey, C. L. McCormick, *J. Macromol. Sci., Rev. Macromol. Chem. Phys.* **30,** 405 (1990)
6. N. Schelosky, T. Röder, T. Baldinger, *Papier (Darmstadt)* **53**, 728 (1999)
7. T. Röder, B. Morgenstern, O.Glatter, *Polymer* **42**, 6765 (2001)
8. A. Potthast, T. Rosenau, R. Buchner, T. Röder, G. Ebner, H. Bruglachner, H. Sixta, P. Kosma, *Cellulose*, in press
9. D. Lehner, H. Lindner, O. Glatter, *Langmuir* **16**, 1689 (2000)
10. H. Schnablegger, O. Glatter, *J. Appl. Opt.* **30-33**, 4889 (1991)
11. W. Burchard, *Adv. Colloid Interface Sci.* **64**, 45 (1996)
12. T. Röder, B. Morgenstern, O. Glatter, *Macromol. Symp.* **162**, 87 (2000)

Macromol. Symp. ***190***, *161–172 (2002)*

Structure and Ordering Kinetics of Micelles in Triblock Copolymer Solutions in Selective Solvents

*R. Bansil** [1], *H. Nie* [1], *Y. Li* [1], *G. Liao* [1], *K. Ludwig* [1], *M. Steinhart* [2], *Č. Koňák* [2], *J. Lal* [3]

[1] Department of Physics, Boston University, Boston, MA 02215. USA
[2] Institute of Macromolecular Chemistry, Academy of Sciences of the Czech Republic, 162 06 Prague 6, Czech Republic
[3] Argonne National Laboratory, Argonne, Illinois, USA

SUMMARY: We have used small-angle x-ray scattering (SAXS), and small-angle neutron scattering (SANS) to study the micelle structure of a polystyrene-*block*-poly(ethene-*co*-butene)-*block*-polystyrene triblock copolymer in dilute - semidilute solutions in solvents selective for either the outer styrene block (dioxane) or for the middle block (heptane or tetradecane). Measurements of equilibrium structure factors showed that micelles were formed in both types of selective solvents. In the case of dioxane, the micelles are isolated whereas in the case of heptane or tetradecane, a bridged micellar structure may be formed at higher copolymer concentrations. In both cases we observed an ordered cubic structure of insoluble domains (micellar cores) at high concentrations (> 8 %). The micellar scattering function was fit to the Percus-Yevick interacting hard-sphere model. The temperature dependence of the core radius, the hard-sphere interaction radius and the volume fraction of hard spheres were obtained. We also used synchrotron-based time-resolved SAXS to examine the kinetics of ordering of the micelles on a cubic lattice for many different temperature jumps into the ordered cubic phase starting from the disordered micellar fluid phase in different solvents at different concentrations. The time evolution of the structure changes was determined by fitting the data with Gaussians to describe the structure factor of the ordered Bragg peaks and the Percus-Yevick structure factor was used to describe the micellar fluid. The time dependence of the peak intensities and widths as well as of the micellar parameters will be presented. The results showing the kinetics of the transformation from the fluid to the ordered phase were analyzed using the Mehl-Johnson-Avrami theory of nucleation.

INTRODUCTION

The structure and properties of block copolymers have attracted much interest recently.[1,2] It is well known that, in a selective solvent which preferentially solubilizes one of the chemical components, the insoluble polymer blocks of both di- and triblock copolymers form micelles.[2] These micelles are also known to exhibit a transition between a liquid-like state and an ordered solid-like state at higher copolymer concentrations.[3-5] In the liquid-like structure, the

 CCC 1022-1360/00/$ 17,50+.50/0

positions of micelles do not exhibit any long-range spatial order, but show a similar short-range order to that in a liquid.

The thermodynamics of micelle formation and structures in diblock copolymer is well understood.[6-10] Recently a few studies have investigated the structure of micelles and their ordering in triblock copolymer solutions.[3,11-13] In the case of symmetric triblocks (ABA), the morphology of the micelles depends on whether the solvent solubilizes the outer (A) or the inner (B) block, leading to the formation of either isolated or bridged micelles, respectively.[12] Lodge et al[13] have used small-angle x-ray scattering (SAXS) to investigate the structure of polystyrene-*block*-polyisoprene-*block*-polystyrene triblock copolymer in a selective solvent good for the outer block, while Raspaud et al[11] have used small-angle neutron scattering (SANS) to study a similar triblock copolymer in a selective solvent good for the middle block. Raspaud et al[11] found that the network formed by the outer blocks has a cubic-ordered structure at high concentrations. The cubic structure was also observed on the polystyrene-*block*-poly(ethene-*co*-butene)-*block*-polystyrene copolymer in a selective solvent good for the middle block by Kleppinger et al.[3] Understanding of the kinetics of micelle formation and of micellar ordering is not as advanced as of the equilibrium copolymer structures, despite the obvious fundamental and practical interest in this question. A few detailed studies on the kinetics of the ordering process in the diblock copolymers have been reported.[4,14-17] Temperature ramp measurements on triblock copolymers have been reported, along with some measurements of temperature jump kinetics. No detailed analysis of kinetics in these systems has been reported.

In this paper we present the results of SANS and a time-resolved SAXS study of the micelle structure and kinetics of the ordering process in the triblock copolymer polystyrene (S)-*block*-poly(ethene-*co*-butene) (EB)-*block*-polystyrene in the selective solvents, tetradecane, heptane and dioxane. Tetradecane and heptane are good solvents for the middle EB block, while dioxane is good for the outer S blocks.

EXPERIMENTAL

Block copolymer

The sample of S-EB-S block copolymer (Shell Chemicals, Kraton G1650) was fractionated and the middle fraction corresponding to 15 wt % of the raw sample, was used in this study. The molecular weight M_w of the fraction was 70,000 with polydispersity M_w/M_n= 1.05, and the styrene block content was 28 wt %.[18] This sample fraction was used in an earlier room temperature SAXS, light scattering and rheological studies.[18,19] Solutions ranging in concentration from 4 to 7 wt % were prepared by direct dissolution in the solvent around 90 °C. Higher concentrations were prepared by evaporating the solvent from the 7-% sample. Two kinds of deuterated solvents were used in the SANS study: heptane-d_{16}, and dioxane-d_8.

Small-angle X-ray scattering (SAXS)

The copolymer solutions were sealed between two thin flat Kapton windows glued to aluminum plates separated by an O-ring of diameter 1 cm. The SAXS measurements were performed at the X-20C beamline of the National Synchrotron Light Source at Brookhaven National Laboratory using a high-flux multilayer monochromator and a position-sensitive detector. The photon energy was 6.9 keV, which corresponds to a wavelength λ = 1.8 Å. The detector was positioned approximately 100 cm away from the sample to obtain a scattering vector range of 0.005 - 0.08 $Å^{-1}$. The absolute incident beam intensity was monitored with a helium-filled ion chamber. The background scattering from an empty Kapton cell was subtracted from the total scattering and the scattering intensities were normalized to an absolute scale per monomer unit using the ion-chamber current to correct for any change of the incident beam intensity or the sample absorption.

Small-angle neutron scattering (SANS)

Neutron scattering experiments were conducted at the small-angle neutron diffractometers at the intense pulsed neutron source, Argonne National Laboratory[20]. Neutrons with wavelengths 0.9-14 Å were used to cover the scattering vector range q = 0.005-0.35 $Å^{-1}$. The block copolymer solutions were maintained at constant temperature in a custom-designed demountable quartz cells with a path length of 1 mm. Further details of the time-of-flight instrument used can be found in literature.[20]

THEORETICAL MODEL FOR MICELLE SOLUTIONS

For a monodisperse system of particles, the scattering function can be written as a product of the single-particle form factor $P(q)$ and a structure factor $S(q)$, describing the interparticle interference:

$$I(q) = KNP(q)S(q) \qquad (1)$$

where K is the contrast factor and N is the number density of scatterers. The form factor of a hard sphere with radius R_c can be written as

$$P(q) = v_0^2 \Phi^2(qR_c) \qquad (2)$$

where $v_0 = (4/3)\pi R_c^3$, is the volume of the hard sphere and

$$\Phi(x) = \frac{3}{x^3}(\sin x - x\cos x). \tag{3}$$

The Percus-Yevick hard-sphere structure factor is given by [21, 22]

$$S(q) = \frac{1}{1 + 24\phi G(2qR_{hs})/(2qR_{hs})}. \tag{4}$$

Here $\phi = 4/3\pi R_{hs}^3 N$ is the hard-sphere volume fraction, R_{hs} being the hard-sphere interaction radius and

$$G(x) = \frac{\alpha}{x^2}(\sin x - x\cos x) + \frac{\beta}{x^3}[2x\sin x + (2 - x^2)\cos x - 2]$$
$$+ \frac{\gamma}{x^5}\{-x^4\cos x + 4[(3x^2 - 6)\cos x + (x^3 - 6x)\sin x + 6]\} \tag{5}$$

Thus the scattering intensity is a function of the hard-sphere volume fraction ϕ, the core radius R_c and the hard-sphere interaction radius R_{hs}. These parameters can be obtained by the least-square fitting of the scattering intensity data to Eqs. 1-5.

RESULTS AND DISCUSSION

Equilibrium structure of semidilute triblock copolymer solutions

Static scattering measurements were made with both SANS and SAXS techniques, as a function of temperature, concentration and solvent. Typical results are discussed below.

Temperature dependence

Figure 1 shows the scattering intensity from 7-wt % S-EB-S solution in tetradecane at different temperatures. At 90° C, the scattering intensity $I(q)$ is featureless, implying that micelles are not formed at this temperature. Below 80 °C, $I(q)$ shows one well-formed maximum in the structure factor at low q values near 0.018 Å^{-1}. In addition, a shoulder is discernible near 0.03 Å^{-1}, which becomes more pronounced as the temperature decreases. This scattering behavior originates from the short-range ordering of micelle cores formed by the insoluble S blocks and can be described by the Percus-Yevick hard-sphere model.[21,22] Similar scattering behavior has been observed in many block copolymer micellar systems.[3,5,11,13,17,21] The solid lines in Fig. 1 show the fitted result. The fit works well around the peak at $q \approx 0.018$ Å. For the values beyond the shoulder, significant deviation from the hard-sphere approximation is observed. This may indicate that, although the hard-sphere interaction radius R_{hs} is well defined, the micelles do not have sharp surfaces. Moreover, the data at higher

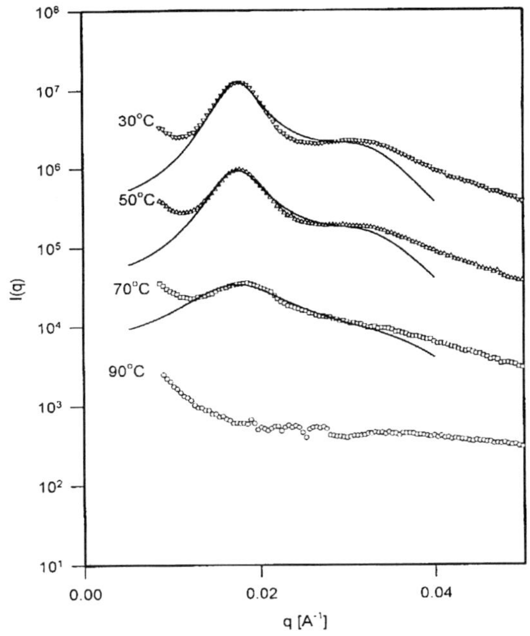

Figure 1. The scattering intensity $I(q)$ versus q from 7-wt % solution in tetradecane at different temperatures. The data are offset by factors of 10 for clarity. The solid lines are fit to the Percus-Yevick hard-sphere model.

q's can be better described by introducing a Gaussian distribution of R_c[23] to reflect the polydispersity in droplet size distribution. For the fitting parameters presented below, we have restricted fitting to the q range from $q_0/2$ to $2q_0$ and polydispersity effects of R_c are not included in the fit.

From the fits we obtained information on the micellar core radius, hard-sphere interaction radius and hard-sphere volume fraction as a function of temperature. The results are shown in Fig. 2. The micellar core radius R_c and the hard-sphere radius R_{hs} both increase as the temperature decreases below 80 °C. This probably reflects changes in the aggregation number of micelles but can also be partly due to the change in micelle size due to conformational changes of the polymer chains. Below 40 °C, both R_c and R_{hs} appear to saturate, and show a very small temperature dependence. Similar temperature dependences of R_c and R_{hs} were observed in other triblock copolymer solutions.[5] The core radius increases from approximately 76 Å at 80 °C to 86 Å at 40 °C; the hard-sphere interaction radius increases from 160 Å at 80 °C to 188 Å at 40 °C.

Since tetradecane is a poor solvent for the outer S blocks and good solvent for the middle EB blocks, two possible micellar structures can be formed. One possibility is that the micelles formed by the end S blocks are bridged by the middle EB blocks so that the system behaves as

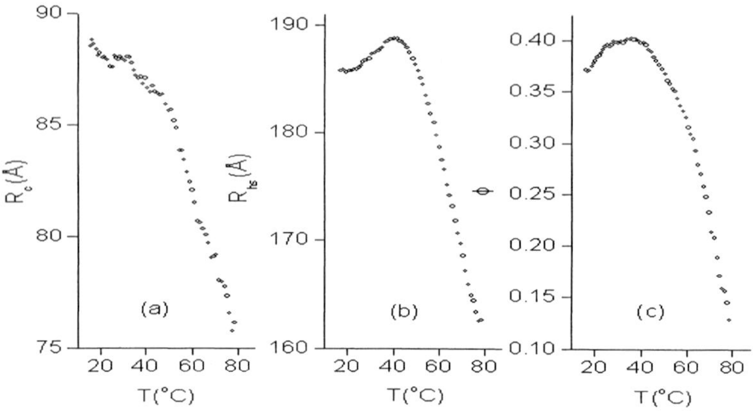

Figure 2. The temperature dependences of the micellar parameters for 7-wt% S-EB-S in tetradecane: (a) the core radius, R_c, (b) the hard-sphere interaction radius, R_{hs} and (c) the hard-sphere volume fraction, ϕ.

a polymer gel. The other possibility is that the middle blocks fold up in one micelle forming isolated flower-like micelles. Although the exact relations between R_c, R_{hs}, the segment length b, and the degree of polymerization N of polymer chains are unknown, in the case of bridged micelles we can estimate R_{hs} as half of the sum of the S and EB chain length. Taking the segment length b of the EB polymer and S polymer as 4.9 Å and 6.8 Å, respectively[24], and assuming the polymer chain is described as a random walk, we obtain that the length of EB chain in the good solvent (self-avoiding random walk) is $bN^{0.6}$ = 294 Å and the length of the S chain in the poor solvent (Gaussian chain) is $bN^{0.5}$ = 66 Å using the degree of polymerizations, N, for the EB chain 918 and for the S chain 94. This gives R_{hs} as 180 Å, consistent with the value obtained by fitting the data. We cannot say which structure is formed in the sample by our SAXS measurements. However, the interconnected structure was found to be favored in the rheological study on the same sample.[19]

Compared with R_c and R_{hs}, the volume fraction of hard spheres is more sensitive to the change of temperature as shown in Fig. 2c. The volume fraction ϕ increases from 0.12 at 80 °C to 0.4 at 40 °C. The increase in ϕ, R_{hs} and R_c is due to the fact that the outer S chains become less soluble with the decrease in temperature. Insoluble chains can be added to the existing micelles, causing the increase in hard-sphere radius R_c and thus R_{hs}, and/or the formation of new micelles, increasing the number of micelles per unit volume. At temperatures higher than 60 °C, the increase in the number of micelles dominates the increase in the hard-sphere volume fraction, since the increase in the volume of one hard sphere $v_{hs} = 4/3\pi R_{hs}^3$ is small compared with the increase in ϕ as easily seen in Fig. 2c. However, in the temperature region between 40 °C and 60 °C, the increase in R_{hs} is the main contributor to the increase in ϕ. Below 40 °C, ϕ saturates. This saturated value $\phi_0 \approx 0.4$ may correspond to the situation that all S chains are aggregated in micelles. In this situation, the aggregation number of S chains N_m in a micelle can be calculated from ϕ_0 and the corresponding R_{hs} by the relation[5]

$$N_m \phi_0 /(4/3\pi R_{hs}^3) = 2cN_A / M_w \qquad (6)$$

where c is the concentration of the sample, N_A is Avogadro number and M_w is the molecular weight of the copolymer. The factor 2 is due to the fact that there are two S chains in a triblock copolymer. Taking R_{hs} = 188 Å, ϕ_0 = 0.4 at 40 °C, c = 0.07 g/cm^3, M_w = 70,000, we obtain N = 84.

The aggregation number N_m can also be obtained from the core radius R_c. If we assume the micelle core is considered as a condensed sphere consisting of only S chains, then

$$N_m V_S = (4/3)\pi R_c^3 \qquad (7)$$

where $V_S = M_S /(\rho_S N_A)$ is the volume of a S chain, M_S = 9,800 is the molecular weight of S chain of our sample and ρ_S= 1.05 g/cm^3 is the mass density of S. This leads to N_m ranging from approximately 120 at 75 °C to 185 at 40 °C, which is twice as large as that obtained from the previous calculation, implying that the micelle cannot be a condensed sphere formed by only S chains. The presence of some solvent in the micelle core was also suggested in previous light scattering studies of the same sample.[25]

Similar results were obtained for the 7-wt % heptane solution of the sample. We did not observe an ordered structure of micelles either in the 7-wt % heptane or in 7-wt % tetradecane solution of the sample. Mortensen et al[26] found that ϕ_{hs} = 0.47 is the lower boundary for the existence of an ordered cubic arrangement of spherical microdomains. In both samples, the maximum value for ϕ we obtained is 0.4, lower than 0.47. This may explain why no ordered structure was observed in these samples. However, in the 15-wt % sample solution (discussed later), we obtained ϕ larger than 0.47 and observed a clear cubic structure.

Concentration and solvent dependence

Samples with different concentrations (4 - 15%) were studied at different temperatures from 25 °C to 90 °C in both the SAXS and the SANS experiments. Shown in Fig. 3 are SANS results at T = 25 °C for two concentrations in both the solvents.

A comparison of the two solvents shows that the interaction between micelles, characterized by the main peak in the scattering data, is stronger in heptane samples (bridged or flower-like micelles) than in the dioxane samples (isolated micelles). In the 4% sample, the micelles are not bridged, in either solvent. Details of fitting of these data to hard-sphere as well as core-shell models will be discussed elsewhere. In the 15% sample, the SAXS data clearly show the presence of Bragg peaks, indicating a cubic-ordered structure. The somewhat poorer resolution of the SANS data, makes it harder to see these features clearly from the SANS data, shown in Fig. 3.

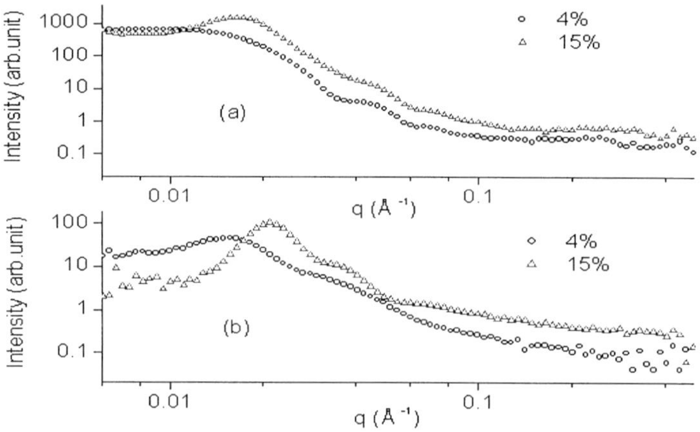

Figure 3. The SANS scattering intensity $I(q)$ versus q for 4-wt% and 15-wt % solution (a) in dioxane-d_8; (b) in heptane-d_{16} at 25 °C.

Ordering kinetics of the micelles in semidilute solutions selective for the middle block

The kinetics of the formation of the ordered micellar phase in a 15% S-EB-S solution in heptane was examined by SAXS measurements performed in a time-resolved way by averaging the scattered intensity on a position-sensitive detector camera for short durations (1 s or less) and repeating the measurements for periods up to 1 h.

Shown in Fig. 4 is the time evolution of the scattered intensity following a quench from 90 °C to 40 °C. The temperature became stable at 45 s after the quench. The scattering patterns during the first 45 s are characteristic of scattering from interacting hard-sphere and can be described by the Percus-Yevick model. After 45 s, well defined sharp peaks developed, signaling the formation of a long-range order. The relative positions of four observed peaks are approximately $1 : \sqrt{2} : \sqrt{3} : 2$, clearly revealing a cubic structure, either a simple cubic (sc) or a body-centered cubic (bcc) structure. The scattering from these Bragg peaks can be represented

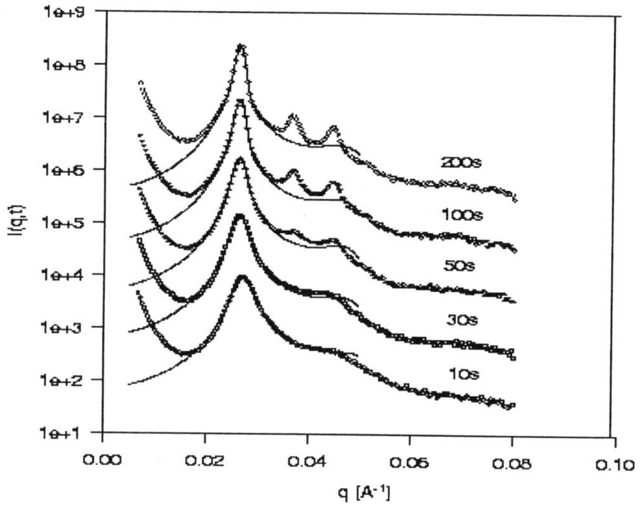

Figure 4. The scattering intensity from a 15% S-EB-S solution in heptane at different times following a quench from 90 °C to 40 ° C. The data are offset by factors of 10 for clarity. The scattering data around the first Bragg peak was fit by Eq. 8 (only i =1 term was included). The q region chosen for the fit was from 0.02 $Å^{-1}$ to 0.035 $Å^{-1}$. The curves calculated from the fitted parameters are plotted as the solid lines.

by a Gaussian. The observed scattering $I(q,t)$ is then made up of two components:

$$I(q,t) = I_{\text{hs}}(q,t) + \sum_i A_i(t) \mathrm{e}^{-\frac{1}{2}(\frac{q-b_i}{\sigma_i})^2} \tag{8}$$

where $I_{\text{hs}}(q,t)$, given by Eq. 1, is the scattering due to the hard spheres, and A_i , b_i , σ_i are the amplitude, position and peak width of the i-th Bragg peak, respectively. The scattering data around the first Bragg peak was fit by Eq. 8 (only i =1 term was included). The q region chosen for the fit was from 0.02 $Å^{-1}$ to 0.035 $Å^{-1}$. The curves calculated from the fitted parameters are plotted as the solid lines in Fig. 4. Full details of fits with all three Gaussians included will be discussed in future publications.

Figure 5 shows the peak intensity of the three Bragg peaks of Fig. 4. The change of temperature with time is also included in Fig. 5. The appearance of the second Bragg peak indicates that the system starts an ordering process. While the micelle formation starts rapidly as the temperature is lowered, the ordering process occurs at a much later time. The same behavior was observed in a diblock copolymer in the selective solvent (Fig. 2 in [4]), suggesting that the kinetics of the ordering process in the triblock copolymer is similar to that in the diblock copolymer.

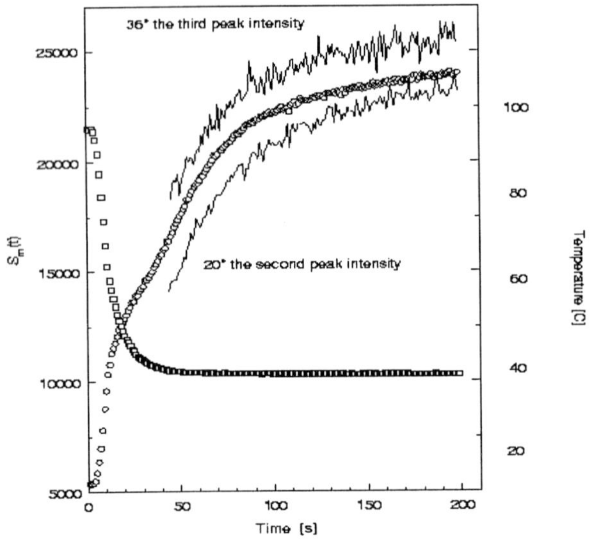

Figure 5. The peak intensities of the three Bragg peaks and temperature versus time. The appearance of the second Bragg peak indicates that the system starts an ordering process.

Figure 6 shows the fitted intensity of first-order Bragg peak, A_1(t) as a function of time following the quench. The first-order Bragg peak $A_1(t)$ is proportional to the fraction of material in the cubic-ordered state. Thus we can compare our experimental result with the nucleation theory prediction of the relation between the fraction of the assembly transformed as a function of time at constant temperature.

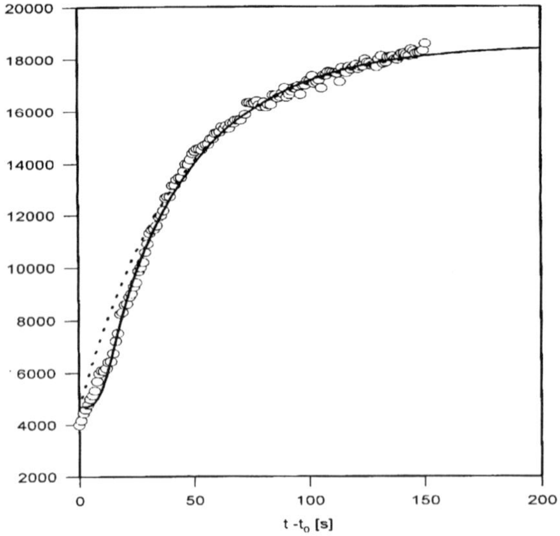

Figure 6. The fitted intensity of first-order Bragg peak, $A_1(t)$ as a function of time following the quench. The dotted line is fit to Eq. 9 and the solid line to Eq. 10.

For the nucleation and growth reaction, Johnson, Mehl and Avrami developed a transformation theory.[27] The theory predicts that for a three-dimensional nucleation and growth process, the volume fraction of the transformed phase η, should be

$$\eta = 1 - e^{-kt^{n}} \tag{9}$$

where $3 < n < 4$, k is a constant. Letting $A_1(t) = A_1(0) + [A_1(\infty) - A_1(0)]\eta$, we fit the first Gaussian intensity to Eq. 9. The dotted line in Fig. 6 is the result of the fit, giving $n \approx 1$. Cahn[27,28] derived a theory which describes nucleation occurring on two-dimensional defects. Cahn's theory includes the effect of impingement from different boundaries and assumes that the planar boundaries are randomly distributed in space. According to Cahn's theory, the volume fraction of the transformed phase can be written:

$$\eta = 1 - \exp(-bf(t - t_0)) \tag{10}$$

where $$f(t) = (t/a)\int_0^1 \{1 - \exp[-\frac{\pi}{3}(t/a)^3(1 - 3\xi^2 - 2\xi^3)]\}\,d\xi \tag{11}$$

and a and b are time constants related with the isotropic growth rates and nucleation rates. Both rates are assumed to be constant. In the early time, $f(t) \sim t^4$; at later times, $f(t) \sim t$. The direct fit of this model to the data with variable parameters a, b, I_0 and I_∞ is shown by the solid line in Fig. 6. The model fits the data better than Eq. 9. From Eqs. 10 and 11, it can be seen that the quantity $\tau = a/b$ plays the role of an effective time constant at the late time limit.[4,14] In the quench shown in Fig. 6, τ = 42.4 s. We do not know the physical origin of two-dimensional defects inherent in Cahn's treatment of the ordering process. Singh et al [14] argued that the likeliest source appears to be due to anisotropic, solid impurities incorporated into the samples.

CONCLUSION

In conclusion, we performed SANS and SAXS study on the micelle structure and the kinetics of the ordering process of S-EB-S triblock copolymer in semidilute solutions of selective solvents. Our results showed that in heptane or tetradecane, a solvent selective for the middle block, the scattering function can be fit by the Percus-Yevick hard-sphere model. The core radius, the hard-sphere interaction radius and the volume fraction of hard spheres increase and then saturate as the temperature decreases. The bridged micellar structure may be formed. In the opposite case of S-EB-S in dioxane, a solvent good for the outer S blocks, only isolated micelles were formed. At higher concentrations, we were able to observe a simple cubic structure of cores. The kinetics of the disorder-order transition following a quench of the temperature from 90 °C to 40 °C was measured, and qualitatively similar results to those observed in previous studies with diblock copolymer solutions were obtained. The results could be described by the classic nucleation and growth theory.

ACKNOWLEDGEMENT

This work was funded by the NSF U.S.-Czech Collaborative Research Grants (NSF INT-9600679 and ME CR ES 044). R.B. acknowledges the support of NSF, Division of Materials Research (NSF-DMR No. 9618467). C.K. and M.S. acknowledge the financial support of the Grant Agency of the Academy of Sciences of the Czech Republic (No. A1050201) and of the Grant Agency of the Czech Republic (No. 203/00/1317).

REFERENCES

[1] Bates, F.S.; Fredrickson, G.H. *Annu. Rev. Phys. Chem.* **1990**, *41*, 525.
[2] Tuzar, Z.; Kratochvil, P. *Surf. Colloid Sci.* **1992**, *15*, 1.
[3] Kleppinger, R.; Reynders, K.; Mischenko, N.; Overbergh, N.; John, M.H.J.; Mortensen K.; Reynaers, H. *Macromolecules* **1997**, *30*, 7008.
[4] Harkless, C.R.; Singh, M.A.; Nagler, S.E.; Stephenson, G.B.; Jordan-Sweet, J.L. *Phys. Rev. Lett.* **1990**, *64*, 2285.
[5] Mortensen, K.; Pedersen, J.S. *Macromolecules* **1993**, *26*, 805.
[6] Leibler, L. *Macromolecules* **1980**, *13*, 1602.
[7] Fredrickson, G.H.; Helfand, E. *J. Chem. Phys.* **1987**, *87*, 697.
[8] Shibayama, M.; Hashimoto, T.; Kawai, H. *Macromolecules* **1983**, *16*, 16.
[9] Stuhn, B.; Mutter, R.; Albrecht, T. *Europhys. Lett.* **1992**, *18*, 427.
[10] Almdal, K.; Bates, F.S. *J. Chem. Phys.* **1992**, *96*, 9122.
[11] Raspaud, E.; Lairez, D.; Adam, M.; Carton, J.-P. *Macromolecules* **1994**, *27*, 2956 *Macromolecules* **1996**, *29*, 1269.
[12] Balsara, N.P.; Tirrell, M.; Lodge, T.P. *Macromolecules* **1992**, *24*, 1973.
[13] Lodge, T.P.; Xu, X.; Ryu, C.Y.; Hamley, I.W.; Fairclough, J.R.A.; Ryan, A.J.; Pedersen, J.S. *Macromolecules* **1996**, *29*, 5955
[14] Singh, M.A.; Harkless, C.R.; Nagler, S.E.; Shannon, R.F.; Ghosh, S.S. *Phys. Rev. B* **1993**, *47*, 8425.
[15] Hashimoto, T.; Sakamoto, N.; Koga, T. *Phys. Rev. E* **1996**, *54*, 5832.
[16] Hashimoto, T.; Sakamoto, T. *Macromolecules* **1995**, *28*, 4779.
[17] Adams, J.L.; Quiram, D.J.; Graessley, W.W.; Register, R.A.; Marchand, G.R. *Macromolecules* **1996**, *29*, 2929.
[18] Plestil, J.; Hlavata, D.; Hrouz J.; Tuzar, Z. *Polymer* **1990**, *31*, 2112.
[19] Tuzar, Z.; Konak, C.; Stepanek, P.; Plestil, J.; Kratochvil, P. *Polymer* **1990**, *31*, 2119.
[20] Thiyagarajan, P.; Epperson, J.E.; Crawford, R.K.; Carpenter, J.M.; Klippert, T.E.; Wozniak, D.G. *J. Appl. Crystallogr.* **1997**, *30*, 280.
[21] Kinning, D.J.; Thomas, E.L. *Macromolecules* **1984**, *17*, 1712.
[22] Percus, J.K.; Yevick, G.J. *Phys. Rev.* **1958**, *110*, 1.
[23] Hashimoto, T.; Fujimura, M.; Kawai, H. *Macromolecules* **1980**, *13*, 1660.
[24] Kroschwitz, J.I. *Concise Encyclopedia of Polymer Science and Engineering* , J. Wiley, New York 1990.
[25] Konak, C; Fleischer, G.; Tuzar, Z.; Bansil, R. *J. Polym. Sci B: Polym. Phys.* **2000**, *38*, 1312.
[26] Mortensen, K. *Europhys. Lett.* **1992**, *92*, 599.
[27] Christian, J.W. *The Theory of Transformations in Metals and Alloys*. Part I, Pergamon, New York 1981.
[28] Cahn, J.W. *Acta Metall.* **1956**, *4*, 449.

Time-Resolved SAXS Studies on Crystallization Behavior of Poly(ethylene isophthalate-*co*-ethylene terephthalate)s

*B. Lee, T. J. Shin, S. W. Lee, J.W. Lee, M. Ree**

Pohang University of Science and Technology, Department of Chemistry, Center for Integrated Molecular Systems, BK21 Functional Polymer Thin Group, Polymer Research Institute and Pohang Accelerator Laboratory
San 31, Hyoja-dong, Pohang 790-784, Republic of Korea

SUMMARY: Small-angle X-ray scattering measurements using synchrotron radiation were carried out for poly(ethylene terephthalate) and poly(ethylene isophthalate-*co*-ethylene terephthalate)s. In addition, differential scanning calorimetric measurements were conducted. Measurements were made both on polymers undergoing isothermal crystallization and during subsequent remelting. The primary and secondary crystallization behaviors are examined. Isophthalate units were found to be excluded from the crystals into amorphous layers during crystallization. No crystal thickening was observed during isothermal crystallization, which may be due to the relatively high chain rigidity. Secondary crystallization, detected predominantly at the later stages of crystallization, causes densification and shrinkage of the amorphous layer. Considering the results, it is proposed that secondary crystallization involves the formation of *short-range molecular order* in the amorphous layers of a lamellar stack as well as in the amorphous regions between lamellar stacks. This short-range-ordered phase has a lower density than the lamellar crystal formed by primary crystallization.

Introduction

A representative of semicrystalline polymers is poly(ethylene terephthalate) (PET), which is widely used as an engineering plastic material. The much investigated primary crystallization of PET was found to involve nucleation and growth.[1-7] In recent years, the secondary crystallization behavior of PET has also been studied.[1-17] As a result of such research efforts, several models have been discussed, particularly with respect to the secondary crystallization of PET polymer: (i) the thin lamellar stack insertion mechanism,[4,9-11] (ii) the single thin lamellar insertion mechanism,[8,12-14] (iii) the fringed micelle formation,[15-17] (iv) perfection of the crystals formed in the primary crystallization,[5,6] and (v) the positive edge-on branched lamellar stack formation mechanism.[2] Nevertheless, the secondary crystallization behavior of PET is still not fully understood; this subject needs more research and debate of researchers in both academia and industry. In contrast, the crystal growth of PET during primary crystallization has been studied in detail, providing an extensive database related to this process. However, there is lack of experimental data on the nucleation and growth of PET at the early stages of crystallization, the subject that needs further investigation.

 CCC 1022-1360/00/$ 17,50+.50/0

In the present work, small-angle X-ray scattering (SAXS) studies of such complex crystallization behavior are extended to poly(ethylene isophthalate-*co*-ethylene terephthalate) copolymers. Differential scanning calorimetry (DSC) investigations were also performed. The results of these studies are analyzed together with those for PET.

Experimental

Materials and sample preparation. Poly(ethylene terephthalate) and two poly(ethylene isophthalate-*co*-terephthalate) samples rich in ethylene terephthalate units synthesized in our laboratory were chosen for study, which are referred to as PET, 5IPT (4.9 mol-% isophthalate units), and 10IPT (9.8 mol-% isophthalate units) (see Fig. 1).[7]

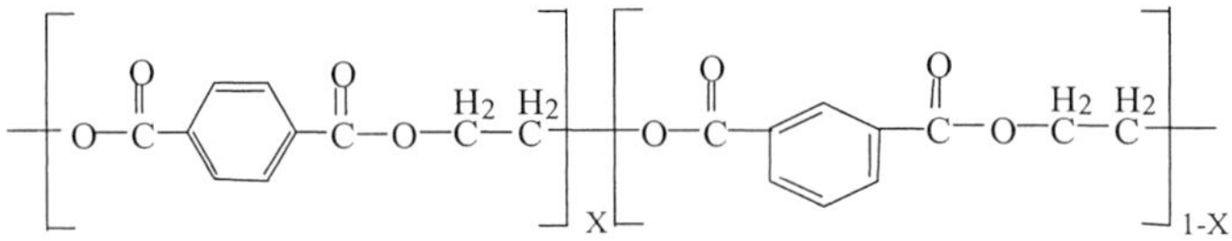

Fig. 1. Chemical structure of poly(ethylene isophthalate-*co*-ethylene terephthalate).

The molecular weight (M_w) was 36 000 for PET, 37 000 for 5IPT, and 36 000 for 10IPT.[7] The polymers were melt-molded under compression in a nitrogen atmosphere and cooled to room temperature, giving 2-mm thick sheets. These polymer sheets were cut either into disks of diameter 4 mm for use in X-ray scattering measurements, or into tiny pieces for use in DSC measurements.

Time-resolved SAXS measurements. SAXS measurements were conducted at the 4C1 SAXS beam line[18] of the Pohang Light Source facility with 2.5 GeV power at the Pohang University of Science and Technology, Korea. The X-ray radiation source has a wavelength of λ = 1.608 Å, and the beam size at the sample position was 0.6 x 0.6 mm. SAXS patterns were measured using a one-dimensional silicon-photodiode array detector (model X/PDA-2048, Princeton Instruments) consisting of 2048 pixels. A jumping hot-stage consisting of two independent chambers was employed. Each specimen was first melted for 5 min in the top chamber and then quickly transferred to the bottom chamber, which was held at the chosen crystallization temperature (T_c). The PET, 5IPT and 10IPT specimens were melted at 285 °C, 275 °C and 265 °C, respectively. The temperature of each chamber was controlled by a Eurotherm controller with a K-type thermocouple. The specimen chambers had nitrogen-blowing holes to protect the sample against thermal oxidation at high temperatures. To correctly monitor the sample temperature, a third Eurotherm controller was used that had a K-type thin-wire thermocouple contacted directly to the specimen. The distance between sample and detector was 1.0 m. The scattering angle was calibrated with a collagen standard prepared from chicken tendon. SAXS measurements were carried out during isothermal crystallization over 170-240 °C and subsequent remelting at 3.0 °C/min. Each measurement was collected for 10 s. Each SAXS intensity profile was normalized to the incident X-ray beam intensity which was monitored by an ionization chamber placed in front of the sample, and corrected further for the background run.

The Lorentz-corrected SAXS profile was obtained as follows. The measured SAXS profile was extrapolated to $q = 4\ \mathrm{nm}^{-1}$ by nonlinear least-square fitting of Porod's law,[19,20] giving the Porod constant and the constant scattering from density fluctuations. The SAXS profile was then treated by subtraction of the constant scattering from density fluctuations and corrected by multiplying by q^2, giving the Lorentz-corrected SAXS profile. The long period L (i.e., the mean distance between adjacent lamellae) was determined from the position (q_{max}) of the peak maximum in the Lorentz-corrected SAXS profile; q is given by $q = (4\pi/\lambda)\sin\theta$, where λ is the wavelength of the X-ray source and 2θ is the scattering angle. The Lorentz-corrected SAXS profile was inverse-cosine-Fourier-transformed to a one-dimensional correlation function $\gamma_1(z)$[21-23] where z is the direction perpendicular to the layer faces in the stack. Morphological parameters such as the lamellar crystal thickness d_c and the amorphous layer thickness d_a were determined from the normalized correlation function $\gamma_1(z)$.

In general, correlation function analysis can give the long period as well as the thickness (l_1) of one layer type (either the lamellar crystal layer or amorphous layer), according to the Babinet reciprocity. Thus, the other layer thickness (l_2) can be obtained from the long period and l_1. In the present study, l_1 is always smaller than l_2. We assign the thickness l_1 to the lamellar crystal thickness (d_c), and l_2 to the amorphous layer thickness (d_a). These assignments are in contradiction to those reported in the literature.[4,11,24] The reasons for this contradiction are as follows. First, the value of l_1 is larger for PET than for the copolymers crystallized under the same degree of supercooling. Instead, the value of l_2 is smaller for PET than for the copolymers. It is well known that noncrystallizable units incorporated into the polymer backbone have a strong tendency to be excluded from lamellar crystals. Thus, the copolymers should have thicker amorphous layers than PET.

In addition, invariant Q, which represents the total integrated scattering,[13,19,21] was obtained from the corrected SAXS profile. Q is related to the difference between the electron densities of the crystalline and noncrystalline regions (ρ_c-ρ_a), the volume fraction of lamellar stacks (φ_s), and the relative volume fraction of lamellae (φ_c) within the lamellar stack:[4,9,13]

DSC measurements. DSC thermograms were measured using a Seiko calorimeter calibrated with indium and tin standards. In the DSC measurements, all polymer samples were first melted and then heated to the chosen T_c, as described above for SAXS measurements. After isothermal crystallization, the crystallized samples were remelted using a heating rate of 3.0 °C/min. All measurements were made in nitrogen atmosphere.

Results and Discussion

Time-resolved SAXS measurements were conducted on PET, 5IPT and 10IPT during isothermal crystallization over a temperature range of 170-240 °C from the melt. Figures 2 and 3 show typical SAXS patterns measured during isothermal crystallization of 5IPT at a chosen temperature and subsequent remelting, respectively.

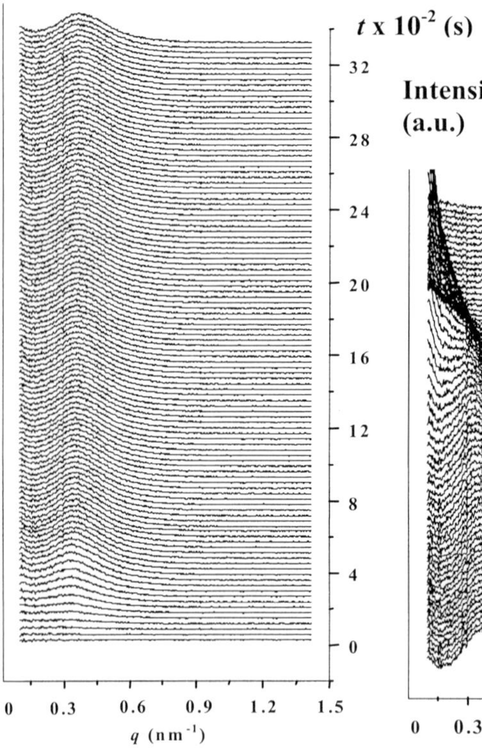

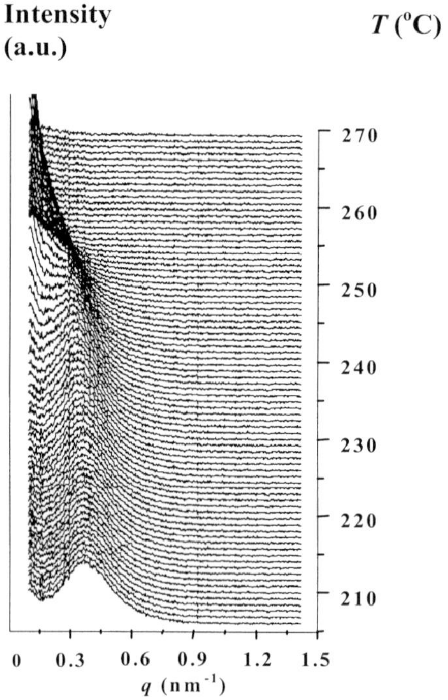

Fig. 2. Time-resolved SAXS patterns of 5IPT copolymer measured during isothermal crystallization at 216 °C for 60 min.

Fig. 3. Time-resolved SAXS patterns of 5IPT copolymer measured during heating at 3.0 °C/min after isothermal crystallization at 216 °C for 60 min.

As can be observed on a typical example in Fig. 2, in all the cases, the SAXS pattern is not detected in the beginning, but develops with the structural evolution associated with crystallization. With increasing crystallization time, the SAXS peak increases in intensity and its maximum shifts to the high-q region, remaining finally unchanged with a further increase in the crystallization time. In this study, the SAXS results are divided into two regimes according to the crystallization time: the primary and secondary crystallization regimes. The two regimes are divided approximately using the crystallization time ($t_{Q\max}$) at which the invariant Q reaches its maximum. The periods before and after this reference time are taken as the primary and secondary crystallization regimes, respectively.

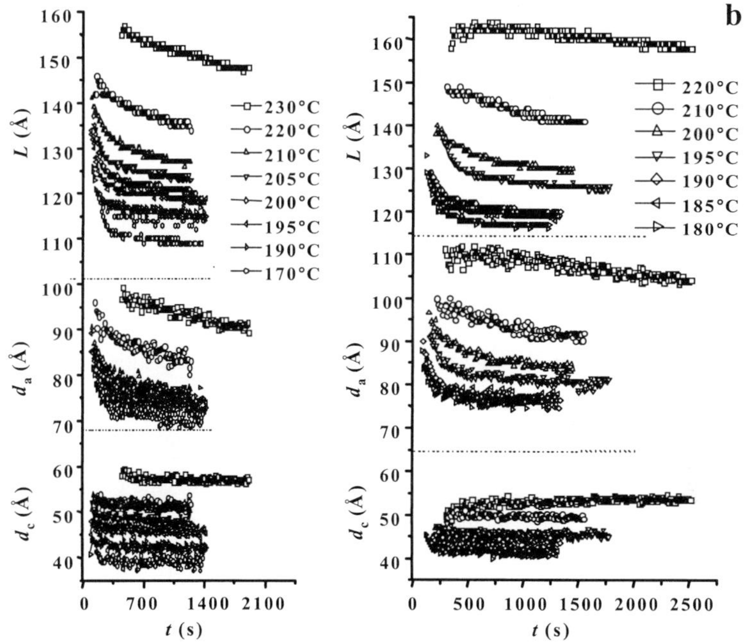

Fig. 4. Variation of morphological parameters (long period L, amorphous layer thickness d_a, and lamellar crystal thickness d_c) with time obtained from the time-resolved SAXS patterns measured during isothermal crystallization at various temperatures. (a) PET, (b) 5IPT.

Primary crystallization. Figure 4 shows the time dependence of morphological parameters (long period L, amorphous layer thickness d_a and lamellar crystal thickness d_c) obtained from the time-resolved SAXS patterns measured during isothermal crystallization of PET and 5IPT at various temperatures.

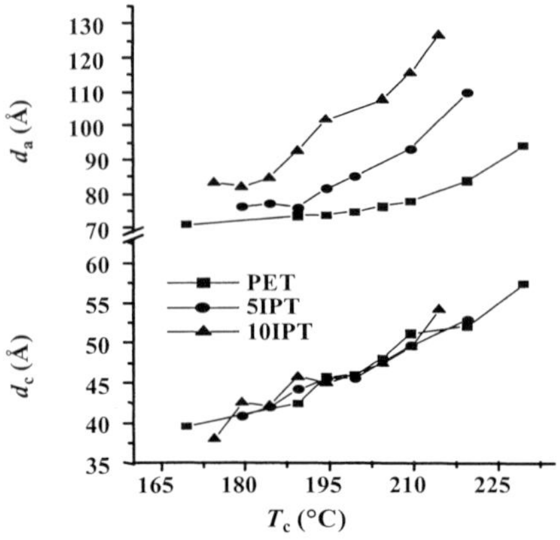

Fig. 5. Variations of amorphous layer thickness d_a and lamellar crystal thickness d_c with crystallization temperature in PET, 5IPT, and 10IPT. For each polymer, the morphological parameters were obtained from the SAXS pattern measured at $t = (t_{Q\max} + 100\ \text{s})$, where $t_{Q\max}$ is the time at which the invariant Q of the SAXS pattern reaches its maximum during crystallization.

Figure 5 illustrates the variations of d_c and d_a with isothermal crystallization temperature T_c, measured for PET and 5IPT crystallized for the time required for the invariant Q to reach a maximum ($t_{Q\max}$).

The lamellar crystal thickness d_c varies little with time in the crystallization of PET, indicating that no thickening occurred in the isothermal crystallization. A similar trend is observed for the crystallization of the copolymers. Furthermore, d_c varies little with the content of isophthalate units for systems crystallized at the same temperature. One can imagine that the kinked isophthalate units are included in the crystal formation at the early stages of crystallization and then excluded from the crystal at a later stage. If this were the case, the exclusion process would cause an increase in the density of the lamellar crystal and a decrease in the density of amorphous layer, leading to a large density difference between these layers. This density difference should cause an increase in the SAXS intensity profile and invariance at the later stages of crystallization. However, the SAXS results observed in this study show no such behavior. We therefore conclude that in the crystallization, lamellar crystals form and grow laterally without thickening and isophthalate units are excluded during this process.

The question remains as to why both the homopolymer and copolymers form lamellar crystals with almost the same value of d_c when crystallized at the same crystallization temperature. In fact, the copolymers have lower equilibrium melting temperatures T_m° than the

PET homopolymer.[7] Thus, taking into account only the degree of supercooling, one might expect the copolymers to make thicker lamellar crystals than the homopolymer. However, this behavior was not observed. Therefore, it is likely that the measured results are due to the restriction of crystallization by the kinked and relatively rigid isophthalate units.

In contrast, the value of d_a increases with increasing content of isophthalate units, even though the systems with different amounts of isophthalate crystallize at the same temperature. In addition, the difference between the d_a values of PET and its copolymer is large at higher isothermal crystallization temperatures. The increase in the thickness of the amorphous layers leads to a reduction in the crystallinity. This increase may be due to the exclusion of the kinked isophthalate units from lamellar crystals into the amorphous region.

In the isothermal crystallization runs the value of d_a decreases with time, regardless of polymer composition. Furthermore, d_a decreases more rapidly with time in the PET homopolymer than in the copolymers for systems crystallizing under the same degree of supercooling. There are two possible major factors that could account for the decrease in d_a during crystallization. The first factor is the distribution of the d_a values and its variation with time. The number of lamellae increases with time during stack building, which reduces the standard deviation in the d_a value and narrows its distribution. This has the effect of shifting the peak maximum of the SAXS profile into the high-q region.[25] The second factor is the relaxation of polymer chains in the amorphous layers. When lamellar stacks form, the polymer chains in the amorphous layers may be in a strained state. The strained polymer chains relax with time, which leads to a reduction in d_a.

Secondary crystallization. The change in the invariant Q with crystallization time was determined from the SAXS pattern measured during isothermal crystallization. A selection of the Q values obtained from PET and 5IPT is presented in Fig. 6.

In the results presented here, the value of Q is normalized by its maximum value. For PET, the value of Q is small and constant for an initial induction period; then increases rapidly with time up to a maximum at $t_{Q\max}$, after which it decreases slowly. This decrease in Q appears more clearly at the later stages of crystallization when the PET is quenched to a higher degree of supercooling. Similar Q variations are observed for the copolymers. The decrease in Q after $t_{Q\max}$ might be due to secondary crystallization taking place in addition to the primary crystallization. If secondary crystallization did not take place, Q would be expected to increase continuously with crystallization time because the continuation of primary crystallization would cause the density of the amorphous layers to decrease with time.

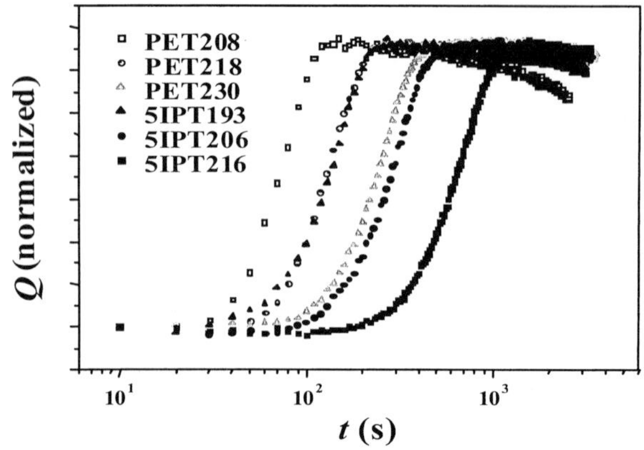

Fig. 6. Time dependence of invariant Q obtained from SAXS profiles measured during isothermal crystallizations of PET polymer and 5IPT copolymer at various temperatures (the codes indicate the polymer and its crystallization temperature).

To investigate the melting behavior of PET, 5IPT, and 10IPT, SAXS and DSC measurements were performed during the heating run at a rate of 3.0 °C/min after isothermal crystallization for 60 min at a temperature in the range 190-230 °C. In these experiments, 60 min was sufficient for both primary and secondary crystallization to occur. Representative results are shown in Fig. 7: the lamellar crystal thickness d_c and amorphous layer thickness d_a determined from the measured SAXS patterns are plotted together with the measured DSC thermograms.

In the DSC thermogram in Fig. 7a, the PET crystallized at 230 °C displays two major melting peaks: one peak in the high-temperature region with a very large heat of melting, and another peak in the low-temperature region. In general, lamellar crystals formed by primary crystallization are larger than those formed by secondary crystallization, causing them to melt in the high-temperature region. Thus, the peak in the high-temperature region can be assigned to the melting of crystals formed by primary crystallization. However, this melting endotherm may overlap either in part or fully with the exothermic peak of the recrystallization that may occur during heating.[2] The peak in the low-temperature region is assigned to the melting of the crystals formed by secondary crystallization. Similar melting behavior is observed for the 5IPT sample crystallized at 216 °C (see Fig. 7c).

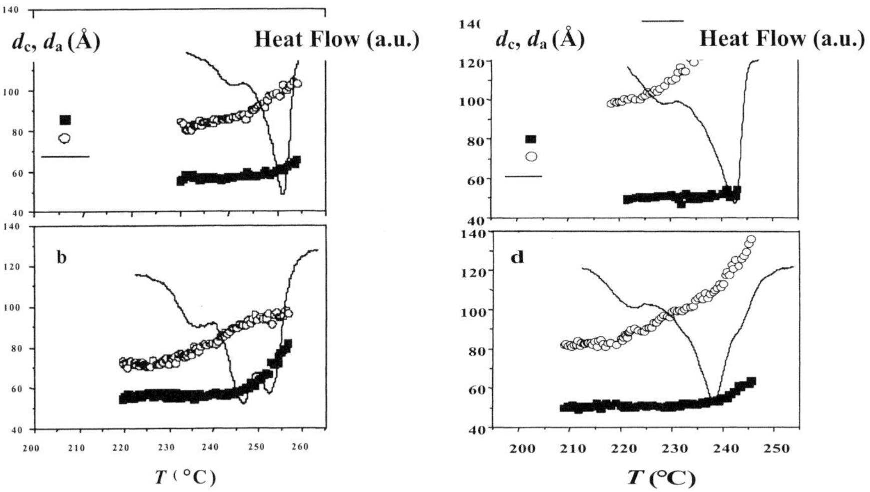

Fig. 7. DSC thermograms, as well as variations of amorphous layer thickness d_a and lamellar crystal thickness d_c with temperature, measured during heating at 3.0 °C/min after isothermal crystallization of a polymer for 60 min at various temperatures.
a PET230, *b* PET219, *c* 5IPT216, *d* 5IPT206 (the codes indicate the polymer and its crystallization temperature); d_c, d_a, heat flow.

In contrast to the results presented above, the PET crystallized at 219 °C shows three major melting peaks (see Fig. 7b). This polymer exhibits two melting peaks in the high-temperature region, rather than the single melting peak seen for PET crystallized at 230 °C. The peaks in the high-temperature region appear at around 247 °C and 253 °C, which fall on either side of the peak due to the lamellar crystals formed by primary crystallization at 230 °C. It is well known that thinner lamellar crystals form at lower crystallization temperatures, and that these crystals exhibit low melting points. We therefore assign the peak at around 247 °C to the melting of the lamellar crystals formed by the primary crystallization, and the peak at around 253 °C is assigned to the melting of crystals that recrystallized during the heating run. The melting point of the crystals from the secondary crystallization is also detected in this sample. The PET samples crystallized at 210-220 °C show a similar behavior to that described above, as do the 5IPT and 10IPT polymers crystallized at 190-210 °C (see an example in Fig. 7d).

From the DSC results presented above we find that the multiple melting points in the isothermally crystallized PET and its copolymers originate from the melting of the crystals formed by primary and secondary crystallization. In addition, it is noteworthy that the melting point of the recrystallized crystals is observed for PET and its copolymers crystallized at relatively high degrees of supercooling.

As displayed in Fig. 7, the value of d_c (which was determined from the measured SAXS pattern) in the heating run varies very little with temperature through the melting of the crystals formed by secondary crystallization, and continues to show little variation up to the temperature of the peak maximum corresponding to the melting of primarily formed lamellar crystals. This behavior was observed regardless of polymer composition and crystallization temperature. At higher temperatures, d_c increases rapidly with increasing temperature through the melting of the primarily formed and recrystallized crystals. The increase in d_c with temperature is particularly marked through the melting of the crystals formed by recrystallization during the heating run. This behavior is detected in samples crystallized with high degrees of supercooling.

In contrast, the value of d_a initially increases with temperature due to melting in the low-temperature region of the crystals formed by secondary crystallization. This behavior is observed regardless of polymer composition and crystallization temperature. With further increase in temperature, the increase in d_a becomes steeper due to the melting in the high-temperature region of the crystals formed by primary crystallization and recrystallization. The increase in d_a is larger in the copolymer than the homopolymer. Overall, d_a shows greater variations in the heating run than d_c. The relatively large variations in d_a are attributed to two main factors. First, the amorphous layer has a larger expansivity than the lamellar crystals, which makes a positive contribution to the expansion of d_a. The second factor is that the secondarily formed crystals melt in the low-temperature region, increasing the amorphous material proportion and causing the increase in d_a. The melting of the crystals formed by secondary crystallization contributes significantly to the expansion of d_a, even below the value of T_m for the crystals formed by primary crystallization. From these results we conclude that secondary crystallization causes PET and its copolymers to undergo densification.

Of the models mentioned earlier in the Introduction, both the thick and the thin lamellar stack insertion model[4,9-11] as well as the single thin lamellar insertion model[8,12-14] have been widely investigated. Continuing this trend, we consider the first two models and evaluate their validity for the secondary crystallization of PET and its copolymers investigated in this study. First, the thin lamellar stack insertion model predicts: (a) the SAXS profile to broaden and shift to the high-q region as crystallization proceeds and (b) Q to increase throughout the crystallization process, which is due to the overall crystallinity increased continuously by the formation of the thin lamellar stacks through secondary crystallization. Second, the single thin lamellar insertion model also predicts Q to always increase with crystallization time, which is due to the overall crystallinity increased by the formation of single thin lamellae, which have the same electron density as the primarily formed lamellae, through the secondary crystallization.

However, the SAXS profiles and their Q variations found in the present study, as described above, are quite different from these predictions. Therefore, taking into consideration the results presented above, we propose a *new structural model* describing the amorphous layers in a lamellar stack and the amorphous regions between lamellar stacks, which become dense and shrink due to secondary crystallization. The proposed model involves a *short-range-ordered structure* that has a lower density than the lamellar crystals formed by primary crystallization. The *short-range-ordered structure* can consist of either single thin lamellae or a fringed micelle-like molecular order, which must have a lower density than the lamellar crystals formed by primary crystallization.

Conclusion

For 5IPT and 10IPT, the isophthalate units are excluded from the lamellar crystals into amorphous layers through the crystallization process. This exclusion leads to an increase in the amorphous layer thickness. However, PET and the copolymer systems crystallized at the same temperature form lamellar crystals with a similar thickness. This behavior might be due to the restriction of crystallization by the isophthalate units, which possess a kinked nature with some rigidity. The presence of isophthalate units lowers both the crystallization rate and crystallinity of the resulting polymer.

The crystal thickness is highly dependent on the crystallization temperature, showing that the degree of supercooling is the major driving force determining the thickness of lamellar crystals. However, no thickening occurs in the isothermal crystallizations, regardless of composition. Instead, the lamellar crystal undergoes perfectioning and lateral growth during the crystallization process, causing its melting temperature to shift to the high-temperature region. On the other hand, in the isothermal crystallization, the amorphous layer thickness d_a decreases with time, regardless of polymer composition. The reasons for the decrease in d_a have been considered, taking into account several factors possibly influencing the crystallization process.

Evidence of secondary crystallization is present in all the SAXS and DSC measurements for both the homopolymer and copolymers. Secondary crystallization causes densification of the amorphous layer, leading to a reduction in the amorphous layer thickness. This is strong evidence that secondary crystallization occurs mainly in amorphous layers. The melting point of the secondary crystals shifts to the high-temperature region with increasing crystallization time. On the basis of these results, it is proposed that *the crystals formed by secondary crystallization form a short-range-ordered phase*. This short-range-ordered phase has a lower density than the lamellar crystals formed by primary crystallization.

Acknowledgements

This study was supported by the Korean Ministry of Science and Technology (MOST) (KISTEP - Basic Research Grant of Nuclear Energy) and by the Center for Integrated Molecular Systems (KOSEF). The synchrotron SAXS measurements were supported by MOST and POSCO.

References

1. S. W. Lee, B. Lee, M. Ree, *Macromol. Phys. Chem.*, **201**, 453 (2000).
2. F. J. Medellin-Rodriguez, P. J. Phillips, J. S. Lin, *Macromolecules*, **29**, 7491 (1996).
3. G. Hauser, J. Schmidtke, G. Strobl, *Macromolecules*, **31**, 6250 (1998).
4. Z.-G. Wang, B. S. Hsiao, B. B. Sauer, W. G. Kampert, *Polymer*, **40**, 4615 (1999).
5. P. J. Holdsworth, A. Turner-Jones, *Polymer*, **12**, 195 (1971).
6. U. W. Gedde, in *Polymer Physics*, Chapman & Hall, London, 1995, Chap. 7.
7. S. W. Lee, M. Ree, C. E. Park, Y. K. Jung, C.-S. Park, Y. S. Jin, D. C. Bae, *Polymer*, **40**, 7137 (1999).
8. A. M. Jonas, T. P. Russell, D. Y. Yoon, *Colloid Polym. Sci.*, **272**, 1344 (1994).
9. R. Verma, H. Marand, B. Hsiao, *Macromolecules*, **29**, 7767 (1996).
10. W. Wang, J. M. Schultz, B. S. Hsiao, *J. Macromol. Sci.: Phys.*, **37**, 667 (1998).
11. B. S. Hsiao, Z.-G. Wang, F. Yeh, Y. Gao, K. C. Z. Sheth, *Polymer*, **40**, 3515 (1999).
12. K.-N. Kruger, H. G. Zachmann, *Macromolecules*, **26**, 1756 (1993).
13. A. M. Jonas, T. P. Russell, D. Y. Yoon, *Macromolecules*, **28**, 8491 (1995).
14. C. Fougnies, P. Damman, D. Vollers, M. Dosiere, M. H. J. Koch, *Macromolecules*, **30**, 1385 (1997).
15. H. Marand, A. Alizadeh, R. Farmer, R. Desai, V. Velikov, *Macromolecules*, **33**, 3392 (2000).
16. B. Crist, E. S. Claudio, *Macromolecules*, **32**, 8945 (1999).
17. A. Alizadeh, L. Richardson, J. Xu, S. McCartney, H. Marand, Y. W. Cheung, S. Chum, *Macromolecules*, **32**, 6221 (1999).
18. J. Bolze, J. Kim, B. Lee, T. J. Shin, J.-Y. Huang, S. Rah, H. S. Youn, M. Ree, *Korea Polym. J.*, **9**, 450 (2001).
19. W. Ruland, *J. Appl. Crystallogr.*, **4**, 70 (1971).
20. J. T. Koberstein, R. S. Stein, *J. Polym. Sci.: Polym. Phys. Ed.*, **21**, 2181 (1983).
21. C. G. Vonk, G. Kortleve, *Colloid Polym. Sci.*, **220**, 19 (1967).
22. B. Goderis, H. Reynaers, M. H. J. Koch, V. B. F. Mathot, *J. Polym. Sci.: Polym. Phys. Ed.*, **37**, 1715 (1999).
23. G. R. Strobl, M. Schneider, *J. Polym. Sci.: Polym. Phys. Ed.*, **18**, 1343 (1980).
24. H. G. Zachmann, H. A. Stuart, *Makromol. Chem.*, **41**, 148 (1960).
25. R. Vignaud, J. M. Schultz, *Polymer*, **27**, 651 (1986).

Progress in SANS Studies of Polymer Systems

(Panel Discussion)

Panel Members:

G. D. Wignall (Oak Ridge National Laboratory, USA, Chair),
H. Benoît (Institut Charles Sadron, France),
T. Hashimoto (Kyoto University, Japan),
J.S. Higgins (Imperial College, UK),
S. King (Rutherford Appleton Laboratory, UK),
T.P. Lodge (University of Minnesota, USA),
K. Mortensen (Risø National Laboratory),
A.J. Ryan (Sheffield University, UK)

Availability of Reactor-Based SANS Instrumentation

The aim of the Panel Discussion was to discuss cutting edge research in progress, in addition to predictions and prospects for future developments in the field. However, the first requirement of current and future research is to have access to small-angle neutron scattering (SANS) facilities, many of which are currently located on 20-40 year old reactors, and may become unavailable in the next 1-2 decades. Table 1 lists some of the main facilities that have been used for SANS research. The first instrument (1,2) suitable for the study of polymers was built in the late 1960s at the FRJ2 reactor at the Kernforschungsanlage (Forschungszentrum), Jülich, Germany, and pioneered the use of long-wavelength neutrons and large overall instrument length (~ 40 m). This was a direct consequence of the low source brilliance of neutron sources, which are orders of magnitude below that of x-ray sources. In order to compensate for this difference, it is necessary to use large sample areas (1 - 20 cm^2), which means that the overall size of the instrument must be large in order to maintain resolution in the range 10-2000 Å.

Table 1. SANS facilities, past, present and future.

SANS FACILITIES PAST, PRESENT AND FUTURE

1972-80	FRJ2-JULICH	ILL-D11	
	HARWELL*	ORR-10M*	SACLAY–EL3*
1980-99	HFIR-30M*	BNL*	RISO*
	NIST-8M, 30M	ORPHEE-LLB	ILL-D17, D22
	IPNS-SAD, SAND+	LANSCE+	ISIS-RAL+
	KEK-TSUKUBA+	PSI-VILLIGEN+	GKSS ECN*
	FRJ2-KWS1, 2	JAERI-JRR3	BENSC-HMI
2000-10	FRM	SNS+	KOREA-HANARO
(FUNDED)	TRR11-TAIWAN	ANSTO	ORNL-SANS1, 2
2000-10	ESS+ CNF	AUSTRON+	KEK-JAERI+ ISIS-2+
(PROPOSED)			

*facility no longer in operation
+pulsed source (all others reactor-based)

 CCC 1022-1360/00/$ 17,50+.50/0

The FRJ2 SANS facility (1,2) was the first to employ a large overall instrument size and to boost the flux of the long-wavelength ($\lambda > 4$ Å) component of the Maxwellian spectrum by moderating the neutrons to a lower temperature by means of a cold source containing a small volume of liquid hydrogen at $T \sim 20$ K. This gives flux gains of over an order of magnitude at $\lambda \sim 10$ Å, and it was on this instrument that the first SANS experiments on polymers were performed. The D11 facility, built during the early 1970s on the high-flux reactor (HFR) at the Institut Laue-Langevin (ILL), Grenoble, France, incorporated many of the features of the FRJ2 instrument, including a cold source and long (~ 80 m) dimensions (3). Both the FRJ2 and HFR instrumentation made use of neutron guide tubes as proposed by Maier-Leibnitz and Springer (4) and have both been subsequently upgraded and expanded to be among the most productive SANS facilities worldwide.

Subsequently, over twenty other SANS facilities have been constructed worldwide, though several of these instruments are no longer operational, as indicated in Table 1. This trend can be expected to continue as many currently available facilities were constructed on reactors that were built in the 1960s and 1970s. A recent forward survey (5) estimated that over the next two decades, the installed capacity of neutron beams for research could decrease to a level below one third of the present capacity. Fortunately, the decline in the availability of reactor-based SANS instruments, as exemplified by the recent shutdown of the Brookhaven and Risø National Laboratory reactors, has been offset by two competing trends. Firstly, several new reactors are under construction worldwide (Table 1), along with upgrades to existing sources (e.g., at the ILL in the mid-90s, and ORNL during 2000-2003).

SANS Instrumentation on Pulsed Sources

In addition, a range of accelerator-based SANS instruments have been developed over the past 15 years, and, in particular, a spallation neutron source is under construction at Oak Ridge (6). In Europe, the planned European spallation source (7) and/or a second target station at ISIS (8) would do much to offset and even reverse predicted decline in the availability of SANS facilities. Thus, it seems likely that pulsed sources will make a much greater contribution to SANS studies of polymers in future than they have in the past. Also, reactors have been optimized over the past several decades, and the flux of instruments planned on new or upgraded reactor sources will either be less than or equal to the current state of the art instruments (e.g. at D22 at the ILL). However, this is not the case for pulsed facilities, which have not yet begun to reach their full potential so we can still expect order-of-magnitude gains over the current facilities, via the ESS,

SNS, ISIS-2, etc. Some of the opportunities created by this development were described by S. King during the Discussion and are explored in more detail in the Appendix.

Inelastic Effects in SANS

Since the 1970s, the treatment of SANS data has been based on the assumption that the scattering is predominantly elastic and the data may be integrated over all energies to give the time-averaged structure of the system. However, few studies have been performed to check whether inelastic effects may be safely neglected, as currently assumed, and those that have been undertaken indicate that this may not always be the case. For example, critical scattering from binary mixtures of metallic alloys near the phase demixing temperature (9), with an incident wavelength of $\lambda_0 \sim 5$ Å, showed a strong inelastic component at Q values that are typical of many SANS studies (0.01 - 0.4 $Å^{-1}$), as shown in Fig. 1. At higher Q values, which are still within the range of current (e.g. D22 at the ILL) and planned (10) facilities, the elastic peak *virtually disappears from the spectrum* (Fig. 2).

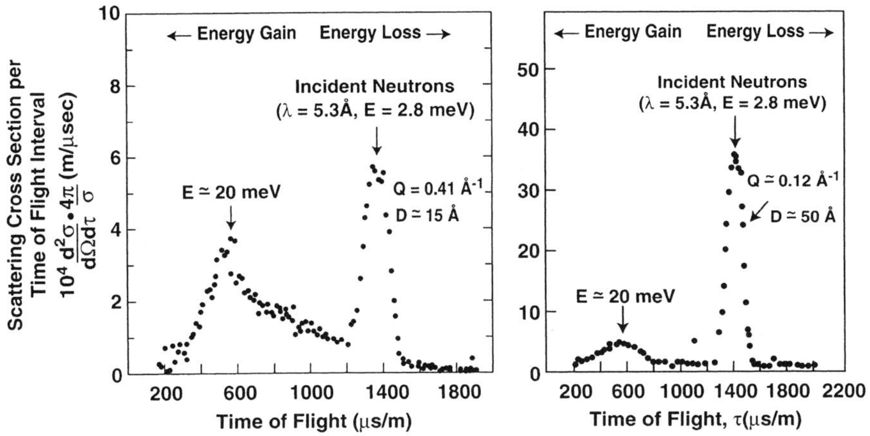

Figure 1: Inelastic spectrum from Bi-Zn solutions (646 °C) near the phase boundary for demixing [*J. Phys. C*, 1, 1088 (1968)].

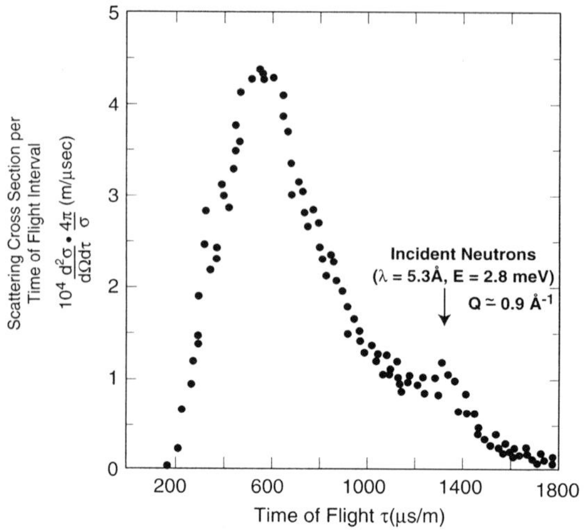

Figure 2: Inelastic spectrum from Bi-Zn solutions (646 °C) at higher Q ($Q_{\text{elastic}} \sim 0.9$ Å^{-1}).

For "soft matter" such as polymers and longer incident wavelengths, inelastic effects can be large even at low Q values, as shown by the analysis of the energy spectrum of a beam of 12 Å neutrons, scattered from a variety of solvents and polymer solutions (11). The data indicate that less than half of the neutrons were scattered elastically as assumed by conventional SANS methodology. Strong inelastic effects have also been observed (12) on pulsed facilities (Fig. 3), where they are particularly important as the assignment of the correct Q values is based on the time of flight, which is strongly affected by changes in the neutron velocity on scattering. New detectors such as those being installed on the ORNL SANS facilities (10) will permit time-stamping and these instruments, along with the other new facilities that are planned (e.g. FRM, ISIS-2, SNS, ESS), should have sufficient flux to add an energy analysis capability to the scattered beam. This will permit the option of restricting the measured data to only those events which correspond strictly to elastic scattering, and allow systematic studies of inelastic effects in SANS to delineate more precisely the types of experiments where the "standard" methodology is appropriate.

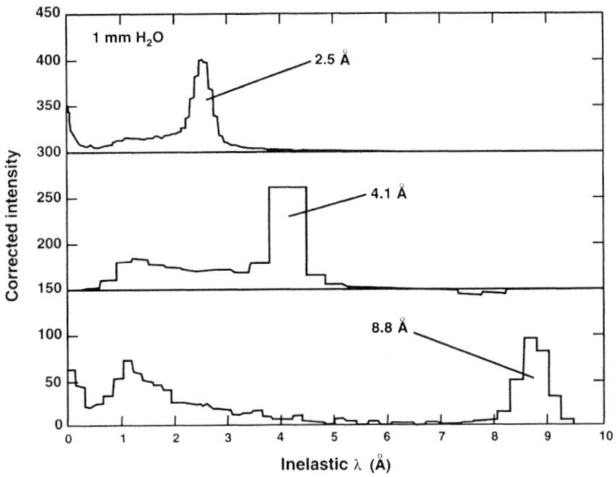

Figure 3: Inelastic spectrum for H_2O recorded on ISIS-SANS with different incident wavelengths [*J. Appl. Crystallogr.* 30, 2240 (1997)].

Time-Resolved Experiments

New high-flux sources will also provide the polymer community with an opportunity to design novel classes of non-equilibrium time-resolved studies and shear-induced phenomena. Processing is a key issue in enabling new materials and technology to reach the market place, and also an area that lags behind synthesis and morphology. Deformation leads to changes in morphology that must be studied under actual conditions, and flow instabilities in polymer extrusion are often the most vexing problem in the commercialization of new products. The need for analyzing systems close to actual processing conditions has been recognized previously (13), and neutrons have a unique advantage since they can penetrate macroscopic extruders and flow devices. Thus, SANS is the technique of choice for exploring the effects of flow in industrially relevant equipment.

Overlapping SANS and Light Scattering

Other instrumental developments have shown promise of lifting the upper resolution limit for pinhole SANS and SAXS instruments, which has traditionally been ~ 10^3 Å. Improvements in the field of ultrasmall-angle neutron scattering (USANS) should make it possible to routinely decrease $Q_{\min}$ by an order of magnitude, thus permitting overlap with light scattering techniques. In the mid-1960s, Bonse and Hart (14) pioneered the use of ultrahigh resolution small-angle x-ray scattering (USAXS) techniques, which used double perfect single crystals to Bragg-reflect the beam with an extremely small angular divergence. By using channel-cut crystals, the wings of the rocking curve (which determine the ultimate resolution) can be dramatically reduced. The critical parameter characterizing the sensitivity is the width of the rocking curve of the analyzer crystal (i.e., the intensity suppression at an angle of ~10", relative to the zero-angle intensity) and the best value of this parameter for USAXS diffractometers is ~ 10^{-5} at λ ~ 1 Å.

Attempts to apply the double-crystal technique for neutron diffraction have been less successful, as the rocking curve parameters (~ 5 x 10^{-3}) have been much higher than for USAXS diffractometers. It was eventually discovered (15) that the wings of the rocking curve had been contaminated by a parasitic neutron flux, propagating inside the single crystal. Channel-cut crystals that have been modified to include neutron absorbing components (Cd) to prevent the propagation of neutrons inside the crystal, have been shown to give rocking curve parameters similar to USAXS, and thus dramatically improve the sensitivity (Fig. 4). In addition, etching the crystals to remove surface imperfections further enhances the sensitivity by another order of magnitude (Fig. 4).

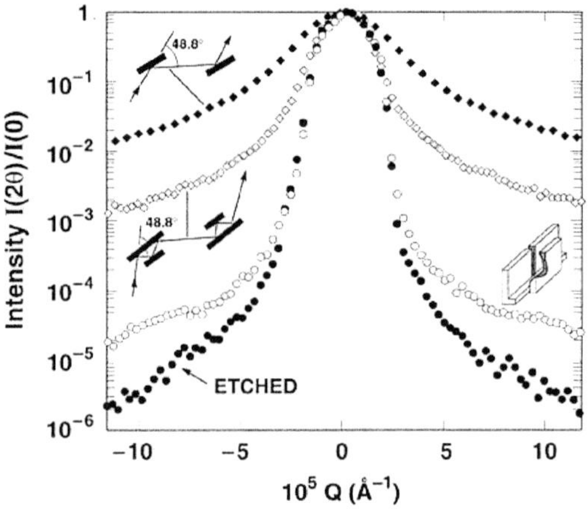

Figure 4: Rocking curves for single- and double-bounce crystals. A Cd-absorber prevents propagation inside the Si walls and improves the USANS sensitivity. Etching removes surface imperfections and further enhances the sensitivity.

These improvements in the signal-to-noise allow the study of particles with dimensions up to 10-30 μm, thus overlapping with light scattering techniques, and an upgraded USANS instrument has recently been installed at the ILL, containing an "Agamalian cut" in the multiple-bounce crystals (16).

However, Bonse-Hart instruments do not measure a two-dimensional pattern and the data are slit-smeared. Alefeld and co-workers (17,18) have proposed an alternative design using focusing toroidal mirrors (FTM) and a demonstration instrument will shortly be commissioned at the FRJ2 reactor in Jülich (19). The advantages of this design are that the FTM-SANS instrument is quite compact and so count rates remain high. Furthermore it also measures a 2D scattering pattern. Provided the technological problems of mirror fabrication and the need for a high-count-rate detector with millimeter resolution can be overcome, this type of instrument would be an excellent choice for a pulsed source and would open up a whole new area of science at the neutron/light scattering interface.

USANS is useful not only in exploring large length scale structures but also in analyzing line profiles of sharp diffraction peaks which appear in a small-angle region. The latter utilizes the sharpness of the rocking curves in USANS apparatus and will yield important information on various elastic moduli for thermally stimulated modes of deformation of the structures associated with the diffractions. USAXS is also useful in the same sense as described above, some application of which have been reported (20) for investigation of order-disorder transition in block copolymers.

Complementary SANS and SAXS Structural Investigations

Professor T. Hashimoto (Kyoto) introduced this subject and gave examples from his research on block copolymers, which are produced by joining two or more chemically distinct polymer blocks, that may be thermodynamically incompatible. Segregation on the molecular scale (10-1000 Å) produces complex nanostructures by self-assembly. The simplest molecular architecture is obtained by connecting a block of A units with a block of B units to make an A-B diblock copolymer, and more complex molecules can be created by linking together three or more monomer unit types.

Two competing effects govern the structure and at high temperatures (T), the chains having UCST-type segmental interactions, for example, are mixed homogeneously. As T is reduced, incompatible blocks tend to segregate due to the different chemical affinities of the components, though the bond between polymer blocks constrains intrachain block separation to a length-scale on the order of the overall chain radius of gyration. In recent years, the rich variety of ordered microdomain morphology of triblock (e.g. ABC) terpolymers has attracted the interest of many researchers (21). So far, most of the copolymers studied have been in the strong segregation regime (where there is little mixing of the A, B and C blocks), and their morphologies have been examined mainly by transmission electron microscopy. However, if one can control the various transitions (e.g. between order and disorder as a function of T) among various ordered phases of multiblock copolymers, it will further enrich potential applications for advanced materials or devices. Scattering techniques are vital for in-situ characterization of the statics and dynamics of such phase transitions as discussed above, and as these systems have characteristics of multicomponents and multiphases, one has to use contrast variation techniques in order to fully characterize the ordered multiphase morphologies.

An example of such a study (22) was used to illustrate the combined use of SANS and SAXS, coupled with deuterium labeling for exploring phase transitions in a triblock terpolymer, polyisoprene (PI)-*block*-poly(deuterated styrene) (DPS)-*block*-poly(methyl vinyl ether) (PMVE), with number-average molecular weights of the PI, DPS and PMVE blocks of 5,000, 23,500 and 35,000, respectively. The temperature dependence of the SAXS and SANS profiles exhibited intriguing features, as shown in Fig. 5, where the D spacing estimated from the scattering vector Q_m of the first-order maximum is plotted as a function of the inverse temperature (K) for SAXS (open squares) and SANS (filled squares). At $T < 140$ °C, the SAXS and SANS D values have a similar temperature dependence, while at $T > 140$ °C, the temperature variation of D for SAXS and SANS are completely opposite. Moreover, the SAXS profile showed a double maximum and spacing, though the SANS profile did not. Noting that the SANS contrasts of PI and PMVE are about equal but much less than that of DPS, and that the SAXS contrasts of DPS and PMVE are similar but greater than PI, the striking temperature dependence of D can be interpreted as follows.

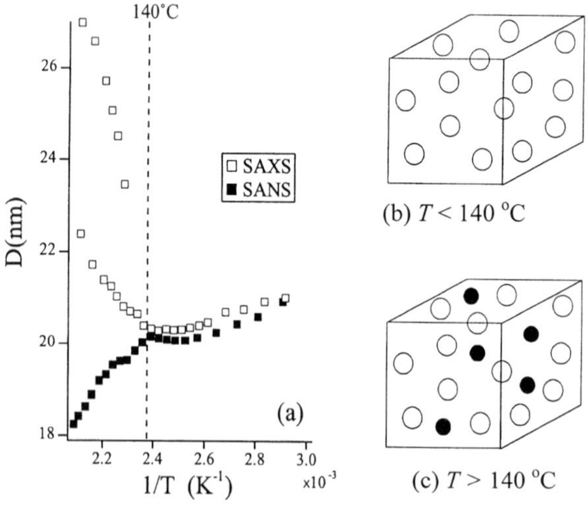

Figure 5: (a) Temperature dependence of the domain spacing D as calculated from the first-order scattering maxima for SAXS (open squares) and SANS (filled squares) from a PI-*b*-DPS-*b*-PMVE triblock terpolymer. Schematic model for the domain structure at (b) $T < 140$ °C and at (c) $T > 140$ °C. The spheres shown by open and filled circles represent, respectively, spheres composed of PI and PMVE blocks.

At temperatures lower than 140 °C, the spherical microdomains of PI block are dispersed in the matrix of mixed DPS and PMVE blocks. Conversely, at temperatures higher than 140 °C, the PMVE blocks undergo microphase separation from the matrix of mixed DPS and PMVE, so that the spheres of PI and those of PMVE coexist. A plausible explanation of the striking differences between SAXS and SANS for $T > 140$ °C is that SAXS "sees" only the PI spheres, while the SANS "sees" both the PI and PMVE. Moreover, microphase separation of PMVE spheres would occur as a result of segregation of PMVE blocks from the DPS chains anchored by the PI spheres. This phenomenon may cause fluctuations in the association number of PMVE blocks in PMVE spheres, which, in turn, may cause a large-length-scale heterogeneities in number density of PI spheres (22). Thus, the complementary information provided by SANS and SAXS helps to characterize the structure and transitions in a way that neither technique alone could do, and is a possible argument in favor of locating neutron and x-ray scattering sources on the same site (e.g. Grenoble).

Retrospective on "Neutron Scattering in the Nineties"

In order to assess the utility of forecasts of future developments in the field, as attempted in this article, it may be instructive to review previous attempts to do so. Then as now, the availability of SANS facilities was a concern and it was hoped that there would be an approximate doubling of the number of reactor neutrons incident on samples "over the next decade" (23). This prediction has come closer to realization in Europe, where the commissioning of new instruments at the ILL, Jülich, Hahn-Meitner, and Paul Scherrer Institutes has more than offset the loss of the Risø reactor. However, in the U.S., the cancellation of the Advanced Neutron Source and the Missouri University Research Reactor scattering program, and the permanent and temporary shutdowns of the Brookhaven and LANSCE sources have largely negated the forecast that polymer scientists will be more prone to use the technique with increased beam time availability (24). NIST has remained the main "workhorse" for polymer-SANS research and projections that "increased flux will lead to a proliferation of kinetic and time-resolved studies" will have to await new and upgraded facilities before they are fully realized.

On the other hand, predictions that absolute calibration will become routine, that the gap between SANS and light scattering will be closed, that equipment will be more user-friendly as changes in sample-detector distance and beam collimation will be more easily accomplished, and faster computers will permit real-time analysis and comparison with theory have been largely authenticated. In addition, the forecast that solid-state detectors will be developed for better *Q*-resolution and/or more compact equipment has been partially fulfilled, as new area detectors have been constructed for the FRJ2 SANS instruments. Further developments along these lines could be of even greater importance for the future of pulsed SANS facilities.

Finally, it may be worth re-emphasizing a point made at the "Neutron Scattering in the Nineties" Symposium (24) as follows:

"The greatest limitation for SANS experimentalists is the securing of suitable samples. To take full advantage of the power of SANS, samples should be selectively deuterated in designated places. The investments in SANS experiments is great enough that samples should be well characterized for MW, tacticity, etc."

In view of the planned large investments in new sources, upgrades and instrumentation, the commitment of a small fraction of this amount to a complementary synthetic program could dramatically increase the overall impact and productivity of future research on polymers.

Acknowledgments: The research at Oak Ridge is supported by the U.S. Department of Energy under contract DE-AC05-00OR22725 with the Oak Ridge National Laboratory, managed by UT-Battelle, LLC.

References

(1) J. Schelten in S. H. Chen, B. Chu, and R. Nossal, Eds., *Scattering Techniques Applied to Supramolecular and Nonequilibrium Systems*, Plenum Press, NATO Advanced Study Series 73, 75 (1981).
(2) J. Schelten, *Kerntechnik* 4, 86 (1972).
(3) K. Ibel, *J. Appl. Crystallogr.* 9, 196 (1976).
(4) H. Maier-Leibnitz and T. Springer, *Annu. Rev. Nucl. Sci.* 16, 207 (1966).
(5) ESF-OECD Technical Report, *A Twenty Years Look Forward at Neutron Scattering Facilities*, http//www.oecd.org/dsti/sti/s_t/ms/prod/scattering.htm (1998).
(6) http://www.sns.gov/
(7) ESS; A Next Generation Neutron Source for Europe, http://www.isis.rl.ac.uk/ess
(8) http://www.isis.rl.ac.uk/targetstation2/
(9) G.D. Wignall and P.A. Egelstaff, *J. Phys. C* 1, 1088 (1968).
(10) http://neutrons.ornl.gov/NSatHFIR/Upgrades/HFIRUpFac.HTML
(11) R.E. Ghosh and A.R. Rennie, *Inst. Phys. Conf. Ser.*, No 107, Chap. 5, 233 (1990).
(12) R.K. Heenan, J. Penfold and S.M. King, *J. Appl. Crystallogr.* 30, 1140 (1997).
(13) Report of the European Science Foundation Workshop, *Scientific Prospects for Neutron Scattering with Present and Future Sources*, Autrans (France) (1996).
(14) U. Bonse and M. Hart, *Appl. Phys. Lett.* 7, 238 (1965).
(15) M. M. Agamalian, G. D. Wignall and R. Triolo, *J. Appl. Crystallogr.* 30, 345 (1997).
(16) ILL Annual Report for 2000, p. 95.
(17) B. Alefeld, D. Schwahn, and T. Springer, *Nucl. Instrum. Meth. Phys. Res., Sect. A* 274, 210 (1989).
(18) B. Alefeld, C. Hayes, F. Mezei, D. Richter and T. Springer, *Physica B* 236, 1052 (1998).

(19) http://www.kfa-juelich.de/iff/Institute/ism/pictures/poster.jpg
(20) T. Koga, T. Koga, K. Kimishima, and T. Hashimoto, *Phys. Rev. E* 60, R3501 (1999); T. Koga, T. Koga, and T. Hashimoto, *J. Chem. Phys.* 110, 11076 (1999).
(21) F.S. Bates and G.H. Fredrickson, *Phys. Today* 52, 32 (1999).
(22) K. Yamauchi, H. Hasegawa, and T. Hashimoto, in preparation.
(23) R. M. Moon, *Neutron Scattering in the Nineties*, IAEA, Vienna, CN-46, 119 (1985).
(24) R. S. Stein, *Neutron Scattering in the Nineties*, IAEA, Vienna, CN-46, 335 (1985).

APPENDIX: NEW SOURCES - NEW OPPORTUNITIES

Stephen King, ISIS Facility, Rutherford Appleton Laboratory, Didcot, OX11 0QX, UK

Historical Perspective

Most neutron scatterers would probably agree that the high-flux reactor at the Institute Laue-Langevin (ILL) in Grenoble represents the best steady-state neutron source in the world today (A1). Commissioned in 1971, but refurbished in the mid-1990s, and with an ongoing program of technical and instrument development, it will no doubt continue to play a significant part in the research careers of many a polymer scientist for at least a couple more decades.

However, as our scientific curiosity develops, so do the demands we place on neutron facilities and their instrumentation. We want to do experiments faster, or with better precision, or with higher resolution, or indeed with any combination of the above. This is of course a trend that is not unique to neutron scattering, but because the facilities involved are large and expensive they must be planned years in advance and designed in such a way that they are utilized to the fullest possible extent. There are currently something like 41 neutron sources around the world operating as "user facilities" (A2). Of these 24 are in Europe (including Russia), 9 are in North America (including Canada), 5 are on the Pacific Rim, 2 are on the Indian sub-continent, and 1 is in North Africa. And there are probably a few others not counted in the above. Of these sources, 36 are reactor facilities, but two-thirds of them were commissioned more than 30 years ago and, consequently, now have increasingly finite lifetimes. New sources are therefore needed. Not just to replace what will be an inevitable decline in measuring capacity, but to ensure that the instrumentation at the disposal of the experimentalist is "state of the art". Building new reactors though is unlikely to be the answer.

To improve on the performance of the ILL reactor (core flux 1.5×10^{15} n cm^{-2} s^{-1}; 58 MW) would be difficult. Difficult technologically (because we are at the limit of our ability to remove heat from such compact reactor cores even though commercial power reactors are typically 1000 MW), difficult politically (because a "next-generation" ILL would need enriched fuel which is subject to international treaties), and difficult environmentally (because the mood of the public has changed and reactors are no longer seen as desirable). Nevertheless, a handful of new reactors have come online (e.g. HMI, JAERI) or have been refurbished in the last 10-20 years (e.g. FRJ2 in the mid-80s, ILL in the mid-90s; ORNL during 2000-3) and a new facility (FRM-2), comparable to ILL in flux terms, is under construction in Munich, Germany (A3).

Spallation Sources

The alternative to building reactors is to use particle accelerators to produce neutrons. Four spallation (or pulsed) sources have been in operation since the 1980s (ISIS in the United Kingdom (A4), KENS in Japan (A5), and IPNS (A6) and LANSCE (A7) in the United States), whilst a continuous variant came online at the Paul Scherrer Institute (PSI) in Switzerland in 1995 (A8). At the time of writing (mid-2001), the US is constructing a "next-generation" spallation neutron source (simply called the SNS) at the Oak Ridge National Laboratory (A9), whilst members of the European Uni-n are considering a proposal for an even more powerful facility (the European Spallation Source or ESS (A10)).

The spallation process involves bombarding a heavy metal (e.g. Ta, W or even Hg) target with relativistic (high-energy) protons. Protons hitting nuclei trigger an intranuclear cascade, placing those nuclei into a highly excited state. These lose energy by "evaporating" nucleons (mainly neutrons). In the case of a tungsten target, each proton delivered results in the production of approximately 15 neutrons but, importantly, with only modest heat production. In most instances the protons are accelerated in pulses and so neutron production also occurs in pulses. This allows the neutron instruments to employ time-of-flight (TOF) techniques.

Essentially, shorter-wavelength neutrons travel faster, and so arrive at a detector earlier, than longer-wavelength neutrons. There is thus no need to employ a velocity selector to monochromatize the incident beam. Wavelength resolution (and hence Q resolution, where Q is the scattering vector) is then largely determined by the precision with which the neutrons can be timed and is, consequently, very good. For a pulsed-source SANS instrument, a figure of 5 % would not be uncommon, compared to 12-30 % for its reactor counterpart (A11). Allied to this is the fact that the count rate at a detector varies as the inverse fourth power of the Q resolution (A12) and so a pulsed-source SANS instrument will generally outperform a reactor-SANS instrument at intermediate Q values anyway.

The other benefit of the TOF approach is that any given point on a detector corresponds to several different Q values, determined by the wavelength of the neutrons arriving there. Hence the broader the incident wavelength band, the greater the range of Q values that can be measured with any given configuration of the instrument. Pulsed-source SANS instruments therefore have much greater dynamic range in Q than reactor-source instruments. As an example, the SANS instrument at ISIS [called LOQ (A13)] can simultaneously measure the range $0.06 < Q\ (\mathrm{nm}^{-1}) < 16$ (though to do so it must operate at half of the source frequency and thus pays a flux penalty). Not having to make measurements (or re-do calibrations) at multiple sample-detector distances and/or at different wavelengths is also a significant experimental advantage of pulsed-source SANS which should not be overlooked lightly. Indeed, polymeric systems often exhibit structure on a range of length scales from phase dimensions to radii-of-gyration to persistence lengths, which require a wide dynamic range.

One important point that should be borne in mind is that none of the operational spallation sources were designed for cold (long-wavelength) neutron production - the source frequency is typically higher than is optimal (e.g. 50 Hz at ISIS) and the cold moderators are decoupled (i.e., separated from the target by a layer of neutron absorber) - so no one has yet truly experienced the best that pulsed-source SANS potentially has to offer. This will change once the new sources come onstream and the existing ones upgrade.

To put these sources in perspective, the most powerful at present is ISIS with a beam power of 0.24 MW (A14). By contrast, the SNS will be a 2-MW source when it comes online in 2006, and the ESS will have a 5-MW target station. For obvious reasons, these beam power levels cannot be directly compared with the thermal power levels of reactors. Comparisons must instead be made on the basis of the neutron fluxes from the moderators (A15,16). On this basis, and comparing the known performance of the (decoupled) ISIS cold moderator to that of the original ILL cold source, the time-averaged flux from a coupled cold moderator on a 4-MW spallation source is fully expected to be about equivalent to that of the ILL second cold source, which supplies the D22 SANS instrument (A17,18). This would represent something like a factor 50 improvement over the present ISIS cold moderator. The proposed low-frequency second target station at ISIS (TS-2) will have a coupled cold moderator and flux gains of at least a factor 10 are projected (A19). If the TS-2 project is approved, then an entirely new SANS instrument will be constructed on it to take advantage of this.

New Opportunities

Despite some initial skepticism at the outset, cold neutron instruments on pulsed sources have more than proved their worth over the years (A20). The next generation of pulsed SANS instruments will offer polymer scientists new experimental opportunities in several areas, some due to improvements in technology, and others due to the complementary nature of the TOF approach, some aspects of which have been discussed above. Present reactor-based instrumentation can certainly be taken as a measure of what the new pulsed-source SANS instruments will be capable of achieving but, beyond that, comparisons and inferences are at best arbitrary because the different instruments utilize their neutrons differently.

Pulsed-source SANS data are always statistically limited at the extremes of the Q range, but the data in between are generally of very good quality (and always of better resolution) compared to a reactor instrument. The experimentalists will therefore benefit most if they adopt a data treatment that takes this into account, such as model-fitting or inversion, rather than persisting with (albeit perhaps more familiar) treatments such as Guinier plots which are by definition restricted to a narrow range of applicability.

To conclude this contribution, it is worth examining the key aspects of future pulsed-source SANS and how these will benefit different areas of polymer science. Arguably the first, and most important aspect is that the experimentalist will have access to instruments with higher neutron fluxes and higher count rates. At a pulsed source with more than one target station, this gain will principally come through utilizing a higher source frequency but in combination with a reduced wavelength band. In other words, there will be a trade-off between flux and Q range. Two categories of measurement will be possible; those that use the higher flux with acceptable statistics to push the time resolution of experiments, and those that use the higher flux to dramatically improve data quality. In the former category, one can envisage time-resolved and kinetic measurements such as in-situ polymerizations, polymer processing, electrochemistry (including the possibility of gating to the neutron pulse), and parametric studies. In the other category are studies of (even more) dilute systems and inelasticity effects. The more routine use of smaller sample volumes or smaller beam sizes (e.g., to a couple of millimetres) will also be possible with the higher fluxes.

Alternatively, by utilizing a low-frequency target in combination with a broad wavelength band, the experimentalists will have a very wide (and simultaneous) Q range at their disposal. This will be important for experiments where the system is responding to a physical or chemical stimulus on many different length scales, or indeed, where it is not at equilibrium.

Examples include polymers under flow, extrusion, or shear, and studies of complex multicomponent systems such as ternary blends and polymer-surfactant complexes. This aspect will also improve the investigation of anisotropic scatterers like liquid crystals. Though, as has already been discussed, improved Q resolution is an inherent property of pulsed-source SANS, further enhancements will further assist studies of interfacial structures, membranes and other ordered systems. And since most proteins interact with biological membranes, there is likely to be a significant increase in the number of experiments looking at membrane interactions in the post-genomic era. In any event, improvements in Q resolution will allow more value to be placed on deuterium contrast variation studies.

Lastly, advances in other areas of scientific equipment design have generally resulted in more compact apparatuses whilst retaining, or even improving, resolution, etc. Without wishing to gaze too deeply into the crystal ball, in the future there will really not be any reason why an experimentalist should not simultaneously measure the UV/Vis or IR signature of a sample, or even its small-angle X-ray scattering, at the same time as the SANS data are recorded. Exploiting the complementarity of different techniques in this way will prove significant in areas such as biopolymers and polymer crystallization, to name but two.

References

A1. http://www.ill.fr
A2. T. Riste, *Neutron News* 6(4), 32 (1995).
A3. http://www.frm2.tu-muenchen.de/index.html
A4. http://www.isis.rl.ac.uk/
A5. http://neutron-www.kek.jp/
A6. http://www.pns.anl.gov/
A7. http://www.lansce.lanl.gov/index_ext.htm
A8. http://sinq.web.psi.ch/
A9. http://www.sns.gov/
A10. http://www.ess-europe.de/ess_js/index.html
A11. These figures allow for the different shapes of the wavelength distributions; the effective width of a Gaussian velocity selector distribution in terms of a rectangular (time) bin is $[\sqrt{12}/\sqrt{(8\ln 2)}]$ x FWHM.
A12. D.F.R. Mildner, J.M. Carpenter, *J. Appl. Crystallogr.* 17, 249 (1984).
A13. http://www.isis.rl.ac.uk/LargeScale/LOQ/loq.htm
A14. Beam power (MW) = 10^{-6} x proton current (μA) x proton energy (MeV). ISIS will produce 300 μA of protons at 800 MeV from March 2002.
A15. In the case of a spallation source, flux is roughly linear in beam power.
A16. In any single pulse, ISIS is at least as bright – in terms of neutrons produced - as the ILL, but the time-averaged output would obviously be lower. This subtlety complicates comparisons of the different types of neutron source. However, provided the pulsed source and its instruments are designed in such a way that all the neutrons in every pulse

can be utilized, it is legitimate to compare the peak flux of a pulsed source with the mean flux of a reactor. A more detailed discussion is presented by F. Mezei, *Neutron News* 5(3), 2 (1994).

A17. R.K. Heenan, J. Penfold, S.M. King, *J. Appl. Crystallogr.* 30, 1140 (1997).

A18. R.K. Heenan, R. Cubitt, K. Mortensen, D. Schwahn, A. Wiedenmann, F. Mezei in *Performance of a Suite of Generic Instruments on ESS*, Report of the ESS Instrumentation Groups at the ESS Science Advisory Committee Workshop, May 2001, ISSN 1433-559X (http://www.ess-europe.de/documentation/ESS_Instr_SAC.pdf).

A19. http://www.isis.rl.ac.uk/targetstation2/

A20. C.J. Carlile, J. Penfold, *Neutron News* 6(2), 5 (1995).

Springer Series in Materials Science

Volume 310

The Springer Series in Materials Science covers the complete spectrum of materials research and technology, including fundamental principles, physical properties, materials theory and design. Recognizing the increasing importance of materials science in future device technologies, the book titles in this series reflect the state-of-the-art in understanding and controlling the structure and properties of all important classes of materials.

More information about this series at http://www.springer.com/series/856

Valeriy V. Ginzburg · Lisa M. Hall
Editors

Theory and Modeling of Polymer Nanocomposites

Springer

Editors
Valeriy V. Ginzburg
Dow Chemical Company (United States)
Midland, MI, USA

Lisa M. Hall
The Ohio State University
Columbus, OH, USA

ISSN 0933-033X ISSN 2196-2812 (electronic)
Springer Series in Materials Science
ISBN 978-3-030-60442-4 ISBN 978-3-030-60443-1 (eBook)
https://doi.org/10.1007/978-3-030-60443-1

This Springer imprint is published by the registered company Springer Nature Switzerland AG
The registered company address is: Gewerbestrasse 11, 6330 Cham, Switzerland

Preface

Polymer nanocomposites (PNCs) are multiphase, multicomponent materials in which one or more components come in the form of particulates with the size less than 100 nm in at least one spatial dimension. The particulates (dispersed phase or fillers) can be metal, ceramic, carbon, or even polymer nanocrystals; they could have spherical, cylindrical, platelet, or other shapes; they can possess various chemical, mechanical, optical, or electrical properties. The polymer matrix, in turn, can be thermoplastic or thermoset; the polymer can be partially crystalline, rubbery, or glassy. One example of a PNC is cross-linked styrene-butadiene rubber with nanoscopic silica or carbon black filler particles, along with other additives, included in varying amounts to create car tires with good traction and acceptable rolling resistance and wear properties. Another example is a typical paint, where sub-100-nm pigments can be added to a liquid formulation, which then undergoes the processes of drying and film formation to create a nanocomposite coating seen on the walls of our houses and on the bodies of our vehicles. In these and other cases where PNCs have already found significant use, including in consumer products such as shoe soles, packaging, and membranes, even minor adjustments to the formulation to improve the material properties can have significant practical implications. It is thus important to develop the Formulation→Structure→Property→Performance roadmap, in which both experimental and theoretical/modeling studies are utilized.

Nanocomposite modeling has made significant progress in recent decades, and many excellent reviews highlighting various specific topics have been published in recent years. Yet, with some notable exceptions (such as "Modeling and Prediction of Polymer Nanocomposite Properties," edited by Vikas Mittal, Wiley, 2012), there have been very few books surveying multiple theoretical methods and analyzing multiple topics under the same cover. Indeed, in order to successfully develop, say, a nanocomposite gas separation membrane, one needs to consider the question of the nanoparticle design (length and density of grafted ligands), their dispersion in the polymer matrix, the processability of the material, and lastly, the gas permeability and selectivity. This design requires multiscale, multiphysics, interdisciplinary modeling that does not come naturally even to the best researchers.

In putting this book together, we hope to stimulate such efforts and highlight some successful examples.

The structure of the book can be described as follows. The first three chapters describe molecular or mesoscale approaches to predict the structure, dispersion, and nanoscale morphologies of nanocomposite materials, given the nanoparticle shape and loading and, in some cases, the length and density of grafted ligands. The next two chapters discuss the dynamics of nanocomposites, concentrating specifically on the behavior of polymer chains at and near the particle surfaces. The remaining six chapters deal with the prediction of mechanical, transport, electrical, and optical properties of nanocomposites, and the implications for product design. We now proceed to describe each chapter more specifically.

In Chap. 1, Arthi Jayaraman discusses the use of the polymer reference interaction site model (PRISM) theory, which is generally applied at the coarse-grained level and includes monomer-scale packing effects along with overall chain and nanoparticle statistical behavior. PRISM requires information about the polymer conformations as an input, and this can be either assumed or calculated self-consistently from molecular simulations as is briefly discussed. A recently developed open-source code pyPRISM is described that allows for easy use of these methods.

In Chap. 2, Valeriy V. Ginzburg describes the application of density functional theory (DFT) to PNCs, which is also generally applied at the coarse-grained level. This approach, based on classical statistical mechanical functionals that calculate the free energy from the density profile, has been particularly useful for describing monomer and nanoparticle packing in confinement or near surfaces. The chapter first details several examples of standard DFT methods, including calculations of the polymer-mediated interactions between polymer-grafted spherical, rod, and plate-shaped particles. The combination of DFT with self-consistent field theory (SCFT) is then discussed. As is made clear by several examples, this is especially attractive to describe nanoparticle behavior in systems with long-range microphase separation such as in block copolymers (where SCFT has found widespread application for systems without nanoparticles).

In Chap. 3, Robert A. Riggleman and co-workers review their newly developed coarse-grained method of simulating nanocomposite morphologies, polymer nanocomposite field theory (PNC-FT). Based on the state-of-the-art polymer field theory, PNC-FT incorporates the nanoparticles via the excluded volume interactions (“cavity functions”) and enables the evaluation of the free energy of various nanoscale structures. The authors discuss the application of their modeling approach to predict the distribution of “hairy” nanospheres in drying polymer films; calculate the effective interactions between nanoparticles with multicomponent ligands; and evaluate the interaction between nanorods in a cylinder-forming block copolymer film. Similar to SCFT-DFT, PNC-FT is now widely used to predict nanocomposite equilibrium structures and morphologies.

In Chap. 4, Argyrios V. Karatrantos and Nigel Clarke describe the dynamics of polymers near the nanofiller surfaces, and specifically the impact of the nanoparticle size and the particle–polymer interaction on the relaxation time, the self-diffusion coefficient, and the glass transition temperature. They utilize coarse-grained molecular dynamics (CG-MD) to show that the mobility of the polymer chains near a strongly attracting particle surface is severely depressed compared to the bulk polymer; near a weakly attracting or repulsive surface, on the other hand, the mobility seems to be slightly enhanced. Overall, CG-MD provides a very useful insight into the interfacial dynamics of PNCs.

In Chap. 5, Jack F. Douglas, Francis W. Starr, and co-workers attack a very similar problem using a combination of CG-MD and analytical scaling theories. They specifically investigate the relationship between the glass transition temperature shift in nanocomposite materials and the density changes induced by the addition of nanoparticles. The authors discuss the relationship between the mobility gradient scale and the density gradient scale, and also comment on the similarities and differences between nanocomposites and thin films in terms of their glass transition and glassy dynamics.

In Chap. 6, Alexey V. Lyulin and co-workers describe the use of multiscale modeling to predict the transport properties of polyelectrolyte membranes (PEMs). Such membranes are used in fuel cells and batteries. The authors discuss how the addition of nanoparticles (such as graphene oxide or silicon dioxide) can influence the proton conductivity through the membrane by impacting the phase separation between the water-rich and polymer-rich domains, as well as the diffusion coefficient and mobility of water near the particle surfaces. Several relevant simulation techniques, from ab initio simulations of electronic structure to classical atomistic molecular dynamics to coarse-grained dynamic density functional theory (DDFT), are employed in this multiscale modeling study.

In Chap. 7, Nitin K. Hansoge and Sinan Keten describe the use of multiscale modeling to predict mechanical properties of matrix-free (or "one-component nanocomposites") assemblies of "hairy nanoparticles." They start by discussing the "energy renormalization" procedure to go from atomistic to coarse-grained molecular dynamics. Using CG-MD, they calculate the morphology of the nanocomposite (i.e., the crystalline structure of the rigid nanoparticle cores separated by the soft ligands). They then used CG-MD to calculate mechanical properties (Young's modulus and toughness) and generate Ashby plots and Pareto fronts to come up with the optimal designs to balance the two properties. Finally, an approach is described whereby the conformational degrees of freedom for the ligands can be integrated out, allowing for computationally inexpensive mesoscale models involving only the particle cores interacting via effective pairwise potential of mean force (PMF).

In Chap. 8, Pavan K. Valavala and Gregory M. Odegard review various approaches to predict mechanical properties of composites and nanocomposites. They start by differentiating between discrete-medium (like MD) and continuous-medium models, and provide an extensive overview of the latter. Within the continuous-medium framework, the authors distinguish between analytical micromechanical theories and computational finite element analysis (FEA), both of which are based on linear elasticity and enable the prediction of linear elastic moduli (Young's modulus and shear modulus for the case of isotropic materials). Multiscale modeling is also discussed for those cases where a number of disparate length scales is involved.

In Chap. 9, Valeriy V. Ginzburg and Jian Yang describe theoretical and modeling approaches to the prediction of thermal conductivity (critically important in many applications, including electronics and transportation). Similar to the previous chapter, the authors discuss the use of analytical micromechanical models and FEA. The role of interfacial thermal resistance—often leading to thermal conductivity being significantly lower than expected—is also reviewed.

In Chap. 10, Michael J. A. Hore discusses ways to predict optical and electrical properties of PNCs. Calculating those properties requires solving Maxwell's equations in complex media, at either very high frequencies (optical) or low frequencies (electrical). Optical properties of interest include scattering and absorption cross section, UV–Vis spectra, and plasmonic resonances. As the author shows, the calculation techniques here range from analytical theories (Rayleigh and Mie scattering) to more computationally expensive finite difference, time domain (FDTD) simulations. For the case of electrical properties (static electrical conductivity), the author emphasizes the importance of percolation (infinite clusters of conducting bonds) and simple methods to predict it (resistor network models). The dependence of electrical conductivity on the conducting filler loading, shape, and alignment is also discussed.

Finally, in Chap. 11, Craig Burkhart, L. Catherine Brinson, and co-workers describe the application of both first-principles and data-driven approaches to the design of nanocomposite materials for automotive tires. The material considered is carbon-black (CB)-filled styrene-butadiene rubber (SBR). The authors discuss the importance of experimental characterization (microscopy) to precisely determine the morphology of the CB network in the SBR matrix; once this morphology is determined, it can serve as input into FEA simulations analyzing the viscoelastic behavior of the material. The predicted linear viscoelastic (LVE) response can be then compared to the "optimal" response known to the industry. Given the complexity of the process, one also looks to simplify the search for best compositions by employing machine learning (ML) tools, and the authors highlight their roadmap in this quest.

In summary, PNC material designers seeking to take advantage of many potential options for polymer and nanoparticles, including their chemical types and amounts, length/size, and architecture (e.g., whether and to what extent chains are cross-linked or grafted to the nanoparticle surface), can turn to modeling approaches to understand how PNC component properties lead to desirable overall material behavior. While the chapters of this book, in order, are meant to present a coherent picture of the range of current PNC modeling techniques from small to large scales and from structure to dynamics to specific material properties, each chapter is written to be independent and readable by itself. We hope the reader will take away an understanding of which modeling methods can be effective and how they can be used, depending on the system and properties of interest, to closely connect synthetically controllable polymer and particle variables to the resulting overall PNC structure and/or dynamics, and allow for rational design of future materials.

Midland, USA — Valeriy V. Ginzburg
Columbus, USA — Lisa M. Hall

Acknowledgments We thank the staff at Springer Nature, and specifically Sara Kate Heukerott, Zachary Evenson, Viradasarani Natarajan, Priyadharshini Subramani, and Adelheid Duhm for their help and encouragement in making this book possible. We would like to thank our families for their patience and support.

Contents

1 Polymer Reference Interaction Site Model (PRISM) Theory and Molecular Simulation Studies of Polymer Nanocomposites 1
Arthi Jayaraman
1.1 Introduction 1
1.2 Relevant Past Experimental and Computational Studies on PNCs 2
1.3 PRISM Theory 5
1.3.1 Basics 5
1.3.2 Linking to Molecular Simulation 9
1.3.3 Open-Source Package *pyPRISM* 12
1.4 Past, Present, and Future of PRISM Theory-Based PNC Studies 13
1.5 Limitations of PRISM Theory 17
1.6 Conclusions 18
References 18

2 Density Functional Theory-Based Modeling of Polymer Nanocomposites 23
Valeriy V. Ginzburg
2.1 Introduction 23
2.2 DFT Overview 25
2.3 Applications of DFT to PNCs 26
2.3.1 Polymers and Nanospheres 26
2.3.2 Polymers and Nanorods 28
2.3.3 Polymers and Nanoplatelets 29
2.4 The SCF-DFT Approach 33
2.4.1 Nanoparticles in Block Copolymers—Bulk and Films 35
2.4.2 "Giant Surfactants"—Single-Ligand Nanoparticles 36

2.4.3 "Hairy Nanoparticles"—Single-Component Assemblies and Mixtures with Polymers 38
2.5 Summary and Outlook 39
References 40

3 Coarse-Grained Modeling of Polymer Nanocomposites: Field-Theoretic Simulations 45
Jason P. Koski, Huikuan Chao, Christian Tabedzki, and Robert A. Riggleman
3.1 Introduction 45
3.2 Standard Field Theory—Compressible Diblock with Homopolymer 47
3.3 Hybrid Particle-Field Theory— Explicit Nanospheres 51
3.4 Field-Based Nanospheres and Finite-Segment Polymers 53
3.4.1 Recent Results 56
3.5 Field-Based Nanorods 56
3.5.1 Recent Results 58
3.6 Homopolymer Grafted Nanoparticles 58
3.7 Complex Grafted Nanoparticles 61
3.7.1 Recent Applications 63
3.8 General Interparticle Potentials 64
3.8.1 Adsorbing Diblock Copolymer 66
3.9 Theoretically-Informed Langevin Dynamics (TILD) 69
3.9.1 Recent Applications 71
3.10 Comparison of the Different Methods 73
3.11 Challenges and Area for Development 75
References 77

4 Polymer Dynamics in Polymer-Nanoparticle Interface 81
Argyrios V. Karatrantos and Nigel Clarke
4.1 Introduction 81
4.2 Polymer Dynamics Around Spherical or Anisotropic Nanoparticles 82
4.2.1 Experiments 82
4.2.2 Simulations 86
4.3 Polymer Dynamics Around a Flat Surface 89
4.3.1 Experiments 89
4.3.2 Simulations 90
4.4 Conclusions 94
References 94

5 The Interfacial Layers Around Nanoparticle and Its Impact on Structural Relaxation and Glass Transition in Model Polymer Nanocomposites 101
Wengang Zhang, Hamed Emamy, Fernando Vargas-Lara, Beatriz A. Pazmiño Betancourt, Dong Meng, Francis W. Starr, and Jack F. Douglas
5.1 Introduction 102
5.2 Modeling and Simulations 104
5.3 Interfacial Zone Around Nanoparticles in Nanocomposites and Near Substrate in Thin Polymer Films 106
5.3.1 Qualitative Description of the Influence of Interfacial Interactions on Mobility Near Interfaces 106
5.3.2 Mobility Gradient Near Interfaces 108
5.3.3 Density Gradient Near Interfaces and Its Scale ξ_ρ 114
5.3.4 Comparison Between Interfacial Mobility and Density Scales 117
5.4 Effects of Bound Interfacial Layer Around NP and Near Substrate on Polymer Relaxation and Overall T_g 118
5.4.1 Influence of Interface Interaction on Polymer Relaxation 119
5.4.2 Effect of Interfacial Interaction Strength on T_g Estimates from Thermodynamic and Dynamic Methods 122
5.5 Conclusions 125
References 127

6 Multiscale Modeling Examples: New Polyelectrolyte Nanocomposite Membranes for Perspective Fuel Cells and Flow Batteries 133
Soumyadipta Sengupta, Alexey V. Lyulin, Georgios Kritikos, Konstantinos Karatasos, Arun Venkatnathan, Rakesh Pant, and Pavel V. Komarov
6.1 Introduction 134
6.2 Multiscale Modeling of Polyelectrolyte Membranes and Their Transport Properties 140
6.3 Confined Film Model of a Nanocomposite Membrane 142
6.4 Atomistic Simulations of Nafion/Graphene Oxide Membranes 149
6.5 Modeling of Doped Non-humidified Membranes 156
6.6 Mesoscopic Simulations and Simulated Example: SPEEK Membrane 160
6.7 Summary 168
References 169

7 Explorations into the Mechanics of Hairy Nanoparticle Assemblies with Molecular Dynamics . . . 179
Nitin K. Hansoge and Sinan Keten
7.1 Introduction . . . 180
7.2 Energy Renormalization (ER) Approach to Coarse-Graining . . . 181
7.2.1 ER Methodology . . . 181
7.2.2 Validation of ER Approach . . . 184
7.3 Modeling of Hairy Nanoparticle Assemblies . . . 185
7.3.1 System Setup . . . 186
7.3.2 Simulation Protocols . . . 187
7.3.3 Mechanical Property Characterization . . . 188
7.3.4 Evaluating Polymer Conformational Behavior . . . 191
7.4 Mesoscale Modeling of aHNPs . . . 193
7.4.1 PMF-Based Approach . . . 194
7.4.2 Functional Form of Effective Interactions . . . 195
7.5 Conclusion . . . 197
References . . . 198

8 Predicting Mechanical Properties Using Continuum Mechanics-Based Approach: Micro-mechanics and Finite Element Analysis . . . 203
Pavan K. Valavala and Gregory M. Odegard
8.1 Introduction . . . 203
8.2 Discrete-Medium Modeling Tools . . . 204
8.3 Continuous-Medium Modeling Tools . . . 205
8.3.1 Analytical Modeling: Micro-mechanics . . . 206
8.3.2 Comparison of the Different Analytical Micro-mechanical Models . . . 211
8.3.3 Laminate Plate Theory . . . 212
8.4 Computational Modeling: Finite Element Analysis (FEA) . . . 215
8.4.1 Computational Micro-mechanics . . . 217
8.5 Multiscale Modeling Concepts . . . 219
8.5.1 Bottom-Up Approach: Equivalent Continuum . . . 220
8.5.2 Top-Down Approach . . . 228
8.5.3 Concurrent Modeling . . . 229
8.6 Summary . . . 230
References . . . 230

9 Modeling the Thermal Conductivity of Polymer-Inorganic Nanocomposites . . . 235
Valeriy V. Ginzburg and Jian Yang
9.1 Introduction . . . 235

9.2 Thermal Conductivities of Typical Polymers and Fillers 237
9.2.1 Polymers 237
9.2.2 Fillers 239
9.3 Analytical Theories and Micromechanical Models 240
9.3.1 The Parallel and Series Models as the Upper and Lower Limit Estimates 240
9.3.2 EMA for Spherical Fillers 241
9.3.3 Non-spherical Fillers 243
9.3.4 Bruggeman Model 244
9.3.5 Combined Approaches 245
9.4 Computer Simulations 245
9.4.1 Finite Element Modeling 246
9.4.2 Interfacial Thermal Resistance Modeling 251
9.4.3 Multiscale Modeling 252
9.5 Summary and Outlook 254
References 255

10 Predicting the Optical and Electrical Properties of Polymer Nanocomposites 259
Michael J. A. Hore
10.1 Introduction 259
10.2 Calculation of Optical Properties 260
10.2.1 Mie Theory Applied to Polymer Nanocomposites 261
10.2.2 The Discrete Dipole Approximation (DDA) 263
10.2.3 Finite-Difference Time-Domain (FDTD) Method 269
10.3 Calculation of Electrical Properties 273
10.3.1 Resistor Network Models 273
10.3.2 The Poisson-Nernst-Planck (PNP) Formalism 276
10.4 Summary and Outlook 278
References 279

11 Data-Driven Multiscale Science for Tire Compounding: Methods and Future Directions 281
Hongyi Xu, Richard J. Sheridan, L. Catherine Brinson, Wei Chen, Bing Jiang, George Papakonstantopoulos, Patrycja Polinska, and Craig Burkhart
11.1 Introduction 282
11.1.1 Early Laboratory Compound Design Concepts 283
11.1.2 Chapter Scope 285
11.2 Methods Framework 285
11.2.1 Simple Representations of Filled Soft Composites 287
11.2.2 Filler Microstructure Statistical Characterization and Stochastic Reconstruction 289

11.2.3 Nanocomposite Properties and Statistical Study of the Representative Volume Element 300
11.2.4 Summary 304
11.3 Future Directions 306
11.3.1 Material Informatics and Cheminformatics 306
11.3.2 Machine Learning 308
11.3.3 Summary 309
References 309

Index 313

Contributors

Beatriz A. Pazmiño Betancourt Materials Science and Engineering Division, National Institute of Standards and Technology, Maryland, USA;
Department of Physics, Wesleyan University, Connecticut, USA

Craig Burkhart Global Materials Science, and Tire and Performance Science Divisions, The Goodyear Tire and Rubber Company, Akron, OH, USA

L. Catherine Brinson Department of Materials Science and Engineering, Duke University, Durham, NC, USA

Huikuan Chao Core Research and Development, The Dow Chemical Company, Midland, USA

Wei Chen Department of Mechanical Engineering, Northwestern University, Evanston, IL, USA

Nigel Clarke Department of Physics and Astronomy, University of Sheffield, Sheffield, UK

Jack F. Douglas Materials Science and Engineering Division, National Institute of Standards and Technology, Maryland, USA

Hamed Emamy Department of Physics, Wesleyan University, Connecticut, USA; Department of Chemical Engineering, Columbia University, New York, NY, USA

Valeriy V. Ginzburg Core Research and Development, The Dow Chemical Company, Midland, MI, USA

Nitin K. Hansoge Department of Mechanical Engineering, Center for Hierarchical Materials Design, Northwestern University, Evanston, IL, USA

Michael J. A. Hore Department of Macromolecular Science and Engineering, Case Western Reserve University, Cleveland, OH, USA

Arthi Jayaraman Department of Chemical and Biomolecular Engineering, Colburn Laboratory, Department of Materials Science and Engineering, University of Delaware, Newark, DE, USA

Bing Jiang Global Materials Science, and Tire and Performance Science Divisions, The Goodyear Tire and Rubber Company, Akron, OH, USA

Konstantinos Karatasos Department of Chemical Engineering, Aristotle University of Thessaloniki, Thessaloniki, Greece

Argyrios V. Karatrantos Materials Research and Technology, Luxembourg Institute of Science and Technology, Esch-sur-Alzette, Luxembourg

Sinan Keten Department of Mechanical Engineering, Center for Hierarchical Materials Design, Department of Civil & Environmental Engineering, Northwestern University, Evanston, IL, USA

Pavel V. Komarov A.N. Nesmeyanov Institute of Organoelement Compounds of Russian Academy of Sciences, Moscow, Russia;
Tver State University, Tver, Russia

Jason P. Koski Sandia National Laboratories, Albuquerque, NM, USA

Georgios Kritikos Department of Chemical Engineering, Aristotle University of Thessaloniki, Thessaloniki, Greece

Alexey V. Lyulin Group Theory of Polymers and Soft Matter and Center for Computational Energy Research, Department of Applied Physics, Technische Universiteit Eindhoven, Eindhoven, The Netherlands

Dong Meng Department of Chemical Engineering, Columbia University, New York, NY, USA

Gregory M. Odegard Mechanical Engineering—Engineering Mechanics, Michigan Technological University, Houghton, India

Rakesh Pant Department of Chemistry and Center for Energy Science, Indian Institute of Science Education and Research, Pashan, Pune, Maharashtra, India

George Papakonstantopoulos Global Materials Science, and Tire and Performance Science Divisions, The Goodyear Tire and Rubber Company, Akron, OH, USA

Patrycja Polinska Global Materials Science, and Tire and Performance Science Divisions, The Goodyear Tire and Rubber Company, Colmar-Berg, Luxembourg

Robert A. Riggleman Chemical and Biomolecular Engineering, University of Pennsylvania, Philadelphia, USA

Soumyadipta Sengupta Group Theory of Polymers and Soft Matter and Center for Computational Energy Research, Department of Applied Physics, Technische Universiteit Eindhoven, Eindhoven, The Netherlands

Richard J. Sheridan Department of Materials Science and Engineering, Duke University, Durham, NC, USA

Francis W. Starr Department of Physics, Wesleyan University, Connecticut, USA

Christian Tabedzki Chemical and Biomolecular Engineering, University of Pennsylvania, Philadelphia, USA

Pavan K. Valavala Research and Development, Packaging and Specialty Plastics, Dow Inc., Midland, USA

Fernando Vargas-Lara Materials Science and Engineering Division, National Institute of Standards and Technology, Maryland, USA;
Department of Physics, Wesleyan University, Connecticut, USA

Arun Venkatnathan Department of Chemistry and Center for Energy Science, Indian Institute of Science Education and Research, Pashan, Pune, Maharashtra, India

Hongyi Xu Department of Mechanical Engineering, The University of Connecticut, Storrs, CT, USA

Jian Yang PE Product Research, The Dow Chemical Company, Lake Jackson, TX, USA

Wengang Zhang Materials Science and Engineering Division, National Institute of Standards and Technology, Maryland, USA;
Department of Physics, Wesleyan University, Connecticut, USA

Chapter 1
Polymer Reference Interaction Site Model (PRISM) Theory and Molecular Simulation Studies of Polymer Nanocomposites

Arthi Jayaraman

Abstract This chapter is focused on Polymer Reference Interaction Site Model (PRISM) theory and its use along with molecular simulation techniques for studying polymer nanocomposites (PNCs). In the first section of this chapter, we summarize key experimental and computational studies on PNCs from the literature to show the reader the types of fundamental questions that these studies have tackled. These are the types of questions that one could also use PRISM theory to answer. Then, we describe the basics of PRISM theory with relevant equations and show how PRISM theory is linked to molecular simulations to obtain meaningful results pertaining to PNC structure and thermodynamics. We also bring to the readers' attention the open-source package, *pyPRISM*, developed for both expert and novice computational researchers to easily incorporate PRISM theory into their PNC studies. We then discuss briefly past, present, and potential new directions of PNC studies using PRISM theory and conclude the chapter highlighting some of the limitations of PRISM theory.

1.1 Introduction

Polymer nanocomposites (PNCs) are a specific class of macromolecular soft materials that have garnered significant attention in the past two to three decades from researchers around the world. This is evident from the many review articles and perspectives on this topic [1–17]. As the name suggests, a polymer nanocomposite is a mixture of the polymer (matrix) as the majority component and nanoscale additives (fillers) as the minority component. The chemical and physical nature of the fillers and matrix and their relative composition in the PNC is selected based on the eventual

A. Jayaraman (✉)
Department of Chemical and Biomolecular Engineering, Colburn Laboratory,
University of Delaware, 150 Academy Street, Newark, DE 19716, USA
e-mail: arthij@udel.edu

Department of Materials Science and Engineering, University of Delaware, Newark, DE 19716, USA

V. V. Ginzburg and L. M. Hall (eds.), *Theory and Modeling of Polymer Nanocomposites*, Springer Series in Materials Science 310,
https://doi.org/10.1007/978-3-030-60443-1_1

application of the PNC. There is a broad range of applications where PNCs have been used, such as in automobile parts (e.g., car bumpers, tires) [18], aerospace industry (e.g., components of thermal resistance aircraft body parts) [19], in separations (e.g., membranes for gas transport and absorption) [20], in electronics (e.g., organic/hybrid solar cells) [21], and in biomaterials (e.g., tissue mimics, drug delivery vehicles) [22]. To achieve the desired function or macroscopic property of the PNC needed for these applications, one has to select the components of the PNCs based on their individual properties (e.g., biocompatible matrix and filler when used in biomedical applications, electron donating/accepting matrix and filler in organic/hybrid electronics) as well as on how they interact and spatially organize within the PNC (i.e., structure/morphology). Additionally, the ease during processing (e.g., response to shear, temperature) of the PNC, a critical feature for large-scale manufacturing, is another variable to consider during the material selection process. The non-trivial relationship between the structure/morphology, dynamics, and thermodynamics for a selected filler and matrix at a specific composition within PNCs has motivated exciting research aimed at unraveling the complex interplay between PNC design and its macroscopic behavior. In particular, computational tools (e.g., coarse-grained models, theory, and simulation) have become valuable tools in this quest to both predict PNC behavior for a variety of PNC design parameters as well as to explain less understood experimental observations in PNCs.

This chapter specifically focuses on how Polymer Reference Interaction Site Model (PRISM) theory and molecular simulation methods have been used to tackle questions about PNC *structure and thermodynamics.* To help the reader appreciate the value of these tools, we first review in Sect. 1.2 relevant past experimental, computational, and theoretical work on PNC structure and thermodynamics to describe the types of fundamental questions that researchers working with PNCs tackle and that PRISM theory can be used for. Then, in Sect. 1.3, the details of PRISM theory, the open-source package for readers wishing to use this method, and complementary molecular simulation methods used along with PRISM theory are discussed. In Sects. 1.4 and 1.5, the past, present, and potential new directions for PRISM theory-based studies of PNCs are described along with the key limitations of PRISM theory.

1.2 Relevant Past Experimental and Computational Studies on PNCs

Past studies focused on PNCs (see references in review articles [1–13, 15, 17]) have shown that any experimental or computational tool that is used to study PNC structure and morphology should be able to tackle/probe all relevant length scales including the size of the monomer, the polymer chain, the nanoscale (bare or unmodified) particle or filler, and if phase separated, size of the domains of fillers and matrix chains. As the nanoscale dimension of the fillers decreases (e.g., curvature of spherical nanoparticle increases), the total interfacial area between filler and matrix in the PNC increases, which in turn affects the polymer chain conformation (e.g., radius of gyration) as compared to its (zero-filler or neat polymer) melt-state ideal chain conformation. The chain conformation also alters the *effective* filler-matrix interaction. If one sets aside the polymer and filler chemistry and the resulting enthalpic driving forces from consideration for a moment, they will realize that the effective filler-matrix interaction is purely a function of the entropic driving forces determined by the filler shape and size, the matrix polymer molecular weight and architecture, and their relative volume fractions in the PNC. The effective filler-matrix or filler-filler interactions dictate whether the filler and the polymer matrix remain mixed (i.e., dispersed filler morphology) or demixed (i.e., aggregated filler morphology) within the PNC. If one also takes into consideration the actual chemistries of the (unmodified) filler and matrix polymer(s), the enthalpic driving forces arising from the filler-filler, matrix-matrix, and filler-matrix (direct) interactions will cooperate/compete with entropic driving forces to alter/retain the chain conformations and the effective filler-matrix interactions that lead to dispersed/aggregated filler morphologies.

To go beyond the (limited) design space of PNCs with *unmodified* or *bare* additives and incorporate additional tuning parameters that can alter the energetic and entropic driving forces for the PNC morphology and dynamics, one can functionalize the filler surfaces. The many advances in synthetic schemes in the past two decades have paved the way for controllably engineering functionalized/grafted/tethered nanoparticles at the molecular level with a variety of ligand—surfactants, polymers, proteins, and nucleic acids—at a desired grafting density, dispersity, and with desired chemistries. PNCs comprised of polymer-grafted nanoparticles (PGNs) have garnered much interest within the PNC community, both from computational and experimental researchers. References in review articles [10–13, 15–17] have established the link between the molecular design of PGNs and the macroscale structure of PGNs within a polymer matrix for linear (architecture) graft and matrix polymers. At high grafting density or the brush-limit, if the chemistry of the graft and matrix polymer is the same or similar, the system is driven purely by entropic driving forces. Much focus has been placed on tailoring the PGNs such that the resulting entropic driving forces favor mixing of PGNs and matrix chains. The extent to which the matrix chains mix with and penetrate the grafted layer is termed extent of 'wetting.' As wetting increases, the matrix chains and PGNs mix, and the extent of filler dispersion within the PNC increases. The key design rules for tuning entropic driving forces to favor wetting are: (a) use matrix polymers of a smaller molecular weight than that of the graft polymer, with the curvature of the filler particle and the value of grafting density dictating how much smaller the matrix chains need to be compared to the graft chain to achieve wetting; (b) include a high dispersity in molecular masses of the graft polymers so as to include few long graft polymers and many short polymers within the grafted layer; and (c) if grafting density and particle size are fixed, choose graft and matrix polymer chemistries that are less flexible and have higher persistence lengths. If the matrix and graft polymer chemistries are dissimilar, then the Flory-Huggins χ parameter between the matrix and graft polymer chemistry together with the other factors that govern the entropic driving forces dictateS whether the PGNs and the matrix will demix or mix at a given temperature. Using a negative χ graft–matrix pair, one can improve grafted layer wetting and push the limits of PGN volume fraction that exhibit PNC dispersion. Recent studies [23–25] have shown that the first-order dispersion to aggregation transition is not synonymous with the wetting to dewetting transition; the latter occurs continuously as the PNC transitions from dispersed to aggregated states. For linear graft and matrix polymers, at low grafting density, the filler surface may not be completely shielded by the grafted chains from the matrix chains and then the energetic driving forces based on the direct filler-graft, filler-filler, and filler-matrix interactions along with the entropic driving forces decide the extent of mixing/demixing of the PGNs and the matrix.

Moving away from a traditional definition of PNCs that has a matrix and a filler component mixed together, one could also consider a melt of PGNs in the absence of a matrix making an effectively one-component PNC. These systems are termed 'matrix-free PGNs' and have been receiving attention in the past few years (see for example studies in references [20, 26–29]). One advantage of using these matrix-free PGNs is the ease of processing without dealing with the hassles of creating a homogeneously mixed PNC. Second advantage is the regularity of particle spacing in a polymer melt accomplished through the grafted polymer segments extending from the nanoparticle core. This regularity in interparticle spacing can be exploited in applications like optics, photonics, separation of small molecules, etc. In the absence of *free* matrix chains, at high grafting density, the grafted chains create a dense region around the core extended outward. Depending on the molecular weight of the grafted chains and the grafting density, one could observe interpenetration between grafted chains on adjacent particles or isolated 'soft'-grafted particles arranged in an array. Instead of melts, if one created a dense solution of matrix-free PGNs, one can create porous polymer material that can be used for separations [20, 29, 30]. In a recent study conducted on a *dried* matrix-free PGN solution, upon exposure to a solvent, a combination of small angle x-ray and neutron scattering results showed that the small molecule solvent uniformly distributes around the nanoparticle surface [29]. Going beyond homopolymers as grafts, one could also use block copolymer grafts in such matrix-free PGN melts/dense solutions to achieve matrix-free PNCs with nanoparticles at precise periodicities within microphase-separated block copolymer domains. For example, in a recent experimental study done on poly(styrene-block-(n-butyl acrylate))-grafted silica nanoparticles, the authors found that at higher grafting density, there was better microphase separation of the block copolymers and more uniform filler (core) dispersion in the covalently linked matrix polymer than at lower grafting density with similar polymer chain length and composition [31].

In thin films of PGNs with and without matrix, additional factors including the competing PGN interaction with the free surface, substrate and matrix will affect structure and dynamics. Kim and Green [32], for instance, showed the three regimes of structural organization—PGNs phase separating from the matrix and residing at the free surface and bottom substrate (unstable film is regime I and stable film is regime II) or PGNs being miscible within the film (regime III). They also find that the phase separation in regime I is analogous to thin film of polymer blends (specifically micelles in homopolymers) while the phase separation in regime II is like that seen in hard spheres in an athermal homopolymer melt. Going beyond homogeneous surfaces, recent experiments on topographically patterned substrate [33] show that the segregation of PGNs can be achieved by varying the relative confinement entropy of grafted versus matrix chains. In studies of matrix-free PGNs adsorbed on a surface, it has been found that increasing the monomer–surface adsorption strength leads to extended polymer chain conformations and adsorption on the surface and adoption of a ‘canopy’ shape. The grafted chains from neighboring ‘canopies’ can interpenetrate at low-moderate graft density and low surface attraction strength. In contrast, at high graft density and/or high surface attraction strengths, the spacing between particles increases and interparticle entanglements reduce. Self-assembly within monolayers of PGNs on surfaces as a function of core–core and core–polymer interactions has also been shown using simulations [34]. These simulations show the relative values of these interactions where one can observe dispersed PGNs, single PGN wide strings of PGNs, stripes of PGNs with width equal to two or three PGNs and clusters of PGNs.

PRISM theory has played an important role in some of the above studies. Before describing these studies in Sect. 1.4, it is useful to first go into the basics of PRISM theory, how it links to simulations and ways one could conduct PRISM theory calculations, in the next section.

1.3 PRISM Theory

1.3.1 Basics

PRISM theory was developed by Schweizer and Curro [35, 36] from the Reference Interaction Site Model (RISM) theory [37–40] to study chain molecules (i.e., polymers). Its formalism is similar to the Ornstein–Zernike integral equation [40, 41] with the total site–site intermolecular pair correlation function, $h_{ij}(r)$, being related to the intermolecular direct correlation function, $c_{ij}(r)$, and intramolecular pair correlation function, $\omega_{ij}(r)$. In Fourier space, PRISM theory takes on the form:

$$\widehat{H}(k) = \widehat{\Omega}(k)\widehat{C}(k)\left[\widehat{\Omega}(k) + \widehat{H}(k)\right] \tag{1.1}$$

where each term is a matrix of size $N \times N$ for a system that can be modeled with N types of interaction sites. For example, in a PNC comprised of homopolymer matrix chains of chemistry A with bare nanoscale fillers of chemistry B, if all repeat units along each matrix chain can be treated as equivalent sites of type A and each nanofiller treated as a single site or collection of sites of type B, N would be equal to 2. If the PNC has PGNs, then the graft polymer site (A), matrix polymer site (B) and particle site (C) make it a 3-site system (see Fig. 1.1), and thus, the matrices would be 3×3. In that 3-site system, the components of the matrices in (1.1) will be:

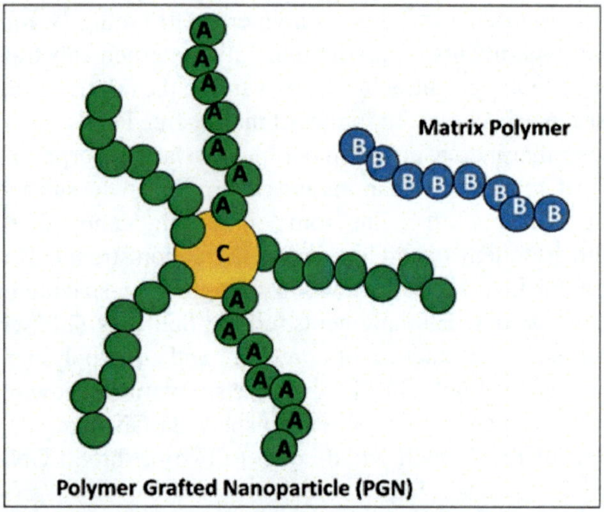

Fig. 1.1 Schematic of a polymer-grafted nanoparticle (PGN) and a matrix polymer chain, and the definition of the three possible types of sites A, B, and C. The reader should note that in this figure, for the grafted chains, the tethered end segment, the intermediate segments and the free end segment are treated equally with the same type of site (type A). Similarly, for the matrix chain, the free ends and intermediate segments are treating equally (with type B site). In principle, this could be relaxed by using different types of sites representing various parts of the chains. As the number of sites used to represent/model, the system increases so does the size of the matrices in (1.2a–c) and the corresponding numerical complexity for solving the PRISM equations

$$\widehat{H}(k) = \begin{bmatrix} \rho_{AA}^{pair}\hat{h}_{AA}(k) & \rho_{AB}^{pair}\hat{h}_{AB}(k) & \rho_{AC}^{pair}\hat{h}_{AC}(k) \\ \rho_{BA}^{pair}\hat{h}_{BA}(k) & \rho_{BB}^{pair}\hat{h}_{BB}(k) & \rho_{BC}^{pair}\hat{h}_{BC}(k) \\ \rho_{CA}^{pair}\hat{h}_{CA}(k) & \rho_{CB}^{pair}\hat{h}_{CB}(k) & \rho_{CC}^{pair}\hat{h}_{CC}(k) \end{bmatrix} \quad (1.2a)$$

in which $\rho_{\alpha\beta}^{pair} = \rho_\alpha \rho_\beta$ and ρ_α and ρ_β correspond to the site number densities of site types α and β,

$$\widehat{\Omega}(k) = \begin{bmatrix} \rho_{AA}^{site}\widehat{\omega}_{AA}(k) & \rho_{AB}^{site}\widehat{\omega}_{AB}(k) & \rho_{AC}^{site}\widehat{\omega}_{AC}(k) \\ \rho_{BA}^{site}\widehat{\omega}_{BA}(k) & \rho_{BB}^{site}\widehat{\omega}_{BB}(k) & \rho_{BC}^{site}\widehat{\omega}_{BC}(k) \\ \rho_{CA}^{site}\widehat{\omega}_{CA}(k) & \rho_{CB}^{site}\widehat{\omega}_{CB}(k) & \rho_{CC}^{site}\widehat{\omega}_{CC}(k) \end{bmatrix} \quad (1.2b)$$

where $\rho_{\alpha\beta}^{site} = (\rho_\alpha + \rho_\beta)$ if $\alpha \neq \beta$ otherwise $\rho_{\alpha\beta}^{site} = \rho_\alpha$, and

$$\widehat{C}(k) = \begin{bmatrix} \hat{c}_{AA}(k) & \hat{c}_{AB}(k) & \hat{c}_{AC}(k) \\ \hat{c}_{BA}(k) & \hat{c}_{BB}(k) & \hat{c}_{BC}(k) \\ \hat{c}_{CA}(k) & \hat{c}_{CB}(k) & \hat{c}_{CC}(k) \end{bmatrix} \quad (1.2c)$$

where $c_{\alpha\beta} = c_{\beta\alpha}$. The design parameters of the PNC, for example, the filler composition (or volume fraction of filler), grafting density, and graft and/or matrix molecular masses (i.e., chain lengths of the polymer) dictate the ρ_i of the sites of type i.

The matrix elements of $\widehat{\Omega}(k)$ are the intramolecular pair correlation functions in inverse space (or wave vector k) between sites i and j within a certain molecule; these are known and given as an input to the (1.1). For some systems, these functions can be derived analytically (e.g., ideal chain conformations in a polymer melt). The analytical expression for a Gaussian chain conformation is

$$\widehat{\omega}_{\alpha\beta}(k) = \frac{1 - f^2 - \frac{2f}{n} + \frac{2f^{n+1}}{n}}{(1-f)^2} \tag{1.3}$$

where f is defined as $f = \exp\left(-k^2\sigma^2/6\right)$, n is the number of monomers in the Gaussian chain, and σ is the characteristic distance unit [42]. Similarly, (1.3) can be used to model a freely jointed chain (FJC) with $f = \sin(kl)/kl$, where l is the bond length between sites along the chain [42]. For PGNs, analytic expressions for the particle–graft and graft–graft $\widehat{\omega}_{\alpha\beta}(k)$ have been derived when the number of grafted chains on the surface is small [43–46]. The scattering community has also derived many analytical forms of $\widehat{\omega}_{\alpha\beta}(k)$ as 'form factors' for fitting the scattering data of commonly studied polymer systems. However, in many cases, analytical expressions for $\widehat{\omega}_{\alpha\beta}(k)$ for the PNC of interest does not exist. Rather than deriving a new analytical expression, one can simulate the PNC and sample configurations of the molecule of interest to calculate the $\widehat{\omega}_{\alpha\beta}(k)$ (see description in Sect. 1.3.2.1).

The matrix elements of $\widehat{H}(k)$ and $\widehat{C}(k)$ are usually unknown and are to be solved for. Given two unknown matrices and one PRISM equation (1.1), we need one more independent equation to solve for the two unknowns. This additional equation is called the closure relation. Closure relations connect the real-space direct pair correlation functions $c_{ij}(r)$, total intermolecular pair correlation function $h_{ij}(r)$, and pair-wise interaction potentials, $U_{ij}(r)$. Since the closures include pair-wise interaction potentials, $U_{ij}(r)$, it is through these closures that the chemical details of the PNC are specified. Examples of closures include Percus–Yevick (PY), Hypernetted Chain (HNC), Mean-Spherical Approximation (MSA), Martynov–Sarkisov (MS), and Laria–Wu–Chandler (LWC) [47, 48]. The choice of the type of closure used to solve (1.1) is critical as this choice dictates how realistic or correct the PRISM theory predictions are. We discuss this again in the limitations of this method (Sect. 1.5). Therefore, the right closure must be found or developed for the system being studied. For PNCs, based on agreement between PRISM theory and molecular simulations, for mixtures of nanoparticles and homopolymers, many studies [43–46, 49–54] have used the PY closure for polymer–polymer pair and polymer–particle pair, and the HNC closure for particle–particle pair. If σ_{ij} is the distance of closest approach between sites i and j, (e.g., $\sigma_{ij} = d$ for monomer–monomer pairs and $\sigma_{ij} = (D + d)/2$ for particle–monomer pairs, where d is the size of the monomer site and D is the size of the particle) the impenetrability condition applies inside the hard core.

$$g_{ij}(r) = 0 \quad r < \sigma_{ij} \tag{1.4a}$$

Outside the hard core, the PY approximation for *ij* pair of sites is written as

$$c_{ij}(r) = \left(1 - e^{\beta U_{ij}(r)}\right) g_{ij}(r) \quad r > \sigma_{ij} \tag{1.4b}$$

and the HNC closure, often used for the particle–particle pair (hence the subscript CC based on type of sites as shown in Fig. 1.1), is described as

$$c_{CC}(r) = h_{CC}(r) - \ln g_{CC}(r) - \beta U_{CC}(r) \qquad r > D \tag{1.4c}$$

Past studies by Schweizer and Yethiraj have suggested that molecular closures [47, 55, 56] could prove more accurate for systems with strongly segregating polymers or monomer species (e.g., strongly segregating copolymer matrix polymer or copolymers grafts, or strongly segregating graft and matrix homopolymers). For this statement to be confirmed for the PNC system at hand, one needs to conduct systematic comparisons of results from PRISM theory using atomic and molecular closures against results from experiments and/or molecules simulation results.

After the closures have been selected, one can solve the above system of coupled nonlinear integral (PRISM theory) equation and closure(s). Analytical solutions only exist for atomic systems while numerical solutions are found for complex polymer systems. In principle, one could select any method from existing numerical methods used commonly to solve integral equations [57, 58]. Recent studies have employed the KINSOL algorithm [59] with the line search optimization strategy which minimizes the objective function along an optimum descent direction. KINSOL exhibits convergence for complex nonlinear integral equations easier and faster as compared to the Picard technique, which has also been used in some PRISM theory studies [7, 58]. Irrespective of the numerical algorithm one chooses to use, upon solving the PRISM equation and closure(s), one will obtain the total intermolecular pair correlation functions, $h_{ij}(r)$, which can be related to the $g_{ij}(r)$ ($=h_{ij}(r) + 1$) and the partial collective structure factors, S_{ij}(k). The partial structure factors are the elements of the $\widehat{S}(k)$ matrix which are related to the direct pair correlation functions and total pair correlation functions as

$$\widehat{S}(k) = \widehat{C}(k) + \widehat{H}(k) \tag{1.5a}$$

These pair-wise structure factors can be compared to the corresponding pair correlation functions from simulations or scattering results from experiments. By obtaining the $g_{ij}(r)$, one can calculate other properties of interest as well. For example, the potential of mean force (PMF) between particles, $W_{CC}(r)$, can be calculated, from the particle–particle pair correlation function as follows.

$$W_{CC}(r) = -kT \ln g_{CC}(r) \tag{1.5b}$$

Similarly, the second virial coefficient can be calculated from $g_{cc}(r)$ for PNCs at infinitely dilute concentrations. The pair-wise structure factors and the intermolecular direct correlation function together also provide an estimate of the induced solvation potential that one component (e.g., PGN) feels because of the other component (e.g., matrix chains); this is described in Sect. 1.3.2.2.

For some PNC design parameters, (e.g., stronger attractive interaction strengths, large site size ratios, *D*/*d*, etc.), the numerical method selected may not yield any solutions. This could be an issue with numerical convergence due to a poor initial guess [57] or due to deviations from liquid-like behavior due to (micro/macro) phase separation within the system. This is discussed further in the limitations described in Sect. 1.5.

1.3.2 Linking to Molecular Simulation

PRISM theory calculations can be complemented by molecular simulations in multiple ways and this section describes some of these synergies.

1.3.2.1 Providing Realistic Intramolecular Pair Correlation Function (i.e., Shapes of Molecules)

In many past PRISM theory studies of PNCs (e.g., bare fillers in homopolymer matrix or lightly grafted PGNs in homopolymer matrix) [43–46, 49–54] the intramolecular pair correlation, Ω_{ij}, was calculated by assuming *ideal* conformations of the matrix chains. But, assuming ideal chain conformations for the polymer could be far from reality in some systems. For example, the chains on a densely grafted PGN or a copolymer-grafted particle have been shown to adopt non-ideal conformations, and therefore, using ideal conformations for the grafted chains is not valid. Similarly, the grafted chains in chemically dissimilar graft–matrix systems could adopt extended (collapsed) configurations to initiate (deter) formation of energetically favorable (unfavorable) contacts with matrix chains. For such cases where an off-the-shelf analytical expression of the intramolecular pair correlation, Ω_{ij}, is not available or existing analytical intramolecular pair correlation functions are not valid, molecular simulations serve as a valuable tool for providing realistic intramolecular pair correlation, Ω_{ij}. For example, one can use molecular dynamics simulations or Monte Carlo (MC) simulation techniques (e.g., see review articles on simulations of polymers and PNCs [60–64]) to sample a single homopolymer-grafted nanoparticle in a polymer matrix. These simulations provide the ensemble average graft monomer–graft monomer, graft monomer–particle, matrix monomer–matrix monomer intramolecular pair correlation functions. These correlation functions are calculated using the Debye scattering relation [65, 66]

$$\widehat{\omega}_{\alpha\beta}(k) = \left\langle \frac{1}{N^{\text{total}}} \sum_{i}^{N_\alpha} \sum_{j}^{N_\beta} \frac{\sin(k \cdot r_{ij})}{k \cdot r_{ij}} \right\rangle \tag{1.6}$$

in which α and β represent site types, N_α is the total number of sites of type α in each molecule, N_β is the total number of sites of type β in each molecule, $N^{\text{total}} = (N_\alpha + N_\beta)$ if $\alpha \neq \beta$ otherwise $N^{\text{total}} = N_\alpha$, r_{ij} is the distance between sites i and j, and the angle brackets represent ensemble averaging over uncorrelated snapshots in a simulation trajectory.

In some cases, to reduce the computational intensity of having to simulate the PGN(s) in a melt-like matrix, one could simulate separately a single matrix polymer chain in vacuum and a single PGN in vacuum to obtain the pair-wise intramolecular correlations. However, these vacuum simulations do not capture the effects of the matrix chains on the grafted polymer configurations or vice versa, and thus, are approximate intramolecular pair correlations.

1.3.2.2 Self-consistent PRISM Theory-Simulation Loop

PRISM theory and molecular simulations can also be linked together in a self-consistent loop where the intramolecular pair correlation functions (e.g., chain conformations) that are input to PRISM theory are provided by simulations of a single PGN in an external medium-induced solvation potential that is obtained from PRISM theory. The interdependence of the chain conformations and the medium-induced solvation potential gives rise to the self-consistent loop (Fig. 1.2). The steps involved in this self-consistent loop are described next.

First, the pair-wise decomposed medium-induced solvation potential, $\Delta\psi_{ij}(r)$, is obtained from the PRISM theory output. This pair-wise function describes the interaction between any two sites i and j as mediated by all the other sites in the system. The form of the solvation potential depends on the approximation used in its derivation [67–72]. The PY and HNC forms of the solvation potential are as follows:

$$\Delta\psi_{ij}^{PY}(r) = -k_B T \ln\left[1 + c_{ik}(r) * s_{kk'}(r) * c_{k'j}(r)\right] \tag{1.7a}$$

$$\Delta\psi_{ij}^{HNC}(r) = -k_B T\, c_{ik}(r) * s_{kk'}(r) * c_{k'j}(r) \tag{1.7b}$$

where '*' in (1.7a) and (1.7b) denote a convolution integral in spatial coordinates and k_B is the Boltzmann constant and T is the temperature. The terms on the right-hand side of (1.7a) and (1.7b) are the real-space pair-wise direct pair correlation function and real-space analog of the structure factors. The solvation potential, $\Delta\psi_{ij}(r)$, is then input to a molecular simulation of a single PGN and/or a single matrix chain. In the simulation, the total interaction potential between sites i and j separated by a distance r, $U_{ij}^{\text{tot}}(r)$ is defined as the sum of $U_{ij}(r)$ and the solvation potential, $\Delta\psi_{ij}(r)$, obtained from the preceding PRISM step. To ensure the sites in the PRISM theory

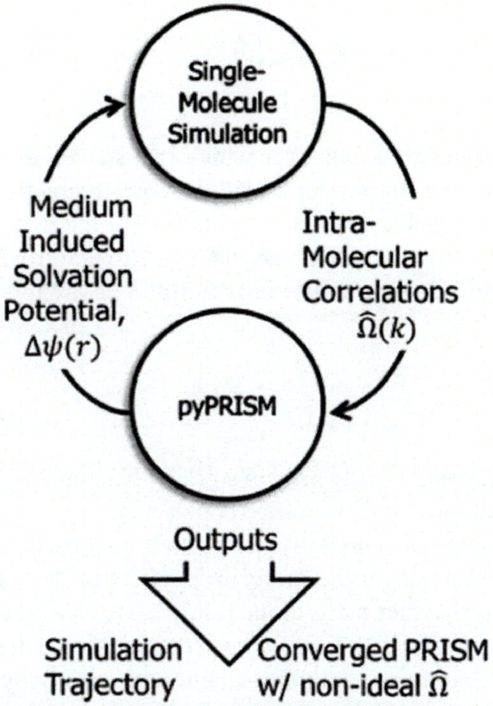

Fig. 1.2 Flowchart of self-consistent PRISM—molecular simulation approach. 'pyPRISM' is the open-source package which the reader could use to conduct this calculation. This package is described in Sect. 1.3.3. This figure is reprinted with permission from [58] (2018) American Chemical Society

and molecular simulations represent the same thing, it is best to keep the model chosen to represent the PGN and the matrix polymer to be the same in both PRISM theory and molecular simulations.

Next, the PRISM theory calculation and molecular simulation steps are repeated one after the other. In PNCs with PGNs, one could either alternately simulate a single PGN or a single matrix chain with the newest set of pair-wise solvation potentials from PRISM calculations or simultaneously simulate the single PGN and a single matrix chain (in parallel) with the same set of solvation potentials calculated from the previous PRISM step. Irrespective of the simulation method, the intramolecular structure factors between site pairs are sampled from uncorrelated configurations and the ensemble average of the intramolecular structure factors is calculated to serve as the new input for the following iteration of PRISM theory calculations.

The self-consistent PRISM theory—molecular simulation loop iterations are continued until convergence is achieved. One convergence criterion is the sum of squared errors (SSE) of $\Delta\psi_{ij}(r)$ between iterations n and $n+1$:

$$SSE_{n \to n+1} = \sum_{i,j} \sum_{m=1}^{N_r} \left[\Delta\psi_{ij}^{n+1}(r_m) - \Delta\psi_{ij}^{n}(r_m) \right]^2 \tag{1.7a}$$

where N_r is the number of points over which real space has been discretized. The subscripts i and j denote those pairs of sites whose interactions are relevant to the simulation. The convergence counter, n, is set to 0 for the first iteration. The *SSE* for the transition $n \to n+1$ ($n \geq 1$) is expressed relative to that from $0 \to 1$. Convergence is attained when the following criterion is fulfilled in say three or five consecutive iterations:

$$\frac{SSE_{n \to n+1}}{SSE_{0 \to 1}} \leq 0.01 \qquad (n \geq 1) \tag{1.7b}$$

This approach was specifically used in work by Nair and Jayaraman [73] to tackle non-ideal conformations along the grafted copolymers and matrix polymers that were neglected in the studies of homopolymer-grafted nanoparticles [43–46] preceding Nair and Jayaraman's work. Furthermore, in contrast to older self-consistent PRISM-MC studies on homopolymer melts alone [68, 70, 71, 74–76] or work on bare particles in a homopolymer melt [77, 78] where the self-consistent loop involved MC simulations of only one matrix polymer chain, in the study by Nair and Jayaraman [73], alternate self-consistent loops for a single copolymer-grafted particle and a single matrix chain were used. This approach accounts for non-idealities in both the grafted and matrix chain conformations. A similar self-consistent PRISM-MC approach was previously applied to a blend of two polymers where the self-consistent loops involved MC simulations of each polymer [79, 80].

1.3.3 Open-Source Package pyPRISM

Having discussed the above basics features of PRISM theory and how to link it to molecular simulations, we now present briefly some details of an open-source package that one could use to conduct PRISM theory calculations on PNCs. Recently, we developed *pyPRISM,* a Python-based open-source package, to ease the implementation and use of PRISM theory by expert and novice researchers who study PNCs and other soft materials. *pyPRISM* provides a user-friendly interface for setting up PRISM theory calculations for the problem at hand, along with underlying data structures that simplify the mathematics involved in setting up and solving PRISM theory. The documentation and knowledge base materials are hosted on the repository website [81–83] and maintained by Dr. Tyler Martin and coworkers at National Institute of Standards and Technology (NIST). This online resource helps users to produce correct results with PRISM theory and easily contribute, create, and modify closures, numerical solution algorithms, and analyses for other users. Along with the details of this package, we published a detailed pedagogical review [58] describing

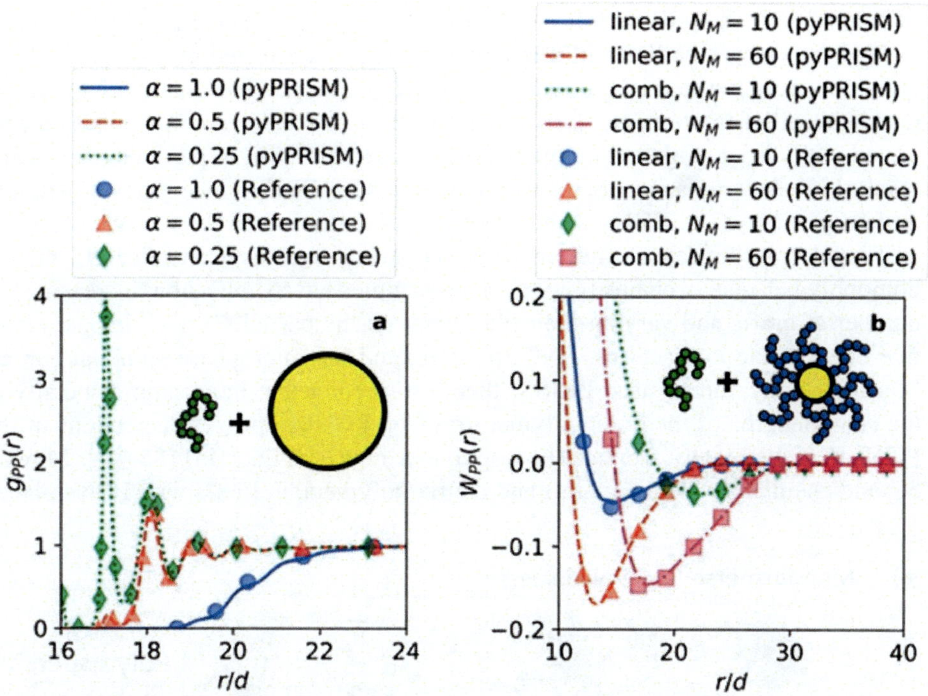

Fig. 1.3 **a** Particle–particle site–site pair correlations, $g_{PP}(r)$, versus reduced separation distance, r/d, in a PNC at a total packing fraction, $\eta = 0.4$, with a freely jointed chain matrix of length, $N = 100$, with attractive polymer–particle interactions at an attraction strength, $\varepsilon = 1.0$, and with varying interaction range, α, as described in Fig. 1.3 of [49]. **b** Particle–particle potential of mean force, $w_{PP}(r)$, versus reduced separation distance, r/d, between polymer grafted particles in a linear polymer matrix at a total packing fraction of $\eta = 0.35$ for varying graft architecture and matrix length, N_M, as discussed in Fig. 9 of [84]. The lines are the predictions from pyPRISM and the symbols are data extracted from the corresponding referenced literature. In both subplots, $d = 1.0$ is the characteristic length scale of the system equal to the monomer site diameter. The details of how these data are generated using pyPRISM can be found in the companion pyPRISM tutorial [83]. Caption and figure were reprinted with permission from [58] (2018) American Chemical Society

how PRISM theory has been applied to a large range of soft matter studies along with results for a few case studies, including PNCs. The results of the PNC case studies calculated using *pyPRISM* were compared to the original studies (see Fig. 1.3), some of which we describe next.

1.4 Past, Present, and Future of PRISM Theory-Based PNC Studies

PRISM theory has been a valuable tool for predicting of how PNC design parameters impact the PNC morphology and phase behavior. Early studies by Hooper and Schweizer as well as Hall and Schweizer have established how the structure of PNCs with bare particles in free polymer matrix is affected by particle–particle attraction, polymer–particle attraction, particle size to polymer matrix chain length ratio, PNC filler fraction, etc. [49–54]. Some of these results have also been corroborated with direct structural comparisons with experiments, in particular scattering results [52]. Jayaraman and Schweizer have extended PRISM theory to study lightly grafted particles (i.e., PGNs with homopolymers tethered on the particles at a low grafting density) in dense solutions without matrix as well as in chemically identical homopolymer matrix chains [43–46]. These studies have shown how increasing the number of grafts and varying their placement on the particle surface impact effective interparticle interactions, PNC structure, and the microphase spinodal curves. Martin and Jayaraman used PRISM theory to predict that increasing dispersity in the chain lengths of the homopolymer grafts on PGNs improves dispersions of the PGNs in a chemically identical homopolymer matrix (Fig. 1.4a) [85–87]. Moving beyond chemically identical graft and matrix polymers, in PNCs with PGNs placed

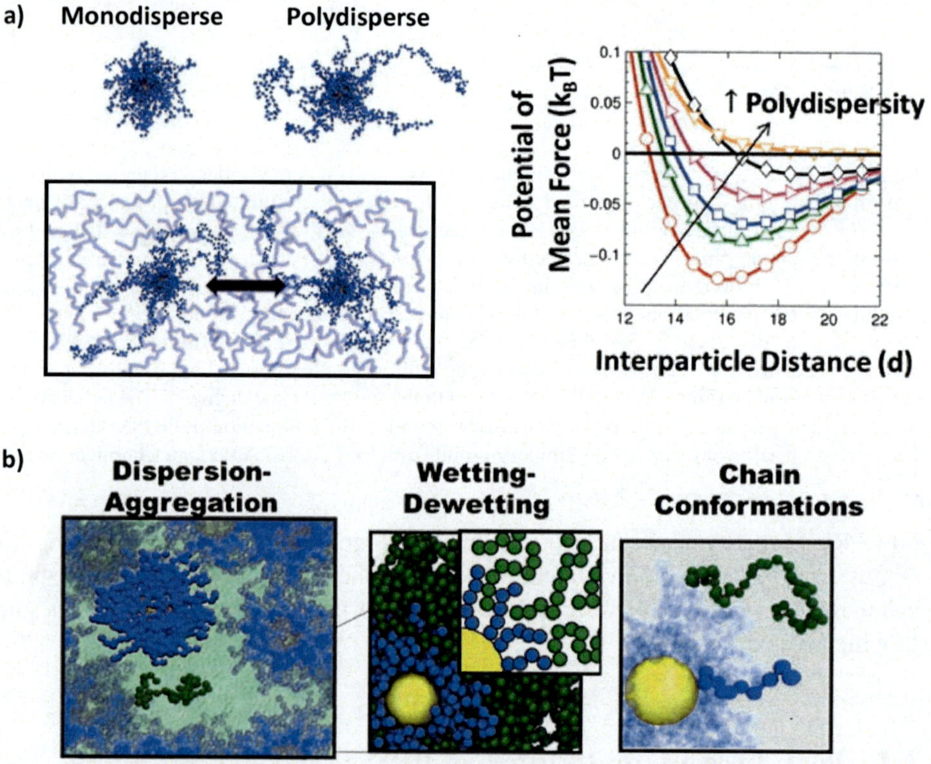

Fig. 1.4 **a** PGNs with disperse polymer grafts increase the particle–particle repulsion in a polymer nanocomposite. **b** PNCs with chemically distinct graft and matrix polymers exhibit gradual change in wetting with temperature in contrast to a first-order particle dispersion to aggregation transition. Figures in part a are reproduced from [85] with permission from Royal Society of Chemistry. Figure in part b has been reprinted with permission from [25] (2016) American Chemical Society

in a chemically dissimilar polymer matrix (Fig. 1.4b), PRISM theory along with molecular simulations has shown that the grafted layer wetting–dewetting transition is a second-order transition and distinct from the PGN dispersion–aggregation (first-order) transition [23–25]. Calculation of the effective χ parameter and second virial coefficient underlying these wetting–dewetting and dispersion–aggregation transitions provided additional insight into the dispersion–aggregation transition occurring when the extent of grafted layer wetting matched that of the analogous athermal PNC.

The above PRISM theory-based studies have focused on linear polymers, both in graft and matrix chains, and polymer chemistries that can be modeled using isotropic intersite interaction potentials. One can expect to observe unique phase behavior, not seen previously, by moving toward nonlinear polymer architectures (e.g., bottlebrushes, comb/graft polymers, cyclic/ring polymers) and/or polymer chemistries with directional and specific interactions (e.g., hydrogen bonding, pi-pi stacking).

Given the many advances in polymer synthesis and techniques for grafting polymers to or from particle surfaces, there is a strong motivation to understand potential impacts of nonlinear polymer architecture in graft and/or matrix chains on PNC structure and dynamics. For example, bottlebrush polymer is one specific class of nonlinear architectures that has received significant attention in many recent synthetic, computational, and theoretical studies [88–94]. Bottlebrush polymers are comprised of a linear backbone with side chains emanating from the backbone. These polymers can be synthesized in a controlled manner with varying side chain chemistry, molecular weight, and grafting density along the backbone. In the context of PNCs, the complexity brought about by grafting such bottlebrush polymers onto particle surface is worth exploring. The increased crowding between the sidechains on the bottlebrush and between the bottlebrushes on the nanoparticle surface should significantly increase the conformational entropy loss of the (linear or nonlinear) matrix polymer chain penetrating the grafted layer. By dialing down the grafting density on the particle surface, one may approach the same wetting–dewetting and/or dispersion–aggregation physics as the densely grafted linear polymer containing PGNs. Two PRISM theory-based studies have tackled some of the above questions. One study by Modica et al. using PRISM theory and Langevin dynamics compared and contrasted comb polymers (i.e., bottlebrush with low grafting density along the polymer backbone) grafts versus linear grafts in PNCs with linear matrix chains [84]. They found that compared to linear polymers, the comb polymers exhibit a stronger effective attraction between the grafted particles in both small molecule solvent and polymer matrix due to the increased crowding in the grafted layer from the comb polymer side chains. They also found that the effective interactions (i.e., PMF) between the grafted particles are more sensitive to the comb polymer design (i.e., side chain length and spacing) in a small molecule solvent than in a *polymer* matrix. Using nonlinear polymer architecture in the matrix can also alter the tendency for particle dispersion/aggregation as compared linear polymer matrix. In another PRISM theory study [95], the use of comb polymer architecture in the matrix leads to reduced (bare) nanoparticle aggregation for moderate nanoparticle–monomer attraction strength and increased bridging of nanoparticles at relatively large nanoparticle–monomer attraction. These effects brought about by using comb polymer architecture versus linear polymer architecture in the matrix can be subdued/enhanced through the choice of comb polymer grafting density, side chain to backbone length ratio, etc. Gradients in sidechain length and crowding along bottlebrushes could be exploited, say with increasing crowding from the tethering point outward to create polymer grafted particles with uniform monomer density radially. Alternately, gradient brushes could be designed to engineer layers of varying monomer densities in the grafted layer. These are potential new topics for PRISM theory-based studies.

The subject of how associating polymer chemistries (e.g., hydrogen bonding or π-π stacking between graft and matrix polymers) alter the grafted chain conformation, matrix-graft interpenetration, grafted layer structure, effective interactions, PNC morphology has received far less attention than other simpler polymer chemistries modeled using isotropic interactions. These directional interactions have been studied in detail in polymer blends (i.e., mixture of homopolymers or copolymers with hydrogen bonding chemistries) where hydrogen bonding has been shown to alter the structure (e.g., conformations associated with intrachain vs. interchain hydrogen bonding) and chain dynamics. Past experimental studies [96–99] have shown that polymer blends with intermolecular hydrogen bonds are characterized by a single glass transition temperature indicating miscibility between blend components. Hydrogen bonded polymer blends show coupling of segmental relaxations (leading to a single alpha relaxation) in blend components and also suppressed concentration fluctuations [100]. The inter- and intra-associations between hydrogen bonding moieties have also been shown to affect the dynamic moduli (G' and G") in the terminal regime [101]. The potential of similar striking effects in PNCs motivates investigations in this avenue. Motivated by these experiments, Kulshreshtha et al. developed new coarse-grained models to study PNCs with a single polymer grafted nanoparticle with (linear) graft polymers placed in a linear polymer matrix with graft and matrix polymers having directional and specific interactions [102]. Interestingly, even if one observed equivalent grafted layer wetting in PNCs with isotropic interactions and PNCs with directional graft–matrix interactions, there were some major differences in the way graft and matrix chains interacted. Each graft polymer in the PNC with directional graft–matrix interactions interacts with fewer matrix chains and has much lower free volume than in the PNCs with isotropic graft–matrix interactions. This difference could have major implications on the particle–particle effective interactions, glass transition temperature of the PNC, as well as the mechanical and rheological properties for these PNCs. Incorporation of this new coarse-grained model [102] within the PRISM theory framework and studies of PNCs with directional and specific interactions between the grafts and matrix polymers at finite filler fractions is another new direction for a PRISM theory-based study.

1.5 Limitations of PRISM Theory

Despite the many successful applications of PRISM theory for PNC studies, it is important to highlight some limitations associated with the use of PRISM theory especially for studying PNCs.

The first limitation is the strong dependence of PRISM theory's results on the closures chosen. For example, past work has shown that by using *atomic* closures to describe the phase behavior of polymer blends results leads to an incorrect scaling dependence of the spinodal temperature on the polymer chain length [103]. To produce the correct scaling, one could use thermodynamic perturbation theory [104] or *molecular closures* [47, 105]. These issues are also relevant for PNCs with a mixed polymer blend as matrix or with chemically dissimilar graft and matrix polymers. As stated above in Sect. 1.3.3, the *pyPRISM* package allows users to assess the effect of the closure itself on the results by making an easier user-interface to switch between closures and/or implement new closures.

PRISM theory's foundation as a liquid-state theory implies that it is only applicable for systems with *liquid-like* structural correlations [106]. One could interpret this as an inability of PRISM theory to directly predict correlations for the glassy or crystalline states of polymer matrix or grafted layer in the PNC. Also, PRISM theory cannot directly predict the structure of an ordered, macrophase-separated, or microphase-separated phases in PNCs. However, modern thermodynamic density functional theories that can be used to study spatially inhomogeneous systems need liquid-state correlations as input, and PRISM theory could be useful for this task [107–110]. Furthermore, PRISM theory study of thin films of PNCs or PNCs near surfaces/interfaces is not as straightforward as studies of bulk PNCs (i.e., away from surface effects) and needs a fairly complex treatment (see, for example, [111]).

Another major challenge is simply (numerically) solving the PRISM theory equations. In particular, the choice of the algorithm, (e.g., Picard, Inexact Newton), the chosen initial guess of the solution, and the convergence criteria for the iterations within the numerical solution method are among the many considerations that impact convergence of the PRISM theory equations to a solution. These choices are not trivial and require multiple tests before a lack of solution for the given system, can be attributed to the system being structurally inhomogeneous rather than a numerical issue.

Lastly, unlike stochastic and deterministic molecular simulation methods like molecular dynamics or MC, PRISM theory does not produce a trajectory of coordinates of the 'sites' in the system. The output from PRISM theory is purely in the form of pair correlations in real/inverse space, and inferences about the structure and phase behavior come from these pair correlations. For computational or experimental users of PRISM theory, the lack of a visual interpretation of their results (e.g., in the form of simulation snapshots) is often disappointing. Interestingly, self-consistent PRISM-simulation methods *partially* mitigate this problem by providing trajectories of single molecules in a mean-field, medium-induced solvation potential. Lastly, unlike molecular dynamics, PRISM theory cannot provide any information about the *dynamics* within the PNCs.

1.6 Conclusions

This chapter presented the basics of PRISM theory and the key considerations that one needs to make in order to obtain meaningful results from PRISM theory studies of PNC structure and phase behavior. We presented a review of past studies on PNCs using experiments, theory, and simulations to describe the main fundamental questions that have been posed and tackled in this field. PRISM theory's successes were also highlighted to demonstrate the many valuable research contributions to the PNC literature made possible by use of PRISM theory. We also briefly mentioned *pyPRISM,* an open-source package developed by Martin, Jayaraman, and coworkers and hosted by NIST to encourage readers to use PRISM theory without concerning themselves with the tedious computational implementation. We concluded the chapter with a few key limitations of PRISM theory to caution the readers of the potential challenges they could encounter when using PRISM theory for studying PNCs.

References

1. R.A. Vaia, E.P. Giannelis, Polymer nanocomposites: status and opportunities. MRS Bull. **26**(5), 394–401 (2001)
2. J. Jordan et al., Experimental trends in polymer nanocomposites—a review. Mater. Sci. Eng. A **393**(1–2), 1–11 (2005)
3. A.J. Crosby, J.Y. Lee, Polymer nanocomposites: the "nano" effect on mechanical properties. Polym. Rev. **47**(2), 217–229 (2007)
4. R. Krishnamoorti, R.A. Vaia, Polymer nanocomposites. J. Polym. Sci. Part B: Polym. Phys. **45**(24), 3252–3256 (2007)
5. K.I. Winey, R.A. Vaia, Polymer nanocomposites. MRS Bull. **32**(4), 314–322 (2007)
6. V. Ganesan, C.J. Ellison, V. Pryamitsyn, Mean-field models of structure and dispersion of polymer-nanoparticle mixtures. Soft Matter **6**(17), 4010–4025 (2010)
7. L.M. Hall, A. Jayaraman, K.S. Schweizer, Molecular theories of polymer nanocomposites. Curr. Opinion Solid State Mater. Sci. **14**(2), 38–48 (2010)
8. J. Jancar et al., Current issues in research on structure–property relationships in polymer nanocomposites. Polymer **51**(15), 3321–3343 (2010)
9. S.K. Kumar, R. Krishnamoorti, Nanocomposites: structure, phase behavior, and properties. Ann. Rev. Chem. Biomol. Eng. **1**, 37–58 (2010)
10. P.F. Green, The structure of chain end-grafted nanoparticle/homopolymer nanocomposites. Soft Matter **7**(18), 7914–7926 (2011)
11. A. Jayaraman, Polymer grafted nanoparticles: effect of chemical and physical heterogeneity in polymer grafts on particle assembly and dispersion. J. Polym. Sci. Part B—Polym. Phys. **51**(7), 524–534 (2013)
12. S.K. Kumar et al., Nanocomposites with polymer grafted nanoparticles. Macromolecules **46**(9), 3199–3214 (2013)

13. V. Ganesan, A. Jayaraman, Theory and simulation studies of effective interactions, phase behavior and morphology in polymer nanocomposites. Soft Matter **10**(1), 13–38 (2014)
14. S. Srivastava et al., 25th anniversary article: polymer–particle composites: phase stability and applications in electrochemical energy storage. Adv. Mater. **26**(2), 201–234 (2014)
15. V.V. Ginzburg, *In: Problems of Nonlinear Mechanics and Physics of Materials* Springer, (2019)
16. M.J. Hore, Polymers on nanoparticles: structure & dynamics. Soft Matter **15**(6), 1120–1134 (2019)
17. S.K. Kumar et al., 50th anniversary perspective: are polymer nanocomposites practical for applications? Macromolecules **50**(3), 714–731 (2017)
18. J.M. Garcés et al., Polymeric nanocomposites for automotive applications. Adv. Mater. **12**(23), 1835–1839 (2000)
19. V.T. Rathod, J.S. Kumar, A. Jain, Polymer and ceramic nanocomposites for aerospace applications. Appl. Nanosci. **7**(8), 519–548 (2017)
20. J.W. Barnett, S.K. Kumar, Modeling gas transport in polymer-grafted nanoparticle membranes. Soft Matter **15**(3), 424–432 (2019)
21. S.R. Silva, et al., *Hybrid and Nanocomposite Materials for Flexible Organic Electronics Applications* (2015), pp. 57–84
22. I. Armentano et al., Biodegradable polymer matrix nanocomposites for tissue engineering: a review. Polym. Degrad. Stab. **95**(11), 2126–2146 (2010)
23. T.B. Martin, A. Jayaraman, Tuning the wetting–dewetting and dispersion–aggregation transitions in polymer nanocomposites using composition of graft and matrix polymers. Mater. Res. Express **3**(3), 034001 (2016)
24. T.B. Martin, A. Jayaraman, Using theory and simulations to calculate effective interactions in polymer nanocomposites with polymer-grafted nanoparticles. Macromolecules **49**(24), 9684–9692 (2016)
25. T.B. Martin et al., Wetting-dewetting and dispersion aggregation transitions are distinct for polymer grafted nanoparticles in chemically dissimilar polymer matrix. J. Am. Chem. Soc. **137**(33), 10624–10631 (2015)
26. N.J. Fernandes et al., Hairy nanoparticle assemblies as one-component functional polymer nanocomposites: opportunities and challenges. MRS Commun. **3**(1), 13–29 (2013)
27. G. Hou et al., Designing superlattice structure via self-assembly of one-component polymer-grafted nanoparticles. J. Phys. Chem. B **123**(9), 2157–2168 (2019)
28. C.R. Bilchak et al., High-frequency mechanical behavior of pure polymer-grafted nanoparticle constructs. ACS Macro Lett. **8**(3), 294–298 (2019)
29. E. Buenning et al., Location of imbibed solvent in polymer-grafted nanoparticle membranes. ACS Macro Lett. **7**(9), 1051–1055 (2018)
30. C.R. Bilchak et al., Polymer-grafted nanoparticle membranes with controllable free volume. Macromolecules **50**(18), 7111–7120 (2017)
31. Y. Huang et al., Matrix-free polymer nanocomposite thermoplastic elastomers. Macromolecules **50**(12), 4742–4753 (2017)
32. J. Kim, P.F. Green, Phase behavior of thin film brush-coated nanoparticles/homopolymer mixtures. Macromolecules **43**(3), 1524–1529 (2010)
33. R. Zhang et al., Entropy-driven segregation of polymer-grafted nanoparticles under confinement. Proc. Natl. Acad. Sci. **114**(10), 2462 (2017)
34. T. Lafitte, S.K. Kumar, A.Z. Panagiotopoulos, Self-assembly of polymer-grafted nanoparticles in thin films. Soft Matter **10**(5), 786–794 (2014)
35. K.S. Schweizer, J.G. Curro, Integral-equation theory of the structure of polymer melts. Phys. Rev. Lett. **58**(3), 246–249 (1987)
36. J.G. Curro, K.S. Schweizer, Theory of polymer melts—an integral-equation approach. Macromolecules **20**(8), 1928–1934 (1987)
37. C.S. Hsu, D. Chandler, RISM calculation of the structure of liquid chloroform. Mol. Phys. **37**(1), 299–301 (1979)

38. C.S. Hsu, D. Chandler, RISM calculation of structure of liquid acetonitrile. Mol. Phys. **36**(1), 215–224 (1978)
39. C.S. Hsu, D. Chandler, L.J. Lowden, Applications of RISM equation to diatomic fluids—liquids nitrogen, oxygen and bromine. Chem. Phys. **14**(2), 213–228 (1976)
40. L.J. Lowden, D. Chandler, Solution of a new integral-equation for pair correlation-functions in molecular liquids. J. Chem. Phys. **59**(12), 6587–6595 (1973)
41. L.S. Ornstein, F. Zernike, Acculental deviations of density and opalescence at the critical point of a simple substance. Proc. Koninklijke Akademie Van Wetenschappen Te Amsterdam **17**, 793–806 (1914)
42. K.S. Schweizer, J.G. Curro, Integral-equation theory of polymer melts—intramolecular structure, local order, and the correlation hole. Macromolecules **21**(10), 3070–3081 (1988)
43. A. Jayaraman, K.S. Schweizer, Effective interactions and self-assembly of hybrid polymer grafted nanoparticles in a homopolymer matrix. Macromolecules **42**(21), 8423–8434 (2009)
44. A. Jayaraman, K.S. Schweizer, Effective interactions, structure, and phase behavior of lightly tethered nanoparticles in polymer melts. Macromolecules **41**(23), 9430–9438 (2008)
45. A. Jayaraman, K.S. Schweizer, Effect of the number and placement of polymer tethers on the structure of concentrated solutions and melts of hybrid nanoparticles. Langmuir **24**(19), 11119–11130 (2008)
46. A. Jayaraman, K.S. Schweizer, Structure and assembly of dense solutions and melts of single tethered nanoparticles. J. Chem. Phys. **128**(16) (2008)
47. K.S. Schweizer, A. Yethiraj, Polymer reference interaction site model theory: new molecular closures for phase separating fluids and alloys. J. Chem. Phys. **98**, 9053–9079 (1993)
48. A. Yethiraj, K.S. Schweizer, Self-consistent polymer integral-equation theory—comparisons with monte-carlo simulations and alternative closure approximations. J. Chem. Phys. **97**(2), 1455–1464 (1992)
49. J.B. Hooper, K.S. Schweizer, Contact aggregation, bridging, and steric stabilization in dense polymer-particle mixtures. Macromolecules **38**, 8858–8869 (2005)
50. J.B. Hooper, K.S. Schweizer, Theory of phase separation in polymer nanocomposites. Macromolecules **39**, 5133–5142 (2006)
51. J.B. Hooper et al., Structure, surface excess and effective interactions in polymer nanocomposite melts and concentrated solutions. J. Chem. Phys. **121**(14), 6986–6997 (2004)
52. L.M. Hall et al., Concentration fluctuations, local order, and the collective structure of polymer nanocomposites. Macromolecules **42**(21), 8435–8442 (2009)
53. L.M. Hall, K.S. Schweizer, Many body effects on the phase separation and structure of dense polymer-particle melts. J. Chem. Phys. **128**, 234901 (2008)
54. L.M. Hall, K.S. Schweizer, Structure, scattering patterns and phase behavior of polymer nanocomposites with nonspherical fillers. Soft Matter **6**(5), 1015–1025 (2010)
55. K.S. Schweizer, Analytic RISM theory of polymer alloys—molecular closure predictions for structurally symmetrical blends. Macromolecules **26**(22), 6033–6049 (1993)
56. A. Yethiraj, K.S. Schweizer, Integral equation theory of polymer blends: numerical investigation of molecular closure approximations. J. Chem. Phys. **98**, 9080–9093 (1993)
57. I. Collins, W. George, *Fundamental Numerical Methods and Data Analysis* (Harvard University Press, 2003)
58. T.B. Martin et al., pyPRISM: a computational tool for liquid-state theory calculations of macromolecular materials. Macromolecules **51**(8), 2906–2922 (2018)
59. A.C. Hindmarsh et al., SUNDIALS: suite of nonlinear and differential/algebraic equation solvers. ACM Trans. Math. Softw. **31**(3), 363–396 (2005)
60. M. Kotelyanskii, D.N. Theodorou, *Simulation Methods for Polymers* (Marcel Dekker, New York, 2004)
61. A. Karatrantos, N. Clarke, M. Kroger, Modeling of polymer structure and conformations in polymer nanocomposites from atomistic to mesoscale: a review. Polym. Rev. **56**(3), 385–428 (2016)
62. A. Gooneie, S. Schuschnigg, C. Holzer, A review of multiscale computational methods in polymeric materials. Polymers **9**(1), 16 (2017)

63. G.G. Vogiatzis, D.N. Theodorou, Multiscale molecular simulations of polymer-matrix nanocomposites: or what molecular simulations have taught us about the fascinating nanoworld. Arch. Comput. Methods Eng.: State Art Rev. **25**(3), 591–645 (2018)
64. T.E. Gartner, A. Jayaraman, Modeling and simulations of polymers: a roadmap. Macromolecules **52**(3), 755–786 (2019)
65. J.D. McCoy et al., Single-chain structure in model polyethylene melts. Macromolecules **25**(19), 4905–4910 (1992)
66. P.J. Flory, *Statistical Mechanics of Chain Molecules* (John Wiley & Sons, Inc, 1969)
67. K.S. Schweizer, J.G. Curro, Integral equation theory of polymer melts: intramolecular structurem local order, and the correlation hole. Macromolecules **21**, 3070–3081 (1988)
68. J. Melenkevitz, K.S. Schweizer, J.G. Curro, Self consistent integral equation theory for the equilibrium properties of polymer solutions. Macromolecules **26**, 6190–6196 (1993)
69. C.J. Grayce, K.S. Schweizer, Solvation potentials for macromolecules. J. Chem. Phys. **100**, 6846–6856 (1994)
70. C.J. Grayce, A. Yethiraj, K.S. Schweizer, Liquid state theory of the density dependent conformation of nonpolar linear polymers. J. Chem. Phys. **100**, 6857–6872 (1994)
71. P.G. Khalatur, A.R. Khokhlov, Hybrid MC/RISM technique for simulation of polymer solutions: MC + RISM integral eqns. Mol. Phys. **93**, 555–572 (1998)
72. K.S. Schweizer, J.G. Curro, PRISM theory of the structure, thermodynamics, and phase-transitions of polymer liquids and alloys. Adv. Polym. Sci. **116**, 319–377 (1994)
73. N. Nair, A. Jayaraman, Self-consistent PRISM theory—Monte Carlo simulation studies of copolymer grafted nanoparticles in a homopolymer matrix. Macromolecules **43**(19), 8251–8263 (2010)
74. S. Mendez et al., An integral equation theory for polymer solutions: explicit inclusion of the solvent molecules. J. Chem. Phys. **115**, 5669–5678 (2001)
75. M. Putz, J.G. Curro, G.S. Grest, Self-consistent integral equation theory for polyolefins: comparison to molecular dynamics simulations and x-ray scattering. J. Chem. Phys. **114**, 2847–2860 (2001)
76. A. Yethiraj, C.K. Hall, Monte Carlo simulations and integral equation theory for microscopic correlations in polymeric fluids. J. Chem. Phys. **96**, 797–807 (1991)
77. A.L. Frischknecht, E.S. McGarrity, M.E. Mackay, Expanded chain dimensions in polymer melts with nanoparticle fillers. J. Chem. Phys. **132**, 204901 (2010)
78. P.G. Khalatur, L.V. Zherenkova, A.R. Khokhlov, Entropy-driven polymer collapse: application of the hybrid MC/RISM method to the study of conformational transitions in macromolecules interacting with hard colloidal particles. Eur. Phys. J. B **5**(4), 881–897 (1998)
79. D.R. Heine, G.S. Grest, J.G. Curro, Structure of polymer melts and blends: comparison of integral equation theory and computer simulations. Adv. Comput. Simul. Approaches Soft Matter Sci. **I**(173), 209–249 (2005)
80. D. Heine et al., Role of intramolecular energy on polyolefin miscibility: Isotactic polypropylene/polyethylene blends. J. Chem. Phys. **118**(2), 914–924 (2003)
81. T.B. Martin, *typyPRISM Tutorial Package*. Available from: http://github.com/usnistgov/typyPRISM_tutorial
82. T.B. Martin, *typyPRISM Package*. Available from: http://github.com/usnistgov/typyPRISM
83. T.B. Martin, T.E. Gartner III, *pyPRISM Tutorial*. Available from: http://pyprism.readthedocs.io/en/latest/tutorial/tutorial.html
84. K.J. Modica, T.B. Martin, A. Jayaraman, Effect of polymer architecture on the structure and interactions of polymer grafted particles: theory and simulations. Macromolecules **50**(12), 4854–4866 (2017)
85. T.B. Martin, A. Jayaraman, Polydisperse homopolymer grafts stabilize dispersions of nanoparticles in a chemically identical homopolymer matrix: an integrated theory and simulation study. Soft Matter **9**(29), 6876–6889 (2013)
86. T.B. Martin, A. Jayaraman, Identifying the ideal characteristics of the grafted polymer chain length distribution for maximizing dispersion of polymer grafted nanoparticles in a polymer matrix. Macromolecules **46**(22), 9144–9150 (2013)

87. T.B. Martin, P.M. Dodd, A. Jayaraman, Polydispersity for tuning the potential of mean force between polymer grafted nanoparticles in a polymer matrix. Phys. Rev. Lett. **110**(1), 018301 (2013)
88. J.N. Albert, T.H. Epps III, Self-assembly of block copolymer thin films. Mater. Today **13**(6), 24–33 (2010)
89. P.M. Claesson et al., Bottle-brush polymers: adsorption at surfaces and interactions with surfactants. Adv. Coll. Interface. Sci. **155**(1–2), 50–57 (2010)
90. S. Peleshanko, V.V. Tsukruk, The architectures and surface behavior of highly branched molecules. Prog. Polym. Sci. **33**(5), 523–580 (2008)
91. R. Verduzco et al., Structure, function, self-assembly, and applications of bottlebrush copolymers. Chem. Soc. Rev. **44**, 2405–2420 (2015)
92. G. Xie et al., Molecular bottlebrushes as novel materials. Biomacromol **20**(1), 27–54 (2018)
93. G.E. Stein, T.S. Laws, R. Verduzco, Tailoring the attraction of polymers toward surfaces. Macromolecules **52**(13), 4787–4802 (2019)
94. D.J. Walsh, D. Guironnet, Macromolecules with programmable shape, size, and chemistry. Proc. Natl. Acad. Sci. **116**(5), 1538–1542 (2019)
95. Q. Xu et al., Structure and effective interactions of comb polymer nanocomposite melts. J. Chem. Phys. **141**(20), 204901 (2014)
96. E.J. Moskala et al., On the role of intermolecular hydrogen bonding in miscible polymer blends. Macromolecules **17**(9), 1671–1678 (1984)
97. S. Viswanathan, M.D. Dadmun, Guidelines to creating a true molecular composite: inducing miscibility in blends by optimizing intermolecular hydrogen bonding. Macromolecules **35**(13), 5049–5060 (2002)
98. S.-W. Kuo, C.-L. Lin, F.-C. Chang, The study of hydrogen bonding and miscibility in poly (vinylpyridines) with phenolic resin. Polymer **43**(14), 3943–3949 (2002)
99. Y. He, B. Zhu, Y. Inoue, Hydrogen bonds in polymer blends. Prog. Polym. Sci. **29**(10), 1021–1051 (2004)
100. S. Zhang, P.C. Painter, J. Runt, Coupling of component segmental relaxations in a polymer blend containing intermolecular hydrogen bonds. Macromolecules **35**(25), 9403–9413 (2002)
101. Z. Yang, C.D. Han, Rheology of miscible polymer blends with hydrogen bonding. Macromolecules **41**(6), 2104–2118 (2008)
102. A. Kulshreshtha, K.J. Modica, A. Jayaraman, Impact of hydrogen bonding interactions on graft-matrix wetting and structure in polymer nanocomposites. Macromolecules **52**(7), 2725–2735 (2019)
103. H.P. Deutsch, K. Binder, Evidence against the integral-equation theory of polymer blends. Europhys. Lett. **17**(8BIS), 697–702 (1992)
104. J.J. Rajasekaran, J.G. Curro, J.D. Honeycutt, Theory for the phase-behavior of polyolefin blends—application to the polyethylene/isotactic polypropylene blend. Macromolecules **28**(20), 6843–6853 (1995)
105. A. Yethiraj, K.S. Schweizer, Integral-equation theory of polymer blends—numerical investigation of molecular closure approximations. J. Chem. Phys. **98**(11), 9080–9093 (1993)
106. J.-P. Hansen, I.R. McDonald, Chapter 4—distribution function theories, *Theory of Simple Liquids*, 4th edn. (Academic Press, Oxford, 2013), pp. 105–147
107. C.N. Patra, A. Yethiraj, Density functional theory for nonuniform polymers: accurate treatment of the effect of attractive interactions. J. Chem. Phys. **118**(10), 4702–4706 (2003)
108. A.L. Frischknecht, et al., Density functional theory for inhomogeneous polymer systems. I. Numerical methods. J. Chem. Phys. **117**(22), 10385–10397 (2002)
109. A.L. Frischknecht, J.G. Curro, L.J.D. Frink, Density functional theory for inhomogeneous polymer systems. II. Application to block copolymer thin films. J. Chem. Phys. **117**(22), 10398–10411 (2002)
110. J.P. Donley et al., Microscopic approach to inhomogeneous polymeric liquids. J. Chem. Phys. **103**(12), 5061–5069 (1995)
111. J.G. Curro et al., The structure of amorphous polymers near surfaces: athermal systems. Comput. Theor. Polym. Sci. **8**(1), 159–168 (1998)

Chapter 2
Density Functional Theory-Based Modeling of Polymer Nanocomposites

Valeriy V. Ginzburg

Abstract Density functional theory (DFT) is a powerful approach utilized successfully in both quantum and classical theoretical and computational physics. Since the 1970s, DFT has been applied to predict the phase behavior of simple fluids, including the liquid-to-crystal transition in hard-sphere and Lennard–Jones fluids. Beginning in the 1990s, it was recognized that DFT can be adapted to describe the equilibrium morphologies of polymer-based nanocomposites (PNC). Here, we review various examples where DFT is applied to PNCs, from polymer–clay and polymer-nanotube mixtures to one-component hairy nanoparticle assemblies. We also discuss hybrid approaches where DFT is combined with other coarse-grained field theories, in particular, the Self-Consistent Field/Density Functional Theory (SCF-DFT) method and its applications.

2.1 Introduction

The performance of polymer nanocomposites depends crucially on the "morphology" or "nanostructure". The nanofillers can be organized into crystal-like patterns or dispersed randomly within a polymer matrix or aggregated into large clusters. The polymers can also form a homogeneous phase or, alternatively, undergo a micro- or macro- phase separation. The overall nanocomposite morphology can also be impacted by confinement (e.g., thin films vs. bulk). Understanding and predicting the morphology as a function of the nanocomposite formulation, filler type and geometry, and processing conditions is the biggest challenge in expanding the use of nanocomposite materials to new application areas. Multiple approaches are utilized to address this challenge [1–22]. In this chapter, we mainly concentrate on one approach to predicting the equilibrium nanocomposite morphology—the coarse-grained Density Functional Theory (DFT) based modeling. While DFT—like other equilibrium models—does not describe the dynamics of ordering, phase separation,

V. V. Ginzburg (✉)
Core R&D, The Dow Chemical Company, Midland, MI 48674, USA
e-mail: vvg851966@gmail.com

V. V. Ginzburg and L. M. Hall (eds.), *Theory and Modeling of Polymer Nanocomposites*, Springer Series in Materials Science 310,
https://doi.org/10.1007/978-3-030-60443-1_2

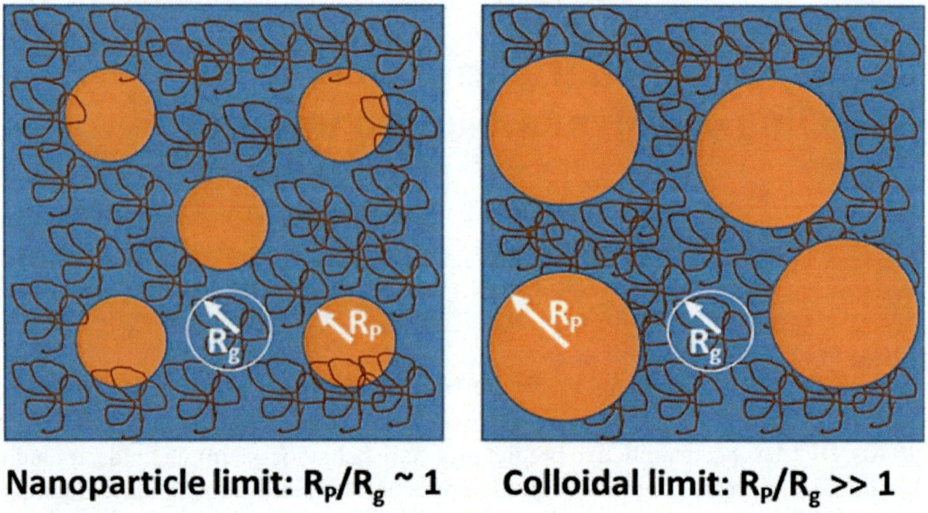

Fig. 2.1 Two classes of polymer nanocomposites

or alignment, it can predict the morphologies that are the most advantageous in a sense that they correspond to the free energy minima, either global or local.

Before discussing DFT in-depth, we would like to introduce a "mini-classification", dividing all nanocomposites into two classes (Fig. 2.1). In the first class, "Nanoparticle limit", the polymer chains and the nanofillers have comparable dimensions; in the second class, "Colloidal limit", the nanofillers are significantly larger than the polymers (at least in one dimension).

The implication of this distinction, in our view, is as follows. For the first class of nanocomposites, the equilibration occurs simultaneously for both polymers and nanofillers. Thus, one needs to construct the free energy as a functional of the single-particle distribution functions ("densities") of both polymers and fillers, and then minimize it appropriately to obtain the equilibrium morphologies. For the second class, however, the equilibration of the nanofillers is much slower than the equilibration of the polymers. Thus, it is necessary to construct the free energy for the polymers and minimize it subject to the "frozen" positions of the fillers; usually, only two filler particles are used for this initial minimization. The positions of the two particles are then updated, and the polymer-free energy is calculated for each inter-particle separation. The result of this process is the so-called potential of mean force (PMF) for the particles due to their interactions with the polymers. Once the PMF is known, the particles are equilibrated next. Thus, the approach used to model the morphology and phase behavior of nanocomposites depends crucially on the nanoparticle size. It is clear that spherical nanoparticles with $R_P < 2$ nm belongs to the first class, while most nanorods, nanoplatelets, and nanospheres with $R_P > 10$ nm belong to the second class. However, for nanospheres with 2 nm $< R_P <$ 10 nm, the classification might be more ambiguous, especially if the fillers and polymers are polydisperse.

The rest of the chapter is structured as follows. In Sect. 2.2, we review the history of the density functional theory and its applications to polymers, liquid crystals, and colloids. In Sect. 2.3, the applications of the DFT to polymer nanocomposites are discussed. In Sect. 2.4, we describe a special "hybrid" approach to nanocomposite modeling, combining the "standard" DFT for spherical particles with the polymer self-consistent field theory (SCFT). Lastly, Sect. 2.5 contains the summary and outlook for future theory developments.

2.2 DFT Overview

Density functional theory originated in the 1960s with the pioneering work of Hohenberg and Kohn [23] and Kohn and Sham [24], outlining a radically new method of calculating electronic structures of complex atoms. This work was subsequently recognized by 1998 Nobel Prize in Chemistry awarded to Walter Kohn [25]. The main idea of the DFT approach is that the energy and other characteristics of a multiparticle system can be fully described by a functional of a single-particle density, provided that one correctly captures correlations and "exchange" interactions. The triumph of DFT in quantum chemistry also helped promote its applications in classical condensed matter. Ramakrishnan and Yussouff [26] first formulated a classical DFT-like model to describe the first-order liquid–solid transition ("freezing") in simple liquids. To parameterize the non-ideal portion of the free energy functional, several approaches (local density approximation [LDA] [27, 28], weighted density approximation [WDA] [29–33], and fundamental measure theory [FMT] [34–39]) have been developed and refined over the course of the last thirty years (see reviews [40, 41] for a more detailed discussion). The original analysis centered on the hard-sphere and Lennard–Jones sphere freezing; subsequently, the method was also extended to describe phase transitions in anisotropic, liquid-crystalline particles [42–50]. Equally momentous, in the 1980s and 1990s DFT was extended to describe flexible polymer chains [51–62]; for more details on those polymer DFTs, see reviews [63, 64]. Example applications of polymer DFT include bulk crystallization [65–67] and ordering near hard surfaces [59, 68–71].

The general idea of DFT in the context of classical fluids is as follows. Let us consider, as an example, a multicomponent polymer melt. For each monomer type, labeled α, one can define the single-particle density, $\rho_\alpha(\mathbf{r})$, and write down the free energy of the overall system in the following form (here, we use the grand-canonical ensemble, with the chemical potential of the α-polymer denoted μ_α),

$$\begin{aligned}\Omega[\rho_\alpha(\mathbf{r})] &\equiv F[\rho_\alpha(\mathbf{r})] + \sum_\alpha \int \mathbf{dr}\rho_\alpha(\mathbf{r})[V_\alpha(\mathbf{r}) - \mu_\alpha] \\ &= F_{id}[\rho_\alpha(\mathbf{r})] + F_{ex}[\rho_\alpha(\mathbf{r})] + F_{\mathrm{ch}}[\rho_\alpha(\mathbf{r})] \\ &\quad + F_{\mathrm{int}}[\rho_\alpha(\mathbf{r})] + \sum_\alpha \int \mathbf{dr}\rho_\alpha(\mathbf{r})[V_\alpha(\mathbf{r}) - \mu_\alpha]\end{aligned} \tag{2.1}$$

Here, the first term in the right-hand side, F_{id}, is the ideal-gas contribution, given by,

$$F_{id}[\rho_\alpha(\mathbf{r})] = k_B T \sum_\alpha \int \mathbf{dr} \rho_\alpha(\mathbf{r}) \left[\ln\left(\Lambda_\alpha^3 \rho_\alpha(\mathbf{r})\right) - 1\right] \tag{2.2}$$

The second term, F_{ex}, describes the hard-sphere excess free energy contribution. As discussed above, there are multiple ways to capture those terms in an approximate fashion, from weighted density approximation [WDA] to fundamental measure theory [FMT]. The third term, F_{ch}, evaluates the effects of connecting hard-spheres into polymeric chains; it is generally based on the Wertheim perturbation scheme [72, 73]. The fourth term, F_{int}, estimates enthalpic contributions to the free energy, e.g., from van-der-Waals forces. Finally, the last term considers the role of external interactions (such as hard walls) via an imposed potential $V_\alpha(r)$. To calculate the morphology of the overall system, one needs to minimize the free energy (2.1) with respect to all the densities $\rho_\alpha(\mathbf{r})$,

$$\frac{\delta F[\rho_\alpha(\mathbf{r})]}{\delta \rho_\alpha(\mathbf{r})} = \mu_\alpha - V_\alpha(\mathbf{r}) \tag{2.3}$$

The resulting self-consistency equations generally are quite complex and need to be solved numerically in an iterative fashion. In particular, F_{ex} generally is a complex nonlinear and nonlocal functional of the single-particle densities. Thus, the minimization (2.3) leads to fairly complicated integral equations, requiring a fairly fine spatial discretization and leading to slow convergence of the iterative process. For more details on the numerical algorithms used, we refer the readers to papers by Frischknecht and co-workers [61, 62]. We now proceed to the discussion of DFT application to nanocomposites.

2.3 Applications of DFT to PNCs

2.3.1 *Polymers and Nanospheres*

As a first example of the application of DFT to nanocomposites, one can look at the problem of the interaction between hard spheres of radius R_{P} and homopolymers (melt or solution). A simple treatment of this problem for the solution case was proposed in the early 2000s by Schmidt and co-workers [74–77] who used the Asakura-Oosawa (AO) [78, 79] approximation. Within the AO analysis, the polymer chains are treated as impermeable spheres of radius R_{g}, where R_{g} is the polymer radius of gyration and depends on the polymer molecular weight and the solvent quality (good or poor) [80]. The AO analysis and subsequent theoretical work [81, 82] famously demonstrated that the colloidal particles experience so-called "depletion attraction", as the polymer chains resist confinement in the narrow spaces between the colloidal particles. In the cases where $q = R_g/R_P < 0.15$ (the "colloidal limit" we discussed earlier), the polymeric degrees of freedom can be integrated out to yield an effective pair potential between the colloids. Schmidt and co-workers built on that earlier body of work and developed a new DFT-based approach to calculate phase diagrams of AO mixtures as a function of the size ratio q, the polymer volume fraction, and the colloid volume fraction. The phase diagrams included colloid-poor phase ("Gas"), colloid-rich liquid phase ("Liquid"), and colloidal crystal phase ("Crystal"). Depending on the size ratio q, the demixing can occur either via "Gas–Liquid" or via "Gas–Solid" coexistence. Furthermore, DFT was applied to describe the particle and polymer density profiles at the "Gas–Liquid" interfaces or near a hard wall.

The above analysis, while straightforward and accurate in the limit of large nanospheres, is expected to run into difficulties for the cases where $q = R_g/R_P \sim 1$, and representing polymers as simple spheres is no longer tenable. In the 2000s, new approaches have been developed based on polymer DFTs. Some examples are shown in Figs. 2.2 and 2.3.

In Fig. 2.2, the density profiles of polymers and nanospheres near a hard wall are plotted as a function of the distance from the wall, as computed by McGarrity, Frischknecht, and Mackay [83, 84]. They considered an athermal mixture (no van-der-Waals interactions), used an FMT approach for the excess sphere free energy, and the Tripathy-Chapman association term [85, 86]. The calculations are done for $R_P = 2\,\sigma$, where σ is the monomer Kuhn length; the chain length $N = 100$. At low particle densities, the polymer expels the particles from the wall (Fig. 2.2a), while at high particle densities, the particles segregate to the surface layer (Fig. 2.2b).

Additional complexity can be introduced by adding van-der-Waals interactions and considering block copolymers instead of homopolymers. Cao and Wu [87] modeled a solution of AB diblock copolymers and nanoparticles near a hard wall; the total film thickness was set to 20 σ. The A-block is attractive to the particles, while the B-block is either neutral (Fig. 2.3a, b) or slightly repulsive (Fig. 2.3c, d) to the particles. The calculations show that as the interaction between the particles and the B-block become less favorable, the system undergoes a transition from a "multilayer film" to a "monolayer". The "multilayer" morphology is likely due to the three-dimensional ordering of the particles into soft colloidal crystals (even though the simulations are one-dimensional); the "monolayer" morphology represents the block copolymer templating, with the particles segregating into the favorable domain.

One last example in this section describes the melt of "hairy nanoparticles" or "nanoparticle-organic hybrid materials" (NOHM) [11, 88–94]. Here, DFT was used to compute the transition between the liquid and the face-centered cubic (FCC) solid (Fig. 2.4). We will return to this same system below when discussing the hybrid SCF-DFT technique.

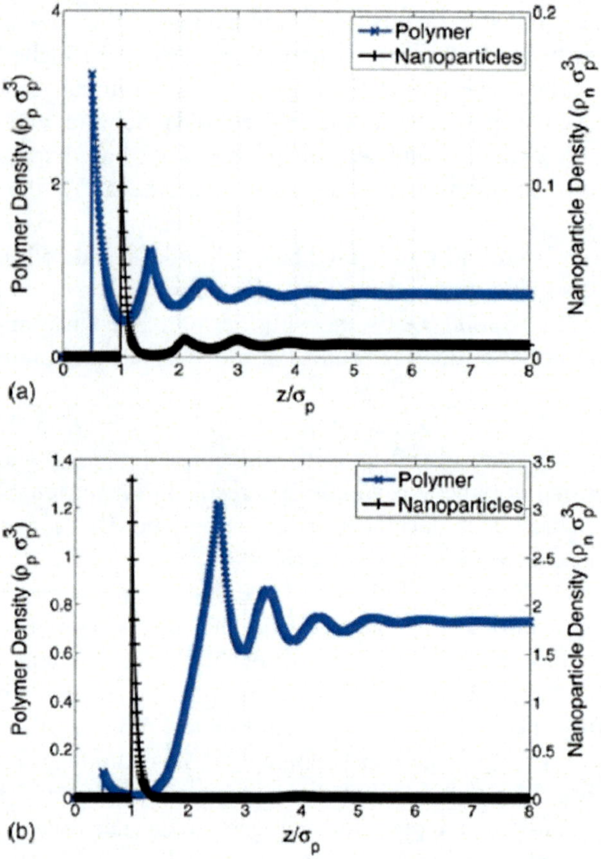

Fig. 2.2 Polymer/nanoparticle mixture near a hard wall. As the particle density is increased, the layer near the wall changes from polymer-rich to particle-rich. **a** Particle (black) and polymer (blue) density profiles for the polymer-rich structure; **b** Same for the particle-rich structure. Reproduced with permission from [83]. Copyright (2009) American Institute of Physics

2.3.2 *Polymers and Nanorods*

Our next example includes rodlike nanoparticles in a polymer melt, as shown in Fig. 2.5, left panel. The nanorods are assumed to have a high aspect ratio and well-aligned. The surfaces of the rods are covered with grafted oligomers of chain length *N*, and they are immersed in a homopolymer melt with chains of length *P*. The grafting density σ_{gr} is obviously one important parameter governing the dispersion of the rods; another parameter is the ratio of the free chain length to the grafted chain length, $\alpha = P/N$. The DFT calculations by Frischknecht, Hore, Composto, and co-workers [95–97] determined the effective interaction potential between the nanorods and evaluated whether it was attractive (leading to the nanorod aggregation) or repulsive (leading to their dispersion). The results are shown in Fig. 2.5, right panel. For the moderate grafting densities, the phase behavior was found to be sensitive to

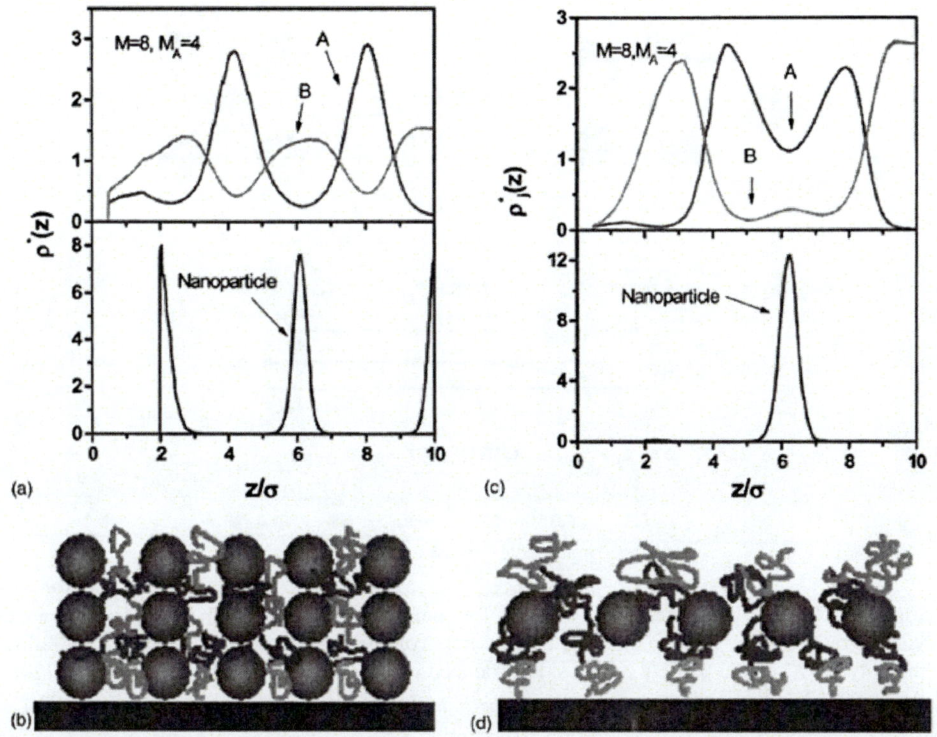

Fig. 2.3 Self-organization of nanoparticles (with diameter $\sigma_P = 4\sigma$) by short-chain block copolymers consisting of A/B segments (chain length $M = 8$). **a** and **c** give the reduced density profiles of the polymeric segments and of the nanoparticles confined within a thin film of thickness $H =$ 20 s. **b** and **d** Visualize the corresponding structures. In (a) and (b), B segments are neutral to the nanoparticles, while in (c) and (d), B segments repel the particles. In both cases, the packing fraction of the polymer in the bulk is 0.01 and that of the particles in the bulk is 0.1. Reproduced with permission from [87]. Copyright (2007) American Institute of Physics

the ratio of the rod radius to the polymer R_g; for the high grafting densities, this dependence has disappeared. The results were compared with experiments, and a good agreement was found.

2.3.3 Polymers and Nanoplatelets

Finally, we consider the case of polymers and nanoplates. Balazs, Ginzburg, and co-workers developed a two-step approach in analyzing the structure and dispersion of such systems. First, polymer self-consistent field theory (SCFT) is used to compute the effective polymer-mediated interaction potential between the platelets

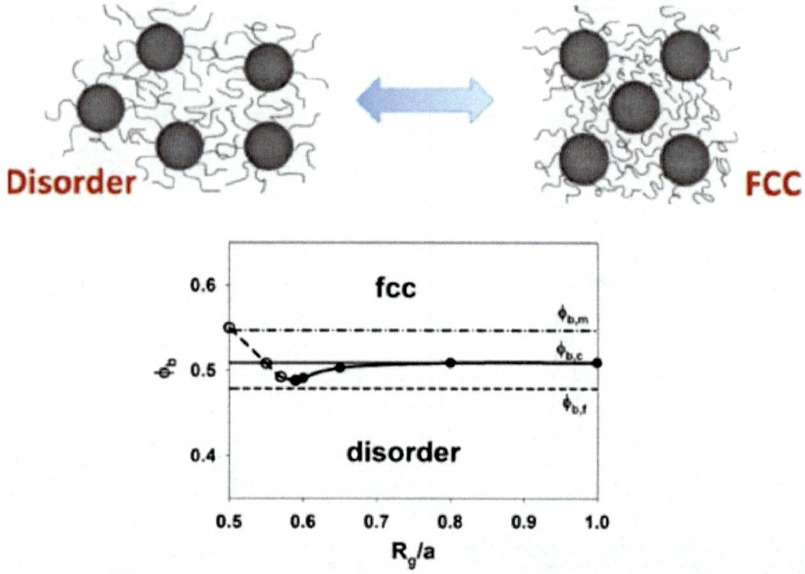

Fig. 2.4 The DFT-calculated phase diagram for "hairy nanoparticles" with the particle/monomer volume ratio of 106. The Y-axis is the nanoparticle volume fraction, and the X-axis is the ratio of the polymer R_g to the particle radius a. Reproduced with permission from [90]. Copyright (2013) American Chemical Society

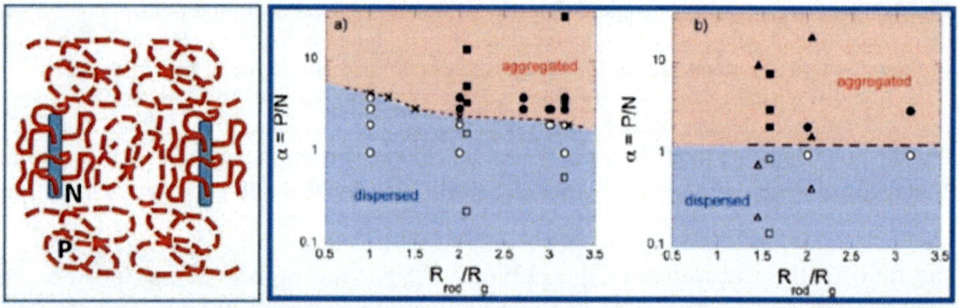

Fig. 2.5 Polymer-grafted nanorods in polymer melt. The cartoon on the left shows the schematics, with N being the length of the grafted ligands, and P being the length of the free chains. The phase diagrams on the right show the regions corresponding to rods being aggregated (red) or well-dispersed (blue); triangles and squares are experimental data, and circles represent DFT calculations. **a** High grafting density; **b** Moderate grafting density. Reproduced with permission from [95]. Copyright (2013) American Chemical Society

[98, 99]. Then, DFT is constructed to model the anisotropic assemblies of oblate ellipsoids with long-range interactions due to the SCFT-calculated potentials [6, 7, 100–103]. Trial single-particle densities are constructed to model various morphologies (Fig. 2.6).

As shown in Fig. 2.7, for nanoplatelets with relatively low aspect ratio $D/L = 30$, the phase diagram is quite nontrivial and complex. In the example here, the platelets

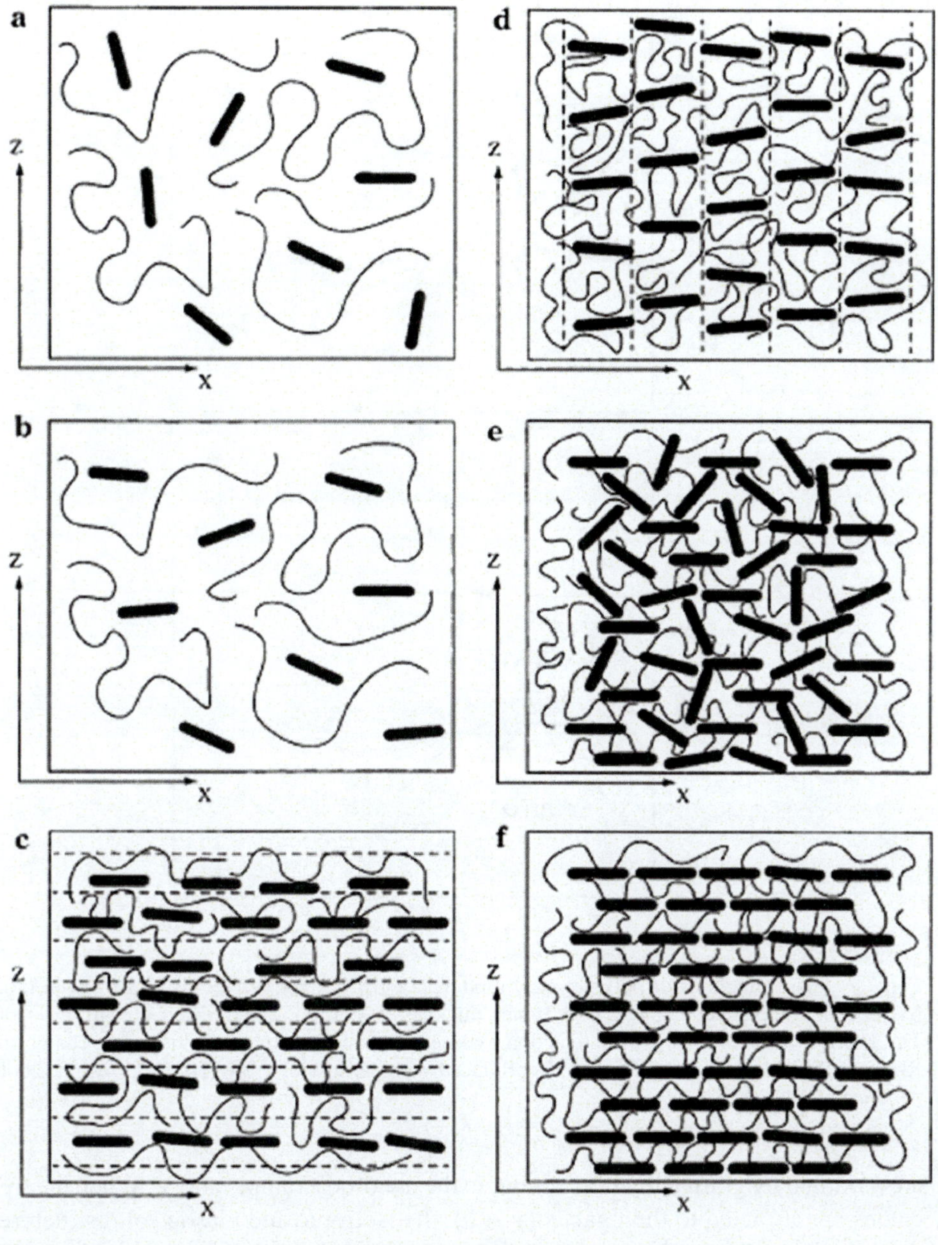

Fig. 2.6 The morphologies of oblate ellipsoid fluids: **a** Isotropic (I); **b** Nematic (N); **c** Smectic (Sm); **d** Columnar (Col); **e** Plastic Solid ("House of Cards") (PS); **f** Crystal (Cr). Reproduced with permission from [103]. Copyright (2000) American Chemical Society

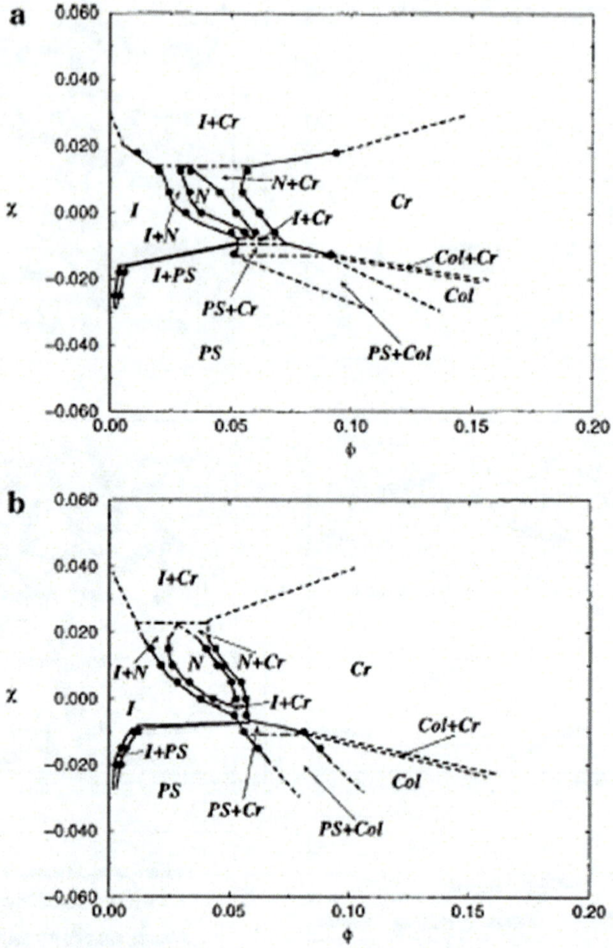

Fig. 2.7 Phase diagrams of polymer–clay mixtures: **a** $\rho_{gr} = 0.04$, $N_{gr} = 50$, and **b** $\rho_{gr} = 0.02$, $N_{gr} = 100$. The platelet diameter $D = 30$ nm, thickness $L = 1$ nm. The free chain length $P = 300$. The Y-axis is the Flory–Huggins interaction parameter between the ligands and the free chains, the X-axis is the platelet volume fraction. Reproduced with permission from [103]. Copyright (2000) American Chemical Society

are modified by grafted ligands, similar to the nanorod example above. When the free chains are attracted to the ligands ($\chi < 0$), the isotropic and plastic solid structures prevail at low filler loadings; eventually, the platelets and polymers arrange into a 3D crystal structure. When the free chains are repelled from the ligands ($\chi > 0$), a wide isotropic-crystal coexistence is observed, indicating aggregation or poor dispersion. Note that in the calculations, we assume that the fillers interact via weak van der Waals attractive forces.

As the platelet diameter is increased, the phase diagram becomes simpler. The isotropic, plastic solid, and columnar phases become squeezed to the limit of extremely low concentrations. It should become possible to consider a simpler model

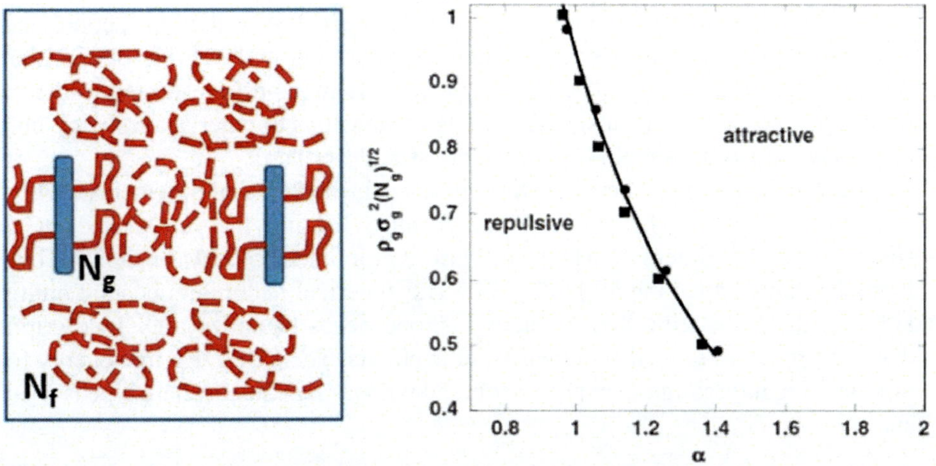

Fig. 2.8 The cartoon on the left shows the schematics, with N_g being the length of the grafted ligands, and N_f being the length of the free chains. The plot on the right shows the locus of the critical values of $\alpha = N_f/N_g$ at which the interaction force between the grafted monolayers becomes attractive for fixed bulk-free polymer density, $\rho_f \sigma_f{}^3 = 0.6$. The chain lengths of the grafted polymers are $N_g = 101$ (squares) and $N_g = 151$ (circles). $\sigma_g = \sigma_f = \sigma$ Reproduced with permission from [105]. Copyright (2009) American Institute of Physics

where the nanoplatelets are well-aligned and ask only whether the effective potential between them is attractive (causing aggregation) or repulsive (causing dispersion). This simplified problem was studied quite extensively (see, e.g., reviews [6, 7, 104]). In one example, Jain et al. [105] used DFT to estimate the interaction potential between two parallel plates in a polymer melt (Fig. 2.8). It was shown that the boundary between the aggregation and dispersion depends strongly on both the grafting density and the ratio of the ligand to free polymer chain lengths, as one indeed can expect.

Overall, the use of DFT in predicting nanocomposite morphology has been quite extensive. However, as examples above show, most DFT applications in the nanocomposite field either utilize the "colloidal approximation" (integrating out the polymeric degrees of freedom) or are restricted to lower-dimensionality morphologies (usually 1D, sometimes 2D) and/or shorter polymers. This is because all polymeric DFTs are still computationally expensive. In the next section, we discuss a simplified approach that aims to reduce the computation time and enable the calculation of more complex morphologies.

2.4 The SCF-DFT Approach

The hybrid Self-Consistent Field/Density Functional Theory (SCF-DFT) method was originally proposed by Thompson, Ginzburg, Matsen, and Balazs [106, 107] to describe the bulk morphologies of mixtures of nanospheres and diblock copolymers. It was subsequently used by many other investigators and applied to various problems, some of which are highlighted below. So far, the method is restricted to the case where the particles are nanospheres, though extensions to other shapes (e.g., oblate or prolate ellipsoids) should be reasonably straightforward.

The starting point for the SCF-DFT approach is the SCFT of polymers (see books and reviews [108–111] for more details). Within SCFT, polymers are treated as Gaussian chains and the interaction between various monomers is local and captured by the Flory-Huggins lattice model [112–114]. The spherical nanofillers occupy multiple lattice sites; to describe their interactions, one needs to utilize a DFT description that captures long-ranged excluded volume effects. The SCF-DFT free energy for a mixture of a multicomponent polymer blend with a single nanosphere type is written as,

$$
\begin{aligned}
\frac{FNv}{k_B T V} &= -\sum_j \frac{\psi_j}{\alpha_j} \ln\left(\frac{Q_j}{\psi_j V}\right) \\
&+ \frac{1}{V}\int \mathbf{dr}\left[\frac{1}{2}\sum_{\alpha,\beta}(\chi N)_{\alpha\beta}\{\varphi_\alpha(\mathbf{r})-\phi_\alpha\}\{\varphi_\beta(\mathbf{r})-\phi_\beta\}\right. \\
&\left. -\xi(\mathbf{r})\left(1-\sum_\alpha \varphi_\alpha(\mathbf{r})\right)\right] \\
&- \frac{1}{V}\int \mathbf{dr}\left[w_P(\mathbf{r})\rho_P(\mathbf{r})+\sum_\alpha w_\alpha(\mathbf{r})\varphi_\alpha(\mathbf{r})\right] \\
&+ \frac{1}{V}\int \mathbf{dr}\left[\rho_P(\mathbf{r})\Psi_{HS}\left(\overline{\overline{\varphi}}_P(\mathbf{r})\right)\right]
\end{aligned}
\tag{2.4}
$$

Here, V is the total volume, T is temperature, N is the chain length of the first (reference) polymer chain, ψ_j is the volume fraction of the j-th component, Q_j is the partition function of the j-th component, and α_j is the volume ratio of the j-th component and the first (reference) component. In addition, $\varphi_\alpha(\mathbf{r})$ is the position-dependent volume fraction of the α-th chemical species, ϕ_α is the overall volume fraction of the α-th chemical species, and $w_\alpha(\mathbf{r})$ is the conjugate field. Finally, $\xi(\mathbf{r})$ is the pressure-like field, and the matrix $(\chi N)_{\alpha\beta}$ describes all the pairwise Flory–Huggins interactions. For the particles, one uses the DFT-like description based on the center probability distribution, $\rho_P(\mathbf{r})$, from which the position-dependent volume fraction is then computed via,

$$
\varphi_P(\mathbf{r}) = \frac{1}{Nv}\int \mathbf{dr}'\rho_P(\mathbf{r}')\Theta\left(R_P - |\mathbf{r}-\mathbf{r}'|\right) \tag{2.5}
$$

Here, Θ is the standard Heaviside function. Also, consistent with DFT for hard spheres, one estimates the excess non-ideal free energy using a Weighted-Density Approximation (WDA) [27]. The Carnahan-Starling [115] excess term and a weighted density function are utilized,

$$\Psi_{HS}(x) = \frac{4x - 3x^2}{(1 - x)^2} \tag{2.6}$$

$$\overline{\overline{\varphi}}_P(\mathbf{r}) = \frac{1}{2^3 N v} \int \mathbf{dr}' \rho_P(\mathbf{r} + \mathbf{r}') \Big[w_0(\mathbf{r}') + w_1(\mathbf{r}') \{\rho_P(\mathbf{r}')\} + w_2(\mathbf{r}') \{\rho_P(\mathbf{r}')\}^2 \Big] \tag{2.7}$$

For simplicity, in the expansion (2.7), w_1 and w_2 are both set to zero, and $w_0(\mathbf{r}) = \Theta(2R_P - |\mathbf{r}|)$. Lastly, the partition functions for the particles and the polymers are given by,

$$Q_P = \int \mathbf{dr} \exp[-w_P(\mathbf{r})] \tag{2.8a}$$

$$Q_{j \neq P} = \int \mathbf{dr} q(\mathbf{r}, \alpha_j) \tag{2.8b}$$

In (2.8b), the propagator q is evaluated in a standard SCFT fashion [106–111]. The free energy (2.4) needs to be minimized with respect to densities, conjugate fields, and pressure; the resulting equations are solved iteratively. The solutions would then describe the morphologies of nanocomposites, including the distribution of nanoparticles within various polymer domains.

2.4.1 Nanoparticles in Block Copolymers—Bulk and Films

Historically, the first example of SCF-DFT was the case where spherical nanoparticles were admixed with an AB-diblock copolymer; the nanoparticles were assumed to be strongly preferential to the A-block. Figure 2.9 shows the density profiles and particle center distributions for the lamellar morphology [106].

The model shows that the particles segregate into the A-lamellae; furthermore, larger particles prefer to be located at the center of the lamella (Fig. 2.9a, b), while the smaller ones tend to push closer to the edges (Fig. 2.9c). This result agrees qualitatively with experimental data of Bockstaller et al. [116] and other authors [117–120]. Similar particle segregation was calculated for other morphologies (cylindrical and spherical) [107]. Subsequent work by Balazs and co-workers discussed the influence of mixed spheres, as well as the behavior of the block copolymer/particle system in thin films [121–128]. It was shown, in particular, that under some conditions, the particles can segregate towards the surfaces and provide "self-healing" effect.

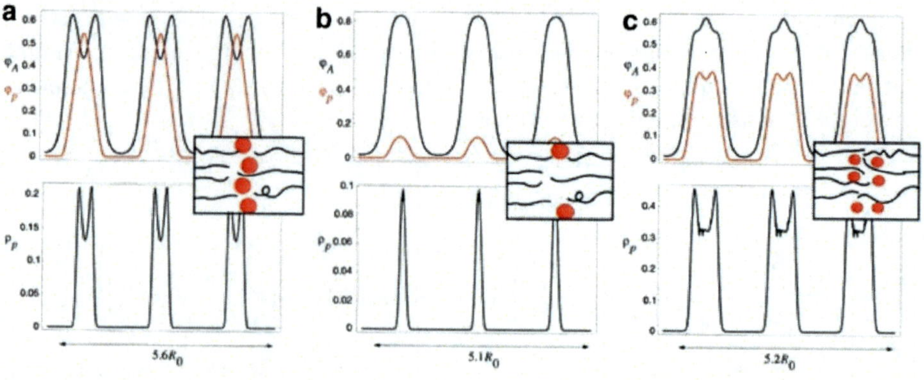

Fig. 2.9 Concentration profiles of diblock-particle systems [$\varphi_A(x)$, density distribution of A blocks; $\varphi_P(x)$, and $\rho_P(x)$, distributions of particles and particle centers, respectively]. **a** Large particle ($R_P = 0.74R_g$), particle volume fraction $\phi_P = 0.15$. **b** Large particle ($R_P = 0.74R_g$), $\phi_P = 0.03$. **c** Small particle ($R_P = 0.5R_g$), $\phi_P = 0.15$. B blocks are not depicted. Three periods are shown to better illustrate the morphologies. Sketches showing the structure of each system accompany the profiles. Reproduced with permission from [106]. Copyright (2001) American Association for Advancement of Science

2.4.2 "Giant Surfactants"—Single-Ligand Nanoparticles

The above theory was formulated assuming that the fillers were "hard spheres". In reality, however, many nanoparticles have oligomeric ligands grafted to their surfaces ("giant surfactants" [19, 129–131], and the theory needs to take this effect into account. One of the first examples of the SCF-DFT extension to account for the effects of grafting was the study by Lee et al. [127], in which they calculated phases of the amphiphilic "tadpole" sphere-coil copolymers. It was shown that the spheres can organize into hexagonal closed-pack lattice, with polymers filling the spaces between them. This analysis was later refined and extended by other authors [132]. Similar arrangements were subsequently observed in Brownian Dynamics and other particle-based simulations [18, 20, 133].

Several recent studies utilized SCF-DFT to predict the morphology and phase behavior of single- or double-tethered nanoparticles with block-copolymer (rather than homopolymer) ligand [134–136]. In one example, Wang and co-workers investigated the phase behavior of ABP, APB, and PAB triblocks, where A and B are flexible chains, and P is a nanoparticle; the A-block is favorable to the particles. Figure 2.10 shows the calculated phase diagrams, based on 2D SCF-DFT simulations. The Y-axis is the strength of attraction between the A-block and the P-particle, and the X-axis is the A-block volume fraction. It can be seen that the morphology and the phase behavior are sensitive to the particle placement relative to the A-block.

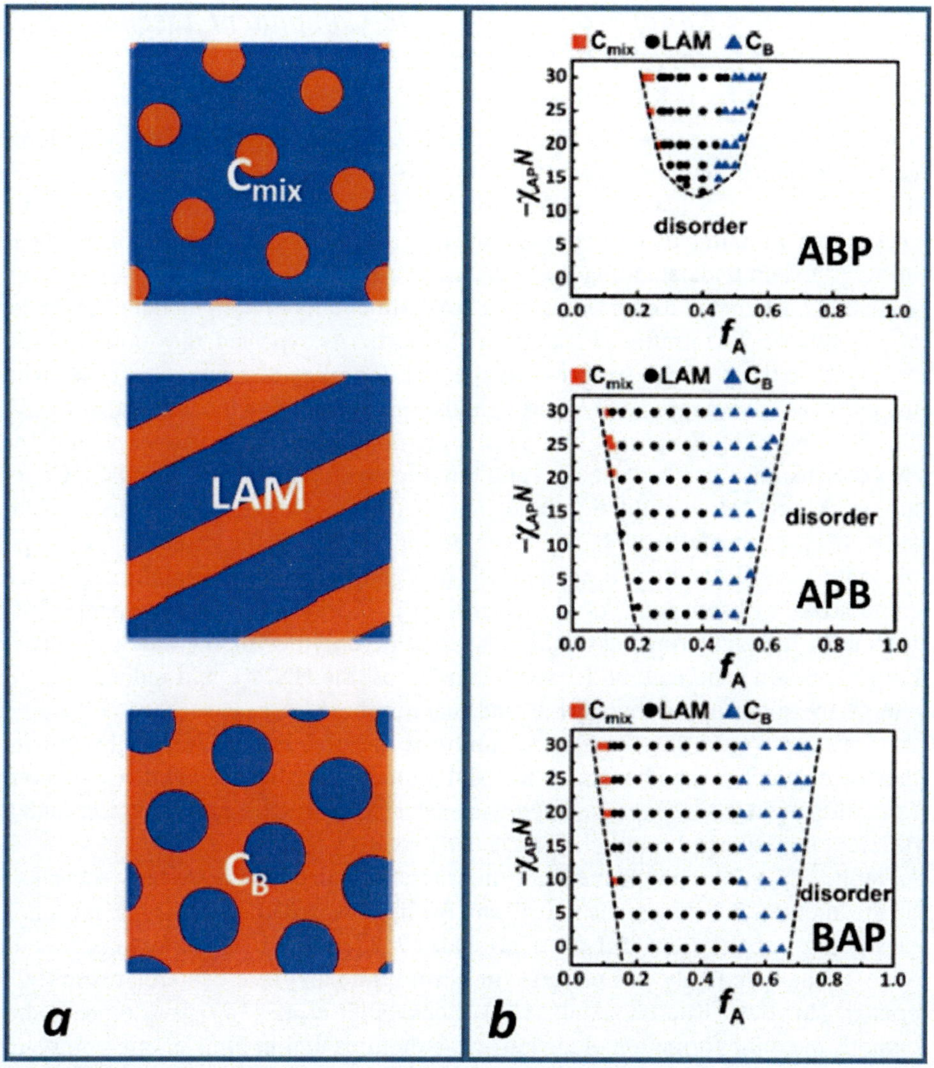

Fig. 2.10 **a** 2D ordered structures self-assembled by various giant surfactants: (Top) hexagonally packed cylinders formed by mixed A block and nanoparticles (NP) (C_{mix}); (Middle) parallel lamellae formed by mixed A block and NP and pure B block (LAM); (Bottom) hexagonally packed cylinders formed by pure B block (C_{B}). The orange color denotes the regions formed by mixed A block and NP, and the blue color denotes the regions formed by pure B block. **b** Phase diagrams for (Top) ABP, (Middle) APB, and (Bottom) BAP giant surfactants as a function of the chain length ratio of A block (f_{A}) and the interaction parameter between A block and NP ($-\chi_{\mathrm{AP}}N$). Reproduced with permission from [137]. Copyright (2019) American Chemical Society

2.4.3 "Hairy Nanoparticles"—Single-Component Assemblies and Mixtures with Polymers

The last example we consider is related to the application of SCF-DFT to polymer-grafted or "hairy" nanoparticles (PGNP or HNP) [11, 17, 137]. This topic was briefly discussed in Sect. 3.1 when we described the use of DFT to estimate the liquid-to-crystal transition in bulk one-component HNP assemblies. We now highlight the use of SCF-DFT in understanding these systems.

In 2009, Akcora and co-workers [138] experimentally investigated the dispersions of polystyrene (PS)-grafted silica-NP in PS matrix. As expected, they observed some cases where the particles were well-dispersed (long ligands) and some cases where the particles were aggregated (short ligands), in agreement with earlier theories [12, 21, 22, 139–141]. Surprisingly, under some conditions, the particles would form two-dimensional arrays ("sheets") and one-dimensional chains ("strings"). Coarse-grained Monte Carlo simulations and analytical strong-segregation theory modeling [138, 142] successfully explained those morphologies for the case of nanocomposites with low NP loading; the combined modeling and experimental insights were summarized by Kumar and co-workers in their 2013 Macromolecules paper [17]. The reason for the anisotropic assembly was that the polymer shell ("canopy") did not have to remain spherical; under some conditions, the HNPs could reduce their free energy by re-arranging the ligands and making their surfaces somewhat "patchy"; the particles would then be able to minimize the free energy further by "fusing" the "bald" patches together. In an SCF-DFT study, Ginzburg [143] investigated the formation of sheets, strings, and aggregates at different NP volume fractions and for different ligand lengths and grafting densities.

In the HNP-homopolymer binary mixtures, the anisotropic assembly was caused by the interaction between the matrix and the ligands. Are the anisotropic morphologies going to be present if there is no matrix? The DFT analysis was discussed in Sect. 3.1 included only two phases—uniform liquid and face-centered cubic (FCC) crystal. However, Koerner et al. [144], Bockstaller et al. [145], and other authors demonstrated the formation of anisotropic structures in thin films of one-component HNPs under conditions of "moderate" and "sparse" grafting. To describe the phase behavior of these systems, Ginzburg [146] applied a slightly modified version of SCF-DFT; the results are shown in Figs. 2.11 and 2.12.

The phase map in Fig. 2.11 shows the equilibrium morphologies for one-component nanocomposites having the core radius $R_P = 3.05$ nm as the grafting density and the grafted chain length is varied. The structures predicted in simulations include FCC crystal and isotropic liquid, but also lamellar (sheets) and cylindrical (strings) morphologies. It was further hypothesized that the phase map can be re-plotted in scaled coordinates, R_P/R_g and φ_P (Fig. 2.12), and a reasonable agreement with experiments was obtained, although more studies are needed.

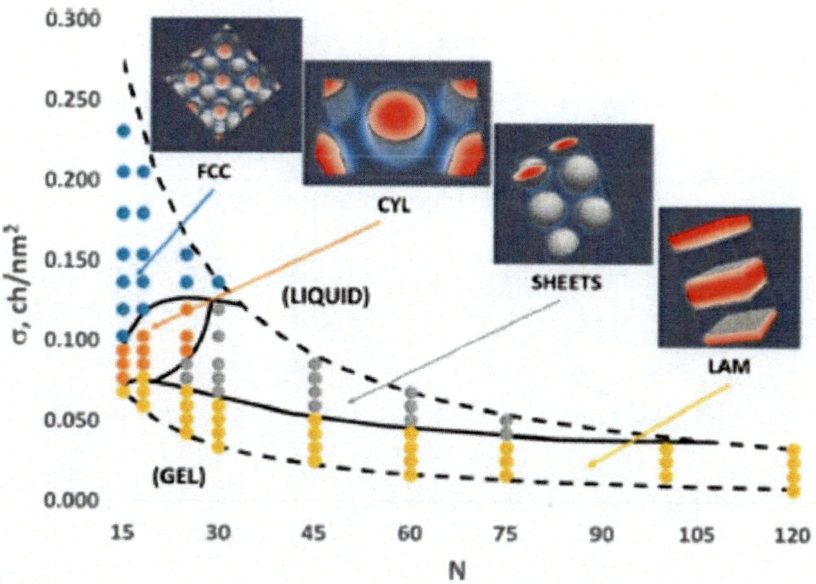

Fig. 2.11 Phase map for the HNP melt. Here, N is ligand length and σ is grafting density. Symbols represent simulation results, with equilibrium (minimum free energy) morphologies shown as follows: yellow = lamellar, gray = sheets, orange = cylindrical, and blue = FCC spherical. Dashed lines are boundaries of the region studied in these simulations, while solid lines are approximate locations of transitions between various phases. Reproduced with permission from [147]. Copyright (2017) American Chemical Society

2.5 Summary and Outlook

Density functional theory (DFT) is a powerful technique to study the phase behavior of complex fluids. Over the past four decades, DFT has been extensively parameterized for various systems, from hard spheres to Lennard–Jones spheres to hard ellipsoids and spherocylinders to flexible polymer chains. In particular, DFT is generally well-suited to model the behavior of a mixture comprised of hard spheres of different sizes; one can then combine one or more sphere types into polymer chains via the Wertheim perturbation procedure. At the same time, numerical solution of the DFT equations often remains challenging due to long-range nature of the non-ideal free energy terms in either WDA or FMT approximations; thus, most DFT examples for nanocomposites so far have been describing either 1D or 2D structures.

Hybrid field-theory approaches, such as SCF-DFT, can often accelerate the calculations and allow for fast exploration of phase maps and various morphologies. The use of SCFT instead of DFT for polymer components neglects some intermolecular correlations but significantly increases the computation speed. Under some conditions, the use of compressible SCFT [104, 105, 148] can help improve the agreement with experiment and/or approximate the DFT results. The SCF-DFT approach, as originally formulated by Thompson et al. [106, 107], and subsequently refined and extended by other authors [127, 136, 146, 149], enables fast and straightforward

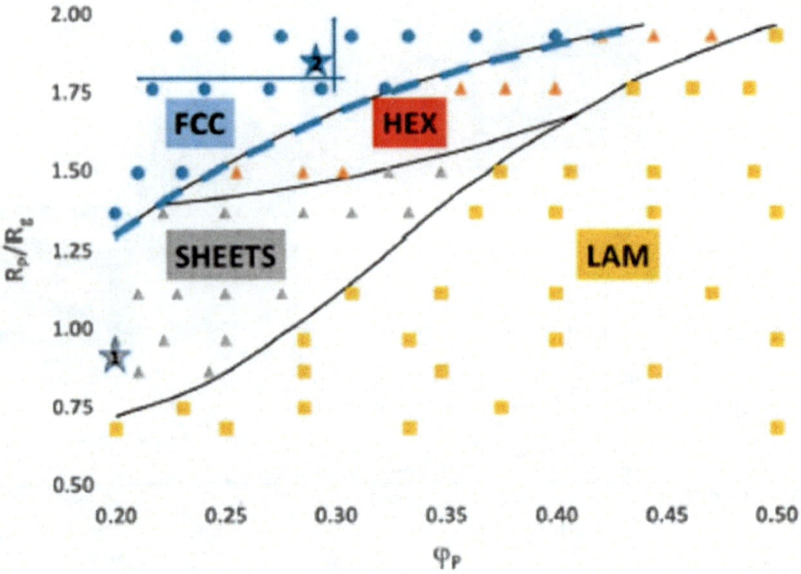

Fig. 2.12 Phase map of the HNP one-component systems expressed in terms of φ_P and R_P/R_g. Symbols are labeled as in Fig. 2.11. Solid lines are guided to the eye depicting approximate phase boundaries. Stars with numbers represent the phase map locations of HNP systems from [144] and [147] (see text for more details). The dashed blue line is the predicted FCC melting transition. Reproduced with permission from [147]. Copyright (2017) American Chemical Society

calculation of the phase behavior for variety of nanocomposite systems ("bald" and "hairy" nanoparticles in homopolymers, homopolymer blends, and block copolymers; melts or solutions; bulk or thin films). It is feasible to suggest that in the future, SCF-DFT can be used for molecular design of nanocomposites in the same way as today, SCFT is used for molecular design of pure block copolymers [109, 150–154].

Acknowledgements I am thankful to Dow for supporting this work and gratefully acknowledge Drs. Jeffrey Weinhold and Jian Yang for helpful discussions.

References

1. Q. Zeng, A. Yu, G. Lu, Prog. Polym. Sci. **33**, 191 (2008)
2. L.-T. Yan, X.-M. Xie, Prog. Polym. Sci. **38**, 369 (2013)
3. J. Yan, M. R. Bockstaller, K. Matyjaszewski, Progr. Polym. Sci., 101180 (2019)
4. G. Allegra, G. Raos, M. Vacatello, Prog. Polym. Sci. **33**, 683 (2008)
5. V. V. Ginzburg, in *Problems of Nonlinear Mechanics and Physics of Materials* (Springer, 2019), p. 205.
6. A. C. Balazs, J. Bicerano, V. V. Ginzburg, in *Polyolefin Composites* (Wiley, 2007), p. 415.
7. A.C. Balazs, V.V. Ginzburg, Y. Lyatskaya, C. Singh, E. Zhulina, in *Polymer-Clay Nanocomposites*. ed. by T. Pinnavaia, G.W. Beall (Wiley, New York, 2000), p. 281
8. A. Karatrantos, N. Clarke, M. Kröger, Polym. Rev. **56**, 385 (2016)

9. J.J. de Pablo, Annu Rev Phys Chem **62**, 555 (2011)
10. J. Feng, H. Heinz, Adv. Polyolefin Nanocompos., 205 (2010)
11. N.J. Fernandes, H. Koerner, E.P. Giannelis, R.A. Vaia, MRS Commun. **3**, 13 (2013)
12. V. Ganesan, A. Jayaraman, Soft Matter **10**, 13 (2014)
13. S.C. Glotzer, M.J. Solomon, Nat. Mater. **6**, 557 (2007)
14. H. Heinz, S.S. Patnaik, R.B. Pandey, B.L. Farmer, Modeli. Simul. Polym., 37 (2010).
15. S.K. Kumar, B.C. Benicewicz, R.A. Vaia, K.I. Winey, Macromolecules **50**, 714 (2017)
16. S.K. Kumar, V. Ganesan, R.A. Riggleman, J. Chem. Phys. **147**, 020901 (2017)
17. S.K. Kumar, N. Jouault, B. Benicewicz, T. Neely, Macromolecules **46**, 3199 (2013)
18. R.L. Marson, T.D. Nguyen, S.C. Glotzer, MRS Commun. **5**, 397 (2015)
19. W.-B. Zhang et al., Macromolecules **47**, 1221 (2014)
20. Z.L. Zhang, M.A. Horsch, M.H. Lamm, S.C. Glotzer, Nano Lett **3**, 1341 (2003)
21. A. Jayaraman, in *Theory and Modeling of Polymer Nanocomposites*, ed. by V. Ginzburg, L. Hall (Springer, 2020)
22. L.M. Hall, A. Jayaraman, K.S. Schweizer, Curr. Opin. Solid State Mater. Sci. **14**, 38 (2010)
23. P. Hohenberg, W. Kohn, Phys. Rev. **136**, B864 (1964)
24. W. Kohn, L.J. Sham, Phys. Rev. **140**, A1133 (1965)
25. W. Kohn, Rev. Mod. Phys. **71**, 1253 (1999)
26. T. Ramakrishnan, M. Yussouff, Phys. Rev. B **19**, 2775 (1979)
27. P. Tarazona, Phys. Rev. A **31**, 2672 (1985)
28. P. Tarazona, L. Vicente, Mol. Phys. **56**, 557 (1985)
29. W. Curtin, N. Ashcroft, Phys. Rev. A **32**, 2909 (1985)
30. A. Denton, N. Ashcroft, Phys. Rev. A **39**, 4701 (1989)
31. R. Leidl, H. Wagner, J. Chem. Phys. **98**, 4142 (1993)
32. J.F. Lutsko, M. Baus, Phys. Rev. A **41**, 6647 (1990)
33. J.F. Lutsko, M. Baus, Phys. Rev. Lett. **64**, 761 (1990)
34. Y. Rosenfeld, Phys. Rev. E **50**, R3318 (1994)
35. Y. Rosenfeld, J. Phys.: Condens. Matter **8**, 9289 (1996)
36. Y. Rosenfeld, M. Schmidt, H. Löwen, P. Tarazona, Phys. Rev. E **55**, 4245 (1997)
37. Y. Rosenfeld, P. Tarazona, Mol. Phys. **95**, 141 (1998)
38. Y. Rosenfeld, J. Phys.: Condens. Matter **14**, 9141 (2002)
39. P. Tarazona, Phys. A **306**, 243 (2002)
40. P. Tarazona, J.A. Cuesta, Y. Martínez-Ratón, in *Theory and Simulation of Hard-Sphere Fluids and Related Systems* (Springer, 2008), pp. 247.
41. M. Baus, J.F. Lutsko, Phys. A **176**, 28 (1991)
42. A. Somoza, P. Tarazona, J. Chem. Phys. **91**, 517 (1989)
43. A. Somoza, P. Tarazona, Phys. Rev. A **41**, 965 (1990)
44. E. Velasco, A. Somoza, L. Mederos, J. Chem. Phys. **102**, 8107 (1995)
45. E. Velasco, L. Mederos, J. Chem. Phys. **109**, 2361 (1998)
46. E. Velasco, L. Mederos, D. Sullivan, Phys. Rev. E **62**, 3708 (2000)
47. E. Velasco, L. Mederos, D. Sullivan, Phys. Rev. E **66**, 021708 (2002)
48. G.J. Vroege, H.N. Lekkerkerker, Rep. Prog. Phys. **55**, 1241 (1992)
49. V.V. Ginzburg, M.A. Glaser, N.A. Clark, Liq. Cryst. **23**, 227 (1997)
50. H. Graf, H. Löwen, J. Phys.: Condens. Matter **11**, 1435 (1999)
51. A. Yethiraj, C.E. Woodward, J. Chem. Phys. **102**, 5499 (1995)
52. A. Yethiraj, J. Chem. Phys. **109**, 3269 (1998)
53. C.E. Woodward, J. Chem. Phys. **94**, 3183 (1991)
54. C.E. Woodward, A. Yethiraj, J. Chem. Phys. **100**, 3181 (1994)
55. D. Chandler, J.D. McCoy, S.J. Singer, J. Chem. Phys. **85**, 5971 (1986)
56. D. Chandler, J.D. McCoy, S.J. Singer, J. Chem. Phys. **85**, 5977 (1986)
57. E. Kierlik, M. Rosinberg, J. Chem. Phys. **97**, 9222 (1992)
58. E. Kierlik, M. Rosinberg, J. Chem. Phys. **99**, 3950 (1993)
59. S.K. Nath, P.F. Nealey, J.J. de Pablo, J. Chem. Phys. **110**, 7483 (1999)
60. J.B. Hooper, J.D. McCoy, J.G. Curro, J. Chem. Phys. **112**, 3090 (2000)

61. L. Frink, A. Salinger, M. Sears, J. Weinhold, A. Frischknecht, J. Phys.: Condens. Matter **14**, 12167 (2002)
62. A.L. Frischknecht, J.D. Weinhold, A.G. Salinger, J.G. Curro, L.J. Douglas Frink, J.D. McCoy, J. Chem. Phys. **117**, 10385 (2002).
63. J. Wu, AIChE J. **52**, 1169 (2006)
64. C.P. Emborsky, Z. Feng, K.R. Cox, W.G. Chapman, Fluid Phase Equilib. **306**, 15 (2011)
65. J.D. McCoy, K.G. Honnell, K.S. Schweizer, J.G. Curro, J. Chem. Phys. **95**, 9348 (1991)
66. J.D. McCoy, K.G. Honnell, K.S. Schweizer, J.G. Curro, Chem. Phys. Lett. **179**, 374 (1991)
67. N. Sushko, P. van der Schoot, M. Michels, J. Chem. Phys. **115**, 7744 (2001)
68. T. Cherepanova, A. Stekolnikov, Chem. Phys. **154**, 41 (1991)
69. J.B. Hooper, J.D. McCoy, J.G. Curro, F. van Swol, J. Chem. Phys. **113**, 2021 (2000)
70. A. Yethiraj, Adv. Chem. Phys. **121**, 89 (2002)
71. X. Xu, D. Cao, J. Wu, Soft Matter **6**, 4631 (2010)
72. M. Wertheim, J. Stat. Phys. **35**, 19 (1984)
73. M. Wertheim, J. Stat. Phys. **35**, 35 (1984)
74. M. Schmidt, H. Löwen, J.M. Brader, R. Evans, Phys. Rev. Lett. **85**, 1934 (2000)
75. J. Brader, R. Evans, M. Schmidt, H. Löwen, J. Phys.: Condens. Matter **14**, L1 (2001)
76. R. Evans, J. Brader, R. Roth, M. Dijkstra, M. Schmidt, H. Löwen, Philos. Trans. R. Soc. Lond. Ser. A: Math. Phys. Eng. Sci. **359**, 961 (2001)
77. J.M. Brader, R. Evans, M. Schmidt, Mol. Phys. **101**, 3349 (2003)
78. S. Asakura, F. Oosawa, J. Chem. Phys. **22**, 1255 (1954)
79. S. Asakura, F. Oosawa, J. Polym. Sci., A: Polym. Chem. **33**, 183 (1958)
80. M. Rubinstein, R. H. Colby, *Polymer Physics*, vol. 23 (Oxford University Press, New York, 2003)
81. A. Gast, C. Hall, W. Russel, J. Colloid Interface Sci. **96**, 251 (1983)
82. H.N. Lekkerkerker, W.-K. Poon, P.N. Pusey, A. Stroobants, P. Warren, EPL (Europhys. Lett.) **20**, 559 (1992)
83. E. McGarrity, A. Frischknecht, M. Mackay, J. Chem. Phys. **128**, 154904 (2008)
84. E. McGarrity, A. Frischknecht, L. Frink, M. Mackay, Phys. Rev. Lett. **99**, 238302 (2007)
85. S. Tripathi, W.G. Chapman, J. Chem. Phys. **122**, 094506 (2005)
86. S. Tripathi, W.G. Chapman, Phys. Rev. Lett. **94**, 087801 (2005)
87. D. Cao, J. Wu, J. Chem. Phys. **126**, 144912 (2007)
88. H.-Y. Yu, D.L. Koch, Langmuir **26**, 16801 (2010)
89. A. Chremos, A.Z. Panagiotopoulos, H.-Y. Yu, D.L. Koch, J. Chem. Phys. **135**, 114901 (2011)
90. H.-Y. Yu, D.L. Koch, Langmuir **29**, 8197 (2013)
91. H.-Y. Yu, S. Srivastava, L.A. Archer, D.L. Koch, Soft Matter **10**, 9120 (2014)
92. A.B. Bourlinos, R. Herrera, N. Chalkias, D.D. Jiang, Q. Zhang, L.A. Archer, E.P. Giannelis, Adv. Mater. **17**, 234 (2005)
93. A.B. Bourlinos, S. Ray Chowdhury, R. Herrera, D.D. Jiang, Q. Zhang, L.A. Archer, E.P. Giannelis, Adv. Funct. Mater. **15**, 1285 (2005).
94. S. Srivastava, P. Agarwal, L.A. Archer, Langmuir **28**, 6276 (2012)
95. A.L. Frischknecht, M.J. Hore, J. Ford, R.J. Composto, Macromolecules **46**, 2856 (2013)
96. A.L. Frischknecht, J. Chem. Phys. **128**, 224902 (2008)
97. M.J.A. Hore, R.J. Composto, Macromolecules **47**, 875 (2014)
98. A.C. Balazs, C. Singh, E. Zhulina, Macromolecules **31**, 8370 (1998)
99. A.C. Balazs, C. Singh, E. Zhulina, Y. Lyatskaya, Accounts Chem. Res. **32**, 651 (1999)
100. Y. Lyatskaya, A.C. Balazs, Macromolecules **31**, 6676 (1998)
101. V.V. Ginzburg, A.C. Balazs, Macromolecules **32**, 5681 (1999)
102. V. Ginzburg, A. Balazs, Adv. Mater. **12**, 1805 (2000)
103. V.V. Ginzburg, C. Singh, A.C. Balazs, Macromolecules **33**, 1089 (2000)
104. P.K. Jog, V.V. Ginzburg, R. Srivastava, J.D. Weinhold, S. Jain, W.G. Chapman, in *Advances in Chemical Engineering* (Elsevier, 2010), p. 131
105. S. Jain, V.V. Ginzburg, P. Jog, J. Weinhold, R. Srivastava, W.G. Chapman, J. Chem. Phys. **131**, 044908 (2009)

106. R.B. Thompson, V.V. Ginzburg, M.W. Matsen, A.C. Balazs, Science **292**, 2469 (2001)
107. R.B. Thompson, V.V. Ginzburg, M.W. Matsen, A.C. Balazs, Macromolecules **35**, 1060 (2002)
108. G. Fredrickson, *The Equilibrium Theory of Inhomogeneous Polymers* (Oxford University Press, 2006)
109. M.W. Matsen, F.S. Bates, Macromolecules **29**, 1091 (1996)
110. A.C. Shi, in *Developments in Block Copolymer Science and Technology*, ed. by I.W. Hamley (Wiley-VCH, Weinheim, 2004)
111. F. Schmid, J Phys-Condens Mat **10**, 8105 (1998)
112. S.F. Edwards, Proc. Phys. Soc. **85**, 613 (1965)
113. P.J. Flory, J. Chem. Phys. **9**, 660 (1941)
114. M.L. Huggins, J. Chem. Phys. **9**, 440 (1941)
115. N.F. Carnahan, K.E. Starling, J. Chem. Phys. **51**, 635 (1969)
116. M.R. Bockstaller, Y. Lapetnikov, S. Margel, E.L. Thomas, J. Am. Chem. Soc. **125**, 5276 (2003)
117. B.J. Kim, J.J. Chiu, L. Bang, C.J. Hawker, D.J. Pine, E.J. Kramer, Abstr. Pap. Am. Chem. S. **230**, U1066 (2005)
118. B.J. Kim, J.J. Chiu, G.R. Yi, D.J. Pine, E.J. Kramer, Adv. Mater. **17**, 2618 (2005)
119. B.J. Kim, G.H. Fredrickson, C.J. Hawker, E.J. Kramer, Langmuir **23**, 7804 (2007)
120. B.J. Kim, G.H. Fredrickson, E.J. Kramer, Macromolecules **41**, 436 (2008)
121. J.Y. Lee, R.B. Thompson, D. Jasnow, A.C. Balazs, Macromolecules **35**, 4855 (2002)
122. J.-Y. Lee, R.B. Thompson, D. Jasnow, A.C. Balazs, Phys. Rev. Lett. **89**, 155503 (2002)
123. R.B. Thompson, J.Y. Lee, D. Jasnow, A.C. Balazs, Phys. Rev. E **66**, 031801 (2002)
124. G.A. Buxton, J.Y. Lee, A.C. Balazs, Macromolecules **36**, 9631 (2003)
125. J. Y. Lee, Z. Shou, and A. C. Balazs, Phys. Rev. Lett. **91** (2003)
126. J.Y. Lee, Z.Y. Shou, A.C. Balazs, Macromolecules **36**, 7730 (2003)
127. J.Y. Lee, A.C. Balazs, R.B. Thompson, R.M. Hill, Macromolecules **37**, 3536 (2004)
128. S. Tyagi, J.Y. Lee, G.A. Buxton, A.C. Balazs, Macromolecules **37**, 9160 (2004)
129. X. Yu et al., Proc. Natl. Acad. Sci. **110**, 10078 (2013)
130. B. Ni, X.-H. Dong, Z. Chen, Z. Lin, Y. Li, M. Huang, Q. Fu, S.Z. Cheng, W.-B. Zhang, Polym. Chem. **5**, 3588 (2014)
131. W. Zhang, Y. Liu, J. Huang, T. Liu, W. Xu, S. Cheng, X. Dong, Soft Matter **15**, 7108 (2019)
132. T. Zhang, C. Fu, Y. Yang, F. Qiu, J. Chem. Phys. **146**, 054902 (2017)
133. S.C. Glotzer, M.A. Horsch, C.R. Iacovella, Z. Zhang, E.R. Chan, X. Zhang, Curr. Opin. Colloid in **10**, 287 (2005)
134. B. Zhang, X. Ye, B.J. Edwards, J. Chem. Phys. **139**, 244909 (2013)
135. T. Zhang, C. Fu, Y. Yang, F. Qiu, Macromolecules **51**, 1654 (2018)
136. Y. Wang, J. Cui, Y. Han, W. Jiang, Langmuir **35**, 468 (2019)
137. M.G. Moffitt, J. Phys. Chem. Lett. **4**, 3654 (2013)
138. P. Akcora et al., Nat. Mater. **8**, 354 (2009)
139. I. Borukhov, L. Leibler, Macromolecules **35**, 5171 (2002)
140. A. Jayaraman, J. Polym. Sci. B: Polym. Phys. **51**, 524 (2013)
141. V. Ganesan, C.J. Ellison, V. Pryamitsyn, Soft Matter **6**, 4010 (2010)
142. V. Pryamtisyn, V. Ganesan, A.Z. Panagiotopoulos, H. Liu, S.K. Kumar, J. Chem. Phys. **131**, 221102 (2009)
143. V.V. Ginzburg, Macromolecules **46**, 9798 (2013)
144. H. Koerner, L. Drummy, B. Benicewicz, L. Yu, R. Vaia, Acs Macro Lett. **2**, 670 (2013)
145. J. Lee, Z. Wang, J. Zhang, J. Yan, T. Deng, Y. Zhao, K. Matyjaszewski, M.R. Bockstaller, Macromolecules (2020)
146. V.V. Ginzburg, Macromolecules **50**, 9445 (2017)
147. V. Goel, J. Pietrasik, H.C. Dong, J. Sharma, K. Matyjaszewski, R. Krishnamoorti, Macromolecules **44**, 8129 (2011)
148. K. Hong, J. Noolandi, Macromolecules **14**, 1229 (1981)
149. K. Chen, H.-S. Li, B.-K. Zhang, J. Li, W.-D. Tian, Sci. Rep.-Uk **6**, 20355 (2016)

150. G. Fredrickson, *The Equilibrium Theory of Inhomogeneous Polymers*, vol. 134 (Oxford University Press on Demand, 2006)
151. V. Ginzburg, J. Weinhold, P. Hustad, P. Trefonas, B. Kim, N. Laachi, and G. Fredrickson, in *Directed Self-assembly of Block Co-polymers for Nano-manufacturing* (Elsevier, 2015), p. 67
152. A. Arora, D.C. Morse, F.S. Bates, K.D. Dorfman, J. Chem. Phys. **146**, 244902 (2017)
153. A. Arora, J. Qin, D.C. Morse, K.T. Delaney, G.H. Fredrickson, F.S. Bates, K.D. Dorfman, Macromolecules **49**, 4675 (2016)
154. F.S. Bates, G.H. Fredrickson, Phys. Today **52**, 32 (1999)

Chapter 3
Coarse-Grained Modeling of Polymer Nanocomposites: Field-Theoretic Simulations

Jason P. Koski, Huikuan Chao, Christian Tabedzki, and Robert A. Riggleman

Abstract For several decades polymer field theory has been an important computational and theoretical tool for predicting and interpreting the phase behavior of polymers. Many emerging applications of polymer materials involve blending polymers with nanoparticles to improve performance; however, the thermodynamics of these systems remains relatively poorly understood. In this chapter, we summarize numerous recent advances in polymer field theory with a focus on our approach to extend polymer field theory to describe polymer nanocomposites. We discuss recent applications of the methods as well as the mutual advantages and disadvantages of various implementations and conclude with a discussion of the challenges and ongoing development of the methods.

3.1 Introduction

One of the intrinsic challenges with modeling of polymer nanocomposites is the inherent multiple length scales involved with the materials, and phenomena on each length scale can affect the resulting properties [1–3]. On the most local level, one must consider the Coulombic, van der Waals, and in some cases Hydrogen-bonding interactions between all species; on the polymer segmental level, packing effects begin to come into play. On yet larger length scales, polymer conformations and

J. P. Koski
Sandia National Laboratories, P.O. Box 5800, Albuquerque, NM 87185, USA
e-mail: jkoski@sandia.gov

H. Chao
Core Research and Development,The Dow Chemical Company, Midland, USA
e-mail: hchao@dow.com

C. Tabedzki · R. A. Riggleman (✉)
Chemical and Biomolecular Engineering,University of Pennsylvania, Philadelphia, USA
e-mail: rrig@seas.upenn.edu

C. Tabedzki
e-mail: tabedzki@seas.upenn.edu

V. V. Ginzburg and L. M. Hall (eds.), *Theory and Modeling of Polymer Nanocomposites*, Springer Series in Materials Science 310,
https://doi.org/10.1007/978-3-030-60443-1_3

nanoparticle sizes and shapes exist, and finally on the largest scales one must account for the bulk structure of the composite, which could in general involve multiple phases coexisting. While there are established strategies for capturing the behavior on relatively small ($\lesssim$ 20 nm) and continuum mechanics works well on the largest length scales, only a handful of approaches are able to predict the structure and phase behavior of composites on larger, mesoscale length scales.

In addition to molecular simulations, two other statistical mechanical approaches are commonly used: polymer reference interaction site model (PRISM) theory and field-theoretic approaches, the latter of which is the primary subject of this chapter. As PRISM theory and background of this approach has been nicely reviewed recently by Martin et al. [4], the details of the theory will only be presented at a high level here in order to focus on the results generated for polymer nanocomposites. Briefly, a set of nonlinear matrix integral equations that are based on the microscopic interactions are employed to calculate the thermodynamic properties of the fluid of interest, and typically the focus is on the static structure factor between the various components, $S(k)$. A challenge of standard PRISM theory is the apparent difficulty in incorporating densely grafted nanoparticles into the theory, although the hybrid Monte-Carlo/PRISM theory approach of Jayaraman et al. provides an elegant solution to this issue [5, 6]. PRISM theory has played an instrumental role in developing theoretical phase diagrams for polymer nanocomposites [7, 8].

Separately, polymer field theory has a rich history that gives it some advantages over other approaches. From the early work of Edwards showing that one can estimate the statistics of Gaussian polymer chains with excluded volume interactions [9] to ubiquitous application to various block copolymer structures [10–15], field theory has a record of generally agreeing well with experiments. Experimental strategies for estimating the parameters of common versions of the theory are well-established and tabulated, and the theory lends itself to both analytic and numerical analysis, depending on the level of approximations employed [10]. Numerical implementations of field theory as either field-theoretic simulations (FTS) or as self-consistent field theory (SCFT) also allow the study of inhomogeneous (phase separated) states.

Two primary strategies have emerged for the study of polymer nanocomposites using field-theoretic simulations. First among them was the seminal work by Thompson et al. [16] combining density functional theory with self-consistent field theory (SCFT/DFT) to predict the distribution of nanoparticles in block copolymer matrices. This method has been employed in a variety of applications with great success at capturing the behaviors observed in experiments [17, 18]. As this method is discussed at length in Chap. 2 of this text, we will not focus on an extended discussion of this approach. Rather, we focus on the hybrid particle/field theory, wherein explicit nanoparticles are retained in the partition function describing the system [19], and its subsequent generalization by many of us [20].

In this chapter, we present a detailed derivation and discussion of the extensions of polymer field theory that allow for the inclusion of excluded volume correlations (see Fig. 3.1) and the associated study of polymer nanocomposites using field-based simulations. To most clearly demonstrate the extensions of conventional methods, we first present a standard field theory derivation of a compressible diblock melt

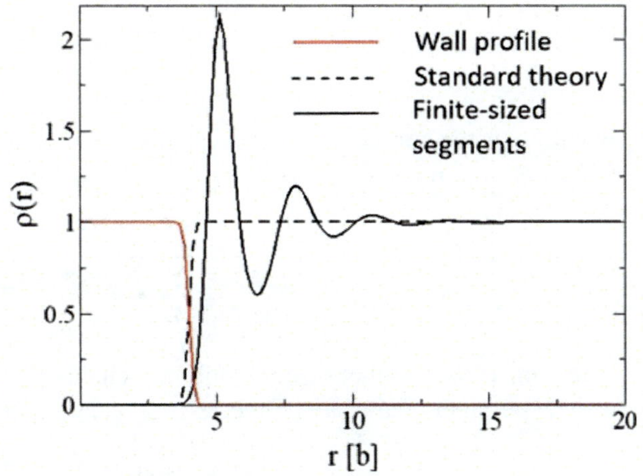

Fig. 3.1 Segment density near a polymer wall for the standard field theory using delta function densities and for the theory using finite-sized segments. The latter gives rise to liquid-like layering near the wall

blended with a homopolymer. Here, we demonstrate the particle-to-field transformation and derive the corresponding density operators for the polymer chains. Second, we present the framework developed by Sides and coworkers [19] to include explicit particles and establish a hybrid particle-field theory (HPFT) technique used to study polymer nanocomposites. We then present extensions of these frameworks to include field-based nanoparticles that allow for rapid calculation of nanoparticle density distributions in bulk systems while also providing a method to trivially include field fluctuations through complex Langevin (CL) simulations. These extensions include field-based nanospheres, nanorods, simple grafted nanoparticles (homopolymer grafting), complex grafted nanoparticles (diblock, mixed, and Janus grafting), and the incorporation of advanced pair-potentials in a field-based formulation. Next, we provide a non-equilibrium dynamic mean-field theory that goes beyond the aforementioned equilibrium methods and allows the integration and study of dynamic processes. Finally, we present a short section on the numerical implementation of the various methods and highlight some recent results obtained using the methods presented herein before concluding by summarizing areas for future development and outstanding challenges.

3.2 Standard Field Theory—Compressible Diblock with Homopolymer

We first present a standard field theory formalism of a compressible diblock copolymer melt with homopolymer. The polymers are modeled as discrete Gaussian chains with N_D and N_H segments for the diblock and homopolymer chains, respectively. Here, $N_D = N_A + N_B$ where N_A and N_B are the number of segments in the A and B block of the copolymer, respectively. The microscopic density of the diblock copolymer is given by

$$\hat{\rho}_K(\mathbf{r}) = \sum_i^{n_D} \sum_j^{N_K} \delta(\mathbf{r} - \mathbf{r}_{i,j}), \tag{3.1}$$

where n_D is the number of diblock chains and K denotes either the A or B block. The homopolymer microscopic density is given by

$$\hat{\rho}_H(\mathbf{r}) = \sum_i^{n_H} \sum_j^{N_H} \delta(\mathbf{r} - \mathbf{r}_{i,j}), \tag{3.2}$$

where n_H is the number of homopolymer chains.

The polymer chains are connected via a Gaussian bonding potential

$$\beta U_0 = \sum_i^{n_T} \sum_j^{N_T - 1} \frac{3|\mathbf{r}_{i,j} - \mathbf{r}_{i,j+1}|^2}{2b^2}, \tag{3.3}$$

where b is the statistical segment length of the polymer and T is either D or H to denote the type of polymer chain. Here, we assume that $b_A = b_B = b_H = b$. A Helfand compressibility [21] potential is used to enforce an energetic penalty for deviations away from the total system density, ρ_0, given by

$$\beta U_1 = \frac{\kappa}{2\rho_0} \int d\mathbf{r} \, [\hat{\rho}_+(\mathbf{r}) - \rho_0]^2, \tag{3.4}$$

where $\hat{\rho}_+ = \hat{\rho}_A + \hat{\rho}_B + \hat{\rho}_H$ is the spatially varying total microscopic density and κ controls the strength of the density fluctuations. Finally, a repulsive Flory-like potential is used to include enthalpic repulsions between dissimilar components

$$\beta U_2 = \sum_{I<J} \frac{\chi_{IJ}}{\rho_0} \int d\mathbf{r} \, \hat{\rho}_I(\mathbf{r}) \hat{\rho}_J(\mathbf{r}), \tag{3.5}$$

where $\hat{\rho}_I$ and $\hat{\rho}_J$ are the microscopic densities of components I and J, respectively, and χ_{IJ} is the Flory parameter governing the strength of the interaction between components I and J. The sum is taken over distinct combinations of A, B, and H.

The total partition function for the particle model can now be written as

$$\mathcal{Z} = z_0 \int d\mathbf{r}^{n_H N_H} \int d\mathbf{r}^{n_D N_D} \, e^{-\beta U_0 - \beta U_1 - \beta U_2}, \tag{3.6}$$

where z_0 contains the numerical prefactors such as the thermal de Broglie wavelengths. Hubbard-Stratonovich transformations are then used to decouple the particle coordinates and transform the microscopic densities into conjugated chemical potential fields [10, 22]. In order to proceed through the Hubbard-Stratonovich transformation for the nonbonded potential in (3.5), it is necessary to write the equation in a quadratic form. This can be done by re-writing (3.5) as the difference of two quadratic terms,

$$\beta U_2 = \sum_{I \neq J} \frac{\chi_{IJ}}{4\rho_0} \int d\mathbf{r}\, [\hat{\rho}_{IJ}^{(+)}(\mathbf{r})]^2 - \frac{\chi_{IJ}}{4\rho_0} \int d\mathbf{r}\, [\hat{\rho}_{IJ}^{(-)}(\mathbf{r})]^2, \tag{3.7}$$

where $\hat{\rho}_{IJ}^{(\pm)}(\mathbf{r}) = \hat{\rho}_I(\mathbf{r}) \pm \hat{\rho}_J(\mathbf{r})$. As a result, for each pair of enthalpically distinct components, two energetic terms arise in the particle-to-field transformation from (3.7). Note that for a two component *incompressible* system, $\hat{\rho}_I(\mathbf{r}) + \hat{\rho}_J(\mathbf{r})$ must equal ρ_0 at every point such that $\int d\mathbf{r}\, [\hat{\rho}_{IJ}^{(+)}(\mathbf{r})]^2 = \int d\mathbf{r}\, \rho_0^2 = \rho_0(n_I N_I + n_J N_J)$ and this term can be lumped into z_0. In this case, only a single energetic term arises from (3.7). However, for a compressible system or systems where there are more than two enthalpically distinct components, $\hat{\rho}_I(\mathbf{r}) + \hat{\rho}_J(\mathbf{r})$ is not restricted to equal ρ_0 at every point and may be spatially varying in which the previously described simplification of (3.7) is not applicable.

Here, the most general result of the particle-to-field transformation is the field-theoretic partition function,

$$\mathcal{Z} = z_1 \int \mathcal{D}w_+ \prod_{I \neq J} \int \mathcal{D}w_{IJ}^{(+)} \int \mathcal{D}w_{IJ}^{(-)}\, e^{-\mathcal{H}[\{w\}]}, \tag{3.8}$$

where z_1 is the re-defined numerical prefactor that differs from z_0 in that it contains the normalization constants from the Gaussian functional integrals used to de-couple the particle interactions, $\mathcal{H}[\{w\}]$ is the effective Hamiltonian of the system, and $\{w\}$ is the set of chemical potential fields for our system, w_+, $w_{AB}^{(+)}$, $w_{AB}^{(-)}$, $w_{AH}^{(+)}$, $w_{AH}^{(-)}$, $w_{BH}^{(+)}$, $w_{BH}^{(-)}$. We note that all of the integrals in (3.8) are along the real axis. The effective Hamiltonian is given by

$$\begin{aligned}\mathcal{H}[\{w\}] = {} & \frac{\rho_0}{2\kappa} \int d\mathbf{r}\, w_+(\mathbf{r})^2 - i\rho_0 \int d\mathbf{r}\, w_+(\mathbf{r}) \\ & + \sum_{I \neq J} \frac{\rho_0}{\chi_{IJ}} \int d\mathbf{r}\, w_{IJ}^{(+)}(\mathbf{r})^2 + \sum_{I \neq J} \frac{\rho_0}{\chi_{IJ}} \int d\mathbf{r}\, w_{IJ}^{(-)}(\mathbf{r})^2 \\ & - n_D \ln Q_D[w_A, w_B] - n_H \ln Q_H[w_H],\end{aligned} \tag{3.9}$$

where Q_D and Q_H are the partition functions of a single diblock copolymer and homopolymer chain, respectively. Q_D is calculated from the chain propagator, $q_D(j, \mathbf{r})$,

$$Q_D[w_A, w_B] = \frac{1}{V} \int d\mathbf{r}\, q_D(N_D, \mathbf{r}), \tag{3.10}$$

which is in turn constructed by iterating a Chapman-Kolmogorov equation

$$q_D(j+1,\mathbf{r}) = e^{-w_K(\mathbf{r})} \int d\mathbf{r}'\ \Phi(\mathbf{r}-\mathbf{r}')\, q_D(j,\mathbf{r}), \tag{3.11}$$

where w_K is either w_A or w_B if segment $j+1$ is an A or B segment, respectively. $\Phi(\mathbf{r}-\mathbf{r}')$ represents the transitional probability density describing the conditional probability of the next discrete polymer segment existing at $\mathbf{r}$ given the previous segment's location at $\mathbf{r}'$; for Gaussian spring models, the Fourier transform of transitional probability density is $\hat{\Phi}(\mathbf{k}) = \exp(-b^2|\mathbf{k}|^2/6)$. The initial condition is taken from the A end of the block copolymer and is given as

$$q_D(1,\mathbf{r}) = e^{-w_A(\mathbf{r})}. \tag{3.12}$$

In (3.10)–(3.12) we have suppressed the functional dependence of q on w_A and w_B for notational simplicity. For a homopolymer, w_K and w_A are replaced with w_H and q_D is replaced with q_H in (3.10)–(3.12). Note that in the case of $N_H = 1$, the homopolymer is represented as an explicit solvent and the single molecule partition function simply becomes $Q_H[w_H; N=1] = \frac{1}{V}\int d\mathbf{r}\, e^{-w_H(\mathbf{r})}$ and that no iteration of the Chapman-Kolmogorov equation is necessary.

The fields w_A, w_B, and w_H are each given by

$$\begin{aligned} w_A &= i[w_+ + w_{AB}^{(+)} + w_{AH}^{(+)}] - w_{AB}^{(-)} - w_{AH}^{(-)}, \\ w_B &= i[w_+ + w_{AB}^{(+)} + w_{BH}^{(+)}] + w_{AB}^{(-)} - w_{BH}^{(-)}, \\ w_H &= i[w_+ + w_{AH}^{(+)} + w_{BH}^{(+)}] + w_{AH}^{(-)} + w_{BH}^{(-)}. \end{aligned} \tag{3.13}$$

The above equations (and all similar equations in this chapter) assume non-negative values for all χ_{IJ}. In cases where χ_{IJ} is negative, $w_{IJ}^{(-)}$ carries the imaginary unit rather than $w_{IJ}^{(+)}$ [23, 24].

The density operators of the polymer segments are given by

$$\tilde{\rho}_A(\mathbf{r}) = \frac{n_D}{VQ_D}\sum_{j=1}^{N_A} q_D(j,\mathbf{r})\, e^{w_A(\mathbf{r})}\, q_D^{\dagger}(N_D-j+1,\mathbf{r}), \tag{3.14}$$

$$\tilde{\rho}_B(\mathbf{r}) = \frac{n_D}{VQ_D}\sum_{j=N_A+1}^{N_D} q_D(j,\mathbf{r})\, e^{w_B(\mathbf{r})}\, q_D^{\dagger}(N_D-j+1,\mathbf{r}), \tag{3.15}$$

$$\tilde{\rho}_H(\mathbf{r}) = \frac{n_H}{VQ_H}\sum_{j=1}^{N_H} q_H(j,\mathbf{r})\, e^{w_H(\mathbf{r})}\, q_H(N_H-j+1,\mathbf{r}). \tag{3.16}$$

The complementary propagator, $q_D^{\dagger}(j,\mathbf{r})$, is determined by iterating the same Chapman-Kolmogorov equation described above in (3.11) starting from the B end of the diblock copolymer.

At this point, a mean-field approximation is often used to simplify (3.8) such that it is assumed that a single solution of $\{w\}$, denoted as $\{w^*\}$, dominates the partition function [10]. This leads to $\mathcal{Z} \approx z_1\, e^{-\mathcal{H}[\{w^*\}]} = e^{-\beta F}$ and a simple steepest descent method can be used to determine $\{w^*\}$. Recently, it has been demonstrated that complex Langevin (CL) sampling can be used to sample the fully fluctuating field theory rather than using a mean-field approximation [10, 25–28]. While the CL method is expected to exactly sample the underlying particle model, numerical implementation of CL sampling leads to larger computational expense and one loses direct access to the free energy [29]. Furthermore, as discussed below, there seem to be cases where the CL method does not converge.

An additional consideration of CL simulations is the presence of ultraviolet (UV) divergences [10, 30]. UV divergences arise from the issue of resolving a delta function, which becomes increasingly problematic with increased lattice resolution. The presence of UV divergences causes thermodynamic parameters such as the free energy, chemical potential, and the osmotic pressure to diverge as the resolution of the simulation grid increases. This can be shown analytically where integrands arise as a function of the wavenumber, $\mathbf{k}$, which do not converge without choosing a wavenumber cutoff. The issue of UV divergences is not present at the mean-field level.

Furthermore, representing both the microscopic densities and the interaction potentials as delta functions in the standard field theory formalism results in the absence of excluded volume correlations. The conventional field theory formalism is thus unable to model systems where excluded volume correlations are essential, such as polymer nanocomposites. This is critical in a system with nanoparticles where a nanoparticle needs to interact with a second nanoparticle at distances greater than 0. The excluded volume correlations are what give rise to aggregated nanoparticles forming ordered lattice structures, which cannot be observed with using delta functions to represent both the microscopic densities and the interaction potentials. In Sect. 4 below, we demonstrate how to resolve this issue.

3.3 Hybrid Particle-Field Theory— Explicit Nanospheres

In an effort to extend the field theory framework, Sides and coworkers [19] developed a hybrid particle-field theory (HPFT) method to model polymer nanocomposites. Within this framework, the polymers are treated as they are in a typical field theory formalism while the nanoparticles retain their explicit particle positions in the Hamiltonian. The microscopic density of the nanoparticles are given by

$$\hat{\rho}_P(\mathbf{r}) = \sum_{i}^{n_P} \Gamma(\mathbf{r} - \mathbf{r}_i), \tag{3.17}$$

where n_P is the number of nanoparticles and $\Gamma(\mathbf{r})$, which describes the NP shape, is defined as

$$\Gamma(\mathbf{r}) = \frac{\rho_0}{2}\,\mathrm{erfc}\left(\frac{|\mathbf{r}| - R_P}{\xi}\right). \tag{3.18}$$

Here, R_P is the radius of the particle and ξ is the length scale in which the particle density goes from ρ_0 inside the core to 0 outside the core. For a system where the NPs have neutral interactions with all polymeric species, we introduce the nanoparticle density into the $\hat{\rho}_+$ term in (3.4) and keep the explicit particle coordinates of the nanoparticles intact during the particle-to-field transformation. As a result, (3.8) and (3.9) become

$$\mathcal{Z} = z_1 \int d\mathbf{r}^{n_P} \int \mathcal{D}w_+ \prod_{I \neq J} \int \mathcal{D}w_{IJ}^{(+)} \int \mathcal{D}w_{IJ}^{(-)}\, e^{-\mathcal{H}[\mathbf{r}^{n_P},\{w\}]}, \tag{3.19}$$

and

$$\begin{aligned}\mathcal{H}[\mathbf{r}^{n_P}, \{w\}] = & \frac{\rho_0}{2\kappa} \int d\mathbf{r}\; w_+(\mathbf{r})^2 - i \int d\mathbf{r}\; w_+(\mathbf{r}) \left(\rho_0 - \hat{\rho}_P(\mathbf{r})\right) \\ & + \sum_{I \neq J} \frac{\rho_0}{\chi_{IJ}} \int d\mathbf{r}\; w_{IJ}^{(+)}(\mathbf{r})^2 + \frac{\rho_0}{\chi_{IJ}} \int d\mathbf{r}\; w_{IJ}^{(-)}(\mathbf{r})^2 \\ & - n_D \ln Q_D[w_A, w_B] - n_H \ln Q_H[w_H],\end{aligned} \tag{3.20}$$

respectively.

The cavity functions, $\Gamma(\mathbf{r})$, introduced in (3.17) are now coupled to the w_+ field such that there is an energetic penalty, governed by κ, for the polymer chains penetrating the core of the nanoparticles. Sides and coworkers [19] demonstrated that Brownian dynamics or force-bias Monte Carlo schemes could be used to update the particle positions where the force on each particle can be computed by

$$\mathbf{F}_j = -\frac{\partial H}{\partial \mathbf{r}_j} = -i\rho_0 \int d\mathbf{r}\, \mathbf{g}(\mathbf{r}_j - \mathbf{r}) w_+(\mathbf{r}), \tag{3.21}$$

where $\mathbf{g}(\mathbf{r}) = \frac{1}{r}\frac{d\Gamma(r)}{dr}\mathbf{r}$.

Unfortunately, spanning a large nanoparticle parameter space is difficult using these methods, particularly in dense polymer melts and/or systems with several nanoparticles. As a result, the HPFT method is often used to model just one or two nanoparticles where the positions of the nanoparticles are fixed. This method is useful to see how polymers or solvent respond or interact with a particle where at the mean-field level, the free energy, chain entropy, or the polymer/solvent distribution near the particle can rapidly be calculated. This approach has been used in numerous applications to model an explicit particle sitting at various defects in a diblock copolymer [31], comparing vesicle and cluster configurations in a diblock copolymer solution [32], or to generate potential of mean force curves by varying the interparticle distance of two explicit particles for an array of polymer melt conditions [33]. While the method of using one or two explicit particles with fixed positions is useful in these situations, the utility of this technique relies on either biasing the particle(s) in certain positions and the polymers equilibrating around the particle(s), or being able to sufficiently sample the particle coordinates. Consequently, the HPFT approach overall is limited in characterizing bulk polymer nanocomposite behavior, particularly in 3D. Lastly, relaxing the mean-field approximation through complex Langevin sampling is a non-trivial task in the HPFT approach. In short, the complex nature of the Hamiltonian would involve forces that are complex-valued, which would push the nanoparticles out of the real plane, and it is not clear how to treat certain aspects of the simulation, such as periodic boundary conditions.

3.4 Field-Based Nanospheres and Finite-Segment Polymers

In this section, we present a generalization of the HPFT approach to incorporate nanoparticles into the field theory framework as field-based particles rather explicit particles [20]. This can be done by using the same microscopic density operator in (3.17) and rather than keeping the nanoparticle coordinates intact, the particle coordinates can be exactly integrated out of the partition function, where (3.20) becomes (Fig. 3.2)

$$\begin{aligned}\mathcal{H}[\{w\}] = {} & \frac{\rho_0}{2\kappa}\int d\mathbf{r}\, w_+(\mathbf{r})^2 - i\rho_0 \int d\mathbf{r}\, w_+(\mathbf{r}) \\ & + \sum_{I \neq J} \frac{\rho_0}{\chi_{IJ}} \int d\mathbf{r}\, w_{IJ}^{(+)}(\mathbf{r})^2 + \frac{\rho_0}{\chi_{IJ}} \int d\mathbf{r}\, w_{IJ}^{(-)}(\mathbf{r})^2 \\ & - n_D \ln Q_D[w_A, w_B] - n_H \ln Q_H[w_H] - n_P \ln Q_P[\Gamma * w_P], \end{aligned} \tag{3.22}$$

where Q_P is the partition function of a single nanoparticle. Here we use $*$ as the shorthand for a convolution operation $f * g = \int_{-\infty}^{\infty} dx' f(x - x') g(x')$. In a similar manner, we can redefine the microscopic densities of the polymers in (3.1) and (3.2) as an array of unit Gaussians rather than delta functions such that

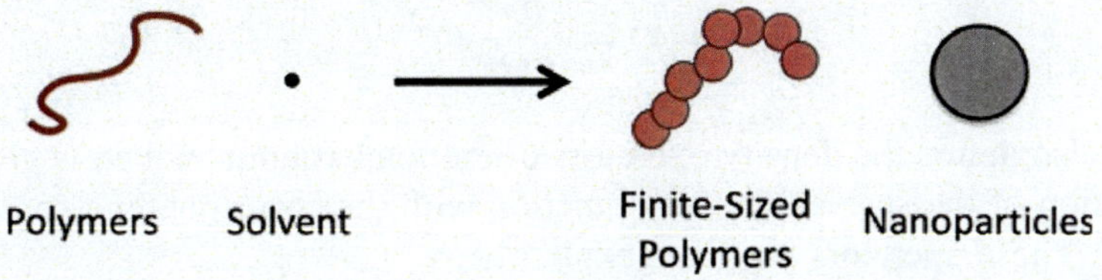

Fig. 3.2 Schematic illustrating the introduced physics in this section; polymers and solvent to finite-sized polymers and nanoparticles

$$\hat{\rho}_K(\mathbf{r}) = \sum_i^{n_D} \sum_j^{N_K} h(\mathbf{r} - \mathbf{r}_{i,j}), \tag{3.23}$$

$$\hat{\rho}_H(\mathbf{r}) = \sum_i^{n_H} \sum_j^{N_H} h(\mathbf{r} - \mathbf{r}_{i,j}), \tag{3.24}$$

where

$$h(\mathbf{r}) = \left(\frac{1}{2\pi a^2}\right)^{\mathbb{D}/2} \exp\left(-\frac{\mathbf{r}^2}{2a^2}\right). \tag{3.25}$$

Here, $\mathbb{D}$ is the dimension of the problem and a is the effective monomer size. We note explicitly that in defining (3.17), (3.23), and (3.24), that these representations are a simple convolution of the corresponding density function (i.e. $\Gamma(\mathbf{r})$ for the nanospheres or $h(\mathbf{r})$ for the polymers) with the array of delta functions used in the standard field theory formulation presented in Sect. 2.2. Tracing these steps through the particle-to-field transformation leads to the effective Hamiltonian

$$\begin{aligned}\mathcal{H}[\{w\}] = {} & \frac{\rho_0}{2\kappa} \int d\mathbf{r}\; w_+(\mathbf{r})^2 - i\rho_0 \int d\mathbf{r}\; w_+(\mathbf{r}) \\ & + \sum_{I \neq J} \frac{\rho_0}{\chi_{IJ}} \int d\mathbf{r}\; w_{IJ}^{(+)}(\mathbf{r})^2 + \frac{\rho_0}{\chi_{IJ}} \int d\mathbf{r}\; w_{IJ}^{(-)}(\mathbf{r})^2 \\ & - n_D \ln Q_D[h * w_A, h * w_B] - n_H \ln Q_H[h * w_H] \\ & - n_P \ln Q_P[\Gamma * w_P]. \end{aligned} \tag{3.26}$$

Now, we have two sets of density operators where the density operators presented in (3.15) and (3.16) within Sect. 2.2 denote the diblock and homopolymer segment center density, respectively. These operators correspond to the densities that would typically be used to calculate a pair distribution function in a molecular dynamics simulation, for example. The nanoparticle center density takes the form

$$\tilde{\rho}_P(\mathbf{r}) = \frac{n_P}{V Q_P} e^{-w_P(\mathbf{r})}, \tag{3.27}$$

We can also define the density operators of the total volume, which is given by the convolution of the species' density function with the corresponding center density operator. These operators are given explicitly as

$$\breve{\rho}_K(\mathbf{r}) = (h * \tilde{\rho}_K)(r), \tag{3.28}$$

$$\breve{\rho}_H(\mathbf{r}) = (h * \tilde{\rho}_H)(r), \tag{3.29}$$

and

$$\breve{\rho}_P(\mathbf{r}) = (\Gamma * \tilde{\rho}_P)(r). \tag{3.30}$$

Here, we point out that the Hamiltonian (and associated partition function) are once again strictly a function of chemical potential fields and all particle coordinates have been integrated out of the partition function. As the Hamiltonian is now solely a function of chemical potential fields, introducing field fluctuations through the CL method is straightforward. Additionally, the density functions address the issue of UV divergences in the fully fluctuating field theory of the standard field theory formalism by regularizing the theory at $\mathbf{r} = 0$. Furthermore, the introduction of the field-based nanospheres does not change how the computational expense scales. The sum over I and J can be further generalized to distinct combinations of A, B, H, and P. In many of our published examples [20, 34–36], we commonly treat the nanoparticles as A particles where $w_A = w_P$ or as neutral nanoparticles ($\chi_{KP} = 0$) so as to reduce the parameter space and the number of fields to equilibrate and/or sample. However, the most general definitions of the fields w_A, w_B, w_H, and w_P are now given by

$$\begin{aligned}
w_A &= i[w_+ + w_{AB}^{(+)} + w_{AH}^{(+)} + w_{AP}^{(+)}] - w_{AB}^{(-)} - w_{AH}^{(-)} - w_{AP}^{(-)}, \\
w_B &= i[w_+ + w_{AB}^{(+)} + w_{BH}^{(+)} + w_{BP}^{(+)}] + w_{AB}^{(-)} - w_{BH}^{(-)} - w_{BP}^{(-)}, \\
w_H &= i[w_+ + w_{AH}^{(+)} + w_{BH}^{(+)} + w_{HP}^{(+)}] + w_{AH}^{(-)} + w_{BH}^{(-)} - w_{HP}^{(-)}, \\
w_P &= i[w_+ + w_{AP}^{(+)} + w_{BP}^{(+)} + w_{HP}^{(+)}] + w_{AP}^{(-)} + w_{BP}^{(-)} + w_{HP}^{(-)}.
\end{aligned} \tag{3.31}$$

An additional advantage of this generalization of the HPFT method is that both explicit and field-based particles can be introduced into the same system. This is demonstrated by representing (3.17) as

$$\hat{\rho}_P(\mathbf{r}) = \sum_i^{n_P} \Gamma(\mathbf{r} - \mathbf{r}_i) = \sum_i^{n_{P,\mathrm{exp}}} \Gamma(\mathbf{r} - \mathbf{r}_i) + \sum_j^{n_{P,\mathrm{field}}} \Gamma(\mathbf{r} - \mathbf{r}_j), \tag{3.32}$$

which leads to

$$\begin{aligned}
\mathcal{H}[\mathbf{r}^{n_{P,\mathrm{exp}}}, \{w\}] = {} & \frac{\rho_0}{2\kappa} \int d\mathbf{r}\, w_+(\mathbf{r})^2 - i \int d\mathbf{r}\, w_+(\mathbf{r}) \left(\rho_0 - \hat{\rho}_{P,\mathrm{exp}}(\mathbf{r})\right) \\
& + \sum_{I \neq J} \frac{\rho_0}{\chi_{IJ}} \int d\mathbf{r}\, w_{IJ}^{(+)}(\mathbf{r})^2 + \frac{\rho_0}{\chi_{IJ}} \int d\mathbf{r}\, w_{IJ}^{(-)}(\mathbf{r})^2 \\
& - n_D \ln Q_D[h * w_A, h * w_B] - n_H \ln Q_S[h * w_H] \\
& - n_{P,\mathrm{field}} \ln Q_P[\Gamma * w_P],
\end{aligned} \tag{3.33}$$

where $n_{P,\mathrm{exp}}$ and $n_{P,\mathrm{field}}$ are the number of explicit and field based nanoparticles, respectively. The incorporation of both explicit and field-based particles leverages the advantages of both representations. The explicit particles allow for exact control and knowledge of a particle's position while the field-based particles lead to rapid calculations of the probability distribution of several particles around the explicit particle. The combination of both these approaches allows for efficient calculation of the particle distribution around a single explicit particle, which can be used to calculate the radial distribution function or potential of mean force. This approach effectively captures many-body effects and the capability to study the effect of field fluctuations to go beyond the mean-field approximation.

3.4.1 *Recent Results*

As an example where we employ a mix of explicit and field-based nanoparticles, in Fig. 3.3 we show recent results characterizing the interactions between spherical nanoparticles embedded in a cylinder-forming block copolymer [36]. A single, explicit nanoparticle is placed in the center of the box, and field-based nanoparticles are allowed to equilibrate around the explicit particle (Fig. 3.3a, b). We next calculate the distribution of the field-based particle centers along the length of the cylinder as a function of distance from the particle surface, $\rho(\Delta r - D_P)$, which is then converted into a free energy (potential of mean force) through $\Delta F/k_BT = -\log[\rho(\Delta r - D_P)/\rho(\infty)]$, where $\rho(\infty)$ is the far-field density. In all cases, the free energy profile is non-monotonic as a function of the particle separation. A barrier emerges when the particles are $\sim 3R_g$ apart, which is due to deformations of the interfaces between the A and B phases. A free energy well of approximately $4k_BT$ emerges when the particles are in contact as the entropic penalty of chains stretching to fill the gap between the nanoparticles is relaxed.

3.5 Field-Based Nanorods

We can extend this approach further to incorporate anistropic particles [20] where we focus our discussion here on nanorods (Fig. 3.4).

To do so, we first define an orientation vector, **u**, along the length of the nanorod where the nanorod density function is given by

$$\Gamma_{NR}(\mathbf{r}, \mathbf{u}) = \frac{\rho_0}{4} \operatorname{erfc}\left(\frac{|\mathbf{u} \cdot \mathbf{r}| - \mathrm{L_P}/2}{\xi}\right) \cdot \operatorname{erfc}\left(\frac{|\mathbf{u} \times \mathbf{r}| - \mathrm{R_P}}{\xi}\right), \tag{3.34}$$

and L_P is the length of the nanorod, R_P the radius, and ξ again plays the role of a parameter that controls the length scale over which the nanorod density decays from ρ_0 to 0. The Hamiltonian and partition function are represented in the same form as they are for the nanospheres except that the convolution of the nanoparticle density function with its chemical potential field is now taken with respect to both position and orientation. This additional complication only changes the nanoparticle partition function, Q_P, and the corresponding center and total density operators, $\tilde{\rho}(\mathbf{r},\mathbf{u})$ and $\breve{\rho}(\mathbf{r},\mathbf{u})$, respectively. These quantities now take the form

$$Q_P[w_P] = \frac{1}{4\pi V}\int d\mathbf{r}\int d\mathbf{u}\,\exp\left[-(\Gamma_{NR} * w_P)(\mathbf{r},\mathbf{u})\right], \tag{3.35}$$

$$\tilde{\rho}_P(\mathbf{r},\mathbf{u}) = \frac{n_P}{4\pi V Q_P} e^{-(\Gamma_{NR} * w_P)(\mathbf{r},\mathbf{u})}, \tag{3.36}$$

and

$$\breve{\rho}_P(\mathbf{r}) = \int d\mathbf{u}\,(\Gamma_{NR} * \tilde{\rho}_P)(\mathbf{r},\mathbf{u}). \tag{3.37}$$

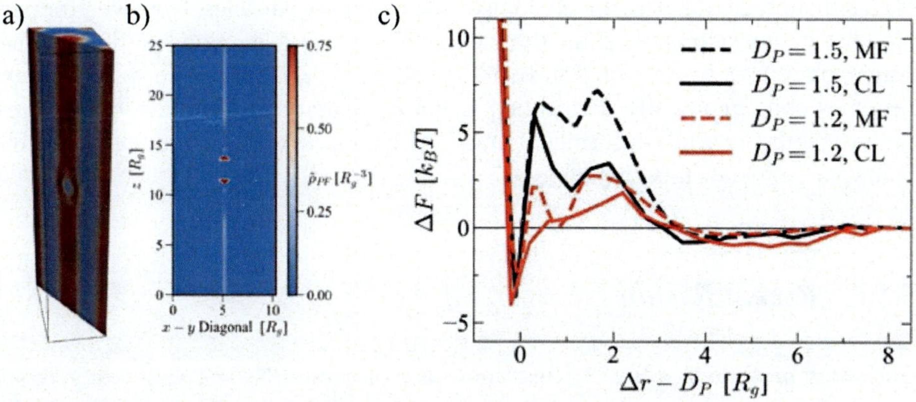

Fig. 3.3 Comparison of the complex Langevin and mean-field distributions of spherical nanoparticles around a central, explicit spherical nanoparticle embedded in the cylindrical phase of a block copolymer. **a** Slice through the simulation box showing the explicit spherical nanoparticle embedded in the minority component of the cylinder phase. **b** Distribution of field-based spherical particles around the central explicit particle in the plane visible in panel a. **c** Potential of mean force along the length of the cylinder for nanoparticles with diameter $D_P = 1.5R_g$ or $1.2R_g$ obtained using complex Langevin simulations (CL) or under the mean-field (MF) approximation. Figure adapted from [36]

Fig. 3.4 Schematic illustrating the introduced physics in this section; nanoparticles to nanorods

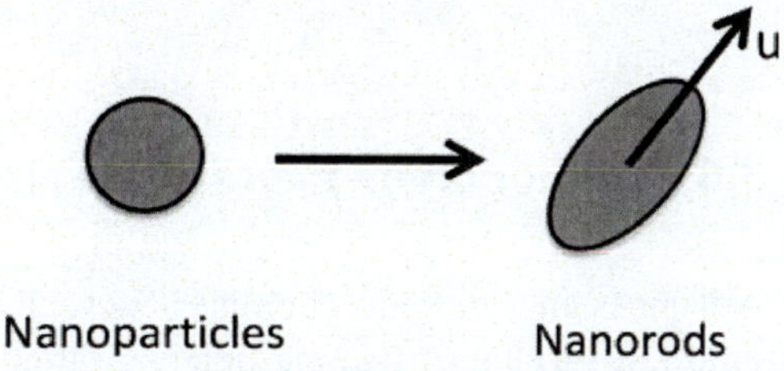

Note that the 4π in the prefactor of in (3.35) and (3.36) is for 3D calculations while it is 2π in 2D calculations. The orientation of the nanorods can be quantified by defining a tensor order parameter

$$S_{\alpha,\beta}(\mathbf{r}) = \frac{\int d\mathbf{u}(u_\alpha u_\beta - \frac{1}{3}\delta_{\alpha,\beta}) \exp(-(\Gamma_{NR} * w_P)(\mathbf{r}, \mathbf{u}))}{\int d\mathbf{u} \exp(-(\Gamma_{NR} * w_P)(\mathbf{r}, \mathbf{u})}, \tag{3.38}$$

where α and β refer to the Cartesian directions and $\delta_{\alpha,\beta}$ is the Kronecker delta function. The exponential terms represent the Boltzmann weight of the particle center density, and the eigenvector associated with the largest eigenvalue denotes the primary direction of alignment.

The introduction of the orientation vector, **u**, is implemented through a discretization of the unit-sphere in spherical coordinates using Gaussian-Legendre weights to efficiently calculate the spatial coordinates associated with **u**. The discretization of the unit-sphere allows for our method to characterize an array of anisotropic particles such as nanorods, Janus particles, Janus rods, or patchy particles. However, for particles that are not symmetric about the azemuthal angle (e.g. diamond particles or most patchy particles), two orientation vectors need to be defined and the system quickly becomes computationally demanding if not computationally intractable. Interestingly, the theoretically-informed Langevin dynamics (TILD) method, described below, circumvents this challenge.

3.5.1 Recent Results

The study presented in Fig. 3.5 is a showcase of using PNC-FT to investigate spatial distribution of nanorods in cylindrical phased PS-b-P2VP thin films [37, 38]. The field theoretic simulation and experimental results (Figs. 3.5a, b) suggest that nanorods segregated to the base of P2VP cylinder when the substrate wets the cylinder and induces defects at the base. The segregation of nanorods to the defect region relieves chain stretching and becomes thermodynamically favored. Figure 3.5c quantifies the segregation by showing the fraction of nanorod vertically oriented to the substrate as a functions of the wetting strength. When the nanorod length becomes comparable to the film thickness, the study shows the fraction saturates and becomes invariant to the wetting strength.

3.6 Homopolymer Grafted Nanoparticles

Advances in synthetic chemistry have led to the ability to graft polymer chains onto the surface of nanoparticles. Grafted chains serve as an additional degree of freedom in controlling the spatial distribution of particles within the polymer matrix.

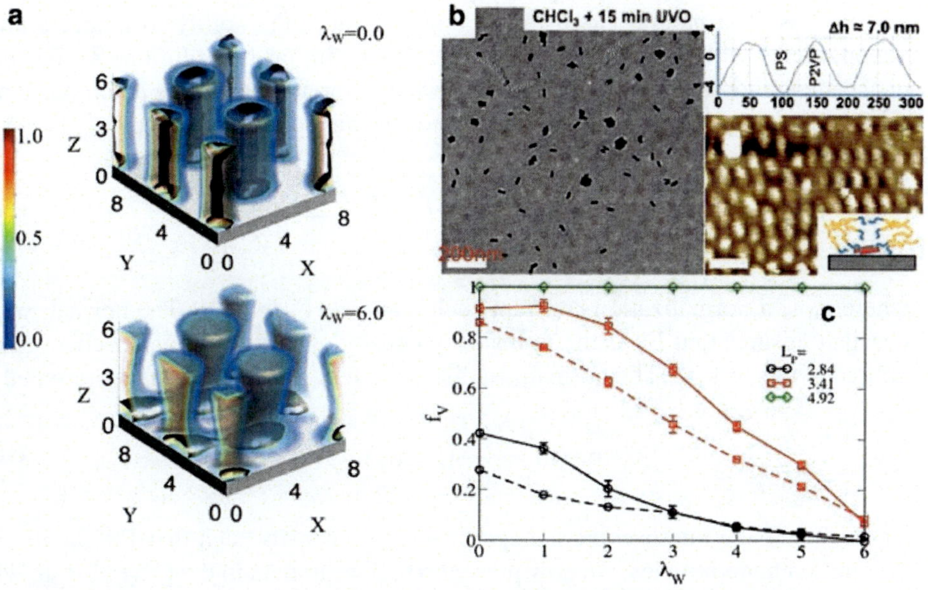

Fig. 3.5 **a** Density field maps containing isosurfaces that enclose regions where the A-block (minority component) local volume fraction is given by the colors on the colorbar with $\lambda W = 0.0$ (top) and $\lambda W = 6.0$ (bottom). **b** TEM (left) and AFM (right) topography images of a P2VP-grafted gold nanorod thin film. Majority of the nanorods are showing horizontal orientation to the substrate in the P2VP cylinder. **c** Fraction of vertically oriented nanorods f_v as a function of the strength of the bottom substrate selectivity λW in films with hard confinement (solid) or soft confinement (dashed). Panels a, c adapted from [38], and panel b reproduced from [37] with permission from The Royal Society of Chemistry

Fig. 3.6 Schematic illustrating the introduced physics in this section; nanoparticles to homopolymer grafted nanoparticles

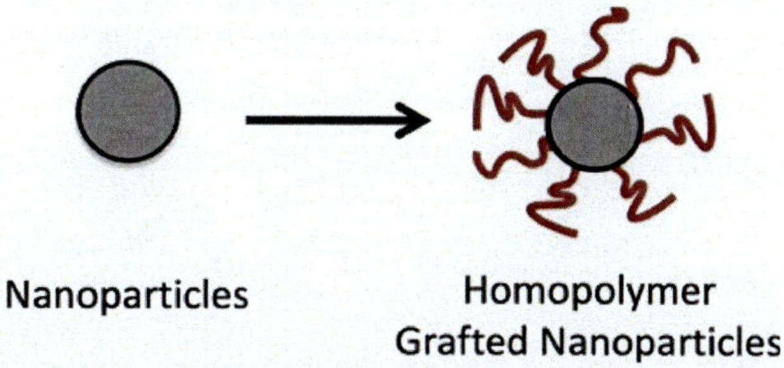

Specifically, the grafted chains can induce extra steric hindrance as particles approach one another and effectively prevent the particle aggregation that would arise due to attractive particle-particle interactions. Significant research has been done in the realm of grafted nanoparticles, particularly in varying the relative lengths of the matrix and grafted chains (Fig. 3.6).

As a result, a natural extension to our method is the ability to model grafted nanoparticles within our framework [39]. Using similar convolutions as above to distribute the mass of a nanoparticle about its center, we can define a homogeneous distribution of grafting sites at the surface of the nanoparticle given by

$$\Gamma_\sigma(\mathbf{r}) = \frac{1}{\sigma_0} \exp\left[-\left(\frac{|\mathbf{r}| - R_P - \xi}{\xi}\right)^2\right], \tag{3.39}$$

where σ_0 is a normalization constant such that $\int d\mathbf{r}\, \Gamma_\sigma(\mathbf{r}) = 1$. The normalization constant ensures that Γ_σ is the probability density of a *single* polymer chain being grafted to the surface. The normalized distribution of grafting sites is represented as

$$\sigma_G(\mathbf{r}) = \frac{1}{n_P} \left(\Gamma_\sigma * \hat{\rho}_{P,c}\right)(\mathbf{r}), \tag{3.40}$$

where $\hat{\rho}_{P,c}$ is the microscopic nanoparticle center density such that $\Gamma * \hat{\rho}_{P,c} = \hat{\rho}_P$. For grafted nanoparticles, we now present our Hamiltonian in the general form with explicit and field-based grafted nanoparticles

$$\begin{aligned}
\mathcal{H}[\mathbf{r}^{n_{P,\mathrm{exp}}}, \{w\}] = {} & \frac{\rho_0}{2\kappa} \int d\mathbf{r}\, w_+(\mathbf{r})^2 - i \int d\mathbf{r}\, w_+(\mathbf{r}) \left(\rho_0 - \hat{\rho}_{P,\mathrm{exp}}(\mathbf{r})\right) \\
& + \sum_{I \neq J} \frac{\rho_0}{\chi_{IJ}} \int d\mathbf{r}\, w_{IJ}^{(+)}(\mathbf{r})^2 + \frac{\rho_0}{\chi_{IJ}} \int d\mathbf{r}\, w_{IJ}^{(-)}(\mathbf{r})^2 \\
& - n_{G,\mathrm{exp}} \int d\mathbf{r}_\perp\, \sigma_{G,\mathrm{exp}}(\mathbf{r}_\perp) \ln q_{G,\mathrm{exp}}[r_\perp; h * w_G] \\
& - n_D \ln Q_D[h * w_A, h * w_B] - n_H \ln Q_H[h * w_H] \\
& - n_{P,\mathrm{field}} \ln Q_P[\mu_P],
\end{aligned} \tag{3.41}$$

where

$$\mu_{P,\mathrm{Homo}}(\mathbf{r}) = (\Gamma * w_P)(\mathbf{r}) - \frac{n_{G,\mathrm{field}}}{n_{P,\mathrm{field}}} \left(\Gamma_\sigma * \ln q_{G,\mathrm{field}}\right)(\mathbf{r}). \tag{3.42}$$

Here, $\mathbf{r}_\perp$ are the positions of the grafting sites for the explicit particles, n_G is the number of grafted chains, and $q_G(\mathbf{r})$ is the partition function of a single grafted chain propagated from the free end and terminating at $\mathbf{r}$. At this point, the derivation of (3.41) and (3.42) warrants additional detail. To simplify the derivation, lets first assume there are no explicit particles in the system. In the step after the particle-to-field transformation, the contribution from the grafted chains is given by

$$- n_G \int d\mathbf{r}_G\, \sigma_G(\mathbf{r}_G) \ln q_G(\mathbf{r}_G), \tag{3.43}$$

where $\mathbf{r}_G$ are the positions of the grafting sites for all of the nanoparticles. If we plug in (3.40) for σ_G above, we now get

$$\begin{aligned}&-\frac{n_G}{n_P}\int d\mathbf{r}_G\,(\Gamma_\sigma * \hat{\rho}_{P,c})(\mathbf{r}_G)\ln q_G(\mathbf{r}_G)\\&= -\frac{n_G}{n_P}\int d\mathbf{r}_G\int d\mathbf{r}'\,\Gamma_\sigma(\mathbf{r}_G-\mathbf{r}')\hat{\rho}_{P,c}(\mathbf{r}')\ln q_G(\mathbf{r}_G).\end{aligned} \tag{3.44}$$

Now, to go to the next step, we first point out that the following relation, $\int d\mathbf{r}\,h(\mathbf{r})\int d\mathbf{r}' f(\mathbf{r}-\mathbf{r}')g(\mathbf{r}') = \int d\mathbf{r}' g(\mathbf{r}')\int d\mathbf{r} f(\mathbf{r}'-\mathbf{r})h(\mathbf{r})$, is only valid when f is even. Since, Γ_σ is even, we can rewrite (3.44) as

$$\begin{aligned}&-\frac{n_G}{n_P}\int d\mathbf{r}_G\int d\mathbf{r}'\,\Gamma_\sigma(\mathbf{r}'-\mathbf{r}_G)\hat{\rho}_{P,c}(\mathbf{r}')\ln q_G(\mathbf{r}_G)\\&= -\frac{n_G}{n_P}\int d\mathbf{r}'\,\hat{\rho}_{P,c}(\mathbf{r}')(\Gamma_\sigma * \ln q_G)(\mathbf{r}'),\end{aligned} \tag{3.45}$$

where we can now trivially reduce the explicit particle center coordinates such that $-\frac{n_G}{n_P}(\Gamma_\sigma * \ln q_G)(\mathbf{r}')$ arises as a second term in the single nanoparticle partition function, Q_P, as shown in (3.42).

The first term in (3.42) encompasses the interactions felt by the nanoparticle core while the second term represents the free energy of a grafted chain terminating at the surface of the nanoparticle. Without the second term, the nanoparticle distribution would be unaffected by the grafted chains. It should be noted explicitly that (3.41) is generalized for a given μ_P and that (3.42) is the correct representation for a homopolymer grafted nanoparticle. We next proceed to generalize the result for diblock, mixed, and Janus grafted nanoparticles.

3.7 Complex Grafted Nanoparticles

As described in the last section, we have now introduced the main components to include complex grafted nanoparticles into our framework [35] (Fig. 3.7).

Following the homopolymer grafted nanoparticles, diblock grafted nanoparticles are implemented in the same manner where

$$\begin{aligned}&\mu_{P,Diblock}(\mathbf{r}) = (\Gamma * w_P)(\mathbf{r}) -\\&\frac{n_{GD,\text{field}}}{n_P}\left(\Gamma_\sigma * \ln q_{GD,\text{field}}\right)(\mathbf{r}),\end{aligned} \tag{3.46}$$

and n_{GD} is the number of grafted diblocks. The only fundamental difference between (3.42) and (3.46) is that q_{GD} is made up of the propagators that form a diblock rather than a homopolymer. To incorporate an A/B mixed homopolymer brush grafted particle, μ_P is altered to

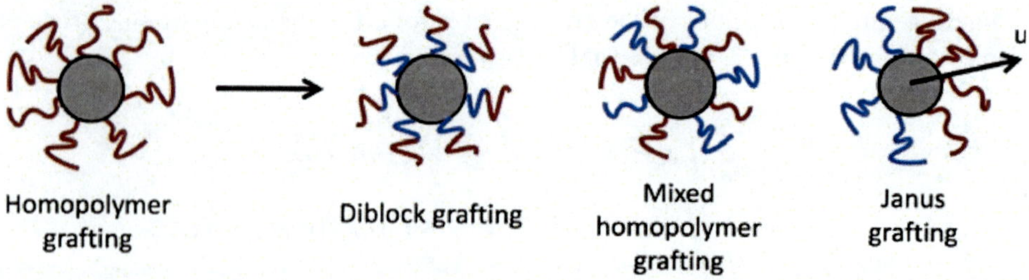

Fig. 3.7 Schematic illustrating the introduced physics in this section; homopolymer grafted nanoparticles to diblock grafted, mixed homopolymer grafted, and Janus grafted nanoparticles

$$\mu_{P,\text{Mixed}}(\mathbf{r}) = (\Gamma * w_P)(\mathbf{r}) - \frac{n_{G,\text{field}}}{n_P}\left[\Gamma_\sigma * \left(f_A \ln q_{GA,\text{field}} + (1-f_A)\ln q_{GB,\text{field}}\right)\right](\mathbf{r}), \tag{3.47}$$

where f_A is the fraction of grafted A homopolymer chains in the grafted brush. Lastly, to include a Janus grafted nanoparticle, we require an orientation-dependent grafting site density. We first present the final form of μ_P given by

$$\mu_{P,\text{Janus}}(\mathbf{r},\mathbf{u}) = (\Gamma * w_P)(\mathbf{r}) - \frac{n_{G,\text{field}}}{n_P}\left[f_A(\tilde{\Gamma}_{\sigma A} * \ln q_{GA,\text{field}})(\mathbf{r},\mathbf{u}) + (1-f_A)(\tilde{\Gamma}_{\sigma B} * \ln q_{GB,\text{field}})(\mathbf{r},\mathbf{u})\right], \tag{3.48}$$

where $\tilde{\Gamma}_{\sigma A}$ and $\tilde{\Gamma}_{\sigma B}$ are orientation dependent grafting site distributions. We define $\tilde{\Gamma}_{\sigma K}$ such that $\tilde{\Gamma}_{\sigma K}(\mathbf{r},\mathbf{u}) = \Gamma_{\sigma K}(-\mathbf{r},\mathbf{u})$. This definition is necessary as $\Gamma_{\sigma K}$ is not even, which is necessary to switch the convolution from the particle centers to the single grafted chain partition function, as discussed in the previous section in going from (3.43) to (3.45). $\Gamma_{\sigma A}$ and $\Gamma_{\sigma B}$ are given by

$$\Gamma_{\sigma A}(\mathbf{r},\mathbf{u}) = \frac{1}{\sigma_0}\exp\left[-\left(\frac{|\mathbf{r}| - R_P - \xi}{\xi}\right)^2\right] \operatorname{erfc}\left(\frac{1 - 2f_A - \mathbf{r}\cdot\mathbf{u}/|\mathbf{r}|}{\xi_{\mathbf{u}}}\right), \tag{3.49}$$

$$\Gamma_{\sigma B}(\mathbf{r},\mathbf{u}) = \frac{1}{\sigma_0}\exp\left[-\left(\frac{|\mathbf{r}| - R_P - \xi}{\xi}\right)^2\right] \operatorname{erfc}\left(\frac{\mathbf{r}\cdot\mathbf{u}/|\mathbf{r}| - 1 + 2f_A}{\xi_{\mathbf{u}}}\right), \tag{3.50}$$

where the complimentary error function term confines the grafting site distribution to a fraction of the nanoparticle surface area. Specifically, $\Gamma_{\sigma A}$ takes up f_A of the surface

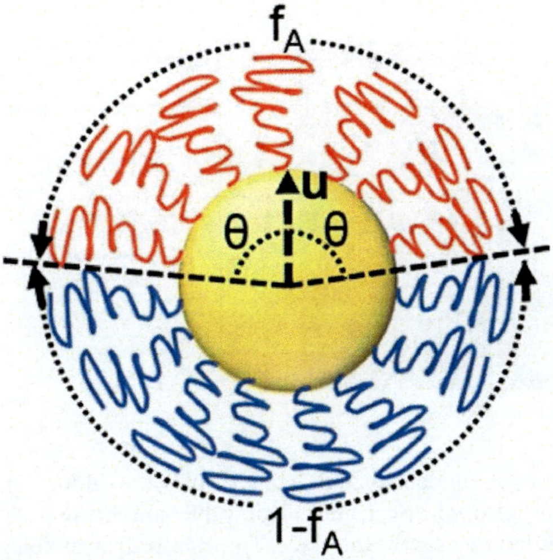

Fig. 3.8 Schematic for Janus grafted nanoparticle showing the setup for the orientation **u**, the fraction of grafting area for each type of chains, and the angle θ between **u** and the directions long the grafting area boundaries. Based on our definition, $\theta = \cos^{-1}(1 - 2f_A)$. Reproduced from [35] with permission from The Royal Society of Chemistry

area while $\Gamma_{\sigma B}$ takes up $1 - f_A$ of the surface area. The difference between (3.47) and (3.48) is that the mixed brush grafting function is not orientation dependent and the *A* and *B* chains are evenly distributed on the surface of the nanoparticle. However, the grafting functions in the Janus particle case are orientation dependent and effectively confines the *A* grafted chains on f_A of the nanoparticle surface while the remaining surface consists of *B* grafted chains (Fig. 3.8).

As is the case with homopolymer grafted particles, the inclusion of both explicit and field-based complex grafted particles is trivial. Furthermore, the framework allows inclusion of a combination of one or more grafted nanoparticle types.

3.7.1 Recent Applications

Figure 3.9 highlights an application of PNC-FT to study the structure and interfacial activity of complex grafted nanoparticles in an immisible polymer blends consisted of homopolymers (denoted as A and B polymers below) [35]. When the radius of gyration of grafted polymer is larger to the size of nanoparticle core ($R_g/R_P = 2.5$), the study shows that the grafted diblocks close to the A–B interface stretch into the A phase while the grafted chains far away from the interface are instead forced to locally phase separate on the side of the particle that resides in the B phase. For the mixed brush case, the grafted chains in their corresponding unfavorable phase pack close to the particle and locally phase separate while the grafted chains in their preferred phase extend into the bulk phase. In the case of Janus gratfing, the chains do not have to sacrifice their entropy to avoid unfavored enthalpic contacts with the bulk phase. The graft structure observed here plays an important role to control the

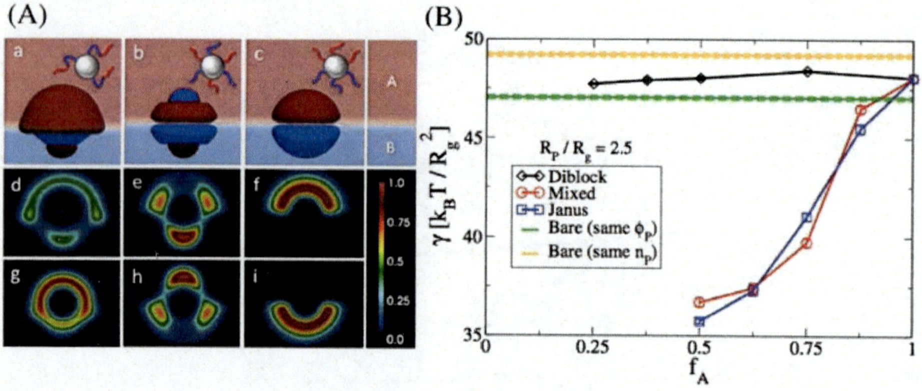

Fig. 3.9 **a** Local grafted chain density.**a**, **d**, and **g** display the diblock grafted particles, **b**, **e**, and **h** display mixed brush particles, while **c**, **f**, and **i** display the Janus brush particles. **a–c** Isosurfaces of the explicit grafted particle sitting near the interface. The red isosurfaces denote the A grafted monomers while the blue isosurfaces represent the B grafted monomers. The red and blue background show the A and B matrix chain density, respectively. **d–f** Color maps indicating the A grafted chain density. **g–i** Color maps indicating the B grafted chain density. The colormaps are taken from slices through the particle center. The color bar denotes the volume fraction of the corresponding component. **b** Interfacial tension as a function of the A component of the grafted layer,f_A. Reproduced from [35] with permission from The Royal Society of Chemistry

interfacial property. The interfacial tension in the cases of mixed and Janus grafting is found significantly lower than both the bare particle and diblock grafted particles given that the overall A component fraction (f_A) is about 0.5.

3.8 General Interparticle Potentials

Interestingly, the implementation of the density functions via convolutions allows for the incorporation of more realistic pair potentials in an exchange field theory formulation. Specifically, to go from a particle-based to a field-based model, we utilize Gaussian functional integrals that either take the form

$$\frac{\int Dw \exp[-(1/2)\int \mathrm{d}r \int \mathrm{d}r' w(r)u(r,r')w(r') + \int \mathrm{d}r \hat{\rho}(r)w(r)]}{\int Dw \exp[-(1/2)\int \mathrm{d}r \int \mathrm{d}r' w(r)u(r,r')w(r')} = \exp\left(\frac{1}{2}\int \mathrm{d}r \int \mathrm{d}r' \hat{\rho}(r)u^{-1}(r,r')\hat{\rho}(r')\right), \tag{3.51}$$

or

$$\frac{\int Dw \exp[-(1/2)\int \mathrm{d}r \int \mathrm{d}r' w(r)u(r,r')w(r') + i\int \mathrm{d}r \hat{\rho}(r)w(r)]}{\int Dw \exp[-(1/2)\int \mathrm{d}r \int dr' w(r)u(r,r')w(r')} = \exp\left(-\frac{1}{2}\int \mathrm{d}r \int \mathrm{d}r' \hat{\rho}(r)u^{-1}(r,r')\hat{\rho}(r')\right), \tag{3.52}$$

where u and u^{-1} are the functional inverse of each other, defined such that

$$\int \mathrm{d}r' u(r, r') u^{-1}(r', r'') = \delta(r - r''). \tag{3.53}$$

In order for the Gaussian integrals in (3.51) and (3.52) to be applied, u must be real, symmetric, and positive definite. In the standard field theories described above in Sect. 3.2, one typically writes the interaction potential as purely local

$$\beta U = C \int \mathrm{d}r\, \hat{\rho}(r)^2 = C \int \mathrm{d}r \int \mathrm{d}r'\, \hat{\rho}(r)\delta(r - r')\hat{\rho}(r'). \tag{3.54}$$

Here, C is a constant and the delta function is a convenient choice for the interaction potential as the functional inverse of a delta function is simply another delta function. If instead we employ the density-smearing approach, we define $\hat{\rho} = \Gamma_i * \hat{\rho}_c$ and (3.54) can be re-written in two equivalent forms

$$\beta U = C \int \mathrm{d}r \int \mathrm{d}r'\, (\Gamma_i * \hat{\rho}_c(r))\delta(r - r')(\Gamma_i * \hat{\rho}_c(r')) \tag{3.55}$$

$$= C \int \mathrm{d}r \int \mathrm{d}r'\, \hat{\rho}_c(r)(\Gamma_i * \Gamma_i)\hat{\rho}_c(r'). \tag{3.56}$$

Tracing through the particle-to-field transformation in the first form (3.55) is similar to that described above in Sect. 3.4 on field-based nanospheres: the effective chemical potential fields would simply be convolved with $\Gamma_i(r)$. In contrast, the second form has the particle centers interacting through an effective pair potential, $(\Gamma_i * \Gamma_i)(r)$. Shown in Figure 3.10 is an example of a pair potential of two nanoparticles described by the equation below

$$u(|\mathbf{r} - \mathbf{r}'|) = \frac{\kappa}{2\rho_0} \int d\mathbf{r}''\, \Gamma(\mathbf{r}'' - \mathbf{r})\, \Gamma(\mathbf{r}'' - \mathbf{r}'). \tag{3.57}$$

While the effective pair potential in (3.56) would need to be positive definite to use the Gaussian integrals, the first form in (3.55) *does not have this requirement*, and this opens a path to the use of more complex effective interactions. For example, in the case where nanoparticles of species i and j have two distinct shapes, (3.55) and (3.56) can be further generalized to

$$\begin{aligned}\beta U &= C \int \mathrm{d}r\, \hat{\rho}_i(r)\hat{\rho}_j(r) \\ &= C \int \mathrm{d}r \int \mathrm{d}r'\, (\Gamma_i * \hat{\rho}_c(r))\delta(r - r')(\Gamma_j * \hat{\rho}_c(r')) \\ &= C \int \mathrm{d}r \int \mathrm{d}r'\, \hat{\rho}_c(r)(\Gamma_i * \Gamma_j)\hat{\rho}_c(r').\end{aligned} \tag{3.58}$$

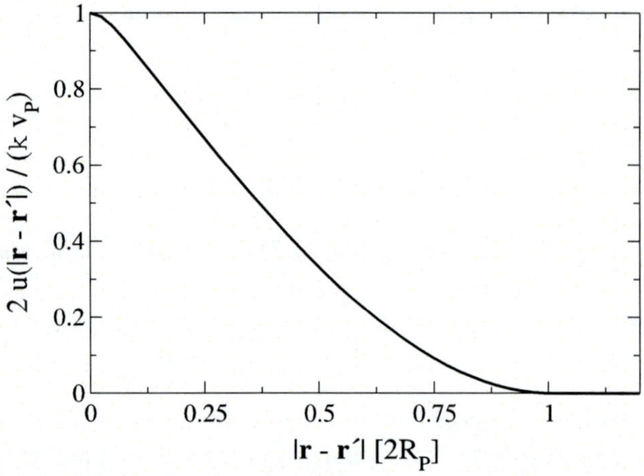

Fig. 3.10 Effective pair potential between two nanoparticle centers given by (3.57) with $\xi = 0.05R_P$

This realization allows for this framework to incorporate more realistic non-local potentials through the microscopic density functions rather than the pair potential, u. However, as Γ_i and Γ_j still arise as convolutions in the Hamiltonian, they should be well behaved in Fourier space to efficiently evaluate the convolution.

Interesting applications of this methodology could be to incorporate van der Waals-like particle-particle attractions or to study adsorbing polymers to the surface of nanoparticles by defining an attractive well at the surface of the nanoparticles. In each of these cases, the interaction potential is not limited by having to be real, symmetric, or positive definite if one employs a Weeks-Chandler-Andersen splitting [40] of the potential into its attractive and repulsive terms. The derivation and implementation of such a form for adsorbing diblock copolymers is provided in the next section.

3.8.1 Adsorbing Diblock Copolymer

In this subsection, we derive the field-theoretic representation of a system where nanoparticles have a purely repulsive interaction with the B blocks, but a local surface attraction with the A blocks of a diblock copolymer. The repulsive part of the interaction for both species is equal, and we continue to capture it through the Helfand compressibility potential in (3.4). We expand the potential to directly see the pair-wise interactions between the species as

$$\begin{aligned}\beta U_1 &= \frac{\kappa}{2\rho_0}\int d\mathbf{r}\,[\hat{\rho}_+(\mathbf{r}) - \rho_0]^2 \\ &= \frac{\kappa}{2\rho_0}\int d\mathbf{r}\,[\hat{\rho}_A(\mathbf{r})^2 + \hat{\rho}_B(\mathbf{r})^2 + \hat{\rho}_P(\mathbf{r})^2 + \rho_0^2 \\ &\quad + 2\hat{\rho}_A\hat{\rho}_B + 2\hat{\rho}_A\hat{\rho}_P + 2\hat{\rho}_B\hat{\rho}_P \\ &\quad - 2\hat{\rho}_A\rho_0 - 2\hat{\rho}_B\rho_0 - 2\hat{\rho}_P\rho_0].\end{aligned} \tag{3.59}$$

Here, we can more clearly see that the A block of the diblock has a repulsive interaction with the nanoparticle cores of the form

$$\begin{aligned}\beta U_{AP,\text{repuls}} &= \frac{\kappa}{\rho_0}\int d\mathbf{r}\hat{\rho}_A\hat{\rho}_P \\ &= \frac{\kappa}{\rho_0}\int d\mathbf{r}\int d\mathbf{r}'(h * \hat{\rho}_{A,c})(\Gamma * \hat{\rho}_{P,c}) \\ &= \frac{\kappa}{\rho_0}\int d\mathbf{r}\int d\mathbf{r}'\hat{\rho}_{A,c}(h * \Gamma)\hat{\rho}_{P,c} \\ &= \int d\mathbf{r}\int d\mathbf{r}'\hat{\rho}_{A,c}(\mathbf{r})u_{AP,\text{repuls}}(|\mathbf{r}-\mathbf{r}'|)\hat{\rho}_{P,c}(\mathbf{r}),\end{aligned} \tag{3.60}$$

where the repulsive pair potential is

$$u_{AP,\text{repuls}}(|\mathbf{r}-\mathbf{r}'|) = \frac{\kappa}{\rho_0}\int d\mathbf{r}''h(\mathbf{r}''-\mathbf{r})\Gamma(\mathbf{r}''-\mathbf{r}'). \tag{3.61}$$

Now, we can define an attractive interaction between the A block and the nanoparticle surface by smearing the particle center density with a *negative Gaussian ring* $-\Gamma_S$ that lies just off the particle surface. This adds a potential of the form

$$\begin{aligned}\beta U_{AP,\text{attract}} &= -\frac{\epsilon}{\rho_0}\int d\mathbf{r}\hat{\rho}_A\hat{\rho}_S \\ &= -\frac{\epsilon}{\rho_0}\int d\mathbf{r}\int d\mathbf{r}'(h * \hat{\rho}_{A,c})(\Gamma_S * \hat{\rho}_{P,c}) \\ &= -\frac{\epsilon}{\rho_0}\int d\mathbf{r}\int d\mathbf{r}'\hat{\rho}_{A,c}(h * \Gamma_S)\hat{\rho}_{P,c} \\ &= \int d\mathbf{r}\int d\mathbf{r}'\hat{\rho}_{A,c}(\mathbf{r})u_{AP,\text{attract}}(|\mathbf{r}-\mathbf{r}'|)\hat{\rho}_{P,c}(\mathbf{r}),\end{aligned} \tag{3.62}$$

where ϵ controls the strength of the attractive surface and the attractive pair potential is

$$u_{AP,\text{attract}}(|\mathbf{r}-\mathbf{r}'|) = -\frac{\epsilon}{\rho_0}\int d\mathbf{r}''h(\mathbf{r}''-\mathbf{r})\Gamma_S(\mathbf{r}''-\mathbf{r}'), \tag{3.63}$$

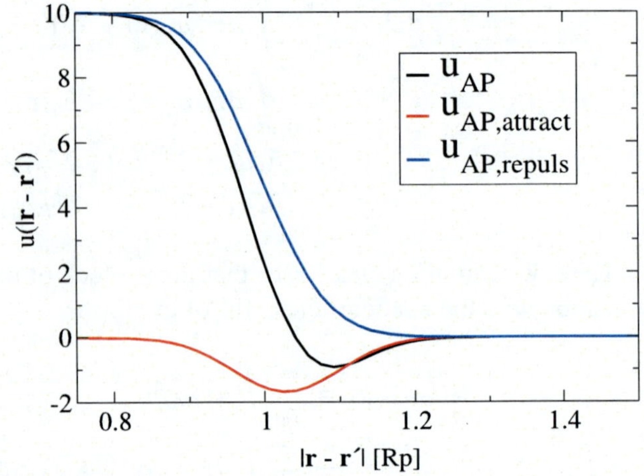

Fig. 3.11 Effective pair potential between an adsorbing polymer and a nanoparticle center (black) given by the summation of an attractive (red) and repulsive (blue) contribution based on (3.63) and (3.61), respectively. These pair potentials are for $\kappa = 10$, $\epsilon = 8$, $a = 0.067R_P$, and $\xi = 0.033R_P$

and Γ_S is defined by

$$\Gamma_S(\mathbf{r}) = \rho_0 \exp\left[-\left(\frac{|\mathbf{r}| - R_P - \xi}{\xi}\right)^2\right]. \tag{3.64}$$

Assuming the pair potentials are additive, the effective pair potential between the A block of the diblock and the nanoparticles is $u_{AP} = u_{AP,\text{repuls}} + u_{AP,\text{attract}}$, which is plotted in Fig. 3.11.

It is clear that despite using an ϵ value of 8, the attractive well is only ≈ 1. In this particular case, it is more appropriate to think of the depth of the well as the strength of attraction between the A block and the nanoparticle surface rather than the absolute value of ϵ. Furthermore, the forms of the density functions are a choice, and one can easily imagine using more realistic interaction potentials, such as a regularized version of colloid potentials [41].

Finally, we present the full set of equations for a field theory that uses the attractive potential described above. Incorporating the potentials defined in (3.3)–(3.5) and introducing (3.63) into a system of AB diblocks and neutral nanoparticles ($n_H = 0$) leads to a Hamiltonian given by

$$\begin{aligned}\mathcal{H}[\{w\}] = {} & \frac{\rho_0}{2\kappa}\int d\mathbf{r}\, w_+(\mathbf{r})^2 - i\rho_0 \int d\mathbf{r}\, w_+(\mathbf{r}) \\ & + \frac{\rho_0}{\chi_{AB}}\int d\mathbf{r}\, w_{AB}^{(+)}(\mathbf{r})^2 + \frac{\rho_0}{\chi_{AB}}\int d\mathbf{r}\, w_{AB}^{(-)}(\mathbf{r})^2 \\ & + \frac{\rho_0}{\epsilon}\int d\mathbf{r}\, w_{AS}^{(+)}(\mathbf{r})^2 + \frac{\rho_0}{\epsilon}\int d\mathbf{r}\, w_{AS}^{(-)}(\mathbf{r})^2 \\ & - n_D \ln Q_D[h * w_A, h * w_B] \\ & - n_P \ln Q_P[\mu_P],\end{aligned} \tag{3.65}$$

where

$$\mu_P(\mathbf{r}) = (\Gamma * w_P)(\mathbf{r}) + (\Gamma_S * w_S)(\mathbf{r}), \tag{3.66}$$

and

$$\begin{aligned} w_A &= i[w_+ + w_{AB}^{(+)} - w_{AS}^{(-)}] - w_{AB}^{(-)} + w_{AS}^{(+)}, \\ w_B &= i[w_+ + w_{AB}^{(+)}] + w_{AB}^{(-)}, \\ w_P &= iw_+, \\ w_S &= iw_{AS}^{(-)} + w_{AS}^{(+)}. \end{aligned} \tag{3.67}$$

Here, the term involving $w_{AS}^{(+)}$ is real while the term involving $w_{AS}^{(-)}$ carries an imaginary unit i because the surface potential in (3.62) is attractive. Furthermore, w_S does not have a contribution from the w_+ field as the purpose of Γ_S in this context is to represent a region in which the A block of the copolymer has a positive affinity towards rather than giving the surface an excluded volume. As a result, $\Gamma_S * \hat{\rho}_{P,c}$ is not included in $\hat{\rho}_+$.

3.9 Theoretically-Informed Langevin Dynamics (TILD)

The TILD method is a particle-based implementation of the models used in FTS, with one significant caveat: the non-bonded interactions are taken between continuous density fields instead of summing up pairs of interactions. Briefly, for a Gaussian-regularized Edwards model (GREM) for n linear homopolymers of length N in an implicit solvent, one can consider the total potential as being broken up into its bonded (b) and non-bonded (nb) components,

$$\beta U_{tot} = \beta U_b(\mathbf{r}^{nN}) + \beta U_{nb}[\rho]. \tag{3.68}$$

The bonded interactions are taken directly from the particle coordinates and do not involve any inter-molecular interactions. As such, the expense of calculating U_b is expected to scale as $\mathcal{O}(nN)$.

The non-bonded contribution to the potential is calculated by mapping the explicit particle densities onto a continuous density field,

$$\rho(\mathbf{r}) = \sum_i^{nN} \delta(\mathbf{r} - \mathbf{r}_i) \approx \frac{1}{\delta V} \sum_i W(\mathbf{r} - \mathbf{r}_i), \tag{3.69}$$

where $W(\mathbf{r} - \mathbf{r}_i)$ is the weight from particle i given to the grid point at $\mathbf{r}$ and δV is the volume of the grid employed; the notation for this mapping closely follows that commonly used in the particle-to-mesh Ewald summation technique in particle-based simulations of charged species [42]. Equation (3.69) assumes a regular grid where each point is associated with the same grid volume, though this is not strictly required. With this continuous density field, the non-bonded potential for the GREM becomes

$$2\beta U[\rho] = \int d\mathbf{r} \int d\mathbf{r}' \, \rho(\mathbf{r}) \, u_G(\mathbf{r} - \mathbf{r}') \, \rho(\mathbf{r}') - \frac{nNu_G(0)}{2}, \tag{3.70}$$

$$u_G(\mathbf{r}) = u_0 \left(2\pi a^2\right)^{-\mathbb{D}/2} e^{-|\mathbf{r}|^2/2a^2}. \tag{3.71}$$

Here, $\mathbb{D}$ is the dimensionality of the system and a is the range of the interaction potential, which is closely related to the monomer smearing size described above (Fig. 3.12).

To use this approach in a Langevin dynamics (LD) scheme, for each particle one needs to obtain the conservative part of the force that arises from the potential energy, $\mathbf{f}_i = -\frac{\partial U_{tot}}{\partial \mathbf{r}_i}$. Similar to the potential energy, contributions from the bonded term will arise naturally as in a particle-based simulation, and these can be easily analytically implemented for most common potentials. The contribution to the force from the non-bonded term is most efficiently handled on the grid by first calculating the continuous (grid-based) non-bonded force between points $\mathbf{r}$ and $\mathbf{r}'$ as

$$\mathbf{f}_{nb}(\mathbf{r}) = -\int d\mathbf{r}' \nabla u_G(\mathbf{r} - \mathbf{r}') \, \rho(\mathbf{r}'), \tag{3.72}$$

which can be handled as $\mathbb{D}$ convolution operations of the gradient of the pair potential with the density field. The continuous force is then mapped onto the particles as

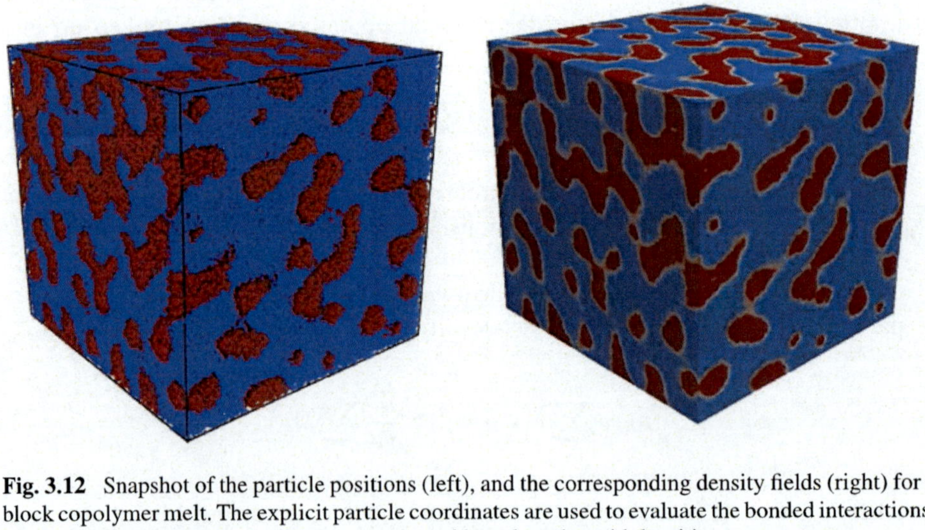

Fig. 3.12 Snapshot of the particle positions (left), and the corresponding density fields (right) for a block copolymer melt. The explicit particle coordinates are used to evaluate the bonded interactions, while the non-bonded interactions are evaluated based on the grid densities

$$\mathbf{f}_i^{(nb)} = \int d\mathbf{r}\, \delta(\mathbf{r} - \mathbf{r}_i)\, \mathbf{f}_{nb}(\mathbf{r}) \tag{3.73}$$

$$\approx \sum_{\mu} W(\mathbf{r}_\mu - \mathbf{r}_i)\, \mathbf{f}_{nb}(\mathbf{r}_\mu), \tag{3.74}$$

where the sum over μ is over the grid points with finite support ($W \neq 0$) for particle i, and $\mathbf{r}_\mu$ indicates a position associated with grid point μ.

An important feature of the TILD method is that it is even more straightforward to introduce more complex pair potentials between the components without the need to use the smearing function approach described in the previous sections. As the potential will be evaluated in Fourier space when evaluating the convolutions in (3.72), there is still a need to employ a regularized potential that is continuous even at overlap (i.e., $u(r = 0)$ should be finite, even if large), but the WCA-like splitting into attractive and repulsive contributions is not necessary.

Our group initially began implementing and using this dynamic implementation of field theories after the theoretical developments presented by Fredrickson and Orland [43] and related work by Grzetic et al. [44] that is well-grounded in the Martin-Siggia-Rose (MSR) path integral formulation of classical mechanics [45]. Our early implementations of the method were referred to as a "dynamic mean-field theory" (DMFT) to highlight the connection to the MSR formalism; however, recently we have realized that our implementation is very closely related to the "theoretically-informed Monte Carlo" (TIMC) methods developed by de Pablo et al. [46]. The primary differences are that the TIMC methods are typically based on Monte Carlo sampling the system, while we employ dynamic integration schemes. There are some common implementation choices that are distinct from our usual applications, but the connection between the methods is sufficiently deep that we have chosen to adopt the name theoretically-informed Langevin dynamics (TILD) to not generate yet another acronym for sampling the models commonly used in field-based simulations.

3.9.1 Recent Applications

Figure 3.13 shows highlights from two recent applications of TILD. First, Fig. 3.13a–c show how solvent annealing conditions can be used to tune the distribution of nanoparticles in block copolymer thin films. In dry films, the nanoparticles were found to be surface-active; the particles adsorbed to the free (top) interface. Upon exposure to solvent, if the nanoparticles had a high grafting density, the solvent would displace the particles from the surface, dispersing the nanoparticles throughout the film. The change in the dispersion of the particles with the presence of solvent allows us to use the drying process to control the distribution of nanoparticles in the block copolymer film by varying the drying rate. By defining a Péclet number as the ratio of the film retraction rate relative to the nanoparticle diffusion time, we show

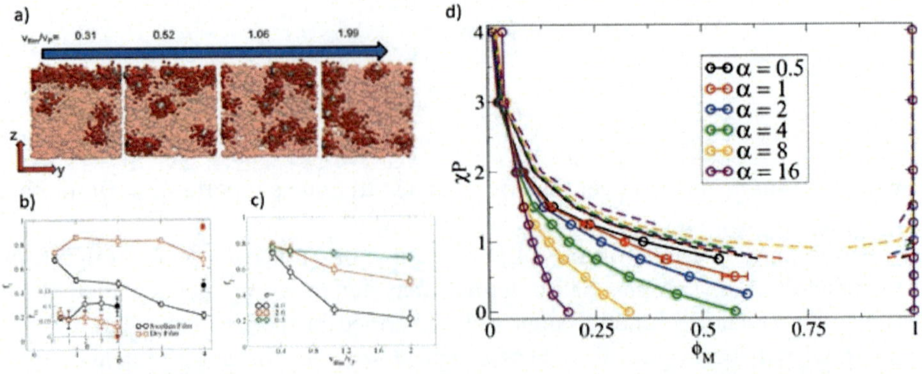

Fig. 3.13 **a** Visualization of the effect of drying rate on the distribution of polymer-grafted nanoparticles in the thin film, where the drying rate is scaled by an effective particle diffusion time. **b** Influence of grafting density σ on the final fraction of nanoparticles that reside at the free surface f_I. **c** Quantification of the effect of drying rate on f_I. Figures a, b, and c reproduced from [47] with permission from The Royal Society of Chemistry. **d** Phase diagram for polymer-grafted nanoparticles dispersed in a homopolymer matrix as a function of the relative lengths of the polymer chains, $\alpha = P/N$ and χP. Panel d is adapted from [48]

visually in Fig. 3.13a, c how the fraction of nanoparticles at the interface f_I can be controlled after solvent evaporation [47].

The second example shows the calculation of phase diagrams for polymer-grafted nanoparticles dispersed in homopolymer matrices [48]. In short, we found that the thermodynamics of these systems depend primarily on two important factors: the repulsion between the grafted polymer chains and the matrix, as governed by χ, and the ratio of the length of the matrix polymers P to the grafted polymers N, $\alpha = P/N$. As shown in Fig. 3.13d, when the matrix is shorter than the grafted chains ($\alpha = 0.5$), the phase separation is primarily controlled by χ, and phase separation is not observed until $\chi P \approx 0.75$. In contrast, when the matrix chains are significantly longer than the grafts ($\alpha \geq 4$), phase separation is observed even in the athermal limit ($\chi = 0$). These results agree qualitatively very well with experiments in the cases where direct comparisons are possible.

The final example presented in this section focuses on the brush structure of mixed brush particles in solution [49]. In Fig. 3.14, the top two rows use HPFT with a mean-field approximation while the bottom row uses TILD where fluctuations are inherent in the model. As these particles are in the presence of solvent, fluctuations are prominent and they are necessary to include realistic descriptions of the mixed brush phases when the radius of the nanoparticle to the grafted chain radius of gyration ratio, R_P/R_g, is ≈ 1 or greater. As R_P/R_g is increased, the mean-field approach predicts increasingly striped/patchy phases. However, the incorporation of fluctuations via the TILD method demonstrates that these structures form significant defects relative to the mean-field predicted structures. Interestingly, when R_P/R_g is small, the Janus structure predicted from the mean-field model is also predicted from the TILD approach.

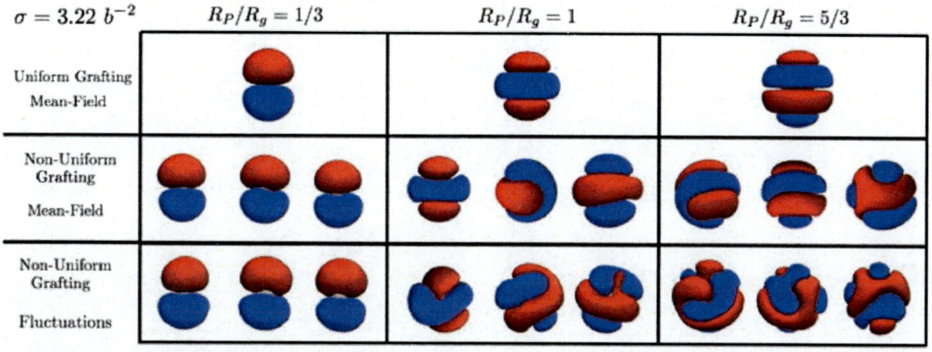

Fig. 3.14 Visualizations of single-particle mixed brush structures in solution. The red isosurface denotes the *A* chains while the blue isosurface denotes the *B* chains. Each row uses specific approximations while each column represents the ratio of the nanoparticle radius to the grafted chains' radius of gyration ratio. The top two rows use HPFT with a mean-field approximation while the bottom row uses TILD where fluctuations are inherent in the model. This figure is adapted from [49]

3.10 Comparison of the Different Methods

Each of the methods discussed herein have their associated advantages and disadvantages. Mean-field solutions are attractive because they enable rapid solutions and well-behaved algorithms that are generally not severely limited by stability, but it remains unclear how accurate mean-field solutions can be for polymer nanocomposites. The mean-field approximation would seem to be particularly suspect at high nanoparticle loading, where one expects a strongly correlated fluid structure. A promising strategy to circumvent this is to employ a hybrid particle-field (HPF) theory with a single explicit nanoparticle surrounded by field-based particles, but evaluated under the mean-field approximation. When applied to a 2D system containing only nanoparticles, Fig. 3.15 shows that this method lead to signatures of hexagonal packing as long as the particle repulsions are sufficiently strong [20].

The complex Langevin field-theoretic simulations (CL-FTS) relax the mean-field approximation and allows for sampling the fully-fluctuating version of the model, assuming it converges to a steady state [10, 50, 51]. While initial implementations of CL-FTS were of limited numerical stability, recent advances in numerical algorithms have significantly improved this situation [26, 30, 52], and for polymer solutions and melts CL-FTS remains a very attractive approach due to its ability to rapidly equilibrate simulations. CL-FTS is particularly attractive in the high molecular weight limit with well-resolved chain discretization where it rapidly equilibrates, while techniques such as TILD will require explicitly simulating the dynamics of the polymer molecules. A challenge with CL-FTS is the difficulty in generating stable trajectories in some regimes of parameter space. For the data presented in Fig. 3.13 (right) where we were interested in the phase behavior of grafted nanoparticles in a polymer matrix, we were unable to generate stable trajectories in the limit where $\chi \to 0$ and

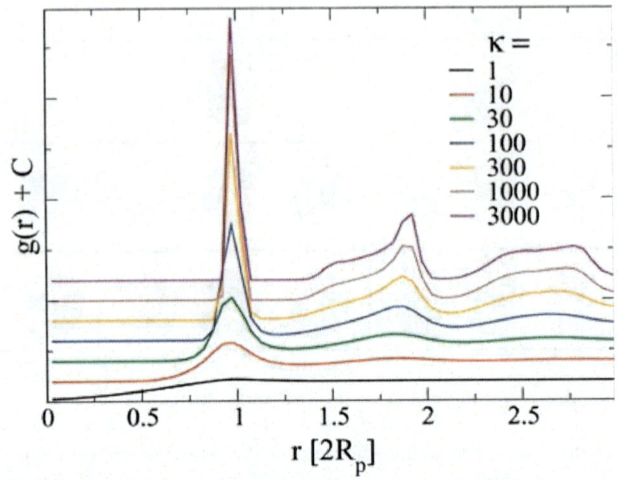

Fig. 3.15 Pair distribution function between particles that interact through the convolution of their shape functions as in (3.57). The prefactor κ increases the magnitude of the repulsion between the particles, and the $g(r)$ curves are vertically offset for clarity. Peaks consistent with hexagonal packing emerge in the large κ limit. Figure adapted from [20]

large P/N. The trajectories generated at low χ would frequently crash, and what data we could generate was inconsistent with the expectation that the systems would macrophase separate. This does not indicate a general failure of the CL method, but it does challenge the interpretation of results since it is not straightforward to anticipate *a priori* which systems will present problems.

Monte Carlo field-theoretic simulations (MC-FTS) have also recently emerged as a viable technique for sampling block copolymer field theories, though it has not yet been applied to PNC-FT models [53–55]. In the most successful implementations of MC-FTS, fields that are purely imaginary are evaluated under the mean-field approximation. For example, in a two-field implementation of a diblock copolymer, the partition function is approximated as

$$\mathcal{Z} = z_0 \int \mathcal{D}w_+ \mathcal{D}w_- \, e^{-\mathcal{H}[w_+, w_-]} \approx z_0 \int \mathcal{D}w_- \, e^{-\mathcal{H}[w_+^*, w_-]}, \tag{3.75}$$

where $w_+^*(r)$ is the mean-field configuration of w_+ for a given w_-. This approximation renders the weights of the microstates positive definite and allows Metropolis Monte Carlo to be performed on the field theoretic model. While not justified analytically, this approximation has shown to capture fluctuation corrections to the phase diagram of block copolymers [53]. However, the systems where this approach can be used will be limited to those where the mean-field approximation on the imaginary fields is justified and of no consequence. In PNCs or other systems where strong repulsive interactions play a dominant role, or in systems where electrostatics play a key role, this approximation is likely to severely limit the accuracy of the model.

Theoretically-informed Langevin dynamics (TILD) and related methods retain many implementation details of the underlying particle model while increasing the sampling efficiency relative to the naive implementation of the particle model. A drawback of this method is that in the best case scenario, the polymer chains will exhibit Rouse dynamics, and so the relaxation time of a chain will increase with the number of discretization sites on the chain as $\tau \sim N^2$. However, the numerical stability of the method combined with the ability to reproduce fluctuation effects is a very attractive feature. In addition, the TILD methods can easily treat more complicated intramolecular interactions, such as the potentials required to model semiflexible polymers. Semiflexibility can be important in many conjugated polymers, or in cases where a fine chain discretization is used and the backbone flexibility of the real polymer may play a role. Spakowitz and co-workers have developed a discrete, shearable, stretchable worm-like chain model [56, 57] that is particularly attractive for smoothly spanning from scales where the persistence length matters to more coarse grained levels where the chains behave more like Gaussian polymers. Polydispersity effects are also more naturally incorporated into the TILD framework, because each chain or nanoparticle can have a distinct size without the need to calculate a distinct molecular partition function.

3.11 Challenges and Area for Development

There are numerous challenges to be resolved and further opportunities for future extensions of the methods presented herein. One notable challenge is the implementation of hydrodynamic interactions in the dynamic methods. In polymeric systems, hydrodynamic interactions play a important role, and the interplay with viscous dissipation such that local fluctuations can propagate and induce long range correlations. For example, dynamic phase separation can be induced by applied shear in polymer solutions at micrometer scales [58, 59]. Another example of hydrodynamic effects in manufacture of polymeric materials is the nonsolvent driven phase inversion in processing porous polymer films. Hydrodynamic effects due to nonsolvent from the bath entering and dewetting the films contribute to the formation of porous microstructures in the processed films [60–62]. Currently there only a few established field-based simulation approach taking into account hydrodynamic interactions at molecular level. In the hydrodynamic self-consistent field theory developed by Fredrickson and coworkers, the "two-fluid" formalism proposed by Doi and Onuki [63] is extended with constitutive relations from continuum mechanics to incorporate hydrodynamics effect [64–66]. This method, while retaining computation efficiency over particle-based method due to field-based nature, requires input of constitutive relations between local stress and densities/velocity fields. On the other hand the form of free energy input in the method is typically from mean field approximation under equilibrium.

Another promising route to address the challenge of including hydrodynamic interactions is to extend the TILD method to capture hydrodynamic effects. Our current efforts along these lines [67] involves extending the framework to employ a grid-based version of the dissipative particle dynamics (DPD) algorithm. In the standard particle-based implementation of DPD, hydrodynamic effects are captured because the method locally conserves momentum; both the stochastic forces and the friction from the missing degrees of freedom are taken as pair-wise additive. In our grid-based implementation, we apply this idea at the level of the grid used to resolve the density fields, and we have shown that we obtain hydrodynamic coarsening of domains during spinodal decomposition [67].

There are additionally numerous possibilities for combining field theoretic approaches with particle based approaches that have not been explored. One interesting example of this could be in the context of Gibbs ensemble simulations [27, 68] where two non-interacting simulation boxes each model phases at coexistence. In many potentially interesting applications of FTS to study phase coexistence, the resulting phases are such that the FTS approach is the method of choice for one phase (e.g., a dense polymer melt) while at a disadvantage for simulating the other phase, such as a supernatant solvent phase or a highly correlated nanoparticle packing. Since in the Gibbs ensemble the particle-to-field transformation proceeds independently in the two simulation boxes [27], one box could remain a particle-based simulation while the second is a field-theoretic simulation.

Finally, as alluded to above in Sect. 3.10 in the discussion of the inability to capture some macrophase separation using the CL-FTS method, there is a need to assess *a priori* when the CL method will be the method of choice. We have additionally encountered these "silent failures" of the CL method in attempts by Riggleman to extend the CL method to the density-explicit formulation of a field theory. In such a field theory for a single component system, the partition function and effective Hamiltonian of the system take the form

$$\mathcal{Z} = z \int \mathcal{D}\rho \int \mathcal{D}w \; e^{-\mathcal{H}[\rho,w]}, \tag{3.76}$$

$$\begin{aligned} \mathcal{H}[\rho, w] = \frac{1}{2k_B T} \int d\mathbf{r} \int d\mathbf{r}' \, \rho(\mathbf{r}) \, u(|\mathbf{r}-\mathbf{r}'|) \, \rho(\mathbf{r}') - i \int d\mathbf{r} \; w(\mathbf{r})\rho(\mathbf{r}) \\ - n \log Q[iw]. \end{aligned} \tag{3.77}$$

This form of a field theory is advantageous because it places fewer restrictions on the form of the interaction potential $u(\mathbf{r})$. The potential does not have to be invertible for the application of this theory, and furthermore higher-order interaction terms, such as those cubic in the density field $\rho(\mathbf{r})$, can be included. Two numerical implementations of the CL method have been attempted: an explicit Euler integration of the Langevin equations that arise from the functional derivatives of $\mathcal{H}$ in (3.76) and an integration scheme where the effective forces are modified by a non-Hermitian mobility matrix [69]. In the former case, we were unable to obtain stable trajectories, while in the second case a steady state solution gave results that disagree with other simulation methods. While we cannot rule out a possible programming error, the algorithm behaves as expected with the noise terms disabled, and this remains an open area for development.

Acknowledgements Sandia National Laboratories is a multimission laboratory managed and operated by National Technology and Engineering Solutions of Sandia, LLC, a wholly owened subsidiary of Honeywell International Inc., for the U.S. Department of Energy's National Nuclear Security Administration under contract DE-NA0003525. Financial support from the National Science Foundation through awards NSF DMR-1410246, NSF PIRE-OISE-1545884, MRSEC-DMR-1720530 as well as computational support from XSEDE award TG-DMR150034 are gratefully acknowledged.

References

1. S.K. Kumar, V. Ganesan, R.A. Riggleman, Perspective: outstanding theoretical questions in polymer-nanoparticle hybrids. J. Chem. Phys. **147**(2), 020901 (2017)
2. V. Ganesan, A. Jayaraman, Theory and simulation studies of effective interactions, phase behavior and morphology in polymer nanocomposites. Soft Matter **10**(1), 13–38 (2014)
3. K.I. Winey, R.A. Vaia, Polymer nanocomposites. MRS Bull. **32**(4), 314–322 (2007)
4. T.B. Martin, T.E. Gartner III, R.L. Jones, C.R. Snyder, A. Jayaraman, pyprism: a computational tool for liquid-state theory calculations of macromolecular materials. Macromolecules **51**(8), 2906–2922 (2018)
5. A. Jayaraman, N. Nair, Integrating prism theory and monte carlo simulation to study polymer-functionalised particles and polymer nanocomposites. Mol. Simul. **38**(8–9), 751–761 (2012)
6. N. Nair, A. Jayaraman, Self-consistent prism theory-monte carlo simulation studies of copolymer grafted nanoparticles in a homopolymer matrix. Macromolecules **43**(19), 8251–8263 (2010)
7. L.M. Hall, A. Jayaraman, K.S. Schweizer, Molecular theories of polymer nanocomposites. Curr. Opinion Solid State Mater. Sci. **14**(2), 38–48 (2010)
8. J.B. Hooper, K.S. Schweizer, Theory of phase separation in polymer nanocomposites. Macromolecules **39**(15), 5133–5142 (2006)
9. S.F. Edwards, The statistical mechanics of polymers with excluded volume. Proc. Phys. Soc. **85**(4), 613 (1965)
10. G. Fredrickson, *The Equilibrium Theory of Inhomogeneous Polymers*, vol. 134 (Oxford University Press on Demand, 2006)
11. W. Zheng, Z.-G. Wang, Morphology of abc triblock copolymers. Macromolecules **28**(21), 7215–7223 (1995)
12. M.W. Matsen, M. Schick, Stable and unstable phases of a diblock copolymer melt. Phys. Rev. Lett. **72**(16), 2660 (1994)
13. N. Xie, W. Li, F. Qiu, A.-C. Shi, Sigma phase formed in conformationally asymmetric ab-type block copolymers. Acs Macro Lett. **3**(9), 906–910 (2014)
14. L. Leibler, Theory of microphase separation in block copolymers. Macromolecules **13**(6), 1602–1617 (1980)
15. K.T. Delaney, G.H. Fredrickson, Recent developments in fully fluctuating field-theoretic simulations of polymer melts and solutions. J. Phys. Chem. B **120**(31), 7615–7634 (2016)
16. R.B. Thompson, V.V. Ginzburg, M.W. Matsen, A.C. Balazs, Predicting the mesophases of copolymer-nanoparticle composites. Science **292**(5526), 2469–2472 (2001)
17. T.N. Hoheisel, K. Hur, U.B. Wiesner, Block copolymer-nanoparticle hybrid self-assembly. Progr. Polym. Sci. **40**, 3–32 (2015)
18. V.V. Ginzburg, Polymer-grafted nanoparticles in polymer melts: modeling using the combined scft-dft approach. Macromolecules **46**(24), 9798–9805 (2013)
19. S.W. Sides, B.J. Kim, E.J. Kramer, G.H. Fredrickson, Hybrid particle-field simulations of polymer nanocomposites. Phys. Rev. Lett. **96**(25), 250601 (2006)
20. J. Koski, H. Chao, R.A. Riggleman, Field theoretic simulations of polymer nanocomposites. J. Chem. Phys. **139**(24), 244911 (2013)

21. E. Helfand, Theory of inhomogeneous polymers: fundamentals of the Gaussian random-walk model. J. Chem. Phys. **62**(3), 999–1005 (1975)
22. P.M. Chaikin, T.C. Lubensky, T.A. Witten, *Principles of Condensed Matter Physics*, vol. 10 (Cambridge university press Cambridge, 1995)
23. M.J. Hore, R.J. Composto, Using miscible polymer blends to control depletion-attraction forces between au nanorods in nanocomposite films. Macromolecules **45**(15), 6078–6086 (2012)
24. N.M. Krook, C. Tabedzki, K.C. Elbert, K.G. Yager, C.B. Murray, R.A. Riggleman, R.J. Composto, Experiments and simulations probing local domain bulge and string assembly of aligned nanoplates in a lamellar diblock copolymer. Macromolecules **52**(22), 8989–8999 (2019)
25. G.H. Fredrickson, V. Ganesan, F. Drolet, Field-theoretic computer simulation methods for polymers and complex fluids. Macromolecules **35**(1), 16–39 (2002)
26. D.J. Audus, K.T. Delaney, H.D. Ceniceros, G.H. Fredrickson, Comparison of pseudospectral algorithms for field-theoretic simulations of polymers. Macromolecules **46**(20), 8383–8391 (2013)
27. R.A. Riggleman, G.H. Fredrickson, Field-theoretic simulations in the gibbs ensemble. J. Chem. Phys. **132**(2), 024104 (2010)
28. R.A. Riggleman, R. Kumar, G.H. Fredrickson, Investigation of the interfacial tension of complex coacervates using field-theoretic simulations. J. Chem. Phys. **136**(2), 024903 (2012)
29. E.M. Lennon, K. Katsov, G.H. Fredrickson, Free energy evaluation in field-theoretic polymer simulations. Phys. Rev. Lett. **101**(13), 138302 (2008)
30. M.C. Villet, G.H. Fredrickson, Efficient field-theoretic simulation of polymer solutions. J. Chem. Phys. **141**(22), 224115 (2014)
31. Y. Kim, H. Chen, A. Alexander-Katz, Free energy landscape and localization of nanoparticles at block copolymer model defects. Soft Matter **10**(18), 3284–3291 (2014)
32. R.J. Hickey, J. Koski, X. Meng, R.A. Riggleman, P. Zhang, S.J. Park, Size-controlled self-assembly of superparamagnetic polymersomes. ACS nano **8**(1), 495–502 (2014)
33. D.M. Trombly, V. Ganesan, Curvature effects upon interactions of polymer-grafted nanoparticles in chemically identical polymer matrices. J. Chem. Phys. **133**(15), 154904 (2010)
34. H. Chao, B.A. Hagberg, R.A. Riggleman, The distribution of homogeneously grafted nanoparticle s in polymer thin films and blends. Soft Matter **10**(40), 8083–8094 (2014)
35. J. Koski, H. Chao, R.A. Riggleman, Predicting the structure and interfacial activity of diblock brush, mixed brush, and Janus-grafted nanoparticles. Chem. Commun. **51**, 5440–5443 (2015)
36. B.J. Lindsay, R.J. Composto, R.A. Riggleman, Equilibrium field theoretic study of nanoparticle interactions in diblock copolymer melts. J. Phys. Chem. B **123**(44), 9466–9480 (2019)
37. B. Rasin, H. Chao, G. Jiang, D. Wang, R.A. Riggleman, R.J. Composto, Dispersion and alignment of nanorods in cylindrical block copolymer thin films. Soft Matter **12**, 2177–2185 (2016)
38. H. Chao, B.J. Lindsay, R.A. Riggleman, Field-theoretic simulations of the distribution of nanorods in diblock copolymer thin films. J. Phys. Chem. B **121**(49), 11198–11209 (2017)
39. H. Chao, B.A. Hagberg, R.A. Riggleman, The distribution of homogeneously grafted nanoparticles in polymer thin films and blends. Soft Matter **10**(40), 8083–8094 (2014)
40. J.D. Weeks, D. Chandler, H.C. Andersen, Role of repulsive forces in determining the equilibrium structure of simple liquids. J. Chem. Phys. **54**(12), 5237–5247 (1971)
41. R. Everaers, M. Ejtehadi, Interaction potentials for soft and hard ellipsoids. Phys. Rev. E **67**(4), 041710 (2003)
42. M. Deserno, C. Holm, How to mesh up ewald sums. i. a theoretical and numerical comparison of various particle mesh routines. J. Chem. Phys. **109**(18), 7678–7693 (1998)
43. G.H. Fredrickson, H. Orland, Dynamics of polymers: a mean-field theory. J. Chem. Phys. **140**(8), 084902 (2014)
44. D.J. Grzetic, R.A. Wickham, A.-C. Shi, Statistical dynamics of classical systems: a self-consistent field approach. J. Chem. Phys. **140**(24), 244907 (2014)
45. P.C. Martin, E. Siggia, H. Rose, Statistical dynamics of classical systems. Phys. Rev. A **8**(1), 423 (1973)
46. F.A. Detcheverry, H. Kang, K.C. Daoulas, M. Müller, P.F. Nealey, J.J. de Pablo, Monte carlo simulations of a coarse grain model for block copolymers and nanocomposites. Macromolecules **41**(13), 4989–5001 (2008)

47. H. Chao, J. Koski, R.A. Riggleman, Solvent vapor annealing in block copolymer nanocomposite films: a dynamic mean field approach. Soft Matter **13**(1), 239–249 (2017)
48. J.P. Koski, N.M. Krook, J. Ford, Y. Yahata, K. Ohno, C.B. Murray, A.L. Frischknecht, R.J. Composto, R.A. Riggleman, Phase behavior of grafted polymer nanocomposites from field-based simulations. Macromolecules (2019)
49. J.P. Koski, A.L. Frischknecht, Fluctuation effects on the brush structure of mixed brush nanoparticles in solution. ACS Nano **12**(2), 1664–1672 (2018)
50. J.R. Klauder, A Langevin approach to fermion and quantum spin correlation functions. J. Phys. A: Math General **16**(10), L317 (1983)
51. G. Parisi, J. Klauder, W. Petersen, J. Ambjorn, S. Yang, F. Karsch, H. Wyld, On complex probabilities. Stochastic Quant. **131**, 381 (1988)
52. E.M. Lennon, G.O. Mohler, H.D. Ceniceros, C.J. García-Cervera, G.H. Fredrickson, Numerical solutions of the complex Langevin equations in polymer field theory. Multiscale Model. Simulation **6**(4), 1347–1370 (2008)
53. B. Vorselaars, P. Stasiak, M.W. Matsen, Field-theoretic simulation of block copolymers at experimentally relevant molecular weights. Macromolecules **48**(24), 9071–9080 (2015)
54. R.K. Spencer, M.W. Matsen, Field-theoretic simulations of bottlebrush copolymers. J. Chem. Phys. **149**(18), 184901 (2018)
55. R.K. Spencer, M.W. Matsen, Critical point of symmetric binary homopolymer blends. Macromolecules **49**(16), 6116–6125 (2016)
56. E.F. Koslover, A.J. Spakowitz, Systematic coarse-graining of microscale polymer models as effective elastic chains. Macromolecules **46**(5), 2003–2014 (2013)
57. E.F. Koslover, A.J. Spakowitz, Discretizing elastic chains for coarse-grained polymer models. Soft Matter **9**(29), 7016–7027 (2013)
58. G. Helfand, G.H. Fredrickson, Large fluctuations in polymer solution under shear. Phys. Rev. Lett. **62**, 2468–2471 (1989)
59. S.T. Milner, Dynamical theory of concentration fluctuations in polymer solutions under shear. Phys. Rev. E **48**, 3674–3691 (1993)
60. C.A. Smolders, A.J. Reuvers, R.M. Boom, I.M. Wienk, Microstructures in phase-inversion membranes. Part 1. Formation of macrovoids. J. Membrane Sci. **73**, 259–275 (1992)
61. P. Van De Witte, P.J. Dijkstra, J.W.A. Van Den Berg, J. Feijen, Phase separation processes in polymer solutions in relation to membrane formation. J. Membrane Sci. **117**, 1–31 (1996)
62. I.M. Wienk, R.M. Boom, M.A.M. Beerlage, A.M.W. Bulte, C.A. Smolders, H. Strathmann, Recent advances in the formation of phase inversion membranes made from amorphous or semi-crystalline polymers. J. Membrane Sci. **113**, 361–371 (1996)
63. M. Doi, A. Onuki, Dynamic coupling between stress and composition in polymer solutions and blends. Journal de Physique II **2**, 1631–1656 (1992)
64. D.M. Hall, T. Lookman, G.H. Fredrickson, S. Banerjee, Hydrodynamic self-consistent field theory for inhomogeneous polymer melts. Phys. Rev. Lett. **97**, 114501 (2006)
65. D.M. Hall, T. Lookman, G.H. Fredrickson, S. Banerjee, Numerical method for hydrodynamic transport of inhomogeneous polymer melts. J. Comput. Phys. **224**, 681–698 (2007)
66. D.R. Tree, K.T. Delaney, H.D. Ceniceros, T. Iwama, G.H. Fredrickson, A multi-fluid model for microstructure formation in polymer membranes. Soft Matter **13**, 3013–3030 (2017)
67. H. Chao, T. Zhang, P. Español, *R.A. Riggleman, A Particle-to-Mesh Implementation of Dissipative Particle Dynamics* (in preparation)
68. A.Z. Panagiotopoulos, N. Quirke, M. Stapleton, D. Tildesley, Phase equilibria by simulation in the Gibbs ensemble: alternative derivation, generalization and application to mixture and membrane equilibria. Mol. Phys. **63**(4), 527–545 (1988)
69. X. Man, K.T. Delaney, M.C. Villet, H. Orland, G.H. Fredrickson, Coherent states formulation of polymer field theory. J. Chem. Phys. **140**(2), 024905 (2014)

Chapter 4
Polymer Dynamics in Polymer-Nanoparticle Interface

Argyrios V. Karatrantos and Nigel Clarke

Abstract In this chapter, we focus on the polymer nanoparticle (either spherical, anisotropic or free surface) interface and specifically on polymer dynamics considering attractive interaction with the nanoparticles. We concisely report the main experimental and computer molecular simulation studies regarding the polymer mobility at the interface. We show how changes of the glass transition (T_g) are correlated with segmental dynamics and relaxation at the interface.

4.1 Introduction

By dispersing spherical [1–9] or anisotropic nanoparticles [10–18] into a dense polymer matrix several material properties can be enhanced, whether mechanical [19, 20], electrical, optical [21] or plasmonic [22]. Although there is a plethora of experimental studies into these kinds of properties or regarding the polymer structure [23–33], there is still controversy and much less focus on the dynamics of polymers [34, 35] in the polymer-nanoparticle interface. Some research studies claim that there is a "bound" polymer layer around the nanoparticle surface, which is immobile [36–38], whereas other research efforts dispute this result insisting that the polymer around the nanoparticle has a certain translational and segmental dynamics [39, 40]. It is not only the packing of chains in the interfacial regions that contributes to the material property enhancements in nanocomposites [42, 43] but also the "bound" polymer layer, which is formed on nanoparticles interacting favorably, or even the chemical heterogeneity of interfacial layers [39, 44] around the nanoparticles contributes largely to the mechanical reinforcement that is observed in nanoparticle-filled elas-

A. V. Karatrantos
Materials Research and Technology, Luxembourg Institute of Science and Technology, 5, Avenue des Hauts-Fourneaux, 4362 Esch-sur-Alzette, Luxembourg
e-mail: argyrioskaratrantos@gmail.com

N. Clarke (✉)
Department of Physics and Astronomy, University of Sheffield, Sheffield S3 7RH, UK
e-mail: n.clarke@sheffield.ac.uk

V. V. Ginzburg and L. M. Hall (eds.), *Theory and Modeling of Polymer Nanocomposites*, Springer Series in Materials Science 310,
https://doi.org/10.1007/978-3-030-60443-1_4

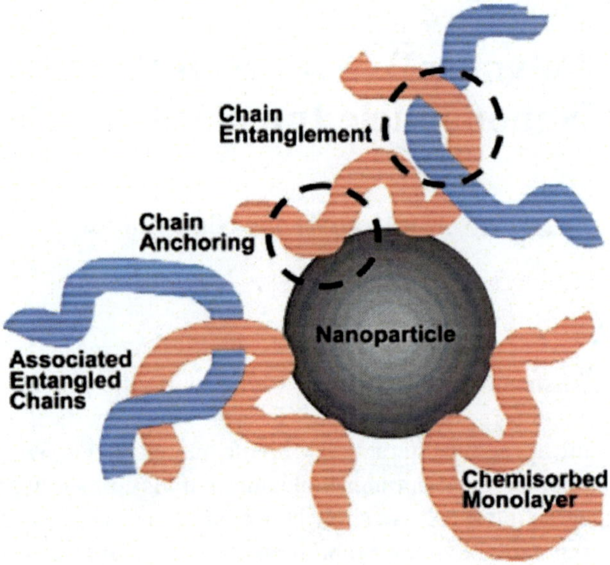

Fig. 4.1 Schematic representation of the interface region between a filler and a polymer matrix. Reprinted with permission from [41]. Copyright American Chemical Society

tomers. Polymer mobility near nanoparticle surfaces has been extensively discussed qualitatively, however, a direct experimental measurement of polymer mobility in the nanocomposites is not easy. In this chapter, we do not aim to report on and discuss studies that investigate the dynamics of polymers in melt [45, 46] or under confinement since there are excellent articles [47, 47–50, 50, 51] and reviews [45, 52] available. Neither do we aim to report, in detail, experimental techniques (results) [24] nor different molecular simulation techniques for measuring polymer dynamics in the interface. In this chapter, we provide a review of theoretical and experimental research that has been undertaken, in order to shed light into chain interfacial dynamics. We report studies that investigate this interface not only using experimental methods but also with computer simulations (in particular coarse grained molecular dynamics or Monte Carlo methods are incorporated) of both polymer–spherical nanoparticle and also polymer–surface substrate (Fig. 4.1).

4.2 Polymer Dynamics Around Spherical or Anisotropic Nanoparticles

4.2.1 Experiments

From the 1970s on, there were efforts to carry out nuclear magnetic resonance (NMR) experiments that identified an immobilized region of rubber around carbon black particles [53, 54]. It was speculated that this was a "bound" rubber layer around the carbon black [53]. The amount of immobilized polymer was estimated, using solid-state NMR, and a correlation with the mechanical reinforcement was established, identifying a direct, strongly nonlinear dependence on the immobilized polymer fraction. The observation of a temperature–independent filler percolation threshold suggested that immobilized polymer fractions may not necessarily form contiguous layers around the filler particles but could only reside in highly confined regions between closely packed filler particles, where they dominated the bending modulus of aggregated nanoparticles [55]. In addition to that immobilized region around carbon black, there was a relatively mobile region, but its dynamics was much slower than that in bulk polymer melt [53, 54, 56]. A nanosilica network formed in industrial styrene–butadiene rubber (SBR) nanocomposites contributed to the storage modulus at different frequencies and temperatures and demonstrated that it was viscoelastic in nature [55]. Moreover, in mixtures such as nitrile butadiene rubber (NBR)/nanosilicas[38] or crosslinked poly(ethyl acrylate)/nanosilicas [57], regions with different segmental dynamics were found [57]. In particular, a "glassy" (immobilized) layer region was observed [38, 57]. In addition, in other NMR studies of poly(ethylene glycol)/SiO_2 mixtures, the existence of a "glassy" layer [58, 59] was also noted. It has been observed that in nanocomposites which contain attractive polymer–nanoparticle interaction, such as nanosilica/poly(ethylene glycol) (PEG) or poly(butylene oxide) (PBO) mixtures [60], polymer chain dynamics was reduced by the nanoparticle surface [61] with the effect being greater where nanoparticles loading was larger than the percolation threshold, since the interface would increase, meaning more monomers existed into the interfacial area [62]. Thus, the mechanism of such polymer diffusion reduction was due to the nanoparticle interfacial area [62–64].

The small angle neutron scattering (SANS) technique was used to probe the temporal persistence of a layer of poly(2-vinylpyridine) (P2VP) that was "bound" with nanosilica [40, 65, 66] at two characteristic temperatures [67]. In particular, it was observed that there is almost no long-term reorganization at 150 °C (which corresponds to $T_{g,\mathrm{P2VP}} + 50\,°\mathrm{C}$), but instead, a reduction in the "bound" layer thickness at 175 °C (Fig. 4.2) was found. It was claimed that this strong temperature dependence arose from the polyvalence of the binding of a single P2VP chain to the nanosilica. The adsorption–desorption process of polymer segments was an activated procedure that occurred over a broad temperature range [67]. The interfacial dynamics of P2VP/silica nanocomposites and matrix-free P2VP grafted from nanosilicas has been shown [40, 68] to be a significant influence, at an intermediate polymer molecular weight, due to the impact of the chain attachment mechanism on segmental dynamics, as depicted in Fig. 4.3. Similar behavior was observed on segmental dynamics for poly(ethyl acrylate)/silica nanocomposites [69].

Furthermore, poly(ethylene oxide) nanocomposites with spherical nanosilicas have been studied by proton NMR spectroscopy, in order to identify and characterize reduced-mobility components, which may arise from either room-temperature lateral adsorption or end-group mediated high-temperature bonding to the surface of nanosilica [70]. In that work, neutron scattering spectroscopy showed that the thickness of a layer, characterized by significant internal mobility, resembling backbone rotation, ranged from 2 nm, for longer (20 K) chains adsorbed on 42 nm diameter nanoparticles, to 0.5 nm and below for shorter (2 K) chains adsorbed on 13 nm

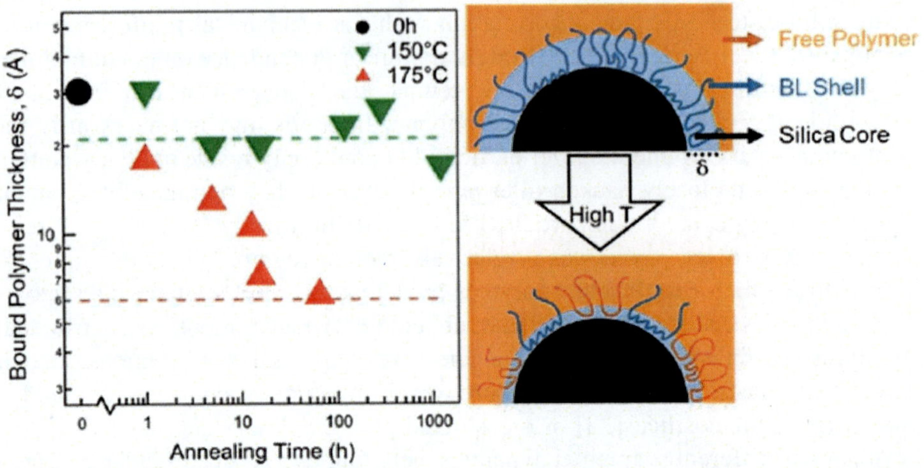

Fig. 4.2 Schematic of the polymer-nanoparticle interface. The silica core (black) has a well–defined "bound" layer (blue, d3–P2VP) and is well-dispersed in a polymer matrix (orange, h–P2VP). Average shell thickness, δ, versus annealing time at 150 °C (green) and 175 °C (red), starting from an average preannealed thickness of ≈3 nm (black). Green and red dashed lines are drawn as guides to the eye. Reprinted with permission from [67]. Copyright American Chemical Society

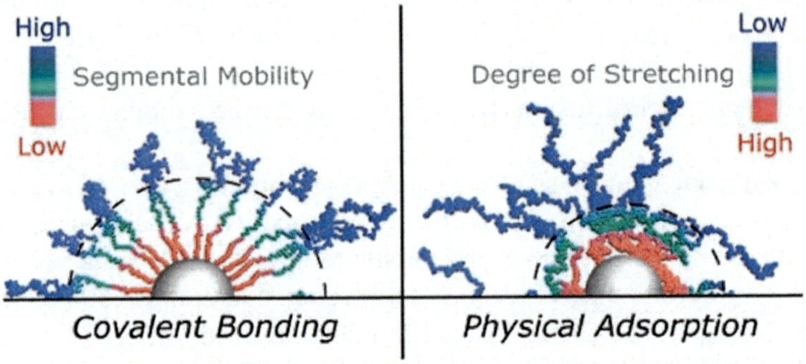

Fig. 4.3 Cartoon based on the self consistent field theory (SCFT) results that illustrates how chain stretching can cause the suppression of segmental mobility at the interface in polymer-grafted nanoparticles (left) and polymer nanocomposites (right). Reprinted with permission from [68]. Copyright American Chemical Society

nanoparticles. In the same type of mixture, by using proton time-domain NMR, three polymer layers were inferred with a different mobility: a strongly adsorbed layer ("glassy" layer) which does not change with temperature, a solid-like layer with intermediate relaxation times, and a highly mobile polymer layer [59]. The fraction of polymers in the glassy state was independent of the molecular weight of the polymer [59]. In order to investigate further the hypothesized "glassy" layer at the nanosilica surface, local dynamics was studied by time of flight (TOF) spectroscopy. It was concluded that PEO chains underwent fast picosecond dynamics which was very similar to that in the bulk, thus there was no indication for the hypothesized "glassy" layer [59]. In a mixture of polydimethylsiloxane (PDMS) chains adsorbed onto nanosilica particles, with PDMS molecular weight below the entanglement molecular weight, interfacial PDMS mobility showed a suppression of dynamics dominated by the PDMS–nanosilicas contacts as observed by broadband dielectric spectroscopy (BDS) and thermally stimulated depolarization currents (TSDC) in a system of [71].

As the nanosilica loading increases, the average particle spacing decreases and approaches the distance at which "bound" polymer layers interact, and there is a strong increase in the network polymer formation. This onset of network polymer formation alters the rheology of the mixture. For adequate polymer nanoparticle interaction strength, the nanoparticles result in mechanical reinforcement [59]. When acrylate polymers are adsorbed to nanosilica, increasing the temperature and adding water can induce a softening associated with a T_g gradient, whereas in poly(ethylene oxide) systems, temperature and water do not lead to significant changes in T_g [70].

Another neutron scattering investigation on the structure (by SANS) and dynamics (by neutron spin echo) of a single component nanocomposite consisted of polyisoprene chains grafted on to silica nanoparticles [72] showing the brush crowding, due to the mutual penetration of the coronas, which led to the topological confinement of the polymer chain dynamics. In particular, on a small scale, the segmental dynamics was unchanged compared to the reference melt, whereas on a large scale, the dynamics appears to have slowed down. By performing a mode analysis in terms of end-fixed Rouse chains, it was shown that the slower dynamics was tracked to topological confinement by the adjacent grafts. If 50% of matrix chains were added, the topological confinement of the grafted chain decreased thus chain motion was accelerated. A crossover from pure Rouse motion at short times to topological confined motion beyond the time at which the segmental mean squared displacement reached the distance to the next graft was also observed [72]. In a strongly attractive poly(vinyl acetate)s (PVAc)–nanosilica mixture, no "dead" layer was observed for any molecular weight used [73]. For that particular mixture, the thickness of the interfacial layer with a hindered segmental relaxation decreased as the molecular weight of polymer increased, in contrast to the theoretical predictions [74–76]. It is worth noting that in (PVAc)–nanosilica mixtures [73], much higher polymer molecular weights were used than in PEG–nanosilica mixtures [59]. Furthermore, no "dead" layer was also observed, after quasi elastic neutron scattering (QENS), in PDMS chains confined between anodic aluminum oxide pores [77]. In nanocomposites containing nanosilicas tethered with poly(ethylene glycol) chains in a poly(methyl methacrylate) (PMMA) matrix, PMMA dynamics underwent a transition from a bulk–like behavior at low nanosilica loading to a confinement behavior at intermediate loading and eventually to glassy behavior at high loading [78]. The dynamics of free polystyrene (PS) and (PMMA) chains around bare or PS grafted nanosilica particles showed that the α–relaxation dynamics of the PS matrix was not affected by nanosilica loading, however it decreased in PMMA nanocomposites due to the attractive PMMA-nanosilica interfacial interaction [79]. In the case of PS grafted nanosilica, the mobility of the interfacial layer was enhanced. However, the α'–relaxation time in the vicinity of T_g of the polymer matrix increased when the temperature was increased [79]. Interfacial layers around nanoparticles, consisting of both adsorbed and free matrix chains, could alter the rheological behavior of nanocomposites. Low glass-transition temperature composites, consisting of PVAc coated nanosilicas in PEO and poly(methyl acrylate) (PMA) matrices and of PMMA coated silica nanoparticles in a PMA matrix have been probed using rheology and X-ray photon correlation spectroscopy. It was shown that the miscibility between adsorbed and matrix chains in the interfacial layers led to the reinforcement of the material [42]. This large dynamical asymmetry between the miscible matrix and surface-bound polymers, and their interfaces has been observed using X-ray photon correlation spectroscopy [80]. The polymer mobility changed in the interfaces of nanocomposites, as extracted from the length-scale-dependent slow particle motion [80]. The local mobility gradient was signified by an unprecedented increase in the relaxation time at length scales on the order of the polymer radius of gyration. This effect was accompanied by a transition from diffusive to sub–diffusive behavior.

4.2.2 Simulations

Using a hybrid model comprised of an atomistic representation of a single wall carbon nanotube (SWCNT) and the coarse grained polymer Kremer–Grest model, it has been shown that the polymer diffusivity depended on the enthalpic interaction between the monomers and SWCNT, as depicted in Fig. 4.4. When the interaction energy between SWCNT and polymers was k_BT, the diffusivity of the first layer, next to SWCNT, slowed down [35] by 80% in comparison to bulk polymer diffusivity [34, 81], as depicted in Fig. 4.4. At distance $\approx 2R_g$ from the SWNCT surface, the polymer diffusivity regained its bulk value. Similar effect was observed in a model nanocomposite containing spherical attractive nanoparticles, where polymer dynamics were slowed dramatically [82]. In the case of repulsive polymer nanoparticle interaction, interfacial dynamics appeared to be faster than in the bulk polymer melt [82–87]. In a coarse grained molecular investigation of a nanosilica (either bare or grafted with PS chains) embedded in a coarse grained PS melt, the PS chain dynamics was affected not only by the nanosilica surface but also by the polymer matrix length and the grafting density [88]. The impact of the nanoparticle decreased with its distance from the surface [89]. Moreover, the polymer matrix molecular weight could control the dynamics in the grafted corona [90]. In particular, short polymer matrix molecular weights were able to "wet" the grafted chains and induce a faster relaxation to those chains. The higher grafting density slowed down the polymer chain dynamics [88].

The interfacial zone alters not only the monomer segmental packing but also the polymer mobility [92]. The scale for density perturbations around the interface decreases with cooling, which means that the interface becomes more sharply defined during cooling [92]. However, the interfacial mobility scale ξ for both nanoparticles and supported films increases with cooling and is of the order of a few nanometers, independent of the polymer-interfacial interaction. There is a distinct relaxation, which is slower than the α-relaxation due to the "bound" polymer on the nanoparticle

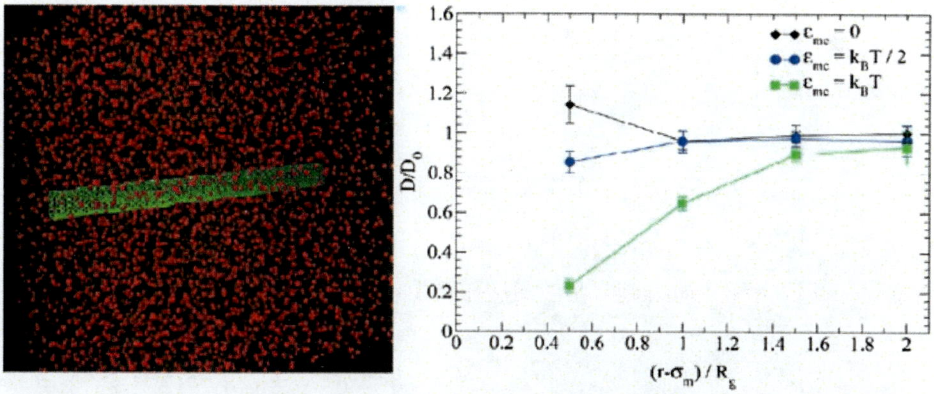

Fig. 4.4 Self diffusion coefficient of polymer chains ($N = 60$) in a monodisperse polymer melt with a SWCNT ($r_{\mathrm{SWVNT}} = 0.66$) normalized by its melt value as a function of the distance r from the SWCNT surface to the polymer center of mass: (i) no attractive interaction between polymers and SWCNT (diamonds), (ii) attractive ($k_{\mathrm{B}}\mathrm{T}/2$) interaction between polymers and SWCNT (circles), and (iii) attractive ($k_{\mathrm{B}}\mathrm{T}$) interaction between polymers and SWCNT (squares). The solid lines have been added to guide the eye. Reprinted with permission from [34]. Copyright American Chemical Society

surface [91, 93]. That "bound" layer "cloaks" the nanoparticle surface [91] at large nanoparticle surface attraction, thus the polymer matrix dynamics is unaffected. The T_{g} of a polymer nanocomposite depends on the local polymer segmental dynamics; however, it is insensitive to the polymer-nanoparticle surface interaction strength when it is stronger than the polymer-polymer interactions (Fig. 4.5).

By using molecular dynamics on model polymer nanocomposites, it is been observed that there is a dynamic heterogeneity [82] of polymer dynamics depending on the nanoparticle volume fraction, with its maximum at the nanoparticle volume fraction in which the Kuhn length of polymers is equal to the average distance of the nanoparticle. Such heterogeneity is due to the decoupling between polymer chain dynamics and nanoparticles [94] as can be depicted in self-intermediate scattering function which measures the correlation of the position of monomers at a given length scale $1/q$ (Fig. 4.6). In addition, polymer dynamics in the interface was affected less by decreasing the nanoparticle size. However, the T_{g} changed more substantially where the nanoparticle size was extremely small [95], due to the smaller nanoparticle spacing on decreasing the particle size [95]. It has also been seen that for larger sized nanoparticles, interfacial relaxations were substantially slower than the matrix, thus the "bound" polymer dynamically decoupled from the polymer matrix, and there were small changes in the T_{g} relative to those of the bulk polymer for large nanoparticles [95]. The monomer relaxation time, as a function of distance from the nanoparticle surface for different nanoparticle sizes, was determined by fitting the self-intermediate scattering function $F_{\mathrm{self}}(q, t, r)$ to (4.1) (see Fig. 4.7).

$$F_{\mathrm{self}}(q, t, r) = [1 - A(r)]e^{(-t/\tau_s)^{3/2}} + A(r)e^{(-t/\tau_\alpha(r))^{\beta(r)}} \tag{4.1}$$

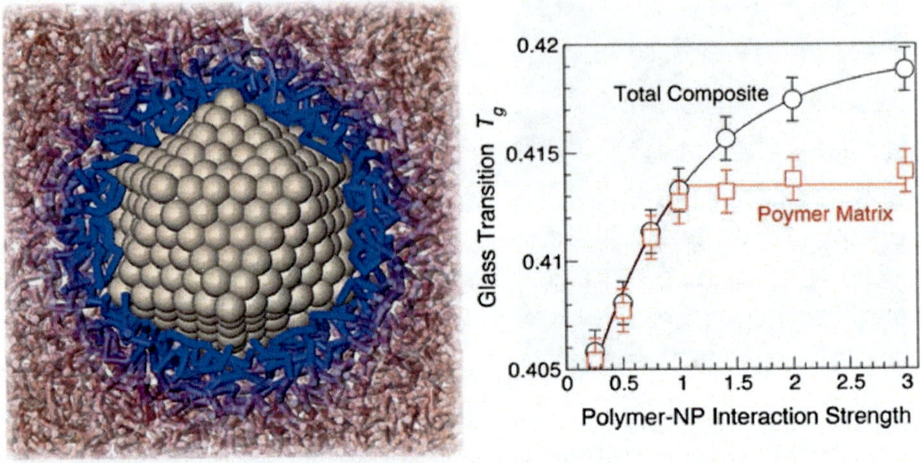

Fig. 4.5 Dynamical definition of T_g , namely, $\tau(T_g) = 10^3$ (in Lennard Jones units). Black symbols define T_g for the overall composite relaxation, τ, while red symbols define T_g from the matrix component of relaxation, $\tau(\alpha)$. Like the thermodynamic T_g, the matrix component plateaus when "bound" polymer develops. The right-hand axis indicates T_g normalized by the corresponding bulk material, highlighting the small amplitude of T_g changes. Reprinted with permission from [91]. Copyright American Chemical Society

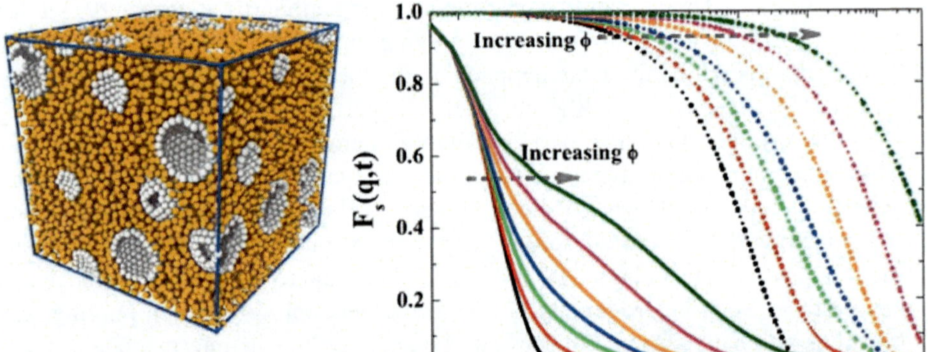

Fig. 4.6 Typical snapshot of simulated polymer nanocomposite. The volume fraction of nanoparticles, ϕ, is 0.25, and the radius of nanoparticles is 4.0σ. The self-intermediate scattering function $F_s(q, t)$ of polymer beads in PNCs with $R = 2.0$ as a function of time. The solid lines ($q = 7.0$) and dotted lines ($q = 0.4$) correspond to different k values. From left to right, the volume fractions are 0.00, 0.10, 0.18, 0.25, 0.36, 0.44, and 0.50, respectively. Reprinted with permission from [94]. Copyright American Institute of Physics

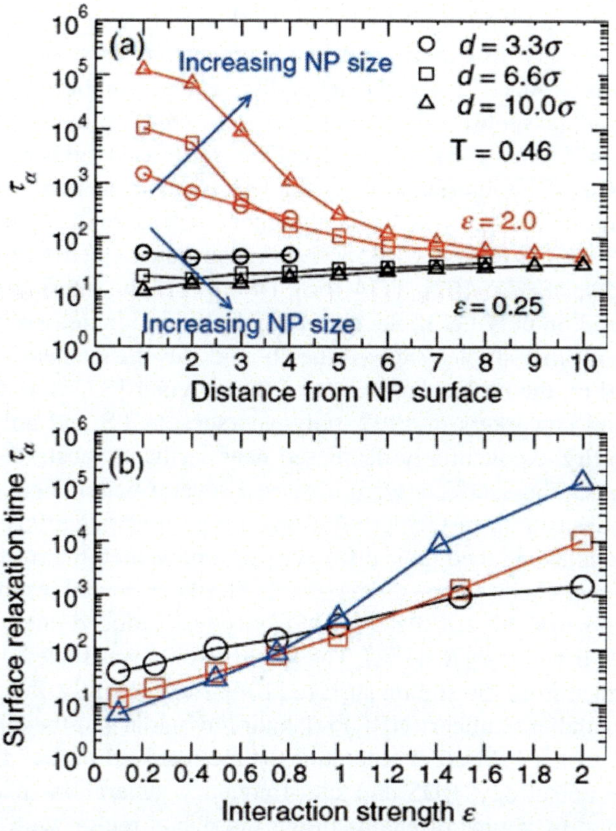

Fig. 4.7 **a** The monomer relaxation time as a function of distance from the nanoparticle surface for different nanoparticle sizes determined by fitting $F_{\text{self}}(q, t, r)$ using (4.1) ($\tau_s = 0.29$). The first term of the equation is vibrational relaxation and the second term represents the α relaxation. For a weak nanoparticle interaction strength $\epsilon_{ps} = 0.25$ (black lines), the monomer relaxation is enhanced approaching the nanoparticle surface. For a strong nanoparticle interaction strength $\epsilon_{ps} = 2.0$ (red lines), the monomer relaxation is significantly slowed approaching the nanoparticle surface. **b** The relaxation time nearest to the nanoparticle surface as a function of interaction strength for different nanoparticle sizes. The data show that the effect of interactions on surface relaxation diminishes as nanoparticle size decreases. Reprinted with permission from [95]. Copyright American Physical Society

4.3 Polymer Dynamics Around a Flat Surface

4.3.1 Experiments

In a polymer-substrate interface, experiments have indicated the existence of a "dead" layer, permanently adsorbed on the substrate, for example entangled PS on an aluminum [96] or a gold [97] substrate. By using the resonance enhanced dynamic light scattering (REDLS), polybutadiene (PB) dynamics were found to slow down at an interface, with a 25 nm thick gold substrate [98]. Two different interfacial regions have been identified by NMR experiments in a polymer-film substrate. First a "glassy", or in other words, a "dead" layer and secondly, another immobilized layer which had reduced segmental mobility [58]. However, some differential scanning calorimetry (DSC) studies [61, 99–101] did not reveal any significant "glassy" layer in different nanocomposites. The interfacial layer thickness was measured in the range of 1.5–5 nm by using various experimental techniques, such as small angle X-Ray scattering (SAXS) [61], SANS [67], NMR [58, 59], transmission electron microscopy (TEM) [66], BDS [24, 40, 71, 102], TGA [66], DSC [71]. The interfacial layer thickness did not depend on polymer molecular weight [39] and increased upon cooling [24, 39]. However, other studies showed the "bound" layer thickness [103] and dynamics depended on the molecular weight of the polymer [73, 104, 105]. As reported by Zheng and coworkers in 1997, polymer (such as PS and poly(vinyl pyridine) (PVP)) mobility exponentially decreased near a silicon substrate due to enhanced entanglements. These effects were observed several nanometers into the polymer film from the substrate surface [106]. Long–range substrate effects on the adsorbed PS or PVP chains reduced their diffusivity and prevented them from diffusing into the free polymer chains above them [106, 107]. Experimental results by dynamic ion mass spectrometry showed that PS diffusivity was reduced due to the silicon substrate up to a distance $\approx 4R_g$ [108]. The PS diffusivity was scaled down to $N^{-2.5}$ at a certain distance from the silicon surface [108]. Very recently, the polyisoprene (PI) and nitrile butadiene rubber (NBR) relaxation dynamics at the silica interface was investigated using 3 different techniques (time-resolved evanescent wave-induced fluorescence anisotropy, BDS and sum-frequency generation spectroscopy) [109]. The segmental dynamics of chains in the interfacial region with the silica surface was dependent on whether they were loosely or strongly adsorbed at the surface. In particular, the segmental relaxation of chains in the strongly adsorbed layer at the interface could be slower than the dynamics of bulk chains by more than 10 orders of magnitude [109].

4.3.2 Simulations

A molecular dynamics simulation study of a chemically realistic model of 1,4-PB chains, between graphite walls [110–112], showed a slowing down of polymer chain dynamics in accordance with experiments by Koga et al. [97]. In addition, atomistic molecular dynamics simulations of a chemically realistic model of atactic short-chain PS on gold surfaces were performed and proved that the dynamics of the film is reduced compared to bulk polymer dynamics [113]. Moreover, the mobility of polyethylene (PE) melt near a graphite substrate was highly anisotropic. In particular, it was dramatically reduced in the direction perpendicular to graphite whilst it was unaltered in the direction parallel to the substrate [114]. The PE diffusivity was heterogeneous perpendicular to the substrate, and reached its bulk value approximately at a distance $6R_g$ [114]. Such dynamics depended on the interfacial type of polymer–

nanoparticle interaction. It was generally believed that an attractive polymer-surface interaction could slow relaxation in thin supported polymer films [115, 116] and polymer nanocomposites [34, 87]. Instead, a repulsive polymer-surface interaction accelerated the polymer dynamics with respect to the bulk [82, 117]. Recent research on nanocomposites has shown that this slowing of polymer chain dynamics [34, 118] occurred more strongly near a highly attractive nanoparticle surface where a "bound" polymer layer [119] of a much lower mobility could be formed, strongly influencing the dynamics of the film. By performing coarse grained simulations, a similar "bound" interfacial layer in thin supported polymer films was observed when the polymer-polymer attraction was stronger than the polymer-polymer interaction strength [120] (Fig. 4.8). This "bound" polymer layer insulated the remainder of the film from the strong substrate interfacial interaction. This layer gave rise to an additional relaxation process in the self-intermediate scattering function, which was not observable in the bulk material, and eventually slowed down the relaxation time of the film.

Simulations have been performed for both unentangled and entangled polymers, considering the film thicknesses to the bulk state [121, 122]. The relaxation time of confined polymer chains was measured through the end-to-end vector correlation functions. It has been shown that there is a minimum in the relaxation time of entangled chains when decreasing the film thickness, due to the disentanglement induced by the confinement [121]. For a large substrate attraction, the "bound" polymer layer effectively "cloaked" the substrate, reducing the effect of the polymer-surface interaction on T_g [92]. Very recently, a fully atomistic model of poly(vinyl–alcohol)(PVA)-silica nanocomposite was simulated by molecular dynamics [123], where the effect of the polymer–nanoparticle interaction on T_g was investigated. The potential of mean force and segmental dynamics indicated that the strong binding interaction between the hydroxylated-silica surface and PVA induced an increase in the T_g of the nanocomposite in comparison to bulk PVA [123]. While it was expected that the T_g of the system would increase with an increasing amount of hydrogen bonds arising from the increasing surface hydroxylation, the trend of increasing T_g reached a maximum value when the surface was about 75% hydroxylated. Beyond that value, a drop in the T_g was observed. It was found that the competition between inter PVA-silanol and intra silanol-silanol interactions was the main reason that contributed to the drop of T_g. Thus, the number and strength of the different kinds of hydrogen bonds in such nanocomposites could be tuned to enable the optimization of the T_g changes that are desired for specific applications [123]. The T_g of PMMA/graphene model systems was found to depend strongly on tacticity in agreement with experiments as was investigated by atomistic simulations [124]. The local PMMA dynamics was studied by analyzing torsional autocorrelation functions for various dihedral angles. A spectrum of different correlation times and activation energies was observed for the motions of different parts of PMMA chains. The dynamic heterogeneity of the PMMA chains has been studied in detail for the different stereo-chemistries via the temperature dependence of the stretching exponent [124]. In particular, the backbone dynamics, the ester group and α-methyl motion were affected by tacticity. The

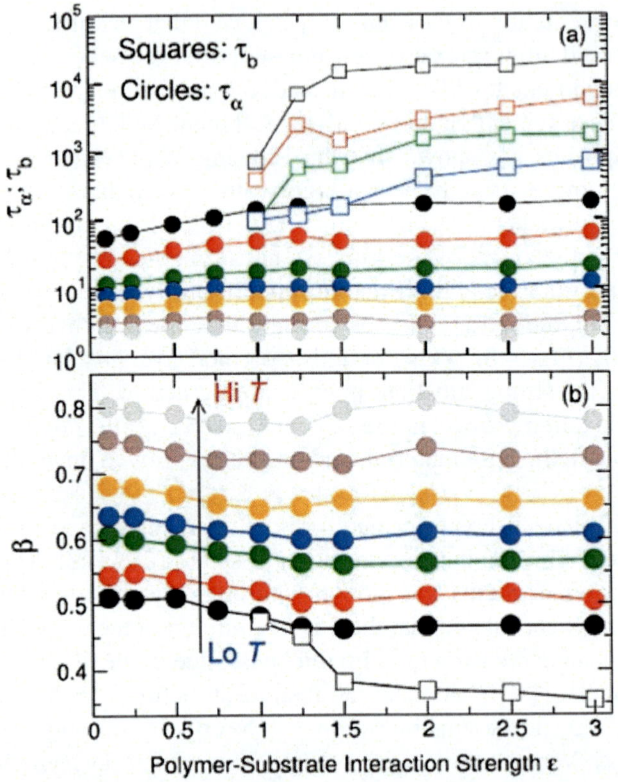

Fig. 4.8 **a** The relaxation time of unbound film τ_α and "bound" layer τ_b. From top to bottom, the temperatures are 0.45 (black), 0.47 (red), 0.50 (green), 0.52 (blue), 0.55 (yellow), 0.60 (brown), and 0.65 (grey). For $\epsilon_{ps} \geq 1$ and $T \leq 0.52$ it is distinguished the bound relaxation, where τ_b is roughly 2 orders of magnitude larger than τ_α. **b** The value of the exponent β for the stretched-exponential fit of the α-relaxation in (5) (circles); β is only weakly dependent on ϵ_{ps}. The squares show the results of fitting the relaxation without an additional "bound" layer (4.1), which indicates that the "bound" layer has the effect of decreasing the fit value of β if it is not accounted for. Such a decrease of β may provide a useful indicator that an additional process to describe relaxation data is necessary. Reprinted with permission from [119]. Copyright American Institute of Physics

ester-methyl group and the end-chain monomers had higher tendency to adsorb on the graphene. The adsorbed layer presented reduced mobility [124].

The dynamic behavior of polymer chains adsorbed on an attractive, homogeneous substrate surface has been studied by dynamic Monte Carlo simulations [125]. The translational diffusion coefficient D_{xy} parallel to the substrate decreased if either the intra-polymer attraction strength, ϵ_{pp}, or the polymer-surface attraction strength, ϵ_{ps}, increased. The rotational relaxation time increased with the increase of polymer-surface attraction strength, but the dependence of this relaxation time on intra-polymer attraction strength was also dependent on the adsorption state of the polymer. It has been found that this relaxation time decreased with increasing intra-polymer attraction strength for a partially adsorbed polymer and increased for a fully adsorbed

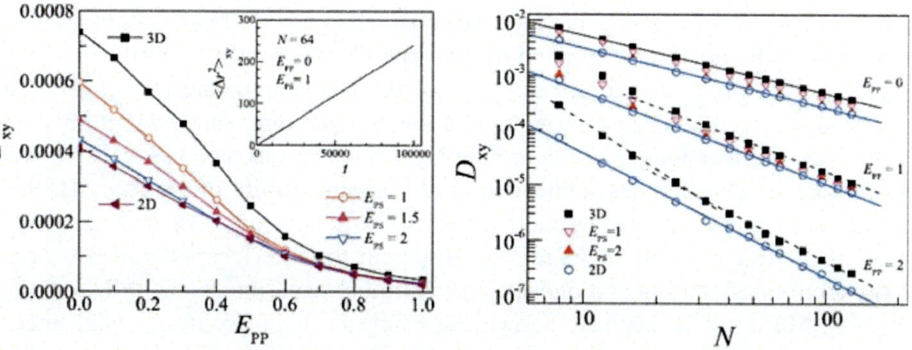

Fig. 4.9 Plot of the translational diffusion coefficient D_{xy} versus the intra-polymer attraction ϵ_{pp} for polymer in 3D solution, on attractive surfaces at attraction strengths $\epsilon_{ps} = 1, 1.5, 2$, and on 2D surface. Polymer length $N = 64$. The insert shows the dependence of $\Delta r^2(xy)$ on time t for $N = 64$ at $\epsilon_{pp} = 0$ and $\epsilon_{ps} = 1$. Log-log plot of the translational diffusion coefficient D_{xy} versus the polymer chain length N for polymers in 3D solution, on attractive surfaces with attractions $\epsilon_{ps} = 1, 1.5, 2$ and on 2D surface. Three data sets with intra-polymer interactions $\epsilon_{pp} = 0, 1, 2$ are presented. Solid lines are linear fit of simulation data, while dashed lines are guides for the eye. Reprinted with permission from [125]. Copyright Springer Nature

polymer [125]. The scaling relation $D_{xy} \approx N^{-a}$ held and the rotational relaxation time was proportional to $\approx N^{\beta}$ (Fig. 4.9). The scaling exponent α was not dependent on the polymer-substrate attraction. The scaling parameter $\beta \approx 2.7$ was not dependent on polymer-substrate attraction for the adsorbed polymer at $\epsilon_{pp} = 0$, but increased with polymer-substrate attraction when $\epsilon_{pp} > 0$. The β always decreased with increasing ϵ_{pp}. When ϵ_{ps} k_BT polymer dynamics slowed down and for ϵ_{ps} $4k_B$T it approached glassy dynamics [86, 126–129]. In another study by molecular dynamics, the polymer diffusion of a single polymer adsorbed on surfaces with a different roughness was simulated. The simulations demonstrated that when there was a strong adsorption and under a strong confinement, the scaling exponent β of the polymer diffusivity on the chain length exhibited three cases depending on the confinement size: a) a Rouse plateau with $\beta \approx 1$ (the lateral motion of the polymer chains was free), a reptation–like plateau with $\beta \approx 1.5$ (the polymer chains could move, through the confinement, in the perpendicular direction) and a transition from the Rouse plateau to the reptation– like plateau with $1.5 < \beta < 1$ appeared (confinement hindered the lateral motion of the polymer chains). By decreasing the confinement (increasing the distance between obstacles) the polymer lateral motion was allowed, which resulted in the higher plateau [130].

Using molecular dynamics simulations, the dynamical properties of a single polymer chain dissolved in an explicit solvent and was strongly adsorbed at the solid–liquid interface. This showed that on analytically smooth interfaces, in accordance with the slip boundary condition, the motions of the polymer chain and the surrounding solvent were hydrodynamically coupled. This led to the chain diffusion coefficient, D, scaling with the chain degree of polymerization N as $D \approx N^{-3/4}$, consistent with the Zimm dynamics for strongly adsorbed chains. The incorporation of transverse forces at the interface resulted in a loss of correlation between the motion of the polymer chain and the solvent. Consequently, the diffusion scaled as $D \approx N^{-1}$, which was a characteristic of the Rouse dynamics [131]. In addition, a bead-spring model for oligomers at a low temperature, showed that only a weak temperature dependence of the mobile layer thickness existed [132]. Moreover, the interfacial layer thickness depended on the semiflexibility of the chain [133, 134]. The segmental dynamics was slower for stiffer chains and there was a correlation between structure and segmental dynamics in the interface as was also observed by neutron reflectivity and dielectric permittivity experiments of PMMA thin films [135]. Moreover it has been shown, by using the Kremer-Grest model, that there was no "glassy" layer, and the dynamic slowing down of stiffer chains was due not only to the densification of the adsorbed layer, but also to segment layering [133]. The segmental dynamics was less sensitive to chain stiffness [133]. Swollen "bound" hydrogenated PB chains exhibited the collective dynamics (the so-called breathing mode) at polymer concentrations (c) below and above the overlap polymer concentration c^* as observed by SANS and neutron spin echo techniques [136]. Furthermore, molecular dynamics simulations showed that polymer matrix chains, with a length equal to that of the "bound" polymer, could be accommodated in the "bound" polymer layer effectively, while the longer polymer matrix chains partially penetrated the "bound" chains and their diffusion behavior was hardly affected compared to their behavior in the bulk region [136].

4.4 Conclusions

In this chapter, we have briefly outlined the main studies (experimental and simulations) that deal with the problem of polymer mobility in the interface, near the nanoparticle (or flat surface). The first studies of rubber around carbon black denoted a "bound" polymer layer at the carbon black. Moreover, other studies of polystyrene on metallic substrates (such as aluminum or gold) denoted the existence of a "dead" layer due to the very high adsorption to the substrate. However, a large amount of experimental studies in a variety of systems, with attractive surfaces, dispute the appearance of such a layer with no mobility at temperatures above T_g. Furthermore, coarse grained simulations verify this observation, denoting interfacial and segmental dynamics even at high attractive interaction strengths.

References

1. M.E. Mackay, A. Tuteja, P.M. Duxbury, C.J. Hawker, B. Van Horn, Z. Guan, G.H. Chen, R.S. Krishnan, General strategies for nanoparticle dispersion. Science **311**, 1740 (2006)
2. M. Supova, G.S. Martynkova, K. Barabasyova, Effect of nanofillers dispersion in polymer matrices: a review. Sc. Adv. Mater. **3**, 1–25 (2011)
3. L. Hall, A. Jayaraman, K.S. Schweizer, Molecular theories of polymer nanocomposites. Current Opin. Solid State Mater. Sci. **14**, 38–48 (2010)

4. V.V. Ginzburg, C. Singh, A.C. Balazs, Theoretical phase diagrams of polymer/clay composites: the role of grafted organic modifiers. Macromolecules **33**(3), 1089–1099 (2000)
5. J.A. Pomposo, A. Ruiz de Luzuriage, A. Etxeberria, J. Rodriguez, Key role of entropy in nanoparticle dispersion: polystyrene-nanoparticle/linear-polystyrene nanocomposites as a model system. Phys. Chem. Chem. Phys. **10**, 650–651 (2008)
6. N. Jouault, F. Dalmas, F. Boue, J. Jestin, Multiscale characterization of filler dispersion and origins of mechanical reinforcement in model nanocomposites. Polymer **53**, 761 (2012)
7. C. Chevigny, F. Dalmas, E. Di Cola, D. Gigmes, D. Bertin, F. Boué, J. Jestin, Polymer-grafted-nanoparticles nanocomposites: dispersion, grafted chain conformation, and rheological behavior. Macromolecules **44**, 122–133 (2011)
8. G.D. Smith, D. Bedrov, Dispersing nanoparticles in a polymer matrix: are long, dense polymer tethers really necessary? Langmuir **25**, 11239–11243 (2009)
9. A. Karatrantos, Y. Koutsawa, P. Dubois, N. Clarke, M. Kröger, Miscibility and diffusion in ionic nanocomposites. Polymers **10**, 1010 (2018)
10. A.A. Golosova, J. Adelsberger, A. Sepe, M.A. Niedermeier, P. Lindner, S.S. Funari, R. Jordan, C.M. Papadakis, Dispersions of polymer-modified carbon nanotubes: a small-angle scattering investigation. J. Phys. Chem. C **116**, 15765–15774 (2012)
11. Y. Zare, H. Garmabi, Attempts to simulate the modulus of polymer/carbon nanotube nanocomposites and future trends. Polym. Rev. **54**, 377–400 (2014)
12. M.J.A. Hore, R.J. Composto, Strategies for dispersing, assembling, and orienting nanorods in polymers. Curr. Opin. Chem. Eng. **2**, 95–102 (2013)
13. A. Karatrantos, R.J. Composto, K.I. Winey, N. Clarke, Nanorod diffusion in polymer nanocomposites by molecular dynamics simulations. Macromolecules **52(6)**, 2513–2520 (2019a)
14. Y. Gao, J. Liu, J. Shen, L. Zhang, D. Cao, Influence of various nanoparticle shapes on the interfacial chain mobility: a molecular dynamics simulation. Phys. Chem. Chem. Phys. **16**, 21372 (2014)
15. A. Karatrantos, N. Clarke, R.J. Composto, K.I. Winey, Topological entanglement length in polymer melts and nanocomposites by a DPD polymer model. Soft Matter **9**, 3877–3884 (2013)
16. K.S. Khare, R. Khare, Effect of carbon nanotube dispersion on glass transition in cross-linked epoxy-carbon nanotube nanocomposites: role of interfacial interactions. J. Phys. Chem. B **117**, 7444–7454 (2013)
17. A.L. Frischknecht, M.J.A. Hore, J. Ford, R.J. Composto, Dispersion of polymer-grafted nanorods in homopolymer films: theory and experiment. Macromolecules **46**, 2856–2869 (2013)
18. U.K. Sankar, M. Tripathy, Dispersion, depletion, and bridging of athermal and attractive nanorods in polymer melt. Macromolecules **48**, 432–442 (2015)
19. G.A. Buxton, A.C. Balazs, Predicting the mechanical and electrical properties of nanocomposites formed from polymer blends and nanorods. Molec. Simul. **30**, 249–257 (2004)
20. L. Tang, Christoph Weder, Cellulose whisker/epoxy resin nanocomposites. ACS Appl. Mater. Interf. **2**, 1073–1080 (2010)
21. M.J.A. Hore, A.L. Frischknecht, R.J. Composto, Nanorod assemblies in polymer films and their dispersion-dependent optical properties. ACS Macro Lett. **1**, 115–121 (2012)
22. R.C. Ferrier, H.-S. Lee, M.J.A. Hore, M. Caporizzo, D.M. Eckmann, R.J. Composto, Gold nanorod linking to control plasmonic properties in solution and polymer nanocomposites. Langmuir **30**, 1906–1914 (2014)
23. B. O'Shaughnessy, D. Vavylonis, Non-equilibrium in adsorbed polymer layers. J. Phys.: Condens. Matter **17**, R63–R99 (2005)
24. S. Cheng, B. Carroll, V. Borachova, J.M. Carrillo, B. Sumpter, A.P. Sokolov, Focus: structure and dynamics of the interfacial layer in polymer nanocomposites with attractive interactions. J. Chem. Phys. **146**, 203201 (2017)
25. A. Karatrantos, N. Clarke, M. Kröger, Modeling of polymer structure and conformations in polymer nanocomposites from atomistic to mesoscale: a review. Polym. Rev. **56**, 385–428 (2016a)

26. M.J.A. Hore, Polymers on nanoparticles: structure & dynamics. Soft Matter **15**, 1120–1134 (2019)
27. V. Ganesan, A. Jayaraman, Theory and simulation studies of effective interactions, phase behavior and morphology in polymer nanocomposites. Soft Matter **10**, 13–38 (2014)
28. J.G. Ethier, L.M. Hall, Structure and entanglement network of model polymer-grafted nanoparticle monolayers. Macromolecules **51**(23), 9878–9889 (2018a)
29. J.G. Ethier, L.M. Hall, Modeling individual and pairs of adsorbed polymer-grafted nanoparticles: structure and entanglements. Soft Matter **14**(4), 643–652 (2018b)
30. A. Karatrantos, N. Clarke, R.J. Composto, K.I. Winey, Polymer conformations in polymer nanocomposites containing spherical nanoparticles. Soft Matter **11**, 382 (2015)
31. A. Karatrantos, N. Clarke, R.J. Composto, K.I. Winey, Entanglements in polymer nanocomposites containing spherical nanoparticles. Soft Matter **12**, 2567 (2016b)
32. A. Karatrantos, N. Clarke, R.J. Composto, K.I. Winey, Structure, entanglements and dynamics of polymer nanocomposites containing spherical nanoparticles. IOP Conf. Series: Mat. Sci. Eng **64**, 012041 (2014)
33. E.N. Skountzos, P.G. Mermigkis, V.G. Mavrantzas, Molecular dynamics study of an atactic poly(methyl methacrylate)-carbon nanotube nanocomposite. J. Phys. Chem. B **122**, 9007–9021 (2018)
34. A. Karatrantos, R.J. Composto, K.I. Winey, M. Kröger, N. Clarke, Entanglements and dynamics of polymer melts near a SWCNT. Macromolecules **45**, 7274 (2012a)
35. A. Karatrantos, N. Clarke, A theoretical model for the prediction of diffusion in polymer/SWCNT nanocomposites. Soft Matter **7**, 7334–7341 (2011)
36. A. Serbescu, K. Saalwachter, Particle-induced network formation in linear pdms filled with silica. Polymer **50**, 5434–5442 (2009)
37. A. Wurm, M. Ismail, B. Kretzschmar, D. Pospiech, C. Schick, Retarded crystallization in polyamide/layered silicates nanocomposites caused by an immobilized interphase. Macromolecules **43**, 1480–1487 (2010)
38. H. Xu, Y. Song, E. Jia, Q. Zheng, Dynamics heterogeneity in silica-filled nitrile butadiene rubber. J. Appl. Polym. Sci. **135**, 63–67 (2018)
39. I. Popov, B. Carroll, V. Bocharova, A.C. Genix, S. Cheng, A. Khamzin, A. Kisliuk, A.P. Sokolov, Strong reduction in amplitude of the interfacial segmental dynamics in polymer nanocomposites. Macromolecules **53**, 4126–4135 (2020)
40. S. Gong, Q. Chen, J.F. Moll, S.K. Kumar, R.H. Colby, Segmental dynamics of polymer melts with spherical nanoparticles. ACS Macro Letters **3**, 773–777 (2014)
41. D. Ciprari, K. Jacob, T. Tannenbaum, Characterization of polymer nanocomposite interphase and its impact on mechanical properties. Macromolecules **39**, 6565–6573 (2006)
42. S. Yang, S. Liu, S. Narayanan, C. Zhang, P. Akcora, Chemical heterogeneity in interfacial layers of polymer nanocomposites. Soft Matter **14**, 4784–4791 (2018)
43. E.J. Bailey, K.I. Winey, Dynamics of polymer segments, polymer chains, and nanoparticles in polymer nanocomposite melts: a review. Progress Polym. Sci. **105**, 101242 (2020)
44. A.C. Genix, V. Bocharova, B. Carroll, M. Lehmann, T. Saito, S. Krueger, L. He, P. Dieudonné-George, A.P. Sokolov, J. Oberdisse, Understanding the static interfacial polymer layer by exploring the dispersion states of nanocomposites. ACS Appl. Mater. Inter. **11**, 17863–17872 (2019)
45. A. Karatrantos, R.J. Composto, K.I. Winey, M. Kröger, N. Clarke, Modeling of entangled polymer diffusion in melts and nanocomposites: a review. Polymers **11**(5), 876 (2019b)
46. D. Richter, L. Willner, A. Zirkel, B. Farago, L.J. Fetters, J.S. Huang, Polymer motion at the crossover from rouse to reptation dynamics. Macromolecules **27**, 7437–7446 (1994)
47. Z.E. Dell, M. Muthukumar, Anomalous packing and dynamics of a polymer chain confined in a static porous environment. J. Chem. Phys. **149**, 174902 (2018)
48. M. Krutyeva, A. Wischnewski, D. Richter, Polymer dynamics in nanoconfinement: interfaces and interphases. EPJ Web of Conf. **83**, 02009 (2015)
49. G.J. Schneider, K. Nusser, L. Willner, P. Falus, D. Richter, Dynamics of entangled chains in polymer nanocomposites. Macromolecules **44**, 5857–5860 (2011)

50. G.S. Schneider, K. Nusser, S. Neueder, M. Brodeck, L. Willner, B. Farago, O. Holderer, W.J. Briels, D. Richter, Anomalous chain diffusion in unentangled model polymer nanocomposites. Soft Matter **9**, 4336–4348 (2013)
51. M. Lungova, M. Krutyeva, W. Pyckhout-Hintzen, A. Wischnewski, M. Monkenbusch, J. Allgaier, M. Ohl, M. Sharp, D. Richter, Nanoscale motion of soft nanoparticles in unentangled and entangled polymer matrices. Phys. Rev. Lett. **117**, 147803 (2016)
52. D. Richter, M. Krutyeva, Polymer dynamics under confinement. Soft Matter **15**, 7316–7349 (2019)
53. S. Kaufman, W.P. Slichter, D.D. Davis, Nuclear magnetic resonance study of rubber–carbon black interactions. J. Polym. Sci. Part A-2 **9**, 829 (1971)
54. G.R. Hamed, S. Hatfield, On the role of bound rubber in carbon black reinforcement. Rubber Chem. Technolog. **62**, 143–156 (1989)
55. A. Mujtaba, M. Keller, S. Ilisch, H.J. Radusch, M. Beiner, T. Thurn-Albrecht, K. Saalwachter, Detection of surface-immobilized components and their role in viscoelastic reinforcement of rubber silica nanocomposites. ACS Macro Lett. **3**, 481–485 (2014)
56. A.J. Trazkovich, T.H. Akyuz, L.M. Hall, Effects of copolymer sequence on adsorption and dynamics near nanoparticle surfaces in simulated polymer nanocomposites. Tire Sci. Technol. **48**, 62–75 (2018)
57. J. Berriot, H. Montes, F. Lequeux, D. Long, P. Sotta, Macromolecules **35**, 9756–9762 (2002a)
58. A. Papon, H. Montes, F. Lequeux, J. Oberdisse, K. Saalwachter, L. Guy, Solid particles in an elastomer matrix: impact of colloid dispersion and polymer mobility modification on the mechanical properties. Soft Matter **8**(15), 4090–4096 (2012)
59. S.Y. Kim, H.W. Meyer, K. Saalwachter, C.F. Zukowski, Polymer dynamics in peg-silica nanocomposites: effects of polymer molecular weight, temperature and solvent dilution. Macromolecules **45**(10), 4225–4237 (2012)
60. T. Glomann, A. Hamm, J. Allgaier, E.G. Hubner, A. Radulescu, B. Farago, G.J. Schneider, A microscopic view on the large scale chain dynamics in nanocomposites with attractive interactions. Soft Matter **9**, 10559 (2013)
61. A.P. Holt, P.J. Griffin, V. Bocharova, A.L. Agapov, A.E. Imel, M.D. Dadmun, J.R. Sangoro, A.P. Sokolov, Dynamics at the polymer/nanoparticle interface in poly(2-vinylpyridine)/silica nanocomposites. Macromolecules **47**, 1837–1843 (2014)
62. G.N. Toepperwein, R.A. Riggleman, J.J. de Pablo, Dynamics and deformation response of rod-containing nanocomposites. Macromolecules **45**, 543–554 (2011)
63. R. Casalini, C.M. Roland, Local and global dynamics in polypropylene glycol/silica composites. Macromolecules **49**, 3919–3924 (2016)
64. J.M. Kropka, V.G. Sakai, P.F. Green, Local polymer dynamics in polymer-c60 mixtures. Nano Lett. **8**, 1061–1065 (2008)
65. P.J. Griffin, V. Bocharova, L.R. Middleton, R.J. Composto, N. Clarke, K.S. Schweizer, K.I. Winey, Influence of the bound polymer layer on nanoparticle diffusion in polymer melts. ACS Macro Lett. **5**, 1141 (2016)
66. N. Jouault, J.F. Moll, D. Meng, K. Windsor, S. Ramcharan, C. Kearney, S.K. Kumar, Bound polymer layers in nanocomposites. ACS Macro Lett. **2**, 371–374 (2013)
67. A.M. Jimenez, D. Zhao, K. Misquitta, J. Jestin, S.K. Kumar, Exchange lifetimes of the bound polymer layer on silica nanoparticles. ACS Macro Lett. **8**, 166–171 (2019)
68. A.P. Holt, V. Bocharova, S. Cheng, M. Kisliuk, B.T. White, T. Saito, D. Uhrig, J.P. Mahalik, R. Kumar, A.E. Imel, T. Etampawala, H. Martin, N. Sikes, B.G. Sumpter, M.D. Dadmun, A.P. Sokolov, Controlling interfacial dynamics: covalent bonding versus physical adsorption in polymer nanocomposites. ACS Nano **10**, 6843 (2016)
69. J. Berriot, F. Lequeux, L. Monnerie, H. Montes, D. Long, P. Sotta, Filler-elastomer interaction in model filled rubbers, a ^{1}H NMR study. J. Non-Cryst. Solids **307–310**, 719–724 (2002b)
70. Y. Golitsyn, G.J. Schneider, K. Saalwachter, Reduced-mobility layers with high internal mobility in poly(ethylene oxide)-silica nanocomposites. J. Chem. Phys. **146**, 203303 (2017)
71. P.A. Klonos, K. Kulyk, M.V. Borysenko, V.M. Gun'ko, A. Kyritsis, P. Pissis, Effects of molecular weight below the entanglement threshold on interfacial nanoparticles/polymer dynamics. Macromolecules **49**, 9457–9473 (2016)

72. C. Mark, O. Holderer, J. Allgaier, E. Hubner, W. Pyckhout-Hintzen, M. Zamponi, A. Radulescu, A. Feoktystov, M. Monkenbusch, N. Jalarvo, D. Richter, Polymer chain conformation and dynamical confinement in a model one-component nanocomposite. Phys. Rev. Lett. **119**, 047801 (2017)
73. A.P. Cheng, S. Holt, H. Wang, F. Fan, V. Bocharova, H. Martin, B.T. Etampawala, T. White, N.G. Saito, T. Kang, M.D. Dadmun, J.W. Mays, A.P. Sokolov, Unexpected molecular weight effect in polymer nanocomposites. Phys. Rev. Lett. **116**(3), 038302 (2016)
74. P.G. de Gennes, Polymers at an interface; a simplified view. Adv. Colloid Interface Sci. **27**, 189–209 (1987)
75. J.M.H.M. Scheuthens, G.J. Fleer, Statistical theory of the adsorption of interacting chain molecules. 2. train, loop, and tail size distribution. J. Phys. Chem. **84**, 178 (1980)
76. O. Guiselin, Irreversible adsorption of a concentrated polymer solution. EuroPhys. Lett. **17**, 225 (1992)
77. M. Krutyeva, A. Wischnewski, M. Monkenbusch, L. Willner, J. Maiz, C. Mijangos, A. Arbe, J. Colmenero, A. Radulescu, O. Holderer, M. Ohl, D. Richter, Effect of nanoconfinement on polymer dynamics: surface layers and interphases. Phys. Rev. Lett. **110**, 108303 (2013)
78. R. Mangal, S. Srivastava, L.A. Archer, Multiscale dynamics of polymers in particle-rich nanocomposites. Macromolecules **49**, 5502 (2016)
79. Y. Lin, L. Liu, D. Zhang, Y. Liu, A. Guan, G. Wu, Unexpected segmental dynamics in polystyrene-grafted silica nanocomposites. Soft Matter **12**(41), 8542–8553 (2016)
80. E. Senses, S. Narayanan, A. Faraone, Nanoscale particle motion reveals polymer mobility gradient in nanocomposites. ACS Macro Lett. **8**, 558–562 (2019)
81. A. Karatrantos, R.J. Composto, K.I. Winey, N. Clarke, Primitive path network, structure and dynamics of SWCNT / polymer nanocomposites. IOP Conf. Series: Mat. Sci. Eng. **40**, 012027 (2012b)
82. G.D. Smith, D. Bedrov, L. Li, O. Byutner, A molecular dynamics simulation study of the viscoelastic properties of polymer nanocomposites. J. Chem. Phys. **117**, 9478 (2002)
83. F.W. Starr, J.F. Douglas, Modifying fragility and collective motion in polymer melts with nanoparticles. Phys. Rev. Lett. **106**, 115702 (2011)
84. F.W. Starr, T.B. Schroder, S.C. Glotzer, Effects of a nanoscopic filler on the structure and dynamics of a simulated polymer melt and the relationship to ultrathin films. Phys. Rev. E **64**, 021802 (2001)
85. F.W. Starr, T.B. Schroder, S.C. Glotzer, Molecular dynamics simulation of a polymer melt with a nanoscopic particle. Macromolecules **35**, 4481–4492 (2002)
86. T. Desai, P. Keblinski, S.K. Kumar, Molecular dynamics simulations of polymer transport in nanocomposites. J. Chem. Phys. **122**, 134910 (2005)
87. B.A. Pazmino Betancourt, J.F. Douglas, F.W. Starr, Fragility and cooperative motion in a glass-forming polymer–nanoparticle composite. Soft Matter **9**, 241–254 (2013)
88. A. Ghanbari, M. Rahimi, J. Dehghany, Influence of surface grafted polymers on the polymer dynamics in a silica polystyrene nanocomposite: a coarse-grained molecular dynamics investigation. J. Phys. Chem. C **117**, 25069–25076 (2013)
89. J. Shen, X. Lin, J. Liu, X. Li, Revisiting stress-strain behavior and mechanical reinforcement of polymer nanocomposites from molecular dynamics simulations. Phys. Chem. Chem. Phys. (2020). https://doi.org/10.1039/D0CP02225J
90. X.M. Jia, H.J. Qian, Z.Y. Lu, The interfacial structure and dynamics in a polymer nanocomposite containing small attractive nanoparticles: a full atomistic molecular dynamics simulation study. Phys. Chem. Chem. Phys. **22**, 11400–11408 (2020)
91. F.W. Starr, J.F. Douglas, D. Meng, S.K. Kumar, Bound layers cloak nanoparticles in strongly interacting polymer nanocomposites. ACS Nano **10**, 10960–10965 (2016)
92. H.E. Zhang,, B.A. Pazmino Betancourt, F. Vargas-Lara, F.W. Starr, J.F. Douglas, The interfacial zone in thin polymer films and around nanoparticles in polymer nanocomposites. J. Chem. Phys. **151**, 124705 (2019)
93. G. Kritikos, A.F. Terzis, Variable density self consistent field study on bounded polymer layer around spherical nanoparticles. Eur. Polym. J. **49**, 613–629 (2013)

94. L.J. Dai, C.L. Fu, Y.L. Zhu, Z.Y. Sun, Heterogeneous dynamics of unentangled chains in polymer nanocomposites. J. Chem. Phys. **150**, 184903 (2019)
95. H. Emamy, S.K. Kumar, F.W. Starr, Diminishing interfacial effects with decreasing nanoparticle size in polymer-nanoparticle composites. Phys. Rev. Letters **121**, 207801 (2018)
96. S. Napolitano, M. Wubbenhorst, The lifetime of the deviations from bulk behaviour in polymers confined at the nanoscale. Nat. Commun. **2**, 260 (2011)
97. T. Koga, N. Jiang, P. Gin, M.K Endoh, S. Narayanan, L.B. Lurio, S.K. Sinha, Impact of an irreversibly adsorbed layer on local viscosity of nanoconfined polymer melts. Phys. Rev. Lett. **107**(22), 225901 (2011)
98. F.Y. Lin, W. Steffen, Capillary wave dynamics of thin liquid polymer films. J. Chem. Phys. **141**, 104903 (2014)
99. X. Cheng, K.W. Putz, C.D. Wood, C. Brinson, Characterization of local elastic modulus in confined polymer films via afm indentation. Macromol. Rapid Commun. **36**, 391–397 (2015)
100. G.P. Baeza, C. Dessi, S. Constanzo, D. Zhao, S. Gong, M. Alegria, A. Rubinstein, R.H. Colby, D. Vlassopoulos, S.K. Kumar, Network dynamics in nanofilled polymers. Nat. Commun. **7**, 11368 (2013)
101. G. Kritikos, Transition of the bounded polymer layer to a rigid amorphous phase: a computational and dsc study. Polymer **55**, 4658–4670 (2014)
102. B. Carroll, S. Cheng, A.P. Sokolov, Analyzing the interfacial layer properties in polymer nanocomposites by broadband dielectric spectroscopy. Macromolecules **50**, 6149–6163 (2017)
103. S. Cheng, A.P. Sokolov, Correlation between the temperature evolution of the interfacial region and the growing dynamic cooperativity length scale. J. Chem. Phys. **152**, 094904 (2020)
104. D.N. Voylov, A.P. Holt, B. Doughty, V. Bocharova, H.M. Meyer, S. Cheng, H. Martin, M.D. Dadmun, A. Kisliuk, A.P. Sokolov, Unraveling the molecular weight dependence of interfacial interactions in poly(2-vinylpyridine)/silica nanocomposites. ACS Macro Lett. **6**, 68–72 (2017)
105. G.V. Gordon, R.G. Schmidt, M. Quintero, N.J. Benton, T. Cosgrove, V.J. Krukonis, K. Williams, P.M. Wetmore, Impact of polymer molecular weight on the dynamics of poly(dimethylsiloxane)-polysilicate nanocomposites. Macromolecules **43**, 10132–10142 (2010)
106. X. Zheng, M. Rafailovich, J. Sokolov, Y. Strzhemechny, S. Schwarz, B. Sauer, M. Rubinstein, Long-range effects on polymer diffusion induced by a bounding interface. Phys. Rev. Lett. **79**(2), 241–244 (1997)
107. P.A. Klonos, O.V. Goncharuk, E.M. Pakhlov, D. Sternik, A. Deryło-Marczewska, A. Kyritsis, V.M. Gun'ko, P. Pissis, Morphology, molecular dynamics, and interfacial phenomena in systems based on silica modified by grafting polydimethylsiloxane chains and physically adsorbed polydimethylsiloxane. Macromolecules **52**, 2863–2877 (2019)
108. Y. Pu, M.H. Rafailovich, J. Sokolov, D. Gersappe, T. Peterson, W.-L. Wu, S.A. Schwarz, Mobility of polymer chains confined at a free surface. Phys. Rev. Lett. **87**, 206101 (2001)
109. H.K. Nguyen, S. Sugimoto, A. Konomi, M. Inutsuka, D. Kawaguchi, K. Tanaka, Dynamics gradient of polymer chains near a solid interface. ACS Macro Lett. **8**, 1006–1011 (2019)
110. L. Yelash, P. Virnau, K. Binder, W. Paul, Slow process in confined polymer melts: layer exchange dynamics at a polymer solid interface. Phys. Rev. E **82**, 050801(R) (2010)
111. L. Yelash, P. Virnau, K. Binder, W. Paul, Three-step decay of time correlations at polymer-solid interfaces. EPL **98**, 28006 (2012)
112. M. Solar, L. Yelash, P. Virnau, K. Binder, W. Paul, Polymer dynamics in a polymer-solid interphase: molecular dynamics simulations of 1,4-polybutadiene at a graphite surface. Soft Mater. **12**, S80–S89 (2014)
113. K. Johnston, V. Harmandaris, Properties of short polystyrene chains confined between two gold surfaces through a combined density functional theory and classical molecular dynamics approach. Macromolecules **46**, 5741–5750 (2013)
114. V.A. Harmandaris, K. Ch Daulas, V.G. Mavrantzas, Molecular dynamics simulation of a polymer melt/solid interface: local dynamics and chain mobility in a thin film of polyethylene melt adsorbed on graphite. Macromolecules **38**, 5796–5809 (2005)

115. B. Zhang, X. Cao, G. Zhou, and N. Zhao, Anomalous diffusion of polystyrene from an attractive substrate based on all-atom simulation, Phys. Chem. Chem. Phys. **20**, 25304–25313 (2018)
116. P.Z. Hanakata, J.F. Douglas, F.W. Starr, Interfacial mobility scale determines the scale of collective motion and relaxation rate in polymer films. Nature Commun. **5**, 4163 (2014)
117. F. Varnik, J. Baschnagel, K. Binder, Reduction of the glass transition temperature in polymer films: a molecular-dynamics study. Phys. Rev. E **65**, 021507 (2002)
118. C. Batistakis, A.V. Lyulin, M.A.J. Michels, Slowing down versus acceleration in the dynamics of confined polymer films. Macromolecules **45**, 7282–7292 (2012)
119. W. Zhang, J.F. Douglas, F.W. Starr, Effects of a "bound" substrate layer on the dynamics of supported polymer films. J. Chem. Phys. **147**, 044901 (2017)
120. K. Zhang, S.K. Kumar, Molecular simulations of solute transport in polymer melts. ACS Macro Lett. **6**, 864–868 (2017)
121. Y. Li, D. Wei, Dynamics of polymer melts confined by smooth walls: crossover from nonentangled region to entangled region. J. Chem. Phys. **126**, 204907 (2007)
122. G. Kritikos, A. Sgouros, G.G. Vogiatzis, D.N. Theodorou, Molecular dynamics study of polyethylene under confinement. J. Phys.: Conf. Ser. **738**, 012012 (2016)
123. S. Chakraborty, F.C.H. Lim, J. Ye, Achieving an optimal Tg change by elucidating the polymer-nanoparticle interface—a molecular dynamics simulation study of poly(vinyl alcohol)-silica nanocomposite system. J. Phys. Chem. C **123**(39), 23995–24006 (2019)
124. A.F. Behbahani, S.M.V. Allaei, G.H. Motlagh, H. Eslami, V.A. Harmandaris, Structure, dynamics, and apparent glass transition of stereoregular poly(methyl methacrylate)/graphene interfaces through atomistic simulations. Macromolecules **51**, 7518–7532 (2018)
125. Q.H. Yang, M.B. Luo, Dynamics of adsorbed polymers on attractive homogeneous surfaces. Scientif. Rep. **6**, 37156 (2016)
126. P.Y. Lai, Statics and dynamics of adsorbed polymer chains: a monte carlo simulation. J. Chem. Phys. **103**, 5742–55 (1995)
127. P.Y. Lai, Apparent glassy dynamics of an adsorbed polymer chain in the bond-fluctuation model. Macromol. Theory Simul. **5**, 255–68 (1996)
128. J. Huang, Z. Mao, C. Qian, Dynamic monte carlo study on the polymer chain in random media filled with nanoparticles. Polymer **47**, 2928–2932 (2006)
129. P.J. Dionne, C.R. Picu, R. Ozisik, Adsorption and desorption dynamics of linear polymer chains to spherical nanoparticles: a monte carlo investigation. Macromolecules **39**, 3089 (2006)
130. J. Li, M. Ding, R. Zhang, T. Shi, Effects of surface roughness on self-diffusion dynamics of single polymer. Soft Matter **14**, 3550–3556 (2018)
131. T. Desai, P. Keblinski, S.K. Kumar, Polymer chain dynamics at interfaces: role of boundary conditions at solid interface. J. Chem. Phys. **128**, 044903 (2008)
132. C.N. Lam, Deeper penetration of surface effects on particle mobility than on hopping rate in glassy polymer films. J. Chem. Phys. **149**, 164909 (2018)
133. J.M.Y. Carrillo, S. Cheng, R. Kumar, M. Goswami, A.P. Sokolov, B.G. Sumpter, Untangling the effects of chain rigidity on the structure and dynamics of strongly adsorbed polymer melts. Macromolecules **48**, 4207–4219 (2015)
134. N. Jiang, M. Sen, W. Zeng, Z. Chen, J.M. Cheung, Y. Morimitsu, M.K. Endoh, T. Koga, M. Fukuto, G. Yuan, S.K. Satija, J.M. Carrillo, B. Sumpter, Structure-induced switching of interpolymer adhesion at a solid-polymer melt interface. Soft Matter **14**, 1108 (2018)
135. T. Hayashi, K. Segawa, K. Sadakane, K. Fukao, N.L. Yamada, Interfacial interaction and glassy dynamics in stacked thin films of poly(methyl methacrylate). J. Chem. Phys. **146**, 203305 (2017)
136. T. Koga, D. Barkley, M. Nagao, T. Taniguchi, J.M.Y. Carrillo, B.G. Sumpter, T. Masui, H. Kishimoto, Koga M., J.G. Rudick, M.K. Endoh, Interphase structures and dynamics near nanofiller surfaces in polymer solutions. Macromolecules **51**(23), 9462–9470 (2018)

Chapter 5
The Interfacial Layers Around Nanoparticle and Its Impact on Structural Relaxation and Glass Transition in Model Polymer Nanocomposites

Wengang Zhang, Hamed Emamy, Fernando Vargas-Lara, Beatriz A. Pazmiño Betancourt, Dong Meng, Francis W. Starr, and Jack F. Douglas

Abstract We quantify the properties of the interfacial layers around bare nanoparticles (NPs) with variable NP diameter in model polymer-NP composites and near the solid substrate and free interfaces of supported thin polymer films using molecular simulations. These interfaces alter both the segmental packing and mobility in an interfacial layer. Variable NP diameter allows us to understand the effect of boundary curvature, where the film sets the limiting value of zero curvature. We find that the interfacial mobility scale ξ for both NPs and supported films increases on cooling, reaching a scale on the order of a few nanometers at low temperatures, regardless of the polymer-interfacial interaction strength. Additionally, we consider the related problem of the interfacial mobility scale of supported polymer films, which provides an upper limit for the interfacial scale of polymer-NP composites as the particle increases in size to macroscopic dimensions. We also characterize the interfacial density gradient scale ξ_ρ near the NP surface and find, in contrast to the length scale from dynamics ξ, that ξ_ρ is even smaller and decreases on cooling in all cases simulated. In other words, the interfaces generally become more sharply defined on cooling. On the other hand, the scale of fluctuations in the interfacial density σ_ρ near solid interfaces, which reflects the degree of packing fluctuations,

W. Zhang (✉) · F. Vargas-Lara · B. A. P. Betancourt · J. F. Douglas
Materials Science and Engineering Division, National Institute of Standards and Technology, Gaithersburg, Maryland 20899, USA
e-mail: wzhang01@wesleyan.edu

J. F. Douglas
e-mail: jack.douglas@nist.gov

W. Zhang · H. Emamy · F. Vargas-Lara · B. A. P. Betancourt · F. W. Starr
Department of Physics, Wesleyan University, Middletown, Connecticut 06459, USA
e-mail: fstarr@wesleyan.edu

H. Emamy · D. Meng
Department of Chemical Engineering, Columbia University, New York, NY 10027, USA

V. V. Ginzburg and L. M. Hall (eds.), *Theory and Modeling of Polymer Nanocomposites*, Springer Series in Materials Science 310,
https://doi.org/10.1007/978-3-030-60443-1_5

and thus "packing frustration", grows on cooling and correlates in a near linear fashion with the interfacial mobility scale ξ, although these characteristic thermodynamic and mobility scales have different dependences on the strength of interfacial interaction. Conversely, σ_ρ for the free interface in thin films grows on heating, similar to that of ξ_ρ. Having characterized the thickness of interfacial layers from both structure and dynamics, we also examine their effect on the overall relaxation of materials and the glass transition temperature T_g based on both thermodynamic and dynamic criteria. In particular, when the interfacial interaction strength is greater than the polymer-polymer interaction strength, a "bound" polymer layer forms near the interface (either the nanoparticle interface or film substrate), giving rise to an additional relaxation process in the self-intermediate scattering function. As a result, we find that T_g, defined from the gradient in the structural relaxation time, increases monotonically with polymer-interface interaction strength ε. In contrast, the thermodynamically defined T_g, defined from "kinks" in the temperature dependence of quasi-thermodynamic properties, is found to saturate to a nearly constant value for ε values greater than the polymer-polymer interaction strength since this bound polymer layer effectively "cloaks" the NP or substrate from the unbound polymer from the strong interfacial interactions. Our findings emphasize the quantitative relationship between the interfacial scales in nanocomposites and thin polymer films and show that interfacial effects are rather universal in systems having significantly mobility gradients.

5.1 Introduction

There is growing appreciation that the dynamics of the interfacial region of both crystalline and amorphous glass-forming materials can be greatly altered from the dynamics deep within these materials [1–5]. Alterations in dynamical properties, such as diffusion and the rate of structural relaxation, are especially great in thin films and polymer nanocomposites because the relatively high surface to volume ratio implies a much greater fraction of molecules are within the interfacial region [6–11]. For example, polymer nanocomposites property changes can be attributed in large part to the fact that NPs modify the dynamics near NP. Indeed, the addition of even a small concentration of bare nanoparticles (NPs) to the polymer matrix can significantly enhance the properties of polymer nanoparticle composites including mechanical properties, optical properties, and electrical conductivity [1, 6, 12–17].

The design of new materials, and an understanding of many existing materials in which interfacial properties are crucially important (nanocrystalline metal alloys and presumably semi-crystalline polymer materials, 3-D printed materials, as well as supported polymer films and polymer nanocomposites) requires a better understanding of how to quantify and control interfacial dynamics of condensed materials [1, 6, 18–22]. In particular, the interfacial layer around nanoparticles in nanocomposites is a topic of extensive experimental investigations due to the relevance of this layer in understanding apparent deviations from continuum composite theory. From a practical standpoint, understanding factors influencing the thickness of interfacial layers is useful for material design [9, 10]. Motivated by the evidence of common features in the interfacial dynamics and properties of nanocomposites and thin polymer films [2, 23–25], we study the interaction of interfacial layers in both nanocomposites and thin polymer films (e.g., interfacial zone near nanoparticles, polymer-air interfacial zone, polymer-substrate interfacial zone, and often the interfacial zone of a grafted polymer layer separating the nanoparticle from the surrounding polymer matrix) [20–22, 26, 27]. This is a vast research field, and the present work is mainly limited to a review of our own efforts aimed at understanding interfacial zones as they arise in glass-forming polymer films supported on solid substrates and polymer nanocomposites.

In particular, we compare the polymer matrix interfacial regime near the surface of bare nanoparticles (NPs) and interfacial layers near the substrate and polymer-air interfaces of thin polymer films. We utilize molecular dynamics (MD) simulations on model coarse-grained polymer melts, investigated extensively in the bulk and under confinement, to explore how the interfacial interaction strength, interface type (free polymer-air interface and polymer-solid interface), boundary curvature (nanoparticle size) and temperature influence the thickness of the interfacial layer and changes in both thermodynamic and dynamic properties of the interfacial zone in comparison to the material far from the interface [28–30]. Special emphasis is given to the evaluation of the gradient in the segmental relaxation time, density, and the variance of density fluctuations near the interfacial region of supported polymer films and nanocomposites, and the characteristic scales describing these gradients. In addition to considering local changes and in the thermodynamic and dynamic properties of these materials, we also consider changes in the overall glass transition temperature defined from both thermodynamic and dynamic properties. We also describe how a bound layer forms around the NP with strong polymer-NP interactions, and how this bound layer "cloaks" the polymer matrix from the strongly interacting NP. Consequentially, the glass transition temperature of the nanocomposites or thin films as whole by thermodynamic measurements saturates for strong polymer-NP interaction. These findings offer an explanation for the observations that T_g has small to no changes for strongly interaction NP or substrate with the polymer [2, 20, 24, 31–34].

Before proceeding, we note some of the limitations of our computational study of the interfacial dynamics of polymer films and composite materials. All the simulations described below, unless otherwise stated, are performed under equilibrium conditions, which practically means that the growth of the relaxation time upon cooling is limited in extent in comparison to the range often studied in experimental systems. While it is possible to extrapolate our results to low temperatures to estimate the glass transition temperature T_g as observed experimentally, such extrapolations are inherently uncertain. We nonetheless believe that such extrapolations are physically meaningful from a qualitative standpoint, despite the uncertainties. The coarse-grained model of the polymer chains that we utilize aims only to understand general trends in changes in the thermodynamic and dynamic properties of the interfacial layer rather than results for particular polymers. In the case of composites, we examine only ideal dispersions of NPs, so that we cannot account for effects due to interactions among the NPs. We also emphasize that many common polymer materials such as epoxies and block copolymer and polymer blends are *multicomponent* materials in which large scale composition gradients often arise as result of phase separation. Interphase regions can be expected to exhibit much larger dimensions around particle additives than found for the pure polymer materials investigated in the present work [35].

5.2 Modeling and Simulations

The simulation data presented in this chapter primarily derives from a series of molecular dynamics simulations of model polymer nanocomposites [23, 30, 36–39] and thin supported polymer films [29, 40–42] in our collaborative work over a period of a few years. Here, we briefly summarize the key elements of the model and simulation results. The complete details can be found in the original works cited.

In the case of the polymer nanocomposites, we idealize the NP to be an icosahedral particle surrounded by a dense unentangled coarse-grained polymers with a chain length of 20 monomers. Due to periodic boundary conditions, our simulations are intended to represent uniformly dispersed NPs in the polymer melt. In Sects. 5.3 and 5.4, we consider the NP concentration in the dilute limit, allowing us to investigate the effect of interfacial layer without the complication of the interactions between NPs. We also vary the polymer-NP interaction strength to evaluate its effect on the dynamics and glass transition temperature.

Polymer chains in the polymer-NP composites are modeled by the well-studied Kremer-Grest spring-bead polymer chain [43]. Each chain has $N = 20$ monomers, which interact via the Lennard-Jones (LJ) potential with strength ε, mass m, diameter σ, and a cutoff of interactions beyond $r_c = 2.5\,\sigma$. Bonded monomers along each polymer chain are connected by a finitely extensive nonlinear elastic (FENE) potential with a bond strength $k_b = 30\,\varepsilon/\sigma^2$ and bond distance $R_0 = 1.5\,\sigma$. The NP consists of LJ particles bonded in icosahedral shape. Specifically, LJ beads in NP are identical to the monomers of the polymers. In Sect. 5.3, we study three different NP sizes, consisting of 12, 104, or 356 beads, which corresponds to an inscribing sphere with radius $1.6\,\sigma$, $3.3\,\sigma$, or $5.0\,\sigma$, respectively. In Sect. 5.4, we only consider the largest NP size listed above. The LJ size parameter between monomers and NP is $\sigma_{\mathrm{p-NP}} = 1.0$ throughout this chapter, while the LJ attraction strengths are $\varepsilon_{\mathrm{p-NP}}$ are varying between 0.1 and 3.0. In particular, in Sect. 5.3, we use $\varepsilon_{\mathrm{p-NP}} = 0.1$ to represent a weakly attractive polymer-NP interaction strength, about one tenth of the polymer-polymer interaction strength, and use $\varepsilon_{\mathrm{p-NP}} = 1.5$ for strongly attractive interactions, which is about 1.5 times the strength of attraction compared to the polymer-polymer interaction. In order to study the effect of interfacial interaction on T_g and relaxation time, we use a polymer-NP interaction strength $\varepsilon_{\mathrm{p-NP}}$ ranging from 0.1, 0.25, 0.5, 0.75, 1.0, 1.25, 1.5, 2.0, 2.5 to 3.0. All simulations are performed at constant averaged pressure $\langle P \rangle = 0.1$ over a temperature range $0.42 \leq T \leq 0.80$.

We also analyze data from simulations of polymer films supported by a solid attractive substrate, where the upper interface is free, corresponding to pressure $\langle P \rangle = 0$; thus the pressure in both composite and film is small. We briefly describe the model details, and a complete description of parameters and simulation protocols are provided in the original work [29, 40, 41, 44]. The polymer chains have 10 beads per chain, half that of the chains in the composite. This difference is not significant: in a prior study (See Fig. S4 from the supplementary information of Ref. [29]) and previously we have shown that the structural relaxation time at a segmental scale for chain lengths 10 and 20 are essentially the identical. We use a harmonic spring potential $U_{\text{bond}} = \frac{k_{\text{chain}}}{2}(r - r_0)^2$ to connect nearest-neighbor monomers within a polymer chain using the equilibrium bond length $r_0 = 0.9$ and the spring constant $k_{\text{chain}} = 1111$ [45]. Though we utilize a harmonic spring potential instead of the FENE potential to model polymer chain in thin film for historical reasons, these two potentials give rise to similar results in the low bond stretching regime, which is the case in our simulations [46]. The polymer films have 600 polymer chains, resulting in a film thickness $h \approx 15\,\sigma$ with dimensions $20\,\sigma$ in directions parallel to the substrate, where periodic boundary conditions are employed. The substrate is modeled as 528 particles arranged in a triangular lattice (the 111 face of an FCC crystal). These particles are tethered to their equilibrium (zero force) locations via a harmonic spring potential $V_{\text{sub}}(r) = (k/2)(r - r_0)^2$, where r_0 is the equilibrium position and $k = 50$ is the spring constant. We use Lennard-Jones (LJ) interactions between non-bonded monomers and substrate particles. Here we examine interaction strengths $\varepsilon_{\text{ps}} = 0.1$ (weakly attractive interaction) and $\varepsilon_{\text{ps}} = 1.5$ (strongly attractive interaction) between monomers and substrate particles. As a qualitative way of characterizing the range of interfacial interaction strength, we note that our "weakly attractive" interfacial interaction strength corresponds to a tenth of that of the polymer-polymer interaction strength, while the "strongly attractive" interfacial interaction strength corresponds to a factor of 1.5 times that of the polymer-polymer interaction strength.

For both composites and films we present results in standard LJ reduced units, where ε is the unit of energy, σ is the unit of length, m is the unit of mass, $\sigma\sqrt{m/\varepsilon}$ is the unit of time, and ε/k_{B} is the unit of temperature (where k_{B} is Boltzmann's constant). To approximately translate to real units, we use $\sigma = 1.0$ nm, the average Kuhn segment length for linear polystyrene (PS), and $\varepsilon = 7.7$ kJ/mol, which leads to $T_{\text{g}} \approx 370$ K, so we may loosely view our polymer model as a coarse-grained representation of unentangled PS [47, 48]. Consequentially, the interfacial (polymer-NP and polymer-substrate) interaction strengths in our model map to 0.77 kJ/mol and 11.55 kJ/mol for the weakly and strongly attractive interfaces, respectively. These interaction strengths are within the range shown in [49], where these authors explored the effects of polymer-NP interaction strength ε on mechanical properties of representative nanocomposites. Additionally, our strongly attractive interface corresponds to an atomically smooth interface having an adhesion energy about 100 mJ/m^2 [50].

5.3 Interfacial Zone Around Nanoparticles in Nanocomposites and Near Substrate in Thin Polymer Films

In this section, we characterize the thickness of interfacial layers ξ, defined by the gradient of the relaxation time in the proximity to the NP surface, for a range of NP sizes and compare it to that in thin polymer films. We use two representative polymer-NP/substrate interactions—"strong" and "weak" interfacial interactions. In doing so, we also quantify the interfacial density gradient scale ξ_ρ, since it is required to estimate ξ and is of independent interest. We show that ξ_ρ exhibits a quite distinct temperature dependence from that of the interfacial mobility scale ξ. To avoid the complication of NP aggregation in the relatively high nanoparticle loading regime, we study the case of NP concentration in the dilute limit to establish a basis of the dynamics and local structure of the interfacial layer around NP. Our results provide a universal framework that may potentially be useful to establish connection between interfacial scales and the scales of cooperative rearrangements within the glass-forming polymer materials. Note that we use the terms "thickness of interfacial layers" and "interfacial mobility scale" interchangeably throughout the text.

5.3.1 *Qualitative Description of the Influence of Interfacial Interactions on Mobility Near Interfaces*

Before quantifying the thickness of interfacial mobile layer around NPs and near substrate in thin films, we present a qualitative description of the perturbation of the dynamics near the interface of the NP in nanocomposites and substrate in thin polymer film. To visualize the mobility gradient, we first shows an illustration of the variations of segmental relaxation time of the polymer $\tau_\alpha(r)$ relative to its value far from the NP interface in Fig. 5.1a, b, for thin films, we show the variations of relaxation time $\tau_\alpha(z)$ relative to its value in the interior bulk-like region of the film $\tau_\alpha^{\mathrm{mid}}$ in Fig. 5.1c, d. Specifically, the panel (a) shows that in the strongly attractive NP-polymer interaction case (where $\varepsilon_{\mathrm{p-NP}} > \varepsilon_{\mathrm{pp}}$ and the subscript p is short for polymer), the segmental relaxation time near the NP is significantly increased, and conversely, in panel (b) a weakly attractive NP (where $\varepsilon_{\mathrm{p-NP}} < \varepsilon_{\mathrm{pp}}$) have significantly reduced τ_α near the surface. Similarly, the panel (c) shows that in the case of thin polymer films with strongly attractive polymer-substrate interaction (where $\varepsilon_{\mathrm{ps}} > \varepsilon_{\mathrm{pp}}$ and the subscript s is short for substrate), relaxation time near the substrate is significantly increased, and conversely, while in panel (d) a weakly attractive substrate (where $\varepsilon_{\mathrm{ps}} < \varepsilon_{\mathrm{pp}}$) have significantly reduced τ_α near the substrate. In both case, there is significant increase in the mobility near the free interface. We observe that the relaxation time τ_α approximately changes by a factor of 10–500 near the NP interface (substrate) at the lowest simulated temperature T, i.e., $T = 0.42$. In the equilibrium state (above T_{g}), this relaxation time ratio $\tau_\alpha(r)/\tau_{\mathrm{far}}$ can be expected to

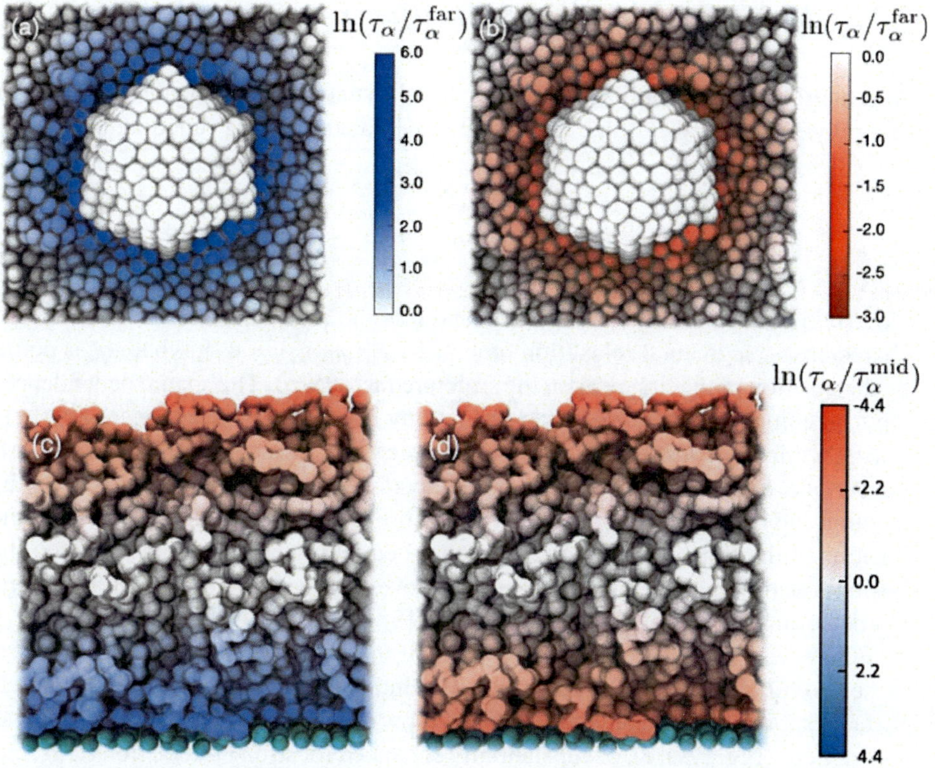

Fig. 5.1 Visualizations of the relaxation time around the NP (panel **a** and **b**) and in thin polymer film (panel **c** and **b**). **a** A strongly attractive polymer-NP interaction slows down the interfacial dynamics considerably, while **b** a weakly attractive polymer-NP interaction accelerates the relaxation in the interfacial region. Similarly **c** a strongly attractive polymer-substrate interaction slows down the interfacial dynamics considerably, while **d** a weakly attractive polymer-substrate interaction accelerates the relaxation near the substrate. In both thin polymer film cases, the free interfacial regions have accelerated dynamics (red color). The diameter of the model NP is 10 nm in laboratory units and the polymer film thickness is about 15 nm

become much larger at lower temperature or in the case of stronger polymer-NPs (polymer-substrate) interaction strength, which results in the formation of bound layer near the NP or substrate [28, 29]. In the glass state, the relaxation ratio decreases on further cooling; in thin polymer film [3, 51–54]. The formation of the bound layer near the interface is due to a separation of time scales for the binding-unbinding transition of the polymer matrix from the interface [28, 29]. It is visually apparent that the width of the interfacial zone extends over several monomer diameters (σ), or a few nanometers in laboratory units. This is in qualitative agreement with recent measurements on the interfacial zone of well-dispersed NPs having an attractive interaction with the matrix to aid in their dispersal [20, 21, 26, 27, 31, 55–57]. Our simulation results thus are reasonable from a qualitative standpoint.

5.3.2 Mobility Gradient Near Interfaces

As a metrology for quantifying the segmental dynamics of the nanocomposites and thin polymer films, we utilize the self-intermediate scattering function,

$$F_{\mathrm{s}}(q,t) \equiv \frac{1}{N}\left\langle \sum_{j=1}^{N} \mathrm{e}^{-i\boldsymbol{q}\cdot[\boldsymbol{r}_j(t)-\boldsymbol{r}_j(0)]} \right\rangle, \tag{5.1}$$

where $\boldsymbol{r}_j$ is the position of monomer j and $\boldsymbol{q}$ is the scattering vector. We define the characteristic segmental relaxation time τ_α by $F_{\mathrm{s}}(q_0, \tau_\alpha) = \mathrm{e}^{-1}$, where q_0 is defined by the position of the first peak in the structure factor $S(q)$. The spatial dependence of the nanocomposites dynamics is quantified by averaging F_{s} over spherical "shells" having a thickness less than a monomer diameter as a function of distance from the NP surface. Similarly, we partition monomers in thin film into layers based on their location and compute F_{s} for each layer. We sort monomers into layers based on their location at the time origin, though we could equally well sort them based on their position at time $t = \tau$, since monomers move (on average) less than a monomer diameter on the scale of the relaxation time [58].

Figure 5.2 shows the expected behavior of the relaxation $F_{\mathrm{s}}(q,t)$ for layers from closest to furthest to the NP (substrate). We find that polymer relaxation near the NP surface (substrate) is substantially slowed for weak polymer-NP (polymer-substrate) interactions (Fig. 5.2a, c), or substantially enhanced for strong interactions (Fig. 5.2b, d). To describe the time dependence of $F_{\mathrm{s}}(q,t)$ for each layer, we fit our data to a phenomenological two-step relaxation functional form,

$$F_{\mathrm{self}}(q,t) = (1-A)\mathrm{e}^{-(t/\tau_{\mathrm{s}})^{3/2}} + A\mathrm{e}^{(-t/\tau_\alpha)^{\beta}}, \tag{5.2}$$

where $\tau_{\mathrm{s}} = 0.29$ is the short vibrational relaxation time scale, whose value is approximately a constant over the range of system parameters studied here [28, 29]; this "stretched Gaussian" dynamics at short time scale reflects dynamics that is between ballistic and Brownian motion.

To better illustrate the variations of monomer relaxation near the NP surface and substrate, Fig. 5.3 shows the layer-resolved relaxation time in nanocomposites and thin films. In nanocomposites, we show the distance r dependence of $\tau_\alpha(r)/\tau_{\mathrm{far}}$ from the center of NP for both strong and weak polymer-NP interfacial interaction strength ε; in thin films, we show the relaxation time as a function of the distance z to the substrate for both interactions. The interfacial relaxation time is much larger or smaller near the NP, depending on the strength of the polymer-NP interactions, as has been documented in many earlier works [28, 36, 37, 39]. The interfacial relaxation increases substantially upon cooling; note the log scale of the relaxation time ratio, $\tau_\alpha(r)/\tau_{\mathrm{far}}$.

Similarly, Fig. 5.3c, d present the distance z dependence of $\tau_\alpha(z)$ from the supporting substrate of the thin polymer films having strongly attractive or weakly attractive

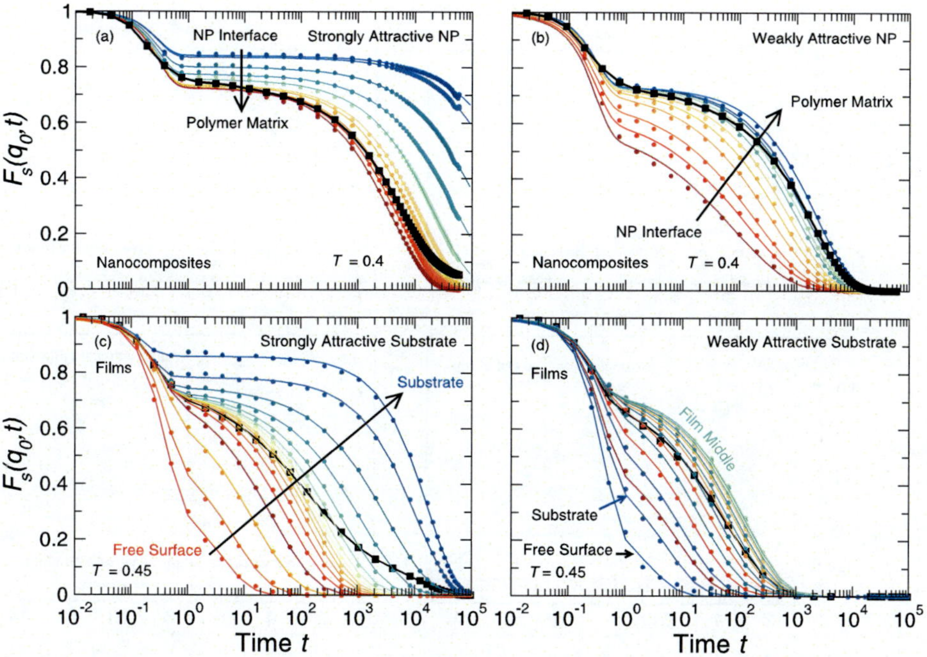

Fig. 5.2 The layer-resolved self-intermediate scatting function $F_s(q, t)$ at a representative temperature for both nanocomposites (panel **a**, **b**) and thin polymer films (panel **c**, **d**) having either strongly (panel **a**, **c**) or weakly (panel **b**, **d**) attractive interfacial interaction. Symbols are simulation data, and lines are the fit defined by (5.2). The black data and line indicates $F_s(q, t)$ for the entire composites or thin film, while colored data are for layers of increasing distance from the NP surface or substrate. Note that for strongly attractive interfacial interaction, where relaxation near the NP or substrate is greatly diminished, the overall $F_s(q, t)$ has qualitatively different behavior at large t, described by (5.7)

interfacial interaction strengths. As in the case of the NP interfacial zone, the relaxation slows down significantly near the strongly attractive substrate and becomes faster near the weakly attractive interface. For both strong and weak substrate interactions, supported films exhibit an enhanced relaxation near the free interface. Accordingly, our methodology to estimate interfacial mobility scale ξ must consider both interfacial regions and their interactions.

Importantly, the interfacial scale defined by the density gradient (ξ_ρ) near a surface is generally quite different from the interfacial mobility scale ξ from dynamics [23, 29, 59]. We next discuss the interfacial density scale ξ_ρ in some detail as we utilize this quantity in the characterization of the interfacial mobility scale ξ, and then quantify the temperature T and polymer-NP and polymer-substrate interaction strength ε variations of ξ to better understand the nature of the mobility interfacial layers from a molecular perspective. Furthermore, as mentioned above, we also characterize the interfacial zones of the thin polymer film since it sets an upper limit for the interfacial mobility scale ξ of NPs.

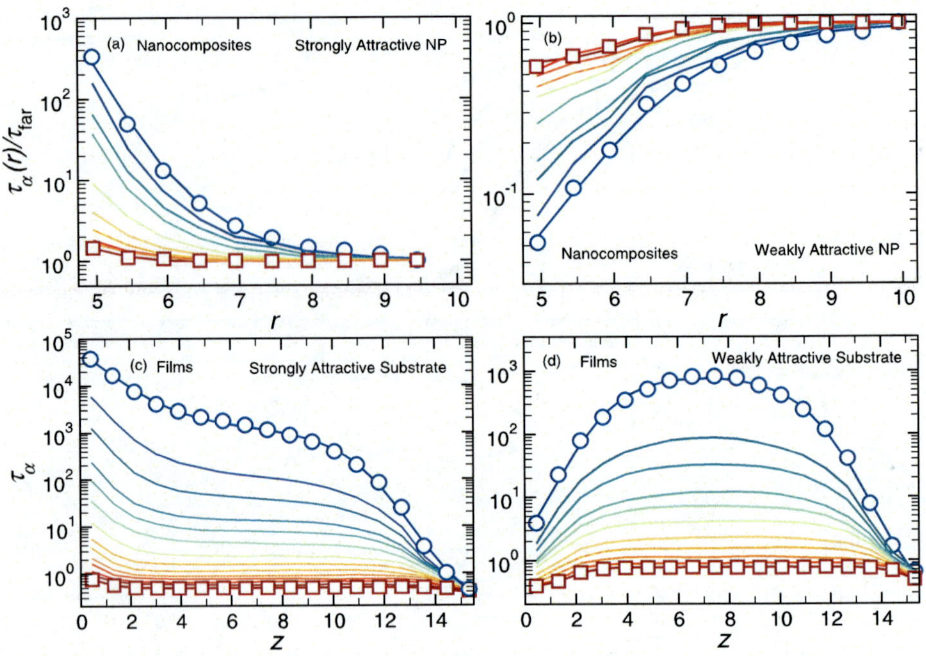

Fig. 5.3 Relaxation profile for both nanocomposites (panel **a**, **b**) and thin polymer film (panel **c**, **d**). For nanocomposites, we show $\tau_\alpha(r)/\tau_{\text{far}}$ as a function of the distance r from the NP center for $r_{\text{NP}} = 5.0$ for **a** strongly and **b** weakly attractive NP-polymer interaction. Similarly, we also show the relaxation profile $\tau(z)$ for thin polymer films with **c** strongly and **d** weakly attractive substrate-polymer interaction strength. We include two representative fits (solid lines) from (5.3) or (5.4) in each panel for the lowest and highest temperatures. The color gradient represents the gradient in T, ranging from blue at low T to red at high T

5.3.2.1 Interfacial Mobility Scale ξ

Having quantified the mobility gradient in both nanocomposites and thin supported polymer films, we next quantitatively extract an interfacial mobility scale. To do so, we look to earlier simulation studies for inspiration. These works were motivated by the possibility of identifying a well-defined growing length scale in cooled glass-forming liquids. Consistent with this expectation, Scheidler, Kob and Binder [60] observed a mobility gradient qualitatively similar to Fig. 5.3c near the "free" interface of a model Lennard-Jones glass-forming film, which they quantified by fitting $\log(\tau_\alpha)$ to an exponential function with a decay length ξ describing the interfacial width. Numerous authors subsequently followed this procedure for estimating the interfacial mobility scale [59, 61], where later evidence correlating this length scale to the activation energy for the film structural relaxation time [62] and the extent of cooperative exchange motion with the film [59] was noted, seemingly confirming the intuition that the mobility interfacial scale might have some relation to the scale of "dynamic heterogeneity" within the film. One difficulty that arises with the approach of Scheidler et al., is that the proximity of the substrate and free interface can potentially cause interference between ξ estimated from the substrate or free surface. Accordingly, it is useful to consider a method that simultaneously incorporates both interfaces. In addition, we should also incorporate the scale of density modifications in the formulation, so that our estimate of ξ takes into account the structural length scale. Indeed, there is evidence that the mobility interfacial layer correlates with the scale ξ_ρ in simulations of proteins in water around room temperature [63–65], suggesting that this scale is somehow related to ξ, even if these quantities are not exactly equal.

Here we use a functional description to determine ξ that addresses these concerns [30]. We first consider a model for the interfacial mobility near the surface of an isolated NP (or a thick polymer film), and then generalize to systems having multiple interfaces. It is natural to postulate a hyperbolic tangent functional form for the mobility profile, by analogy with the known density profile at the polymer-air interface [66], where the distance from the interface is taken with respect to a reference scale that includes a contribution from the density gradient scale ξ_ρ, which helps to determine the boundary of the material. Specifically, we model our mobility gradient for a NP by,

$$\ln \tau_\alpha(r) = \ln \tau_0 + c_0 \left[1 - \tanh\left[\frac{r - (r_{NP} + \xi_\rho)}{\xi}\right]\right], \tag{5.3}$$

where $r_{NP} = a\frac{\sqrt{3}}{12}(3 + \sqrt{5})$ is the radius of a sphere inscribing the icosahedral NP having an edge length a; ξ, τ_0 and c_0 are fitting parameters. ξ_ρ is the interfacial density gradient scale that quantifies the sharpness of an interface, and it is determined from the density profile. To avoid breaking the flow of this section, we describe the quantification of ξ_ρ in detail in the next section. In short, $r_{NP} + \xi_\rho$ determines where the polymer interface starts. We adapt the same form for the mobility profile near a solid substrate where the distance from the planar interface z replaces r in (5.3) and where $r_{NP} = 0$. We fit the layered resolved relaxation time $\tau_\alpha(r)$ using (5.3) with the explicit value of ξ_ρ to obtain the interfacial mobility scale ξ. We treat the more complex situation of a thin film where the interfaces interact (as in Fig. 5.3) using the generalized relation for $\tau_\alpha(z)$,

$$\ln \tau_\alpha(z) = \ln(\tau_0) + c_1\left(1 - \tanh\left[\frac{z - \xi_\rho^{sub}}{\xi_{sub}}\right]\right) + c_2\left(1 - \tanh\left[\frac{(h - \xi_\rho^{int}) - z}{\xi_{int}}\right]\right), \tag{5.4}$$

where ξ_ρ^{sub} and ξ_ρ^{int} are the interfacial density gradient scales of the substrate and free surface, respectively (defined in the next section); ξ_{sub} and ξ_{int} are the interfacial mobility scales of the substrate and free surface, respectively; τ_0 is related to the relaxation time in the middle of the film; h is the temperature dependent film thickness, defined by the height at which density decreases to 0.05 near the interfaces; c_1, and c_2 are fitting parameters. We obtain excellent fits with (5.4) for all cases that we consider. We expect that (5.4) is also applicable to well-dispersed nanocomposites with relatively high concentration of NPs, where the interfacial zones of NPs overlap with each other.

5.3.2.2 Effect of Nanoparticle Size on the Interfacial Zone

Based on our MD simulations, and the analysis to characterize the interfacial mobile layer described in the previous section, we now discuss our results for the interfacial mobility scales around NP in nanocomposites for varying NP size and in thin polymer films. To avoid the complication of NP aggregation in relatively high NP concentration, we focus on the case of well-dispersed NPs in the low NP loading to establish a baseline for the length scale of interfacial mobile layer around NP. In the case of thin polymer film, we study the interfacial mobility scale near both the supporting substrate and free interface. In Fig. 5.4, we show a comparison of ξ estimated for a thin polymer film and our model nanocomposites with variable NP size in the cases of a strongly attractive interface (panel a), and weakly attractive solid interface (panel b). We expect that ξ_{sub} in the film substrate should serve as an approximate limit for ξ of the polymer-NP composite in the limit of large NP size. For the case of strong interfacial interactions, ξ of the composite increases with increasing NP size, and approaches to that of the film substrate. This observation is consistent with that observed for the magnitude of the interfacial effect on the relaxation time $\tau_\alpha(r)$ reported in [38], and supported by measurements [56]. For the case of weakly attractive interfaces, ξ of the composite is only very weakly dependent on NP size, and (like the strongly attractive case) smaller than that of the polymer film substrate. Moreover, the range ξ is diminished for weak interfacial interactions, as compared to the case of strong interactions.

We next consider the behavior of the free interface of the polymer film, a type of interface that does not arise in bulk nanocomposites. Figure 5.4c shows that the interfacial mobile layer near the free interface ξ_{free} increases on cooling, just like

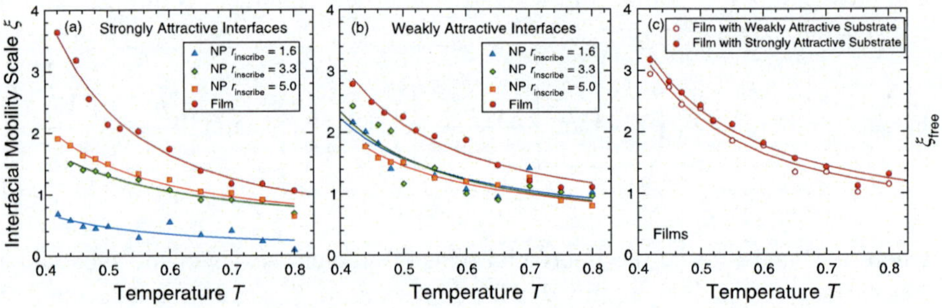

Fig. 5.4 The interfacial mobility scale $\xi(T)$ of thin films and nanocomposites for **a** strongly and **b** weakly attractive interfaces, and **c** the free polymer-air interface. The symbols represent the data and the solid lines are guides for the eye

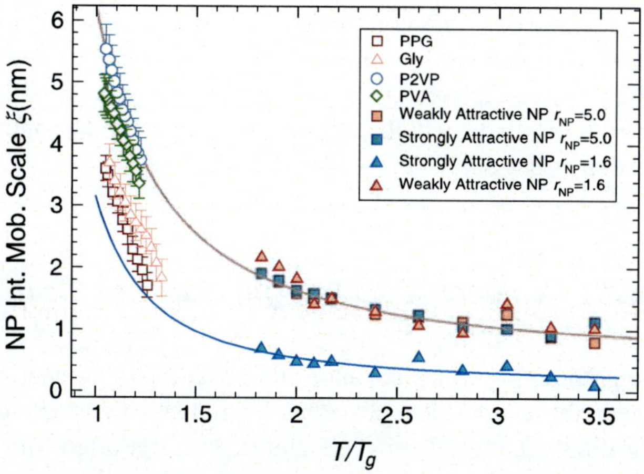

Fig. 5.5 Temperature dependence of the interfacial mobility scale $\xi(T/T_g)$ near the NP from experiments and simulations. The hollow symbols represent the experiments for glycerol (Gly), poly(2-vinylpyridine) (P2VP), polypropylene glycol (PPG), and polyvinyl alcohol (PVA) and the solid symbols represent the simulation data for variable polymer-NP interaction strengths and NP sizes. The experimental data are derived from broadband dielectric spectroscopy and are taken from Cheng et al. [22]. We use the empirical Arrhenius relation to extrapolate the low T behavior of $\xi(T)$ from our simulations

that of the substrate. However, the T dependence of ξ_{free} is nearly *identical* for both substrate interactions. This finding is not surprising, since the free interface is sufficiently far from the substrate to largely avoid its effects. Moreover, the magnitude of ξ_{free} is very similar to that of ξ_{sub} of the weakly interacting substrate. This can be anticipated from the fact that the relaxation profile near the weakly attractive substrate is similar to that of a "free" interface.

We briefly compare our estimates of the interfacial mobility scale ξ in Fig. 5.5 with experimental estimates of the mobility interfacial zone thickness by Sokolov and coworkers for several glass-forming materials (glycerol, poly(2-vinylpyridine), polypropylene glycol, and polyvinyl alcohol) based on broadband dielectric spectroscopy measurements [21, 22]. To extrapolate the low T behavior of $\xi(T)$ from our simulations, we use an empirical Arrhenius relation inspired by the string model [67], $\xi(T) = \xi_0 \exp((\Delta H - T\Delta S)/kT)$ [67, 68] to fit the temperature dependent of ξ from our simulations, where ΔH, ΔS, and ξ_0 are free fitting parameters. This fitting form also assumes that ξ scales linearly with the string length, as found in simulations above T_g [53, 59]. We see that the scale and qualitative temperature dependence of these experimental estimates of ξ are reasonably consistent with our coarse-grained simulations; see Cheng et al. [21, 22] for the description of the materials, measurement methods, and measurements uncertainties etc. It is notable that the order of magnitude of the interfacial scale is not sensitive to whether the glass-former is polymeric or not, and even the strength of interaction between the polymer and the substrate or NP surface seems to be of secondary importance. These experimental estimates are indirect, but later work involving direct imaging of the interfacial region by atomic force microscopy has conformed these observations to a good approximation [27]. The silica NPs used by Sokolov and coworkers [21] in their experimental study has a diameter of 12.5 nm, a value similar to our simulated NPs. Quantitative comparisons with experiment will require estimating the strength of the polymer matrix interaction strength ε [26, 34, 69–72].

5.3.3 *Density Gradient Near Interfaces and Its Scale ξ_ρ*

Similar to the gradient of molecular mobility around NPs or near planar substrates, the density near these interfaces also has a gradient. To show this, we quantify the density profiles around NP in PNC and in the supported thin film in Fig. 5.6. We smooth the packing fluctuations of density profiles by convolving $\rho(z)$ or $\rho(r)$ with a box function. In all cases, the density reaches a well-defined mean value ρ_{mid} far from the interfaces. In the nanocomposites, the density profile $\rho(r)$ is more sharply defined near the strongly attractive NP/substrate than that in the weakly attractive NP/substrate since its density quickly converging to that far away from the NP/substrate. In thin polymer films, the density gradients in the free interface are nearly the same in both types of interfaces, as expected.

To quantify how sharply defined the polymer interface near NP, substrate or air is, we define the interfacial density scale ξ_ρ as the length from the interface at which the density reaches within 5% of the density in the steady region, starting from the interface (NP surface, substrate, or free surface). The starting position of the substrate or free surface is defined by the position at which $\rho = 0.05$; and the starting position of the NP surface is at the radius of sphere inscribing the NP r_{inscribe}.

We show the variation of ξ_ρ for both model NPs and polymer films for strongly attractive (Fig. 5.7a) and weakly attractive (Fig. 5.7b) interfacial interactions. We find that the boundary interaction has a significant influence on ξ_ρ. The interfacial scale ξ_ρ

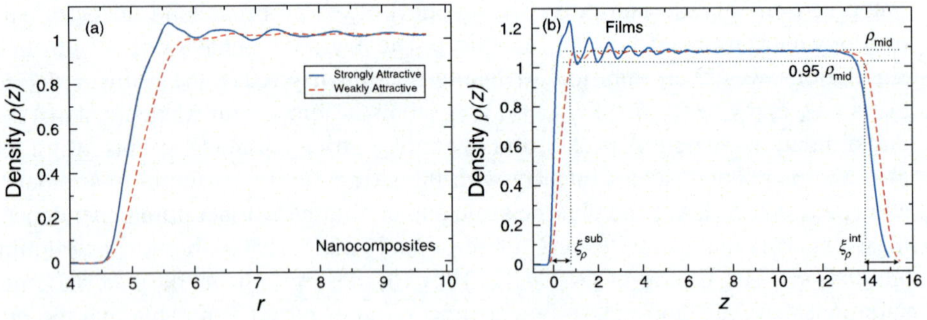

Fig. 5.6 The density profile $\rho(r)$ around **a** NP and $\rho(z)$ for **b** thin films. In panel b, we illustrate the definition of the interfacial density scale ξ_ρ of thin polymer film supported by a substrate with weak substrate-polymer interaction strength $\varepsilon = 0.1$ (black) and strong substrate-polymer interaction strength $\varepsilon = 1.5$ (red) at $T = 0.45$. See the main text for the definition of ξ_ρ

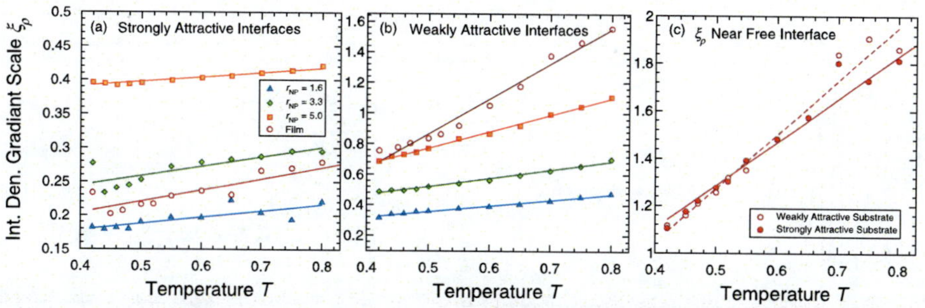

Fig. 5.7 The interfacial density scale ξ_ρ for polymer-NP and polymer-substrate interfaces having either **a** strong and **b** weak attractive interactions, and **c** the free polymer-air interface. ξ_ρ of strongly attractive interfaces in the thin film and NP composite increases only slightly with increasing T. In contrast, ξ_ρ of the weakly attractive interfaces (or free surface) increases much more significantly with increasing T. The solid lines are guides for the eye

is evidently much larger in the case when the interfacial interactions are weak than in the case of the strongly attractive interface; this is expected, since the strongly interacting interface should give rise to a more precisely defined interfacial layer. ξ_ρ also exhibits a much stronger T variation for the weakly attractive interface. Specifically, ξ_ρ grows on heating in the case of weakly attractive interfacial interaction for both the polymer films and polymer-NP composites. Additionally, the interfacial density scale ξ_ρ increases with decreasing curvature of the interface (increasing NP size) for the weakly attractive interfaces since flat substrate is equivalent to zero curvature.

5.3.3.1 Density Variance Near Interfaces and Its the Characteristic Length Scale σ_ρ

The interfacial density scale discussed above quantifies how sharply an interface is defined. Another way to characterize the interfacial density scale is to consider density fluctuations near interfaces, i.e., quantifying the spatial range over which the packing fluctuations near an interface vanish. This scale quantifies the local extent of "packing frustration" and we expect this quantity to have a much more direct relationship to dynamical changes near interfaces than that of the density gradient. Due to molecular packing near a solid interface, the density profile oscillates around approximately the average density of the interior of the material with a period on the order of one monomer distance in the red curve in Fig. 5.8a. This figure shows the density profile of a film with thickness $\approx$15 having polymer-substrate interaction strength $\varepsilon = 1.5$ and for $T = 0.45$. The interfacial density scale from density fluctuationa σ_ρ can thus be defined from the density variance near the interface from its bulk value. Similar to that from [73], we define the depth dependence of density variance,

$$\Delta_\rho^2(z) = \left\langle (\rho(z) - \langle \rho_{\text{mid}} \rangle)^2 \right\rangle, \tag{5.5}$$

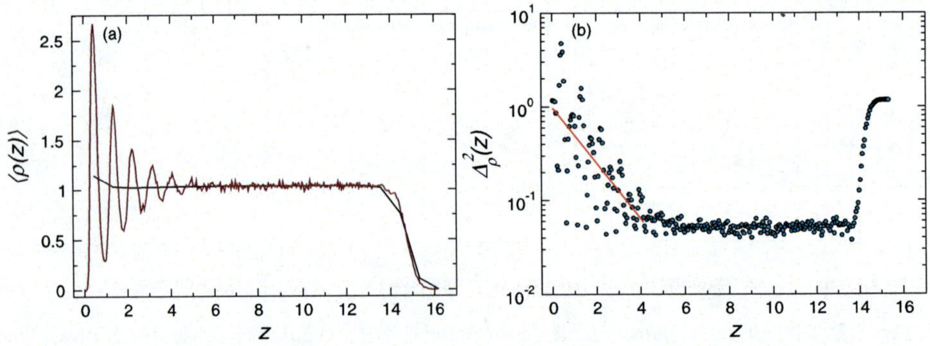

Fig. 5.8 **a** The density profile of a thin polymer film with $\varepsilon = 1.5$ at $T = 0.45$. The black curve is $\rho(z)$ averaged over 1 monomer σ distance; and the dark red curve is the same $\rho(z)$ averaged over $0.05\,\sigma$. **b** Variance of the segment density $\Delta_\rho^2(z)$. The resulting interfacial density scale σ_ρ for this film under these conditions is approximately 2.5 in our reduced units. The red curve is the fit from (5.6), from which we obtain σ_ρ

where ρ_{mid} is the average density in the film middle. Figure 5.8b shows the resulting density variance $\Delta_\rho^2(z)$ for thin film at $T = 0.45$ having $\varepsilon = 1.5$. To obtain the interfacial density variance scale σ_ρ, we can identify σ_ρ as the length scale at which $\Delta_\rho^2(z)$ plateaus. However, the resulting σ_ρ magnitudes obtained from this procedure depend on the degree of noise in the density profile. To avoid this, we fit $\Delta_\rho^2(z)$ near the interface with an exponential function,

$$\Delta_\rho^2(z) = A \exp\left(-\frac{z}{\sigma_\rho/2}\right), \tag{5.6}$$

where A is the fitting parameter, to obtain σ_ρ (Fig. 5.8b red curve). That said, we also evaluate σ_ρ from the plateau value of $\Delta_\rho^2(z)$ and find no qualitative difference with that from the method described above.

The resulting σ_ρ for both strongly and weakly attractive substrates increases on cooling in Fig. 5.9a, a trend that is the opposite to the interfacial density gradient scale ξ_ρ in Fig. 5.7. The interfacial density variance scale quantifies the length scale at which the density variance is nearly the same as that in the film middle. This is in contrast to the quantity ξ_ρ described previously, which quantifies the "sharpness" of the interfaces. Unlike the behavior at the substrate, the interfacial density variance scale $\sigma_\rho^{\text{free}}$ near a free polymer interface increases with increasing temperature in Fig. 5.9b, a trend similar to ξ_ρ near the free interface. We discuss the significance of this finding and its relation to the interfacial mobility scale ξ defined from relaxation profile in the next section. Here, we briefly note that the density variance scale indeed has a much stronger correlation with the interfacial mobility scale ξ than the scale of density gradient ξ_ρ, both for the case of solid interfaces prevalent in nanocomposites and free interfaces prevalent in thin polymer films. These observations were not noted in our previous works.

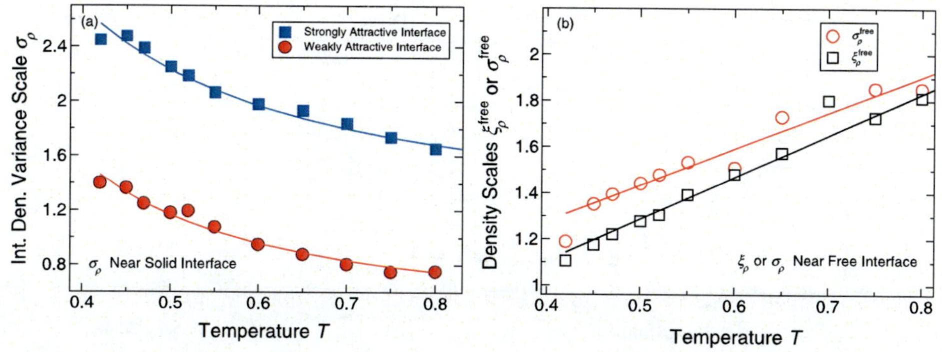

Fig. 5.9 **a** The temperature T dependence of σ_ρ for both the strongly and weakly attractive interface. **b** The interfacial density scale from density gradient ξ_ρ and density variance σ_ρ near the "free" polymer surface in thin polymer film

5.3.4 Comparison Between Interfacial Mobility and Density Scales

We next evaluate how the interfacial interaction strengths affect the interfacial mobility scale ξ by comparing ξ near the interface with strong and weak attraction and ξ near free boundary in Fig. 5.10a. We find that ξ and its temperature dependence are nearly the same for all cases studied (Fig. 5.10a). Having established this, we directly compare the temperature variations of interfacial mobility scale ξ with that from density gradient ξ_ρ and density variance σ_ρ. It is apparent that the interfacial mobility scale ξ and interfacial density gradient scale ξ_ρ have the opposite temperature dependence (Figs. 5.4 and 5.7), while the interfacial density variance scale σ_ρ grows on cooling, similar to $\xi(T)$. To show this relation explicitly, we plot σ_ρ and ξ_ρ parametrically in Fig. 5.10b and show that there is an approximately linear correlation between σ_ρ and ξ near the solid interface. That said, the magnitude of interfacial density variance scale σ_ρ for different interfacial interaction strength is rather different in Fig. 5.9a. In contrast, $\xi(T)$ in Fig. 5.10a we see that σ_ρ that has nearly the same value for all interfaces studied here. We also find that σ_ρ of the free interface of thin polymer film grows on heating, the same as that of ξ_ρ, indicating that the linear correlation between σ_ρ and ξ is only valid for interfacial layers near the solid interfaces (substrate or NP surface). We conclude that while the scale from the density gradient characterizes the "sharpness" of an interface, the interfacial density variance scale σ_ρ quantifies the spatial extent of the gradient in density fluctuation near interfaces. Therefore, these two different density scales characterize different aspects of the mean structure and amplitude of structural fluctuations near an interface, accounting for their opposite temperature dependences. As for the comparison between the density and mobility scales, although the interfacial density variance scale correlates nearly linearly with the interfacial mobility scale, interfacial interaction or free surface affects their magnitudes differently. This further reinforces the idea that the molecular structure and dynamics are distinct near interfaces.

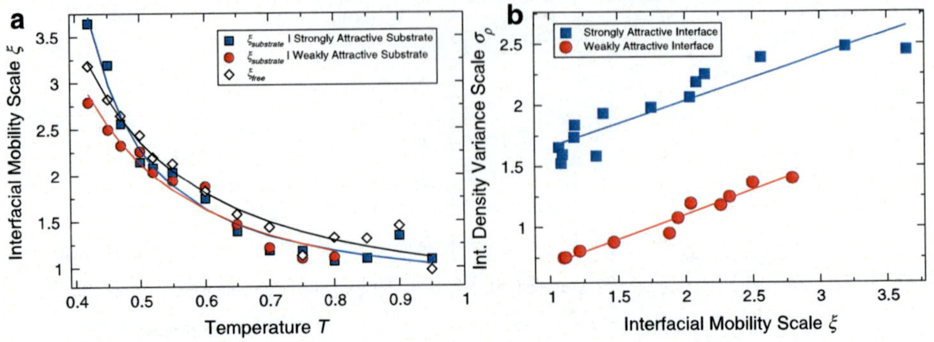

Fig. 5.10 **a** The interfacial mobility scale ξ for polymer-substrate interfaces having either strong and weak attractive interactions, and near the free polymer-air interface. ξ for all interfaces has similar magnitude and increases with increasing T. The solid lines are guides for the eye. **b** Parametric plot for interfacial mobility scale ξ and interfacial density variance scale σ_ρ. We see that there is a correlation between ξ and σ_ρ, but σ_ρ has different magnitudes depending on the interfacial interaction strength

5.4 Effects of Bound Interfacial Layer Around NP and Near Substrate on Polymer Relaxation and Overall T_g

While we have quantified the interfacial mobile layer thickness ξ for varying NP sizes in nanocomposites and near interfaces in thin films and their temperature dependence in the previous section, the influences of these interfacial layers to the overall segmental relaxation time τ_α and the glass transition temperature T_g warrant a separate discussion, since the correspondence between them is not necessarily straightforward. The influence of interfaces on the glass transition temperature T_g of small molecules and polymers has been intensively investigated and discussed [2, 6, 9, 74–77]. The prevailing view is that surface interactions play a critical role in determining the sign and magnitude of the shift in T_g on geometrical confinement, or effective confinement through the addition of NP. In particular, strong, favorable interactions are typically found to increase T_g, while weak or unfavorable interactions decrease T_g. Support for this picture comes from molecular simulations showing that T_g increases as the attractive polymer-surface (polymer-NP or polymer-substrate) interactions become stronger [36, 45, 78–81]. Additionally, the free surface of supported films, which has enhanced dynamics relative to the interior of the film, tends to decrease the overall T_g. As a consequence, the overall change in T_g for nanocomposites or supported films then results from the competition between these interfacial effects, as well as effects due to the intrinsic geometrical confinement. On the other hand, a number of recent experimental investigations report negligible changes in T_g in polymer nanocomposites, even when the polymer-NP interaction is so strong that a large shift might be expected [6, 82, 83]. For example, Kumar and coworkers [2, 31, 32] found that the T_g of three different polymers were essentially unaffected by the presence of varying amounts of strongly interacting silica NPs. Sokolov and coworkers [20, 33] confirmed these findings, and their results suggest that surface chains relax on a time scale that can be several orders of magnitude slower than the bulk chains. Indeed, in the limit of strongly favorable polymer-NP interactions, experimental measurements indicate the presence of an interfacial polymer layer with a substantially slower relaxation [32, 57, 84–86]. In the limit of irreversible adsorbtion, it has been suggested that the T_g changes can be related to the amount of adsorbed polymer [87]. However, the precise effect of this "bound" layer on nanocomposite dynamics, often quantified by shifts in the apparent T_g, remains unclear.

In this section, we investigate the influence of NP interaction in nanocomposites and substrate interaction in thin films on the relaxation and T_g. In particular, we focus on the conditions under which a distinct interfacial bound layer appears, how it influences nanocomposite and supported thin film relaxation, and correspondingly how it impacts thermodynamic and dynamic methods to estimate T_g, which sometimes have been observed to decouple under confinement [24]. We find that when the polymer-NP or polymer-substrate interaction strength ε exceeds the polymer-polymer interaction strength, a distinct relaxation process at very large time develops, associated with the formation of an interfacial bound polymer. The NPs or substrates effectively "cloak" themselves in this bound layer, making standard quasi-thermodynamic and dynamic estimates of T_g, which primarily weight the matrix chains, insensitive to ε. The bound chains also affect the overall relaxation of the nanocomposite or thin films, but care is required to separate the unbound and bound components of the relaxation. These considerations demonstrate that, while the dynamics of the nanocomposites or thin film are indeed greatly altered by strongly interacting NPs or substrate, commonly employed methods of estimating T_g often do not reflect these changes because the unbound portion of the material and the interfacial dynamics effectively decouple.

5.4.1 Influence of Interface Interaction on Polymer Relaxation

Although we examine the layer-resolved relaxation for both nanocomposites and thin films in Figs. 5.2 and 5.3, experimentally, the distance dependence of $F_s(q, t)$ is not readily accessible, so we consider how the substantial variations in local relaxation time are reflected in its overall behavior. Figure 5.11a and c shows the overall $F_s(q, t)$ for various ε at a representative temperature for nanocomposites and thin film, respectively. For $\varepsilon \gtrsim 1$, a qualitative change of behavior becomes apparent; namely, an additional relaxation process in $F_s(q, t)$ appears at very large t, which is most apparent in the double-log representation. This additional relaxation process occurs due to the "bound" polymer with very slow surface relaxation that emerges at large ε. Figure 5.11b and d further shows the emergence of the bound polymer as a function of temperature for a strongly attractive interface (NP or substrate). Evidence of an additional relaxation process in $F_s(q, t)$ was also found in simulations of composites with strongly interacting nanorods [88]. For cases where the bound

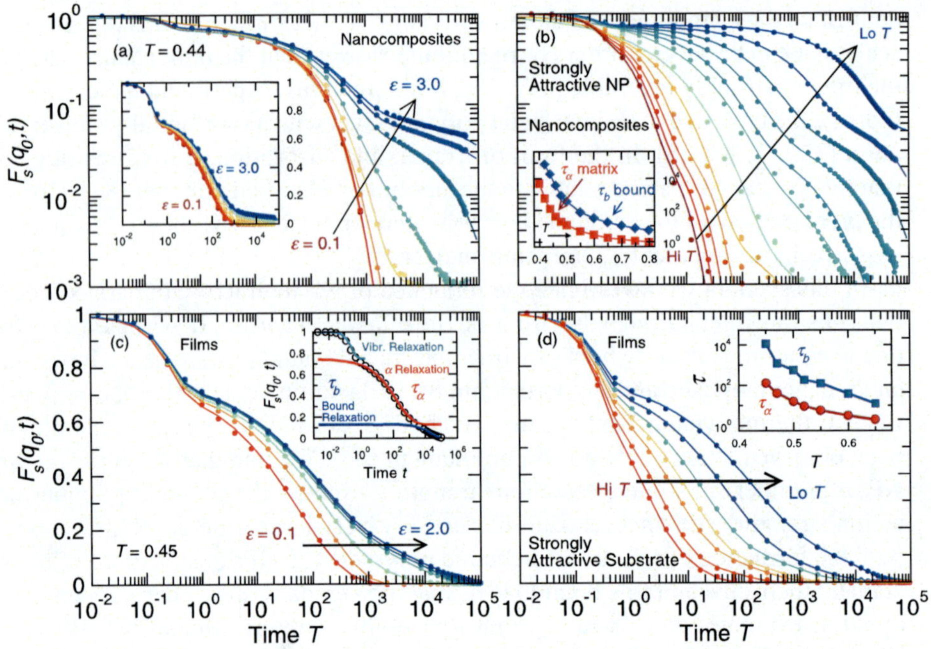

Fig. 5.11 The self-intermediate scattering function $F_s(q_0, t)$ for nanocomposites (panel **a**, **b**) and thin polymer film (panel **c**, **d**). Panels **a** and **b** show $F_s(q_0, t)$ for **a** nanocomposites and **c** thin films at a representative temperature with varying polymer-NP interaction strengths ε. Panel **a** is a log-log representation of $F_s(q_0, t)$, while the inset is a log-linear representation of the same data. For $\varepsilon > 1$, a distinct third relaxation emerges due to slowly relaxing "bound" chains near the NP interface or substrate. For $\varepsilon \leq 1$, the solid lines are the result of the fit to the two-step relaxation model (5.2); for $\varepsilon > 1$ the solid lines are the result of the fit to the three-step bound polymer relaxation model (5.7). The inset of panel **c** shows how each term of (5.7) contributes to the overall relaxation. For clarity, the α and vibrational relaxation terms in (5.7) are shifted upward by A_b and $A_b + A$, respectively. Panels **b** and **d** show $F_s(q_0, t)$ for **b** nanocomposites and **d** thin film having a strongly attractive interfacial interaction strength, where bound polymers play a significant role in relaxation. The inset of panels **b** shows the T dependence of the matrix τ_α and bound τ_b relaxation times for $\varepsilon = 1.4$. Similarly, the inset of panel **d** shows the relaxation time τ_b of bound layer and the primary relaxation τ_α due to the unbound polymer as a function of T

sub-group appears, the relaxation of $F_s(q, t)$ cannot be described by the two-step relaxation of (5.2) in both nanocomposites and thin films. Rather, we can describe the full t dependence of $F_s(q, t)$ by adding a third relaxation process that explicitly defines a time scale τ_b for the bound polymer relaxation,

$$F_s(q, t) = (1 - A)e^{-(t/\tau_s)^{3/2}} + (A - A_b)e^{(-t/\tau_\alpha)^{\beta}} + A_b e^{-(t/\tau_b)^{\beta_b}}, \tag{5.7}$$

where $A_b \approx N_{b/N}$ approximately associated with the fraction of bound polymer; τ_α defines the relaxation of the polymer matrix in nanocomposites or unbound polymer in thin films, which effectively averages over local polymer relaxation away from the NP or substrate. Both the α and bound polymer relaxation processes are described by a stretched exponential with stretching exponents β and β_b (respectively) which are in the range 0.4–0.6 at low T. The solid lines in Fig. 5.11 show that this three-step relaxation function can account for the relaxation over all time scales, spanning decades of relaxation of $F_s(q, t)$. To show these three time scales more clearly, we plot contributions of each term in (5.7) to the overall relaxation in the inset of Fig. 5.11c. The "non-ergodicity parameter" $A \approx 0.75$ is nearly independent of T and ε over the T range investigated. In both nanocomposites and thin films, for $\varepsilon \leq 1$, the bound layer no longer exists, and so $A_b = 0$ within the precision of our results; for ε larger than the polymer-polymer interaction strength, the bound fraction A_b increases weakly in the range 0.05–0.075 in nanocomposite [28], and the same fraction A_b in thin films is approximately ranging from 0.01– 0.13 [29]. Integrating over the monomer density profile as a function of distance from the NP in nanocomposites or distance from the substrate in thin films, this fraction is equivalent to a bound thickness of $\approx$1.5 to 2σ, similar to the interfacial mobility scale that emerges from the fit of (5.3) to the relaxation gradient in Fig. 5.7a. Note that in the regime of unentangled polymer, the thickness of this interfacial mobility layer has no detectable chain length dependence [29]. Consistent with experimental studies of nanocomposites [20], estimates of τ_b in nanocomposites or thin films are typically 1–2 orders of magnitude larger than the segmental relaxation time of the polymer matrix (or unbound polymer) τ_α (insets of Fig. 5.11b, d), and the difference grows with increasing ε. The presence of two structural relaxations—one corresponding to the surface and one to the bulk polymer—has been reported by several experiments [20, 89, 90], including those on polymer films near flat surfaces. The substantial difference in the relaxation of matrix (unbound) and bound polymers causes the overall relaxation, defined by $F_s(q, \tau) = Ae^{-1} \approx 0.28$, to deviate from the matrix (unbound) relaxation time τ_α for $\varepsilon > 1$. As a consequence, T_g, as defined by a fixed relaxation time τ of the material as whole, increases weakly with the polymer-NP (polymer-substrate) attraction ε, even while the matrix (unbound) relaxation is nearly constant. This is the "cloaking" effect of the bound polymer mentioned above, and we shall return to this point when we examine a quasi-thermodynamic T_g definition. Note that the dimensions of the chains in the bound layer are not significantly altered from those of matrix chains, though the interfacial chains tend to align with the NP interface [36].

We did not originally anticipate that a continuous gradient of relaxation times (Fig. 5.2c or d) would give rise to two distinct relaxation processes, though evidence for this has been found theoretically [91]. To confirm consistency between the description of a continuous gradient *versus* effectively distinct relaxations, we checked that the local relaxation averaged over the matrix region (excluding the bound layer) independently recovers the matrix relaxation time τ_α obtained by fitting the overall relaxation to (5.7). In many cases, it may be difficult to measure the relaxation with sufficient precision to estimate the bound polymer thickness and relaxation. Fortunately, the separation of time scales of the bound layer can be utilized to approximate the fraction A_b of bound polymer (and hence its thickness) from (5.7). Specifically, if we treat the bound layer as frozen ($\tau_b \gg \tau_\alpha$), then the intermediate scattering function at the overall relaxation time τ defines the relation,

$$F_s(q,\tau) = Ae^{-1} \approx (A - A_b)e^{-(\tau/\tau_\alpha)^\beta} + A_b, \tag{5.8}$$

where we drop the first term of (5.7) for the fast relaxation, which is essentially zero at time τ. A_b can then be obtained by rearrangement of this equation,

$$A_b = A\frac{e^{1-(\tau/\tau_\alpha)^\beta} - 1}{e^{1-(\tau/\tau_\alpha)^\beta} - e} \approx \frac{A}{e-1}\left[\left(\frac{\tau}{\tau_\alpha}\right)^\beta - 1\right], \tag{5.9}$$

where the final approximate result is obtained by expanding the exponential to first order, and is only valid assuming that $\tau \approx \tau_\alpha$ (i.e. $1 - (\tau/\tau_\alpha)^\beta \ll 1$, as is the case for our data). Perfect NP (or substrate) cloaking would imply that the matrix (unbound) relaxation time τ_α equals that of the pure melt, which should be a good approximation when ε is very large. Thus, for strongly interacting NP or substrate, the bound fraction can be estimated directly from the relaxations of the pure polymer and the composite or thin films.

5.4.2 *Effect of Interfacial Interaction Strength on T_g Estimates from Thermodynamic and Dynamic Methods*

Next, we wish to understand the consequences of distinct bound layer relaxations on the glass transition temperature T_g. Specifically, we evaluate T_g using both a dynamic (relaxation time) approach and commonly used thermodynamic approaches. We first define T_g from thermodynamic quantities. In Fig. 5.12 we show the T dependence of the excess potential energy $\Delta U(T) = U(T) - U_{\text{glass}}(T)$ on heating the glass in nanocomposites (panel a), and thin films (panel b) for the range of polymer-NP and polymer-substrate interactions considered, respectively; note that we include both polymer and NP (substrate) contributions to U, though only the polymer contribution is significant. Generally speaking, in the nanocomposites, the measurement of T_g is complicated by the intrinsic spatial heterogeneity of the NPs. Additionally, in the case of thin films, we also define T_g from the temperature dependence of film thickness, from which we extrapolate T_g in Fig. 5.13c; this procedure is similar to that from the excess potential energy described above [29]. In both cases, glasses were formed by cooling at a rate of 10^{-5}, and are reheated at the same rate. It is noteworthy that these cooling rates are many orders of magnitude faster than in experiments, and, correspondingly, the T_g estimated is substantially higher than would be obtained at an experimental rate. Deep in the glass state ($T < 0.3$), U_{glass} is well described by a quadratic function, equivalent to a linear temperature dependence of the specific heat. Figure 5.12 shows that ΔU grows rapidly in the vicinity of $T \approx 0.4$ at this rate. Analogous to the experimental procedure, we define T_g by the extrapolation of a linear fit in the transition region to $\Delta U \to 0$. We show the resulting T_g defined in this way in Fig. 5.13a, b and T_g varies only weakly with polymer-NP and polymer-

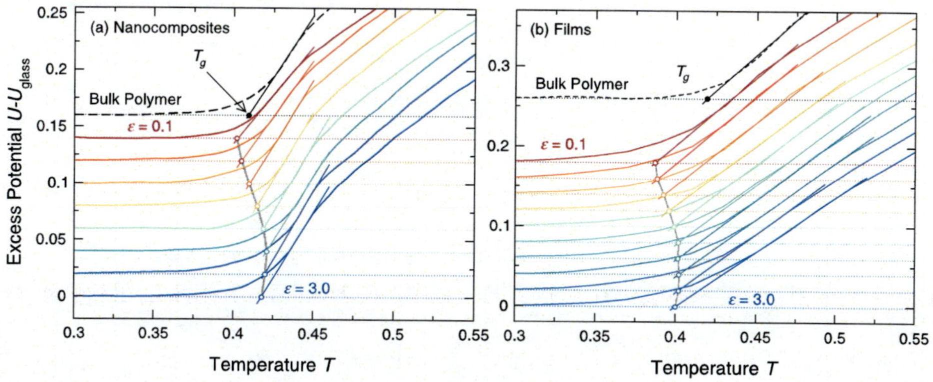

Fig. 5.12 The excess energy $\Delta U(T) = U(T) - U_{\text{glass}}(T)$ at fixed heating rate 10^{-5} for both **a** nanocomposites and **b** thin polymer films. We define the quasi-thermodynamic rate-dependent T_g by a linear extrapolation of the vanishing of ΔU in the transition region. The circles indicate the resulting T_g values, which increase with ε, and then plateau. Data are shifted vertically for clarity. In panel **a** from top to bottom, the curves are the bulk polymer under isobaric condition $\langle P \rangle = 0.1$, $\varepsilon = 0.1$, 0.25, 0.5, 0.75, 1.0, 1.4, 2.0, and 3.0. In panel **b** from top to bottom, curves represent bulk polymer under isobaric condition $\langle P \rangle = 0$, polymer-substrate interaction strength $\varepsilon = 0.1$, 0.25, 0.5, 0.75, 1.0, 1.25, 1.5, 2.0, 2.5, and 3.0

substrate interaction strength. We also considered alternate definitions of T_g based on deviations from liquid-like relaxation at high T, or the extremum in specific heat; all these definitions show the same qualitative increase and saturation of T_g at large ε.

We are now in a position to contrast the behavior of T_g from thermodynamic and dynamic definitions (Fig. 5.13c, d). Dynamically, T_g is commonly defined by a fixed relaxation time that corresponds roughly to accessible cooling and heating rates, and experimentally this is typically 100 s. Since our thermodynamic definition of T_g corresponds to a much faster cooling rate than experiments, we correspondingly define the dynamic T_g by a shorter time scale, specifically as the temperature where $\tau = 10^3$ (in LJ units); our qualitative findings do not change if we choose a larger time scale. There are multiple implications raised by the T_g results. First, the T_g from the quasi-thermodynamic definition is essentially independent of polymer-NP or polymer-substrate interaction strength for $\varepsilon \gtrsim 1.0$ (i.e., greater than polymer-polymer interaction strength) in both nanocomposites and films. In parallel, the behavior of T_g defined from the relaxation of the matrix in nanocomposites and unbound film in thin films τ_α also roughly saturation for $\varepsilon \gtrsim 1.0$. Indeed, Figs. 5.2 and 5.11 confirm that this is precisely the region where a distinct bound polymer relaxation develops. Thus, the saturation of T_g is a consequence of the bound polymer layer cloaking the effects of NP or substrate interactions. On the other hand, T_g defined by the overall relaxation τ (that includes bound chains), increases monotonically with increasing ε, although the rate of increase falls of significantly for $\varepsilon > 1.0$. Apparently, the thermodynamic definition of T_g is insensitive to the very slowly relaxing bound polymer layer, and only reflects the dynamics of the matrix polymer. In other words, the bound layer is thermodynamic "dead layer". Naively, one might

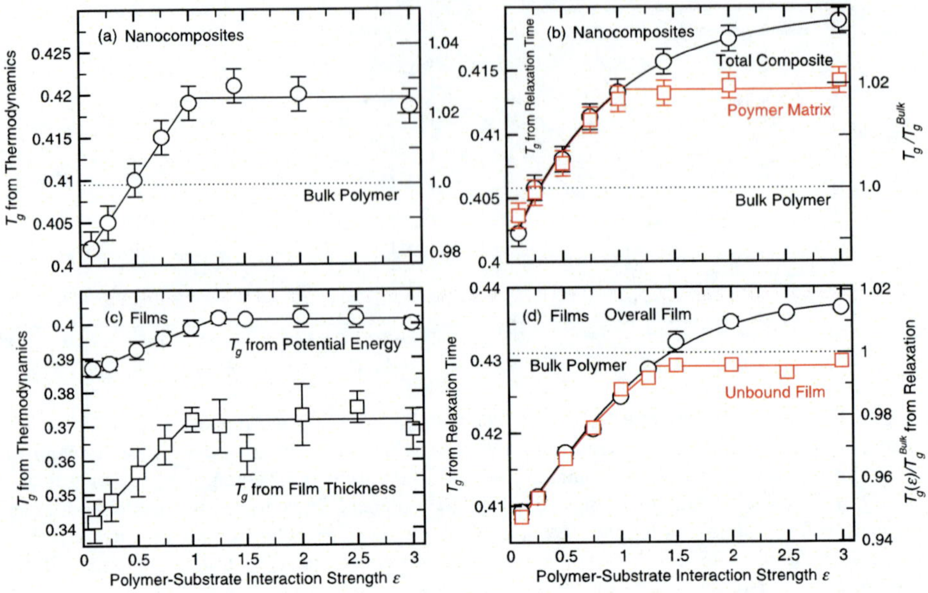

Fig. 5.13 T_g estimates for nanocomposites (panel **a**, **b**) and thin films (panel c, d) defined from thermodynamics (panel **a**, **c**) and relaxation time (panel **b**, **d**). T_g defined from thermodynamic quantities (Fig. 5.12) plateaus when bound polymer emerges for both **a** nanocomposites and **c** thin films. The uncertainty estimates indicate the standard error of the mean of the T_g estimates among 5 independent runs. On the right side of the panel, we normalize $T_g(\varepsilon)$ by the corresponding T_g from the bulk polymer, highlighting the small amplitude of T_g changes. Specifically, in the case of nanocomposites, T_g of the bulk polymer is under isobaric condition with $\langle P \rangle = 0.1$, while in the thin film case, the T_g of bulk polymer is under isobaric condition with $\langle P \rangle = 0$. Panel b and d show T_g defined by a fixed relaxation time for both **b** nanocomposites and **d** thin films. Dynamical definition of T_g, namely $\tau(T_g) = 10^3$. In particular, we examine T_g of the overall and unbound portion of the nanocomposites (polymer matrix) and thin film (unbound film), as well as bulk polymer for comparison. For $\varepsilon > 1.0$, T_g of the unbound portion is nearly an invariant of ε. This indicates the bound polymer "cloaks" the strongly interacting substrate. Like the thermodynamic T_g, the matrix component or unbound film plateaus when bound polymer develops

expect linear growth of T_g from the overall dynamics with increasing ε; however, the overall dynamical T_g increases sub-linearly when bound polymer develops, since the effect of NP/substrate interactions on matrix/unbound chains is cloaked. As a result, even for very strong polymer-NP (polymer-substrate) interactions, the effect on any of these measures of T_g is $\lesssim 5\%$. In the case of nanocomposites, even for larger NP loading fractions (when the fraction of bound polymer must be proportionally larger), the effect on T_g defined by thermodynamic variations or matrix/unbound chain dynamics will be small due to cloaking, consistent with experiments. We have confirmed this effect in simulations at twice the NP concentration as the present data (not shown). This behavior is generally a consequence of the fact that the bulk structural relaxation decouples from the much faster surface relaxation.

5.5 Conclusions

Since nanoparticles have a relatively large interfacial zones in comparison to the size of macro-particle additives in bulk materials, the interfacial zones over which the polymer matrix dynamics are perturbed can be expected to have an appreciable influence on nanocomposite properties [28, 37, 39, 92], despite the relatively small width of this interfacial zone. Motivated by the direct analogy between nanocomposites and thin polymer films, we focus on this problem by quantifying the spatial extent and properties of the interfacial zone around nanoparticles and the closely related problem of the interfaces of supported thin polymer films. Specifically, we characterize the spatial extent to which the polymer structure and dynamics are perturbed near the interface regardless of whether it is the interface around nanoparticle, or the interfacial region near the substrate or the free interface in thin polymer films. Furthermore, we closely examine the effect of the interfacial layer on the overall relaxation of materials and the glass transition temperature T_g from both thermodynamic and dynamic estimates, especially in the case where the interfacial interaction is strongly attractive.

To characterize the scale of spatial perturbation near the interface related to polymer dynamics, we examine the relaxation gradient near nanoparticles and in thin films from which we obtain the interfacial mobility scale ξ. For the scale of spatial perturbation related to the polymer structure, we characterize a length scale from the density gradient ξ_ρ and density variance σ_ρ near an interface separately. In short, the interfacial density gradient scale ξ_ρ characterizes how well-defined the structural interface is, while the interfacial density variance scale σ_ρ describes the spatial extent of density fluctuations near an interface, providing a measure of "packing frustration".

We find that the thickness of the interfacial mobile layer ξ around nanoparticles or near interfaces in thin polymer film is typically a few nanometers when translated to laboratory units appropriate for polystyrene. In all cases, we find that ξ grows progressively upon cooling, but remains on the order of a few nanometers. Notably, the interfacial mobility scale ξ is sensitive to NP size for strong interfacial interactions, but only weakly dependent on NP size for weak substrate interactions. That said, the film appears to offer a limiting value for ξ in case of very large NP. Our findings broadly accord with the early computational study by de Pablo and coworkers [93], and are also in general qualitative accord with measurements of the interfacial zone thickness by Sokolov and coworkers [20–22, 26, 27, 55]. We note that recent X-ray scattering studies have recently provided evidence for a zone of modified solvent density around nanoparticles having a scale extending up to 2 nm beyond the nanoparticle surface [94], another experimental trend in qualitative accord with the present work. Measurements of this kind should be helpful in determining the interaction potential between the nanoparticles and the polymer fluid matrix. In the case when the interfacial interaction strength is above the polymer-polymer interaction strength, we find that it is possible to explicitly separate a distinct fraction of bound polymers with significantly increased relaxation time from the overall relaxation behavior of the materials. Indeed, a bound polymer layer forms near either the nanoparticle near the surface in nanocomposites in this case, or near substrates in thin supported polymer films. We also show that the effective size of the nanoparticles in these systems can be altered by a "bound" layer of polymer. The importance of this layer is appreciated when we consider that the interfacial mobile layer thickness can be comparable to dimensions of the nanoparticle.

Subsequently, we examine the effect of the bound polymer layer on T_g from the relaxation time and thermodynamic estimates for many interfacial interaction strengths ε. We find that T_g defined from relaxation time increases monotonically with ε, and nearly saturates for strongly attractive interaction strength, while the same T_g (defined dynamically) of the unbound portion of the film or polymer matrix in nanocomposites saturates as substrate interaction strength exceeds the polymer-polymer interaction strength. In parallel, we also examine thermodynamic estimates of T_g, which does not capture the signatures of much slower, bound polymer chains near interfaces, and hence yield T_g estimates that mainly reflect the T_g of bulk-like portion of polymer, which corresponds to the unbound film in thin film or polymer matrix in nanocomposites. Specifically, the thermodynamically defined T_g is found to saturate to nearly a constant value for ε values greater than the polymer-polymer interaction strength, independent of the precise thermodynamic features that are used to extract T_g. However, in the thin film case, the saturation value of T_g will generally be thickness dependent. The slowing of dynamics induced by the substrate can exceed the enhancement of the dynamics at the free surface, leading to no change, or even an increase in T_g [40]. Indeed, we do not see significant T_g reductions when the bound layer is present [95]. Similarly, numerous studies on nanocomposites have also reported little to no change in T_g by adding strongly interacting nanoparticles [2, 20, 31–33]. This leads to the conclusion that a bound polymer layer may arise near the strongly attractive interfaces, whether it is substrate in supported polymer films, or nanoparticle surface in nanocomposites. In both cases, the bound polymer layer "cloaks" the film interior or polymer matrix from interfacial interactions, resulting in small shifts in thermodynamic estimates of T_g. As a consequence, the changes in T_g no longer grow in proportion to the interfacial interaction strength when such a bound layer forms. This "cloaking" effect near strongly attractive interfaces and nanoparticles appears to be rather universal in systems having strong interfacial interactions.

For the interfacial scale related to the polymer structure near an interface, we find that the interfacial density gradient scale ξ_ρ grows on heating, an opposite trend compared to that of ξ, while the interfacial density variance scale σ_ρ grows on cooling, exhibiting approximately a linear correlation with that of $\xi(T)$. That said, we find that $\xi(T)$ for different interfaces or interfacial interaction strengths have roughly the same magnitude, while $\sigma_\rho(T)$ for different interfaces have different magnitudes, suggesting that interfacial scale from dynamics and structure have distinct dependence on the interfacial interactions.

Since the interfacial scale ξ defines the range of the mobility gradient, by definition, it is a scale of dynamic heterogeneity. However, it is not clear how the scale of this mobility gradient may (or may not) relate to the scale of spatially and temporally heterogeneous dynamics that occurs both in films and composites, as well as in structurally homogeneous bulk materials. It has been widely argued [39, 67, 92, 96, 97], dating back to the pioneering work of Adam and Gibbs (AG), that the scale of this spatiotemporal heterogeneity can be directly related to the growing activation barrier for molecular rearrangement on cooling. While there has been some support of the suggestion that the interfacial mobility scale grows in a parallel fashion to the extent of cooperative motion [59], simulation has indicated that the extent of particle exchange motion is essentially constant in glass-forming polymer films so that an explanation of mobility gradients in thin films in terms of gradients in the degree of cooperative motion with film depth can be clearly excluded [41]. Recent work has indicated that string-like replacement motions can quantitatively account for the variation in the activation free energy, and thus offer a concrete molecular realization of the abstract "cooperatively rearranging regions" proposed by AG. While the variation of ξ in Fig. 5.4 exhibits the same qualitative trend as the scale of string-like collective motion in our simulated nanocomposites and thin polymer films, further work is required on both nanocomposite and thin film polymer materials to determine if a quantitative relation exists between these quantities.

Acknowledgements We thank S. K. Kumar for valuable discussion. Computer time was provided by Wesleyan University. This work was supported in part by NIST award 70NANB13H202.

References

1. R.D. Priestley, C.J. Ellison, L.J. Broadbelt, J.M. Torkelson, Structural relaxation of polymer glasses at surfaces, interfaces, and in between. Science **309**(5733), 456–459 (2005)
2. A. Bansal, H. Yang, C. Li, K. Cho, B. Benicewicz, S. Kumar, L. Schadler, Quantitative equivalence between polymer nanocomposites and thin polymer films. Nat. Mater. **4**, 693–698 (2005)
3. M.D. Ediger, J.A. Forrest, Dynamics near free surfaces and the glass transition in thin polymer films: a view to the future. Macromolecules **47**(2), 471–478 (2014)
4. X. Wang, X. Tong, H. Zhang, J.F. Douglas, String-like collective motion and diffusion in the interfacial region of ice. J. Chem. Phys. **147**(19), 194508 (2017)
5. W. Xia, S. Keten, Interfacial stiffening of polymer thin films under nanoconfinement. Extreme Mech. Lett. **4**, 89–95 (2015)
6. P. Rittigstein, J.M. Torkelson, Polymer-nanoparticle interfacial interactions in polymer nanocomposites: confinement effects on glass transition temperature and suppression of physical aging. J. Polym. Sci. Part B: Polym. Phys. **44**(20), 2935–2943 (2006)
7. S.C. Tjong, Y.-W. Mai, *Physical Properties and Applications of Polymer Nanocomposites* (Elsevier, 2010)
8. B. Daniel, Y. Da, D. Junyan, O.C.K., Recent progress in high resolution lithography. Polym. Adv. Technol. **17**(2), 94–103 (2006)
9. M. Alcoutlabi, G.B. McKenna, Effects of confinement on material behaviour at the nanometre size scale. J. Phys.: Condens. Matt. **17**(15), R461 (2005)
10. P.A. O'Connell, S.A. Hutcheson, G.B. McKenna, Creep behavior of ultra-thin polymer films. J. Polym. Sci. Part B: Polym. Phys. **46**(18), 1952–1965 (2008)
11. H.-L. Tyan, Y.-C. Liu, K.-H. Wei, Thermally and mechanically enhanced clay/polyimide nanocomposite via reactive organoclay. Chem. Mater. **11**(7), 1942–1947 (1999)
12. J.H. Koo, *Fundamentals, Properties, and Applications of Polymer Nanocomposites* (Cambridge University Press, 2016)

13. R. Gangopadhyay, A. De, Conducting polymer nanocomposites: a brief overview. Chem. Mater. **12**(3), 608–622 (2000)
14. A.C. Balazs, T. Emrick, T.P. Russell, Nanoparticle polymer composites: where two small worlds meet. Science **314**(5802), 1107–1110 (2006)
15. J. Jancar, J. Douglas, F.W. Starr, S. Kumar, P. Cassagnau, A. Lesser, S.S. Sternstein, M. Buehler, Current issues in research on structure-property relationships in polymer nanocomposites. Polymer **51**(15), 3321–3343 (2010)
16. J.F. Moll, P. Akcora, A. Rungta, S. Gong, R.H. Colby, B.C. Benicewicz, S.K. Kumar, Mechanical reinforcement in polymer melts filled with polymer grafted nanoparticles. Macromolecules **44**(18), 7473–7477 (2011)
17. G. Schmidt, M.M. Malwitz, Properties of polymer-nanoparticle composites. Current Opin. Colloid Interface Sci. **8**(1), 103–108 (2003)
18. H. Zhang, P. Kalvapalle, J.F. Douglas, String-like collective atomic motion in the interfacial dynamics of nanoparticles. Soft Matter **6**(23), 5944–5955 (2010)
19. H. Zhang, D.J. Srolovitz, J.F. Douglas, J.A. Warren, Grain boundaries exhibit the dynamics of glass-forming liquids. Proc. Nat. Acad. Sci. **106**(19), 7735–7740 (2009)
20. A.P. Holt, P.J. Griffin, V. Bocharova, A.L. Agapov, A.E. Imel, M.D. Dadmun, J.R. Sangoro, A.P. Sokolov, Dynamics at the polymer/nanoparticle interface in poly (2-vinylpyridine)/silica nanocomposites. Macromolecules **47**(5), 1837–1843 (2014)
21. S. Cheng, S. Mirigian, J.-M.Y. Carrillo, V. Bocharova, B.G. Sumpter, K.S. Schweizer, A.P. Sokolov, Revealing spatially heterogeneous relaxation in a model nanocomposite. J. Chem. Phys. **143**(19), 194704 (2015)
22. S. Cheng, B. Carroll, W. Lu, F. Fan, J.-M.Y. Carrillo, H. Martin, A.P. Holt, N.-G. Kang, V. Bocharova, J.W. Mays et al., Interfacial properties of polymer nanocomposites: role of chain rigidity and dynamic heterogeneity length scale. Macromolecules **50**(6), 2397–2406 (2017)
23. F.W. Starr, T.B. Schrøder, S.C. Glotzer, Effects of a nanoscopic filler on the structure and dynamics of a simulated polymer melt and the relationship to ultrathin films. Phys. Rev. E **64**(2), 021802 (2001)
24. R.D. Priestley, D. Cangialosi, S. Napolitano, On the equivalence between the thermodynamic and dynamic measurements of the glass transition in confined polymers. J. Non-Cryst Solids **407**, 288–295 (2015)
25. S. Sen, Y. Xie, A. Bansal, H. Yang, K. Cho, L. Schadler, S. Kumar, Equivalence between polymer nanocomposites and thin polymer films: effect of processing conditions and molecular origins of observed behavior. Eur. Phys. J. Spec. Top. **141**(1), 161–165 (2007)
26. S. Cheng, S.-J. Xie, J.-M.Y. Carrillo, B. Carroll, H. Martin, P.-F. Cao, M.D. Dadmun, B.G. Sumpter, V.N. Novikov, K.S. Schweizer, A.P. Sokolov, Big effect of small nanoparticles: a shift in paradigm for polymer nanocomposites. ACS Nano **11**(1), 752–759 (2017)
27. S. Cheng, A.P. Holt, H. Wang, F. Fan, V. Bocharova, H. Martin, T. Etampawala, B.T. White, T. Saito, N.-G. Kang, M.D. Dadmun, J.W. Mays, A.P. Sokolov, Unexpected molecular weight effect in polymer nanocomposites. Phys. Rev. Lett. **116**, 038302 (2016)
28. F.W. Starr, J.F. Douglas, D. Meng, S.K. Kumar, Bound layers "cloak" nanoparticles in strongly interacting polymer nanocomposites. ACS Nano **10**(12), 10960–10965 (2016)
29. W. Zhang, J.F. Douglas, F.W. Starr, Effects of a "bound" substrate layer on the dynamics of supported polymer films. J. Chem. Phys. **147**(4), 044901 (2017)
30. W. Zhang, H. Emamy, B.A. Pazmiño Betancourt, F. Vargas-Lara, F.W. Starr, J.F. Douglas, The interfacial zone in thin polymer films and around nanoparticles in polymer nanocomposites. J. Chem. Phys. **151**(12), 124705 (2019)
31. J. Moll S.K. Kumar, Glass transitions in highly attractive highly filled polymer nanocomposites. Macromolecules **45**, 1131–1135 (2012)
32. S.E. Harton, S.K. Kumar, H. Yang, T. Koga, K. Hicks, E. Lee, J. Mijovic, M. Liu, R.S. Vallery, D.W. Gidley, Immobilized polymer layers on spherical nanoparticles. Macromolecules **43**, 3415–3421 (2010)
33. A.P. Holt, J.R. Sangoro, Y. Wang, A.L. Agapov, A.P. Sokolov, Chain and segmental dynamics of poly(2-vinylpyridine) nanocomposites. Macromolecules **46**, 4168–4173 (2013)

34. B. Natarajan, Y. Li, H. Deng, L.C. Brinson, L.S. Schadler, Effect of interfacial energetics on dispersion and glass transition temperature in polymer nanocomposites. Macromolecules **46**(7), 2833–2841 (2013)
35. S. Seethamraju, J. Obrzut, D. Jack F., J.W. Woodcock, J.W. Gilman, Quantifying fluorogenic dye hydration in an epoxy resin by noncontact microwave dielectric spectroscopy. J. Phys. Chem. B **123**(27), 5935–5941, 2020
36. F.W. Starr, T.B. Schrøder, S.C. Glotzer, Molecular dynamics simulation of a polymer melt with a nanoscopic particle. Macromolecules **35**(11), 4481–4492 (2002)
37. B.A.P. Betancourt, J.F. Douglas, F.W. Starr, Fragility and cooperative motion in a glass-forming polymer-nanoparticle composite. Soft Matter **9**(1), 241–254 (2013)
38. H. Emamy, S.K. Kumar, F.W. Starr, Diminishing interfacial effects with decreasing nanoparticle size in polymer-nanoparticle composites. Phys. Rev. Lett. **121**(20), 207801 (2018)
39. F.W. Starr, J.F. Douglas, Modifying fragility and collective motion in polymer melts with nanoparticles. Phys. Rev. Lett. **106**(11), 115702 (2011)
40. W. Zhang, J.F. Douglas, F.W. Starr, Why we need to look beyond the glass transition temperature to characterize the dynamics of thin supported polymer films. Proc. Nat. Acad. Sci. **115**(22), 5641–5646 (2018)
41. W. Zhang, F.W. Starr, J.F. Douglas, Collective motion in the interfacial and interior regions of supported polymer films and its relation to relaxation. J. Phys. Chem. B **123**(27), 5935–5941 (2019)
42. W. Zhang, J.F. Douglas, F.W. Starr, What does the T_g of thin polymer films really tell us? in *AIP Conference Proceedings*, vol. 1981 (AIP Publishing, 2018), p. 020083
43. G.S. Grest, K. Kremer, Molecular dynamics simulation for polymers in the presence of a heat bath. Phys. Rev. A **33**(5), 3628 (1986)
44. W. Zhang, J.F. Douglas, F.W. Starr, What does the instantaneous normal mode spectrum tell us about dynamical heterogeneity in glass-forming fluids? J. Chem. Phys. **151**(18), 184904 (2019)
45. P.Z. Hanakata, B.A. Pazmiño Betancourt, J.F. Douglas, F.W. Starr, A unifying framework to quantify the effects of substrate interactions, stiffness, and roughness on the dynamics of thin supported polymer films. J. Chem. Phys. **142**(23), 234907 (2015)
46. K. Kremer, G.S. Grest, Dynamics of entangled linear polymer melts: a molecular-dynamics simulation. J. Phys. Chem. **92**(8), 5057–5086 (1990)
47. P.G. Santangelo, C.M. Roland, Molecular weight dependence of fragility in polystyrene. Macromolecules **31**(14), 4581–4585 (1998)
48. J.-L. Barrat, J. Baschnagel, A. Lyulin, Molecular dynamics simulations of glassy polymers. Soft Matt. **6**, 3430–3446 (2010)
49. N.K. Hansoge, T. Huang, R. Sinko, W. Xia, W. Chen, S. Keten, Materials by design for stiff and tough hairy nanoparticle assemblies. ACS Nano. **12**(8), 7946–7958 (2018)
50. W. Xia, J. Song, D.D. Hsu, S. Keten, Side-group size effects on interfaces and glass formation in supported polymer thin films. J. Chem. Phys. **146**(20), 203311 (2017)
51. G. Yuan, C. Li, S.K. Satija, A. Karim, J.F. Douglas, C.C. Han, Observation of a characteristic length scale in the healing of glassy polymer interfaces. Soft Matt. **6**(10), 2153–2159 (2010)
52. B. Frieberg, E. Glynos, P.F. Green, Structural relaxations of thin polymer films. Phys. Rev. Lett. **108**(26), 268304 (2012)
53. W. Zhang, F.W. Starr, J.F. Douglas, Reconciling computational and experimental trends in the temperature dependence of the interfacial mobility of polymer films. J. Chem. Phys. **152**(12), 124703 (2020)
54. W. Zhang, J.F. Douglas, F.W. Starr, Dynamical heterogeneity in a vapor-deposited polymer glass. J. Chem. Phys. **146**(20), 203310 (2017)
55. S. Cheng, V. Bocharova, A. Belianinov, S. Xiong, A. Kisliuk, S. Somnath, A.P. Holt, O.S. Ovchinnikova, S. Jesse, H. Martin, T. Etampawala, M. Dadmun, A.P. Sokolov, Unraveling the mechanism of nanoscale mechanical reinforcement in glassy polymer nanocomposites. Nano Lett. **16**(6), 3630–3637 (2016)

56. S. Gong, Q. Chen, J.F. Moll, S.K. Kumar, R.H. Colby, Segmental dynamics of polymer melts with spherical nanoparticles. ACS Macro Lett. **3**(8), 773–777 (2014)
57. N. Jouault, J.F. Moll, D. Meng, K. Windsor, S. Ramcharan, C. Kearney, S.K. Kumar, Bound polymer layer in nanocomposites. ACS Macro Lett. **2**(5), 371–374 (2013)
58. J. Baschnagel, F. Varnik, Computer simulations of supercooled polymer melts in the bulk and in confined geometry. J. Phys.: Condens. Matt. **17**(32), R851 (2005)
59. P.Z. Hanakata, J.F. Douglas, F.W. Starr, Interfacial mobility scale determines the scale of collective motion and relaxation rate in polymer films. Nat. Commun. **5**, 4163 (2014)
60. P. Scheidler, W. Kob, K. Binder, Cooperative motion and growing length scales in supercooled confined liquids. EPL (Europhys. Lett.) **59**(5), 701 (2002)
61. J.E. Pye, K.A. Rohald, E.A. Baker, C.B. Roth, Physical aging in ultrathin polystyrene films: evidence of a gradient in dynamics at the free surface and its connection to the glass transition temperature reductions. Macromolecules **43**(19), 8296–8303 (2010)
62. D.S. Simmons, An emerging unified view of dynamic interphases in polymers. Macromol. Chem. Phys. **217**(2), 137–148 (2016)
63. S.N. Jamadagni, R. Godawat, S. Garde, Hydrophobicity of proteins and interfaces: Insights from density fluctuations. Ann. Rev. Chem. Biomol. Eng. **2**, 147–171 (2011)
64. S. Sarupria, S. Garde, Quantifying water density fluctuations and compressibility of hydration shells of hydrophobic solutes and proteins. Phys. Rev. Lett. **103**(3), 037803 (2009)
65. R. Godawat, S.N. Jamadagni, S. Garde, Characterizing hydrophobicity of interfaces by using cavity formation, solute binding, and water correlations. Proc. Nat. Acad. Sci. **106**(36), 15119–15124 (2009)
66. P.Z. Hanakata, J.F. Douglas, F.W. Starr, Local variation of fragility and glass transition temperature of ultra-thin supported polymer films. J. Chem. Phys. **137**(24), 244901 (2012)
67. B.A. Pazmiño Betancourt, J.F. Douglas, F.W. Starr, String model for the dynamics of glass-forming liquids. J. Chem. Phys. **140**(20), 204509 (2014)
68. J.F. Douglas, J. Dudowicz, K.F. Freed, Does equilibrium polymerization describe the dynamic heterogeneity of glass-forming liquids? J. Chem. Phys. **125**(14), 144907 (2006)
69. X. Qin, W. Xia, R. Sinko, S. Keten, Tuning glass transition in polymer nanocomposites with functionalized cellulose nanocrystals through nanoconfinement. Nano lett. **15**(10), 6738–6744 (2015)
70. W. Xia, D.D. Hsu, S. Keten, Dependence of polymer thin film adhesion energy on cohesive interactions between chains. Macromolecules **47**(15), 5286–5294 (2014)
71. W. Xia, X. Qin, Y. Zhang, R. Sinko, S. Keten, Achieving enhanced interfacial adhesion and dispersion in cellulose nanocomposites via amorphous interfaces. Macromolecules **51**(24), 10304–10311 (2018)
72. S.Y. Kim, C.F. Zukoski, Super-and sub-einstein intrinsic viscosities of spherical nanoparticles in concentrated low molecular weight polymer solutions. Soft Matt. **8**(6), 1801–1810 (2012)
73. J.-M.Y. Carrillo, S. Cheng, R. Kumar, M. Goswami, A.P. Sokolov, B.G. Sumpter, Untangling the effects of chain rigidity on the structure and dynamics of strongly adsorbed polymer melts. Macromolecules **48**(12), 4207–4219 (2015)
74. C.L. Jackson G.B. McKenna, The melting behavior of organic materials confined in porous solids. J. Chem. Phys. **93**, 9002–9011 (1990)
75. J.L. Keddie, R.A.L. Jones, R.A. Cory, Size-dependent depression of the glass-transition temperature in polymer-films. Europhys. Lett. **27**, 59–64 (1994)
76. J.L. Keddie, R.A.L. Jones, R.A. Cory, Interface and surface effects on the glass-transition temperature in thin polymer-films. Faraday Discuss. **98**, 219–230 (1994)
77. P. Rittigstein, R.D. Priestley, L.J. Broadbelt, J.M. Torkelson, Model polymer nanocomposites provide an understanding of confinement effects in real nanocomposites. Nat. Mater. **6**, 278–282 (2007)
78. D.S. Fryer, R.D. Peters, E.J. Kim, J.E. Tomaszewski, J.J. De Pablo, P.F. Nealey, C.C. White, W.-L. Wu, Dependence of the glass transition temperature of polymer films on interfacial energy and thickness. Macromolecules **34**(16), 5627–5634 (2001)

79. J. Torres, P. Nealey, J. De Pablo, Molecular simulation of ultrathin polymeric films near the glass transition. Phys. Rev. Lett. **85**(15), 3221 (2000)
80. S. Peter, H. Meyer, J. Baschnagel, Molecular dynamics simulations of concentrated polymer solutions in thin film geometry. I. equilibrium properties near the glass transition. J. Chem. Phys. **131**(1), 014902 (2009)
81. W.L. Merling, J.B. Mileski, J.F. Douglas, D.S. Simmons, The glass transition of a single macromolecule. Macromolecules **49**(19), 7597–7604 (2016)
82. H. Lu, S. Nutt, Restricted relaxation in polymer nanocomposites near the glass transition. Macromolecules **36**, 4010–4016 (2003)
83. B. Ash, L. Schadler, R. Siegel, Glass transition behavior of alumina/polymethylmethacrylate nanocomposites. Mater. Lett. **55**(1–2), 83–87 (2002)
84. C. Yu, S. Granick, Revisiting polymer surface diffusion in the extreme case of strong adsorption. Langmuir **30**, 14538–14544 (2014)
85. A. Sargsyan, A. Tonoyan, S. Davtyan, C. Schick, The amount of immobilized polymer in pmma sio2 nanocomposites determined from calorimetric data. Eur. Polym. J. **43**(8), 3113–3127 (2007)
86. S. Napolitano, M. Wubbenhorst, Dielectric signature of a dead layer in ultrathin films of a nonpolar polymer. J. Phys. Chem. B **111**, 9197–9199 (2007)
87. M.J. Burroughs, S. Napolitano, D. Cangialosi, R.D. Priestley, Direct measurement of glass transition temperature in exposed and buried adsorbed polymer nanolayers. Macromolecules **49**(12), 4647–4655 (2016)
88. G.N. Toepperwein, R.A. Riggleman, J.J. de Pablo, Dynamics and deformation response of rod-containing nanocomposites. Macromolecules **45**(1), 543–554 (2012)
89. E.C. Glor, Z. Fakhraai, Facilitation of interfacial dynamics in entangled polymer films. J. Chem. Phys. **141**(19), 194505 (2014)
90. K. Paeng, R. Richert, M. Ediger, Molecular mobility in supported thin films of polystyrene, poly (methyl methacrylate), and poly (2-vinyl pyridine) probed by dye reorientation. Soft Matt. **8**(3), 819–826 (2012)
91. S. Mirigian, K.S. Schweizer, Theory of activated glassy relaxation, mobility gradients, surface diffusion, and vitrification in free standing thin films. J. Chem. Phys. **143**(24), 244705 (2015)
92. B.A.P. Betancourt, P.Z. Hanakata, F.W. Starr, J.F. Douglas, Quantitative relations between cooperative motion, emergent elasticity, and free volume in model glass-forming polymer materials. Proc. Nat. Acad. Sci. **112**(10), 2966–2971 (2015)
93. G.J. Papakonstantopoulos, K. Yoshimoto, M. Doxastakis, P.F. Nealey, J.J. de Pablo, Local mechanical properties of polymeric nanocomposites. Phys. Rev. E **72**, 031801 (2005)
94. M. Zobel, R.B. Neder, S.A. Kimber, Universal solvent restructuring induced by colloidal nanoparticles. Science **347**(6219), 292–294 (2015)
95. N.G. Perez-de Eulate, M. Sferrazza, D. Cangialosi, S. Napolitano, Irreversible adsorption erases the free surface effect on the t g of supported films of poly (4-tert-butylstyrene). ACS Macro Lett. **6**(4), 354–358 (2017)
96. G. Adam, J.H. Gibbs, On the temperature dependence of cooperative relaxation properties in glass-forming liquids. J. Chem. Phys. **43**(1), 139–146 (1965)
97. F.W. Starr, J.F. Douglas, S. Sastry, The relationship of dynamical heterogeneity to the adam-gibbs and random first-order transition theories of glass formation. J. Chem. Phys. **138**(12), 12A541 (2013)
98. T. Salez, J. Salez, K. Dalnoki-Veress, E. Raphaël, J.A. Forrest, Cooperative strings and glassy interfaces. Proc. Nat. Acad. Sci. **112**(27), 8227–8231 (2015)

Chapter 6
Multiscale Modeling Examples: New Polyelectrolyte Nanocomposite Membranes for Perspective Fuel Cells and Flow Batteries

Soumyadipta Sengupta, Alexey V. Lyulin, Georgios Kritikos, Konstantinos Karatasos, Arun Venkatnathan, Rakesh Pant, and Pavel V. Komarov

Abstract Renewable energy production from fuel cells and energy storage in flow batteries are becoming increasingly important in the modern energy transition. Both batteries use polyelectrolyte membranes (PEMs) to allow proton transport. In this chapter, both PEMs and PEMs-based nanocomposites have been discussed using various simulational approaches. A coarse-grained model of a Nafion film capped by the substrates with variable wettability has been used to simulate nanocomposites

---

S. Sengupta · A. V. Lyulin (✉)
Group Theory of Polymers and Soft Matter and Center for Computational Energy Research, Department of Applied Physics, Technische Universiteit Eindhoven, 5600 MB Eindhoven, The Netherlands
e-mail: a.v.lyulin@tue.nl

S. Sengupta
e-mail: s.sengupta@tue.nl

G. Kritikos · K. Karatasos
Department of Chemical Engineering, Aristotle University of Thessaloniki, 54124 Thessaloniki, Greece
e-mail: kritikgio@gmail.com

K. Karatasos
e-mail: kkaratas@cheng.auth.gr

A. Venkatnathan · R. Pant
Department of Chemistry and Center for Energy Science, Indian Institute of Science Education and Research, Dr. Homi Bhabha Road, Pashan, Pune, Maharashtra 411008, India
e-mail: arun@iiserpune.ac.in

R. Pant
e-mail: rakesh.pant@students.iiserpune.ac.in

P. V. Komarov
A.N. Nesmeyanov Institute of Organoelement Compounds of Russian Academy of Sciences, Vavilova st. 28, 11999 Moscow, Russia
e-mail: pvkomarov@gmail.com

Tver State University, Sadoviy per. 35, 170002 Tver, Russia

V. V. Ginzburg and L. M. Hall (eds.), *Theory and Modeling of Polymer Nanocomposites*, Springer Series in Materials Science 310,
https://doi.org/10.1007/978-3-030-60443-1_6

of PEMs by classical molecular-dynamics (MD) method. Classical MD modeling results have also been reviewed for a PEM-graphene oxide nanocomposite internal structure and dynamics. Ab-initio simulations have been implemented to describe the proton transfer pathways in anhydrous PEMs. Finally, the large-scale mesoscopic simulations have been introduced to shed light on the water domain features present in the hydrated PEMs. A brief description of polybenzimidazole membrane as electrolyte and Ionic Liquids as dopants for fuel cells is also presented.

6.1 Introduction

In this chapter, the application of the multiscale modeling methods has been discussed for a very special class of the nanocomposites—polymer electrolyte membranes (PEMs), one of the key elements of the modern fuel cells [1, 2] and flow batteries [3]. Low-temperature fuel cells (FCs) based on polymer ion-exchange membranes powered by hydrogen or methanol are often considered as a replacement for traditional power sources in motor transport and household devices [1]. Their main advantages are portability, low operating temperature (<100 °C), and simplicity of the internal design.

The fuel cell design is quite simple, see Fig. 6.1. It consists of two electrodes and a solid electrolyte. The following types of FC are distinguished: alkaline, carbonate, solid oxide, with a polymeric proton-exchange membrane and phosphoric acid. With respect to operating temperature, the FCs are divided into low-temperature FC (operating temperature <100 °C), medium-temperature FC (<220 °C), and high-temperature FC (600–1000 °C).

In the low-temperature hydrogen fuel cells, the role of electrolyte is played by an ion-exchange membrane. In general, PEM represents a thin film of a polymer with ionogenic groups. The role of membrane is to separate the flow of fuel from

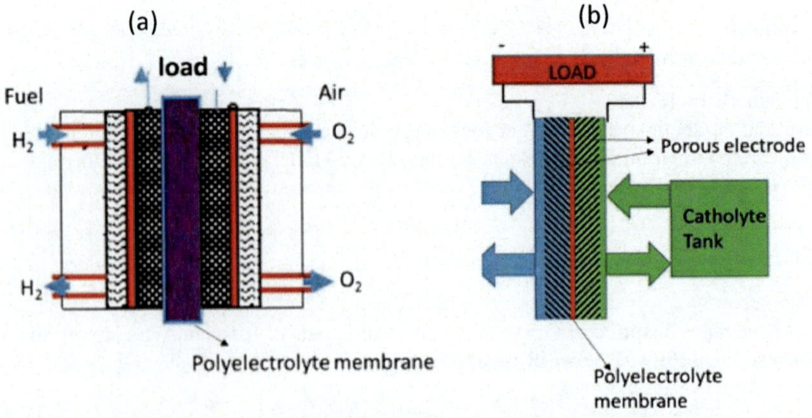

Fig. 6.1 Schematics of the **a** a hydrogen–oxygen fuel cell and **b** a redox flow battery

the oxidizer and to serve as an electron insulator. Hydrogen molecules are oxidized at the anode ($H_2 \rightarrow 2H^+ + 2e^-$) and the formed protons are transferred to the cathode through the membrane [1]. The released electrons pass through an external circuit and are transmitted to the cathode, creating electric current. At the cathode block, in the presence of the catalyst, the reaction $O_2 + 4e^- \rightarrow 2O^{2-}$ occurs. The cycle of the recombination electrochemical transformations is completed with the reaction of oxygen ions with protons; as a result, the water molecules are formed, $O^{2-} + 2H^+ \rightarrow H_2O$. The theoretical value of the electromotive force (EMF) of a hydrogen–oxygen FC at 25 °C is 1.23 V; the no-load voltage is about 1 V. Such FCs have the lowest operating temperature of 40–80 °C, which is both their advantage and disadvantage. The simple design of the low-temperature FCs allows them to be made compact, but, unfortunately, at these temperatures, the platinum catalyst is very sensitive to the degree of the fuel contamination. Therefore, research aimed to improve the efficiency of such FCs is performed in two directions: the development of the highly efficient membranes and the search for chemically resistant and cheap catalysts.

It is assumed that for use in FCs, the membrane conductivity should be above 0.001 S/cm [1–3]. Since PEMs perform the functions of a gas separator and an electrical insulator, it is necessary that these membranes possess low gas permeability and electrical conductivity. In addition, they should have sufficient mechanical strength. This is required for operation at high pressures. Furthermore, the membrane should not be destroyed as a result of electrodes pressing on. Since during the electrochemical reactions peroxide compounds, such as H_2O_2, OH^-, and HOO^- can be formed, the membranes should demonstrate high chemical resistance to oxidation and hydrolysis [1–3].

PEMs are also used in redox flow batteries (RFB) as shown in Fig. 6.1b. The catholyte and the anolyte form a redox couple in which the anolyte gets oxidized and releases electron into the external circuit. The electron travels through the external circuit and reaches the cathodic side to reduce the catholyte. There also exist some protons released during this process which travel toward the cathodic side to maintain the charge balance in the system; the PEM helps in transport of these protons. The speed of charging and discharging is affected by the ease of proton transport across the PEM. The PEM also prevents the catholyte and the anolyte from crossing over and mixing. Such mixing would reduce the efficiency of the battery and could also lead to catastrophic power discharge and fire.

When water is added to the PEM, the microphase separation of the hydrophilic and hydrophobic segments of the polymer chains occurs. It leads to the formation of water domains localized near the hydrophilic groups. As a result, various hydrated proton complexes, i.e., hydronium ions (H_3O^+), Zundel ($H_5O_2{}^+$) and Eigen ($H_9O_4{}^+$) cations are formed. With the increasing amount of water, the water domains merge into a continuous network of water channels that provide proton transport through the membrane volume as shown in Fig. 6.2. At the same time, the hydrophobic domains enhance the mechanical strength of the membranes. They are formed from the fluorinated, aromatic or aliphatic segments of macromolecules. The sulfonic acid groups ($-SO_3H$) and the phosphate groups ($-PO_3H_2$) are the most commonly used

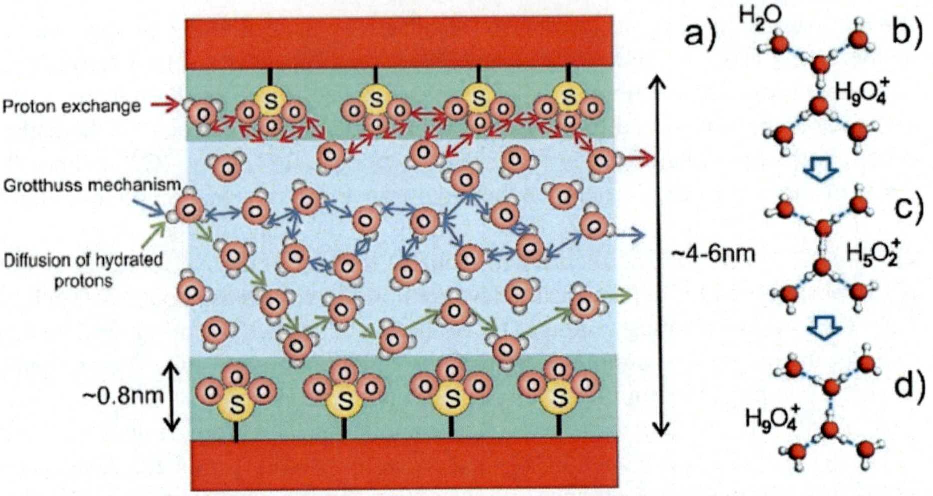

Fig. 6.2 The mechanisms of proton transport: **a** in the ion channel, **b–d** the stages of proton transfer

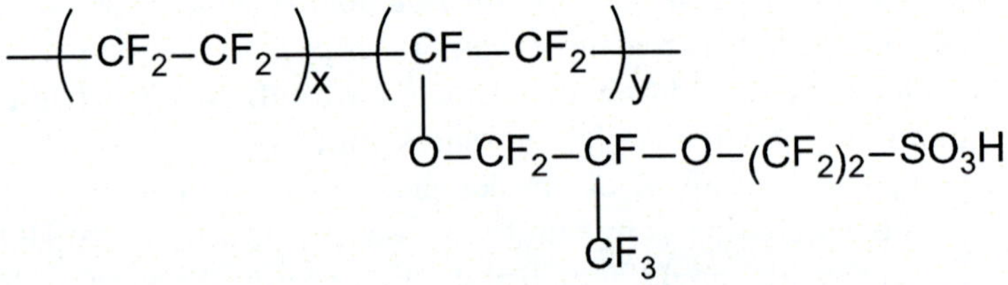

Fig. 6.3 The general structural formula of the Nafion® family polymers

ionic acid-type groups, dissociating with the release of a proton. Sulfonated polymers are more efficient at high degree of hydration, while phosphorylated polymers can function with a low water content and have high heat resistance.

Conventionally, the Nafion derivatives (see Fig. 6.3) are the copolymers of tetrafluoroethylene and perfluorinated vinyl ether side chains with sulfonic acid groups at ends. Membranes based on such polymers are durable, resistant to oxidation and have a record high ionic conductivity (up to 0.1 S/cm at temperatures below 80 °C). The disadvantages of Nafion are the narrow range of the operating temperatures (<90 °C). Limitation of the temperature range leads to the need for utilization of highly purified hydrogen, since even small CO impurities impart a destructive effect on the platinum catalyst. Increasing the operating temperature to above 120 °C allows to shift the equilibrium of the CO and H_2 sorption on the catalyst toward hydrogen. This prevents the poisoning of the catalyst, but it is impossible to raise the temperature above 90 °C. Upon increasing temperature, the water content in the ion channels decreases due to water evaporation. This causes the diameter of the water channels to narrow. If the water content is below the percolation threshold, the network of water channels is disrupted. It is accompanied by a sharp drop in the membrane conductivity. The ionic conductivity of the dry polymer is not sufficient for the operation of FC. The efficient transfer of protons can be ensured only if constant hydrostatic level is maintained. This requires introducing additional structural elements and increases the total cost of the batteries.

In [4, 5] the modification of Nafion by silicon dioxide nanofillers has been studied. Such nanocomposites demonstrate the ability to function effectively at temperatures up to 140 °C, as their ability to retain water at high temperatures increases. In addition to the studies on the modifications of Nafion, research is actively conducted to find alternative solid polyelectrolytes based on mass-produced polymers with an aryl skeleton, such as polycarbonates, polyesters, polystyrenes, and polysulfonic acids [6]. Membranes based on polybenzimidazole (PBI) modified with phosphate groups possess high thermal stability and conductivity in dry conditions. Their operating temperature range is 160–180 °C, which allows their use in medium-temperature FCs. However, upon contact with water, the acid–base complex is destroyed, and the phosphoric acid is washed out into catalytic and gas-diffusion layers. This leads to the blockage of the gas pores, degradation of the electrodes and of the whole gas pipeline system. At the same time, the proton conductivity of the membrane decreases sharply. Another relatively simple method to obtain ion-exchange membranes can be implemented by modifying the polymer matrix with side chains followed by sulfonic acids [7]. This method to obtain ionomers requires lower costs than when using Nafion. The proton conductivity in this case is 0.11 S/cm at 100% hydration and $T = 298$ K. The main disadvantage of such membranes is a high degree of swelling, greatly exceeding similar values for Nafion. There have been also attempts to obtain high-performance membranes based on polysulfonic acids, polyethersulfonic acids, polyimides, and sulfonated poly(ether-ether ketone) (SPEEK) [8–14].

The exact mechanism of the PEM protons transfer is the highly debated subject; the proton transport is described in some detail by Agmon [15]. Two possible mechanisms of the proton transfer through the membrane are discussed. Sulfonic acid groups, which are concentrated mainly on the walls of the channels, dissociate in the presence of water to form ion pairs: $-SO_3H + H_2O \rightarrow -SO_3^- + H_3O^+$. According to the classical Grotthuss mechanism, the protons are transferred through a continuous proton-exchange reaction: $H_2O + H^+ \leftrightarrow H_3O^+$, see Fig. 6.2a. Other mechanisms suggest the protons transfer as a result of the hydronium ions diffusion and hops between adjacent sulfonic acid groups [16]. In this case, the rate of proton exchange depends on the difference in the probabilities of proton capture by water molecules and by sulfonic acid groups.

Figure 6.2b shows the hydronium ion H_3O^+, surrounded by solvated water molecules, forming the Eigen–cation $H_9O_4^+$ [17]. It is hydrated as well, but the hydrogen bonds in the second coordination sphere of H_3O^+ are weaker than in the first sphere. It is assumed that the proton exchange is initiated by a random change in one of the hydrogen bonds between the water molecules forming the second coordination sphere of hydronium. This causes a rapid change in the orientation of water molecules (over a time of the order of 1 ps), which can lead to the breaking of one of the hydrogen bonds. In the remaining cluster, a quick rearrangement of the bond lengths and angles occurs forming a complex $H_2O \cdots H^+ \cdots H_2O$ Zundel cation ($H_5O_2^+$) [18] as shown in Fig. 6.2c. In this complex, in turn, as a result of fluctuations in the bond lengths, the proton can join one of the two water molecules forming a new hydronium ion as shown in Fig. 6.3d. Next, the hydration shell is formed, and the process repeats.

Ab-initio molecular-dynamics (MD) calculations [19–21] confirm that proton transfer is realized through a continuous chain of transformations of Eigen and Zundel cationic complexes. In the empirical valence bonds model, using the example system of a small Nafion/water system (~5000 atoms and 40 excess protons), a detailed study of the Grotthuss mechanism was performed [22, 23]. This model allows the hopping movement of protons from hydronium ions to the nearest water molecules. In a hybrid model using the classical and the quantum–mechanical methods [24], the proton migration was studied for a system containing a single Nafion oligomeric chain with 10 SO_3^- groups, the same number of H_3O^+ ions and 156 water molecules. In [21], using extensive 120-ps-long density-functional theory (DFT)-based simulations of charge migration in the 1200-atom model of the hydrophilic Nafion nanochannel, a bimodality of the van Hove autocorrelation function $G_s(\mathbf{r}, t)$ has been observed. This provides direct evidence for the Grotthuss hopping mechanism being a significant contributor to the proton conductivity.

The PEM proton conductivity strongly depends on the water content in the membrane. The amount of water in the membrane is often characterized by the volume fraction φ and the hydration parameter λ (the number of water molecules per sulfonic acid group). Nafion becomes a proton conductor at $\lambda \approx 2$–5 [21, 25]. In this case, the fraction of hydrated sulfonic acid groups is very small, and they should exist only within segmental clusters.

Hydrophilic membrane domains containing $-SO_3H$ groups have a high affinity to water. Even after air drying, the proton of the sulfonic acid group continues to bind two water molecules. Upon contact with water, the degree of the membrane hydration increases dramatically due to the absorption of water [26–29], which leads to the membrane swelling and to an extensive reformation of its structure.

The presence of well-defined maxima in the intensity of small-angle X-ray scattering (SAXS) data for sulfonic acid membranes shows that their structure must contain ordered water clusters [29, 30]. Based on the analysis of experimental data, Gierke proposed some generalized model of the structure of perfluorinated sulfonic acid membranes [28, 31] as shown in Fig. 6.4. According to this model, the membrane consists of an array of linked clusters connected by narrow channels. On the periphery of the clusters, which in a first approximation have a spherical shape, sulfonic acid groups are located. As a result, the surface energy is minimized and the contact of the hydrophobic segments of the matrix chains with water is limited. The internal volume of the cluster is filled with water molecules and its proton complexes, which are formed during the dissociation of the functional groups. The migration of the proton complexes occurs from cluster to cluster via the extensive network of channels, which can break off with low moisture content. If this happens, the membrane loses its transport properties. Even though this model is widely used, other possible variants of the membrane structure, which also allow the description of the SAXS

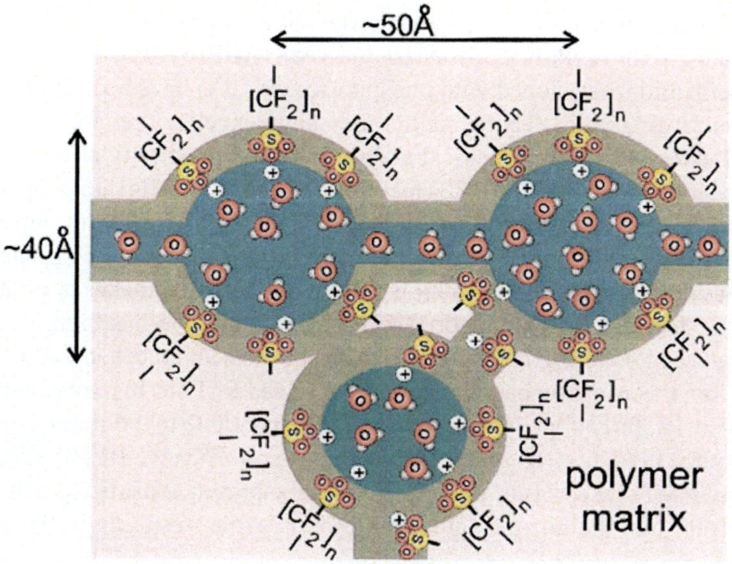

Fig. 6.4 The cluster-channel model of Nafion, proposed by Gierke [28]

results, are present in the literature. One could mention here Fujimura's modified core–shell model [32, 33], Dreyfus's local ordering model [34], Litt's lamellar model [35], and Rubat's rod-like aggregates model [36]. A common feature of these models is adoption of the fact that sulfonic acid groups in a hydrophobic matrix form a penetrating network of linked ionic clusters that swell in a polar solvent and provide ion transport of the membrane. At the same time, each of the models provides its description of the topology of the structure of the ion channels network. In addition to these models, Schmidt–Rohr and Chen [37] proposed parallel cylindrical channels inside a hydrated Nafion membrane. These channels were lined-up with sulfonic acid groups and were filled with water and protons.

Based on the results of X-ray and neutron scattering experiments on samples of Nafion membranes with different water contents, as well as on the basis of energy considerations, Gebel [26] proposed a concept to describe the process of swelling and dissolving ionomers. It is assumed that the dry membrane contains separate isolated ion clusters with a diameter of about 1.5 nm. They swell when water enters the membrane. When the water content in the membrane reaches a percolation threshold λ^*, a linked network of the water clusters forms in the membrane, and the membrane becomes a conductor of ions. With further absorption of water, an inversion of the structure occurs, and, as a result, the system structure takes the form of a network of intertwined rods. A further increase in the water content leads to the complete dissolution of the membrane with the formation of a colloidal dispersion of individual rod-like fibers.

PEMs which require humidification for proton conduction limit the deployment of fuel cells with operating temperatures up to the boiling point of water. The operation of these fuel cells at higher temperature reduces catalyst poisoning, and hence, suitable alternatives as proton carriers have been explored [38, 39]. Aromatic membranes such as Poly (2,5-benzimidazole) (ABPBI), Poly[2,2-(m-phenylene)-5,5-bibenzimidazole) doped with phosphoric acid (PA) have been explored for high-temperature fuel cells. These membranes have excellent properties such as high thermal, chemical and mechanical stability, high conductivity at elevated temperatures and low cost [40, 41]. ABPBI membrane has nitrogen (N) atom on the imidazole ring which serves as a proton acceptor [42] and can interact with dopants. ABPBI shows higher affinity toward PA as compared to PBI, due to the absence of phenyl ring. PA is amphoteric and has a high boiling point, which makes it a suitable proton conductor for fuel cell applications [41]. PBI membrane has high glass-transition temperature (430 °C), excellent chemical resistance, and mechanical strength which stems from the aromatic backbone. The PA-doped PBI has high proton conductivity (0.07 S/cm) in fully doped conditions and is comparable to the state-of-the-art PFSA membranes. In PBI, high conductivity extends up to 200 °C. IR [40, 43, 44], Raman [45], and NMR spectra [46] have confirmed the presence of strong hydrogen bonds between PA and the nitrogen atom on the imidazole, resulting in the formation of $H_2PO_4^-$.

Ionic liquids (ILs) are considered as a promising alternative to water due to their excellent properties such as high thermal stability, low vapor pressure, wide electrochemical window, and high anhydrous ionic conductivity. These properties render ILs promising dopants in electrolytic membranes, and, thus, enabling operation of PEMFC at higher operating temperatures (>100 °C). A protic ionic liquid (PIL) is defined as a combination of a Brønsted base and Brønsted acid, where the base accepts a proton from the acid [47]). The use of PILs as proton carriers has been explored for high-temperature fuel cells [47–49]. A wide range of PILs as a prospective proton conducting material and underlying proton conduction mechanism has been investigated with excess of one constituent, or in PIL-doped membranes [47, 50–54]. Sood et al. [53] demonstrated that IL (triethylammmonium trifluorosulfonate, TFTEA) doping in Nafion (neutralized with triethylamine, TEA) enhances the ionic conductivity anhydrous conditions. The conductivity of IL-doped membrane increases with wt% of IL.

In the next parts of this chapter, computer simulations of some of these PEMs and their nanocomposites are discussed in more detail.

6.2 Multiscale Modeling of Polyelectrolyte Membranes and Their Transport Properties

By now, many theoretical studies on the properties of Nafion have been published [21, 55–57]. However, the studies of other ion-exchange polymers, in particular, sulfonated polyheteroarylenes, are practically absent. Although Nafion and polyheteroarylenes differ significantly in their chemical structure, the formation of the structure of sulfonated polymers must follow general principles. Therefore, the analysis of the accumulated results about Nafion can be useful in the study of other sulfonated acid polymers.

The earliest studies were carried out in the framework of semi-empirical methods with a large number of parameters, adjusted according to experimental data [36, 58–60]. They drew general conclusions regarding the origins of the microphase separation in ionomers into hydrophilic and hydrophobic domains, but they did not provide a detailed description of the membrane structure.

The next group of theoretical approaches includes methods based on atomistic simulations [61–68]. They allow to simulate the microphase separation of polar and non-polar segments of polymer chains and to study the diffusion of water and hydronium. This helps to understand how the local ordering occurs in PEMs, depending on the water content and on the role of the side chains of fluoropolymers containing sulfonic acid groups.

The most rigorous level of consideration was achieved using the quantum–mechanical methods [19, 20, 69–71]; accurate data have been obtained on the properties of different molecular groups, the dissociation of $-SO_3H$, the conformations of fragments of the Nafion chain, and the mechanism of the proton transport.

The experimental and theoretical studies of ion-exchange membranes show that in order to explain the structural features of the water channels of proton-conducting membranes, it is necessary to consider the morphology of ionomers on spatial and temporal scales that significantly exceed the capabilities of atomistic methods. In this case, mesoscopic approaches, such as cellular automaton method [72], MC/PRISM (a combination of Monte Carlo methods and integral equations) [73], the dissipative particle dynamics (DPD) method [74, 75], and a dynamic version of the density-functional theory method (DDFT) [76, 77], can provide the most detailed information. The predictions obtained in the framework of the aforementioned approaches are in a good agreement with pertinent experimental results and with findings from the atomistic modeling. However, neither DPD nor DDFT is suitable for studying the mechanisms of ion transfer.

As has been mentioned before, the Nafion-based nanocomposites fabricated by mixing with nanoparticles are often used in fuel cells and flow batteries to improve proton conductivity, water retention levels and selectivity to the motion of protons only. Classical MD simulation studies of such nanocomposites are appropriate to understand the internal water cluster structure and transport characteristics. Nafion nanocomposites containing highly hydrophilic nanoparticles, like modified silica, have exhibited higher proton conductivity as compared to that of bulk Nafion [78]. Methanol crossover in direct methanol fuel cells and vanadium crossover in flow batteries have been reduced by using such Nafion nanocomposites [79, 80]. In [80], the reduction of methanol crossover and increased proton conduction in a Nafion-modified carbon nanotube (CNT) nanocomposite has been observed. It was proposed that such effects were due to the formation of long hydrophilic pathways along the modified CNTs. All these experiments point toward the need for a better understanding of Nafion–nanoparticle interfaces.

Nafion nanocomposites show a large variation in nanoparticle sizes (5–75 nm) [81, 82]. Therefore, Nafion–nanoparticle interface could be modeled using a coarse-grained description of a Nafion film-flat substrate interface with lateral dimensions in the range of 4–6 nm. Nanocomposites may possess different filler densities which affect the average interparticle distances. The average interparticle distance in a Nafion/titania nanocomposite was found to be 9 nm [81]. Also, previous experiments have sought a more thorough understanding of the structure in Nafion films of thickness less than 10 nm, because this is exactly the range of the interparticle distances commonly met in catalyst layers [83]. Therefore, film thicknesses in the range of 6–11 nm can be chosen to perform the classical coarse-grained MD simulations.

Different types of nanoparticles like silica, zirconia, and modified carbon nanotubes [78, 81] with varying wettabilities have been used in Nafion nanocomposites. In addition, Nafion also exists in catalyst layers between carbon support and platinum nanoparticles [82, 84].

In the next paragraph, a hydrated Nafion film capped by substrates of varying hydrophilicity will be discussed in some detail. Such a model, introduced in [78] initially for the elastomer-based nanocomposites, has been recently used to provide useful insights into the interfacial interactions in Nafion nanocomposites.

6.3 Confined Film Model of a Nanocomposite Membrane

Classical molecular-dynamics (MD) techniques were used for thr simulations. A polymer nanocomposite is comprised by fillers/nanoparticles dispersed inside the polymer matrix as shown in Fig. 6.5a. The representative volume element (RVE) modeled in the simulations is the polymer material present between any two nanoparticles as a confined film. This RVE was modeled by confining 17 Nafion chains along with water molecules and hydronium ions between structureless walls of tuneable hydrophilicity [85, 86] as shown in Fig. 6.5b. A fixed moderate hydration level of $\lambda = 15$ was used for the simulations. The walls represent the nanoparticle surfaces of variable hydrophilicity.

An integrated Lennard–Jones potential [87],

$$E = \epsilon\left(\frac{2}{15}(\sigma/r)^9 - (\sigma/r)^3\right), r < r_c \tag{6.1}$$

has been used to simulate structureless walls at the top and at the bottom of the simulation box [85]. In (6.1) r_c is the cut-off radius, and the interaction energy ϵ of the walls with the hydrophobic parts of the system (i.e., the Nafion molecule except the sulfonic acid groups) was $\varepsilon_{phob} = 0.25$ kcal/mol. Five different values of interaction energy ε_{phyl} of the walls with the hydrophilic part of the system were used in the simulations. Henceforth, walls with $\varepsilon_{phyl} = 0.25, 0.50$ kcal/mol will be referred to as low hydrophilicity (LH) walls and those with $\varepsilon_{phyl} = 1.20, 1.50, 2.00$ kcal/mol will be referred to as high hydrophilicity (HH) walls. In both cases, ϵ_{phob} has been

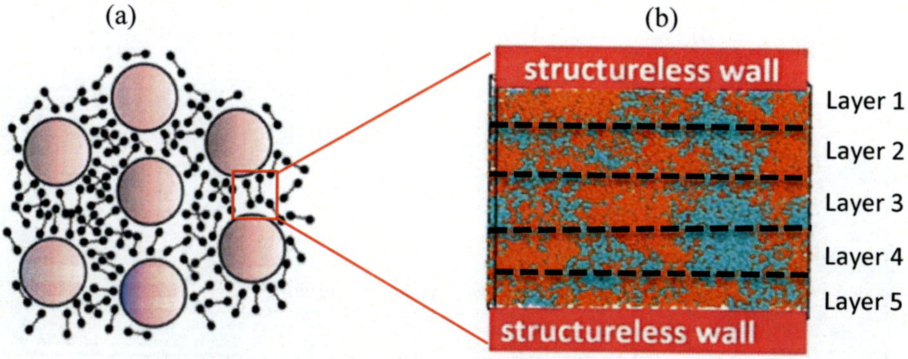

Fig. 6.5 **a** A schematic representation of a nanocomposite membrane. The spheres are the nanoparticles while the black bead-springs are the hydrated Nafion. The hydrated Nafion volume bound between two nanoparticles as shown by the red square is the representative volume element (RVE). This RVE is modeled as a capped film as shown in **b** hydrated Nafion film is capped by two structureless walls. *Z*-axis is the direction perpendicular to the walls and *X*- and *Y*-axes are parallel to the walls. *Z*-direction has fixed boundaries and the film is periodic in *X*- and *Y*-directions. Blue color represents water molecules and hydronium ions; orange color is used for Nafion molecules. Water transport is analyzed in the five different layers of the film [66]

fixed at 0.25 kcal/mol [66, 85, 86]. Three different film thicknesses of 6.3, 8.7, and 11.5 nm were simulated for each of the $\varepsilon_{\text{phyl}}$ values. The film thickness was varied in the *Z*-direction (Fig. 6.5b). The thickness variation modeled the variation of the filler fraction in a nanocomposite, i.e., higher film thickness corresponded to lower filler fraction and vice versa.

Figure 6.6 shows the simulated Nafion film snapshots at the end of the production runs for five different values of wall hydrophilicity. For low values of wall hydrophilicity ($\varepsilon_{\text{phyl}} = 0.25, 0.50$ kcal/mol), there is very little accumulation of water molecules near the walls. However, for the high hydrophilicity walls ($\varepsilon_{\text{phyl}} = 1.20, 1.50, 2.00$ kcal/mol), a noticeable accumulation of water near the walls is observed.

Due to sufficiently high hydration levels, water channels in Nafion form a percolated network [24] which facilitates a proton transport. However, such a percolated network can also allow unwanted crossover of methanol and vanadium ions. Nafion nanocomposites have been shown to reduce the crossover of methanol [80] and vanadium ions [88]. Therefore, water cluster analysis was usually done for the different values of wall hydrophilicity. All the water cluster analysis shown here is for a cut-off distance of 3.7 Å averaged over 3 ns of the simulated physical time [66].

Water cluster distribution at a fixed film thickness and different values of wall hydrophilicity is shown in Fig. 6.7. The cluster distributions for the LH walls ($\varepsilon_{\text{phyl}} = 0.25, 0.50$ kcal/mol) are almost the same as that of the bulk hydrated Nafion ($\lambda = 15$) cluster distribution. The large clusters (cluster size close to 2400) for the HH walls ($\varepsilon_{\text{phyl}} = 1.20, 1.50, 2.00$ kcal/mol) occur less frequently as compared to the LH walls. In addition, there is an emergence of clusters in the 900–1500 size range for the HH walls. This shows that the cluster sizes reduced considerably with increasing

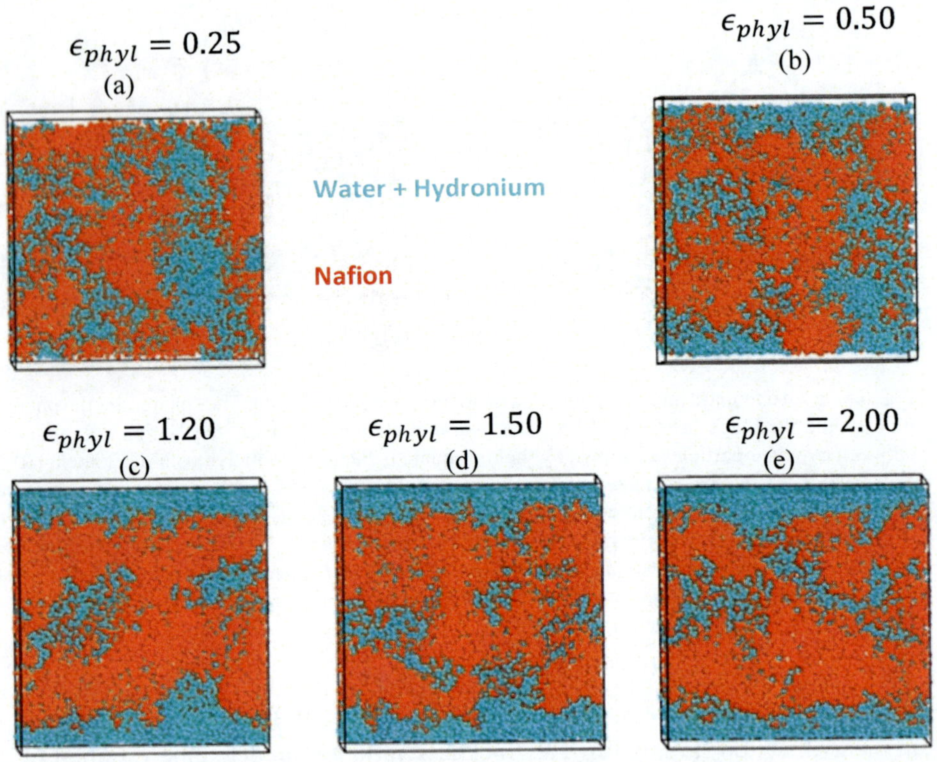

Fig. 6.6 Snapshots for **a** $\varepsilon_{\text{phyl}} = 0.25$ kcal/mol **b** $\varepsilon_{\text{phyl}} = 0.50$ kcal/mol **c** $\varepsilon_{\text{phyl}} = 1.20$ kcal/mol **d** $\varepsilon_{\text{phyl}} = 1.50$ kcal/mol **e** $\varepsilon_{\text{phyl}} = 2.00$ kcal/mol where blue color shows the water molecules and hydronium ions, and orange color shows the Nafion atoms. Reprinted with permission from [66]. © 2018 American Chemical Society

hydrophilicity of the walls for a fixed film thickness of 6.3 nm. Similar effects are also seen for larger films thickness.

Figure 6.8 shows the water cluster count normalized by the bulk Nafion ($\lambda = 15$) water cluster count, for different wall hydrophilicities and film thicknesses. All the normalized cluster counts are larger than 1, meaning that the number of water clusters for all the wall hydrophilicity values and the film thicknesses were greater than that of bulk hydrated samples. This implies a more dispersed water cluster network in the Nafion films, as compared to the bulk hydrated Nafion.

The water cluster count increases for the HH walls as compared to the LH walls for all three different film thickness values as shown in Fig. 6.8. This effect is universal and is weakly dependent on the film thickness. The higher cluster count indicates a more dispersed water cluster network for the HH wall films as compared to the LH wall films, which can also be seen in the inset for the HH wall films as shown in Fig. 6.7. Relevant experiments have shown that unwanted crossover reduces due to the highly hydrophilic nanoparticles like silica, clay, etc. [79] added to Nafion. In fact, the existence of the long-range-oriented pathways along the modified carbon

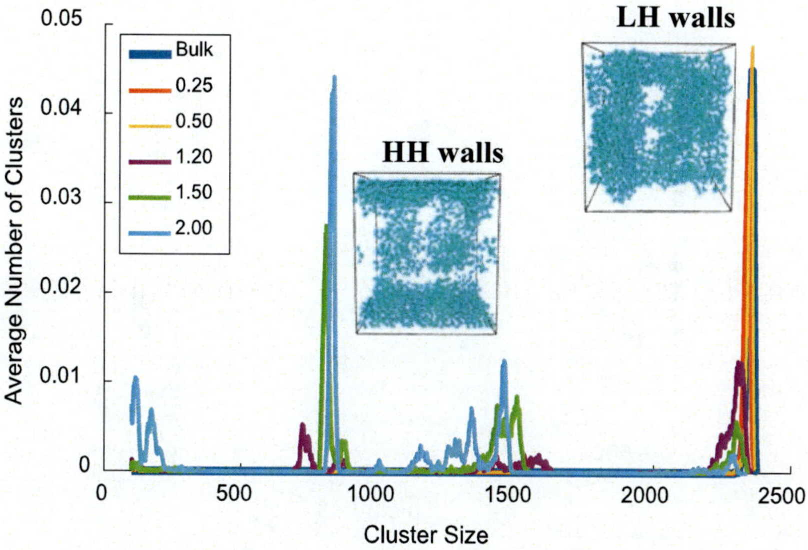

Fig. 6.7 Water cluster distribution for the 6.3 nm film at different wall hydrophilicity values, also for bulk Nafion. The cluster distribution shown is for the cluster sizes ranging from 100 to 2380 water molecules. Reprinted with permission from [66]. © 2018 American Chemical Society

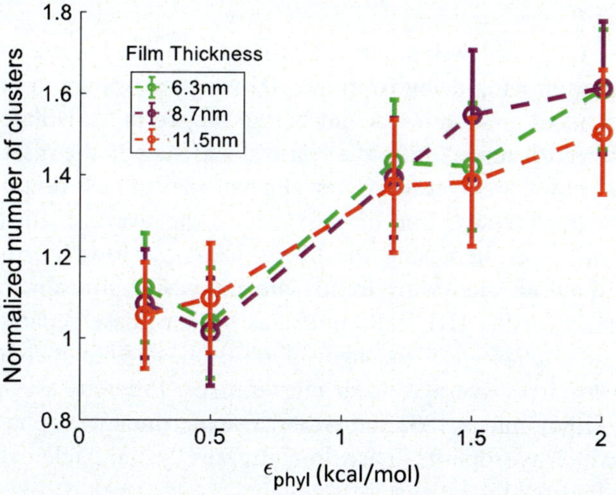

Fig. 6.8 Dependence of water cluster count, normalized by the bulk cluster count, as a function of different wall hydrophilicity (ϵ_{phyl}) for different film thickness [66]

nanotubes was the proposed mechanism for the observed enhanced proton transport and reduced methanol crossover in a Nafion—modified CNT nanocomposite [80]. The MD simulations [66] also show the preferential accumulation of water along the HH walls and a concomitant increase in the water cluster count due to the emergence of a more dispersed water phase and isolated water clusters. It is likely that less polar

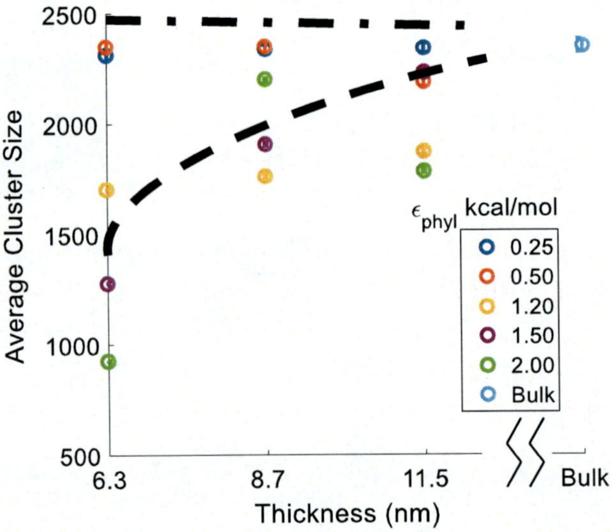

Fig. 6.9 Average water cluster size versus different film thickness for different wall hydrophilicity (ϵ_{phyl}). Average cluster size for bulk Nafion is also shown. The dash-dot line shows the trend for the low hydrophilicity walls and the dash line shows the trend for the high hydrophilicity walls. These lines are not numerical fits. Reprinted with permission from [66]. © 2018 American Chemical Society

molecules, like methanol, will move away from the highly hydrophilic nanoparticles, similar to carbon moving away from the HH walls, as shown in Fig. 6.6. This will increase the chances of such molecules being trapped in the isolated clusters which are found at larger distances from the HH walls as seen in the inset of Fig. 6.7.

Figure 6.9 shows the average water cluster size for different wall hydrophilicities and film thicknesses. For the LH walls, the average cluster sizes remain almost constant upon increasing the film thickness. However, the average water cluster sizes show an increasing trend with increasing film thickness for the HH walls. This trend for the HH walls indicates higher phase separation between the water/hydrophilic phase and the hydrophobic phase with increasing film thickness, leading to larger average water cluster sizes. Previous transmission electron microscopy (TEM) images [83] and GISAXS experiments [89] have shown similar trends in the water/hydrophilic phase domain sizes vs film thickness for Nafion films supported on hydrophilic silica substrates.

Water diffusion plays a key role in the performance of Nafion. The proton attaches itself to the water molecules and diffuses in the membrane. Therefore, water diffusion constants (D_x, D_y and D_z) have been computed in [66] for the X-, Y-, and Z-directions using the Einstein relation for diffusive motion. Water diffusion in the films' XY-plane is studied using the in-plane diffusivity (D),

$$D = (D_x + D_y), \tag{6.2}$$

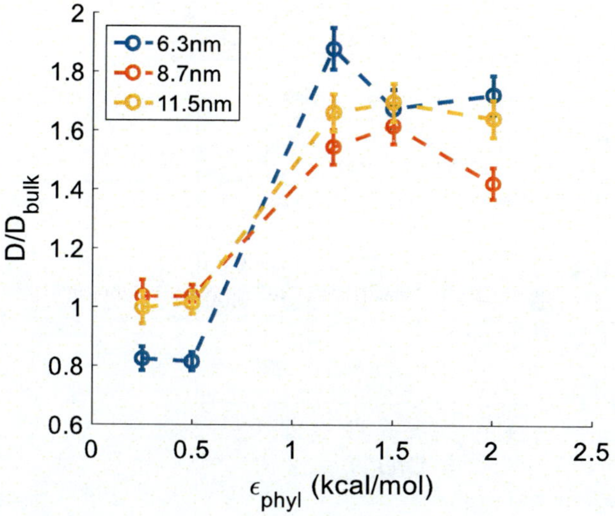

Fig. 6.10 Film averaged in-plane water diffusion constants (D) normalized by the corresponding two-dimensional water diffusion (D_{bulk}) constant at $\lambda = 15$ for bulk Nafion. Reprinted with permission from [66]. © 2018 American Chemical Society

and was compared to the analogous (two-third of the total water diffusion coefficient) values for hydrated Nafion bulk (D_{bulk}),

$$D_{\text{bulk}} = \left(\frac{2}{3}\right) * (D_{\text{bulk}-x} + D_{\text{bulk}-y} + D_{\text{bulk}-z}) \tag{6.3}$$

Henceforth, the diffusion in the XY-plane will be referred to as in-plane water transport. The total water diffusion coefficient ($1.5 * D_{\text{bulk}}$) in bulk Nafion for $\lambda = 15$ at $T = 353$ K was found to be 1.93×10^{-5} cm^2/s [65].

Figure 6.10 shows the in-plane water diffusion normalized by the bulk hydrated Nafion water diffusion constant, for different wall hydrophilicities and different film thicknesses. The in-plane water diffusion is noticeably higher for the HH walls ($\varepsilon_{\text{phyl}} = 1.20, 1.50, 2.00$) as compared to the LH walls ($\varepsilon_{\text{phyl}} = 0.25, 0.50$), for all three different film thicknesses. The cylindrical water channels/micelles in Nafion were found to orient along the hydrophilic substrates and away from the hydrophobic substrates in supported Nafion film experiments [90]. Experimental results further suggested that treated nano-patterned substrates can be used to enhance the directional transport of water within the Nafion membrane since the water transport takes place mostly along the water channels/micelles [90]. A similar enhancement of water transport for the HH walls (substrates) was also observed in the simulations [66].

It is important to keep in mind that the average water cluster sizes decreased significantly for the HH walls. The bulk classical MD simulations of PEMs like Nafion [91], SPEEK [92], PFIA [65] have shown the water diffusion to increase with increasing water cluster sizes and vice versa. However, the capped Nafion films with

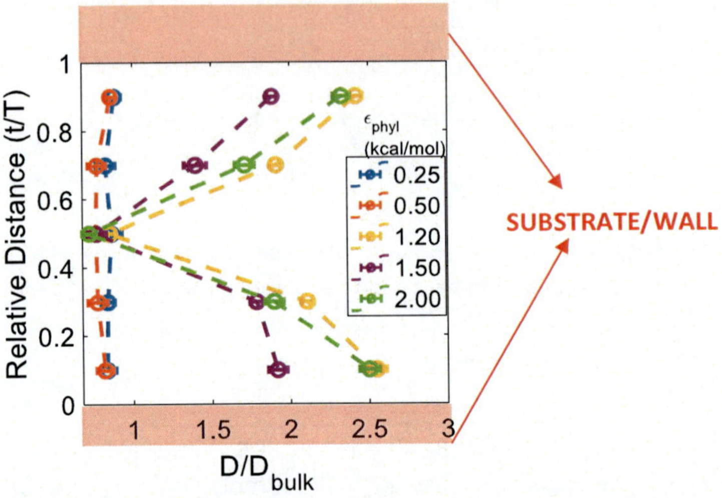

Fig. 6.11 In-plane water diffusion constants (D), normalized by the two-dimensional water diffusion constant (D_{bulk}) at $\lambda = 15$ for bulk Nafion, layer resolved for the 6.3 nm film with varying wall hydrophilicity (ϵ_{phyl}). Relative distance (t/T) is the distance from the bottom of the film (t) normalized by the film thickness (T) [66]

LH walls show much smaller water diffusion than that for the HH walls, despite larger water cluster sizes for the LH walls. This is due to the formation of water channels parallel to the HH walls, in contrast with the long tortuous water channel with bottlenecks in the LH wall films.

The reason behind this enhancement of in-plane water diffusion for the HH walls was further understood by analyzing the water transport in five layers, as shown earlier in Fig. 6.5b. Figure 6.11 shows the layer resolved in-plane water diffusion constant, normalized by the bulk hydrated Nafion two-dimensional water diffusion constant, for the 6.3 nm film. The water diffusion constants for the LH walls are slightly smaller than the bulk hydrated Nafion values. This is probably due to the extreme confinement of the water molecules inside the Nafion film away from the LH walls. For the HH walls, the diffusion constant near the center of the film is close to that for the bulk hydrated Nafion, but the diffusion increases considerably on moving closer to the walls. Similar trends are also observed for thicker films. The presence of such highly mobile water layers near both the HH walls in a capped Nafion film explains the considerably high film averaged in-plane water diffusion constant for the HH walls as compared to the LH walls.

6.4 Atomistic Simulations of Nafion/Graphene Oxide Membranes

Conductivity in polymer electrolyte membranes is significantly affected by water retention. In Nafion and other PFSA-based membranes, ion diffusion takes place at the formed hydrophilic nanochannels [37, 67] and, thus, several studies aiming at the optimization of batteries and fuel cells performance focus on the enhancement of hydrophilicity. In this paragraph, the results have been discussed of the fully atomistic MD simulations of Nafion-based nanocomposite membranes.

The effect of humidification on PFSA membranes has been explored by a variety of experimental techniques such as contact angle measurements [93–95], atomic force microscopy (AFM) [96–99], electrochemical mass-transport measurements [100], and X-ray studies [101]. As was mentioned earlier, it has been observed that at temperatures above 80 °C or at low hydration levels, Nafion membrane conductivity drops significantly. Therefore, the Nafion membrane must be kept sufficiently hydrated in order to retain high proton conductivity levels [37]. In a typical fuel cell, the supplied hydrogen gas is often humidified to enhance proton conductivity. However, the external humidification results in an increase in the size of the fuel cell which is not desirable for portable applications [102]. Another drawback for their use in direct methanol fuel cells (DMFCs) is related to methanol crossover through swelled ionic channels [103].

In order to avoid the need for external humidification and to reduce the methanol crossover effect in DMFCs, self-humidifying membranes doped with inorganic fillers have been examined. Inorganic fillers, such as silica, titania, zirconia, iron oxides, carbon nanotubes, zeolites, and clay, may assist in water retention, enhance proton-conducting properties, and increase mechanical and thermal stability of the membrane [104]. Recently, Nafion/graphene oxide (GO) composites have been explored as potential materials for polymer electrolyte membrane applications [67, 105–109]. The choice of GO stems from properties like large surface area, hydrophilic functional groups, mechanical strength, and chemical stability, which makes it an ideal candidate to form composite membranes [110]. The hydrophilic groups of the GO interact with the protons, which are propagated through the hydrogen bonding network formed with the adsorbed water, allowing for conductivity levels close to 10^{-2} S cm^{-1} [109, 111]. Relevant experiments have shown that the presence of GO modulated the ionic channels of Nafion and decreased the methanol crossover on DMFCs while preserving high ionic conductivity [107].

More detailed studies have demonstrated that the transport of water and other small molecules through graphene-based membranes depends on the interlayer channels and the functional groups of GO [106], like epoxy and hydroxyl groups that may assist water retention due to the formation of a hydrogen bonding network [108, 112]. Investigations of Nafion/GO composites with GOs functionalized with hydrophilic groups, such as $-SO_3H$, –OH, and $-NH_2$ [105], supported the idea that functionalization of GO with such moieties facilitates proton transport and enhances the water retention capability of the Nafion membranes. Pertinent experiments on Nafion/GO composites for DMFC applications [107] attested to their improved properties compared to simple hydrated Nafion membranes. Scanning electron microscopy and transmission electron microscopy studies attributed the enhanced proton conductivity to interactions between the different functional groups in GO and Nafion [108]. The mechanical properties of the Nafion/GO composites as measured by tensile strength experiments [113] were shown to be significantly improved compared to the performance of pristine Nafion membranes, without affecting their swelling properties [108].

GO has also been incorporated in several other composite membranes, like sulfonated polyether ketone [114], PBI [115], polyvinyl alcohol [116], and polyacrylic acid [117, 118] resulting in a marked improvement of their performance. In the case of hydrogels where the water content exceeds 90 wt%, GO may influence in a dramatic manner the physical adsorption of polyelectrolyte chains, the polymer dynamic response at local and global length scales, the charge distributions around the components, and the mobility of the counterions [118]. The vehicular diffusion of water molecules in the presence of GO as studied through MD simulations was found to be slower due to the hydrogen bonding interactions between water molecules and hydroxyl groups of GO [67, 119].

MD simulations, particularly in the fully atomistic representation, allow the study in atomistic detail of structural features and translational dynamics of water and hydronium cations in the formed Nafion channels. The degree of hydration (λ) of the membrane is usually determined by the number of water molecules per side-chain pendant of Nafion. At $\lambda \geq 3$, the sulfonic group loses its hydrogen and so hydronium ions arise [71, 120].

In a recent molecular-dynamics (MD) study [67], the structure and diffusion properties in Nafion/GO systems at three different λ values, i.e., 10, 15, and 20, were simulated. In this work, three different temperatures were examined, i.e., 250, 300, and 350 K at a pressure of 1 bar. The Nafion structure was equilibrated by annealing at 1000 K, while a cooling rate of 50 K/30 ns to the target temperatures was used [67]. More details about the simulation protocol and parameters are given in [67].

In Fig. 6.12, the snapshots of a typical Nafion/GO system are shown, consisting of Nafion, hydronium, water, and GO entities, at $\lambda = 15$, and $T = 300$ K. Differences with respect to the bulk water structure and the formation of water channels can be observed (Fig. 6.12a), a phenomenon that is independent of the presence of the GO in Nafion membranes [65, 67]. As indicated in Fig. 6.12b, the concentration of water onto the GO surface is lower compared to that at distances far from the surface. This results in a lower degree of solvation of the hydronium ions, which are physically adsorbed on the GO flake.

Morphological characteristics related to the spatial arrangement of the different components close to the GO surface were probed by calculating the relevant density profiles, along a direction normal to the GO flake, as shown in Fig. 6.13. It can be observed that the concentration of water on the surface is low, exhibiting a depletion layer. For all λ values examined, a peak is present at a distance close to 8 Å from the surface. It appears that as the hydration level becomes higher, water retention on

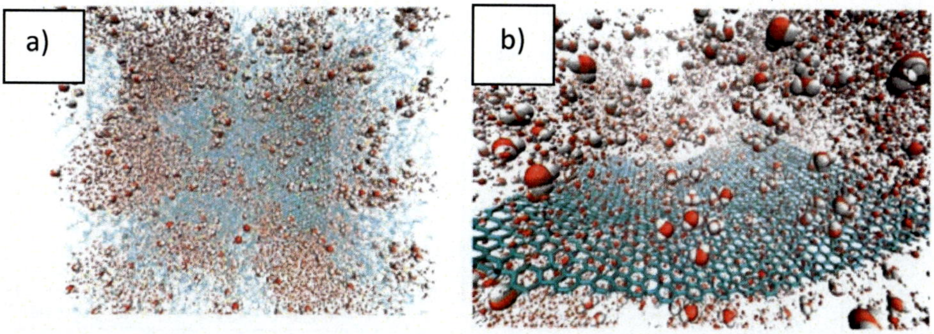

Fig. 6.12 Snapshots [67] of **a** Nafion/hydronium/water/GO and **b** hydronium/water/GO nanocomposite systems. Carbon atoms are shown in dark cyan, oxygen atoms in red, and hydrogen atoms in white

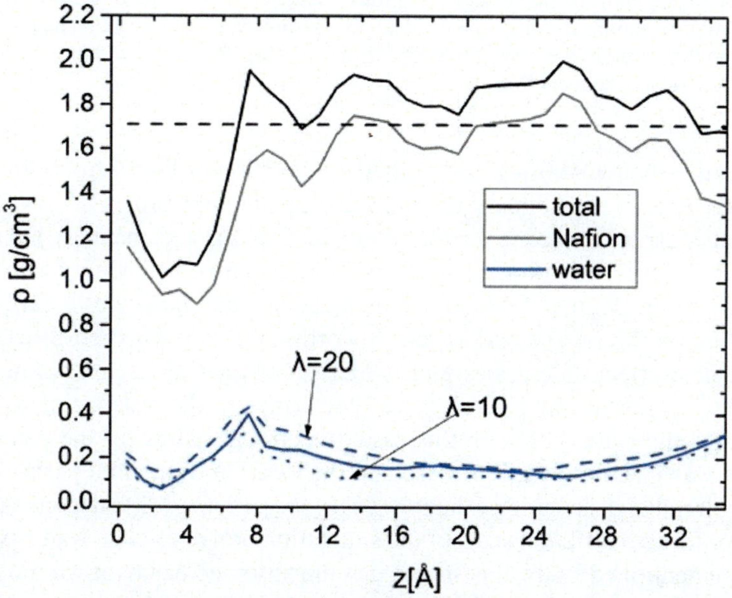

Fig. 6.13 The density profiles along a direction perpendicular to the GO sheet, as a function of the distance from GO surface, for the Nafion/GO systems. The overall density, along with the contributions of Nafion and water, at $T = 300$ K and $\lambda = 15$ is presented. The black dash line depicts the average density of the respective GO-free Nafion system. Water profiles for hydration levels of $\lambda = 10$ (blue dot line) and $\lambda = 20$ (blue dash line), at $T = 300$ K are shown as well. Reprinted with permission from [67]. © 2019 American Chemical Society

the GO tends to increase. On the other hand, the concentration profile of hydronium ions is more uniform (see Fig. 6.3 of [67]). The profile of the Nafion chains shows a reduced concertation close to the GO surface, while the width of the depletion zone of the polymer chains close to GO is commensurate to the size of the side chains [67]. Since the backbone polymer dynamics is expected to be practically frozen at

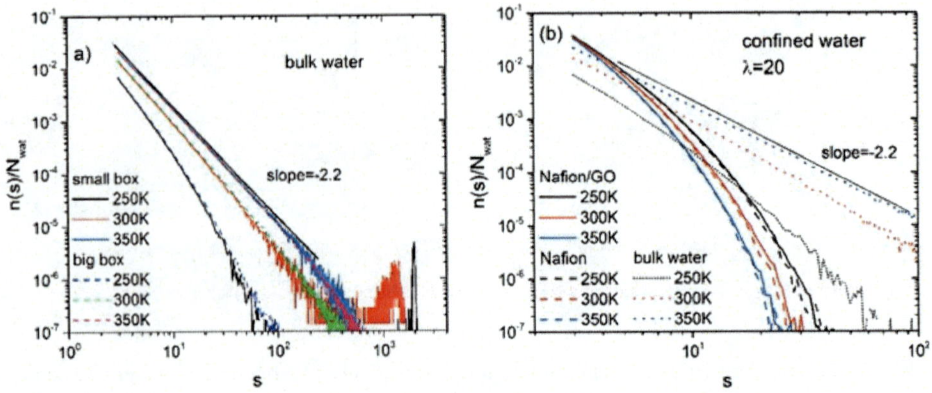

Fig. 6.14 Number of water clusters, $n(s)$, normalized by the total number of water molecules, N_{wat}, as a function of the cluster size, s. **a** Bulk water cluster distributions at the examined temperatures at two different boxes, with sizes of $L \approx 41$ Å (denoted as small box) and $L \approx 126$ Å (denoted as big box) and **b** distributions in Nafion/GO/water and Nafion/water systems are compared with those in bulk water. The straight dash lines indicate a slope of −2.2. Reprinted with permission from [67].

temperatures below 400 K [103], only the relaxation of the hydrophilic side groups is probed by the MD simulations at the examined temperatures.

Another aspect related to the morphology of the formed channels and the conductivity mechanisms in Nafion membranes is the clustering behavior of water molecules [121]. Since the degree of spatial confinement of water molecules is expected to vary depending on the local environment, it worths studying whether different levels of confinement affect its clustering characteristics under different confinement levels becomes of interest. Such a study was performed in the systems of [67] using the DBSCAN algorithm [122]. In this case, the critical radius for the identification of clusters was taken to be 2.8 Å, based on the location of the first peak of the water–water pair radial distribution function (RDF) [67]. Figure 6.14a depicts the clustering behavior for two different sizes of the simulation box in pristine water systems from the aforementioned study. It is shown that the observed behavior was not affected by finite size effects at the examined sizes of the simulation boxes.

Apart from that, it should be noticed that at all temperatures, the highest probability for cluster formation corresponds to a minimum number of three molecules participating in a cluster (s). As the temperature drops, DBSCAN analysis shows that the size of the largest clusters reduces. Moreover, the distribution of the cluster size follows a power law which is in close agreement with the universal exponent of -2.2, as predicted by the Fisher droplet model [123]. The cluster distribution under confined conditions (i.e., in Nafion-based systems) is shown in Fig. 6.14b. Evidently, the bulk water structure is disrupted. The RDF arising from the centers of mass of water molecules in bulk and under confined conditions (not shown here) remains qualitatively the same (in terms of the number and the position of peaks [67]), but

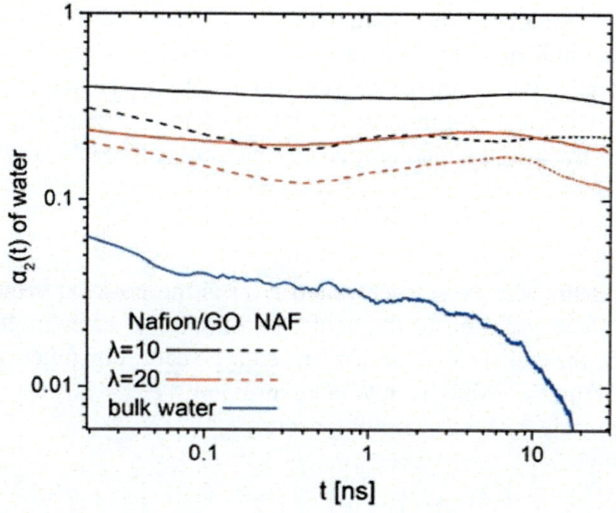

Fig. 6.15 Non-Gaussian parameter [$\alpha_2(t)$] of the water diffusion, in water/Nafion/GO (Nafion/GO) and water/Nafion (NAF) systems (λ equal to 10 and 20) and bulk water, at 300 K. Reprinted with permission from [67]. © 2019 American Chemical Society

the height of the first peak, which relates to the first neighbor shell, becomes higher as the water concertation reduces (i.e., the degree of confinement increases).

Figure 6.15 depicts the non-Gaussian parameter [NGP or α_2 (t)], of the centers of mass water molecules in the examined systems, defined as.

$$\alpha_2(t) = \frac{3\langle \Delta r(t)^4 \rangle}{5\langle \Delta r(t)^2 \rangle^2} - 1, \tag{6.4}$$

where $\Delta r(t)$ is the displacement of the center of mass from the original position, at time t. Square brackets denote time and ensemble average. NGP essentially provides a measure for the degree of dynamic heterogeneities in the diffusional motion of the probed particles. It takes the minimum theoretical value (−0.4) when all centers of mass travel the same distance, a value of 0 when Brownian diffusion is at work and values higher than 0 when the distances traveled after time t by the particles under examination are not Gaussian distributed. The latter case denotes an increased level of heterogeneity in the particles' translational motion. According to this picture, it appears that heterogeneities in the water diffusion increase as λ decreases. This becomes more pronounced following the inclusion of GO in the Nafion membrane. In addition, the $\alpha_2(t)$ parameter under confined conditions appears almost an order of magnitude higher compared to the bulk, unconfined case.

This behavior is consistent with the heterogeneous dynamics of water molecules related to the confined motion within the channels formed by Nafion. In the presence of GO, additional constrictions are imposed close to the GO surface due to specific interactions of water molecules (i.e., hydrogen bonding) with the oxidized carbon groups of GO.

More details regarding the characteristics of water translational motion in the microenvironment formed in the Nafion-based systems can be obtained by studying relevant structural relaxation processes which take place within specific spatial dimensions. Such spatiotemporal resolution in molecular motion can be probed by the self-part of the intermediate scattering function, defined as [124]:

$$F_s(q,t) = \left\langle \exp\left(i\vec{q}\cdot\left[\vec{r}_i(t) - \vec{r}_i(0)\right]\right)\right\rangle \tag{6.5}$$

where $\vec{q}$ is the scattering vector, and $\vec{r}_i(t)$ is the position vector of the examined particle (here pointing to the center of mass of a water molecule) at time t. In order to fit the behavior of the intermediate scattering function, the modified Kohlrausch–Williams–Watts (mKWW) expression [125, 126] was employed for a constant magnitude q of the scattering vector,

$$F_s(t) = \alpha_1 \exp\left[-\frac{t}{\tau_1}\right] + (1-\alpha_1)\exp\left[-\left(\frac{t}{\tau_2}\right)^{\beta}\right]. \tag{6.6}$$

The mKWW fitting function assumes the first term of a simple exponential (i.e., a Debye process) and the second term of a stretched exponential relaxation [125]. The assumption of a Debye relaxation is consistent with the existence of an Arrhenius component of the diffusion in the glassy region [127]. The decorrelation times are calculated, based on the fitting parameters and the expression for the decorrelation time,

$$\tau_c = a_1\tau_1 + (1-a_1)\left(\frac{\tau_2}{\beta}\right)\Gamma\left(\frac{1}{\beta}\right), \tag{6.7}$$

where Γ is the gamma function. In Fig. 6.16, results for the intermediate scattering function are presented, based on the motion of the center of mass of water molecules in pristine water and the two extreme hydration levels studied in the composite systems. As the evaluation of the dynamics close to the Fickian regime, at $T = 250$ K, is rather ambiguous, especially for the Nafion/GO systems, water translational dynamics is only presented for a wave vector representing a length scale commensurate with the dimensions of the simulation box, i.e., $q = 0.22$ Å^{-1} [85].

The results indicate a more sluggish diffusion as the degree of the water confinement increases. The activation energies were calculated to be 7.3, 10.0 and 11.3 kJ/mol for the bulk water and the water in Nafion/GO with $\lambda = 20$ and in Nafion/GO with $\lambda = 10$ systems, respectively. The common interpretation [126] of the increase observed in the activation energy is the presence of heterogeneities, that is an indication of cooperative diffusion, which was implied earlier by the behavior of the NGP (Fig. 6.15). It is noteworthy that the dynamics of water in the confined conditions exhibits an Arrhenius behavior. An interpretation of the apparent Arrhenius behavior under conditions of cooperative diffusion was provided recently, on the basis of an extension of the super-Arrhenius region [127, 128].

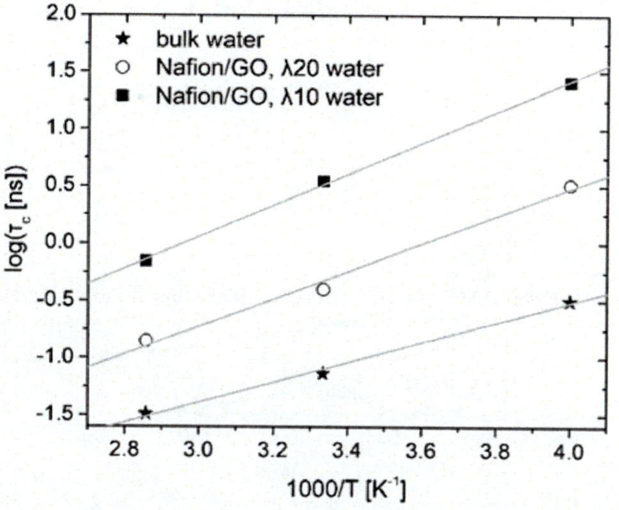

Fig. 6.16 Decorrelation times of the incoherent dynamic structure factor (6.2), $q = 0.22$ Å^{-1}, for water molecules at 300 K. The different symbols refer to the bulk state and to the Nafion/GO systems at $\lambda = 20$ and $\lambda = 10$. Reprinted with permission from [67]. © 2019 American Chemical Society

Apart from the translational dynamics of hydronium ions and the Grotthuss mechanism, the conductivity in the Nafion-based nanocomposite membranes also depends on the adsorption/desorption mechanism of the hydronium ions from the GO surface. This process can be explored by means of a relevant correlation function [117]:

$$h(t) = \frac{\langle g(t)g(0)\rangle}{\langle g^2\rangle} \tag{6.8}$$

where $g(t)$ assumes a value of 1 if an adsorption event is detected at time t and 0 otherwise. Angle brackets denote averaging over all pairs and time origins. As a criterion for adsorption of a hydronium ion onto the GO surface, it was taken that the distance between a hydronium oxygen and a hydroxyl hydrogen of GO at time t was less or equal to a critical distance of 3.7 Å, which corresponds to the first peak of the respective RDF function [67]. The critical distance is practically the same as in the case of the sulfur–hydronium interaction. The results presented in Fig. 6.17 were fitted by the mKWW function (6.6).

The desorption times were evaluated to be 5×10^7, 6×10^6 and 1×10^6 ns for hydration levels of $\lambda = 10$, $\lambda = 15$ and $\lambda = 20$, respectively. The results emphasize the significance of the hydration level on the hydronium interfacial dynamics. At $\lambda = 10$, the desorption time is an order of magnitude larger compared to those at higher hydration levels. Only at $\lambda = 20$, the cations desorb from the GO at times close to 1 ms. It should be noted that the average residence time of the water and hydronium ions on the sulfur group of the Nafion is of the order of ps and ns, respectively [129], although cases of hydronium bound to SO_3^- for longer times (approximately 1 ms)

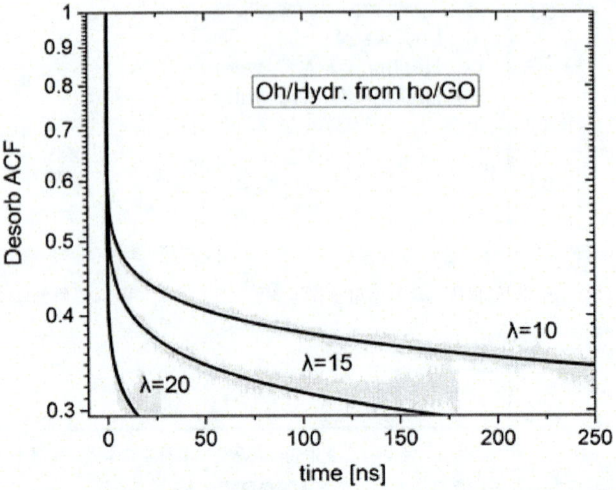

Fig. 6.17 Desorption autocorrelation function (6.8) of the oxygen (Oh) of hydronium ions from the hydrogen (hO) of the GO hydroxyl. Reprinted with permission from [67]. © 2019 American Chemical Society

had also been observed [129]. The deviation in the desorption times between GO and Nafion sulfur groups should be attributed to the low concentration of water near the GO, especially at the hydration levels of $\lambda = 10$ and $\lambda = 15$. Based on the above analysis, it can be concluded that water retention at the Nafion/GO interface appears only at high enough hydration levels of Nafion [67].

6.5 Modeling of Doped Non-humidified Membranes

A few computational studies have been performed on benzimidaozle (BIM)-based membranes. Pahari et al. [130] investigated the structure and dynamics of phosphoric acid-doped PBI membranes at a varying concentration of phosphoric acid using classical MD simulations. The authors observed no phase separation, and phosphoric acid molecules form inter and intramolecular hydrogen bonds at all concentrations of PA. Pahari and Roy [131] employed MD simulations on PA-doped PBI and ABPBI. The authors showed that ABPBI has more affinity toward phosphoric acid as compared to PBI due to the larger number of H-bonds in ABPBI than in PBI. Shirata and Kawauchi [132] employed DFT calculations to examine the interaction of PA with BIM with different configurations. The authors explored several interactions and concluded that *N*-type interactions are the strongest (followed by O–, OH–, and π-type). However, PBI is preferred due to the loss of the mechanical strength of ABPBI with increasing concentration of phosphoric acid. Another drawback of ABPBI membrane is its poor solubility in common solvents employed for membrane casting methods [133]. Also, PA-doped PBI membranes have disadvantages like leaching and condensation of phosphate groups at high temperatures [134]. Hence, alternatives to PA such as ionic liquids (ILs) are potentially interesting.

Iojoiu and coworkers [54] proposed that the triethylamine (TEA) saturated and (TEATF)-doped Nafion membranes consist of nanoaggregates comprised of proton acceptor and proton donor sites. The authors proposed three possible long-range proton transport mechanisms, (a) via cationic clusters, (b) via a concerted interaction between cation and anion clusters, and (c) via direct proton exchange between cationic and anionic clusters. Kumar and Venkatnathan [135] employed quantum chemistry calculations on IL-doped PFSA membrane for high-temperature fuel cells. The authors used DFT method with B3LYP/6–311++G** basis set for all the calculations. The authors explored several proton transport pathways in a triethylammonium-triflate (TEATF) IL-doped Nafion membrane. In TEATF-doped PFSA membranes, both the cation and anion can access suitable sites, simultaneously, on the sulfonic acid end groups. For example, the cation approaches toward the O atom, and the anion approaches the H atom. The anion abstracts the proton from the acid and facilitates the cation to electrostatically bind to the sulfonate end groups of the membranes. Figure 6.18a shows the structure of the resultant complex, a side-chain fragment of a Nafion membrane with a TEATF.

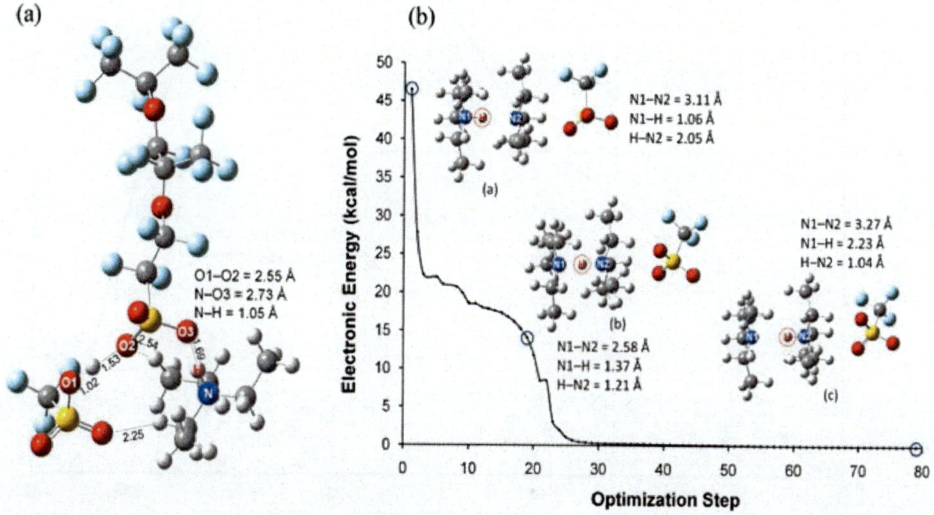

Fig. 6.18 **a** Interaction of a Nafion® side-chain fragment with a TEATF ionic liquid unit. TFA approaches toward hydrogen atom and $TEAH^+$ toward the oxygen atom of the sulfonic acid end group of the membrane fragment. **b** Mechanism of proton transfer in a $TEAH^+ \cdots TEA \cdots TFA^-$ complex (**b**) shows the optimization process of $TEAH^+ \cdots TEA \cdots TFA^-$. The structures shown in the figure are the configurations of the complex at the points denoted by blue circles, and the moving hydrogen atom is denoted by a red circle. A steep decrease in system energy occurs due to TFA^- interaction with the $TEAH^+ \cdots TEA$ complex. The authors showed that the transfer of a proton from a tertiary amine cation to a tertiary amine (see panel b) occurs only on interaction with anion. The interaction of anion with amine increases the basicity of the latter and facilitates the transfer of proton from cation to neutral amine. Reprinted with permission from work of Kumar and Venkatnathan [135]. © American Chemical Society

A high boiling point (256 °C) of imidazole (compared to water) makes them promising materials as proton carriers in high-temperature fuel cells. Hydrogen bonding in imidazole plays a key role in proton conduction and rotation of an imidazole molecule in the process results in the cleavage of hydrogen bonds between molecules. In another quantum chemistry study, Kumar and Venkatnathan [136] explored proton transport and rotation energy barrier in imidazole chains. The authors showed that the propagation of an excess proton along the imidazole chain occurs with energy barriers lower than 1 kcal/mol. The authors calculated the energy barriers for rotation of imidazole molecules in two, three, and four imidazole molecule chains and found that the barrier is equal to the number of hydrogen bonds broken in the process. Figure 6.19 shows the change in structure and potential energy of a chain of three imidazole molecules with the rotation of an imidazole molecule. The rotational barrier in two imidazole chain is 8.0 kcal/mol, in three imidazole chain, it is 17.1 and 20.0 kcal/mol in four imidazole chain.

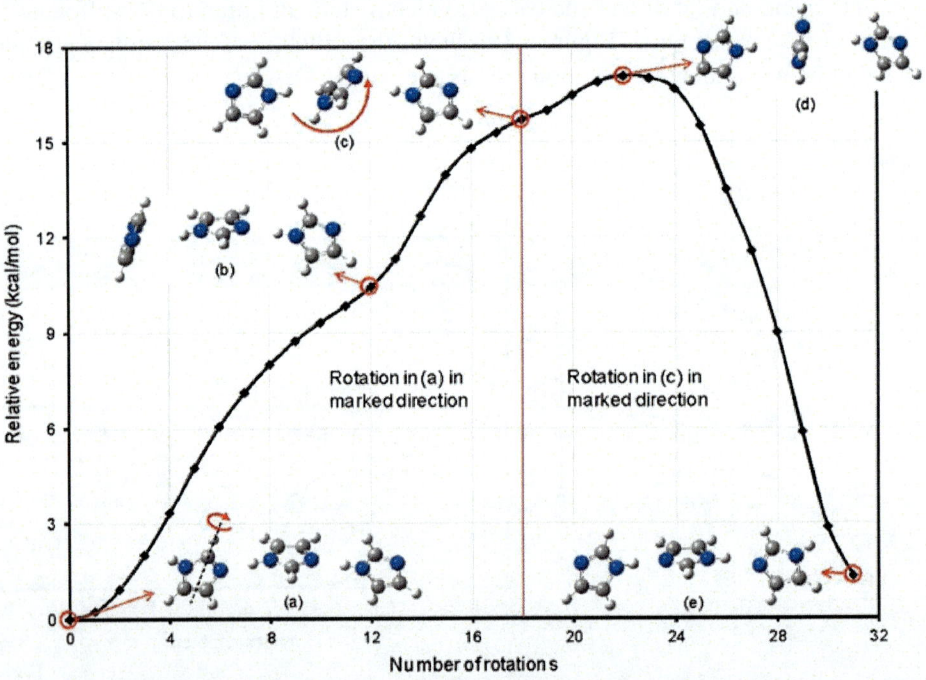

Fig. 6.19 A PES of 3-imidazole-molecule chain. The scan is separated into two parts: left part is due to rotation of first imidazole in structure (**a**); and right part is due to rotation of second imidazole in structure (**c**). The rotation steps, in the marked directions, were chosen as 10° and 9.7°, respectively. Structure (**b**) shows that up to 120° rotation, there is no effect on hydrogen bonding between second and third molecule. Structure (**d**) shows maxima in which the second molecule is showing minimum interaction with other molecules. In structure (**e**), all imidazole molecules are flipped. Reprinted with permission from work of Kumar and Venkatnathan [136]. © American Chemical Society

In a different study, Venkatnathan and coworkers [137] explored proton transport pathways in base-rich imidazolium ionic liquids. The authors suggested proton transport pathways to explain the experimentally observed enhanced conductivity in base-rich PILs. The results showed the barrierless rotation of imidazole in base-rich ionic liquids, could be one of the possible reasons for enhanced conductivity. The figure illustrates the overall mechanism of proton transport in imidazole rich IL. Figure 6.20 illustrates the overall proposed mechanism of proton transport in one such system and suggests barrierless rotation of imidazole molecule to be the reason for the enhanced conductivity of base-rich ILs.

The addition of triflic acid, which exist in its dissociated form (TFA), is known to enhance the efficiency of PA-doped ABPBI membranes. Sunda et al. [138] employed MD simulations to characterize the structure and dynamics of ABPBI + PA as shown in Fig. 6.21, ABPBI + TFA and ABPBI + PA + TFA blends with varying levels of hydration. The structural properties such as RDFs showed that the distance between two adjacent imidazole units on the polymer chain remains unaffected by hydration and the type of blend. The end-to-end polymer chain distance and radius of gyration are also unaffected by hydration and the type of blend, illustrating that the stability of polymer membranes under various hydrated acidic environments remains unaffected. The number of PA, TFA, and water molecules in the cluster around the

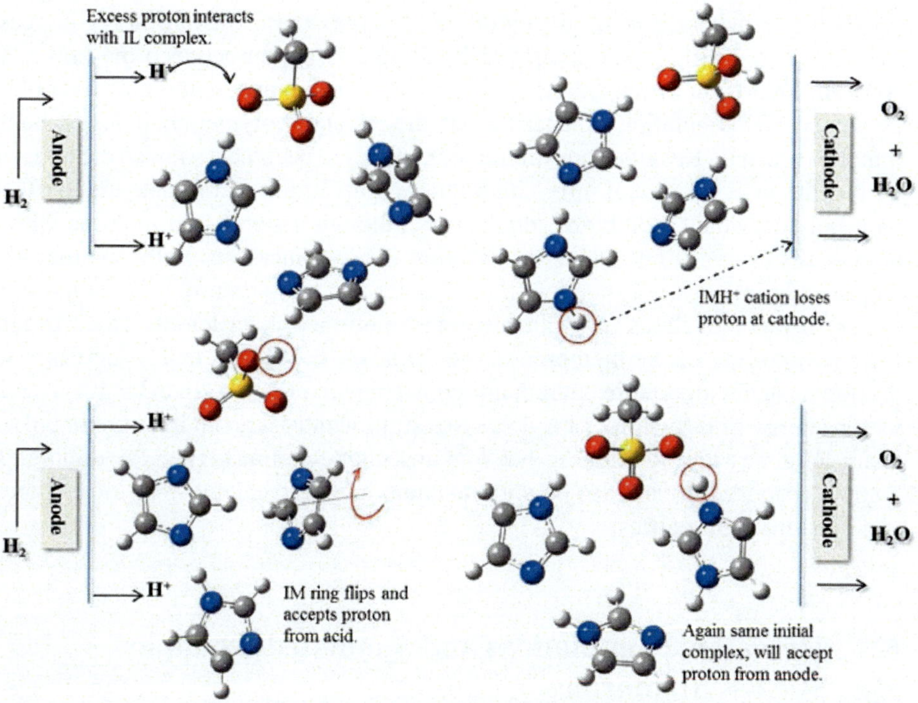

Fig. 6.20 Schematic of proton transport pathway in [MSA]/[IM] = 1:3, i.e., IMMSA with two IM molecules. Reprinted with permission from Venkatnathan and coworkers [137]. © American Chemical Society

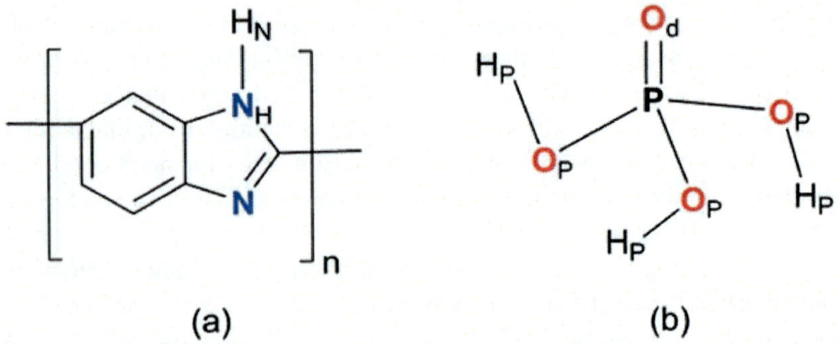

Fig. 6.21 Chemical structure with atom types of **a** ABPBI polymer membrane and **b** phosphoric acid (PA). © Royal Society of Chemistry

polymer membrane (skewed and extended form) is found to depend significantly on the extent of hydration. The lowest water mobility was obtained from the ABPBI + PA + TFA blend, which suggests that this blend could be the most effective in reducing acid leaching from the membrane matrix.

In a different study, Venkatnathan and coworkers [139] performed MD simulations to examine structure and dynamics in neat BIM, phosphoric acid, and PA-BIM mixtures. The authors observed that diffusion coefficients of BIM decrease with increasing phosphoric acid concentration, whereas the diffusion of PA increases. The RDFs showed a strong hydrogen bonding interaction between the imine N of BIM and hydrogen of phosphoric acid. Further Venkatnathan and coworkers [140] employed MD simulations to characterize the effect of polymer chain length using a dimer to 100-mer. Results from simulations (dimer to decamer) showed the following trends: the inter-chain and intra-chain interactions in the membrane are unaffected with polymer chain length and temperature, though a significant increase with PA doping is observed. The radius of gyration linearly increases with polymer chain length and remains unchanged with PA doping and temperature. However, the end-to-end distance deviates from linearity with polymer chain length which suggests increased coiling of the membrane. The diffusion coefficient of PA increases with PA doping and temperature but remains constant with polymer chain length. The activation energy of diffusion of PA decreases significantly with an increase in polymer chain length at low PA doping but remains unaffected at higher PA doping. The authors proposed decamer to be optimal chain length to calculate various structural and dynamical properties.

6.6 Mesoscopic Simulations and Simulated Example: SPEEK Membrane

In this section, two examples of the mesoscopic modeling of polymer ion-exchange-based membranes will be considered. Since the average size of the polymer membrane domains is about 4 nm, it seems rather attractive to use the mesoscopic simulation methods to model the morphology of such systems. These methods allow us to study samples of PEM membranes with a characteristic scale of 20–100 nm over long-time intervals of 1–1000 μs. This is achieved by abandoning atomistic detail when moving to a larger scale using collective variables. To model a SPEEK based membrane with different water content, the mesoscale DDFT simulations (MDDFT) [141] and dissipative particle dynamics [142] methods were used. Although these methods are based on the use of different simulation principles, they are united by a common approach to the construction of coarse-grained models.

MDDFT method is based on a dynamic version of the density-functional theory in combination with the Flory–Huggins model. As collective variables, it uses fields of the number particle densities $\rho_\alpha(\mathbf{r}, t)$. The evolution of density fields can be found by solving a system of diffusion Langevin equations. All details of the method are set out in [76, 141, 143–146].

DPD method is another mesoscale simulation technique proposed initially by Hoogerbrugge and Koelman [142, 147] and generalized later to molecular systems [148]. As in the MD method, the evolution of the coarse-grained particles is described by the system of Newton equations of motion. One can treat as an advantage the fact that both MDDFT and DPD methods use the same strategies for building the mesoscopic models of the molecular systems and evaluating the force parameters through the Flory–Huggins parameters [149]. In MDDFT and DPD method, the simulated system is placed in a cell of a fixed volume V with periodic boundary conditions. All molecular objects are replaced by some equivalent set of beads, called mesoscopic particles (MPs) of various types α. They all have the same volume $v_\alpha \equiv v$. Each of the subsystems corresponds to a selected sequence of structural units of the macromolecule chain (these can be its individual segments), or to a certain number of solvent molecules. The choice of volume v is determined by the parameterization of the system, which, in turn, determines its degree of coarsening. It introduces a unit scale $\sigma = (6v/\pi)^{1/3}$. In MDDFT, polymer chains have the same Gaussian coil conformation, and the spatial distribution of MP can be described by the potential field of local densities $\rho_\alpha(\mathbf{r}, t)$. Unlike MDDFT, the DPD method imposes no restrictions on the conformational mobility of the polymer chains and describes the structure of the polymer systems more accurately.

Consider the process of constructing a coarse-grained model for the sulfonated PEEK, poly(oxy-1,4-phenyleneoxy- 1,4-phenylenecarbonyl- 1,4-phenylene), (SPEEK) [11, 77] (Fig. 6.22). To construct a mesoscopic model of SPEEK based membrane, it is necessary to correctly map the atomistic model of the ionomer chain to the equivalent coarse-grained representation in the form of a sequence of repeating mesoscopic particles and to subsequently calculate the interaction parameters between these MPs based on Flory–Huggins parameters [149]. Although SPEEK can be considered a random block copolymer, for simplicity, the composition and structure of the chains was represented as $[A_nB_m]_N$, where n and m are the number of polar and non-polar beads, N is the number of the chain blocks. The n:m ratio allows to characterize the degree of sulfonic acid (DS) of the polymer. Analysis of the features of the chemical structure of the polymer and the requirement to

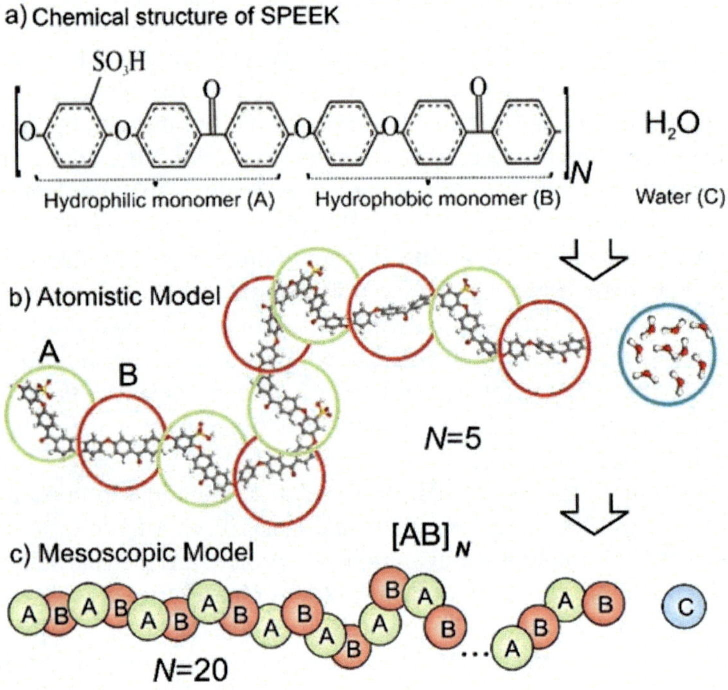

Fig. 6.22 Mapping of a fragment of sulfonated poly(oxy- 1,4-phenyleneoxy- 1,4-phenylenecarbonyl -1,4-phenylene) chain and a cluster of water molecules onto the equivalent mesoscopic representation

comply with the structural characteristics of the model (contour length, rigidity and distance between the ends of the chain) allows to select the size of the statistical segment a of the basic chain,

$$a = C_{\infty} L_{\text{mon}}, \tag{6.9}$$

where $C_{\infty} = 3.36$ is the characteristic ratio (characterizes the rigidity of the chain), L_{mon} is the size of the monomer. Estimates for C_{∞} were done using the Bicerano method [77, 150]. These results indicate a higher degree of conformational mobility of the PEEK chains in comparison with Nafion and its $C_{\infty} = 7.13$ [76]. Aromatic rings were chosen as the basic fragment of the SPEEK chain, which gives $L_{\text{mon}} =$ 5.3 Å (the size of the benzene ring). Based on the L_{mon} and C_{∞} values 6.9 defines the length of the statistical segment $a =$ 17.8 Å, that is close to the length of the sulphonated and non-sulphonated comonomers in SPEEK (13.5 Å). Thus, chains of SPEEK can be matched with mesoscopic particles (MPs or beads) of two types A and B of the same size. The volume of MP can also be estimated by the Bicerano method, which gives $\nu = 250\ \text{cm}^3/\text{mol}$. Mesoscopic particles of the third type (C) are matched to water molecules. Given the value of ν, one MP of water corresponds to about 14 molecules. The mapping of the system to a coarse representation is shown in Fig. 6.22.

To determine the parameters of the intermolecular interactions $\varepsilon_{\alpha\beta}$ used in MDDFT [141] and $a_{\alpha\beta}$ used in DPD [142], it is necessary to calculate the Flory–Huggins parameters $\chi_{\alpha\beta}$ for the A, B and C mesoscopic particles, which are related to the Hildebrand solubility parameters δ_α as

$$\chi_{\alpha\beta} = \frac{\nu(\delta_\alpha - \delta_\beta)^2}{RT} - \chi_s, \tag{6.10}$$

where χ_s is the entropy contribution to the free energy of mixing, R is the gas constant. The contribution of χ_s can be neglected since its estimates for many polymer/solvent blends give the result of 0.34 [151, 152], which is only a small correction to the first term in 6.10. The values of the parameters δ_α can be obtained via the cohesion energy density E_{coh}/V:

$$\delta_\alpha = \sqrt{E_{\text{coh}}^{(\alpha)}/V} \tag{6.11}$$

The estimates of δ_α can be obtained using the Bicerano [150] and Askadsky [153] methods. The MD-based calculations give a more accurate value of δ. The value of $E_{\text{coh}}^{(\alpha)}$ is equal to the change in the potential energy of a single volume of a substance when all intermolecular forces are turned off. It characterizes the intensity of the interatomic and intermolecular interactions. In other words, E_{coh} shows the change in the potential energy of the system during the transition of molecules into the gas phase from the condensed state. The analysis performed in [77] allows us to choose the parameters δ_α and to determine the following values for the Flory–Huggins parameters: $\chi_{AB} = 21.6$, $\chi_{AC} = 1.32$, and $\chi_{BC} = 33.6$.

In the framework of MDDFT, three SPEEK/water systems have been studied in detail with different degree of sulfonation (DS) of the polymer matrix. DS is regulated by the composition of the polymer chain $[A_n B_m]_N$. The length of the chains in all cases has the same value $(n + m)N = 40$. The following compositions are studied: $[A_1 B_1]_{20}$ (DS = 50%), $[A_3 B_2]_8$ (60%), and $[A_3 B_1]_{10}$ (75%). In all calculations, the fixed temperature was used, $T = 298\ K$.

The second main parameter of the calculations is the amount of water in the membrane, which is governed by λ; it was varied from 0.5 to 14 to change the hydration level. The amount of water in the membrane is convenient to control using the water volume fraction φ_C, which corresponds to the ratio of the number of mesoscopic water particles to the total number of particles in the system. The solutions to the system of MDDFT equations are searched on a grid of 32 × 32 × 32 nodes with a step of 1 nm. This gives the modeling cell edge of the size of 32 nm. This step is chosen according to the unit length of 0.92 nm, calculated based on the estimate of the base volume v of the mesoscopic particles.

In Fig. 6.23, the isocontour plot of the distribution of local densities ρ_C of the water MPs at the volume section $V(x, y, 0)$ is performed for three values of DS at

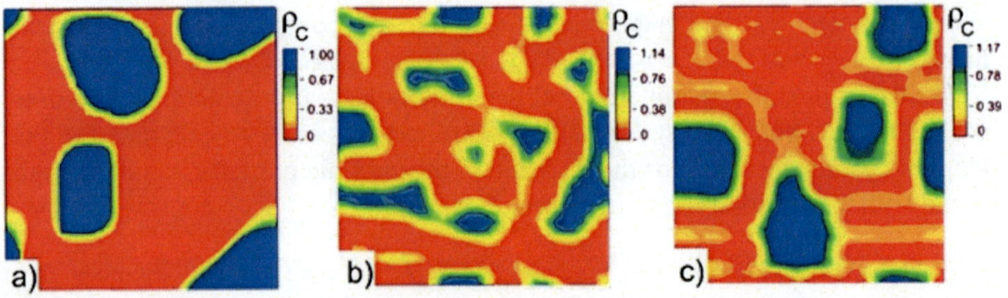

Fig. 6.23 The density distribution $\rho_C(\mathbf{r}, t)$ of the MPs of water on the section plane $V(x, y, 0)$ at the same volume fraction $\varphi_C = 0.29$ of water in the membrane: **a** $[A_1B_1]_8$ ($\lambda = 11$); **b** $[A_3B_2]_8$ ($\lambda = 9.5$); **c** $[A_3B_1]_8$ ($\lambda = 7.6$)

$\varphi_C = 0.29$. In the system with the lowest degree of sulfonation in the case of $[A_1B_1]_8$ chain in the final state, the water forms isolated clusters separated by an average distance of about 19 nm. The shape of the clusters is close to ellipsoidal, and their cross-section is 10 nm. The interface zone has a clear boundary with a thickness of 0.8 nm.

For higher degrees of sulfonation, DS = 60% ($[A_3B_2]_8$) and 75% ($[A_3B_1]_8$), a channel network connecting water clusters is observed in the model membranes. At the same time, the morphologies of the membranes at DS = 60 and 75% are very different. In Fig. 6.23, it is clearly seen that at DS = 60%, the morphology of water channels looks the most preferable. The water forms rather wide channels in which its average field density has high average values—from 0.5 and higher. The average maximum cross-section diameter is about 8 nm. A comparison of the three membrane samples allows us to conclude that the sequence $[A_3B_2]_8$ is optimal, since one can assume that the membrane will have better transport properties.

The volume visualization of the density distribution of water in the SPEEK membrane for $[A_3B_2]_8$ (Fig. 6.24) shows how the percolation of the water subsystem occurs at different threshold densities ρ_C and λ. The three density values are chosen because the concentration of water in the water channels is highly non-uniform. The largest values of the density of the MPs of water are reached in the center of the water channels. The thickness of the interface zone, where the overlap in the density distributions of subsystems A and C is observed, is about 1.6 nm (Fig. 6.23). The choice of the three values for the threshold ρ_C allows to produce a clearer picture of the water channels percolation in the material as shown in Fig. 6.24. It is seen that the connected network of the water channels is already formed with a low water content, $\lambda = 0.5$, due to the presence of the microphase separation of the hydrophobic and hydrophilic segments of the polymer chain in a dry polymer (Fig. 6.23b).

With an increase in the water content to $\lambda > 1.2$, the presence of the water channels domains with densities $\rho_C > 0.2$ is clearly seen. Many elongated and dumbbell-shaped regions are formed in the membrane as a result of the confluence of the small water domains. An estimate of the percolation threshold λ^* for the number density of water particles $\rho_C = 0.2$ indicates that the water domains begin to form a connected network in the range of $1.2 < \lambda^* < 2.3$, that is, when approximately two

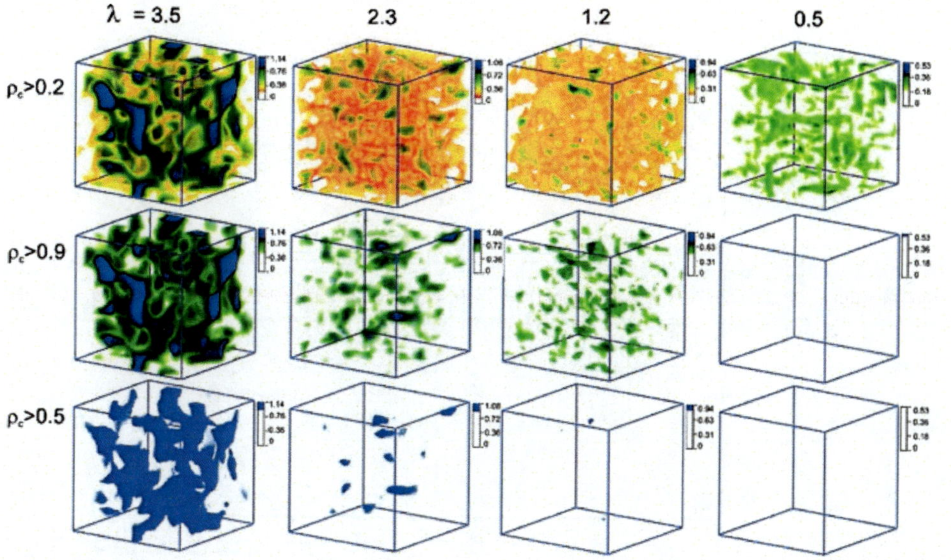

Fig. 6.24 Visualization of the formation of the water channels in the volume of the SPEEK membrane in the case of $[A_3B_2]_8$ monomers sequence, for the different threshold values of ρ_C depending on the water content λ

water molecules correspond to one sulfonic acid group. For other threshold values ρ_C, the following results are obtained: $\lambda^* = 4$ for $\rho_C > 0.5$ and $\lambda^* = 7$ for $\rho_C > 0.9$, which is significantly below the corresponding estimates for Nafion, $\lambda^* = 16$, as obtained in [76].

The low percolation threshold for $[A_3B_2]_8$ chains can be explained by the high degree of sulfonation of the ionomer model chain. The morphology that occurs in the case of $[A_3B_2]_8$ chains is similar to the disordered lamellar phase (Fig. 6.23) for polar and nonpolar domains. Since the volume fraction of the sulfonated MPs in the dry membrane is 60%, the polar domains form a continuous phase. Such a microphase separation is ideal for the fast percolation of the water phase through the membrane at a small λ.

Thus, the small value of the percolation threshold λ^* in the case of $[A_3B_2]_8$ chains is due to the percolation of the polar subsystem. This allows to understand the relatively large value of λ^* for Nafion-1100 in [76]. The degree of sulfonation of Nafion-1100 is 30%, which exactly corresponds to the percolation of randomly distributed homogeneous spheres [154]. However, the spatial distribution of the polymer chain segments strongly depends on their conformational behavior. The percolation threshold for the polar units and, hence, for the water phase may take higher values.

Figure 6.25 shows the dependence of the average maximum diameter D of the channels cross-section on λ. It can be seen that as the volume fraction of water decreases to $\lambda^* \sim 7$ (for areas with $\rho_C > 0.9$), the channels contract only slightly. Thus, the values $\lambda > 7$ should correspond to the stable operation mode for the membrane. However, at $\lambda \leq 7$, a sharp contraction of the cross-section of the channels

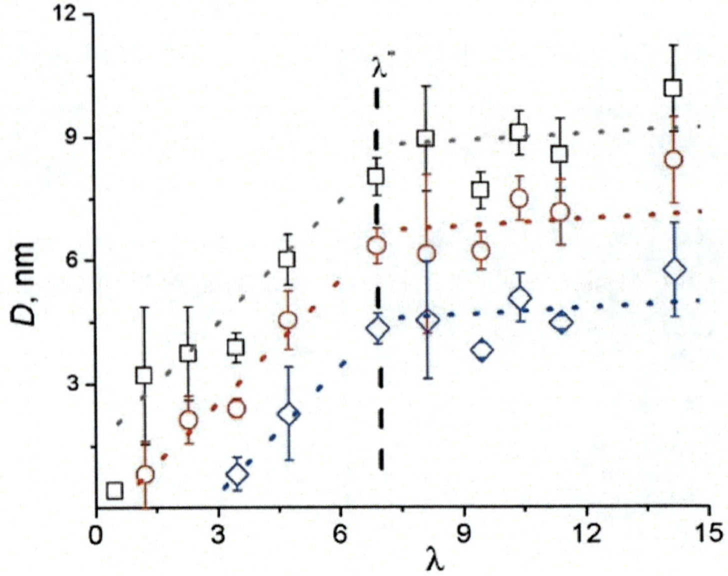

Fig. 6.25 The maximum average diameter D of the cross-section of the water channels as function of the volume fraction of water in case of $[A_3B_2]_8$ chains. The diameters are calculated for the following MP densities: (1) $\rho_C \geq 0.2$; (2) $\rho_C \geq 0.5$; (3) $\rho_C \geq 0.9$. The dotted lines are constructed by linear approximation, λ^* is the percolation threshold

is observed. This behavior is consistent with the behavior of real membranes as a result of water loss at high temperatures. Interesting that at $\lambda \sim 1.5$, when the percolation for water occurs in the system ($\rho_C > 0.2$, see also Fig. 6.24), the maximum cross-section of the channels is $D \sim 2.5$ nm. This fact allows to suppose that SPEEK membrane in the case of $[A_3B_2]_8$ chains can retain its conductive properties even with relatively low water content in the system.

The results of the SPEEK-based ion-exchange membranes simulations show that with a decrease in system hydration, the network of the water channels is broken primarily due to the non-uniform distribution of hydrophilic and hydrophobic domains in the matrix volume. The percolation threshold is higher for more heterogenic domain distribution. It is obvious that the best transport properties can be provided by a material in which, as a result of the microphase separation of polar and non-polar blocks of the polymer chain, the continual distributions of the corresponding domains are formed. In the ideal case, the lowest percolation threshold is demonstrated by ionomers, which in the dry state have the linked network of hydrophilic domains. It can be supposed that the better topology of the structure of the water channels will be demonstrated by ionomers based on diblock copolymers, which contain two types of blocks: (a) susceptible to sulfonation and (b) resistant to H_2SO_4 effect. This statement is confirmed in [155] where the authors changed the degree of sulfonation of the polystyrene block and the total molecular weight of the polymer chain and succeeded to produce the membranes in which hydrophilic domains of various morphologies are observed.

The optimism for the use of diblock copolymers for the production of PEM is due to the extensive experience in the synthesis of such polymers for the production of mesoporous materials [156]. Depending on the composition of the polymer chain, the stable domains are formed as a result of the microphase separation. The domains form various types of supramolecular packaging, such as lamellar, hexagonal, and cubic. The types of cubic packaging: primitive (denoted as *P*), double diamond (DD), and gyroid structures (*G*) are bicontinuous spatial structures. Therefore, the membranes in which such ordering of domains forms as a result of microphase separation, even with low moisture, should have a well-organized network of the water channels. The latter is an important factor for creating high-performance membranes for medium-temperature fuel cells.

To build a model of an ion-exchange membrane, a linear diblock copolymer is used, consisting of *N* MPs of diameter $\sigma = 1$, interacting according to Hooke's law. The model chain includes the polar (hydrophilic *A*) and non-polar (hydrophobic *B*) types of thermodynamically incompatible beads. The chain structure can be written as A_nB_{N-n}, where $N = 24$. The ratio of the hydrophobic and hydrophilic beads is regulated by the parameter *m*, which take values from 1 to 12 ($f = n/(N-n) = 0.04 \div 0.5$). The maximum value of *n* is limited, since for $n > 12$ the polymer becomes water soluble as the polar beads become dominant. As a chemical prototype of monomers of types *A* and *B*, the sulfonated and nonsulfonated SPEEK monomer blocks can be considered. To determine the interaction parameters, the same (as in the DDFT simulations) Flory–Huggins parameters χ_{ij} have been used. To model the membranes, the cubic cell with the edge *L* and periodic boundaries conditions in all three dimensions was used. For a correct description of the hydrodynamic properties of the system, the average number density should be $\rho = 3$ [148], which corresponds to $3L^3$ mesoscopic particles. The total fraction of all MPs in the system is taken as a unity.

For all the simulated systems, the clear inhomogeneous distribution of the mesoscopic particles is observed. An example of such separation for the A_5B_{19}/water system is shown in Fig. 6.26. It is seen that the water MPs form the cylindrical phase. The hydrophilic phase, as shown in Fig. 6.26a, forms some transition zones with a

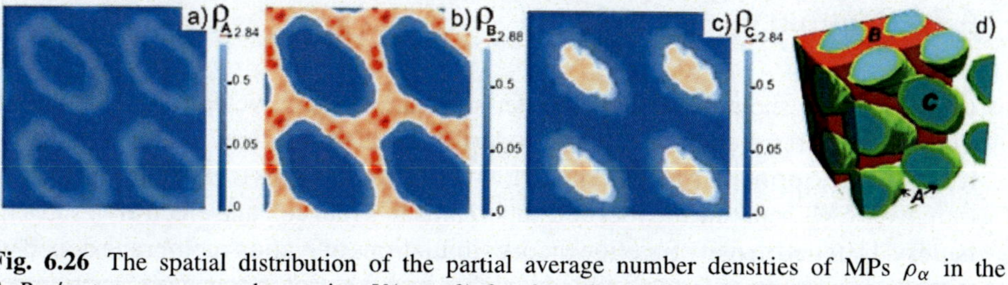

Fig. 6.26 The spatial distribution of the partial average number densities of MPs ρ_α in the A_5B_{19}/water system: at the section $V(x, y, 0)$ for the subsystems: **a** A; **b** B; **c** C. **d** Combined construction for densities $\rho_A \geq 0.5$, $\rho_B \geq 1.75$, $\rho_C \geq 0.75$. $T = 300$ K, $\lambda = 10$, $L = 27\ \sigma$, $t =$ 100,000 DPD steps

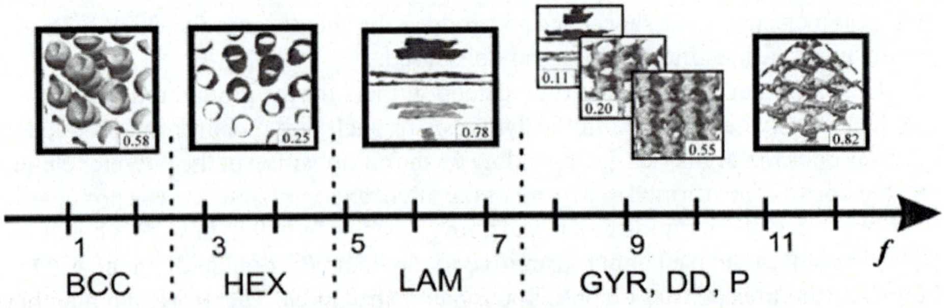

Fig. 6.27 One-dimensional phase diagram of the system $A_nB_{24\text{-}n}$/water ($T = 300$ K, $\lambda = 10$). The vertical lines show the boundaries of the areas of the most frequent occurrence of the following spatial packages: body-centered (bcc), hexagonally packaged cylinders (hex), lamellar (lam), gyroid (G), perforated lamellar (hpl). The examples of snapshots of the simulation cells in the final state demonstrate the boundaries of hydrophobic MPs localization. The isosurfaces are built for $\rho_B = 0.5$

strongly diffused boundary between the subsystems *B* and *C*. From the snapshot as shown in Fig. 6.26d, it is obvious that the water clusters form the cylindrical domains with hexagonal packing; they are embedded in the ionomer matrix and covered with a "shell" of hydrophilic beads.

Figure 6.27 presents the results of the calculation of the one-dimensional phase diagram of the structures that occurs most often during the simulations. The most frequently occurring morphologies are those with the largest probability $\omega(f, L)$. It is defined as the ratio of the number of arisen states with a specific type of symmetry to the number of independent calculations performed. For the accurate identification of the structures found, the structural factors are calculated. Among the identified cubic symmetries, the gyroid structure is characterized by the highest probability $\omega(f, L)$ of occurrence; it forms in the case of the A_9B_{15} chains.

The discussed modeling approaches clearly show that with proper selection of the ratio of the lengths of the hydrophilic and hydrophobic blocks in the ionomer chain, there exists a possibility to create a PEM with a penetrating system of the water channels in the entire volume of the membrane sample.

6.7 Summary

In this chapter, the multiscale computer simulations of polyelectrolyte membranes, the very important components of modern fuel cells and flow batteries, have been reviewed. The primary objective of a PEM is to allow protons to pass through it. Additionally, PEMs separate the fuel and air streams in a fuel cell and electrolytes in a flow battery. Detailed density function theory simulations of Nafion membrane doped with triethylammonium-triflate provided insights into the proton transport reaction coordinates. The classical molecular-dynamics simulations of a Nafion film confined by variable wettability substrates and a Nafion-graphene oxide nanocomposite provided information about the effect of the interfacial interactions in PEM nanocomposites on water clustering and proton diffusion. Finally, the mesoscopic dynamic density function theory simulations of SPEEK membranes revealed the influence of the sulfonation degree on the internal hydrated domain sizes. The MDDFT results on the influence of the amount of PEM hydrophobic and hydrophilic blocks on water cluster morphology have also been discussed.

References

1. L. Carrette, K.A. Friedrich, U. Stimming, Fuel cells—fundamentals and applications. Fuel Cells **1**, 5–39 (2001). https://doi.org/10.1002/1615-6854(200105)1:1%3c5::AID-FUCE5%3e3.0.CO;2-G
2. Wikipedia. Wikipedia (n.d.) https://en.wikipedia.org/wiki/Fuel_cell. Accessed 24 Sept 2019
3. Z. Qi, G.M. Koenig, Review article: flow battery systems with solid electroactive materials. J. Vac. Sci. Technol. B, Nanotechnol. Microelectron. Mater. Process Meas. Phenom. **35**, 040801 (2017). https://doi.org/10.1116/1.4983210
4. P. Antonucci A. Aricò P. Cretì E. Ramunni V. Antonucci, Investigation of a direct methanol fuel cell based on a composite Nafion®-silica electrolyte for high temperature operation. Solid State Ionics **125**, 431–437 (1999). https://doi.org/10.1016/S0167-2738(99)00206-4
5. D.H. Jung, S.Y. Cho, D.H. Peck, D.R. Shin, J.S. Kim, Performance evaluation of a Nafion/silicon oxide hybrid membrane for direct methanol fuel cell. J. Power Sources **106**, 173–177 (2002). https://doi.org/10.1016/S0378-7753(01)01053-9
6. B. Smitha, S. Sridhar, A. Khan, Synthesis and characterization of proton conducting polymer membranes for fuel cells. J. Memb. Sci. **225**, 63–76 (2003). https://doi.org/10.1016/S0376-7388(03)00343-0
7. H. Wang, G.A. Capuano, Behavior of raipore radiation-grafted polymer membranes in H_2/O_2 fuel cells. J. Electrochem. Soc. **145**, 780 (1998). https://doi.org/10.1149/1.1838345
8. R. Nolte, K. Ledjeff, M. Bauer, R. Mülhaupt, Partially sulfonated poly(arylene ether sulfone)—a versatile proton conducting membrane material for modern energy conversion technologies. J. Memb. Sci. **83**, 211–220 (1993). https://doi.org/10.1016/0376-7388(93)85268-2
9. G. Gebel, P. Aldebert, M. Pineri, Swelling study of perfluorosulphonated ionomer membranes. Polymer (Guildf) **34**, 333–339 (1993). https://doi.org/10.1016/0032-3861(93)90086-P
10. J. Kerres, W. Cui, S. Reichle, New sulfonated engineering polymers via the metalation route. I. Sulfonated poly(ethersulfone) PSU Udel® via metalation-sulfination-oxidation. J. Polym. Sci. Part A Polym. Chem. **34**, 2421–2438 (1996). https://doi.org/10.1002/(SICI)1099-0518(19960915)34:12<2421::AID-POLA17>3.0.CO;2-A
11. X. Jin, M.T. Bishop, T.S. Ellis, F.E. Karasz, A sulphonated poly(aryl ether ketone). Br. Polym. J. **17**, 4–10 (1985). https://doi.org/10.1002/pi.4980170102
12. P. Xing, G.P. Robertson, M.D. Guiver, S.D. Mikhailenko, K. Wang, S. Kaliaguine, Synthesis and characterization of sulfonated poly(ether ether ketone) for proton exchange membranes. J. Memb. Sci. **229**, 95–106 (2004). https://doi.org/10.1016/J.MEMSCI.2003.09.019
13. H.-L. Wu, C.-C.M. Ma, C.-H. Li, C.-Y. Chen, Swelling behavior and solubility parameter of sulfonated poly(ether ether ketone). J. Polym. Sci. Part B Polym. Phys. **44**, 3128–3134 (2006). https://doi.org/10.1002/polb.20964
14. R.T.S. Muthu Lakshmi, V. Choudhary, I.K. Varma, Sulphonated poly(ether ether ketone): synthesis and characterisation. J. Mater. Sci. **40**, 629–636 (2005). https://doi.org/10.1007/s10853-005-6300-2

15. N. Agmon, The Grotthuss mechanism. Chem. Phys. Lett. **244**, 456–462 (1995). https://doi.org/10.1016/0009-2614(95)00905-J
16. O. Markovitch, N. Agmon, Structure and energetics of the hydronium hydration shells. J. Phys. Chem. A **111**, 2253–2256 (2007). https://doi.org/10.1021/jp068960g
17. M. Eigen, L. de Maeyer, Self-dissociation and protonic charge transport in water and. Proc. R. Soc. Lond. Ser. A Math. Phys. Sci. **247**, 505–533 (1958). https://doi.org/10.1098/rspa.1958.0208
18. G. Zundel, J. Fritsch, *The Chemical Physics of Solvation*. vol. 2 (Elsevier, 1986). https://doi.org/10.1016/0166-1280(87)85076-5
19. M. Tuckerman, K. Laasonen, M. Sprik, M. Parrinello, *Ab initio* molecular dynamics simulation of the solvation and transport of hydronium and hydroxyl ions in water. J. Chem. Phys. **103**, 150–161 (1995). https://doi.org/10.1063/1.469654
20. M.E. Tuckerman, K. Laasonen, M. Sprik, M. Parrinello, Ab initio simulations of water and water ions. J. Phys. Condens. Matter **6**, A93-100 (1994). https://doi.org/10.1088/0953-8984/6/23A/010
21. P.V. Komarov, P.G. Khalatur, A.R. Khokhlov, Large-scale atomistic and quantum-mechanical simulations of a nafion membrane: morphology, proton solvation and charge transport. Beilstein J. Nanotechnol. **4**, 567–587 (2013). https://doi.org/10.3762/bjnano.4.65
22. M.K. Petersen, G.A. Voth, Characterization of the solvation and transport of the hydrated proton in the perfluorosulfonic acid membrane nafion. J. Phys. Chem. B **110**, 18594–18600 (2006). https://doi.org/10.1021/jp062719k
23. M.K. Petersen, F. Wang, N.P. Blake H. Metiu, G.A. Voth, Excess proton solvation and delocalization in a hydrophilic pocket of the proton conducting polymer membrane Nafion (2005) https://doi.org/10.1021/JP044535G
24. R. Devanathan, A. Venkatnathan, R. Rousseau, M. Dupuis, T. Frigato, W. Gu et al., Atomistic simulation of water percolation and proton hopping in nation fuel cell membrane. J. Phys. Chem. B **114**, 13681–13690 (2010). https://doi.org/10.1021/jp103398b
25. W.Y. Hsu, J.R. Barkley, P. Meakin, Ion percolation and insulator-to-conductor transition in Nafion perfluorosulfonic acid membranes. Macromolecules **13**, 198–200 (1980). https://doi.org/10.1021/ma60073a041
26. G. Gebel, Structural evolution of water swollen perfluorosulfonated ionomers from dry membrane to solution. Polymer (Guildf) **41**, 5829–5838 (2000). https://doi.org/10.1016/S0032-3861(99)00770-3
27. K.A. Mauritz, R.B. Moore, State of understanding of Nafion. Chem. Rev. **104**, 4535–4585 (2004). https://doi.org/10.1021/cr0207123
28. T.D. Gierke, G.E. Munn, F.C. Wilson, The morphology in nafion perfluorinated membrane products, as determined by wide- and small-angle x-ray studies. J. Polym. Sci. Polym. Phys. Ed. **19**, 1687–1704 (1981). https://doi.org/10.1002/pol.1981.180191103
29. C.L. Marx, D.F. Caulfield, S.L. Cooper, Morphology of ionomers. Macromolecules **6**, 344–353 (1973). https://doi.org/10.1021/ma60033a007
30. W.J. Macknight, W.P. Taggart, R.S. Stein, A model for the structure of ionomers. J. Polym. Sci. Polym. Symp. **45**, 113–128 (2007). https://doi.org/10.1002/polc.5070450110
31. W.Y. Hsu, T.D. Gierke, Ion transport and clustering in nafion perfluorinated membranes. J. Memb. Sci. **13**, 307–326 (1983). https://doi.org/10.1016/S0376-7388(00)81563-X
32. M. Fujimura, T. Hashimoto, H. Kawai, Small-angle X-ray scattering study of perfluorinated ionomer membranes. 1. Origin of two scattering maxima. Macromolecules 14, 1309–1315 (1981). https://doi.org/10.1021/ma50006a032
33. M. Fujimura, T. Hashimoto, H. Kawai, Small-angle X-ray scattering study of perfluorinated ionomer membranes. 2. Models for ionic scattering maximum. Macromolecules **15**, 136–144 (1982). https://doi.org/10.1021/ma00229a028
34. B. Dreyfus, G. Gebel, P. Aldebert, M. Pineri, M. Escoubes, M. Thomas, Distribution of the «micelles» in hydrated perfluorinated ionomer membranes from SANS experiments. J. Phys. **51**, 1341–1354 (1990). https://doi.org/10.1051/jphys:0199000510120134100

35. M.H. Litt, Reevaluation of Nafion morphology. Am. Chem. Soc. Polym. Prepr. Div. Polym. Chem. **38**, 80–81 (1997)
36. L. Rubatat, A.L. Rollet, G. Gebel, O. Diat, Evidence of elongated polymeric aggregates in Nafion. Macromolecules **35**, 4050–4055 (2002). https://doi.org/10.1021/ma011578b
37. K. Schmidt-Rohr, Q. Chen, Parallel cylindrical water nanochannels in Nafion fuel-cell membranes. Nat. Mater. **7**, 75–83 (2008). https://doi.org/10.1038/nmat2074
38. J.P. Meyers, J.E. Mcgrath, R. Borup, J. Meyers, B. Pivovar, Y.S. Kim et al., Scientific aspects of polymer electrolyte fuel cell durability and degradation. Chem. Rev. **107**, 3904–3951 (2007). https://doi.org/10.1021/cr050182l
39. M. Kumar, S.J. Paddison, Side-chain degradation of perfluorosulfonic acid membranes: An ab-initio study. J. Mater. Res. 27, 1982–1991 (2012). https://doi.org/10.1557/jmr.2012.191
40. X. Glipa, B. Bonnet, B. Mula, D.J. Jones, J. Rozière, Investigation of the conduction properties of phosphoric and sulfuric acid doped polybenzimidazole. J. Mater. Chem. **9**, 3045–3049 (1999). https://doi.org/10.1039/a906060j
41. Q. Li, J.O. Jensen, R.F. Savinell, N.J. Bjerrum, High temperature proton exchange membranes based on polybenzimidazoles for fuel cells. Prog. Polym. Sci. **34**, 449–477 (2009). https://doi.org/10.1016/j.progpolymsci.2008.12.003
42. D. Rodriguez, C. Jegat, O. Trinquet, J. Grondin, J.C. Lassègues, Proton conduction in poly (acrylamide)-acid blends. Solid State Ionics **61**, 195–202 (1993). https://doi.org/10.1016/0167-2738(93)90354-6
43. P. Musto, F.E. Karasz, W.J. MacKnight, Fourier transform infra-red spectroscopy on the thermo-oxidative degradation of polybenzimidazole and of a polybenzimidazole/polyetherimide blend. Polymer **34**, 2934–2945 (1993). https://doi.org/10.1016/0032-3861(93)90618-K
44. R. Bouchet, E. Siebert, Proton conduction in acid doped polybenzimidazole. Solid State Ionics **118**, 287–299 (1999). https://doi.org/10.1016/S0167-2738(98)00466-4
45. Q. Li, R. He, R.W. Berg, H.A. Hjuler, N.J. Bjerrum, Water uptake and acid doping of polybenzimidazoles as electrolyte membranes for fuel cells. Solid State Ionics **168**, 177–185 (2004). https://doi.org/10.1016/j.ssi.2004.02.013
46. C.E. Hughes, S. Haufe, B. Angerstein, R. Kalim, U. Mähr, A. Reiche et al., Probing structure and dynamics in poly[2,2′-(m-phenylene)-5,5′-bibenzimidazole] fuel cells with magic-angle spinning NMR. J. Phys. Chem. B **108**, 13626–13631 (2004). https://doi.org/10.1021/jp047607c
47. A. Noda, M.A.B. Hasan Susan, K. Kudo, S. Mitsushima, K. Hayamizu, M. Watanabe, Brønsted acid-base ionic liquids as proton-conducting nonaqueous electrolytes. J. Phys. Chem. B **107**, 4024–4033 (2003). https://doi.org/10.1021/jp022347p
48. H. Matsuoka, H. Nakamoto, M.A.B.H. Susan, M. Watanabe, Brønsted acid-base and -polybase complexes as electrolytes for fuel cells under non-humidifying conditions. Electrochim. Acta **50**, 4015–4021 (2005). https://doi.org/10.1016/j.electacta.2005.02.038
49. T.L. Greaves, C.J. Drummond, Protic ionic liquids: properties and applications. Chem. Rev. **108**, 206–237 (2008). https://doi.org/10.1021/cr068040u
50. A. Schechter, R.F. Savinell, Imidazole and 1-methyl imidazole in phosphoric acid doped polybenzimidazole, electrolyte for fuel cells. Solid State Ionics **147**, 181–187 (2002)
51. M.A.B.H. Susan, M. Yoo, H. Nakamoto, M. Watanabe, A novel Brønsted acid–base system as anhydrous proton conductors for fuel cell electrolytes. Chem. Lett. **32**, 836–837 (2003). https://doi.org/10.1246/cl.2003.836
52. M.L. Hoarfrost, M. Tyagi, R.A. Segalman, J.A. Reimer, Proton hopping and long-range transport in the protic ionic liquid [Im][TFSI], probed by pulsed-field gradient NMR and Quasi-elastic neutron scattering. J. Phys. Chem. B **116**, 8201–8209 (2012). https://doi.org/10.1021/jp3044237
53. R. Sood, C. Iojoiu, E. Espuche, F. Gouanvé, G. Gebel, H. Mendil-Jakani et al., Proton conducting ionic liquid doped Nafion membranes: nano-structuration, transport properties and water sorption. J. Phys. Chem. C **116**, 24413–24423 (2012). https://doi.org/10.1021/jp306626y

54. V. Di Noto, M. Piga, G.A. Giffin, S. Lavina, E.S. Smotkin, J.Y. Sanchez et al., Influence of anions on proton-conducting membranes based on neutralized nafion 117, triethylammonium methanesulfonate, and triethylammonium perfluorobutanesulfonate. 2. electrical properties. J. Phys. Chem. C 116, 1370–1379 (2012). https://doi.org/10.1021/jp204242q
55. R. Devanathan, A. Venkatnathan, M. Dupuis, Atomistic simulation of nafion membrane: I. Effect of hydration on membrane nanostructure. J. Phys. Chem. B 111, 8069–8079 (2007). https://doi.org/10.1021/jp0726992
56. A. Venkatnathan, R. Devanathan, M. Dupuis, Atomistic simulations of hydrated nafion and temperature effects on hydronium ion mobility. J. Phys. Chem. B **111**, 7234–7244 (2007). https://doi.org/10.1021/jp0700276
57. C.K. Knox, G.A. Voth, Probing selected morphological models of hydrated Nafion using large-scale molecular dynamics simulations. J. Phys. Chem. B **114**, 3205–3218 (2010). https://doi.org/10.1021/jp9112409
58. K.A. Mauritz, A.J. Hopfinger, Structural properties of membrane ionomers, in *Modern Aspects of Electrochemistry* (Springer US, Boston, MA, 1982), p. 425–508. https://doi.org/10.1007/978-1-4615-7458-3_6
59. H.L. Yeager, A. Steck, Cation and water diffusion in Nafion Ion exchange membranes: influence of polymer structure. J. Electrochem. Soc. **128**, 1880 (1981). https://doi.org/10.1149/1.2127757
60. S.C. Yeo, A. Eisenberg, Physical properties and supermolecular structure of perfluorinated ion-containing (nafion) polymers. J. Appl. Polym. Sci. **21**, 875–898 (1977). https://doi.org/10.1002/app.1977.070210401
61. J.A. Elliott, S. Hanna, A.M.S. Elliott, G.E. Cooley, Atomistic simulation and molecular dynamics of model systems for perfluorinated ionomer membranes. Phys. Chem. Chem. Phys. **1**, 4855–4863 (1999). https://doi.org/10.1039/a905267d
62. A. Vishnyakov, A.V. Neimark, Molecular simulation study of Nafion membrane solvation in water and methanol. J. Phys. Chem. B **104**, 4471–4478 (2000). https://doi.org/10.1021/JP993625W
63. S.S. Jang, V. Molinero, C. Tahir, W.A. Goddard III., Nanophase-segregation and transport in Nafion 117 from molecular dynamics simulations: effect of monomeric sequence. J. Phys. Chem. B **108**, 3149–3157 (2004). https://doi.org/10.1021/jp036842c
64. S.S. Jang, V. Molinero, T. Çagin, W.A. Goddard, Effect of monomeric sequence on nanostructure and water dynamics in Nafion 117. Solid State Ionics **175**, 805–808 (2004). https://doi.org/10.1016/J.SSI.2004.08.039
65. S. Sengupta, R. Pant, P. Komarov, A. Venkatnathan, A.V. Lyulin, Atomistic simulation study of the hydrated structure and transport dynamics of a novel multi acid side chain polyelectrolyte membrane. Int. J. Hydrogen Energy (2017). https://doi.org/10.1016/j.ijhydene.2017.09.078
66. S. Sengupta, A.V. Lyulin, Molecular dynamics simulations of substrate hydrophilicity and confinement effects in Capped Nafion films. J. Phys. Chem. B **122**, 6107–6119 (2018). https://doi.org/10.1021/acs.jpcb.8b03257
67. G. Kritikos, R. Pant, S. Sengupta, K.K. Karatasos, A. Venkatnathan, A.V. Lyulin, Nanostructure and dynamics of humidified Nafion-graphene oxide composites via molecular dynamics simulations. J. Phys. Chem. C **122**, 22864–22875 (2018). https://doi.org/10.1021/acs.jpcc.8b07170
68. S. Sengupta, A.V. Lyulin, Molecular modeling of structure and dynamics of Nafion protonation states. J. Phys. Chem. B **123**, 6882–6891 (2019). https://doi.org/10.1021/acs.jpcb.9b04534
69. S.J. Paddison, T.A. Zawodzinski Jr., Molecular modeling of the pendant chain in Nafion®. Solid State Ionics **113–115**, 333–340 (1998). https://doi.org/10.1016/S0167-2738(98)00298-7
70. S.J. Paddison, L.R. Pratt, T.A. Zawodzinski, Conformations of perfluoroether sulfonic acid side chains for the modeling of Nafion. J. New Mater. Electrochem. Syst. **2**, 183–188 (1999)
71. S.J. Paddison, The modeling of molecular structure and ion transport in sulfonic acid based ionomer membranes. J. New Mater. Electrochem. Syst. **4**, 197–207 (2001)

72. D.A. Mologin, P.G. Khalatur, A.R. Khokhlov, Structural organization of water-containing Nafion: a cellular-automaton-based simulation. Macromol. Theory Simul. **11**, 587 (2002). https://doi.org/10.1002/1521-3919(20020601)11:5%3c587::AID-MATS587%3e3.0.CO;2-P
73. P.G. Khalatur, S.K. Talitskikh, A.R. Khokhlov, Structural organization of water-containing Nafion: the integral equation theory. Macromol. Theory Simul. **11**, 566 (2002). https://doi.org/10.1002/1521-3919(20020601)11:5%3c566::AID-MATS566%3e3.0.CO;2-0
74. S. Yamamoto, S.A. Hyodo, A computer simulation study of the mesoscopic structure of the polyelectrolyte membrane Nafion. Polym. J. **35**, 519–527 (2003). https://doi.org/10.1295/polymj.35.519
75. P.V. Komarov, I.N. Veselov, P.G. Khalatur, Self-organization of amphiphilic block copolymers in the presence of water: A mesoscale simulation. Chem. Phys. Lett. **605–606**, 22–27 (2014). https://doi.org/10.1016/J.CPLETT.2014.05.004
76. J.T. Wescott, Y. Qi, L. Subramanian, C.T. Weston, Mesoscale simulation of morphology in hydrated perfluorosulfonic acid membranes. J. Chem. Phys. **124**, 134702 (2006). https://doi.org/10.1063/1.2177649
77. P.V. Komarov, I.N. Veselov, P.P. Chu, P.G. Khalatur, Mesoscale simulation of polymer electrolyte membranes based on sulfonated poly (ether ether ketone) and Nafion. Soft Matter **6**, 3939 (2010). https://doi.org/10.1039/b921369d
78. B. Muriithi, D. Loy, Proton conductivity of Nafion/ex-situ sulfonic acid-modified Stöber silica nanocomposite membranes as a function of temperature, silica particles size and surface modification. Membranes (Basel) **6**, 12 (2016). https://doi.org/10.3390/membranes6010012
79. R. Gosalawit, S. Chirachanchai, S. Shishatskiy, S.P. Nunes, Krytox–Montmorillonite–Nafion® nanocomposite membrane for effective methanol crossover reduction in DMFCs. Solid State Ionics **178**, 1627–1635 (2007). https://doi.org/10.1016/J.SSI.2007.10.008
80. M.M. Hasani-Sadrabadi, E. Dashtimoghadam, F.S. Majedi, S. Wu, A. Bertsch, H. Moaddel et al., Nafion/Chitosan-wrapped CNT nanocomposite membrane for high-performance direct methanol fuel cells. RSC Adv. **3**, 7337 (2013). https://doi.org/10.1039/c3ra40480c
81. B.R. Matos, R.A. Isidoro, E.I. Santiago, F.C. Fonseca, Performance enhancement of direct ethanol fuel cell using Nafion composites with high volume fraction of titania. J. Power Sources **268**, 706–711 (2014). https://doi.org/10.1016/J.JPOWSOUR.2014.06.097
82. A.Z. Weber, A. Kusoglu, Unexplained transport resistances for low-loaded fuel-cell catalyst layers. J. Mater. Chem. A **2**, 17207–17211 (2014). https://doi.org/10.1039/C4TA02952F
83. M.A. Modestino, D.K. Paul, S. Dishari, S.A. Petrina, F.I. Allen, M.A. Hickner et al., Self-assembly and transport limitations in confined Nafion films. Macromolecules **46**, 867–873 (2013). https://doi.org/10.1021/ma301999a
84. E. Passalacqua, F. Lufrano, G. Squadrito, A. Patti, L. Giorgi, Nafion content in the catalyst layer of polymer electrolyte fuel cells: effects on structure and performance. Electrochim. Acta **46**, 799–805 (2001). https://doi.org/10.1016/S0013-4686(00)00679-4
85. D. Damasceno Borges, A.A. Franco, K. Malek, G. Gebel, S. Mossa, Inhomogeneous transport in model hydrated polymer electrolyte supported ultrathin films. ACS Nano **7**, 6767–6773 (2013). https://doi.org/10.1021/nn401624p
86. D. Damasceno Borges, G. Gebel, A.A. Franco, K. Malek, S. Mossa, Morphology of supported polymer electrolyte ultrathin films: a numerical study. J. Phys. Chem. C **119**, 1201–1216 (2015). https://doi.org/10.1021/jp507598h
87. F.F. Abraham, Y. Singh, The structure of a hard-sphere fluid in contact with a soft repulsive wall. J. Chem. Phys. **67**, 2384 (1977). https://doi.org/10.1063/1.435080
88. Cha S-H. Recent development of nanocomposite membranes for vanadium redox flow batteries. J. Nanomater. **2015**, 1–12 (2015). https://doi.org/10.1155/2015/207525
89. A. Kusoglu, T.J. Dursch, A.Z. Weber, Nanostructure/swelling relationships of bulk and thin-film PFSA ionomers. Adv. Funct. Mater. **26**, 4961–4975 (2016). https://doi.org/10.1002/adfm.201600861
90. M. Bass, A. Berman, A. Singh, O. Konovalov, V. Freger, Surface-induced Micelle orientation in Nafion films. Macromolecules **44**, 2893–2899 (2011). https://doi.org/10.1021/ma102361f

91. S. Cui, J. Liu, M.E. Selvan, D.J. Keffer, B.J. Edwards, W.V. Steele, A molecular dynamics study of a nafion polyelectrolyte membrane and the aqueous phase structure for proton transport. J. Phys. Chem. B **111**, 2208–2218 (2007). https://doi.org/10.1021/jp066388n
92. M. Tripathy, P.B.S. Kumar, A.P. Deshpande, Molecular structuring and percolation transition in hydrated sulfonated poly (ether ether ketone) membranes. J. Phys. Chem. B **121**, 4873–4884 (2017). https://doi.org/10.1021/acs.jpcb.7b01045
93. T.A. Zawodzinski, C. Derouin, S. Radzinski, R.J. Sherman, V.T. Smith, T.E. Springer et al., Water uptake by and transport through Nafion® 117 membranes. J. Electrochem. Soc. **140**, 1041 (1993). https://doi.org/10.1149/1.2056194
94. M. Bass, A. Berman, A. Singh, O. Konovalov, V. Freger, Surface structure of nafion in vapor and liquid. J. Phys. Chem. B **114**, 3784–3790 (2010). https://doi.org/10.1021/jp9113128
95. S. Goswami, S. Klaus, J. Benziger, Wetting and absorption of water drops on nafion films. Langmuir **24**, 8627–8633 (2008). https://doi.org/10.1021/la800799a
96. N.J. Economou, A.M. Barnes, A.J. Wheat, M.S. Schaberg, S.J. Hamrock, S.K. Buratto, Investigation of humidity dependent surface morphology and proton conduction in multi-acid side chain membranes by conductive probe atomic force microscopy. J. Phys. Chem. B **119**, 14280–14287 (2015). https://doi.org/10.1021/acs.jpcb.5b07255
97. H. Noguchi, K. Taneda, H. Minowa, H. Naohara, K. Uosaki, Humidity-dependent structure of surface water on perfluorosulfonated ionomer thin film studied by sum frequency generation spectroscopy. J. Phys. Chem. C **114**, 3958–3961 (2010). https://doi.org/10.1021/jp907194k
98. O. Kwon, Y. Kang, S. Wu, D.M. Zhu, Characteristics of microscopic proton current flow distributions in fresh and aged nafion membranes. J/ Phys/ Chem/ B **114**, 5365–5370 (2010). https://doi.org/10.1021/jp911182q
99. R.S. McLean, M. Doyle, B.B. Sauer, High-resolution imaging of ionic domains and crystal morphology in ionomers using AFM techniques. Macromolecules **33**, 6541–6550 (2000). https://doi.org/10.1021/ma000464h
100. D. Novitski, S. Holdcroft, Determination of O_2 mass transport at the Pt | PFSA ionomer interface under reduced relative humidity. ACS Appl Mater. Interfaces **7**, 27314–27323 (2015). https://doi.org/10.1021/acsami.5b08720
101. J. Tang, W. Yuan, J. Zhang, H. Li, Y. Zhang, Evidence for a crystallite-rich skin on perfluorosulfonate ionomer membranes. RSC Adv. **3**, 8947–8952 (2013). https://doi.org/10.1039/c3ra40430g
102. F.N. Büchi, S. Srinivasa, Operating proton exchange membrane fuel cells without external humidification of the reactant gases. J. Electrochem. Soc. **144**, 2767 (1997). https://doi.org/10.1149/1.1837893
103. H.S. Park, Y.J. Kim, W.H. Hong, Y.S. Choi, H.K. Lee, Influence of morphology on the transport properties of perfluorosulfonate ionomers/polypyrrole composite membrane. Macromolecules **38**, 2289–2295 (2005). https://doi.org/10.1021/ma047650y
104. K. Pourzare, Y. Mansourpanah, S. Farhadi, Advanced nanocomposite membranes for fuel cell applications: a comprehensive review. Biofuel Res. J. **3**, 496–513 (2016). https://doi.org/10.18331/BRJ2016.3.4.4
105. A. Enotiadis, K. Angjeli, N. Baldino, I. Nicotera, D. Gournis, Graphene-based nafion nanocomposite membranes: enhanced proton transport and water retention by novel organo-functionalized graphene oxide nanosheets. Small **8**, 3338–3349 (2012). https://doi.org/10.1002/smll.201200609
106. G. Liu, W. Jin, N. Xu, Graphene-based membranes. Chem. Soc. Rev. **44**, 5016–5030 (2015). https://doi.org/10.1039/c4cs00423j
107. B.G. Choi, Y.S. Huh, Y.C. Park, D.H. Jung, W.H. Hong, H. Park, Enhanced transport properties in polymer electrolyte composite membranes with graphene oxide sheets. Carbon N Y **50**, 5395–5402 (2012). https://doi.org/10.1016/j.carbon.2012.07.025
108. R. Kumar, C. Xu, K. Scott, Graphite oxide/Nafion composite membranes for polymer electrolyte fuel cells. RSC Adv. **2**, 8777 (2012). https://doi.org/10.1039/c2ra20225e
109. H. Zarrin, D. Higgins, Y. Jun, Z. Chen, M. Fowler, Functionalized graphene oxide nanocomposite membrane for low humidity and high temperature proton exchange membrane fuel cells. J. Phys. Chem. C **115**, 20774–20781 (2011). https://doi.org/10.1021/jp204610j

110. D.C. Lee, H.N. Yang, S.H. Park, W.J. Kim, Nafion/graphene oxide composite membranes for low humidifying polymer electrolyte membrane fuel cell. J. Memb. Sci. **452**, 20–28 (2014). https://doi.org/10.1016/j.memsci.2013.10.018
111. M.R. Karim, K. Hatakeyama, T. Matsui, H. Takehira, T. Taniguchi, M. Koinuma et al., Graphene oxide nanosheet with high proton conductivity. J. Am. Chem. Soc. **135**, 8097–8100 (2013). https://doi.org/10.1021/ja401060q
112. T. Bayer, S.R. Bishop, M. Nishihara, K. Sasaki, S.M. Lyth, Characterization of a graphene oxide membrane fuel cell. J. Power Sources **272**, 239–247 (2014). https://doi.org/10.1016/j.jpowsour.2014.08.071
113. L. Wang, J. Kang, J.-D. Nam, J. Suhr, A.K. Prasad, S.G. Advani, Composite membrane based on graphene oxide sheets and Nafion for polymer electrolyte membrane fuel cells. ECS Electrochem. Lett. **4**, F1-4 (2014). https://doi.org/10.1149/2.0021501eel
114. Z. Jiang, X. Zhao, Y. Fu, A. Manthiram, Composite membranes based on sulfonated poly(ether ether ketone) and SDBS-adsorbed graphene oxide for direct methanol fuel cells. J. Mater. Chem. **22**, 24862–24869 (2012). https://doi.org/10.1039/c2jm35571j
115. N. Üregen, K. Pehlivanoğlu, Y. Özdemir, Y. Devrim, Development of polybenzimidazole/graphene oxide composite membranes for high temperature PEM fuel cells. Int. J. Hydrog. Energ **42**, 2636–2647 (2017). https://doi.org/10.1016/j.ijhydene.2016.07.009
116. R. Rudra, V. Kumar, N. Pramanik, P.P. Kundu, Graphite oxide incorporated crosslinked polyvinyl alcohol and sulfonated styrene nanocomposite membrane as separating barrier in single chambered microbial fuel cell. J. Power Sources **341**, 285–293 (2017). https://doi.org/10.1016/j.jpowsour.2016.12.028
117. K. Karatasos, G. Kritikos, Characterization of a graphene oxide/poly(acrylic acid) nanocomposite by means of molecular dynamics simulations. RSC Adv. **6**, 109267–109277 (2016). https://doi.org/10.1039/C6RA22951D
118. K. Karatasos, G. Kritikos, A microscopic view of graphene-oxide/poly(acrylic acid) physical hydrogels: effects of polymer charge and graphene oxide loading. Soft Matter **14**, 614–627 (2018). https://doi.org/10.1039/c7sm02305g
119. R. Devanathan, D. Chase-Woods, Y. Shin, D.W. Gotthold, Molecular dynamics simulations reveal that water diffusion between graphene oxide layers is slow. Sci. Rep. **6**, 1–8 (2016). https://doi.org/10.1038/srep29484
120. S. Feng, G.A. Voth, Proton solvation and transport in hydrated nafion. J. Phys. Chem. B **115**, 5903–5912 (2011). https://doi.org/10.1021/jp2002194
121. R. Devanathan, A. Venkatnathan, R. Rousseau, M. Dupuis, T. Frigato, W. Gu et al., Atomistic simulation of water percolation and proton hopping in Nafion fuel cell membrane. J. Phys. Chem. B **114**, 13681–13690 (2010). https://doi.org/10.1021/jp103398b
122. M. Ester, H.-P. Kriegel, J. Sander, X. Xu, A density-based algorithm for discovering clusters in large spatial databases with noise, in *KDD-96 Proceedings*, vol. 96, p. 226–231 (1996). https://doi.org/10.1103/physicsphysiquefizika.3.255
123. M.E. Fisher, The theory of condensation and the critical point. Phys. Phys. Fiz **3**, 255–283 (1967). https://doi.org/10.1103/PhysicsPhysiqueFizika.3.255
124. W. vanMegen, T.C. Mortensen, J. Müller, S.R. Williams, W. van Megen, T.C. Mortensen et al., Measurement of the self intermediate scattering function of suspensions of hard spherical particles near the glass transition. Phys. Rev. E **58**, 6073–6085 (1998). https://doi.org/10.1103/PhysRevE.58.6073
125. R. Kohlrausch, Theorie des elektrischen Rückstandes in der Leidener Flasche. Ann. Phys. **167**, 179–214 (1854). https://doi.org/10.1002/andp.18541670203
126. K.L. Ngai, *Relaxation and Diffusion in Complex Systems* (Springer, New York, 2011) https://doi.org/10.1007/978-1-4419-7649-9
127. G. Kritikos, K. Karatasos, Temperature dependence of dynamic and mechanical properties in poly(acrylic acid)/graphene oxide nanocomposites. Mater. Today Commun. **13**, 359–366 (2017). https://doi.org/10.1016/j.mtcomm.2017.11.006
128. G. Kritikos, N. Vergadou, I.G. Economou, Molecular dynamics simulation of highly confined Glassy ionic liquids. J. Phys. Chem. C **120**, 1013–1024 (2016). https://doi.org/10.1021/acs.jpcc.5b09947

129. R. Devanathan, A. Venkatnathan, M. Dupuis, Atomistic simulation of Nafion membrane. 2. Dynamics of water molecules and hydronium ions. J. Phys. Chem. B111, 13006–13013 (2007). https://doi.org/10.1021/jp0761057
130. S. Pahari, C.K. Choudhury, P.R. Pandey, M. More, A. Venkatnathan, S. Roy, Molecular dynamics simulation of phosphoric acid doped monomer of polybenzimidazole: a potential component polymer electrolyte membrane of fuel cell. J. Phys. Chem. B **116**, 7357–7366 (2012). https://doi.org/10.1021/jp301117m
131. S. Pahari, S. Roy, Structural and conformational properties of polybenzimidazoles in melt and phosphoric acid solution: a polyelectrolyte membrane for fuel cells. RSC Adv. **6**, 8211–8221 (2016). https://doi.org/10.1039/C5RA22159E
132. K. Shirata, S. Kawauchi, Effect of benzimidazole configuration in polybenzimidazole chain on interaction with phosphoric acid: a DFT study. J. Phys. Chem. B **119**, 592–603 (2015). https://doi.org/10.1021/jp510067n
133. S.C. Kumbharkar, U.K. Kharul, New N-substituted ABPBI: synthesis and evaluation of gas permeation properties. J. Memb. Sci. **360**, 418–425 (2010). https://doi.org/10.1016/j.memsci.2010.05.041
134. M.A. Habib, J.O. Bockris, Adsorption at the solid/solution interface. An FTIR study of phosphoric acid on platinum and gold. J. Electrochem. Soc.132, 108 (1985)
135. M. Kumar, A. Venkatnathan, Mechanism of proton transport in ionic-liquid-doped perfluorosulfonic acid membranes. J. Phys. Chem. B **117**, 14449–14456 (2013). https://doi.org/10.1021/jp408352w
136. M. Kumar, A. Venkatnathan, Quantum chemistry study of proton transport in imidazole chains. J. Phys. Chem. B **119**, 3213–3222 (2015). https://doi.org/10.1021/jp508994c
137. R. Pant, M. Kumar, A. Venkatnathan, Quantum mechanical investigation of proton transport in imidazolium methanesulfonate ionic liquid. J. Phys. Chem. C **121**, 7069–7080 (2017). https://doi.org/10.1021/acs.jpcc.6b11997
138. A.P. Sunda, M. More, A. Venkatnathan, A molecular investigation of the nanostructure and dynamics of phosphoric-triflic acid blends of hydrated ABPBI [poly(2,5-benzimidazole)] polymer electrolyte membranes. Soft Matter **9**, 1122–1132 (2013). https://doi.org/10.1039/c2sm26927a
139. M. More, S. Pahari, S. Roy, A. Venkatnathan, Characterization of the structures and dynamics of phosphoric acid doped benzimidazole mixtures: a molecular dynamics study. J. Mol. Model **19**, 109–118 (2013). https://doi.org/10.1007/s00894-012-1519-8
140. M. More, A.P. Sunda, A. Venkatnathan, Polymer chain length, phosphoric acid doping and temperature dependence on structure and dynamics of an ABPBI [poly(2,5-benzimidazole)] polymer electrolyte membrane. RSC Adv. **4**, 19746–19755 (2014). https://doi.org/10.1039/c4ra01421a
141. J.G.E.M. Fraaije, B.A.C. van Vlimmeren, N.M. Maurits, M. Postma, O.A. Evers, C. Hoffmann et al., The dynamic mean-field density functional method and its application to the mesoscopic dynamics of quenched block copolymer melts. J. Chem. Phys. **106**, 4260–4269 (1997). https://doi.org/10.1063/1.473129
142. P.J. Hoogerbrugge, J.M.V.A. Koelman, Simulating microscopic hydrodynamic phenomena with dissipative particle dynamics. Europhys. Lett. **19**, 155–160 (1992). https://doi.org/10.1209/0295-5075/19/3/001
143. N.M. Maurits, B.A.C. Van Vlimmeren, J.G.E.M. Fraaije, Mesoscopic phase separation dynamics of compressible copolymer melts. (1997)
144. N.M. Maurits, A.V.M. Zvelindovsky, G.J.A. Sevink, B.A.C. Van Vlimmeren, J.G.E.M. Fraaije, N.M. Maurits et al., Hydrodynamic effects in three-dimensional microphase separation of block copolymers). Hydrodynamic effects in three-dimensional microphase separation of block copolymers: Dynamic mean-field density functional approach. J. Chem. Phys. **108**, 9150–9154 (1998). https://doi.org/10.1063/1.476362.
145. A.V.M. Zvelindovsky, G.J.A. Sevink, B.A.C. van Vlimmeren, N. Maurits, J.E.M. Fraaije, Three-dimensional mesoscale dynamics of block copolymers under shear: The dynamic density-functional approach: the dynamic density-functional approach. Phys. Rev. E **57**, R4879–R4882 (1998)

146. J.J. Krueger, P.P. Simon, H.J. Ploehn, Phase behavior and microdomain structure in perfluorosulfonated ionomers via self-consistent mean field theory (2002). https://doi.org/10.1021/MA0020638
147. J.M.V.A. Koelman, P.J. Hoogerbrugge, Dynamic simulations of hard-sphere suspensions under steady shear. Europhys. Lett. **21**, 363–368 (1993). https://doi.org/10.1209/0295-5075/21/3/018
148. R.D. Groot, P.B. Warren, Dissipative particle dynamics: bridging the gap between atomistic and mesoscopic simulation. J. Chem. Phys. **107**, 4423–4435 (1997). https://doi.org/10.1063/1.474784
149. P.J. Flory, M. Volkenstein, Statistical mechanics of chain molecules. Biopolymers **8**, 699–700 (1969). https://doi.org/10.1002/bip.1969.360080514
150. J. Bicerano, *Prediction of Polymer Properties* (Marcel Dekker, 2002)
151. M.L. Huggins, The solubility of nonelectrolytes. By Joel H. Hildebrand and Robert S. Scott. J. Phys. Chem. **55**, 619–620 (1951). https://doi.org/10.1021/j150487a027.
152. P.J. Flory, Fifteenth Spiers Memorial Lecture. Thermodynamics of polymer solutions. Discuss. Faraday Soc. 49, 7 (1970). https://doi.org/10.1039/df9704900007
153. A.A. Askadiskii, V.I. Kondraschenko, Computer material science of polymers. Sci. World **1**, 544 (1999)
154. R. Consiglio, D.R. Baker, G. Paul, H.E. Stanley, Continuum percolation thresholds for mixtures of spheres of different sizes. Phys. A Stat. Mech. Its Appl. **319**, 49–55 (2003). https://doi.org/10.1016/S0378-4371(02)01501-7
155. M.J. Park, K.H. Downing, A. Jackson, E.D. Gomez, A.M. Minor, D. Cookson et al., Increased water retention in polymer electrolyte membranes at elevated temperatures assisted by capillary condensation (2007). https://doi.org/10.1021/NL072617L
156. S. Förster, M. Konrad, From self-organizing polymers to nano- and biomaterials. J. Mater. Chem. **13**, 2671–2688 (2003). https://doi.org/10.1039/B307512P

Chapter 7
Explorations into the Mechanics of Hairy Nanoparticle Assemblies with Molecular Dynamics

Nitin K. Hansoge and Sinan Keten

Abstract This chapter summarizes recent investigations into the mechanical behavior of matrix-free hairy nanoparticles using molecular simulation techniques. Generic as well as systematic coarse-graining techniques are presented, highlighting emerging methods for chemistry-specific modeling of polymeric materials. Applications of coarse-graining to study assembled hairy nanoparticle systems are overviewed, with an emphasis on quantitative structure–property relationships obtained from these investigations. Comparisons of simulations with theoretical scaling relationships such as the Daoud-Cotton theory and experimental data are provided. Methods to accelerate the design space through upscaling techniques and metamodel development are briefly overviewed. Key physical insights obtained from these studies on the effects of grafting density, grafting length, and polymer chain chemistry are mentioned throughout the chapter to illustrate the importance of modeling contributions. A brief outlook into the prospects of using novel mesoscopic approaches such as those based on potentials of mean force is noted. The chapter concludes with a summary and outlook on the state of the art of the computational design of assembled hairy nanoparticles.

N. K. Hansoge · S. Keten (✉)
Department of Mechanical Engineering, Northwestern University, 2145 Sheridan Road, Evanston, IL 60208-3109, USA
e-mail: s-keten@northwestern.edu

N. K. Hansoge
e-mail: NitinHansoge2020@u.northwestern.edu

Center for Hierarchical Materials Design, Northwestern University, 2205 Tech Drive, Evanston, IL 60208-3109, USA

S. Keten
Department of Civil & Environmental Engineering, Northwestern University, 2145 Sheridan Road, Evanston, IL 60208-3109, USA

V. V. Ginzburg and L. M. Hall (eds.), *Theory and Modeling of Polymer Nanocomposites*, Springer Series in Materials Science 310,
https://doi.org/10.1007/978-3-030-60443-1_7

7.1 Introduction

It has been well documented that introducing nanoparticles in a polymer matrix enhances the thermomechanical properties of the nanocomposite [1–5]. The dispersion state of these nanoparticles in the matrix controls the mechanical properties of the material [6–8]. Studies have shown that increasing nanoparticle content beyond a certain weight fraction starts to degrade the mechanical properties [9]. An innovative way to overcome these dispersion issues is to covalently graft polymers onto nanoparticles and get rid of the polymeric matrix [10]. Such self-assembled structures consisting of core–shell systems, called assembled hairy nanoparticles (aHNPs) as shown in Fig. 7.1, provide structural order to the nanocomposite [11]. The grafted polymer chains themselves form the "matrix" phase and provide the necessary balance between steric hindrance and cohesion that leads to nanocomposites with relatively regular spacing between the nanoparticles [12–14]. In these assembled systems, the nanoparticles serve as both a reinforcing phase and tethering points for the polymer chains. The improved structural order helps in tailoring the mechanical properties accurately, thereby making this a promising new approach making new materials with unprecedented mechanical properties [15, 16].

There have been several studies investigating the various properties of aHNPs. Michael Bockstaller and his co-workers have used techniques such as nanoindentation and Brillouin light spectroscopy to characterize the mechanical properties of aHNPs [15–17]. They found out that Young's modulus of polymer-grafted silica nanoparticles is greatly influenced by the cohesive interactions between the polymer chains and their fracture toughness increases substantially beyond a critical degree of polymerization to values that are close to neat glassy polymers. Sanat Kumar and his co-workers have developed theories behind the chain conformation using polymer physics scaling laws [18, 19]. Based on these laws, they found how the particles aggregate based on the molecular weight and grafting density of the polymer chains [20].

These studies have been foundational for explaining structure—mechanical property relationships in aHNPs. Further insights have come from recent coarse-grained simulations that explained key experimental observations [21]. Uncovering the dependence of properties on molecular-scale design parameters remains challenging

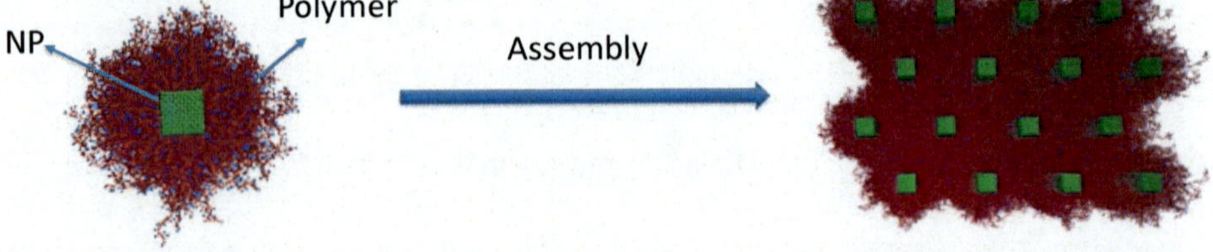

Fig. 7.1 Schematic of matrix-free assembly of polymer-grafted nanoparticles

to address by theory or generic polymer chain models commonly used in molecular dynamics. Atomistic simulations of aHNPs that can reveal these molecular mechanisms while accounting for chemical structure are computationally expensive due to spatiotemporal limitations. Hence, multiscale modeling techniques such as coarse-grained molecular dynamics (CG-MD) simulations are being prevalently used. While there are several approaches to develop a CG model, a temperature transferable energy renormalization approach that preserves the dynamics and mechanical properties across a range of thermodynamic states is typically desired. A recently developed approach to accomplishing this feat will be discussed in the upcoming section.

7.2 Energy Renormalization (ER) Approach to Coarse-Graining

Assessing the complex dynamical and mechanical properties of polymers and polymer nanocomposites using all-atomistic (AA) MD simulations is computationally challenging due to their spatiotemporal limitations. Multiscale modeling techniques such as coarse-graining approaches that are derived from atomistic data can overcome the limitations imposed by AA-MD simulations. In order to improve the computational efficiency of molecular simulations, the "unessential" atomistic features can be removed and thereby reducing the number of degrees of freedom [22]. There have been several CG approaches proposed based on statistical mechanics and theories, including the inverse Boltzmann method (IBM), [23] force matching (also called the multiscale coarse-graining), [24, 25] relative entropy [26] and inverse Monte Carlo methods [27]. Although these methods preserve the static structure of the AA system [e.g., radial distribution function (RDF)], matching the dynamical and mechanical properties across different thermodynamical states has been challenging [28, 29]. By introducing time-scaling factors [30] or non-conservative forces such as frictional and dissipative forces, [31, 32], it is possible to develop the CG model to capture the AA dynamics and thermodynamics. However, the "temperature transferability" of CG modeling remains challenging because of the complex ways that temperature affects molecular friction, configurational entropy, and relaxation behavior of polymers.

7.2.1 ER Methodology

To overcome the issues of temperature transferability, a new coarse-graining strategy, namely the energy renormalization (ER) approach, was proposed for capturing dynamics of polymers and small molecules [33–35]. The inspiration behind this approach is the Adam-Gibbs (AG) theory [36] and the more recent generalized entropy theory (GET) of glass formation [37, 38]. The essential idea borrowed from these theories is the critical role of configurational entropy, s_c, in glass formation. The "entropy-enthalpy compensation effect" [39, 40] states that the effective enthalpy of the system must correspondingly increase as s_c decreases under coarse-graining, to preserve the overall properties of the glass forming material. The monomeric cohesive interaction in these GF systems is often described by the Lennard–Jones (LJ) parameter ε in the nonbonded interactions. The GET predicts that ε has a strong influence on the dynamics and mechanical response of GF liquids through its influence on s_c [41, 42]. Thus by renormalizing ε as a function of temperature, the aim is to "correct" for the activation free energy and thus preserve the dynamics of the CG polymer during glass formation. Hence the name "energy renormalization."

Historically, there have been a few efforts to correlate dynamics of GF systems with Debye Waller Factor $\langle u^2 \rangle$, a physical property that quantifies the molecular free volume and local stiffness at picosecond caging timescale [43, 44]. Considering that the activation "barrier" to molecular motions in a condensed matter should increase with the stiffness of the material that emerges from the mutual interactions of surrounding molecules, Hall and Wolynes (HW) [45] proposed that the structural relaxation time τ of GF liquid should obey the simple relation with Debye Waller factor $\langle u^2 \rangle$ as $\tau \sim \exp(u_0^2/\langle u^2 \rangle)$, where u_0^2 is an adjustable constant. Similar relationships have been argued by Schweizer and co-workers [46–48] which has similar features to the "shoving" model of Dyre and co-workers, [49, 50] where the activation barrier was alternatively identified with the glassy shear modulus of the material.

Starr et al. provided the first simulations to show the qualitative validity of the HW relation in the polymer melt [44, 51]. Leporini and co-workers [52–54] later showed that τ data from many simulations and experiments could be related to $\langle u^2 \rangle$ if the barrier was taken as a power series of $1/\langle u^2 \rangle$, while Simmons and co-workers [55] argued that τ could be described by a localization model (LM) (i.e., an extension of the HW model), $\tau \sim \exp\left[\left(u_0^2/\langle u^2 \rangle\right)^{\alpha/2}\right]$, where α is an exponent related to the shape of the free volume. A problem with these τ and $\langle u^2 \rangle$ correlations is that a number of free parameters involved limit their predictive ability. Thus, Betancourt et al. [56] chose a reduced value of $\langle u^2 \rangle$ at the onset temperature T_A for molecular caging. They fixed the prefactor in the τ and $\langle u^2 \rangle$ relation by the observed τ value τ_A at T_A, leading to the scaling form:

$$\tau(T) = \tau_A \exp\left[\left(u_A^2/\langle u^2(T) \rangle\right)^{\alpha/2} - 1\right] \tag{7.1}$$

This relation has been taken even further by Jack Douglas and co-workers [57] in the case of Cu-Zr metallic glass alloys, where the free volume geometry is expected to be more isotopic, leading to $\alpha \approx 3$. Remarkably, this relation was found to describe the dynamics of Cu-Zr metallic glass over a large range of composition and temperature without any free parameters.

Building upon this LM of relaxation, it is hypothesized that by preserving $\langle u^2 \rangle$ of the AA system via renormalizing ε, the temperature-dependent GF dynamics for the CG modeling can be recovered. To test this idea, the influence of cohesive interaction

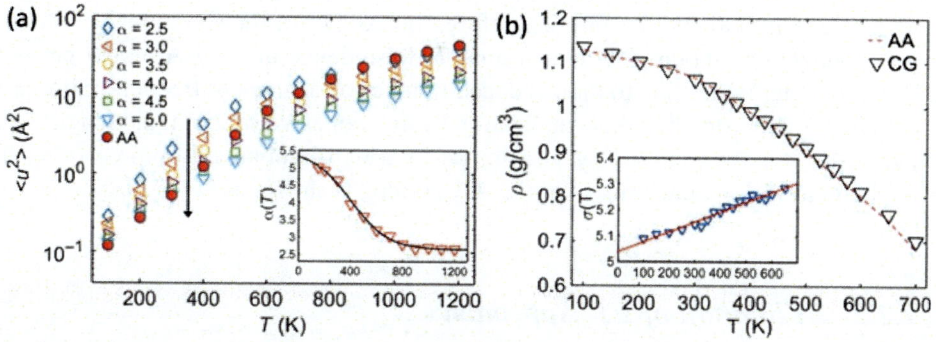

Fig. 7.2 **a** Variation of $\langle u^2 \rangle$ with temperature for polycarbonate with varying $\alpha(T)$. The vertical arrow indicates dataset with increase in α. Inset: The resultant sigmoidal form of $\alpha(T)$ obtained by preserving the temperature dependent $\langle u^2 \rangle$ of the AA model. **b** Variation of ρ with temperature for ortho-ter phenyl (OTP) with varying $\beta(T)$. Inset: Polynomial form for $\beta(T)$. Reprinted with permission from (**a**) [33] © 2019 AAAS and (**b**) [34] © 2018 American Chemical Society

ε on $\langle u^2 \rangle$ is evaluated. Specifically, an ER factor $\alpha(T)$ is introduced where $\varepsilon(T) = \alpha(T)\ \varepsilon_0$ and ε_0 is the constant obtained from the radial distribution function using inverse Boltzmann method. The $\langle u^2 \rangle$ is calculated from the segmental mean-squared displacements (MSD) $\langle r^2(t) \rangle$ at around $t_c \approx 3$ ps, where t_c is estimated from the localized caging effect from simulations. As shown in Fig. 7.2a, the $\langle u^2 \rangle$ increases nonlinearly with T for the AA and CG systems with varying α. The lower $\langle u^2 \rangle$ of the CG model with higher α over T indicates a suppressed segmental mobility due to stronger cohesive interactions. For each α, it is clear that the $\langle u^2 \rangle$ of the CG model intersects with the AA $\langle u^2 \rangle$ at a different temperature as seen from Fig. 7.2a, demonstrating the necessity of rescaling ε at different temperature to preserve the AA $\langle u^2 \rangle$. Accordingly, the $\alpha(T)$ can be determined by preserving the AA $\langle u^2 \rangle$ at each temperature state (inset in Fig. 7.2a), leading to a sigmoidal functional form

$$\alpha(T) = (\alpha_g - \alpha_A)\Phi + \alpha_g \tag{7.2}$$

where α_g and α_A refer to α values at in the low T glassy and high T Arrhenius regimes, respectively; Φ is the two-state crossover function taking the form: $\Phi = 1/[1 + \exp(-k(T - T_T))]$, where k is a parameter related to the temperature breadth of the transition, and T_T describes the crossover point of this sigmoidal function from the Arrhenius to glassy regimes. Note that small changes in the definition of t_c in the picosecond timescale yield nearly the same result for $\alpha(T)$. The $\alpha(T)$ tends to plateau at high-T Arrhenius regimes and low-T glassy regimes and the magnitude of $\alpha(T)$ is higher at lower T. However, qualitatively similar to the picture of how activation energy of relaxation $\Delta G(T)$ changes with T, $\alpha(T)$ is varied strongly with T in the non-Arrhenius regime associated with glass formation. Both GET and AG theory predict that the $\Delta G(T)$ of glass forming liquids increases with decreasing T as s_c decreases upon cooling but saturate at high and low T limits, giving a sigmoidal functional form [36, 58]. Many of these features have been confirmed by recent molecular simulations [56, 59]. Therefore, the $\Delta G(T)$ of the CG model (without ε renormalization) should follow a similar T dependence of the AA model but with different magnitudes due to their reduced number of degrees of freedom. Similar to matching $\langle u^2 \rangle$ to obtain a renormalization for the cohesive strength, the length scale parameter σ is normalized by a factor $\beta(T)$ which is obtained by preserving the temperature-dependent density of the AA system as shown in Fig. 7.2b.

7.2.2 Validation of ER Approach

Once the functional forms of $\alpha(T)$ and $\beta(T)$ is determined, it needs to be validated to ensure the CG model can capture the dynamics of AA system. The segmental dynamics of the system was evaluated by looking at the mean square displacement (MSD) $\langle r^2(t) \rangle$ of the center of mass of monomer for AA and CG models at various temperatures. Figure 7.3a shows the comparison of the $\langle r^2(t) \rangle$ for the AA (lines) and CG (symbols) models. Remarkably, by preserving the $\langle u^2 \rangle$ of atomistic system under coarse-graining through ER, the CG model could capture the entire MSD curve over a wide range of temperature, from low-T glassy regime to high-T Arrhenius regime.

Next the segmental relaxation time, τ, is evaluated by calculating the second order Legendre parameter, P_2, for AA and CG models which is shown in Fig. 7.3b. As a comparison, a constant ER factor of $\alpha = \alpha_A$ (matching high temperature Arrhenius regime) and $\alpha = 1$ (first estimate of ε obtained from IBM by matching

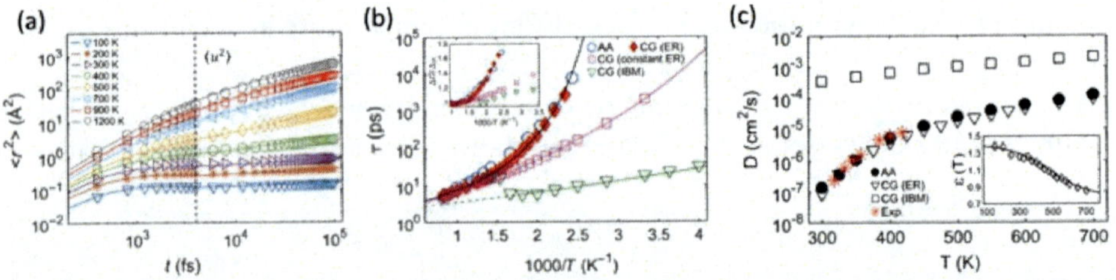

Fig. 7.3 **a** The MSD $\langle r^2 \rangle$ of the center of mass of the monomer versus time for the AA (lines) and CG (symbols) models of polycarbonate over a wide T range. The vertical dashed line marks the time scale (around the "caging" time of 3 ps) when $\langle r^2 \rangle$ is obtained from the $\langle r^2 \rangle$ measurement. **b** T-dependent segmental relaxation time τ for the AA and CG models. As a comparison, the τ estimates from the CG models with constant ER (i.e., $\alpha = \alpha_A$) and derived from the IBM exhibit a growing divergence as lowering T, while the τ estimates from the ER describe the AA τ to a much better approximation. The solid curves show the VFT fits of the τ data. The dashed curve for the CG model from the IBM shows a high T regime where the onset of sample evaporation leads to an increase in τ. Inset shows the activation energies of relaxation ΔG normalized by its value $\Delta\mu$ at high-T Arrhenius regime for the AA and CG models. **c** Self-diffusion coefficient D of OTP molecules for the AA and CG models using both ER and IBM at varying temperatures, and their comparison with the experimental data [60]. (Inset) The cohesive interaction strength $\varepsilon(T)$ (in kcal/mol) for the CG model determined by matching the T-dependent $\langle u^2 \rangle$ of the AA model to the CG model. Reprinted with permission from (**a**) and (**b**) [33] © 2019 AAAS and (**c**) [34] © 2018 American Chemical Society.

the RDF). Apart from high-T Arrhenius regime, just a constant rescaling of ε is not sufficient to describe the dynamics of the atomistic system. Thus, a temperature - dependent application of ER accurately captures the slowing down on dynamics upon approaching the glass transition temperature. The activation energy of relaxation, $\Delta G(T)$, is determined through the relationship: $\Delta G(T) = k_B T \ln[\tau/\tau_0]$, where τ_0 is the vibrational relaxation time on the order of 10^{-12} to 10^{-13} s. The inset of Fig. 7.3b shows the results of ΔG normalized by its value $\Delta\mu$ at the high-T Arrhenius regime. The CG model with ER evidently describes ΔG of the AA model rather well, whereas ΔGwithout ER remains too small at low temperatures.

With the implementation $\alpha(T)$ under the ER approach, the temperature-dependent self-diffusivity of CG models matches the AA models as well as the experimental values over the entire temperature range. For comparison, the diffusivity of CG model obtained from IBM is also shown in Fig. 7.3c and it can be seen that the value is much higher than the AA model. This dramatic speedup of dynamics necessitates ER under coarse-graining to preserve the AA dynamics over a wide T range.

The ER approach is broadly useful for the study of polymeric materials with more truthful representation of dynamical and mechanical properties. Further work needs to be done to understand how the formulation could be used for more complex systems, such as thermosets at varying degrees of crosslinking, and copolymer systems. In the subsequent sections, results from CG-MD simulations that utilized the ER concept will be presented to provide insights into the behavior of hairy nanoparticle assemblies.

7.3 Modeling of Hairy Nanoparticle Assemblies

Much of the experimental work carried out on hairy nanoparticle assemblies involve using a homopolymer, usually polystyrene, [12, 17, 61] and a spherical nanoparticle, usually silica [15, 16]. In the computational field, there have been studies looking at the structure, [62] entanglements [63] and dynamics [64] of solvent-free polymer-grafted nanoparticles. However, assemblies with high-aspect ratio nanoparticles can provide an important blend of anisotropy and improved mechanical properties. As quoted by Christopher Li, "Anisotropic assembly of isotropic nanoparticles is observed in a polymer nanocomposite system and leads to considerable improvements in mechanical properties. One of the approach for achieving controlled particle assembly takes its inspiration from bottom-up nanofabrication techniques and involves encoding nanoparticles with intrinsic anisotropy to synthesize a variety of patchy particles, magnetic particles and nanorods" [65]. An interesting choice of anisotropic nanofiller that has come into spotlight in recent years is the cellulose nanocrystal (CNC) [66]. These eco-friendly bio-compatible crystals can be extracted from renewable and sustainable sources such as wood, bacteria and certain marine animals such as tunicates. These CNCs have a high axial modulus of ~150 GPa which is comparable to Kevlar, material currently used in bullet-proof vests. These transparent crystals have relatively low density for such a high modulus which makes it an ideal candidate for light weight applications. However, the key property that makes it extremely desirable for aHNPs is the presence of reactive sidegroups (–OH) on the CNC surface that allows for easy functionalization and polymer grafting. Some of the common functionalization includes sulfuric acid treatment that provides sulfate esters, TEMPO-mediated hypochlorite oxidation creates carboxylic acids and carboxylic acid halide treatment to create ester linkages [66].

Experimentalists have grafted several polymers onto CNC and made nanocomposites out of it, such as poly(methyl methacrylate), [67] polystyrene[68] and poly(4-vinylpyridine) [69]. Studies have also investigated thermal and mechanical properties of CNC-based nanocomposites [67, 70]. From a modeling perspective, molecular dynamics simulations have been employed to investigate structural [71] and conformational [72] properties and self-assembly [73, 74] of polymer-grafted spherical nanoparticles, but not much is done in understanding the mechanical behavior of polymer-grafted anisotropic nanoparticle assemblies.

7.3.1 System Setup

Hansoge and co-workers have utilized the chemistry-specific CG models to carry out CG-MD simulations and obtain the transverse mechanical properties [75] and understand the conformational properties [76]of the CNC-grafted hairy nanoparticles. The CG schemes of the different polymers and CNC are shown in Fig. 7.4. For polymers with sidegroups, such as polystyrene (PS) and poly(methyl methacrylate) (PMMA), two bead per monomer is used as the CG scheme, one each for the backbone and sidegroup. For linear polymers, the choice of CG bead is based on the functional group in the backbone, i.e., one bead for polybutadiene (PB) and four beads per monomer for polycarbonate (PC) corresponding to phenylene, isopropylidene and carbonate groups. The details of the coarse-graining schemes are described in works carried out by Xia and co-workers [33–35]. While the same principle could have been applied for the CG scheme of CNC as well, the properties of interest do not require extensive parametrization of CNC. Hence, in order to improve the

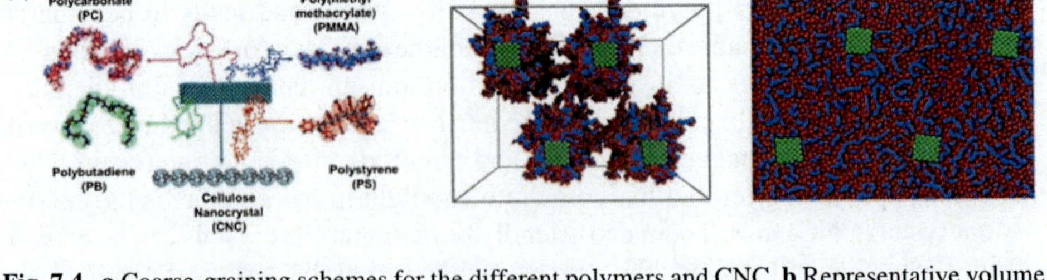

Fig. 7.4 **a** Coarse-graining schemes for the different polymers and CNC. **b** Representative volume element consisting of four polymer-grafted nanoparticles. **c** Snapshot of the representative volume element at the end of the equilibraiton cycle. Reprinted with permission from (**a**) [76] © 2018 American Chemical Society, (**b**) and (**c**) [75] © 2018 American Chemical Society

computational efficiency, two glucose monomers are represented by one CG bead. The entire CNC is modeled as an FCC crystal consisting of beads with high spring constant harmonic bonds ($k = 100$ kcal/mol/Å^2) in the axial direction. Although the surface of CNCs can be quite irregular, the canopy of the grafted polymeric layer should shield these surface features from affecting the mechanical properties. Such a high value is used as the CNC is much stiffer in the axial direction (~150 GPa) than in the transverse direction (38 GPa) [66]. Thus, the non-bonded interactions, which is represented by the standard 12–6 Lennard Jones potential, are parametrized to achieve the experimental value of 38 GPa. CNCs extracted from wood can have an aspect ratio of around 20, whereas those obtained from tunicates can have aspect ratios in several hundreds. Computationally modeling such high-aspect ratio particles will be challenging; hence, a representative volume element of CNC is chosen such that the length is five times the width or height and periodic boundary conditions are employed in the longitudinal direction which effectively creates infinitely long crystals.

The mechanical properties are evaluated using CNC-PMMA nanocomposite as a model. To simulate the CNC-PMMA nanocomposite, a single polymer-grafted CNC is first generated, where the location of graft sites is chosen arbitrarily, and the polymer chain is grown from the designated site using a random walk algorithm as depicted in Fig. 7.4b [77]. The system is then replicated three times (resulting in four total aHNPs) and packed together to create a representative nanocomposite system. The CG-MD simulations are carried out using large-scale atomic/molecular massively parallel simulator (LAMMPS) software [78].

7.3.2 *Simulation Protocols*

In order to simulate the bulk behavior of the nanocomposite, periodic boundary conditions are applied. This also avoids problems with the boundary effects due to finite size. The initial structure generated by random walk algorithm might have some overlapped atoms. Thus, before running the actual equilibration cycle, a soft interatomic cosine potential is applied to push the polymers apart in the NVT ensemble. This soft potential is then replaced with the actual interatomic Lennard–Jones potential to carry out the equilibration. During equilibration, the first step is to remove any residual stresses in the system. This is achieved by annealing at a high temperature of 1000 K for 2 ns. The next step is to ensure that the grafted nanoparticles are packed well for which the system is equilibrated at 600 K and a high pressure of 1000 atm for 2 ns. Finally, the system is cooled down to 300 K (room temperature) and equilibrated for another 2 ns under atmospheric pressure (1 atm). The relaxation of the polymer chains is checked using the second Legendre order parameter (P2) [33] which is an average made over all orientations of the structural units studied in a sample. A P2 values of less than 0.1 indicate that the polymer chains are fully relaxed. The equilibrations steps are carried out under the isothermal-isobaric NPT ensemble. Figure 7.4c shows the snapshot of the nanocomposite system after annealing and equilibration. Since the CG bead encompasses several atoms, an integration timestep of 4 fs is used in these simulations.

The tensile tests are also carried out under the NPT ensemble with zero pressure on the sides so as to not fix the Poisson's ratio. A strain rate of 10^8 s^{-1} is employed, which is typical in such MD simulations. While it might seem that the strain rate is too high as compared to experimental strain rates, the design strategies and main conclusions drawn from this study are not affected as the strain rate dependence of properties is relatively low in glassy regime of polymers. The properties of interest in this study are Young's modulus, which is obtained from the slope of linear region in the stress–strain curve upto 0.2% strain, and toughness, which is the area under the stress–strain curve.

7.3.3 Mechanical Property Characterization

In polymer nanocomposites, there exists a diametric relationship between Young's modulus and toughness of the nanocomposite. Higher nanoparticle content will increase the modulus at the expense of ductility in the system. Effectively sweeping the dependencies of the number of design parameters affecting the mechanical properties is computationally very expensive. Thus, in order to find the sweet spot of achieving maximum stiffness and toughness, Hansoge and co-workers adopted a unique strategy of combining machine learning with molecular dynamics simulations [75]. While coarse-graining does provide some improvement in the computational efficiency, it is still infeasible to sweep the broad parametric range of the design space. Thus, a metamodel-based design optimization (MBDO) [79, 80] framework is adopted to provide design strategies for the optimal mechanical properties of aHNPs.

A multi-response Gaussian process metamodel is trained for this purpose. The input design parameters chosen are chain length (N), grafting density (σ), polymer—NP interaction strength (ε_{pnp}), and NP edge length (l_{np}) and the outputs are Young's modulus and toughness. Initially, 100 CG-MD simulations are run whose parameters span the entire design space. The input–output relationship is then used to train the metamodel. The details of the algorithms used in training can be found in [75]. The metamodel does a great job in replacing the computationally expensive CG-MD simulations as over a million pseudo-simulations were run to populate the entire design space. The outputs from these simulations are plotted in Young's modulus versus toughness Ashby plot as shown in Fig. 7.5. The color in each paned indicated the value of the input parameter. Populating the design space reveals the maximum limit of the properties that can be achieved which is shown in the red line and called the Pareto frontier. For any point on the Pareto frontier, there is no combination of input parameters that can give higher modulus without compromising the toughness and vice versa. The insets in Fig. 7.5 show the distribution of the input parameter that lies on the Pareto frontier.

As seen from Fig. 7.5a, the toughness of the nanocomposite increases with N (as marked by the change from blue to yellow). The main toughening mechanism

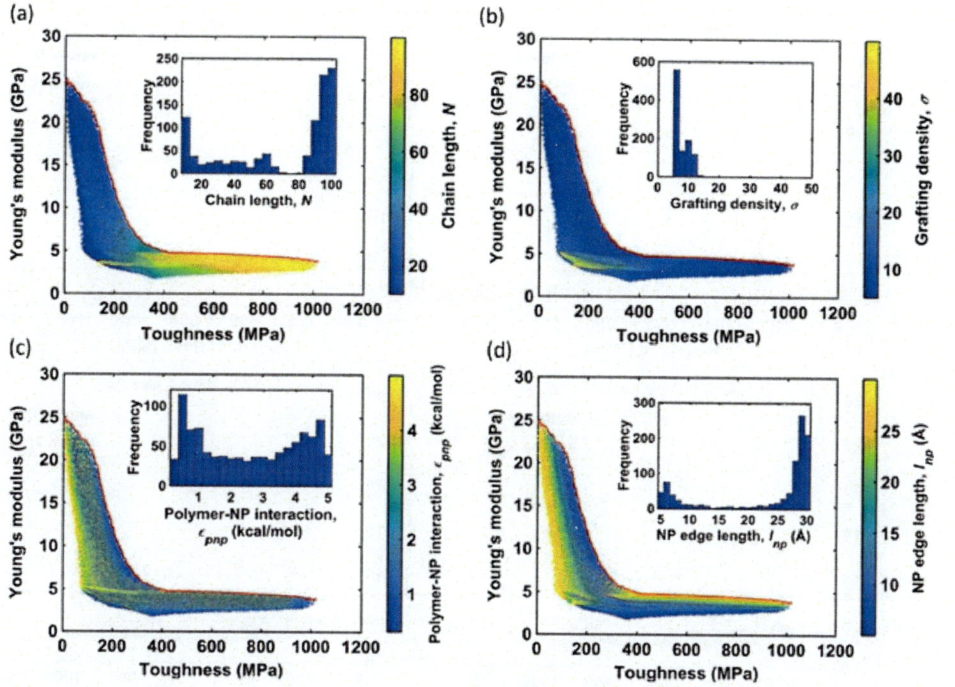

Fig. 7.5 Young's modulus versus toughness for the NP—polymer nanocomposite colored by **a** chain length of polymer N, **b** grafting density σ, **c** polymer—NP interaction strength $\varepsilon_{\mathrm{pnp}}$ and **d** NP edge length l_{np}. The insets show the distribution of each parameter that lies on the Pareto curve. Reprinted with permission from [75] © 2018 American Chemical Society

in these aHNPs is the resistance to sliding of the polymer chains which increases with N as effective cohesive interactions between the polymer chains increases. At lower N, the weight % of nanoparticle is higher, which increases the modulus of the nanocomposite. The effect of σ on the mechanical properties is a bit intricate. Grafting polymers onto the nanoparticle strongly affects the conformations of the chains. A detailed discussion on the conformational behavior of the polymer chains is discussed in Sect. 3.4. Figure 7.5b reveals that as σ is increased, the modulus and toughness both decrease, as evident near the elbow region of the plot (blue to yellow). Thus, it is beneficial to have a lower grafting density to achieve optimal mechanical properties. The presence of covalent bonds between the polymer and nanoparticle limits the influence of $\varepsilon_{\mathrm{pnp}}$ as compared to other input parameters. However, as seen from Fig. 7.5c, there is a higher preference for $\varepsilon_{\mathrm{pnp}}$ to achieve higher modulus when the weight % of nanoparticle is high. With regards to l_{np}, larger length is preferred in the high toughness region, as larger l_{np} provides additional grafting sites that increase the polymer content and thus the toughness as seen from Fig. 7.5d.

The range of mechanical properties that can be achieved at different weight % of nanoparticles provide a deeper insight into the complexity of the dependence of input parameters (Fig. 7.6). The points are also colored based on whether the chain length is above or below the critical chain length, N_{cr}, which marks the transition between

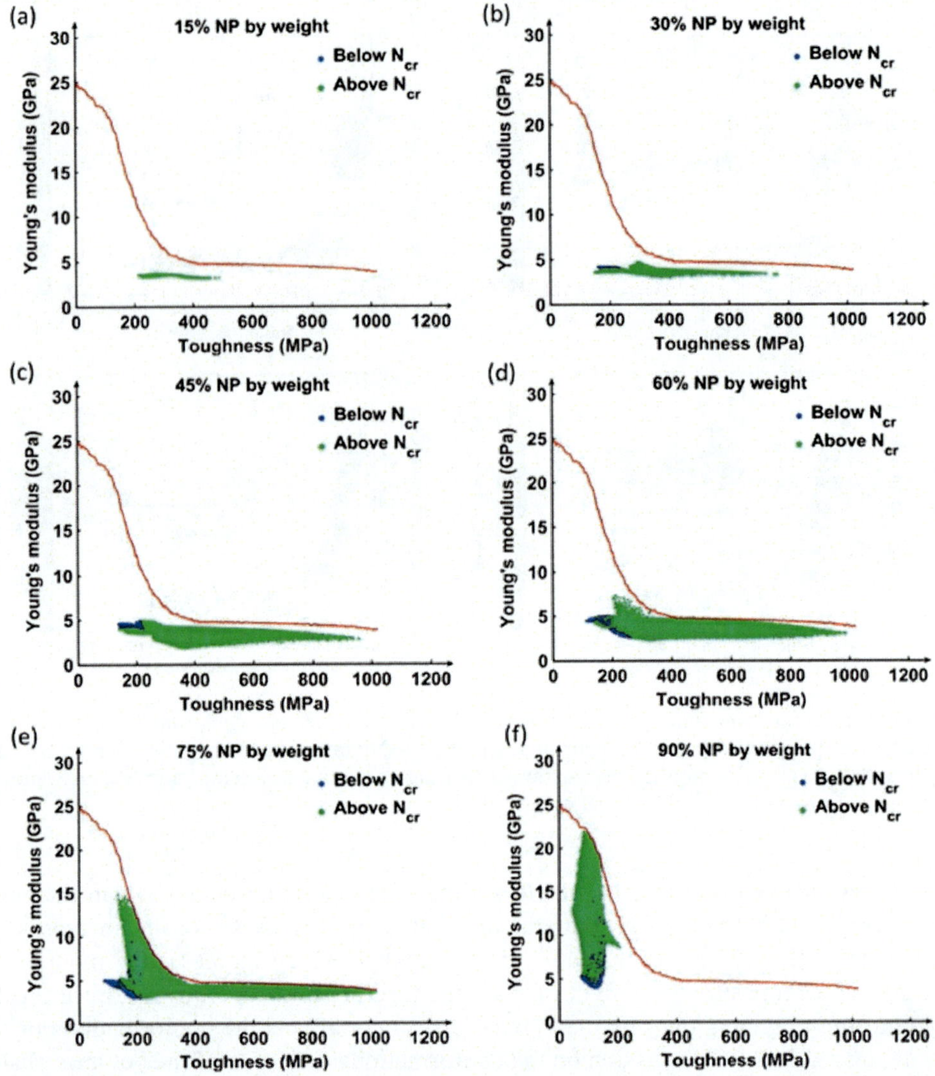

Fig. 7.6 Young's modulus versus toughness of the nanocomposite for different weight percentages of NPs. The shaded region in each plot shows the properties that can be achieved at a given weight %, while the red curve is the Pareto frontier for the overall material system. Reprinted with permission from [75] © 2018 American Chemical Society

semi-dilute and concentrated brush regime (details in Sect. 3.4). The weight % of the nanoparticle can be varied by changing N, σ, and l_{np}. At low weight % (below 45%, Fig. 7.6a–c), the predicted properties do not reach the Pareto frontier as there is not enough NPs to provide stiffness to the nanocomposite. At least 60% NP weight is required (Fig. 7.6d and e) to reach the Pareto frontier. However, at extremely high weight % (Fig. 7.6f), there is not enough polymer content to provide toughness to the nanocomposite beyond 200 MPa. Thus, the ideal range of weight % seems to be between 60 and 90%. Another important point to note is that all the designs that reach the Pareto frontier have chain lengths above the N_{cr}. Traditional polymer nanocomposites cannot achieve such high weight % due to nanoparticle aggregation issues, but since aHNPs can overcome these issues, optimal mechanical properties can be achieved for the nanocomposite.

7.3.4 Evaluating Polymer Conformational Behavior

When σ is increased, the monomer segments near the nanoparticle experience steric hindrance and are pushed away from the nanoparticle. Thus, near the nanoparticle surface, the polymer chains are in the concentrated brush regime (CPB) and gradually relax to a semi-dilute brush regime (SDPB) away from the surface. Looking at the scaling (Fig. 7.7a), the polymer chains are more extended in the direction perpendicular to the nanoparticle surface as $R_g^{\perp} \sim N^{0.6}$ compared to $R_g^{\perp} \sim N^{0.5}$ which is

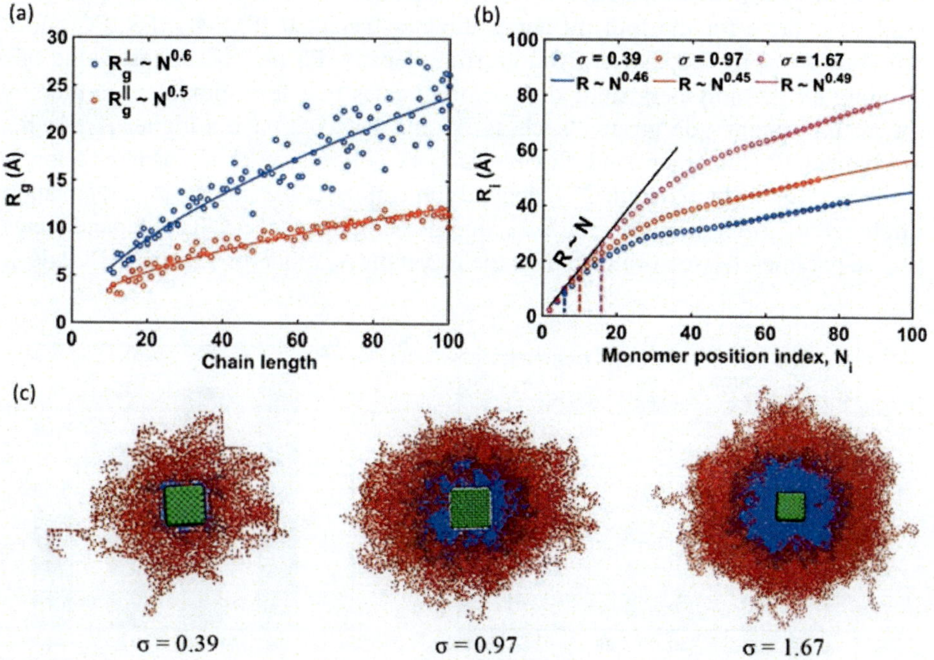

Fig. 7.7 **a** Scaling of radius of gyration in radial $\left(R_g^{\perp}\right)$ and axial $\left(R_g^{\parallel}\right)$ directions with chain length. **b** Scaling relationship in CPB ($R \sim N$) and SDPB ($R \sim N^{0.5}$) regimes. R_i is the distance of ith monomer from the nanoparticle surface, and N_i is the monomer position index in the polymer chain. The dotted line shows the transition point from CPB to SDPB, i.e., N_{cr}. **c** Snapshots of configurations of the hairy nanoparticle showing the onset of the SDPB regime (red) from the CPB regime (blue) with grafting density. All grafting densities are in units of chains/nm^2. Reprinted with permission from [75] © 2018 American Chemical Society

typically observed in polymer melts. In the perpendicular direction, the distance of the monomer from the surface, R, is measured against the monomer position index. It is observed that near the nanoparticle surface, the monomer segments are in the CPB regime, i.e., $R \sim N$. Away from the nanoparticle surface, the segments relax to SPDB regime, $R \sim N^{0.5}$. The transition from CPB to SDPB regime is denoted by the critical chain length, N_{cr}. Theoretical scaling arguments can be used to determine the dependence of the transition on features such as NP radius and grafting density. An extension of theoretical Daoud-Cotton (DC) model [81] was proposed by Ohno et al. [82] where they treated a star polymer as a system consisting of chains grafted onto a spherical nanoparticle. A similar argument is applied to polymer-grafted CNCs to come up with an equation for N_{cr} shown in 7.3

$$N_{cr} = \frac{2r_0}{3a\sigma_0^{*1/2} l_m}\left(r_0^{-3/4}\sigma_0^{*3/4} b^{3/2} - 1\right) \tag{7.3}$$

N_{cr} is an important design parameter for aHNPs. As seen from Fig. 7.6, degree of polymerization needs to be greater than the N_{cr} to obtain optimal mechanical properties [32]. Utilizing the chemistry-specific CG models, a comparison of N_{cr} values is done for the four different polymers selected (PMMA, PS, PC and PB) as shown in Fig. 7.8. For a given grafting density, PS has the largest value of N_{cr} among all the polymers studied. It also indicates that semi-flexible polymers with more prominent side groups, such as PS and PMMA, have a higher N_{cr} in linear polymers. On the other hand, flexible rubbery polymers such as PB have a low N_{cr}.

The critical chain length, N_{cr}, depends on empirical parameters, a and b, nanoparticle radius, r_0, monomer length, l_m, and grafting density, σ^*. The dependencies of these parameters are normalized on the R_i versus N_i plots for various aHNP designs

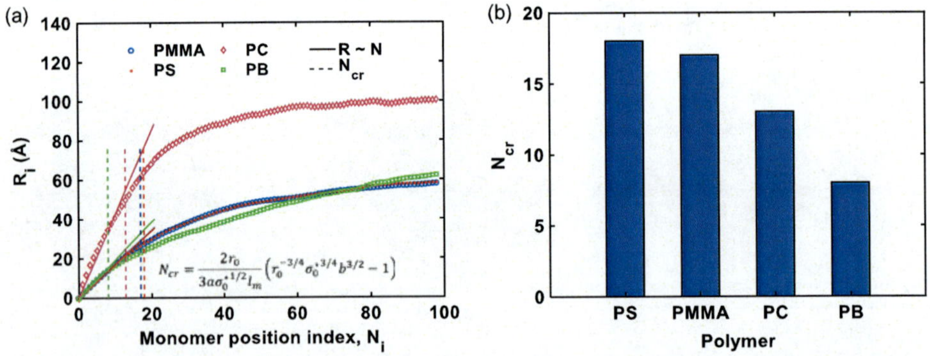

Fig. 7.8 **a** Plot of distance of monomer from the nanoparticle surface versus monomer position index. R_i is distance of ith monomer (N_i) from the nanoparticle surface. The linear solid line shows the scaling relationship in CPB ($R \sim N$) regime and the dotted line shows the transition point from CPB to SDPB $R \sim N^{0.5}$, identified as N_{cr}. The inset shows the equation of N_{cr}. **b** Bar plot showing the comparison of N_{cr} for the different polymers at 30% grafting density. Reprinted with permission from [76] © 2019 American Chemical Society.

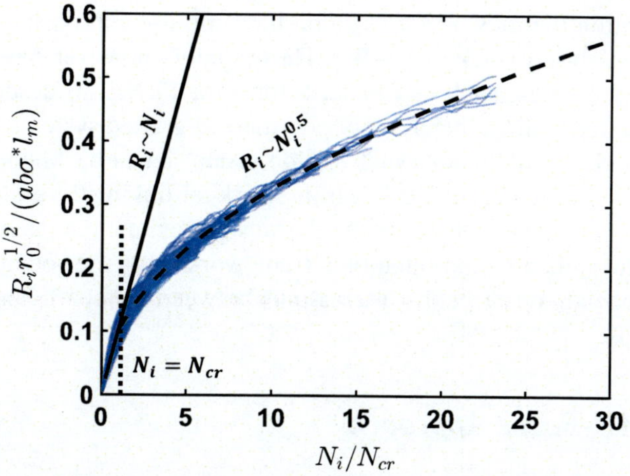

Fig. 7.9 The monomer index N_i is normalized with N_{cr} and the radial distance (R_i) is normalized with material design parameters a, b, r_0, l_{m} and σ^*, shown for 110 different aHNP systems. Reprinted with permission from [76] © 2019 American Chemical Society

and it can be seen that all the curves fall onto a single curve (Fig. 7.9). This consistent collapse of the data suggests the universality of the polymer chain conformations in these hairy nanoparticle assemblies. From this universal relation, one can predict the position of the monomer from the following equations:

$$R_i = \frac{0.1ab\sigma^* l_{\mathrm{m}}}{r_0^{1/2}} \frac{N_i}{N_{\mathrm{cr}}} \mathrm{for} N_i < N_{\mathrm{cr}} \tag{7.4}$$

$$R_i = \frac{0.1ab\sigma^* l_{\mathrm{m}}}{r_0^{1/2}} \left(\frac{N_i}{N_{\mathrm{cr}}}\right)^{0.5} \text{ for } N_i > N_{\mathrm{cr}} \tag{7.5}$$

7.4 Mesoscale Modeling of aHNPs

The length and time scales accessed by CG-MD simulations are in the order of nanometers and nanoseconds. While such spatiotemporal scales are sufficient to evaluate dynamics and structural conformation of polymer chains, predicting size-dependent macroscopic properties such as fracture energy or impact resistance is not possible. A mesoscale model that can access experimental length scales without loss of chemical specificity and accuracy will be beneficial to carry out such analysis. One way to improve the spatiotemporal scales is to implicitly simulate the grafted polymer chains through an effective interaction between the nanoparticles that is function of polymer characteristics such as chain length, grafting density and chemistry. There are several studies in the literature that have developed such effective potentials using integral equation theories, [83–89] umbrella sampling approach, [90–92] and self-consistent field theory (SCFT) [93–95]. These studies, however, have utilized generic interatomic potentials, such as Lennard–Jones or FENE, to describe the polymer chains. Since the confined behavior of polymers in grafted systems strongly depends on molecular chemical features such as side-group size or backbone rigidity, there is a critical need to develop a numerical framework for linking the effective interactions between nanoparticles to these molecular design parameters. Hence, Hansoge and co-workers formulated a computational framework using the potential of mean force (PMF) approach to describe the interactions between nanoparticles in aHNPs.

7.4.1 PMF-Based Approach

The nanoparticle is idealized as a planar surface from which polymers emerge and interdigitate, a model that could be broadly applicable to large, stiff 1D-2D nanoparticles or strongly cohesive assemblies of these particles in a lamellar configuration as shown in Fig. 7.10a. The nanoparticles are grafted with four different polymers (PMMA, PS, PC, PB) with various chain lengths and grafting densities. Tensile and compressive tests are carried out to obtain the PMF between the nanoparticles, which

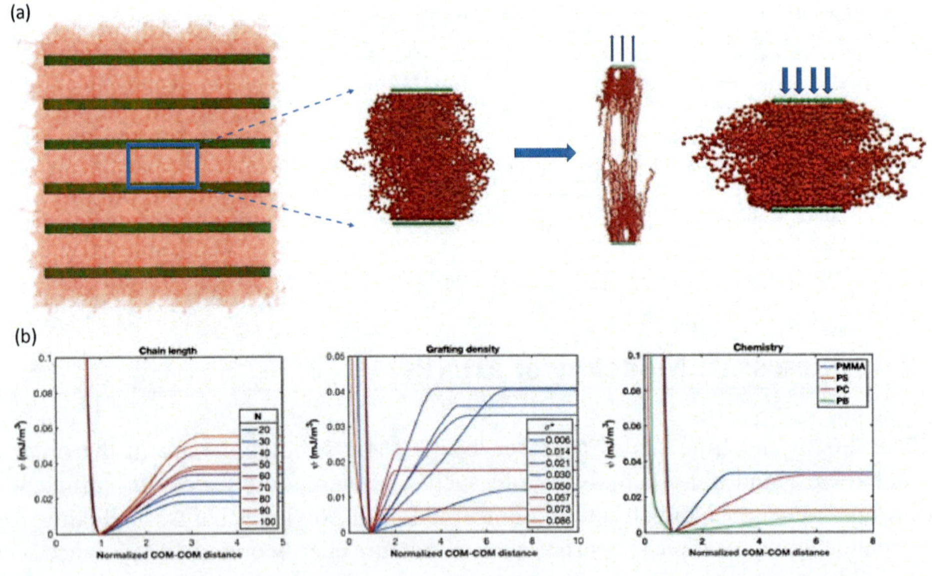

Fig. 7.10 **a** Schematic of the layered polymer grafted to a high-aspect ratio nanoparticle assembly. A representative volume is used to analyze the transverse properties of these assemblies. CG-MD simulations are carried out by fixing the bottom plate and pulling/pushing on the top plate with a constant velocity. **b** Representative plots of potential of mean force per volume (ψ) with respect to center of mass distance between the nanoparticles normalized by their equilibrium distance. The three plots show the variation with chain length, grafting density and polymer chemistry

is the work required to move the plates. The schematic of the tensile and compressive tests is shown in Fig. 7.10b. The sum of forces in the pulling direction is calculated, multiplied with the displacement and divided by the volume to provide the potential of mean force per volume (ψ). The total work required to completely separate the plates starting from their equilibrium position is defined as ψ_{max}, which is the plateau of the PMF curve or depth of the energy well. The simulation protocols are same as mentioned above. An analytical expression is used to describe the PMF behavior and the empirical constants in this expression are linked to the design parameters, i.e., chain length, grafting density and chemistry.

Figure 7.10c shows representative PMF curves for the different design parameters. In general, it is observed that the ψ_{max} increases linearly with chain length. The mechanism behind this is increased in entanglements per chain that increases with chain length. The dependence of ψ_{max} on grafting density is quadratic. The cohesive interaction energy between polymer chains arising from different nanoparticles is the main contributor to ψ_{max}. When the grafting density is low, the polymer chains interdigitate very well and improve with grafting density improves. However, at very high grafting densities, the steric hindrance caused due to chains being very close to each other leads to a decrease in this cohesive energy which in turn decreases ψ_{max}. The chemistry dependence can also be seen from Fig. 7.10c. PB, which is a rubber at room temperature, has the lowest value of ψ_{max}, whereas tough glassy polymers, PMMA and PC have a high value of ψ_{max}. The initial curvature of the attractive part of the PMF curve is related to the modulus of the system. PMMA, which has the highest modulus among the four polymers, has the highest curvature, whereas PB, being a rubber, has the lowest.

7.4.2 *Functional Form of Effective Interactions*

After understanding the role of chain length and grafting density on ψ, the next step is to find an analytical equation that can fit all of the PMF curves and seek to determine a relationship between the empirical constants and the physical parameters. It is that commonly used pairwise potentials such as Morse, Lennard–Jones and even Mei, do not fit the data accurately. Hence, the behavior of the attractive and repulsive portion was separately studied to come up with an analytical function. The repulsive portion of the curve is well described the exponential form shown in 7.6.

$$\psi_r = a_r * \mathrm{e}^{-b_r x} + c_r \tag{7.6}$$

where a_r, b_r and c_r are the empirical constants and x is the center of mass distance between the plates. The parameters a_r and b_r are dependent on N and $\sigma*$ and the parameter c_r, which is a very small number, shifts the potential to zero at equilibrium distance. For the attractive part, it was found that an empirical sigmoid relation shown in 7.7 fits all the PMF curves. The constant $\frac{a_a}{b_a+\mathrm{e}^{-c_a x_0}}$ is added to shift the curves to zero at equilibrium distance ($x = x_0$).

$$\psi_a = \frac{a_a}{b_a + e^{-c_a x}} - \frac{a_a}{b_a + e^{-c_a x_0}} \tag{7.7}$$

The constants a_a and b_a control the plateau of the PMF curve (ψ_{max}), whereas c_a controls the curvature of the curve. The empirical constants are obtained by fitting the data to the above equation. For each polymer chemistry, chain length and grafting density, five separate trials were run to get better statistics. Adding the attractive and repulsive forms gives the final potential form:

$$\psi = a_r * e^{-b_r x} + \frac{a_a}{b_a + e^{-c_a x}} - \frac{a_a}{b_a + e^{-c_a x_0}} \tag{7.8}$$

We note that $\frac{a_a}{b_a+e^{-c_a x_0}} - c_r \sim \frac{a_a}{b_a+e^{-c_a x_0}}$ as c_r is negligible. This five-parameter equation describes the interaction between nanoparticles in all different designs of aHNPs. The goodness of the fit is evaluated based on the mean square error (MSE) and the correlation coefficient (R^2) between the fit and the data. It is ensured that the $R^2 > 0.99$ and the MSE $< 1\%$. Figure 7.11 shows a representative fit of the function to the data obtained from the simulations.

The point where the derivative of the potential (given by 7.9) becomes zero is the equilibrium distance between the nanoparticles, x_0.

$$\psi' = -b_r * a_r * e^{-b_r x} + \frac{a_a * c_a * e^{-c_a x}}{(b_a + e^{-c_a x})^2} \tag{7.9}$$

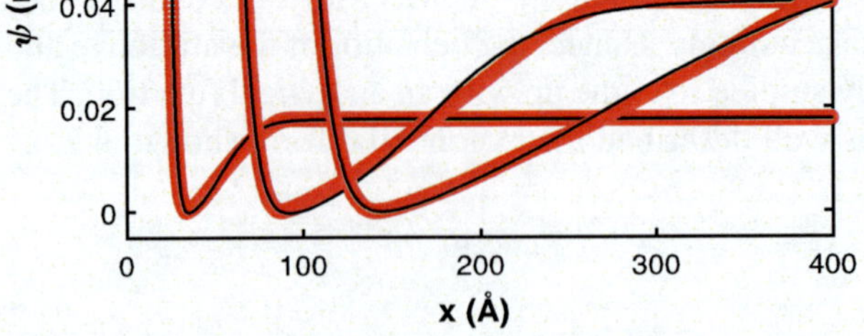

Fig. 7.11 Fitting the functional form to the data obtained from the CG-MD simulations. The figure shows two curves for chain lengths 20, 60 and 100 of PMMA-grafted nanoparticle system

The solution for $\psi' = 0$ cannot be determined analytically as a general expression for all possible values of the empirical constants. However, since the potential is formed of two continuous smooth curves, the derivative can be found numerically for each curve once empirical constants are determined from fitting ψ. The equilibrium distance obtained from the derivative is compared to the equilibrium distance from the simulations and it is observed that the two values are within 2% of each other. This, along with the goodness of fit measures, ensures that the fitting parameters accurately capture the equilibrium position and the shape of the curves.

The PMF methodology now enables treatment of particles at a much coarser level without the need for explicit modeling of polymer chains. Future investigations will need to examine the effects of particle shape, curvature and strain rate effects, as well as temperature on PMF functions to ensure the general form is still applicable. Implementation of this methodology on mesoscale models will pave the way for exciting, large-scale studies of nanocomposites and thin films that approach the scales of experiments.

7.5 Conclusion

In conclusion, this chapter covers various modeling techniques for hairy nanoparticle assemblies. The first section explains the energy renormalization methodology for coarse-graining polymers. This method can capture the dynamics accurately over a wide range of temperature and it requires only a picosecond time scale measure, Debye–Waller factor, for calibration. The second section goes into describing modeling of hairy nanoparticle assemblies with cellulose nanocrystal. A unique methodology of combining machine learning and coarse-grained molecular dynamics simulation is employed to optimize the mechanical properties. A speedup of 6–7 orders of magnitude is observed as compared to all-atomistic simulations. Finally, a mesoscopic model framework is discussed that will extend the spatiotemporal scales of MD simulations to experimental level. An analytical functional form governing the interactions between nanoparticles eliminates the need to model the polymer chain explicitly, thus improving the computational efficiency. These modeling efforts provided important guidelines toward designing assembled polymer-grafted nanoparticle composites to achieve optimal mechanical properties. Looking forward, these approaches will help to accelerate the materials-by-design process and inspire future studies in advancing mechanical performance of composites and other relevant structural materials.

References

1. A.C. Balazs, T. Emrick, T.P. Russell, Nanoparticle polymer composites: where two small worlds meet. Science **314**, (17) (2006).
2. S.C. Tjong, Structural and mechanical properties of polymer nanocomposites. Mater. Sci. Eng.: R: Rep. **53**(3–4), 73–197 (2006)
3. J.R. Potts, D.R. Dreyer, C.W. Bielawski, R.S. Ruoff, Graphene-based polymer nanocomposites. Polymer **52**(1), 5–25 (2011)
4. F. Hussain, M. Hojjati, M. Okamoto, R.E. Gorga, Review article: polymer-matrix nanocomposites, processing, manufacturing, and application: an overview. J. Compos. Mater. **40**(17), 1511–1575 (2006)
5. K.I. Winey, R.A. Vaia, Polymer nanocomposites. MRS Bull. **32**(4), 314–322 (2011)
6. J. Jancar, J.F. Douglas, F.W. Starr, S.K. Kumar, P. Cassagnau, A.J. Lesser, S.S. Sternstein, M.J. Buehler, Current issues in research on structure–property relationships in polymer nanocomposites. Polymer **51**(15), 3321–3343 (2010)
7. R. Krishnamoorti, Strategies for dispersing nanoparticles in polymers. MRS Bull. **32**(4), 341–347 (2011)
8. C. Chevigny, F. Dalmas, E. Di Cola, D. Gigmes, D. Bertin, F. Boué, J Jestin, Polymer-grafted-nanoparticles nanocomposites: dispersion, grafted chain conformation, and rheological behavior. Macromolecules **44**(1), 122–133 (2011).
9. N. Jouault, D. Lee, D. Zhao, S.K. Kumar, Block-copolymer-mediated nanoparticle dispersion and assembly in polymer nanocomposites. Adv. Mater. **26**(24), 4031–4036 (2014)
10. S.K. Kumar, N. Jouault, B. Benicewicz, T. Neely, Nanocomposites with polymer grafted nanoparticles. Macromolecules **46**(9), 3199–3214 (2013)
11. N.J. Fernandes, H. Koerner, E.P. Giannelis, R.A. Vaia, Hairy nanoparticle assemblies as one-component functional polymer nanocomposites: opportunities and challenges. MRS Commun. **3**(1), 13–29 (2013)
12. X.W. Gu, X. Ye, D.M. Koshy, S. Vachhani, P. Hosemann, A.P. Alivisatos, Tolerance to structural disorder and tunable mechanical behavior in self-assembled superlattices of polymer-grafted nanocrystals. Proc. Natl. Acad. Sci. U S A **114**(11), 2836–2841 (2017)
13. D. Maillard, S.K. Kumar, B. Fragneaud, J.W. Kysar, A. Rungta, B.C. Benicewicz, H. Deng, L.C. Brinson, J.F. Douglas, Mechanical properties of thin glassy polymer films filled with spherical polymer-grafted nanoparticles. Nano Lett. **12**(8), 3909–3914 (2012)
14. X. Ye, C. Zhu, P. Ercius, S.N. Raja, B. He, M.R. Jones, M.R. Hauwiller, Y. Liu, T. Xu, A.P. Alivisatos, Structural diversity in binary superlattices self-assembled from polymer-grafted nanocrystals. Na.t Commun. **6**, 10052 (2015)
15. J. Choi, C.M. Hui, J. Pietrasik, H. Dong, K. Matyjaszewski, M.R. Bockstaller, Toughening fragile matter: mechanical properties of particle solids assembled from polymer-grafted hybrid particles synthesized by ATRP. Soft Matter **8**(15) (2012).
16. M. Schmitt, J. Choi, C.M. Hui, B. Chen, E. Korkmaz, J. Yan, S. Margel, O.B. Ozdoganlar, K. Matyjaszewski, M.R. Bockstaller, Processing fragile matter: effect of polymer graft modification on the mechanical properties and processibility of (nano-) particulate solids. Soft Matter **12**(15), 3527–3537 (2016)
17. J. Midya, Y. Cang, S.A. Egorov, K. Matyjaszewski, M.R. Bockstaller, A. Nikoubashman, G. Fytas, Disentangling the role of chain conformation on the mechanics of polymer tethered particle materials. Nano Lett. **19**(4), 2715–2722 (2019)
18. M. Asai, D. Zhao, S.K. Kumar, Role of grafting mechanism on the polymer coverage and self-assembly of Hairy nanoparticles. ACS Nano **11**(7), 7028–7035 (2017)
19. D. Dukes, Y. Li, S. Lewis, B. Benicewicz, L. Schadler, S.K. Kumar, Conformational transitions of spherical polymer brushes: synthesis, characterization, and theory. Macromolecules **43**(3), 1564–1570 (2010)

20. P. Akcora, H. Liu, S.K. Kumar, J. Moll, Y. Li, B.C. Benicewicz, L.S. Schadler, D. Acehan, A.Z. Panagiotopoulos, V. Pryamitsyn, V. Ganesan, J. Ilavsky, P. Thiyagarajan, R.H. Colby, J.F. Douglas, Anisotropic self-assembly of spherical polymer-grafted nanoparticles. Nat. Mater. **8**(4), 354–359 (2009)
21. J.G. Ethier, L.F. Drummy, R.A. Vaia, L.M. Hall, Uniaxial deformation and crazing in glassy polymer-grafted nanoparticle ultrathin films. ACS Nano **13**(11), 12816–12829 (2019)
22. W.G. Noid, Perspective: coarse-grained models for biomolecular systems. J. Chem. Phys. **139**(9), 090901 (2013)
23. M.P. Dirk Reith, Florian Muller-Plathe, deriving effective mesoscale potentials from atomistic simulations. J. Comput. Chem. **24**, 1624–1636 (2003)
24. W.G. Noid, J.W. Chu, G.S. Ayton, V. Krishna, S. Izvekov, G.A. Voth, A. Das, H.C. Andersen, The multiscale coarse-graining method. I. A rigorous bridge between atomistic and coarse-grained models. J. Chem. Phys. **128**(24), 244114 (2008).
25. G.A.V. Sergei Izvekov, A multiscale coarse-graining method for biomolecular systems. J. Phys. Chem. B Lett. **109**, 2469–2473 (2004)
26. M.S. Shell, Coarse-graining with the relative entropy. Adv. Chem. Phys. **161**, 395–441 (2016)
27. A.P. Lyubartsev, A. Laaksonen, Calculation of effective interaction potentials from radial distribution functions: A reverse Monte Carlo approach. Phys. Rev. E Stat. Phys. Plasmas Fluids Relat. Interdiscip. Topics **52**(4), 3730–3737 (1995)
28. D. Fritz, K. Koschke, V.A. Harmandaris, N.F. van der Vegt, K. Kremer, Multiscale modeling of soft matter: scaling of dynamics. Phys. Chem. Chem. Phys. **13**(22), 10412–10420 (2011)
29. H.A. Karimi-Varzaneh, N.F. van der Vegt, F. Muller-Plathe, P. Carbone, How good are coarse-grained polymer models? A comparison for atactic polystyrene. Chemphyschem **13**(15), 3428–3439 (2012).
30. P.K. Depa, J.K. Maranas, Speed up of dynamic observables in coarse-grained molecular-dynamics simulations of unentangled polymers. J. Chem. Phys. **123**(9), 94901 (2005)
31. I.Y. Lyubimov, J. McCarty, A. Clark, M.G. Guenza, Analytical rescaling of polymer dynamics from mesoscale simulations. J. Chem. Phys. **132**(22), 224903 (2010)
32. A. Davtyan, J.F. Dama, G.A. Voth, H.C. Andersen, Dynamic force matching: a method for constructing dynamical coarse-grained models with realistic time dependence. J. Chem. Phys. **142**(15), 154104 (2015)
33. N.K. Hansoge, W. Xia, W.-S. Xu, F.R. Phelan Jr, S. Keten, J.F. Douglas, Energy renormalization for coarse-graining polymers having different segmental structures. Sci. Adv. **5**(4) (2019).
34. W. Xia, J. Song, N.K. Hansoge, F.R. Phelan Jr, S. Keten, J.F. Douglas, Energy renormalization for coarse-graining the dynamics of a model glass-forming liquid. J. Phys. Chem. B **122**(6), 2040–2045 (2018)
35. W. Xia, J. Song, C. Jeong, D.D. Hsu, F.R. Phelan Jr, J.F. Douglas, S. Keten, Energy-renormalization for achieving temperature transferable coarse-graining of polymer dynamics. Macromolecules **50** (2017).
36. G. Adam, J.H. Gibbs, On the temperature dependence of cooperative relaxation properties in glass-forming liquids. J. Chem. Phys. **43**(1), 139–146 (1965)
37. J. Dudowicz, J.F. Douglas, K.F. Freed, Advances in the generalized entropy theory of glass-formation in polymer melts. J. Chem. Phys. **141**(23), 234903 (2014)
38. J.F. Dudowicz, K.F. Douglas, Generalized entropy theory of polymer glass formation. Adv. Chem. Phys. 125–222 (2008).
39. A. Yelon, B. Movaghar, Microscopic explanation of the compensation (Meyer-Neldel) rule. Phys. Rev. Lett. **65**(5), 618–620 (1990)
40. S.R. Rufus Lumry, Enthalpy-entropy compensation phenomena in water solutions of proteins and small molecules: a ubiquitous property of water. Biopolymers **9**, 1125–1227 (1970)
41. W.-S. Xu, J.F. Douglas, K.F. Freed, Influence of cohesive energy on the thermodynamic properties of a model glass-forming polymer melt. Macromolecules **49**(21), 8341–8354 (2016)
42. W.-S. Xu, K.F. Freed, Influence of cohesive energy and chain stiffness on polymer glass formation. Macromolecules **47**(19), 6990–6997 (2014)

43. R.J. Weiss, J.J. DeMarco, G. Weremchuk, An apparent anisotropic Debye-Waller factor in cubic crystals. Acta Crystallogr. **9**, 42–44 (1954).
44. F.W. Starr, S. Sastry, J.F. Douglas, S.C. Glotzer, What do we learn from the local geometry of glass-forming liquids? Phys. Rev. Lett. **89**(12), 125501 (2002)
45. R.W. Hall, P.G. Wolynes, The aperiodic crystal picture and free energy barriers in glasses. J. Chem. Phys. **86**(5), 2943–2948 (1987)
46. S. Mirigian, K.S. Schweizer, Unified theory of activated relaxation in liquids over 14 decades in time. J. Phys. Chem. Lett. **4**(21), 3648–3653 (2013)
47. S. Mirigian, K.S. Schweizer, Elastically cooperative activated barrier hopping theory of relaxation in viscous fluids. II. Thermal liquids. J. Chem. Phys. **140**(19), 194507 (2014).
48. K.S. Schweizer, Relationships between the single particle barrier hopping theory and thermodynamic, disordered media, elastic, and jamming models of glassy systems. J. Chem. Phys. **127**(16), 164506 (2007)
49. J.C. Dyre, Colloquium: the glass transition and elastic models of glass-forming liquids. Rev. Mod. Phys. **78**(3), 953–972 (2006)
50. J.C. Dyre, N.B. Olsen, T. Christensen, Local elastic expansion model for viscous-flow activation energies of glass-forming molecular liquids. Phys. Rev. B **53**(5), 2171–2174 (1996).
51. F.W. Starr, J.F. Douglas, Modifying fragility and collective motion in polymer melts with nanoparticles. Phys. Rev. Lett. **106**(11), 115702 (2011)
52. L. Larini, A. Ottochian, C. De Michele, D. Leporini, Universal scaling between structural relaxation and vibrational dynamics in glass-forming liquids and polymers. Nat. Phys. **4**(1), 42–45 (2007)
53. A. Ottochian, C. De Michele, D. Leporini, Universal divergenceless scaling between structural relaxation and caged dynamics in glass-forming systems. J. Chem. Phys. **131**(22), 224517 (2009)
54. A. Ottochian, D. Leporini, Universal scaling between structural relaxation and caged dynamics in glass-forming systems: free volume and time scales. J. Non-Cryst. Solids **357**(2), 298–301 (2011)
55. D.S. Simmons, M.T. Cicerone, Q. Zhong, M. Tyagi, J.F. Douglas, Generalized localization model of relaxation in glass-forming liquids. Soft Matter **8**(45), 11455–11461 (2012)
56. B.A. Pazmino Betancourt, P.Z. Hanakata, F.W. Starr, J.F. Douglas, Quantitative relations between cooperative motion, emergent elasticity, and free volume in model glass-forming polymer materials. Proc. Natl. Acad. Sci. U S A **112**(10), 2966–2971 (2015)
57. J.F. Douglas, B.A. Pazmino Betancourt, X. Tong, H. Zhang, Localization model description of diffusion and structural relaxation in glass-forming Cu–Zr alloys. J. Stat. Mech.: Theory Experiment **2016** (5) (2016).
58. J. Dudowicz, K.F. Freed, J.F. Douglas, Entropy theory of polymer glass formation revisited. I. General formulation. J. Chem. Phys. **124**(6), 64901 (2006).
59. P.Z. Hanakata, J.F. Douglas, F.W. Starr, Interfacial mobility scale determines the scale of collective motion and relaxation rate in polymer films. Nat. Commun. **5**, 4163 (2014)
60. D.W. McCall, D.C. Douglass, D.R. Falcone, Molecular motion in ortho-terphenyl. J. Chem. Phys. **50**(9), 3839–3843 (1969)
61. Y. Jiao, A. Tibbits, A. Gillman, M.-S. Hsiao, P. Buskohl, L.F. Drummy, R.A. Vaia, Deformation behavior of polystyrene-grafted nanoparticle assemblies with low grafting density. Macromolecules **51**(18), 7257–7265 (2018)
62. A. Chremos, A.Z. Panagiotopoulos, H.Y. Yu, D.L. Koch, Structure of solvent-free grafted nanoparticles: molecular dynamics and density-functional theory. J. Chem. Phys. **135**(11), 114901 (2011)
63. J.G. Ethier, L.M. Hall, Structure and entanglement network of model polymer-grafted nanoparticle monolayers. Macromolecules **51**(23), 9878–9889 (2018)
64. A. Chremos, A.Z. Panagiotopoulos, D.L. Koch, Dynamics of solvent-free grafted nanoparticles. J. Chem. Phys. **136**(4), 044902 (2012)
65. C. Li, Anisotropy unnecessary. Nat. Nanomater. **8** (2009).

66. R.J. Moon, A. Martini, J. Nairn, J. Simonsen, J. Youngblood, Cellulose nanomaterials review: structure, properties and nanocomposites. Chem. Soc. Rev. **40**(7), 3941–3994 (2011)
67. S.A. Kedzior, L. Graham, C. Moorlag, B.M. Dooley, E.D. Cranston, Poly(methyl methacrylate)-grafted cellulose nanocrystals: one-step synthesis, nanocomposite preparation, and characterization. Can. J. Chem. Eng. **94**(5), 811–822 (2016)
68. S. Harrisson, G.L. Drisko, E. Malmstrom, A. Hult, K.L. Wooley, Hybrid rigid/soft and biologic/synthetic materials: polymers grafted onto cellulose microcrystals. Biomacromol **12**(4), 1214–1223 (2011)
69. K.H. Kan, J. Li, K. Wijesekera, E.D. Cranston, Polymer-grafted cellulose nanocrystals as pH-responsive reversible flocculants. Biomacromol **14**(9), 3130–3139 (2013)
70. K.J. De France, K.J. Chan, E.D. Cranston, T. Hoare, Enhanced mechanical properties in cellulose nanocrystal-poly(oligoethylene glycol methacrylate) injectable nanocomposite hydrogels through control of physical and chemical cross-linking. Biomacromol **17**(2), 649–660 (2016)
71. S. Goyal, F.A. Escobedo, Structure and transport properties of polymer grafted nanoparticles. J. Chem. Phys. **135**(18), 184902 (2011)
72. M. Murat, G.S. Grest, Polymers end-grafted onto a cylindrical surface. Macromolecules **24**, 704–706 (1991).
73. S.C. Glotzer, W. Paul, Molecular and mesoscale simulation methods for polymer materials. Annu. Rev. Mater. Res. **32**(1), 401–436 (2002)
74. T.D. Nguyen, Z. Zhang, S.C. Glotzer, Molecular simulation study of self-assembly of tethered V-shaped nanoparticles. J. Chem. Phys. **129**(24), 244903 (2008)
75. N.K. Hansoge, T. Huang, R. Sinko, W. Xia, W. Chen, S. Keten, Materials by design for Stiff and Tough Hairy nanoparticle assemblies. ACS Nano **12**(8), 7946–7958 (2018)
76. N.K. Hansoge, S. Keten, Effect of polymer chemistry on chain conformations in Hairy nanoparticle assemblies. ACS Macro Lett. **8**(10), 1209–1215 (2019)
77. G.S. Grest, K. Kremer, Molecular dynamics simulation for polymers in the presence of a heat bath. Phys. Rev. A Gen. Phys. **33**(5), 3628–3631 (1986)
78. S. Plimpton, Fast parallel algorithms for short-range molecular dynamics. J. Comput. Phys. **117**, 1–19 (1995)
79. H. Fang, M. Rais-Rohani, Z. Liu, M.F. Horstemeyer, A comparative study of metamodeling methods for multiobjective crashworthiness optimization. Comput. Struct. **83**(25–26), 2121–2136 (2005)
80. J.-M. Miao, S.-J. Cheng, S.-J. Wu, Metamodel based design optimization approach in promoting the performance of proton exchange membrane fuel cells. Int. J. Hydrogen Energy **36**(23), 15283–15294 (2011)
81. M. Daoud, J.P. Cotton, Star shaped polymers: a model for the conformation and its concentration dependence. J. De Physique **43**(3), 531–538 (1982)
82. K. Ohno, T. Morinaga, S. Takeno, Y. Tsujii, T. Fukuda, Suspensions of silica particles grafted with concentrated polymer brush- effects of graft chain length on brush layer thickness and colloidal crystallization. Macromolecules **40**, 9143–9150 (2007)
83. A. Jayaraman, K.S. Schweizer, Effective interactions and self-assembly of hybrid polymer grafted nanoparticles in a homopolymer matrix. Macromolecules **42**(21), 8423–8434 (2009)
84. T.B. Martin, A. Jayaraman, Using theory and simulations to calculate effective interactions in polymer nanocomposites with polymer-grafted nanoparticles. Macromolecules **49**(24), 9684–9692 (2016)
85. N. Nair, N. Wentzel, A. Jayaraman, Effect of bidispersity in grafted chain length on grafted chain conformations and potential of mean force between polymer grafted nanoparticles in a homopolymer matrix. J. Chem. Phys. **134**(19), 194906 (2011)
86. M. Doxastakis, Y.L. Chen, J.J. de Pablo, Potential of mean force between two nanometer-scale particles in a polymer solution. J. Chem. Phys. **123**(3), 34901 (2005)
87. M. Doxastakis, Y.L. Chen, O. Guzman, J.J. de Pablo, Polymer-particle mixtures: depletion and packing effects. J. Chem. Phys. **120**(19), 9335–9342 (2004)
88. L. Zhao, Y.G. Li, C. Zhong, J. Mi, Structure and effective interactions in polymer nanocomposite melts: an integral equation theory study. J. Chem. Phys. **124**(14), 144913 (2006)

89. A.J. Clark, J. McCarty, M.G. Guenza, Effective potentials for representing polymers in melts as chains of interacting soft particles. J. Chem. Phys. **139**(12), 124906 (2013)
90. V. Pryamtisyn, V. Ganesan, A.Z. Panagiotopoulos, H. Liu, S.K. Kumar, Modeling the anisotropic self-assembly of spherical polymer-grafted nanoparticles. J. Chem. Phys. **131**(22), 221102 (2009)
91. D. Meng, S.K. Kumar, D. Lane, J. Matthew, G.S. Grest, Effective interactions between grafted nanoparticles in a polymer matrix. Soft Matter **8**(18) (2012).
92. W. You, Z. Tang, C.A. Chang, Potential mean force from umbrella sampling simulations: what can we learn and what is missed? J. Chem. Theory Comput. **15**(4), 2433–2443 (2019)
93. D.M. Trombly, V. Ganesan, Curvature effects upon interactions of polymer-grafted nanoparticles in chemically identical polymer matrices. J. Chem. Phys. **133**(15), 154904 (2010)
94. S.A. Egorov, Interactions between polymer brushes in solvents of variable quality: a density functional theory study. J. Chem. Phys. **129**(6), 064901 (2008)
95. F. Lo Verso, L. Yelash, S.A. Egorov, K. Binder, Interactions between polymer brush-coated spherical nanoparticles: the good solvent case. J Chem Phys **135**(21), 214902 (2011)

Chapter 8
Predicting Mechanical Properties Using Continuum Mechanics-Based Approach: Micro-mechanics and Finite Element Analysis

Pavan K. Valavala and Gregory M. Odegard

Abstract The mechanical properties of nano-structured materials are important field of exploration in the fields of materials science and other engineering disciplines. Thorough understanding of underlying material structure and resulting properties require a variety of tools depending on the length scales of interest. This chapter reviews continuum mechanics-based techniques, with an emphasis on micro-scale modeling techniques: analytical and computational. In addition to micro-mechanics, different approaches to multiscale modeling are presented. With the appropriate choice of techniques, models can be bridged across multiple length scales leading to mechanistic understanding of the mechanics of materials. Some illustrative examples are also discussed that utilize the techniques presented here.

8.1 Introduction

Mechanical properties are of great interest in many applications and can be predicted from the structure of the material. Modeling the mechanics of nanostructured materials can be accomplished using a multitude of techniques that span various length and time scales leading to structure–property relationships (Fig. 8.1). From the nanoscale to sub-micronscale, computational chemistry-based tools provide the best means to capture the mechanics of the material systems. At larger length scales typically micronscale and above, continuum mechanics-based techniques provide a robust toolkit that can be employed to model material behavior with a high degree of accuracy. Each of these techniques provides certain unique advantages that make them

P. K. Valavala (✉)
Research and Development, Packaging and Specialty Plastics, Dow Inc., Lake Jackson, TX 77566, USA
e-mail: PValavala@dow.com

G. M. Odegard
Mechanical Engineering—Engineering Mechanics, Michigan Technological University, Houghton, MI 49931, USA
e-mail: GModegar@mtu.edu

V. V. Ginzburg and L. M. Hall (eds.), *Theory and Modeling of Polymer Nanocomposites*, Springer Series in Materials Science 310,
https://doi.org/10.1007/978-3-030-60443-1_8

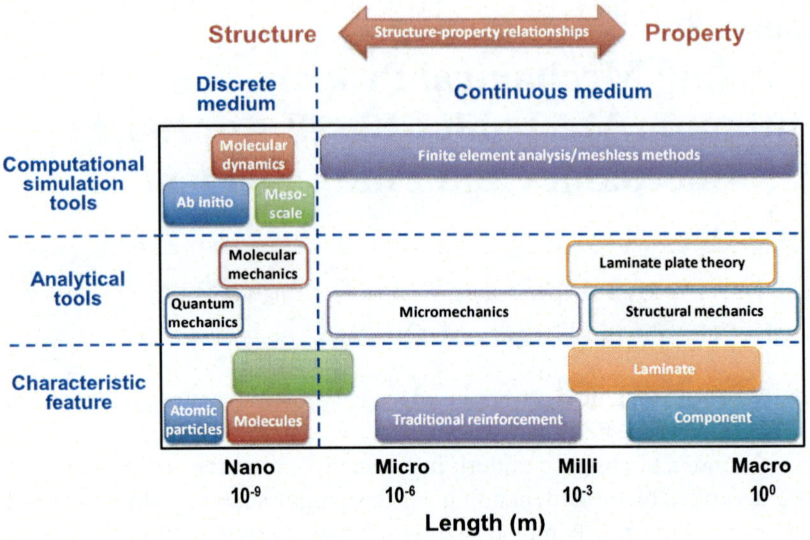

Fig. 8.1 Length scale spectrum associated with modeling mechanical properties of composites. Reproduced with permission from [1]. Copyright (2018) Elsevier

desirable to model the mechanics of nanostructured materials. This chapter will provide a brief overview of the techniques employed at the sub-micron length scales followed by a more detailed overview of the continuum-based techniques generally applied at micronscale and above. Finally, the chapter will conclude with discussion on combining these different techniques to achieve a structurally informed modeling strategy to understand the behavior of nanostructured materials.

8.2 Discrete-Medium Modeling Tools

At the smallest length scales, the physics of the material is accurately modeled as a discrete system with defined structure. Some of the defining features of material systems at this scale are atomic structures, molecular configuration, nanoparticles, etc. The modeling tools that are widely employed at this scale spanning from nano to sub-micronscales are based on quantum mechanics and molecular mechanics, respectively. In quantum mechanics-based techniques, the underlying atomic system is modeled to help elucidate the physics of the material of interest. These models typically solve for the electron density of the atoms for prescribed position of the atomic center [2, 3]. These models are able to capture the influence of external applied forces to the system as required. While providing high degree of accuracy in prediction of properties that depend on electron density, the models are mostly used for a collection of few tens to hundreds of atoms and timescales of the order of femto- to pico-seconds. This significantly hinders their applicability to macro-scale engineering problems. Some of the widely used techniques are based on methods such as the density functional theory, tight-binding approximation, and Hartree–Fock method [2, 3].

At the next larger length scales, molecular mechanics provides an effective method to model systems. These models are able to explore molecules and the collection thereof including their configurational space and the resulting properties [4–6]. Unlike quantum mechanics-based approach, these methods approximate the underlying atomic physics through simplifications to ignore certain electron cloud characteristics in favor of longer scale interactions to enable the modeling up to sub-micron length scales. Atoms are assumed to be hard spheres with interactions that can be described entirely based on the center of the atom. The interatomic interactions are generally captured through semi-empirical analytical expressions, for instance, chemically bonded 2-body interaction can be described by a quadratic expression using a spring constant (K_{Bond}) and an equilibrium bond length parameter (r_{eq}). $E^{\text{Bond}} = K_{\text{Bond}}(r - r_{\text{eq}})^2$, where E^{bond} is the potential energy of the 2-body bonded interaction, K_{bond} is the spring constant, r is the separation distance of two atoms, and r_{eq} is the equilibrium distance of the two atoms. Similar expressions are utilized to describe numerous interactions such as: angular energy (3-body), torsional energy (4-body), electrostatic interactions, and non-bonded interactions [4–14]. Techniques such as molecular dynamics (MD) build upon the framework of molecular mechanics and enable modeling of the mechanics of the material systems [15–19]. Application of these tools enable connecting of sub-micron length scale structure to a variety of physical behavior of the system beyond mechanical properties such as prediction of glass transition temperature to diffusion [15, 20]. But the inherent limitation of techniques like MD are the accessible length and time scales that are governed by the stiffest interactions in the model, i.e., K_{bond}, the time stepping in the model scales with $\Delta t_{\text{Model}} \propto \sqrt{\frac{m_{\text{Atom}}}{K_{\text{Bond}}}}$ where m_{Atom} is the mass of an atom in a bonded interaction. Thus, the longest timescales that are typically accessible for these models are of the order of nanoseconds. Some improvements in length and time scales can be achieved through techniques such as coarse-graining that combine multiple atoms into a heavier collective unit "pseduo-atom" of mass $m_{\text{Atom}}^{\text{pseudo}}$ ($\gg m_{\text{Atom}}$), and to account for the collection of atoms within such a "pseudo-atom" the effective bond stiffness is changed to $K_{\text{Bond}}^{\text{Super-atom}}$ ($\ll K_{\text{Bond}}$) leading to $\Delta t^{\text{Coarse-grain}} \propto \sqrt{\frac{m_{\text{Atom}}^{\text{pseudo}}}{K_{\text{Bond}}^{\text{Super-atom}}}}$. Hence, $\Delta t^{\text{Coarse-grain}} \gg \Delta t_{\text{Model}}$ [21, 22]. However, even with coarse-graining, models typically achieve timescales much less than millisecond. Similar to the atomic modeling techniques, these tools are unable to access larger length and timescales.

8.3 Continuous-Medium Modeling Tools

At lengths starting at around microns, continuum mechanics provide a variety of tools to model the effective mechanical properties of homogenous and multi-phase heterogeneous materials. Unlike the discrete medium modeling tools described in the previous sections, continuum-based approaches neglect underlying structure of the material at lower length scales, the material is assumed continuous at any arbitrary length scale. Also, classical continuum mechanics assumes local interaction, in that stress at a given point depends on the current values and the history of deformation and temperature at that point only. Material points interact with neighboring points through transfer of mass, momentum, energy, and entropy only [23]. Some non-local continuum theories eliminate this restriction by allowing for longer range interactions that resemble discrete systems as described in the previous section [24–27]. The remainder of this chapter will focus only on the classical continuum approaches. For heterogeneous materials with multiple phases, micro-mechanics-based modeling techniques were developed to account for interactions at the constituent level, i.e., inclusion/matrix. These methods were primarily used to predict the properties of traditional composite materials developed during a period when computational resources were expensive. In the past few decades, micro-mechanics-based approaches have evolved from analytical expressions to a more complex and much more computationally rigorous approach. The continuum-based approaches can be broadly classified into these categories: analytical and computational.

8.3.1 Analytical Modeling: Micro-mechanics

The original closed-form solutions to materials with inhomogeneous inclusions was presented by John Eshelby. He proposed a virtual experiment to derive the stress and strain fields for this problem [28]. Consider a linear elastic material with a stiffness (C_{ijkl}) bounded by a surface ($\partial\Omega$) and an enclosed volume (Ω) with a sub-volume (Ω_i) (homogeneous inclusion) of material bounded by ($\partial\Omega_i$). Assume that this sub-volume (Ω_i) was extracted from the material, and subject to an inelastic stress-free transformation, i.e., eigen-strain $\left(\varepsilon_{ij}^*\right)$ as shown in Fig. 8.2. This homogeneous inclusion is subjected to tractions to facilitate returning it back into the original

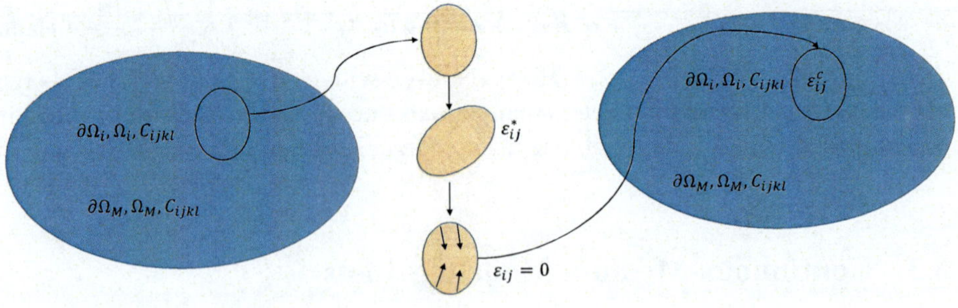

Fig. 8.2 Schematic representation of a thought experiment proposed by Eshelby to compute stresses and strains for a material with homogeneous inclusion

material bounded by surface ($\partial\Omega$). Upon replacement, both the inclusion and matrix experience a localized strain in the vicinity of the replaced inclusion. The resultant strain in the homogeneous inclusion is known as the "constrained strain" $\left(\varepsilon_{ij}^{C}\right)$. The stress and strain fields in the material can be expressed as:

$$\varepsilon_{ij}(x) = \overline{\varepsilon_{ij}} + \varepsilon_{ij}'(x)$$

$$\sigma_{ij}(x) = \overline{\sigma_{ij}} + \sigma_{ij}'(x)$$

where $\varepsilon_{ij}(x), \overline{\varepsilon_{ij}}, \varepsilon_{ij}'(x)$ are the strain at any spatial position described by a coordinate vector 'x', average macroscopic strain and the spatially fluctuating strain respectively. $\sigma_{ij}(x), \overline{\sigma_{ij}}, \sigma_{ij}'(x)$ are the corresponding stress counterparts. The average fields over the domain can be calculated using [29]:

$$\frac{1}{V}\int f(x)\mathrm{d}V = \overline{f}$$

where $f(x)$ is a microscopic field. The stress and strain fields need to satisfy the micro–macro energy equivalence (Hill-Mandel Principle [30, 31]):

$$\frac{1}{V}\int \sigma_{ij}\varepsilon_{ij}\mathrm{d}V = \overline{\sigma_{ij}\varepsilon_{ij}}$$

The constrained strain is related to the eigen-strain through a fourth-order tensor referred to as the Eshelby tensor:

$$\varepsilon_{ij}^{C} = S_{ijkl}\varepsilon_{kl}^{*}$$

The Eshelby tensor has minor symmetry: $S_{ijkl} = S_{jikl} = S_{ijlk} = S_{jilk}$. The Eshelby tensor (hence strain) is constant within the homogeneous inclusion in an infinite matrix of the same material, and depends only on the inclusion shape, and inclusion orientation, and elastic properties of the host [32, 33]. Closed-form solutions have been proposed for some regular shapes such as a spherical inclusion:

$$S_{1111} = S_{2222} = S_{3333} = \frac{7-5\nu}{15(1-\nu)}$$

$$S_{1122} = S_{1133} = S_{2233} = \frac{5\nu-1}{15(1-\nu)}$$

$$S_{1212} = S_{1313} = S_{2323} = \frac{4-5\nu}{15(1-\nu)}$$

where ν is the Poisson's ratio of the matrix material. The stress in the inclusion can be calculated as:

$$\sigma_{mn}^{I} = C_{mnkl}(S_{mnkl} - \delta_{mnkl})\varepsilon_{kl}^{*}$$

where σ_{mn}^{I} is the stress in the inclusion, and δ_{mnkl} is the fourth-order identity tensor.

Through the application of the Eshelby approach, the mechanics of multi-phase materials can be studied. Mean-field homogenization (MFH) provides equivalent properties of a composite material from the constituent phase properties, geometry, and volume fractions [32, 33]. These techniques are generally applicable when there is sufficient scale separation between the inclusion phase and the matrix, typically an order of magnitude in length scales.

8.3.1.1 Mean-Field Homogenization: Linear Elastic Behavior

Following the Eshelby approach, we can calculate the stresses and strain fields in a composite with multiple phases with different properties [28, 32, 33, 34]. Unlike the case presented in the previous section, a heterogeneous inclusion with different material properties $\left(C_{ijkl}^{I}\right)$ than the matrix $\left(C_{ijkl}^{M}\right)$ is shown in Fig. 8.3a. A solution to the inhomogeneous inclusion can be calculated through solving an equivalent homogeneous inclusion that results in the same strain field as the inhomogeneous inclusion. The average strain (or eigen-strain) in the inclusion was provided in the previous section:

$$\overline{\varepsilon_{ij}} = S_{ijkl}\varepsilon_{kl}^{*}$$

The micro-strain in the inclusion can be related to the macro-strain:

$$\varepsilon_{ij}^{I} = A_{ijkl}^{dil,I}\varepsilon_{kl}^{0} \text{ and } \sigma_{ij}^{I} = B_{ijkl}^{dil,I}\sigma_{kl}^{0}$$

where ε_{ij}^{0} and σ_{ij}^{0} are the macro-strain and stress, respectively, and $A_{ijkl}^{dil,I}$ and $B_{ijkl}^{dil,I}$ are the strain and stress concentration tensors, respectively. The strain concentration tensor, also known as the dilute approximation tensor is given by:

$$A_{ijkl}^{dil,I} = \left[\delta_{ijkl} + S_{ijmn}\left(C_{mnpq}^{M}\right)^{-1}\left(C_{pqkl}^{I} - C_{pqkl}^{M}\right)\right]^{-1}$$

Generically, the effective stiffness of the material with multiple inclusions can be expressed as:

$$\overline{C_{ijkl}} = \left[\sum_{\alpha}\left(v_{\alpha}C_{ijmn}^{\alpha}A_{mnkl}^{dil,\alpha}\right)\right]^{-1}$$

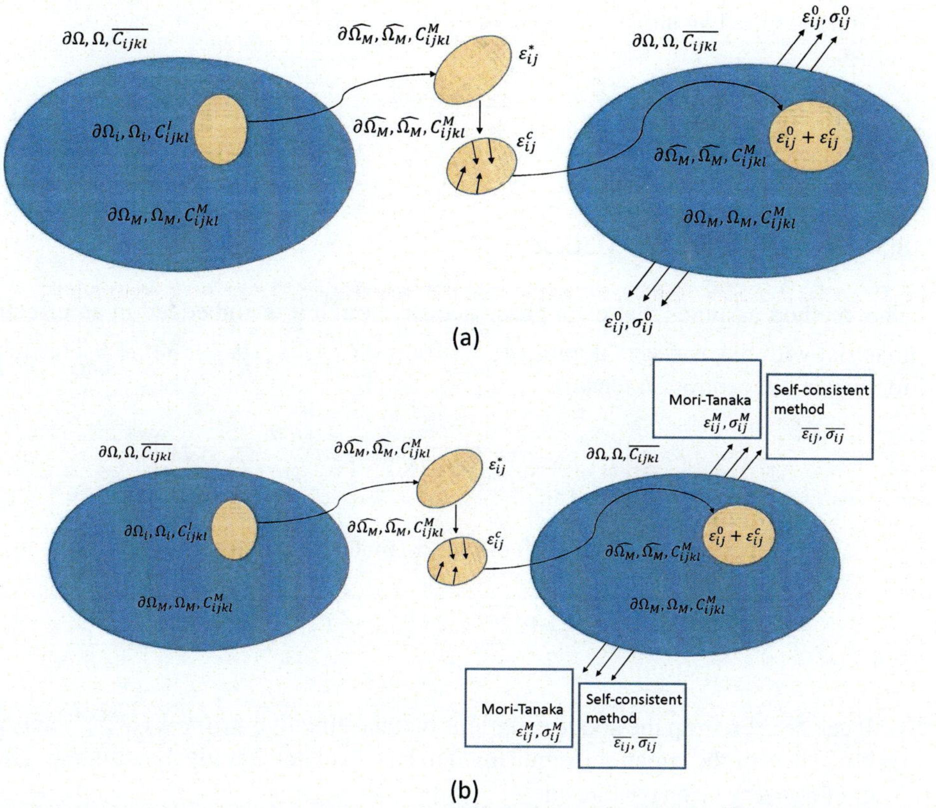

Fig. 8.3 Schematic of a heterogenous inclusion in a matrix. **a** Eshelby approach leading to a dilute approximation. **b** Approximation used to arrive at Mori–Tanaka and self-consistent homogenization

where $v_\alpha = \frac{\Omega_\alpha}{\Omega}$ is the volume fraction of the α phase and $\sum_\alpha v_\alpha = 1$.

8.3.1.2 Mori–Tanaka Method

The Mori-Tanaka approach assumes that each heterogeneous inclusion (α) acts as though it is embedded in an infinite matrix without any interaction with other inclusions, which leads to $\varepsilon_{ij}^{I} = A_{ijkl}^{dil,I,0}\varepsilon_{kl}^{M}$, where $A_{ijkl}^{dil,I,0}$ is the dilute approximation strain concentration tensor [35–37]. The Mori-Tanaka strain concentration tensor for inclusion is given by:

$$A_{ijkl}^{dil,\alpha} = \left[\delta_{ijkl} + S_{ijmn}^{M}\left(C_{mnpq}^{M}\right)^{-1}\left\{C_{pqkl}^{I,\alpha} - C_{pqkl}^{M}\right\}\right]^{-1}$$

$$A_{ijkl}^{MT,\alpha} = A_{ijmn}^{dil,\alpha}\left[v_M\delta_{mnkl} + \sum_\alpha v_\alpha A_{ijkl}^{dil,\alpha}\right]^{-1}$$

And the effective stiffness becomes:

$$\overline{C_{ijkl}^{MT}} = C_{ijkl}^{M} + \sum_{\alpha} v_{\alpha}\left(C_{ijmn}^{I,\alpha} - C_{ijmn}^{M}\right)A_{mnkl}^{MT,\alpha}$$

8.3.1.3 Self-consistent Method

This method assumes that each phase acts as though it is embedded in an effective material with the average "composite" stiffness ($\overline{C_{ijkl}^{SCM}}$) [29, 33, 38, 39]. This leads to a strain concentration tensor:

$$A_{ijkl}^{SCM,\alpha} = \left[\delta_{ijkl} + \overline{S_{ijmn}\left(C_{mnpq}^{CSM}\right)}^{-1}\left\{C_{pqkl}^{I,\alpha} - \overline{C_{pqkl}^{SCM}}\right\}\right]^{-1}$$

And the resulting effective stiffness of the material can be expressed as:

$$\overline{C_{ijkl}^{SCM}} = C_{ijkl}^{M} + \sum_{\alpha} v_{\alpha}\left(C_{ijmn}^{I,\alpha} - C_{ijmn}^{M}\right)A_{mnkl}^{SCM,\alpha}$$

It can be seen from the above equations that the effective stiffness ($\overline{C_{ijkl}^{SCM}}$) appears on both sides of the equations requiring it to be calculated iteratively until a specified level of convergence is achieved.

8.3.1.4 Property Bounds

Each homogenization scheme relies on a set of assumptions to provide an elegant solution to calculate the effective behavior of a composite. It is often difficult to determine which one of these schemes will yield an accurate prediction for a given material system. In addition, all of these techniques are built on the Eshelby approach of an effective single inclusion problem that does not explicitly account for attributes of the inclusion such as distribution of sizes, orientations, etc. To account for such variability one can explore bounds of properties for a material system in addition to the techniques presented in the previous section.

Voigt-Reuss Bounds

An effective method to understand the effect of an inclusion (secondary phase) in a material is through the use of Voigt and Reuss bounds [32, 33, 40, 41]. The resulting bounds for the elastic modulus of a two-phase composite can be expressed using the rule of mixtures (Voigt/iso-strain) and inverse rule of mixtures (Reuss/iso-stress):

$$\overline{E^{Voigt}} = vE^I + (1-v)E^M$$

$$\frac{1}{\overline{E^{Reuss}}} = \frac{v}{E^I} + \frac{1-v}{E^M}$$

where E^I and E^M is the elastic modulus of the inclusion and the matrix, respectively, and v is the volume fraction of the inclusion in the two-phase material system.

Hashin–Shtrikman Bounds

The work of Hashin and Shtrikman was based on the variational principles to derive bounds for the effective stiffness of a two-phase material [32, 33, 42, 43]:

$$\overline{C_{ijkl}^{HS+}} = C_{ijkl}^I + (1-v)\left[\left(C_{ijkl}^M - C_{ijkl}^I\right)^{-1} + vS_{ijmn}^I\left(C_{mnkl}^I\right)^{-1}\right]^{-1}$$

$$\overline{C_{ijkl}^{HS-}} = C_{ijkl}^M + v\left[\left(C_{ijkl}^I - C_{ijkl}^M\right)^{-1} + (1-v)S_{ijmn}^M\left(C_{mnkl}^M\right)^{-1}\right]^{-1}$$

where $\overline{C_{ijkl}^{HS+}}$ and $\overline{C_{ijkl}^{HS-}}$ are the bounds for the effective stiffness, respectively, and v is the volume fraction of the inclusion in the two-phase material system, and C_{ijkl}, and S_{ijkl} are the stiffness tensors of the inclusion and the matrix, respectively, and the superscripts I and M refer to the different phases within the composite.

8.3.2 Comparison of the Different Analytical Micro-mechanical Models

For a composite material with a spherical inclusion of 10× stiffness relative to the matrix material, Fig. 8.4 shows a comparison of predicted modulus with the different micro-mechanics-based estimates [39]. The Voigt-Reuss bounds provide the widest bounds encompassing the predictions from all other homogenization techniques, while Hashin–Shtrikman provides tighter bounds. The Mori-Tanaka and self-consistent approaches are relatively close until 30% volume fraction beyond which they deviate. At low volume fractions (<5% approx.) of the inclusion, most of the approaches are in reasonable agreement with each other. This would fall into the truly dilute approximation where each inclusion behaves as though it is embedded in an infinite matrix. As the inclusion volume fraction increases, the predictions deviate from each other. At higher concentrations, the stress fields from each of the inclusions starts to perturb the stress fields produced by the neighboring inclusion leading to nonlinear behavior. Also, as the inclusion volume fractions increase, the

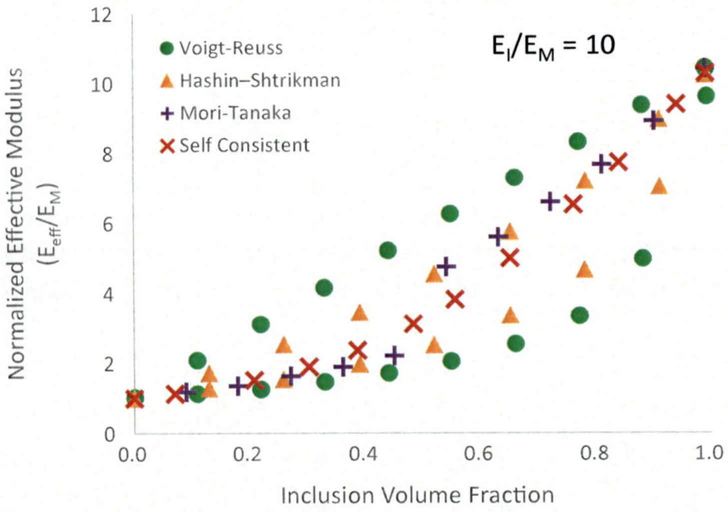

Fig. 8.4 Comparision of stiffness predicted by different homogenization techniques for a two-phase composite

definition of matrix, generally assumed to be the continuous phase, starts to break down. The upper limit is a face-centered cubic (*fcc*) crystal for which the maximum volume fraction is 74%, which leads to spherical inclusion becoming the continuous phase. In general, beyond a certain concentration, the roles of matrix and inclusion have to be rethought as the continuous (matrix) and discontinuous (inclusion) phases may have to be reversed. The predictions of effective properties are also sensitive to the magnitude of difference between the matrix and inclusion stiffness. Typically, as the stiffness discrepancy between the matrix and inclusion is reduced, the different homogenization schemes predict values closer to each other and in the limit case of identical properties of matrix and inclusion, all predictions collapse to a single value.

8.3.3 Laminate Plate Theory

In the previous section, the focus was on understanding the resulting mechanics of a composite from the knowledge of the constituent behavior. However, for many engineering applications, the properties of interest are typically a few length scales larger in magnitude. To address these engineering-scale problems, laminate mechanics are typically employed. A laminate is series of layers of composite materials, typically fiber reinforced, bonded together to form a plate. Each of these constituent layers is referred to as "ply". Most structural laminate applications involve withstanding in-plane stretch loading or bending moments. The mechanics of these structures can be understood through calculation of in-plane stresses/strains and the resulting curvature, respectively [33, 44–46].

Utilizing the symmetry of the Cauchy stress tensor, it can be expressed in Voigt notation as:

$$\sigma_{ij} = \sigma_{ji} \text{ i.e. } \sigma_{ij} = \begin{bmatrix} \sigma_{11} & \sigma_{12} & \sigma_{13} \\ \sigma_{12} & \sigma_{22} & \sigma_{23} \\ \sigma_{13} & \sigma_{23} & \sigma_{33} \end{bmatrix}$$

$$\text{Voigt Notation: } \sigma_i = \begin{bmatrix} \sigma_{11} & \sigma_{22} & \sigma_{33} & \sigma_{23} & \sigma_{13} & \sigma_{12} \end{bmatrix}^{\mathrm{T}} = \begin{bmatrix} \sigma_1 & \sigma_2 & \sigma_3 & \sigma_4 & \sigma_5 & \sigma_6 \end{bmatrix}^{\mathrm{T}}$$

For laminate plate theory, if we assume that the thickness of the plate is very small when compared to the other dimensions of the plate and the plane stress conditions apply, the stresses and the corresponding strains reduce to $\sigma_i = \begin{bmatrix} \sigma_1 & \sigma_2 & \sigma_6 \end{bmatrix}^{\mathrm{T}}$ and $\varepsilon_i = \begin{bmatrix} \varepsilon_1 & \varepsilon_2 & \gamma_6 \end{bmatrix}^{\mathrm{T}}$ where $\gamma_6 = 2\varepsilon_6$ for each ply. The inter-laminar shear can be neglected under this approximation. The stress–strain relationship for a unidirectional fiber-reinforced ply is given by: $\sigma_i = Q_{ij}\varepsilon_j$ where Q_{ij} is the reduced stiffness matrix in the local coordinates of the ply, i.e., 1-direction is aligned to the length of the fiber [44–48].

$$Q_{ij} = \begin{bmatrix} \frac{E_1}{1-\upsilon_{12}\upsilon_{21}} & \frac{\upsilon_{12}E_2}{1-\upsilon_{12}\upsilon_{21}} & 0 \\ \frac{\upsilon_{12}E_2}{1-\upsilon_{12}\upsilon_{21}} & \frac{E_2}{1-\upsilon_{12}\upsilon_{21}} & 0 \\ 0 & 0 & G_{12} \end{bmatrix}$$

where E_1, E_2 and G_{12} are the modulus of the ply in the 1, 2, and 6 directions of the ply, respectively, with the 1-direction aligned to the length of the fiber. For a laminate consisting of multiple plies, the forces and moments are related to the mid-plane strains and curvatures:

$$\begin{Bmatrix} N_x \\ N_y \\ N_{xy} \\ M_x \\ M_y \\ M_{xy} \end{Bmatrix} = \begin{bmatrix} A_{11} & A_{12} & A_{13} & B_{11} & B_{12} & B_{13} \\ A_{12} & A_{22} & A_{23} & B_{12} & B_{22} & B_{23} \\ A_{13} & A_{23} & A_{33} & B_{13} & B_{23} & B_{33} \\ B_{11} & B_{12} & B_{13} & D_{11} & D_{12} & D_{13} \\ B_{12} & B_{22} & B_{23} & D_{12} & D_{22} & D_{23} \\ B_{13} & B_{23} & B_{33} & D_{13} & D_{23} & D_{33} \end{bmatrix} \begin{Bmatrix} \varepsilon_x^0 \\ \varepsilon_y^0 \\ \gamma_{xy}^0 \\ \kappa_x \\ \kappa_y \\ \kappa_{xy} \end{Bmatrix}$$

where $\varepsilon_x^0, \varepsilon_y^0, \gamma_{xy}^0$ and $\kappa_x, \kappa_y, \kappa_{xy}$ are the mid-planes strains and curvatures, respectively, of the laminate. $A_{ij} = \sum_k \overline{Q_{ij}^k} t_k; D_{ij} = \sum_k \overline{Q_{ij}^k}\left(t_k \overline{z_k}^2 + \frac{t_k^3}{12}\right); B_{ij} = \sum_k \overline{Q_{ij}^k} t_k \overline{z_k};$ are the in-plane stiffness matrix, bending stiffness matrix, and the bending-extension coupling matrix, respectively. 'k' is the summation over all plies in the laminate. Q_{ij}^k is the stiffness matrix of the kth ply in the local coordinates, and

$\overline{Q_{ij}^k}$ is the stiffness matrix in the global coordinates obtained through a transformation operation such that $\overline{Q_{ij}^k} = T_{im}^{-1} Q_{mn}^k T_{jn}^{-1}$ and $T_{ij} = \begin{bmatrix} m^2 & n^2 & 2mn \\ n^2 & m^2 & -2mn \\ -mn & mn & m^2 - n^2 \end{bmatrix}$ with $m = \cos\theta$ and $m = \sin\theta$, where θ is the angle between the local fiber axis for the kth ply and the global xyz axes.

The stresses in any ply can be calculated with $\sigma_i^k = \overline{Q_{ij}}\varepsilon_j^k$, where $\varepsilon_i^k = \varepsilon_i^0 + z\kappa_i$ and z is the distance of the kth ply from the mid-plane of the laminate.

8.3.3.1 Failure in Laminate Plate Theory

The previous section describes the calculations for the stress and strain for a laminate from the individual ply material properties, fiber orientation, and ply lay-up. This analysis can be extended to situations where the composite is expected to fail [33, 44–49]. The failure criterion can be expressed simply from the knowledge of stress and strain in the individual plies, and a determination of ply failure can be made. First ply failure does not necessarily mean laminate failure, after the first ply failure, the redistributed load may still be sustained by the laminate. For ply failure, if the maximum stress or strain (i.e., $\sigma_i^k > \left(\sigma_i^k\right)^{Ult.}$ or $\varepsilon_i^k > \left(\varepsilon_i^k\right)^{Ult.}$) is exceeded, in certain cases, it can be assumed that the laminate has failed. These simple failure criteria based on maximum stress and maximum strain implicitly assume there is a clearly dominant stress/strain direction as compared to the others. In situations, where more than one stress component is comparable to the maximum, the maximum stress/strain criterion does not yield very realistic results. Other improved failure criteria have been suggested such as Tsai-Hill, Tsai-Wu, Hasin, and Hashin-Rotem.

The Tsai-Hill criterion addresses the drawback of the failure criterion based on only one of the components of stress (or strain) through use of an approach similar to the von-mises criterion used for isotropic, homogenous systems. The failure envelope is defined as $\frac{\sigma_1^2}{F_1^2} - \frac{\sigma_1\sigma_2}{F_1^2} + \frac{\sigma_2^2}{F_2^2} + \frac{\sigma_6^2}{F_6^2} - 1 = 0$ where F_1, F_2, F_6 are the failure stresses along 1, 2, and 6 local directions of the ply that can be determined either experimentally or through the use of micro-mechanical theories. However, there are certain drawbacks to the Tsai-Hill criterion as it does not distinguish between the tensile and compressive asymmetry of strength of composites particularly in polymer-based materials. The Tsai-Wu criterion addresses this strength asymmetry in the expression for the failure envelope: $f_1\sigma_1 + f_2\sigma_2 + f_{11}\sigma_1^2 + f_{22}\sigma_2^2 + 2f_{12}\sigma_1\sigma_2 + f_{66}\sigma_6^2 - 1 = 0$ where $f_1 = \frac{1}{F_{1t}} - \frac{1}{F_{1c}}$, $f_2 = \frac{1}{F_{2t}} - \frac{1}{F_{2c}}$, $f_{11} = \frac{1}{F_{1t}F_{1c}}$, $f_{22} = \frac{1}{F_{2t}F_{2c}}$, $f_{66} = \frac{1}{F_6^2}$, and $f_{12} = \frac{-1}{F_{1t}^2}$. F_{xt} and F_{xc}, where 'x' is 1, 2, and 6 are the strength of the composite in tension and compression along the respective direction for the ply [44]. Different expressions for f_{12} have been suggested by others [49]. Details of the Hashin [50], Hashin-Rotem [51], and other failure criterion can be found elsewhere [49, 52].

Figure 8.5 shows the comparison of failure criterion for a uniaxial composite (AS4/3501-6 Graphite/Epoxy), where σ_1 and σ_2 are the stresses along the length of

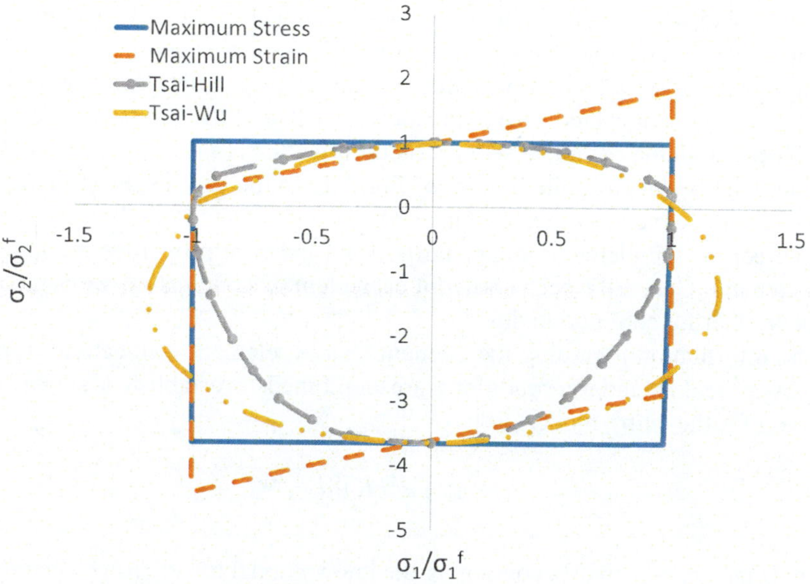

Fig. 8.5 Comparision of the four different failure criterion for a uniaxial fiber composite: Maximum stress, maximum strain, Tsai-Hill, and Tsai Wu

the fiber (1-axis) and the direction perpendicular to it direction (2-axis), respectively, [49, 53]. σ_1^f and σ_2^f are the failure stresses in the 1 and 2 direction of the composite, respectively. The maximum stress criterion ignores any interactions between the stress components hence the failure envelope is just a rectangle. The maximum strain criterion also ignores interaction between the different strains, however, the Poisson's effects lead to a parallelogram for the failure envelope. Tsai-Hill has a quadratic failure envelope in its classical form and does not include tensile and compression asymmetry of strength, however Sun et al. [49] modified it to account for the asymmetry. Tsai-Wu accounts for both linear and quadratic terms.

8.4 Computational Modeling: Finite Element Analysis (FEA)

The primary goal of FEA is to provide an estimated elasticity solution (that is, complete knowledge of the stress and strain fields inside an engineering structure or component) for a complex geometry that does not have a closed-form elasticity solution [54]. This is achieved by discretizing the overall geometry into a series of smaller primitive geometries (elements) that have analytical solutions. For example, consider the truck frame structure shown at the top of Fig. 8.7 represented by a computer-aided design (CAD) image. The structure is clearly very complex and contains welds (pink regions), holes, dimples, curves, and attached parts (brown regions); all of which form a structural model for which there are no known elasticity solutions (try and find an elasticity solution to this complex geometry!). Rather than attempting to solve such a complex problem, the frame geometry has been discretized into a series of connected elements (mesh) such that the entire structure can be reproduced with the complete set of structural features (e.g., welds, holes, dimples, and curves), as shown in the bottom of Fig. 8.6. In the case of the structure shown in Fig. 8.6, the elements types are hexahedrons and tetrahedrons. The FEA software keeps track of the element connectivities, and material properties are assigned to each element. Generally, each corner of each element contains a *node*, with adjacent elements sharing common nodes.

The information regarding the element shapes, element connectivities, material properties, and boundary conditions are combined to establish a single stiffness equation for the entire model [54]:

$$[f] = [K][d]$$

where $[f]$ is a $N \times 1$ matrix containing the known nodal forces (force vector components acting at each node), $[d]$ is a $N \times 1$ matrix containing the known nodal displacements, $[K]$ is the $N \times N$ *stiffness matrix*, and N is the number of degrees of freedom in the model (usually three displacement components associated with each node). Finite element models can contain up to millions of degrees of freedom. The stiffness equation is solved computationally such that all nodal forces and displacements are

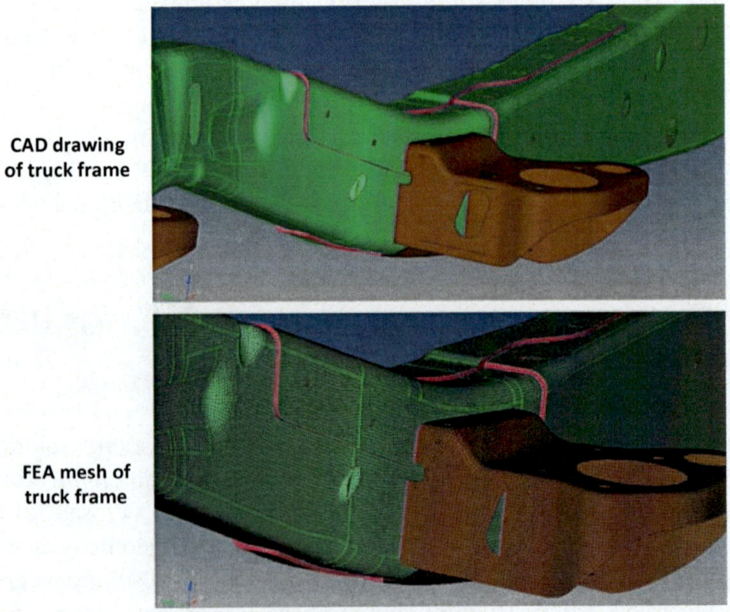

Fig. 8.6 CAD and FEA models of truck frame (Image courtesy Prathamesh Deshpande using HyperWorks modeling software). Reproduced with permission from [1]. Copyright (2018) Elsevier

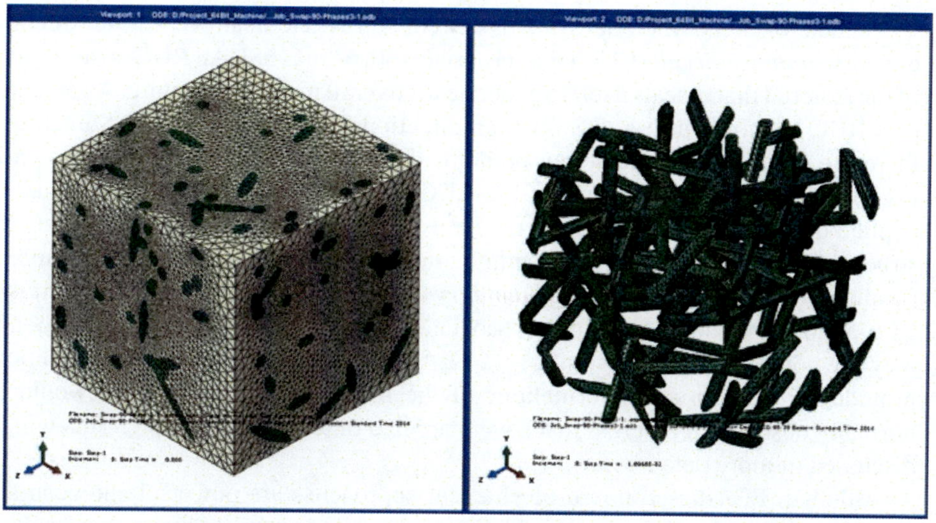

Fig. 8.7 Image showing a three-dimensional RVE of a composite material (left) and the constituent high aspect ratio inclusions (right) [62, 63]

determined, and the corresponding stress and strain fields are calculated thereafter. The stress and strain fields are typically visualized with a *contour plot*. It is important to note that FEA is very powerful and flexible. Not only can it be used to simulate the behavior of engineering structures, it can also be used to predict material properties based on microstructure which will be presented later in this section.

8.4.1 Computational Micro-mechanics

The FEA approach can ultimately provide the best predictions of bulk properties of composites based on the properties and geometry of the constituents relative to other micro-mechanics methods. This is due to the ability of FEA to accurately predict internal stress fields of a composite to a high degree of precision. The required approach with FEA is to simulate a material volume sample [55, 56]. A *representative volume element* (RVE) is a volume of material whose effective behavior is representative of that of the bulk material [32, 33]. For a material system with a stochastic nature (e.g., reinforcing fibers randomly aligned), the RVE must be large enough to represent material symmetries that are the same as those of the bulk (e.g., isotropic, transversely isotropic) [57–61].For example, consider RVE shown in Fig. 8.7. The RVE contains whole small spheres, and symmetric portions of medium and large spheres. Together, these various sizes of spheres represent the bulk material. For a RVE modeled with FEA, each domain within the RVE is meshed with the appropriate material properties applied.

For heterogeneous material systems that contain a periodically repeating structure, a *representative unit cell* (RUC) is appropriate [60, 61, 64–68]. An RUC is the volume of the material that repeats itself to generate the overall material structure. An example of a RUC is shown in Fig. 8.8 for a unidirectional fibrous composite. Sometimes, there are multiple RUCs that can be defined for a material structure, as shown in Fig. 8.8. Regardless of the size of the RUC, special boundary conditions must be applied onto the RUC sides within the FEA framework that effectively simulate the structure as a component inside an infinite medium of repeat RUCs. These boundary conditions are known as *periodic boundary conditions*, which are not to be confused with periodic boundary conditions associated with DFT and MD models (although they are conceptually very similar). In FEA models, periodic boundary conditions provide the appropriate deformations of the model (RUC region) that would be normally observed if external loads were applied onto the bulk sample. This will be discussed in more detail below.

Although FEA-based micro-mechanical approaches are powerful and accurate, they suffer from the same drawback as some of the other micro-mechanical models discussed above, they are not easily accessible to many materials design engineers. The building of the model and the application of the correct boundary conditions is a complex process. An alternative computational approach is available that is accurate, reliable, and relatively easy to implement into a composite material development process. The *generalized method of cells* (GMC) [69] was developed for this purpose and consists of a modeled RUC composed of subcells representing different material phases (e.g., fibers and matrix). A first-order displacement field is assumed to exist within the subcells (higher order displacement fields can be assumed if necessary),

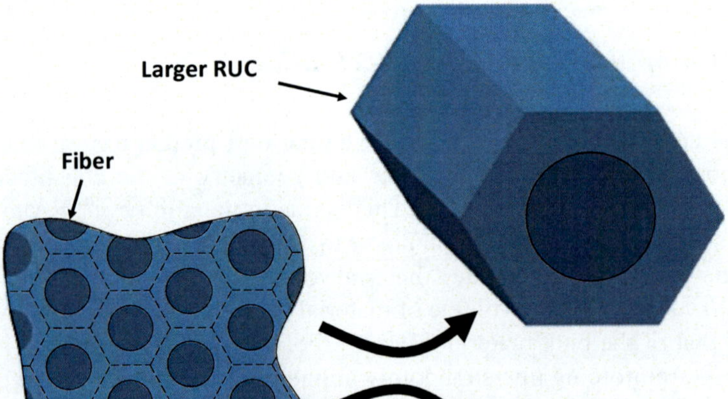

Fig. 8.8 RUC for a fiber composite. Reproduced with permission from [1]. Copyright (2018) Elsevier

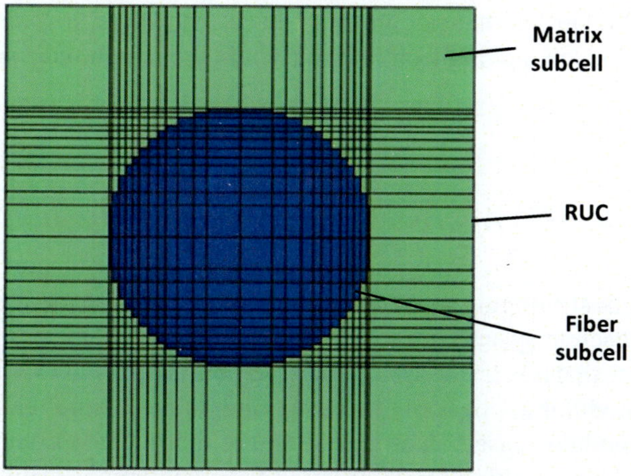

Fig. 8.9 GMC RUC for a fiber composite. Reproduced with permission from [1]. Copyright (2018)

and traction and displacement continuity are imposed between adjacent cells. An example of a GMC RUC is shown in Fig. 8.9 for a fibrous composite.

In this manner, internal fields are simulated within the composite without the need for discretizing the phases into elements, as is required with the FEA method. The GMC method has been implemented in the software program MAC/GMC (Micromechanics Analysis Code base on the generalized Method of Cells). The software is publicly available and easy to learn, making it much more accessible to materials design engineers, with a level of accuracy that is sufficient for most computational material modeling (CMM) efforts. Comprehensive details on this method can be found elsewhere [69].

8.5 Multiscale Modeling Concepts

So far, we have discussed the primary predictive tools available for modeling of composites at various length scales with particular emphasis on techniques applicable at the micronscale. From this discussion, it should be clear that these tools vary greatly in the time and length scales, as well as their assumption of a discrete/continuous framework. As is evident in Fig. 8.1, the prediction of the bulk-level response of an engineering component based on atomic structure requires the use multiple modeling tools across multiple time and length scales. This section discusses some of the important concepts related to the linking of different predictive tools across multiple length and time scales. Multiple options exist for incorporating information from different length scales in computational material modeling (CMM). The first option is to start with a lower-length scale modeling approach, and use the predicted information as input into a higher length scale model. This is known as the *bottom-up approach.* This corresponds to moving from left to right in the length scale spectrum shown in Fig. 8.1.

8.5.1 Bottom-Up Approach: Equivalent Continuum

Perhaps the most difficult conceptual aspect of multiscale modeling is the need to represent a discrete system (all materials are discrete at ~nanometer length scales) using concepts that rely on the assumption of a continuum. For example, the definition of Young's modulus applies only in the continuum mechanics realm. The prediction of Young's modulus based on atomic structure (particularly complex atomic structures) requires a transition from the discrete to continuum realms. We need some governing rules to be able to safely navigate this transition.

An *equivalent continuum* is a material model in the continuum realm that has typical engineering concepts tied to it (e.g., Young's modulus, stress, strain, thermal conductivity, glass transition) that describe a material behavior that is observed (or predicted) in either the discrete realm or a lower-scale continuum realm. The challenge is to find a way to take lower-length-scale-level predictions/observations and establish an equivalent continuum. The first step in this process is typically to establish a RVE (or RUC in the case that the lower-level structure is very simple; we will henceforth just refer to the RVE with the understanding that a RUC can also be used in certain cases). There are multiple factors in selecting an RVE size, including boundary conditions used, crystalline/amorphous nature of the material, desired level of accuracy, and availability of sufficient computational resources.

8.5.1.1 Boundary Conditions

Consider the molecular RVE shown in Fig. 8.10. Suppose that an equivalent volume and shape of the equivalent continuum material are used to represent (overlay) the RVE, denoted as region $\mathcal{R}$ with boundary $\partial\mathcal{R}$. The equivalent continuum model should mimic the behavior of the heterogeneous model as closely as possible under all mechanical loadings. It is important to note that an equivalent continuum model will ideally represent the behavior of an arbitrary volume of the actual material, not just the shape defined by the RVE. The definition of $\mathcal{R}$ is only necessary for relating the mechanical response of the RVE to that of the equivalent material points in the equivalent continuum.

The mechanical response of the RVE is usually established by applying boundary conditions onto the RVE boundary and predicted the corresponding behavior. This is usually performed computationally. There are four types of boundary conditions for RVEs in static equilibrium: *displacement-controlled* boundary conditions (also called *kinematic* or *Dirichlet* boundary conditions), *traction-controlled* boundary conditions (also called *static* or *Neumann* boundary conditions), *periodic* boundary

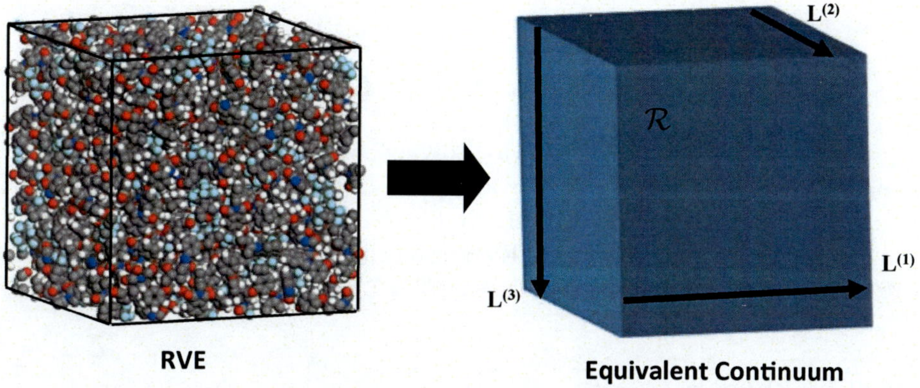

Fig. 8.10 Equivalent continuum modeling. Reproduced with permission from [1]. Copyright (2018) Elsevier

conditions, or a mixture of these three [33]. For displacement-controlled boundary conditions, the components of the prescribed (denoted with overbar) displacement vector are specified everywhere on $\partial\mathcal{R}$

$$\overline{u}_i = u_i \qquad \forall \mathbf{X} \in \partial\mathcal{R}$$

where **u** is the displacement vector and **X** is the material coordinate vector. For traction-controlled boundary conditions, the components of the prescribed traction vector are specified everywhere on $\partial\mathcal{R}$ as

$$\overline{s}_i = S_{ij}N_j \qquad \forall \mathbf{X} \in \partial\mathcal{R}$$

where **S** is the stress tensor and **N** is the surface unit normal vector. For periodic boundary conditions, the prescribed displacements and tractions are given by

$$\overline{u}_i(\mathbf{X}+\mathbf{L}) = u_i(\mathbf{X}) + u_i^{\text{ave}} \qquad \overline{\mathbf{s}}(\mathbf{X}+\mathbf{L}) = -\mathbf{s}(\mathbf{X}) \qquad \forall \mathbf{X} \in \partial\mathcal{R}$$

where $\mathbf{u}^{\text{ave}}$ is the average displacement vector associated with the bulk deformation of the solid material, and **L** is the periodicity vector of the RVE (shown, for example, in Fig. 8.10). For the case of mixed boundary conditions, the boundary of region $\mathcal{R}$ can be divided into three sub-boundaries $\partial\mathcal{R}_{\text{d}}$, $\partial\mathcal{R}_{\text{t}}$, and $\partial\mathcal{R}_{\text{pr}}$ such that

$$\partial\mathcal{R} = \partial\mathcal{R}_d \cup \partial\mathcal{R}_t \cup \partial\mathcal{R}_p \qquad \partial\mathcal{R}_d^\circ \cap \partial\mathcal{R}_t^\circ \cap \partial\mathcal{R}_p^\circ = \emptyset$$

where $\emptyset$ is the null set and the superscript $^\circ$ denotes the relative interior. The corresponding boundary conditions are

$$\overline{u}_i = u_i \qquad \forall \mathbf{X} \in \partial \mathcal{R}_d$$
$$\overline{s}_i = S_{ij} N_j \qquad \forall \mathbf{X} \in \partial \mathcal{R}_t$$
$$\overline{u}_i(\mathbf{X}+\mathbf{L}) = u_i(\mathbf{X}) + u_i^{\mathrm{ave}}; \ \overline{\mathbf{s}}(\mathbf{X}+\mathbf{L}) = -\mathbf{s}(\mathbf{X}) \ \forall \mathbf{X} \in \partial \mathcal{R}_p$$

Although the simplest approach to applying boundary conditions to a RVE (either computationally or analytically) is to use the displacement- or traction-controlled boundary conditions, it has been shown [59, 64] that the application of periodic or mixed boundary conditions to the RVE results in a more realistic predicted material response.

8.5.1.2 RVE Size

It was stated above that the type of boundary conditions used could influence the necessary RVE size. Specifically, the minimum size of an RVE for a periodic material is the minimum size of a possible repeatable structure, as shown in Fig. 8.8 for the smaller RVE of the fiber composite material. However, if periodic boundary conditions are applied to the edges of the RVE, then the minimal size of the RVE is the minimum size necessary to construct the material structure without rigid rotations of the RVE, such as the larger RVE in Fig. 8.8 and the RVE in Fig. 8.10. If periodic boundary conditions are not used, then the computational results will depend on the RVE size. This dependence will depend on the morphology and properties of the material. Generally, the larger the RVE, the more accuracy is obtained with the results. For example, consider the elastic properties of aligned-fiber composites established by Jiang et al. [64]. Traction-controlled, displacement-controlled, and periodic boundary conditions were applied for bulk-level transverse shear (shear in the plane transverse to the fiber direction). Two cases were simulated, one that contained inclusions that were stiffer than the matrix by an order of magnitude and one in which the matrix was stiffer than the inclusions by an order of magnitude. The Young's modulus of the matrix was assumed to be 1 GPa. The corresponding transverse shear modulus was determined using all three boundary conditions for the two composite systems for RVE sizes of $\delta = \delta_0$, $2\delta_0$, $3\delta_0$, and $4\delta_0$ where δ_0 is the RVE size shown in the inset of Figs. 8.11 and 8.12. Figures 8.11 and 8.12 show the calculated shear modulus for the stiff inclusions and matrix, respectively, for the different RVE lengths. The data has been plotted with smooth lines to emphasize the overall trend. Clearly, for the case of stiff inclusions shown in Fig. 8.11, the discrepancy between the predicted modulus from the periodic and displacement boundary conditions is larger than that between the periodic and traction boundary conditions. Both discrepancies decrease as the RVE size increases. For the case of the stiff matrix shown in Fig. 8.12, the predicted shear modulus from the displacement boundary conditions has better agreement with the periodic boundary conditions than does the modulus predicted with the traction boundary conditions. Again, as the RVE size increases, the different sets of boundary conditions predict a more similar shear modulus.

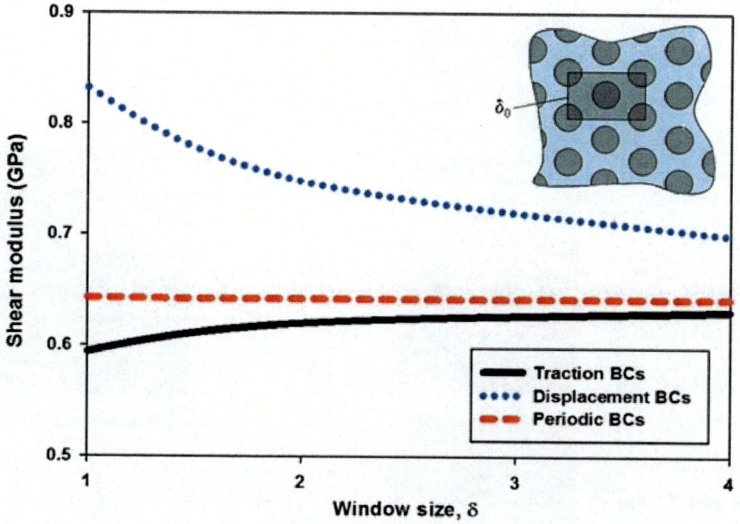

Fig. 8.11 Transverse shear modulus of fiber composite with stiff inclusions. Reproduced with permission from [1]. Copyright (2018) Elsevier

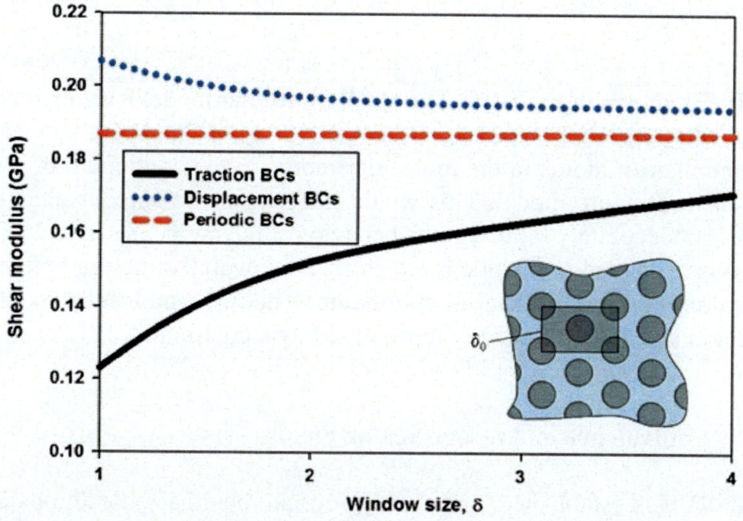

Fig. 8.12 Effect of boundary conditions in equivalent continuum modeling. Reproduced with permission from [1]. Copyright (2018)

For amorphous materials, such as that shown in Fig. 8.10, the minimum size of the RVE is the size that statistically represents the structure of the material. This is a fairly ambiguous statement; however, there is no general agreement of a minimum necessary size of an RVE that produces more accurately predicted bulk-level properties. Of course, as the RVE size increases, the more likely the predicted properties will agree with those measured at the bulk-length scale level. Consider the molecular

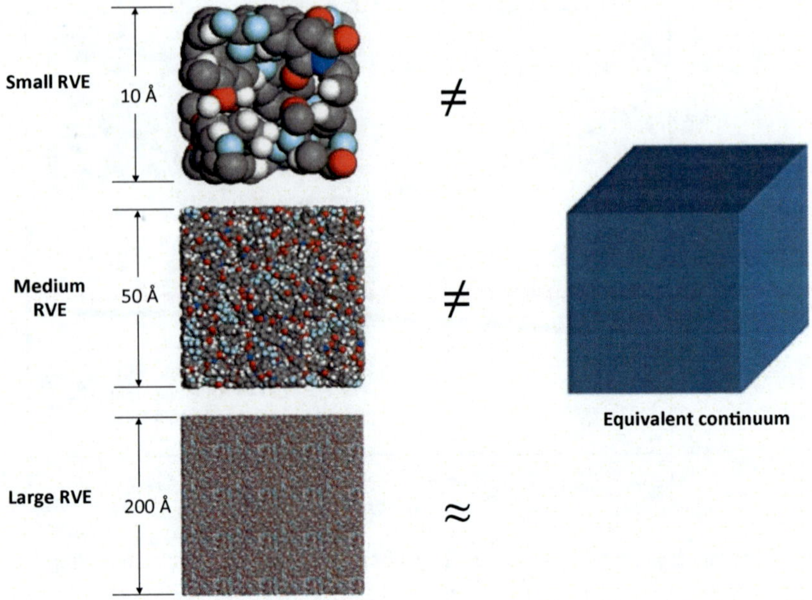

Fig. 8.13 RVE sizes for an equivalent continuum. Reproduced with permission from [1]. Copyright (2018) Elsevier

model RVEs shown in Fig. 8.13. The small molecular model RVE may be too small to effectively represent a continuum, as there is a poor mapping between the very limited number of atoms in the molecular model to the continuum of points in the equivalent continuum model. This would likely result in an equivalent continuum that does not accurately represent the behavior of the molecular model. However, as the RVE is expanded to include more atoms, the overall structure gradually comes closer to representing a material continuum. The large molecular model is likely sufficient in size to be accurately represented by a continuum.

8.5.1.3 Equivalence of Average Scalar Fields

Once an RVE is established, and the appropriate boundary conditions are chosen, the next step is to determine the properties that allow the continuum representation to mimic the RVE behavior. In general, an equivalent continuum must meet two requirements in order to accurately predict the behavior of a particular material:

1. Under identical applied far-field deformations (or loads), the RVE and the equivalent continuum must have identical (or nearly identical) values of one or more scalar fields that are averaged over the volume of the RVE and volume $\mathcal{R}$
2. The material points of the equivalent continuum volume $\mathcal{R}$ must have the same kinematic motion as material points (atoms in some cases) of the heterogeneous RVE at the same locations relative to some defined basis set and origin.

Requirement #1 expresses the need to have one or more scalar parameters; such as the scalar strain energy density or the six scalar components of the stress tensor averaged over the RVE; to have equal values under identical loads applied to the RVE and equivalent continuum models. For example, suppose boundary conditions are applied to the RVE shown in Fig. 8.10. The total potential energy is calculated before and after the application of the boundary conditions using the assigned force field and atomic coordinates. The change in this potential energy correlates to the strain energy of the equivalent continuum that is deformed under the same exact conditions. The material parameters (e.g., Young's modulus) can be adjusted such that the energies of the two models match. It is important to note that many researchers choose to match components of virial and continuum stress tenors (from the RVE and effective continuum models, respectively) under identical conditions. Although this approach makes intuitive sense, it may be unnecessary to do so if Requirement #2 is enforced. Further, it is much more efficient to match a single scalar value (strain energy) than to match six independent components of a symmetric stress tensor. Further details on this requirement can be found elsewhere [70]. Once Requirement #1 is satisfied, then Requirement #2 must be considered.

8.5.1.4 Kinematic Equivalence

Requirement #2 is often referred to as the *Cauchy-Born Rule*, which requires the kinematic motions of the RVE and the equivalent continuum to match for each atom (in the case of a molecular RVE) or each material point (for a heterogeneous micro-scale model) of the RVE. For example, in the case of the RVE shown in Fig. 8.10, the equivalent continuum should deform in the same identical manner on both the RVE surface and in the interior as the atoms in the RVE. This requirement often requires higher-order elasticity theories [24, 26, 27] to be used to accurately match RVE and effective continuum deformations.

Although this requirement can be easily satisfied for the deformations of simple RVEs such as those shown in Figs. 8.11 and 8.12, this rule is usually ignored for more complex RVEs, such as that shown for an amorphous polymer in Fig. 8.10 for one of the following reasons. Either (a) this requirement is over-restrictive and unnecessary given the required predictive accuracy of the effective continuum, or (b) this requirement is extremely difficult to impose on very complex RVEs. An example of point (a) is captured with the heterogeneous RVE of Fig. 8.8. In the composites community, it is often unnecessary to have predictive effective continuum models to predict the point-to-point kinematic mechanical behavior of the microstructure. The relaxation of this requirement has presented few difficulties in the successful design and implementation of most fiber-reinforced composite materials in the last several decades. An example of item (b) is for the amorphous RVE shown in Fig. 8.10. Given the complex atomic interactions that occur on this length scale for a set of atoms that have no local geometric order, the kinematic motion of the atoms is not expected to be uniform. Establishing a higher-order effective continuum model to match a highly non-uniform deformation field would be an exhausting task. Therefore, although Requirement #2 is rigorously followed for simple crystalline (or highly ordered) material systems, it is rarely followed for amorphous or structurally complex materials. Perhaps the most successful use of requirement #2 is for predicting the response of crystalline materials, as is detailed elsewhere [71].

8.5.1.5 An Example: Silica/Polyimide Nanocomposite

Odegard et al. modeled a nanocomposite system with silica reinforcement in a polyimide matrix, to study the effect of the particle size on the effective mechanical properties of the composite [72]. Homogenized properties from Mori–Tanaka technique for this two-phase material system was compared to the equivalent continuum properties of a molecular model using the approaches described in the previous section. At larger silica particle sizes (>100 nm), the models agreed quite well; however, at small silica particle size, the predictions these the two approaches deviate. This can be explained due to the fundamental assumption in continuum based micro-mechanics homogenization techniques such as Mori–Tanaka is the presence of continuum at all length scales techniques. At small enough length scales, the local interactions between the nano-particle and the matrix result in a breakdown of this assumption. The region of localized influence of the reinforcement (silica) on the matrix (polyimide) is generally referred to as an "interphase". Classical Mori–Tanaka does not readily provide options to incorporate the effect of the interphase. The interphase can be regarded as a third phase in this material system in addition to the matrix and the reinforcing particle. In molecular models, the perturbation in the local density due to the introduction of a reinforcing phase serves to validate this presence of interphase. However, as the particle size increases due to changes in the surface to volume ratio of the particle, the interphase contribution to the mechanics becomes negligible and the classical micro-mechanics approaches such as Mori–Tanaka become good approximation for their mechanical behavior. At length scales where interphase is an important contributor to the overall mechanics of the material system [73, 74], modifications to the micro-mechanics approach have been proposed such as the effective interface model that introduces a third phase to account for this localized influence from reinforcing particle, and the stiffness of the composite can be expressed as [72]:

$$\begin{aligned}\overline{C_{ijkl}} = C^M_{ijkl} + \big[(v_I + v_i)\big(C^i_{ijmn} - C^M_{ijmn}\big)A^{Ii}_{mnkl} \\ + v_I\big(C^I_{ijmn} - C^i_{ijmn}\big)A^I_{mnkl}\big]\big[v_I\delta_{ijkl} + (v_I + v_i)A^{Ii}_{ijkl}\big]^{-1}\end{aligned}$$

where $A^I_{ijkl} = \delta_{ijkl} - S^I_{ijmn}\Big[S^I_{mnkl} + \big(C^I_{mnop} - C^M_{mnop}\big)^{-1}C^M_{opkl}\Big]^{-1}$ and

$$A^{Ii}_{mnkl} = \delta_{ijkl} - S^{I}_{ijmn}\left\{\frac{v_I}{v_I + v_i}\left[S^{I}_{mnkl} + \left(C^{I}_{mnop} - C^{M}_{mnop}\right)^{-1}C^{M}_{opkl}\right]^{-1} + \frac{v_i}{v_I + v_i}\left[S^{I}_{mnkl} + \left(C^{i}_{mnop} - C^{M}_{mnop}\right)^{-1}C^{M}_{opkl}\right]^{-1}\right\}$$

In the above equations, A^{I}_{ijkl} and A^{Ii}_{ijkl} are strain concentration tensors, v_I and v_i are the volume fractions of the inclusion and the interphase, respectively. C^{M}_{ijkl}, C^{I}_{ijkl}, C^{i}_{ijkl} and $\overline{C_{ijkl}}$ are the stiffness tensors for the matrix, inclusion, the interphase, and the average for the composite, respectively. S^{I}_{ijkl} is the Eshelby tensor for the inclusion.

The influence of the particle size can be captured using the effective interface model based on micro-mechanics and in conjunction with the equivalent continuum approach can be used for *bottom-up* modeling of nanostructured materials. Figure 8.14 compares the predicted Young's modulus for the silica/polyimide system for different sizes of a functionalized silica particle bonded to a polyimide matrix [72]. The role of interphase is a significant contributor to the mechanics of the materials below certain particles sizes due to the surface interactions between the reinforcement and the matrix, which can be accounted for with appropriate modifications to continuum based approaches.

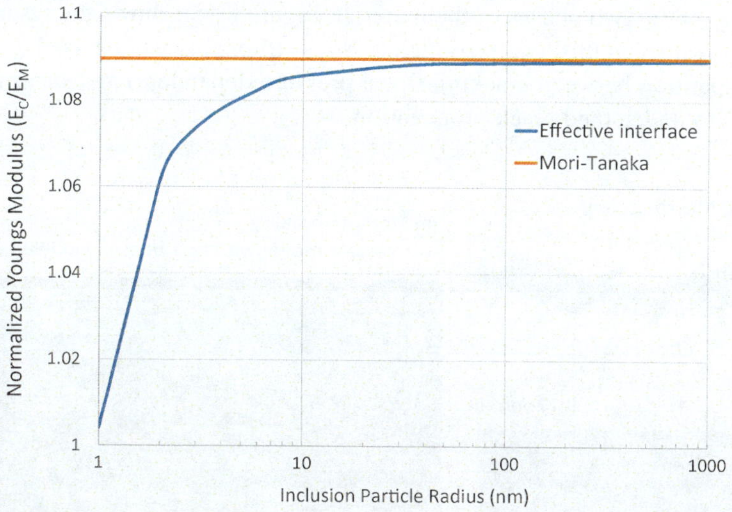

Fig. 8.14 Comparison of Mori–Tanaka and an effective interface micro-mechanical models as a function of the silica particle size for prediction of the modulus of a binary composite of silica functionalized to a polyimide matrix

8.5.2 Top-Down Approach

The second option is to use information from higher length scale model as input into lower-length scale models[55, 56], known as the *top-down approach* as shown in Fig. 8.15, for the specific case of a double notch four point bend (DN4PB) sample subject to impact loading. A DN4PB test is typically used to probe pre-failure damage in multiphase materials [75]. Details of the DN4PB test can be found elsewhere [75]. An FEA model was developed for a stationary crack for a two-phase rubber toughened material [62, 63]. It assumes plane strain conditions and the macro-scale FE model (Fig. 8.15a) are based on an average material behavior calculated using the Mori-Tanaka homogenization technique. The matrix material is assumed to be elastic–plastic with linear hardening and the rubbery inclusion is described using a hyper-elastic material model of arbitrary material constants. Figure 8.15b provides a zoomed in view of the FE mesh and the stress contours for the material showing localization of stresses at the crack tip. The butterfly pattern of the stresses ahead of the crack tip [76] is typical in the DN4PB test. In a localized region ahead of the crack tip (Fig. 8.15b, c), a micro-scale model that explicitly accounts for the two phases can help understand the failure behavior of the two-phase material and the toughening mechanisms. First the macro-scale model is solved with the homogenized two-phase material behavior, and for each loading step the displacement boundary condition in the localized region are passed to the micro-scale model to evolve stresses explicitly in the two phases of the material. Figure 8.15c shows the visualization of stress triaxiality in this localized region. Stress triaxiality can be used to understand the competition between cracking (local pressure dependent) and shearing within a material to understand crack propagation.

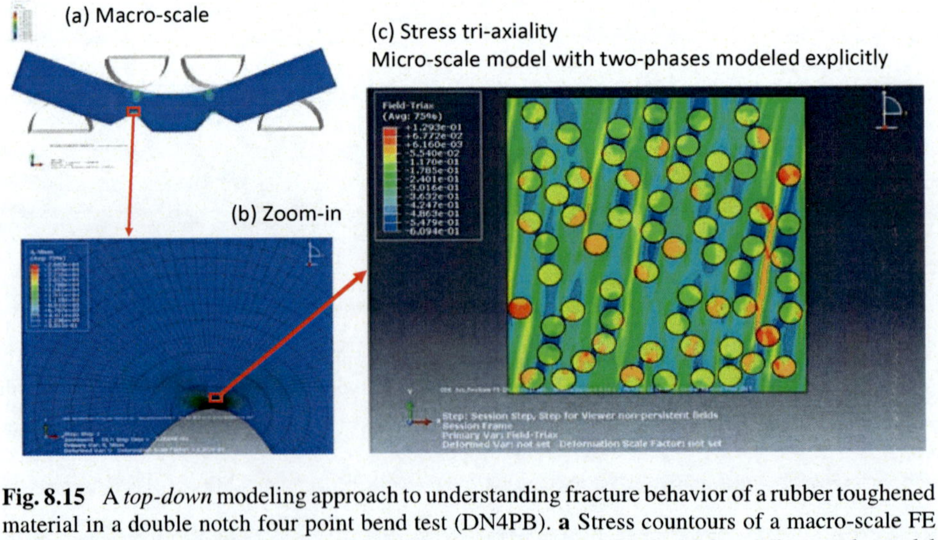

Fig. 8.15 A *top-down* modeling approach to understanding fracture behavior of a rubber toughened material in a double notch four point bend test (DN4PB). **a** Stress countours of a macro-scale FE model. **b** Zoomed-in view of the stresses at the crack tip of the left crack. **c** Micro-scale model showing stress triaxiality with explicit rubber doamins within the matrix ahead of the crack tip

The *top-down* models can be carried out with the macro-scale solution providing the displacement boundary conditions for the micro-scale model with one-way coupling, i.e., macro-FEA model driving the micro-model. However, it is possible to build coupled models, where the two-scales of models can be solved in a coupled fashion to enable information flow both ways [55, 56]. For both *bottom-up* and *top-down* approach, the transition from different modeling schemes is relatively painless. However, these approaches do not provide a direct interaction between modeling schemes.

8.5.3 Concurrent Modeling

A third option is to have two different modeling schemes interact directly such that information can be passed back and forth between the two, and this continuous flow of information influences both simulations. This is known as *concurrent modeling*. This approach makes it possible to simulate the influence of a larger-scale event on the material at a smaller length scale, and vice versa. For example, it is sometime desirable to understand how a large-scale propagating stress wave might affect the atomic behavior of the material. The stress wave is best simulated with a simulation in the continuum realm, while the atomic behavior is best simulated in the discrete realm. For the two different simulations types to "talk to each other", it is necessary to establish an interface in which information can be shared, called a *handshake region*. Figure 8.16 shows a general diagram of the handshake region to interface

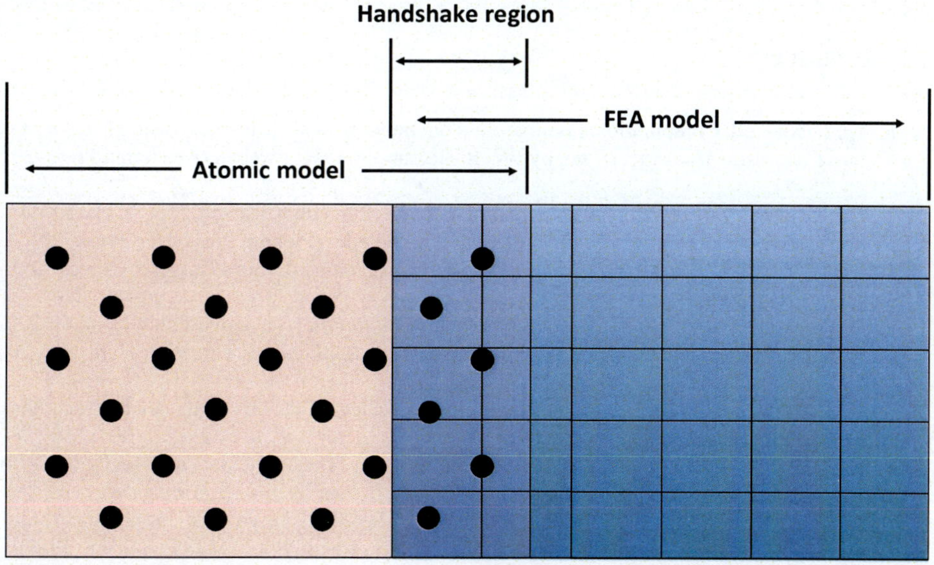

Fig. 8.16 Handshake region between discrete and continuous models. Reproduced with permission from [1]. Copyright (2018) Elsevier

an atomic and FEA model. The atoms don't necessarily correspond directly to FEA nodes. A wide range of methods has been developed to couple concurrent models using the handshake approach. More detailed reviews of these methods can be found elsewhere [71, 77, 78].

8.6 Summary

Mechanical properties of materials can be modeled at various length scales depending on the need for understanding the causal relationship between the structure at a certain length scale and the resulting properties. The current chapter presented a few different techniques of achieving this with particular emphasis on micro-scale modeling techniques: analytical and computational. A few examples demonstrating options for multiscale modeling were discussed that are capable of advancing a mechanism-based understanding of the mechanics of materials. The holy grail for complete elucidation of material physics is a fully coupled atomistic to macro-scale model, which does not quite yet exist. However, stringing together the different modeling techniques at various length scales certainly enhances the study of structure–property relationships.

Acknowledgements The authors would like to acknowledge the help from Valeriy Ginzburg, Mike Cheatham, Jason Brodil and Julia Woertink in preparing this manuscript. Pavan wants to express his gratitude to Dow Inc. for supporting this effort.

References

1. G.M. Odegard, Computational multiscale modeling—nanoscale to macroscale, in *Comprehensive Composite Materials II*, ed. by P.W.R. Beaumont, C.H. Zweben (Academic Press, 2018), pp. 28–51
2. D.J. Griffiths, *Introduction to Quantum Mechanics*, 2nd edn. (Pearson, New York, 2004)
3. R. Shankar, *Principles of Quantum Mechanics*, 2nd edn. (Plenum Press, New York, 2011)
4. M.P. Allen, D.J. Tildesley, *Computer Simulation of Liquids* (Oxford University Press, Oxford, 1987)
5. D. Frenkel, B. Smit, *Understanding Molecular Simulation* (Academic Press, London, 2002)
6. A.R. Leach, *Molecular Modelling: Principles and Applications* (Prentice Hall, New York, 2001)
7. N.L. Allinger et al., Molecular mechanics (MM3) calculations on conjugated hydrocarbons. J. Comput. Chem. **11**, 868–895 (1990)
8. N.L. Allinger, Y.H. Yuh, J.H. Lii, Molecular mechanics. The MM3 force field for hydrocarbons. J. Am. Chem. Soc. **111**, 8551–8566 (1989)
9. J.H. Lii, N.L. Allinger, Molecular mechanics. The MM3 force field for hydrocarbons. 2. vibrational frequencies and thermodynamics. J. Am. Chem. Soc. **111**, 8566–8575 (1989)
10. J.H. Lii, N.L. Allinger, Molecular mechanics. The MM3 force field for hydrocarbons. 3. The van der Waals' potentials and crystal data for aliphatic and aromatic hydrocarbons. J. Am. Chem. Soc. **111**, 8576–8582 (1989)

11. W.L. Jorgensen, D.S. Maxwell, J. Tirado-Rives, Development and testing of the OPLS all-atom force field on conformational energetics and properties of organic liquids. J. Am. Chem. Soc. **117**, 11225–11236 (1996)
12. G.A. Kaminsky et al., Evaluation and reparametrization of the OPLS-AA force field for proteins via comparison with accurate quantum chemical calculations on peptides. J. Phys. Chem. B **105**, 6474–6487 (2001)
13. H. Sun et al., Prediction of shear viscosities using periodic perturbation method and OPLS force field. Fluid Phase Equilib. **260**(2), 212–217 (2007)
14. P.K. Weiner, P.A. Kollman, AMBER—assisted model-building with energy refinement—a general program for modeling molecules and their interactions. J. Comput. Chem. **2**(3), 287–303 (1981)
15. A. Bandyopadhyay et al., Molecular modeling of crosslinked epoxy polymers: the effect of crosslink density on thermomechanical properties. Polymer **52**(11), 2445–2452 (2011)
16. P.K. Valavala et al., Nonlinear multiscale modeling of polymer materials. Int. J. Solids Struct. **44**(3–4), 1161–1179 (2007)
17. P.K. Valavala et al., Multiscale modeling of polymer materials using a statistics-based micromechanics approach. Acta Mater. **57**(2), 525–532 (2009)
18. P.K. Valavala, G.M. Odegard, Modeling techniques for determination of mechanical properties of polymer nanocomposites. Rev. Adv. Mater. Sci. **9**(1), 34–44 (2005)
19. P.K. Valavala, G.M. Odegard, E.C. Aifantis, Influence of representative volume element size on predicted elastic properties of polymer materials. Modell. Simul. Mater. Sci. Eng. **17**(4), 045004 (2009)
20. A. Bandyopadhyay, P.K. Valavala, G.M. Odegard, *Multiscale modeling of epoxy-based composite materials*, in *SAMPE Fall Technical Conference*, Wichita, KS (2009)
21. S.O. Nielsen et al., Coarse grain models and the computer simulation of soft materials. J. Phy.-Condensed Matter **16**(15), R481–R512 (2004)
22. T.C. Clancy, J. Hinkley, *Coarse-Grained and Atomistic Modeling of Polyimides* (National Aeronatics and Space Administration, 2004)
23. Z.P. Bažant, M. Jirásek, Nonlocal integral formulations of plasticity and damage: survey of progress. J. Eng. Mech. **128**(11), 1119–1149 (2002)
24. E.C. Aifantis, On the microstructural origin of certain inelastic models. J. Eng. Mater. Technol.-Trans. ASME **106**(4), 326–330 (1984)
25. B.S. Altan, E.C. Aifantis, On some aspects in the special theory of gradient elasticity. J. Mech. Behav. Mater. **8**(3), 231–282 (1997)
26. A.C. Eringen, *Microcontinuum Field Theories* (Springer-Verlag, New York, 1999)
27. A.C. Eringen, *Nonlocal Continuum Field Theories* (Springer-Verlag, New York, 2002)
28. J.D. Eshelby, The determination of the elastic field of an ellipsoidal inclusion, and related problems. Proc. R. Soc. Lond. Ser. A **241**, 376–396 (1957)
29. R. Hill, *Theory of Mechancial Properties of Fibre-Strengthened Materials—I. Elastic Behavior.* J. Mech. Phys. Solids **12**, 199–212 (1964)
30. R. Hill, On constitutive macro-variables for heterogeneous solids at finite strain. Proc. R. Soc. Lond. A Math. Phys. Sci. **326**(1565), 131–147 (1972)
31. J. Mandel, *Plasticité Classique et Viscoplasticité, in CISM Lecture Notes* (Springer, Udine, 1971)
32. S. Nemat-Nasser, M. Hori, *Micromechanics: Overall Properties of Heterogeneous Materials* (North-Holland, 1993)
33. J. Qu, M. Cherkaoui, *Fundamentals of Micromechanics of Solids* (Wiley, 2006)
34. J. Yu Li, On micromechanics approximation for the effective thermoelastic moduli of multi-phase composite materials. Mech. Mater. **31**, 149–159 (1999)
35. T. Mori, K. Tanaka, Average stress in matrix and average elastic energy of materials with misfitting inclusions. Acta Metall. **21**(5), 571–574 (1973)
36. Y. Benveniste, A new approach to the application of Mori-Tanaka's theory in composite materials. Mech. Mater. **6**(2), 147–157 (1987)

37. R.M. Christensen, H. Schantz, J. Shapiro, On the range of validity of teh Mori-Tanaka method. J. Mech. Phys. Solids **40**, 69–73 (1992)
38. R. Hill, A self-consistent mechanics of composites materials. J. Mech. Phys. Solids **13**, 213–222 (1965)
39. B. Klusemann, B. Svendsen, Homogenization methods for multi-phase elastic composites: comparisons and benchmarks. Technische Mechanik **30**(4), 374–386 (2010)
40. W. Voigt, Theoretische Studien über die Elasticitätsverhältnisse der Krystalle. Abh. Kgl. Ges. Wiss. Göttingen, Math. Kl. **34**, 3–51 (1887)
41. A. Reuss, Berechnung der Fließgrenze von Mischkristallen auf Grund der Plastizitätsbedingung für Einkristalle. J. Appl. Math. Mech. **9**, 49–58 (1929)
42. Z. Hashin, The elastic moduli of hetergeneous materials. J. Appl. Mech. **29**, 143–150 (1962)
43. Z. Hashin, S. Shtrikman, A variational approach to the theory of elastic behaviour of multiphase materials. J. Mech. Phys. Solids **11**, 127–140 (1963)
44. E.J. Barbero, *Introduction to Composite Materials Design* (Taylor and Francis, Philadelphia, PA, 1999)
45. R.M. Christensen, *Mechanics of Composite Materials* (Dover Publications, Inc., Mineola, NY, 2005)
46. D. Hull, T.W. Clyne, *An Introduction to Composite Materials* (Cambridge University Press, Cambridge, 1996)
47. J. Aboudi, *Mechanics of Composite Materials: A Unified Micromechancial Approach* (Elsevier, Amsterdam, 1991)
48. J.N. Reddy, *Mechanics of Laminated Composite Plates: Theory and Applications* (CRC Press, Boca Raton, 1997)
49. C.T. Sun, et al., *Comparative Evaluation of Failure Analysis Methods for Composite Laminates*, F.A.A. U.S. Department of Transportation, Office of Aviation Research, Editor. 1996, U.S. Department of Transportation, Federal Aviation Administration
50. Z. Hashin, Failure criteria for unidirectional fiber composites. J. Appl. Mech. **47**(2), 329–334 (1980)
51. Z. Hashin, A. Rotem, A fatigue failure criterion for fiber reinforced materials. J. Compos. Mater. **7**(4), 448–464 (1973)
52. A. De Luca, F. Caputo, A review on analytical failure criteria for composite materials. AIMS Mater. Sci. **4**(5), 1165–1185 (2017)
53. C.T. Sun, S.G. Zhou, Failure of quasi-isotropic composite laminates with free edges. J. Reinf. Plast. Compos. **7**(6), 515–557 (1988)
54. O.C. Zienkiewicz, *The Finite Element Method in Engineering Science* (McGraw Hill, London, 1971)
55. P.J. Blanco et al., Variational foundations and generalized unified theory of RVE-Based multiscale models. Arch. Comput. Methods Eng. **23**(2), 191–253 (2014)
56. F. Feyel, J.-L. Chaboche, FE2 multiscale approach for modelling the elastoviscoplastic behaviour of long fibre SiC/Ti composite materials. Comput. Methods Appl. Mech. Eng. **183**(3–4), 309–330 (2000)
57. W.J. Drugan, J.R. Willis, A micromechanics-based nonlocal constitutive equation and estimates of representative volume element size for elastic composites. J. Mech. Phys. Solids **44**(4), 497–524 (1996)
58. K. Alzebdeh et al., Fracture of random matrix-inclusion composites: Scale effects and statistics. Int. J. Solids Struct. **35**(19), 2537–2566 (1998)
59. M. Jiang et al., Scale and boundary conditions effects in elastic properties of random composites. Acta Mech. **148**, 63–78 (2001)
60. M. Ostoja-Starzewski, Microstructural randomness versus representative volume element in thermomechanics. J. Appl. Mech.-Trans. ASME **69**(1), 25–35 (2002)
61. M. Ostoja-Starzewski, *Microstructural Randomness and Scaling in Mechanics of Materials* (Chapman & Hall/CRC Press, 2008)
62. Corp, S., *ABAQUS FEA*. 2014, Dassault Systèmes
63. engineering, e.-X., *Digimat*. 2015, MSC Software Corporation, Hexagon

64. M. Jiang, I. Jasiuk, M. Ostoja-Starzewski, Apparent elastic and elastoplastic behavior of periodic composites. Int. J. Solids Struct. **39**, 199–212 (2002)
65. M. Ostoja-Starzewski, Lattice models in micromechanics. Appl. Mech. Rev. **55**(1), 35–60 (2002)
66. S.I. Ranganathan, M. Ostoja-Starzewski, Scaling function, anisotropy and the size of RVE in elastic random polycrystals. J. Mech. Phys. Solids **56**(9), 2773–2791 (2008)
67. S.I. Ranganathan, M. Ostoja-Starzewski, Scale-dependent homogenization of inelastic random polycrystals. J. Appl. Mech.-Trans. ASME **75**(5) (2008)
68. S.I. Ranganathan, M. Ostoja-Starzewski, Mesoscale conductivity and scaling function in aggregates of cubic, trigonal, hexagonal, and tetragonal crystals. Phys. Rev. B **77**(21) (2008)
69. J. Aboudi, S.M. Arnold, B.A. Bednarcyk, *Micromechanics of Composite Materials: A Generalized Multiscale Analysis Approach* (Elsevier Inc., New York, 2013)
70. G.M. Odegard et al., Equivalent-continuum modeling of nano-structured materials. Compos. Sci. Technol. **62**(14), 1869–1880 (2002)
71. W.K. Liu, E.G. Karpov, H.S. Park, *Nano Mechanics and Materials: Theory, Multiscale Methods and Applications* (Wiley, Hoboken, NJ, 2006)
72. G.M. Odegard, T.C. Clancy, T.S. Gates, Modeling of the mechanical properties of nanoparticle/polymer composites. Polymer **46**(2), 553–562 (2005)
73. F.W. Starr, T.B. Schrøder, S.C. Glotzer, Molecular dynamics simulation of a polymer melt with a nanoscopic particle. Macromolecules **35**(11), 4481–4492 (2002)
74. D.A. Jesson, J.F. Watts, The interface and interphase in polymer matrix composites: effect on mechanical properties and methods for identification. Polym. Rev. **52**(3), 321–354 (2012)
75. H.J. Sue, A.F. Yee, Study of fracture mechanisms of multiphase polymers using the double-notch four-point-bending method. J. Mater. Sci. **28**(11), 2975–2980 (1993)
76. T.L. Anderson, *Fracture Mechanics : Fundamentals and Applications*, 2nd edn. (Taylor & Francis, 2005)
77. R.E. Miller, E.B. Tadmor, A unified framework and performance benchmark of fourteen multi-scale atomistic/continuum coupling methods. Modell. Simul. Mater. Sci. Eng. **17**(5), 053001 (2009)
78. H.S. Park, W.K. Liu, An introduction and tutorial on multiple-scale analysis in solids. Comput. Methods Appl. Mech. Eng. **193**(17–20), 1733–1772 (2004)

Chapter 9
Modeling the Thermal Conductivity of Polymer-Inorganic Nanocomposites

Valeriy V. Ginzburg and Jian Yang

Abstract In many applications (LED lighting, consumer electronics, transportation, and others), successfully managing/removing the excess ("waste") heat is crucial to successful performance. This, in turn, requires materials with sufficiently high thermal conductivity values. However, most polymers are characterized by relatively low thermal conductivities (κ < 1 W/m K). Thus, in many applications, it is necessary to design hybrid materials (composites and nanocomposites) with the processability of polymers and thermal conductivity close to that of ceramics or other inorganics (κ ~ 1–10 W/m K). Theory and modeling are widely used to understand the design rules for the development of such composite materials. Here, we review the recent progress in this field, describing methods of predicting thermal conductivity of (nano)composites as function of their composition and the properties of the constitutive materials (matrix and fillers).

9.1 Introduction

In the last several decades, the need for materials combining high thermal conductivity (TC) and good processability has increased dramatically. This need is particularly acute in electronic industry where the miniaturization of devices results in a continuing increase in the waste heat generated during the device operation [1–3]. Thus, it is necessary to direct this waste heat from the high-power elements to the eventual heat sink; otherwise, the high-power elements would overheat quickly and their performance would degrade over time. The heat management system, then,

V. V. Ginzburg (✉)
Core Research and Development, The Dow Chemical Company, Building 1702, Midland, MI 48674, USA
e-mail: vvg851966@gmail.com

J. Yang
PE Product Research, The Dow Chemical Company, 230 Abner Jackson Parkway, Lake Jackson, TX 77566, USA
e-mail: jyang1@dow.com

V. V. Ginzburg and L. M. Hall (eds.), *Theory and Modeling of Polymer Nanocomposites*, Springer Series in Materials Science 310,
https://doi.org/10.1007/978-3-030-60443-1_9

requires various "thermal interface materials" (TIM) such as greases, elastomers, and others [4]. Those TIMs need to combine processability (they need to be sufficiently malleable to fit the spaces between the other elements of the device without leaving any voids) and high thermal conductivity. These attributes, however, are somewhat contradictory. Typical high-TC materials (metals and ceramics) are usually hard and brittle, while most polymers have low TC [5]. Thus, typical TIMs are composites, where high-TC fillers are dispersed in a polymer matrix, either crosslinked or un-crosslinked. The fillers could be ceramic, metal, or carbon-based (such as carbon nanotubes or CNT); they could be nano-sized, micro-sized, or a combination thereof (see Sect. 9.2 for more details).

Given the complexity of the parameter space, the development of theoretical and modeling tools to map it out is critically important. Indeed, prediction of thermal transport in polymer-inorganic composites has been an active field over the past several decades. We refer the reader to a number of reviews on the topic published in recent years [5–10]. Here, we generally follow the discussion from our earlier review, with some updates and modifications.

To begin with, let us define the goal of the study. In a homogeneous material, the heat flux q is proportional to the negative of the temperature gradient ∇T (Fourier's law),

$$q = -\kappa(\nabla T) \tag{9.1}$$

where the proportionality constant κ is the thermal conductivity of the material, measured (in SI units) in W/m K. Fourier's law is valid over large length-scales, and expected to break down at the molecular length-scales. Now, the question we would like to address is as follows—knowing the thermal conductivities of the matrix (κ_m) and the fillers (κ_f), as well as the filler volume fraction ϕ, can we find out the thermal conductivity of the overall composite (κ_c)?

The problem is rather non-trivial, because it contains two separate sub-problems (Fig. 9.1). The first problem is to determine how the fillers are organized in the composite—are they oriented (if anisotropic), aggregated, well-dispersed, etc. The second problem is—assuming the morphology is known, determine the thermal conductivity. In general, most studies concentrate on the second problem, and this is what will be our focus here as well.

Our chapter is structured as follows. First (Sect. 9.2), we briefly discuss the types of fillers and polymers used for thermal management applications and recap the typical values of thermal conductivities for most relevant polymers and fillers. Then, we describe the analytical effective medium approximation (EMA)-type theories (Sect. 9.3) and numerical simulations (Sect. 9.4). Finally, we will discuss the current status of the theory, opportunities, and challenges (Sect. 9.5).

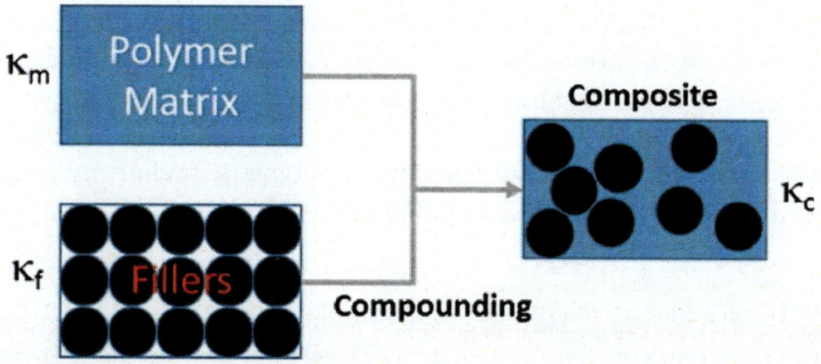

Fig. 9.1 From components to composites. Fillers are compounded into the polymer matrix to result in a nano- or micro-composite material. The final properties of the composite depend not just on the composition, but also on the "morphology" (how the particles are dispersed in the composite), which, in turn, is influenced by the processing (compounding)

9.2 Thermal Conductivities of Typical Polymers and Fillers

9.2.1 Polymers

The polymer matrix TC plays a crucial role in determining the overall TC of thermoplastic composites [11, 12]. In principle, thermal conductivity of a single polymer chain or an "ideal" polymer crystal can be quite high, as predicted in computer simulations [13–16]. Henry and Chen [13] found that the TC of single polyethylene (PE) chains with an extended conformation could be as high as 350 W/(m K) when a chain length is longer than 100 nm (Fig. 9.2). In another example, spider dragline silk with a micrometer-size was recently reported to have exceptionally high TC up to

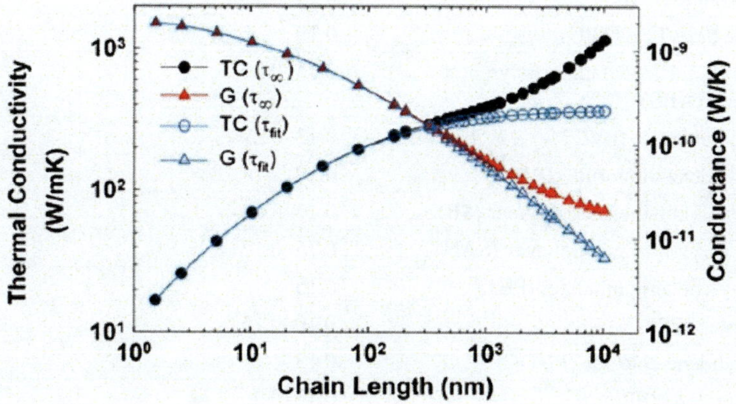

Fig. 9.2 Thermal conductivity (circles) and corresponding thermal conductance (triangles) predictions for a single PE chain. The τ_∞ curves assume that the large-wavelength modes have infinite phonon–phonon relaxation times. The fit curves use extrapolated values for the relaxation times of these modes. Reproduced with permission from [21]. Copyright 2008 American Physical Society

416 W/(m K) [17], which is due to its well-organized, highly crystalline, ordered, and oriented morphology, formed by a near-perfect self-assembling process. However, real-life polymers in practice have much lower values of TC, as can be seen in Table 9.1 (also see [18–20]).

The reason that the thermal conductivity of polymers is relatively low stems from the following. In general, thermal conductivity is given by,

Table 9.1 Thermal conductivities of selected thermoplastic polymers

Thermoplastic polymers	Thermal conductivity at room temperature (W/m K)
High-density polyethylene (HDPE)	0.33–0.53
Ultrahigh molecular weight polyethylene (UHMWPE)	0.41–0.51
Commercial thermotropic liquid crystalline polymers (LCP)	0.30–0.40
Polyoxymethylene (Homo) (POM)	0.30–0.37
Low density polyethylene (LDPE)	0.30–0.34
Poly(ethylene vinyl acetate) (EVA)	0.35
Polyphenylene sulfide (PPS)	0.30
Poly(butylene terephthalate) (PBT)	0.25–0.29
Polytetrafluoroethylene (PTFE)	0.27
Polyamide-6,6(PA66)	0.24–0.33
Polyamide-6 (PA 6)	0.22–0.33
Polyetheretherketone (PEEK)	0.25
Polysulfone (PSU)	0.22
Polymethylmethacrylate (PMMA)	0.16–0.25
Polycarbonate (PC)	0.19–0.21
Urethane base TPE (TPU)	0.19
Poly(acrylonitrile–butadiene–styrene) copolymer (ABS)	0.15–0.20
Polyvinyl-chloride (PVC)	0.13–0.29
Polyvinylidene difluoride (PVDF)	0.19
Styrene/polybutadiene copolymer (SB)	0.17–0.18
Styrene-acrylonitrile copolymer (SAN)	0.15–0.17
Poly(ethylene terephthalate) (PET)	0.15
Polystyrene (PS)	0.10–0.15
Polyvinylidene chloride (PVDC)	0.13
Polysobutylene (PIB)	0.12–0.20
Polypropylene (PP)	0.11–0.17
Polyimide, thermoplastic (PI)	0.11

Reprinted with permission from [5]. Copyright (2016) Elsevier

$$\kappa = \frac{1}{3}\sum_j v_j l_j C_{p,j} \tag{9.2}$$

where the summation is over various phonon wavenumbers and possibly over electron modes (if available—they are significant in metals and some CNT and graphenes, but not in polymers or ceramics); v, C_p, and l are the carrier speed, the contribution to the heat capacity, and the mean free path of the carrier, respectively. For amorphous polymers, the mean free path of the phonons is extremely low, and thus the overall thermal conductivity is also low. In general, most amorphous polymers (thermoplastic and thermoset) have TC ~ 0.1–0.5 W/m K; highly oriented liquid crystalline or crystalline polymers usually have higher TC ~ 0.5–1.0 W/m K. For more detailed discussion, see [5].

9.2.2 *Fillers*

Thermal conductivities of several filler types are summarized in Table 9.2. Note that these generally should be considered "bulk" thermal conductivities. For nanoparticles, especially in the case of metals, the actual thermal conductivity could be lower (see, e.g., [22]).

While thermal conductivity of a filler is important in determining the properties of the overall composite, there are several other factors. First, the primary filler

Table 9.2 Thermal conductivities of selected fillers

Fillers	Category	TC W/(m K)	Electrically conductive?
Aluminum	Metal	234	Yes
Copper	Metal	386	Yes
Silver	Metal	427	Yes
Carbon nanotube (CNT)	Carbon-based	1000–4000	Yes
Carbon fiber	Carbon-based	300–1000	Yes
Graphene	Carbon-based	2000–6000	Yes
Graphite	Carbon-based	100–400	Yes
β-Silicon nitride (β-Si_3N_4)	Ceramics	103–200	No
Hexagonal boron nitride (h-BN)	Ceramics	185–300	No
Aluminum nitride (AIN)	Ceramics	100–300	No
Diamond	Ceramics	1000	No
β-Silicon carbide (β-SiC)	Ceramics	120	
α-Alumina (α-AI_2O_3)	Ceramics	30	No
Beryllium oxide (BeO)	Ceramics	270	No

Reprinted with permission from [5]. Copyright (2016) Elsevier particles could come in different shapes, from high-aspect ratio fibers (like carbon nanotubes) to spheres to "flakes". Second, the primary particles can be weakly or strongly agglomerated. Lastly, the surface treatment (if any) and the interactions between the filler and the matrix can impact the interfacial thermal resistance.

9.3 Analytical Theories and Micromechanical Models

Analytical theories and micromechanical models aim to calculate (in an approximate fashion) the property of a composite (e.g., thermal conductivity, electrical conductivity, Young's modulus, viscosity, etc.) as a function of its composition and the properties of the individual components.

9.3.1 The Parallel and Series Models as the Upper and Lower Limit Estimates

The simplest approach as the first approximation is to consider two limiting cases: linear mixing rule (also known as the parallel model) and inverse mixing rule (often referred to as the series model), see Fig. 9.3. It is generally believed that all linear properties (TC, electrical conductivity, tensile modulus, etc.) have the series model calculating the lower bound and the parallel model predicting the upper bound. The equations for the two models are given by,

$$\kappa_c = \phi\kappa_f + (1-\phi)\kappa_m \quad \text{(Parallel)} \tag{9.3}$$

$$\kappa_c^{-1} = \phi\kappa_f^{-1} + (1-\phi)\kappa_m^{-1} \quad \text{(Series)} \tag{9.4}$$

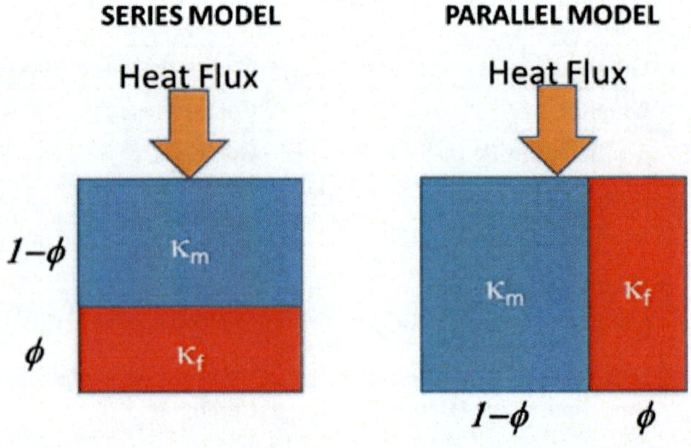

Fig. 9.3 Schematic of the series (left) and parallel (right) composite models

Equations (9.3) and (9.4) could be easily generalized to an arbitrary number of fillers.

Most experimental data fall between these two bounds, with the series model usually closer to the experimental data as compared to the parallel model [23]. Obviously, neither series nor parallel model is expected to be very accurate, given that they do not consider factors like the particle shape, orientation (for the non-spherical particles), or the actual morphology. Those are accounted for in the so-called effective medium approximation (EMA) models. Some of the most widely used EMA approaches are described below.

9.3.2 *EMA for Spherical Fillers*

In the Maxwell-Garnett (MG) model [24], the conductivity of the composite having spherical fillers is given by,

$$\kappa_c = \kappa_m \left(1 + \frac{3\phi(\delta - 1)}{2 + \delta - \phi(\delta - 1)}\right) \tag{9.5}$$

where $\delta = k_f / k_m$, and ϕ is the filler volume fraction. The ϕ-dependence of the relative thermal conductivity of the composite within MG model is shown in Fig. 9.4. At low filler loadings, the dependence of composite TC on filler content is nearly linear.

The MG model equations are generally applicable to composites with low filler volume fractions ("dilute regime") and also disregard the role of the interfacial

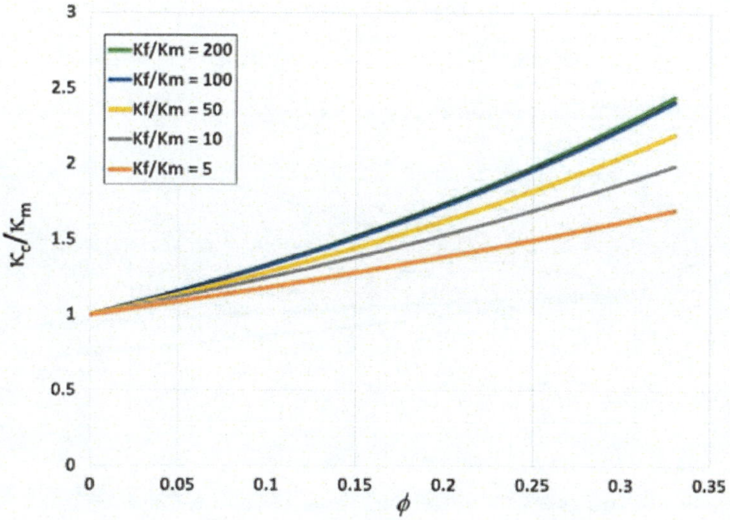

Fig. 9.4 Effective thermal conductivity of composites as a function of filler loading based on MG model prediction. Different curves correspond to different filler thermal conductivities

thermal resistance ("Kapitza resistance" [25]). The latter problem was addressed by Nan et al. [26]. Nan's model can be written as,

$$\kappa_c = \kappa_m\left(1 + \frac{3A\phi}{1 - A\phi}\right) \tag{9.6}$$

$$A = \frac{1 - \left(\frac{\kappa_m}{\kappa_f} + \frac{a_K}{a}\right)}{1 + 2\left(\frac{\kappa_m}{\kappa_f} + \frac{a_K}{a}\right)} \tag{9.7}$$

where $a_K = R\kappa_m$ is the Kapitza radius, R is the interfacial thermal resistance between the matrix and the particle, and a is the particle radius [27]. As the Kapitza resistance is increased, the "effective filler thermal conductivity", $\kappa_f^{\text{eff}} = \left[\kappa_f^{-1} + \frac{R}{a}\right]^{-1}$ decreases correspondingly and the overall composite TC would be reduced as well. The dependence of thermal conductivity on the filler volume fraction based on Nan model is shown in Fig. 9.5; here, we kept the "nominal" ratio $k_f/k_m = 100$ constant but changed the a_K/a. It can be clearly seen that if the Kapitza radius starts to exceed the radius of the filler, the overall composite thermal conductivity would decrease, rather than increase, with the filler loading. In the limit of $a_K/a \to 0$, one recovers the original MG-like behavior (top curve).

The above approaches are, strictly speaking, accurate primarily in the "dilute limit" where the interactions between the particles can be neglected. Ordonez-Miranda and co-workers developed a novel "crowding-factor" model that results in a more accurate description of composites with high filler volume fractions [10, 27–29]. Their approach is based on the ideas originally utilized by Mooney [30] in

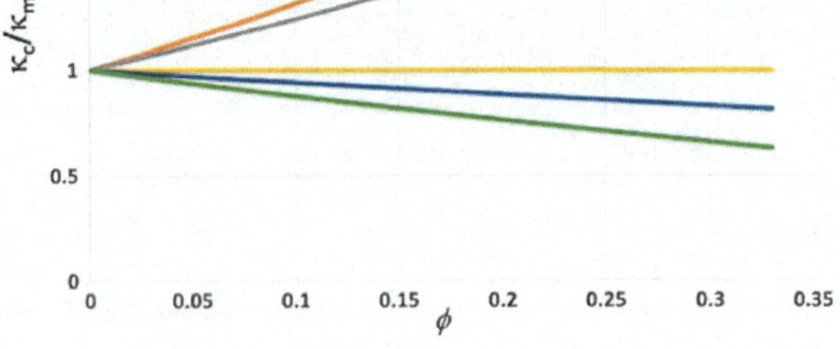

Fig. 9.5 Effective thermal conductivity of composites as a function of filler loading based on Nan model prediction. Different curves correspond to different Kapitza radius values

analyzing the viscosity of concentrated suspensions. After accounting for the "crowding" effects as the filler volume fraction approaches the maximum packing fraction, the Nan (9.6) is replaced by,

$$\kappa_c = \kappa_m \exp\left(\frac{3A\phi}{1 - A\psi\phi}\right) \tag{9.8}$$

$$\psi = 1 + \phi\frac{1 - \phi_0}{\phi_0} \tag{9.9}$$

where ϕ_0 is the maximum packing fraction. While the proposed approach showed a good agreement with experiments in some cases, more validation is still needed. Indeed, at sufficiently high filler loading, thermal conductivity predicted by (9.8) is expected to diverge, rather than smoothly approach κ_f; this is unphysical and thus implies that the model still needs modifications.

9.3.3 Non-spherical Fillers

Fricke [31] generalized the MG model for the case of the spheroidal inclusions. In the dilute limit, suspended particles are considered to be thermally isolated one from another. Fricke's equation can be written in a form resembling the MG (9.5):

$$k_c = k_m\left(1 + \frac{n\phi(\delta - 1)}{n - 1 + \delta - \phi(\delta - 1)}\right) \tag{9.10}$$

where

$$n = \frac{\beta(\delta - 1)}{(\delta - 1) - \beta} \tag{9.11}$$

$$\beta = \frac{1}{3}(\delta - 1)\left[\frac{2}{1 + L_{11}(\delta - 1)} + \frac{1}{1 + L_{33}(\delta - 1)}\right] \tag{9.12}$$

The factors L_{11} and L_{33} depend on the filler aspect ratio r (which is defined as the size divided by thickness and always larger than unity).

$$L_{11} = \frac{r^2}{2(r^2 - 1)} - \frac{r}{2(r^2 - 1)^{3/2}}\cosh^{-1} r$$
$$L_{33} = 1 - L_{11} \tag{9.13}$$

For a sphere, $r = 1$ and the geometrical factors are all equal to 1/3. In this case, the shape factor n becomes equal to 3, and (9.10) reduces to the Maxwell-Garnett formula.

9.3.4 Bruggeman Model

The effective medium approaches described above assume separated particles in the effective medium, disregarding direct contacts between the fillers possibly leading to thermally conductive paths [32]. The so-called Bruggeman model [33] does not suffer from this limitation. It was originally developed to describe thermal or electrical conductivity in polycrystals [34] and can be qualitatively accurate without limitation on the concentration of inclusions. The Bruggeman model yields the following implicit equation for κ_c, for a binary composite containing matrix and spherical fillers,

$$\phi\left(\frac{\kappa_f - \kappa_c}{\kappa_f + 2\kappa_c}\right) + (1 - \phi)\left(\frac{\kappa_p - \kappa_c}{\kappa_p + 2\kappa_c}\right) = 0 \quad (9.14)$$

We demonstrate how the composite TC depends on the matrix TC and the filler loading based on the Bruggeman model in Fig. 9.6. Here, the filler is assumed to have $\kappa_f = 300$ W/m K (similar to, for example, boron nitride), and the matrix TC is varied. Note that at intermediate particle loadings (~20–30 vol%), thermal conductivity exhibits behavior similar to percolation transition. This is indeed a typical characteristic of the Bruggeman model [10], although the determination of the percolation threshold or percolation exponent in 3D is not very accurate (The question of "thermal conductivity percolation" remains very intriguing [35]—it is certainly much less pronounced than for electrical conductivity). Post-percolation, the composite TC becomes relatively independent of the matrix TC, smoothly approaching κ_f as the filler volume fraction approaches 100%.

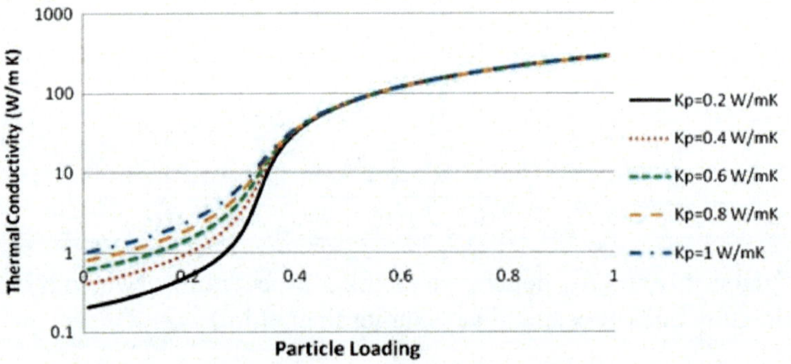

Fig. 9.6 Bruggeman model prediction of composite TC as a function of particle loading (volume fraction) with filler TC (300 W/m K) and various matrix TC (Kp) ranging from 0.2 to 1.0 W/m K. Reproduced with permission from [5]. Copyright (2016) Elsevier

9.3.5 Combined Approaches

The micromechanical models described above are based on the assumption that the fillers are uniformly dispersed and (for the case of anisotropic fillers) randomly oriented. This assumption is often incorrect because of the filler aggregation. Thus, a combined approach, describing the aggregates first, and the overall composite second, are needed. One example of such a combined approach could be found in the Brune-Bicerano micromechanical model for the Young's modulus of partially exfoliated polymer–clay nanocomposites [36]. To describe thermal conductivity in composites where the fillers are partially aggregated, one can use a similar combination approach. First, the Bruggeman model is used to predict the effective TC of particle clusters. Then, one can apply the MG model to approximate the effective TC of composites [37]. The schematics of this approach are shown in Fig. 9.7.

Apart from the above models, some other models [38] include equations by Hashin and Shtrikman [39, 40], Hamilton and Crosser [41, 42], Hatta and Taya [43], Agari [44], Cheng and Vachon [45] as well as by Lewis and Nielsen [46–48]. These models also appear to reasonably fit most of the experimental data for composites based on isotropic particles as well as short fibers and flakes with limited aspect ratio, up to loadings of about 30% in volume. The development of new EMA approaches remains an active field [49].

9.4 Computer Simulations

The following two sections are reproduced with minor changes from our earlier review [5]: "Thermal conductivity of polymer-based composites: Fundamentals and

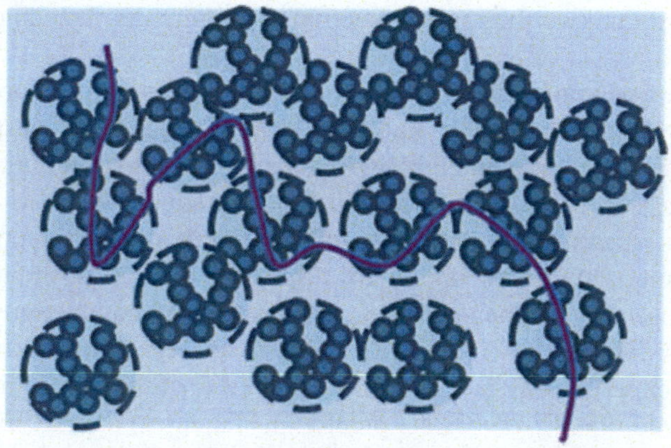

Fig. 9.7 Schematic representation of well-dispersed aggregates in composite materials with a highlighted thermally conductive path. Reproduced with permission from [5]. Copyright (2016) Elsevier

applications", Progress in Polymer Science, 59, 4185 [2016], https://dx.doi.org/10.1016/j.progpolymsci.2016.03.001.

9.4.1 Finite Element Modeling

While the EMA approaches described in the previous section are useful for quick evaluation of the trends expected upon changes in filler loading or filler type, they are limited to specific conditions (e.g., monodisperse spherical fillers in a homogeneous matrix or perfectly aligned ellipsoidal fillers in a homogeneous matrix). In many cases, numerical approaches, such as finite element analysis (FEA) or finite difference (FD) modeling, could be better suited [50–54].

In a typical FEA approach, one considers a realistic composite morphology (based on experimental inputs such as miscroscopy or taken from molecular-level simulations) and attempts to solve the heat transfer equation,

$$\nabla(\kappa(\mathbf{r})\nabla T(\mathbf{r})) = 0 \tag{9.15}$$

It is convenient to use a cubic box and apply a constant heat flux, q, to one of its faces while maintaining the opposite face at constant temperature T_0. The initial condition is $T(r) = T_0$. Once the temperature profile reaches steady state, the average thermal conductivity can be computed as $\kappa_{\text{eff}} = \frac{qX_0}{\langle T(x=0)\rangle - T_0}$. Here, we assumed that the heat flux is applied at the $X = 0$ plane, the length of the cube is X_0, and $\langle\rangle$ refers to averaging in the YZ-plane. If the composite is anisotropic, similar calculations could be done in other directions to compute the principal components of the symmetric tensor k_{ij} (the symmetry of the tensor follows from the Onsager reciprocal relations [55, 56]). In most cases, polymer-based composites are either isotropic (so that only one thermal conductivity number is needed) or having a preferential filler plane orientation (so that two thermal conductivity numbers, in-plane and through-plane, are required).

One of the early numerical studies of thermal conductivity in polymer-inorganic composites was performed by Kumlutas et al. [57]. They investigated composites of high-density polyethylene (HDPE) and spherical aluminum metal particles. The particles were dispersed uniformly and homogeneously in the matrix, and there was no anisotropy or directionality. The authors took electron microscopic images of composites corresponding to various filler loadings, and used finite difference modeling in two dimensions to compute thermal conductivity, assuming that κ(LDPE) = 0.543 W/m K and κ(Al) = 204 W/m K. Numerical results were then compared to experimental data, as well as to number of theoretical models (Fig. 9.8). It was observed that the numerical predictions (squares) and the experimental data (triangles) were in a fairly good agreement at high filler loadings (>15 vol%); at lower filler loadings, numerical predictions were substantially higher than the measured values. There can be a number of explanations for this discrepancy. For example, the model did not consider the role of interfacial resistance [26, 58, 59] (we will

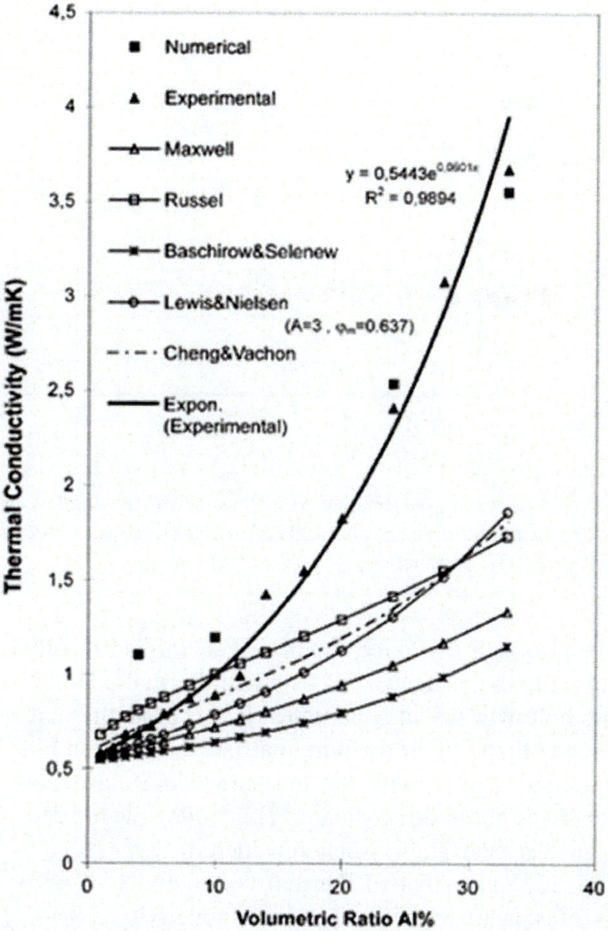

Fig. 9.8 Composite thermal conductivity in HDPE/Al binary mixtures as function of Al filler volume fraction. Experimental data are compared with predictions using various micromechanical models, as well as FEA simulations. Reprinted with permission from [57]. Copyright (2003) Elsevier

address this topic in more detail below). Also, the digitization of the images could have led to incorrect assignment to some pixels in the regions crowded with filler particles. On the other hand, it was shown that constitutive models, such as Cheng and Vachon [45], predict thermal conductivity fairly well at filler loading of less than 10 vol%. This is consistent with the notion that the effective medium theories lose accuracy at higher filler volume fractions as the fillers tend to form large aggregates. Interestingly, though, the authors did not attempt to compare their results with the Bruggeman model which is expected to work better at high filler loadings.

Annapragada et al. [60] studied thermal conductivity of particulate composites with polydisperse spherical fillers. They developed packing algorithms to generate representative volume elements (RVE), both in two and three dimensions. The predicted thermal conductivity was shown to be fairly close to the experimental

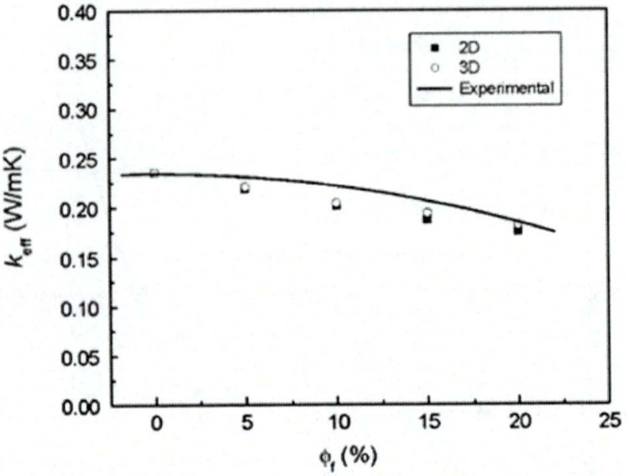

Fig. 9.9 Comparison between experimental data (line) and FEA simulations (2D-filled squares; 3D-open circles) for the polypropylene/hollow glass particle composites. Reprinted with permission from [62]. Copyright (2007) Elsevier

value, and the approach certainly could be generalized to other composites with spherical fillers. Liang and Li [61, 62] studied thermal conductivity of polypropylene filled with hollow glass latex particles; in this case, thermal conductivity of the composite is lower than that of the pure matrix. As shown in Fig. 9.9, the predicted thermal conductivity agrees with the measured data reasonably well, both when two-dimensional and three-dimensional FEA models were used. Again, the role of thermal resistance at interface was not considered.

Nayak et al. [63] investigated thermal conductivity of epoxy/pine wood dust (PWD) composites; again, the filler thermal conductivity (κ(PWD) = 0.068 W/m K) was less than that of the matrix (κ(epoxy) = 0.363 W/m K), so the overall thermal conductivity should decrease as the filler loading is increased. The authors used three-dimensional (3D) sphere-in-cube lattice model, varying the effective sphere radius to change the volume fraction. The problem was solved using ANSYS software. Figure 9.10 shows the dependence of thermal conductivity on the PWD volume fraction; simulations and experiment are in a reasonable agreement, although the non-monotonic behavior of the simulated thermal conductivity is not fully explained and could be an artifact of the lattice model.

The examples above considered composites with spherical fillers, monodisperse, or polydisperse. Anisotropic fillers provide further complexity since both positional and orientational distribution of fillers needs to be considered. In recent years, many studies have been devoted to polymer-carbon nanotube (CNT) composites [18, 64–67]. Indeed, with nominal thermal conductivity of CNTs being in excess of 1000 W/m K, one can expect to produce high-thermal conductivity composites with relatively small CNT volume fractions. It has been shown [11, 64, 68] that for polymer/CNT nanocomposites, their electrical conductivity, σ, increases dramatically (by 5–6 orders of magnitude) as the CNT volume fraction is changed from

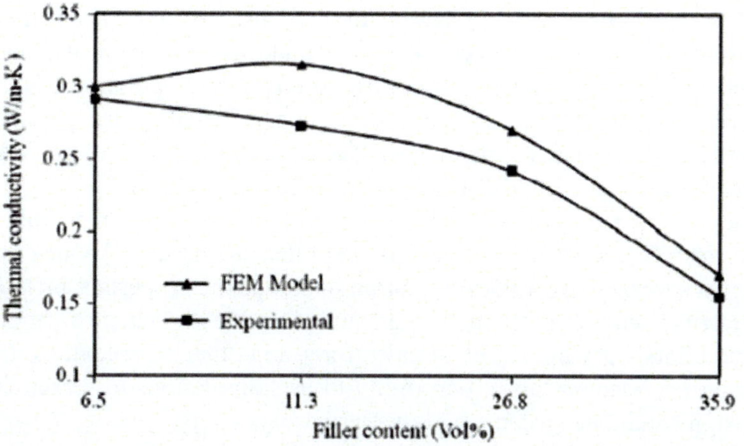

Fig. 9.10 Thermal conductivity of epoxy/pine wood dust (PWD) composites as function of the filler volume fraction. Triangles are FEA simulation results, and squares are experimental data. Reprinted with permission from [63]. Copyright (2010) Elsevier

about 0.2 vol% to about 1 vol%. Indeed, according to percolation theory, high-aspect ratio fillers such as CNT form continuous network at very low concentrations (for the dependence of percolation threshold on aspect ratio, see, for example, review by Bicerano et al. [69]). This percolated network accounts for most of the charge conduction, making polymer/CNT nanocomposites reasonably good conductors, even if not on the same scale as metals. However, thermal conductivity increases in polymer/CNT nanocomposites are not as significant as the electrical conductivity increases, despite the similarity between the two transport properties. Moniruzzaman and Winey [64] review a number of experimental studies on polymer-CNT thermal conductivity, concluding that in general, the composite TC can exhibit about two- to three-fold increase over the matrix TC; this is a substantial increase but nowhere near what is seen for electrical conductivity. Accordingly, several computational and theoretical studies have been performed to explain this discrepancy. For example, Shenogina et al. [35] simulated the heat flow between two nearby nanotubes in a polymer matrix using ABAQUS finite element software. They concluded that the rate of heat flow was fairly independent of the distance between the nanotubes, thus suggesting either absence or weakness of percolation effects. Singh et al. [59] used element-free Galerkin method to numerically compute effective thermal conductivity of polymer/CNT composite as function of CNT volume fraction and interfacial thermal resistance. The nanotubes were assumed to be perfectly aligned, and the calculated thermal conductivity was along the direction of their alignment. The overall TC thermal conductivity was shown to be a strong function of the CNT volume fraction and the interfacial thermal resistance.

More recently, Li et al. [70] used 3D finite element modeling of polymer/single-wall-carbon nanotube (SWCNT) composite with disordered nanotubes, and compared results with experimental data on two model systems, epoxy/SWCNT and polyolefin/SWCNT (Fig. 9.11). The results demonstrated, once again, the importance of interfacial thermal resistance for proper understanding of heat transfer in these materials. Even so, the mechanism of heat transfer in the polymer/CNT composites is not yet fully understood, as more recent experimental studies [71, 72] indicated that the increase in heat conductivity as function of the CNT volume fraction could be indeed consistent with the nanotube percolation.

Other recent applications of FEA modeling include those of polymer/platelet composites [73, 74] and those where mixed fillers are used [75]. One can expect that in the next several years, FEA simulations of composite properties, including thermal conductivity, will be dramatically expanded and will include formulations with multiple fillers, varying degree of anisotropy, and other complexities. Furthermore, as computers become faster, our ability to use simulations to screen formulations and design candidate systems will increase dramatically.

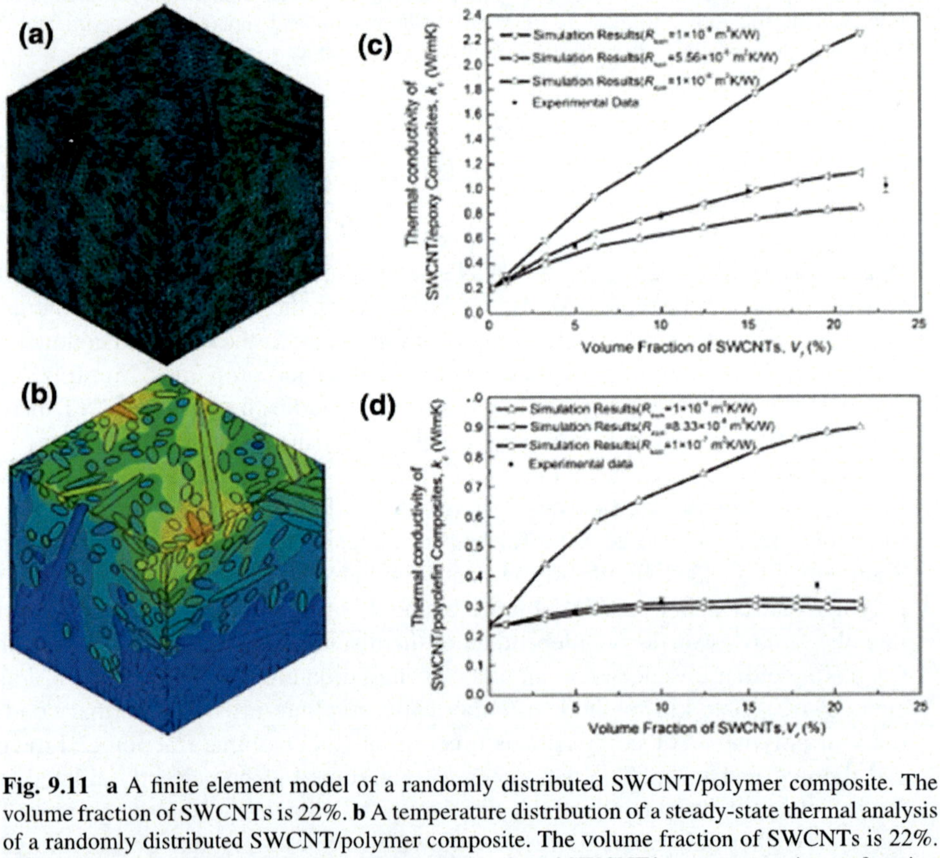

Fig. 9.11 **a** A finite element model of a randomly distributed SWCNT/polymer composite. The volume fraction of SWCNTs is 22%. **b** A temperature distribution of a steady-state thermal analysis of a randomly distributed SWCNT/polymer composite. The volume fraction of SWCNTs is 22%. **c** Calculated and experimental thermal conductivity (k_c) of SWCNT/epoxy composites as function of SWCNT volume fraction. **d** Same as (**c**) but for polyolefin/SWCNT composites. Here, R_{ksm} is the interfacial thermal resistance between SWCNT and matrix. Reprinted with permission from [70]. Copyright (2012) Elsevier

As already mentioned, one critical challenge in ensuring the validity and accuracy of the continuum-level models (whether constitutive or FEA) is description of the heat transfer through interfaces. Below, we briefly review recent progress on this topic.

9.4.2 Interfacial Thermal Resistance Modeling

Interfacial thermal resistance (often called Kapitza resistance following the pioneering work by Kapitza [25]) can be understood as follows. The matrix (such as polymer) and the filler (CNT or an inorganic particle) have different spectra of quantum states for the heat carriers (electrons, holes, and phonons). In most polymers, heat is carried only by phonons, and their density of states is substantially lower than in highly crystalline CNTs or inorganic fillers. As a result, many phonons are back-scattered at the interface between the matrix and the filler. The problem can be exacerbated if the interfacial coupling between the filler and the matrix is weak [76–78]. This results in a temperature discontinuity, ΔT, at the interface; this discontinuity is proportional to the heat flux, J,

$$J = \Lambda(\Delta T). \tag{9.16}$$

The proportionality constant, Λ, is the interfacial thermal conductivity and has units of W/(m^2 K); the inverse of Λ is the interfacial thermal resistance (usually labeled R_k). It is often convenient to define interfacial thickness, h, as follows,

$$h = \frac{\kappa}{\Lambda} \tag{9.17}$$

where κ is the bulk thermal conductivity. Typical values of h are on the order of nanometers [26, 77]; thus, it is often reasonable to disregard the interfacial effects if the filler size is on the order of microns. In many FEA simulations of conventional polymer-inorganic composites discussed above, the assumption of "perfect" interface ($\Delta T = 0$) was thus justified since the typical finite element size was much greater than h. However, if one tried to increase the composite thermal conductivity substantially higher, by increasing the filler loading and/or reducing the filler size, the interparticle distance D_{int} would eventually decrease to the point where it becomes comparable to h. Further increases in the filler volume fraction would then lead to no improvement in the overall TC. As Shenogin et al. [77] emphasize, in these nanocomposites, the fillers could be separated by 1 nm physically, yet from thermal standpoint, their separation could be as large as h ~ 10–20 nm. This insight helps us better understand the results from the FEA and micromechanical models in which the interfacial resistance was varied parametrically (Figs. 9.5 and 9.11)—once the filler volume fraction increased to the point where the interparticle distance becomes comparable to the thermal interfacial thickness, the overall thermal conductivity stops increasing.

So far, we mainly discussed continuum approaches to thermal conductivity, where heat transfer through matrix, filler, and interface was described in a phenomenological manner. In order to predict the interfacial thermal resistance, one needs to turn to molecular models. We already mentioned the study by Shenogin et al. [77] who used Molecular Dynamics to study interfacial heat transport between a carbon nanotube and octane liquid. Alzina et al. [79] used multiscale modeling to study heat transfer in epoxy/glass composites at very low temperatures. They showed that theory and experiment agreed fairly well if one assumed the so-called diffuse mismatch model (DMM), in which all phonons were assumed to diffusely scatter at the interface. Roy et al. [80, 81] demonstrated using Molecular Dynamics the importance of coupling between the carbon nanotube and the polymer matrix. Shin et al. [82] developed a multiscale model combining nonequilibrium Molecular Dynamics and finite element analysis to compute thermal conductivity in SiC/epoxy composites. Similar approaches have been also applied to various other systems [78, 83–85].

One important conclusion from these and other studies was that the surface functionalization of the filler could be crucial to improving thermal conductivity of composites. Within both theories of interfacial resistance (acoustic mismatch model and diffuse mismatch model), the interfacial thermal conductivity is directly proportional to the overlap between vibrational density of states (VDOS) spectra of the matrix and the filler. Wang et al. [78] computed VDOS of pristine or organically modified graphene sheet and compared it with those of a model paraffin (Fig. 9.12).

The calculation shows that accurately matching the VDOS of the filler and the matrix by using the right surface modifier is very important. Even though the overlap integral S increases substantially upon addition of virtually any organic modifier, butyl surfactant results in the highest S-value, leading to the most effective reduction in the interfacial resistance for a given surface coverage (Fig. 9.13). This analysis shows that even modest surface functionalization could lead to substantial improvement in the interfacial thermal transport. Recent experimental studies suggest that surface functionalization can indeed lead to some improvement in thermal conductivity in alumina/nylon-6 composites [86].

Prediction of interfacial thermal properties is still relatively new and rapidly developing field, and this short review does not attempt to be comprehensive. As computers become more powerful and atomistic force fields more accurate, we can expect to see more advances in predicting interfacial thermal resistance for more and more diverse filler/matrix pairs. Creating such a library will help in developing more accurate macroscopic FEA models and thus using modeling as part of new material design.

9.4.3 Multiscale Modeling

As we discussed above, an accurate prediction of thermal conductivity in polymer-inorganic nano- and micro-composites requires the knowledge of structure and dynamics on several length-scales, from atomistic (to determine the electron and phonon spectra and the density of states, both in the bulk and at interfaces) to

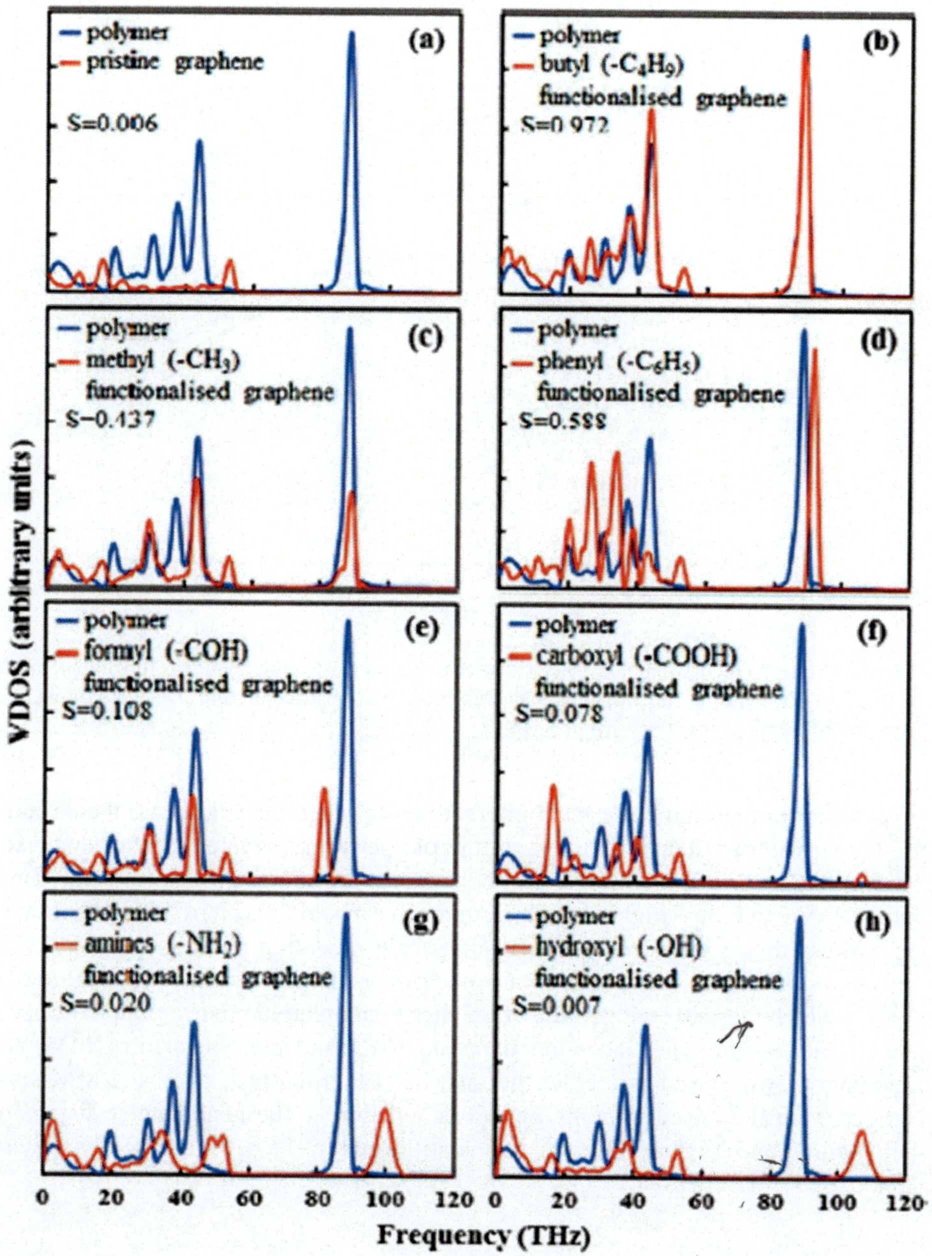

Fig. 9.12 Comparison of VDOS between paraffin (blue) and graphene with various surface functionalities (red). The captions ($S = 0.006$ etc.) describe the calculated VDOS overlap integrals. Reprinted with permission from [78]. Copyright (2015) American Chemical Society

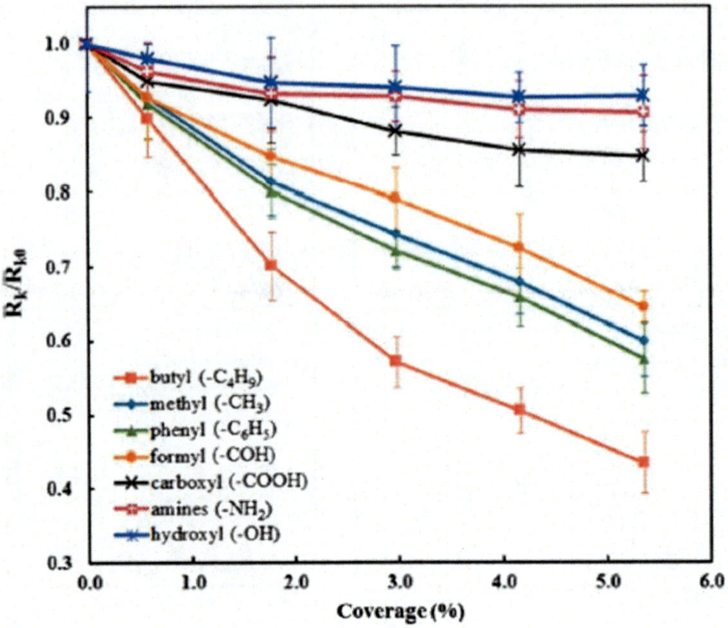

Fig. 9.13 Interfacial thermal resistance (relative to that of pristine graphene) as function of surfactant coverage for various functionalizations. Reprinted with permission from [78]. Copyright (2015) American Chemical Society

nanoscale (to determine the local interactions between the fillers and the structure of the polymer in their proximity) to microscale (to determine the filler dispersion and orientation) to macroscale (to understand the influence of processing conditions). Obviously, no single modeling approach can simulate all of these phenomena; to address them, we need to utilize multiscale modeling paradigm [87–93]. One successful example of such multiscale modeling is the recent work by Odegard et al. [93] in which thermal conductivity of graphene nanoplatelets in cycloaliphatic epoxy was studied by combined molecular dynamics (MD) and micromechanics (MM); MD simulations were used to predict the interfacial thermal resistance and MM simulations were then used to compute TC as function of the nanoplatelet dispersion, orientation, and loading. Overall, this is still a relatively new and rapidly growing field.

9.5 Summary and Outlook

Thermally conductive materials are and will remain critically important for various industries, from electronics to energy to automotive. The use of nano- and micro-composite materials is often required to successfully navigate the balance of properties (thermal conductivity, electrical conductivity or lack thereof, mechanical strength, thermal expansion, viscosity, etc.). The use of modeling is often required to reduce the number of potential experiments, develop new design criteria for the selection of matrix and filler materials, and/or understand the failure mechanisms for existing materials.

Currently, most theoretical and computational research dealing with the prediction and description of thermal conductivity in composites is based on analytical or numerical micromechanical models. Most often, in these models, the composite morphology is simplified, and the fillers are assumed to be uniformly dispersed and—if non-spherical—oriented isotropically. While these approaches often are successful in describing experimental results in commercial composites, the success is typically accomplished by the introduction of fitting parameters (for example, interfacial or Kapitza resistance is often used as one).

In our view, the main challenges for theoretical and computational prediction of thermal conductivity in composite materials are as follows. First, it is necessary to develop robust procedures (either EMA or FEA) to calculate thermal conductivities of composites with multiple fillers having different sizes and different shapes/aspect ratios. Second, it is necessary to understand the relationship between the processing conditions and the composite morphology (aggregation, orientation, and dispersion). Finally, it is necessary to understand the molecular details of heat transport on the nanoscale, both within the fillers and the matrix and through various interfaces.

Addressing these challenges will likely take a long time and will require a significant modification of existing theoretical and computational tools. It is possible, even likely, that new tools (e.g., machine learning) will also be added to the mix. It is our hope that this review will help future researchers as they work on the new approaches to this old but fascinating problem.

Acknowledgements We thank Dow for supporting this work. We are also grateful to Darren Hansen and Huikuan Chao for critical review of the manuscript and many helpful suggestions.

References

1. A.L. Moore, L. Shi, Mater. Today **17**, 163 (2014)
2. X.C. Tong, *Advanced Materials for Thermal Management of Electronic Packaging*, vol. 30, 1st edn. (Springer, New York, 2011)
3. P.K. Schelling, L. Shi, K.E. Goodson, Mater. Today **8**, 30 (2005)
4. F. Sarvar, D.C. Whalley, P.P. Conway, in *2006 1st Electronic Systemintegration Technology Conference* (IEEE, 2006), p. 1292
5. H. Chen, V.V. Ginzburg, J. Yang, Y. Yang, W. Liu, Y. Huang, L. Du, B. Chen, Prog. Polym. Sci. **59**, 41 (2016)
6. P. Keblinski, Modeling of Heat Transport in Polymers and Their Nanocomposites, In: *Handbook of Materials Modeling: Applications: Current and Emerging Materials*, 2-nd edn, editors: W. Andreoni, S. Yip, (Springer Nature, Cham, 2020), p. 975–997 https://doi.org/10.1007/978-3-319-44680-6_31
7. A. Li, C. Zhang, Y.-F. Zhang, Polymers **9**, 437 (2017)

8. X. Yang, C. Liang, T. Ma, Y. Guo, J. Kong, J. Gu, M. Chen, J. Zhu, Adv. Compos. Hybrid Mater. **1**, 207 (2018)
9. C. Huang, X. Qian, R. Yang, Mater. Sci. Eng. R Rep. **132**, 1 (2018)
10. J. Ordonez-Miranda, R. Yang, J.J. Alvarado-Gil, *Nanoscale Thermoelectrics* (Springer, Berlin, 2014), p. 93
11. R. Haggenmueller, C. Guthy, J.R. Lukes, J.E. Fischer, K.I. Winey, Macromolecules **40**, 2417 (2007)
12. I. Krupa, I. Novak, I. Chodak, Synth. Met. **145**, 245 (2004)
13. A. Henry, G. Chen, Phys. Rev. Lett. **101**, 235502/1 (2008)
14. T.F. Luo, K. Esfarjani, J. Shiomi, A. Henry, G. Chen, J. Appl. Phys. **109**, 074321/1 (2011)
15. R.Y. Wang, R.A. Segalman, A. Majumdar, Appl. Phys. Lett. **89**, 173113/1 (2006)
16. Z.H. Wang, J.A. Carter, A. Lagutchev, Y.K. Koh, N.H. Seong, D.G. Cahill, D.D. Dlott, Science **317**, 787 (2007)
17. X.P. Huang, G.Q. Liu, X.W. Wang, Adv. Mater. **24**, 1482 (2012)
18. Z.D. Han, A. Fina, Prog. Polym. Sci. **36**, 914 (2011)
19. https://www.netzsch-thermal-analysis.com/uploads/tx_nxnetzschmedia/files/TPoP_web_mit_Hervorhebung.pdf. Accessed on 19 Jan 2015
20. Y. Yang, in *Physical Properties of Polymers Handbook*, ed. by J.E. Mark (Springer Science & Business Media, New York, 2007), p. 156
21. A. Henry, G. Chen, Phys. Rev. Lett. **101** (2008)
22. Z. Lu, Y. Wang, X. Ruan, J. Appl. Phys. **123**, 074302 (2018)
23. D. Bigg, Adv. Polym. Sci. **119**, 1 (1995)
24. J.C.M. Garnett, Proc. Roy. Soc. Lond. Ser. A **73**, 443 (1904)
25. P.L. Kapitza, J. Phys. **4**, 181 (1941)
26. C.-W. Nan, R. Birringer, D.R. Clarke, H. Gleiter, J. Appl. Phys. **81**, 6692 (1997)
27. J. Ordonez-Miranda, R. Yang, J.J. Alvarado-Gil, J. Appl. Phys. **114**, 064306 (2013)
28. J. Ordonez-Miranda, R. Yang, J.J. Alvarado-Gil, J. Appl. Phys. **111**, 044319 (2012)
29. J. Ordonez-Miranda, J.J. Alvarado-Gil, Compos. Sci. Technol. **72**, 853 (2012)
30. M. Mooney, J. Colloid Sci. **6**, 162 (1951)
31. H. Fricke, Phys. Rev. **24**, 575 (1924)
32. I.H. Tavman, J. Appl. Polym. Sci. **62**, 2161 (1996)
33. D.A.G. Bruggeman, Ann. Phys. Leipzig **24**, 636 (1935)
34. D. Stroud, Superlattices Microstruct. **23**, 567 (1998)
35. N. Shenogina, S. Shenogin, L. Xue, P. Keblinski, Appl. Phys. Lett. **87**, 133106 (2005)
36. D.A. Brune, J. Bicerano, Polymer **43**, 369 (2002)
37. P.E. Gharagozloo, K.E. Goodson, J. Appl. Phys. **108**, 074309/1 (2010)
38. W. Lin, Modeling of thermal conductivity of polymer nanocomposites, in *Modeling and Prediction of Polymer Nanocomposite Properties*, ed. by V. Mittal (Wiley-VCH Verlag GmbH & Co. KGaA, Weinheim, Germany, 2013)
39. Z. Hashin, S. Shtrikman, J. Appl. Phys. **33**, 3125 (1962)
40. Z. Hashin, J. Compos. Mater. **2**, 284 (1968)
41. R.L. Hamilton, O.K. Crosser, Ind. Eng. Chem. Fund. **1** (1962)
42. R. L. Hamilton, *Thermal Conductivity of Heterogeneous Mixtures*, Ph. D. Dissertation, University of Oklahoma (1960)
43. H. Hatta, M. Taya, J. Appl. Phys. **58**, 2478 (1985)
44. Y. Agari, A. Ueda, S. Nagai, J. Appl. Polym. Sci. **49**, 1625 (1993)
45. S. Cheng, R. Vachon, Int. J. Heat Mass Transf. **13**, 537 (1970)
46. L. Nielsen, J. Appl. Phys. **41**, 4626 (1970)
47. L. Nielsen, J. Appl. Polym. Sci. **17**, 3819 (1973)
48. T. Lewis, L. Nielsen, J. Appl. Polym. Sci. **14**, 1449 (1970)
49. A.D. Drozdov, J. deClaville Christiansen, Compos. Sci. Technol. **182**, 107717 (2019)
50. K. Ramani, A. Vaidyanathan, J. Compos. Mater. **29**, 1725 (1995)
51. R.W. Lewis, P. Nithiarasu, K. Seetharamu, *Fundamentals of the Finite Element Method for Heat and Fluid Flow* (Wiley, Hoboken, 2004)

52. S.S. Rao, *The Finite Element Method in Engineering* (Butterworth-Heinemann, UK, 2005)
53. G. Dhatt, E. Lefrançois, G. Touzot, *Finite Element Method* (Wiley, Hoboken, 2012)
54. T.J. Hughes, *The Finite Element Method: Linear Static and Dynamic Finite Element Analysis* (Courier Dover Publications, USA, 2012)
55. L. Onsager, Phys. Rev. **37**, 405 (1931)
56. L. Onsager, Phys. Rev. **38**, 2265 (1931)
57. D. Kumlutas, I.H. Tavman, M.T. Coban, Compos. Sci. Technol. **63**, 113 (2003)
58. R. Peterson, A. Anderson, J. Low Temp. Phys. **11**, 639 (1973)
59. I.V. Singh, M. Tanaka, M. Endo, Int. J. Therm. Sci. **46**, 842 (2007)
60. S.R. Annapragada, D. Sun, S.V. Garimella, Comp. Mater. Sci. **40**, 255 (2007)
61. J.Z. Liang, F.H. Li, Polym. Test. **26**, 1025 (2007)
62. J.Z. Liang, F.H. Li, Polym. Test. **26**, 419 (2007)
63. R. Nayak, P.T. Dora, A. Satapathy, Comp. Mater. Sci. **48**, 576 (2010)
64. M. Moniruzzaman, K.I. Winey, Macromolecules **39**, 5194 (2006)
65. J.M. Wernik, S.A. Meguid, Appl. Mech. Rev. **63** (2010)
66. M. Rahmat, P. Hubert, Compos. Sci. Technol. **72**, 72 (2011)
67. Z. Spitalsky, D. Tasis, K. Papagelis, C. Galiotis, Prog. Polym. Sci. **35**, 357 (2010)
68. F. Du, R.C. Scogna, W. Zhou, S. Brand, J.E. Fischer, K.I. Winey, Macromolecules **37**, 9048 (2004)
69. J. Bicerano, J.F. Douglas, D.A. Brune, J. Macromol. Sci. Rev. Macromol. Chem. Phys. C **39**(4), 561–642 (1999)
70. X.T. Li, X.Y. Fan, Y.D. Zhu, J. Li, J.M. Adams, S. Shen, H.Z. Li, Comp. Mater. Sci. **63**, 207 (2012)
71. B.-W. Kim, S.-H. Park, R. Kapadia, P. Bandaru, Appl. Phys. Lett. **102**, 243105 (2013)
72. S.Y. Kwon, I.M. Kwon, Y.-G. Kim, S. Lee, Y.-S. Seo, Carbon **55**, 285 (2013)
73. R.F. Hill, J.L. Strader, IEEE Trans. Compon. Packag. Technol. **30**, 235 (2007)
74. B. Mortazavi, O. Benzerara, H. Meyer, J. Bardon, S. Ahzi, Carbon **60**, 356 (2013)
75. K. Sanada, Y. Tada, Y. Shindo, Compos. Part A Appl. Sci. Manuf. **40**, 724 (2009)
76. P. Keblinski, J.A. Eastman, D.G. Cahill, Mater. Today **8**, 36 (2005)
77. S. Shenogin, L. Xue, R. Ozisik, P. Keblinski, D.G. Cahill, J. Appl. Phys. **95**, 8136 (2004)
78. Y. Wang, H.F. Zhan, Y. Xiang, C. Yang, C.M. Wang, Y.Y. Zhang, J. Phys. Chem. C **119**, 12731 (2015)
79. A. Alzina, E. Toussaint, A. Béakou, B. Skoczen, Compos. Struct. **74**, 175 (2006)
80. A.K. Roy, B.L. Farmer, V. Varshney, S. Sihn, J. Lee, S. Ganguli, ACS Appl. Mater. Interfaces **4**, 545 (2012)
81. V. Varshney, S.S. Patnaik, A.K. Roy, G. Froudakis, B.L. Farmer, ACS Nano **4**, 1153 (2010)
82. H. Shin, S. Yang, S. Chang, S. Yu, M. Cho, Polymer **54**, 1543 (2013)
83. M.R. Gharib-Zahedi, M. Tafazzoli, M.C. Böhm, M. Alaghemandi, Phys. Chem. Chem. Phys. **17**, 14502 (2015)
84. K. Tanaka, S. Ogata, R. Kobayashi, T. Tamura, T. Kouno, Int. J. Heat Mass Transf. **89**, 714 (2015)
85. D. Konatham, A. Striolo, Appl. Phys. Lett. **95**, 163105 (2009)
86. K. Sato, A. Ijuin, Y. Hotta, Ceram. Int. (2015)
87. T. Gates, G. Odegard, S. Frankland, T. Clancy, Compos. Sci. Technol. **65**, 2416 (2005)
88. T.C. Clancy, S.-J.V. Frankland, J.A. Hinkley, T.S. Gates, Int. J. Therm. Sci. **49**, 1555 (2010)
89. S. Zhai, P. Zhang, Y. Xian, J. Zeng, B. Shi, Int. J. Heat Mass Transf. **117**, 358 (2018)
90. A. Vahedi, M.H.S. Lahidjani, S. Shakhesi, Phys. B **550**, 39 (2018)
91. A.G. Hoekstra, S. Portegies Zwart, P.V. Coveney, Multiscale modelling, simulation and computing: from the desktop to the exascale, Phil. Trans. R. Soc. A. **377**, 20180355 (2019)
92. J.L. Suter, P.V. Coveney, ACS Omega **3**, 6439 (2018)
93. S. Chinkanjanarot, J.M. Tomasi, J.A. King, G.M. Odegard, Carbon **140**, 653 (2018)

Chapter 10
Predicting the Optical and Electrical Properties of Polymer Nanocomposites

Michael J. A. Hore

Abstract Nanoparticles are frequently combined with polymers to create polymer nanocomposite materials with enhanced electrical, optical, or transport properties. In addition to the robust suite of theoretical techniques—including molecular dynamics, Monte Carlo simulations, and polymer field theory—that exists to predict structural information of such polymer nanocomposites, there is also a sizeable selection of techniques available to predict physical properties such as optical scattering/absorption and electrical/ionic conductivity of these materials. In many instances, techniques for predicting structural information may be coupled to those used to predict physical properties. This chapter presents a survey of common approaches to modeling both the optical and electrical properties of polymer nanocomposites, explaining the foundations of each technique, limitations, and highlighting select examples from the literature using each technique.

10.1 Introduction

Polymer nanocomposites result from the combination of a polymer matrix with one or more nanoparticle components. The nanoparticles may be bare or polymer-grafted, and can have shapes ranging from isotropic spheres to complex geometries. A primary motivating factor in adding nanoparticles to polymer materials is to enhance the physical properties of the resulting polymer nanocomposite. These physical properties include the mechanical properties, optical properties, and electrical properties among others [1–3]. For example, carbon nanotubes may enhance the electrical conductivity and/or the mechanical properties of the nanocomposite, whereas noble metal nanoparticles can impart useful optical properties. Key to understanding the resulting physical properties of the nanocomposite is having a robust understanding of the nanoparticle distribution, orientation, or other morphological information.

M. J. A. Hore (✉)
Department of Macromolecular Science and Engineering, Case Western Reserve University, Euclid Avenue, Cleveland, OH 10900, USA
e-mail: hore@case.edu

V. V. Ginzburg and L. M. Hall (eds.), *Theory and Modeling of Polymer Nanocomposites*, Springer Series in Materials Science 310,
https://doi.org/10.1007/978-3-030-60443-1_10

As discussed elsewhere in this book, there are numerous computational approaches, including field theoretic calculations and particle-based simulations, to extract such information.

The focus of this chapter is prediction of nanocomposite properties as a function of nanoparticle distribution, material morphology, and so on. The first section of this chapter contains an extensive discussion of calculating optical absorption and scattering by a polymer nanocomposite that contains metallic nanoparticles, such as Au nanospheres. Mie theory, the discrete dipole approximation (DDA), and finite-difference time-domain (FDTD) calculations can all be applied to the study of polymer nanocomposites. The second section of this chapter discusses determinating the electrical properties of a nanocomposite using resistor network models. Although not typically applied to polymer nanocomposite systems, this chapter closes with an introduction to the Poisson-Nernst-Planck formalism, which can be applied in the future to understand ion transport in polymer nanocomposites.

10.2 Calculation of Optical Properties

Noble metal nanoparticles have attracted interest from the polymer nanocomposite community because of the manner in which they interact with light. Nanoparticles synthesized from elements such as Au or Ag interact with light to preferentially scatter or absorb certain wavelengths. The combination of scattering and absorption is referred to as extinction, and the ability to tune the wavelength where the majority of the extinction occurs is attractive for many applications.

For nanoparticles composed of Au, Ag, and other noble metal nanoparticles, incident light creates collective oscillations of electrons along the surface of the nanoparticle ("surface plasmons"). This process is distinct from those that give rise to optical phenomena in quantum dot materials, such as CdSe. The result of the excitation of surface plasmons is one or more characteristic absorptions at wavelengths that depend on the nanoparticle size, shape, dispersion, and other factors. Photons with wavelengths equal to the surface plasmon resonance (SPR) wavelength experience strong absorption and scattering by the particles. When particles are present at high volume fractions (i.e., are close to one another spatially), the relative interparticle spacing affects the extinction of the nanocomposite. Generally, this "plasmonic coupling" results in a red-shifting of the SPR for spherical nanoparticles. For anisotropic nanoparticles, the polarization of the incident light, relative to the orientation of the nanoparticle, becomes an important determining factor, as does the spacing and relative orientation of the particles with respect to their neighbors. In all instances, the refractive indices of both the nanoparticle and surrounding matrix must be taken into account if an accurate prediction of the SPR wavelength is desired. The prediction of optical absorption or scattering by nanoparticles may be validated experimentally by UV-visible spectrophotometry (UV-vis), dark field optical microscopy, and Fourier-transform infrared spectroscopy to name a few. For the interested reader, Maier provides an extensive review of the fundamentals of plasmonics [4].

10.2.1 *Mie Theory Applied to Polymer Nanocomposites*

Mie theory is the result of solving Maxwell's equations analytically. In 1908, Gustav Mie published the result of solving Maxwell's equation to describe the scattering of light by colloidal (spherical) metal particles, which provided a mechanism behind the colors of colloidal gold solutions. The extinction and scattering cross sections, C_{ext} and C_{sca}, respectively, are typically expressed in terms of their efficiencies. Mie theory gives the following expressions for the extinction and scattering efficiencies:

$$Q_{ext} = \frac{C_{ext}}{\pi r^2} = \frac{2}{x^2}\sum_{n=0}^{\infty}(2n+1)\mathrm{Re}(a_n + b_n) \tag{10.1}$$

$$Q_{sca} = \frac{C_{sca}}{\pi r^2} = \frac{2}{x^2}\sum_{n=0}^{\infty}(2n+1)(|a_n| + |b_n|) \tag{10.2}$$

where the scattering amplitudes a_n and b_n are defined in terms of Riccati-Bessel functions. The mathematical expressions for these amplitudes are complicated, and the interested reader can find their expressions in the work of Bohren and Huffman [5]. In practice, for particles of size r, the summations in (10.1) and (10.2) can be truncated at a value $n_{max} = x + 4x^{1/3} + 2$, with $x = kr = 2\pi r/\lambda$. The absorption characteristics can be analyzed in terms of the absorption cross section $C_{abs} = C_{ext} - C_{sca}$. In addition, analysis with Mie theory can be extended to more complex particle architectures, such as spherical core-shell particles and others. However, analytical solutions for non-spherical shapes have proven elusive to date. Nevertheless, examples of approximate Mie-type theories for other geometries can be found in the literature [6, 7].

The application of Mie theory to predict the optical properties of nanocomposites is relatively straightforward, and consists primarily of calculating the extinction and scattering cross sections at wavelengths of interest. Fortunately, spherical nanoparticles are a common component of polymer nanocomposites, meaning that Mie theory is applicable to a number of nanocomposite systems [8, 9]. However, a limitation of this approach is that the calculations largely ignore the internal structure of the nanocomposite by assuming dispersed nanoparticles, and instead rely on characteristics of the components, such as the size of the nanoparticles and their refractive indices. In light of this, an important aspect of applying Mie theory is correctly representing the refractive indices of the material. As one example, this can be illustrated in recent work by Bockstaller et al. on dispersion of poly(styrene-*r*-acrylonitrile) (PSAN)-grafted SiO_2 nanoparticles in a poly(methyl methacrylate) (PMMA) matrix [10]. At a wavelength of $\lambda = 532$ nm, the refractive indices of SiO_2, PSAN, and PMMA are 1.458, 1.573, and 1.489, respectively. Because of this, the authors were able to index match the PSAN-grafted SiO_2 to the PMMA matrix—a condition that maximizes transmittance. The dielectric constant of the nanoparticles, which consist of a core and a shell, can be approximated using the Maxwell-Garnett effective medium model [11]:

$$\varepsilon_{\text{eff}} = \varepsilon_{\text{shell}} \left[1 + \frac{3fx}{(1 - fx)} \right] \tag{10.3}$$

where f is the volume fraction of the core (relative to the volume of the entire nanoparticle), and

$$x = \frac{(\varepsilon_{\text{core}} - \varepsilon_{\text{shell}})}{(2\varepsilon_{\text{core}} + \varepsilon_{\text{shell}})} \tag{10.4}$$

The dielectric constant (ε) is related to the refractive index (n) as $\varepsilon = n^2$. In practice, this approximation would be tabulated as a function of wavelength, and the extinction and scattering cross sections calculated. Using this approach, the authors found excellent *quantitative* agreement between C_{sca} measured experimentally, and that calculated using Mie theory, as shown in Fig. 10.1. A similar approach was taken by Incel et al. in predicting the scattering and extinction of spherical CeO_2 nanoparticles coated with SiO_2 when incorporated into a polystyrene film [12]. The supporting information of [12] provides computer code demonstrating the calculation of extinction and scattering according to (10.1) and (10.2).

Of the techniques discussed in this chapter for predicting the optical properties of nanocomposites, Mie theory is perhaps the most limiting in that it is restricted to spherically-symmetry scatterers, and cannot describe more complicated situations such as self-assembled structures, for example. In addition, while it has been shown to accurately describe the scattering and extinction cross sections, it is unable to produce detailed features such as the local electric field or the scattering cross section due to plasmonic coupling. To accurately tie the internal structure of a polymer nanocomposite to the resulting macroscopic optical behavior, more detailed approaches are needed.

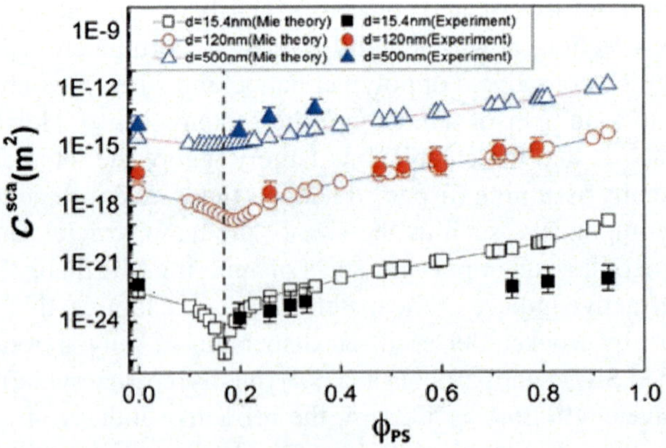

Fig. 10.1 Scattering cross section C_{sca} of polystyrene (PS)-grafted SiO_2 nanoparticles in toluene as a function of the weight fraction of PS, ϕ_{PS}. Open points were calculated using Mie theory, and filled points are experimental measurements. Squares, circles, and triangles correspond to nanoparticles with diameters of 16, 120, and 500 nm, respectively. Reproduced with permission from [8], Copyright ©2014 American Chemical Society

10.2.2 *The Discrete Dipole Approximation (DDA)*

While exact analytical solutions of Maxwell's equations are only possible in certain limiting cases (e.g., spherical nanoparticles), certain approximations have been extremely successful in modeling a variety of systems. The Discrete Dipole Approximation (DDA) can be attributed to work produced by Purcell and Pennypacker in 1973 [13]. Originally developed to describe the scattering and absorption of light by interstellar dust, it models a continuous particle as a collection of discrete dipole particles. The electromagnetic response of the dipoles is calculated for an incident electromagnetic field $\mathbf{E}_{\text{inc}}$. Most often, software packages such as DDSCAT and ADDA are used to calculate the response. Recently, with the advent of general-purpose computing on graphical processing units (GPUs), codes have been developed that take advantage of the highly parallel nature of GPU architectures to increase the speed of DDA calculations. DDA-GPU is one such program, although other examples exist.

10.2.2.1 Theoretical Background

Traditionally, within DDA the nanoparticle ("target") is composed of N dipoles arranged on a cubic lattice. In the original DDA formulation, the polarizibility of each dipole i was calculated from the Clausius-Mossotti relationship,

$$\alpha_i = \frac{3d^3}{4\pi}\left(\frac{\varepsilon_i - 1}{\varepsilon_i + 2}\right) \tag{10.5}$$

where ε_i is the dielectric function of dipole i, and d is the lattice constant of the cubic array. Draine and Flatau noted that this expression is only exact for $kd = 0$, where $k = \omega/c$, with ω representing the angular frequency of the light and c the speed of light in a vacuum. If the wavelength of the light is much larger than the lattice constant of the dipoles, then the polarizibilty can be approximated as

$$\alpha_i^{DG} \approx \frac{\alpha_i}{1 + \left(\alpha_i/d^3\right)\left[(b_1 + n^2 b_2 + n^2 b_3 S)(kd^2) - (2/3)ikd^3\right]} \tag{10.6}$$

In (10.6), $i = \sqrt{-1}$, S is a real number that describes the polarization of the incident radiation, and n is the refractive index. The coefficients $b_1 = -1.891531$, $b_2 = 0.1648469$, and $b_3 = -1.7700004$. Draine and Flatau note that this correction is most accurate for $|n|kd < 1$. For this reason, DDA calculations typically require a large number of dipoles to ensure that d is sufficiently small and the polarizibility is correctly determined.

Because DDA can be used to define nanoparticles of arbitrary geometries, it is convenient to characterize a target by an effective radius a_{eff}. In fact, some software packages such as DDSCAT require this value as a parameter to the calculation. The effective radius of a target is calculated by setting the actual volume of a target equal to the volume of an equivalent sphere, resulting in an expression

$$a_{\text{eff}} = \left(\frac{3V}{4\pi}\right)^{\frac{1}{3}} = \left(\frac{3N}{4\pi}\right)^{\frac{1}{3}} d \qquad (10.7)$$

where N is the number of dipoles that the target contains, V is the physical volume of the target, and d is the spacing of the dipoles. For a spherical nanoparticle with radius r_s, this expression simplifies greatly, and the effective radius can be written,

$$a_{\text{eff}} = r_s = d\left(\frac{3N}{4\pi}\right)^{\frac{1}{3}} \qquad (10.8)$$

While it is not surprising that the effective radius of a sphere would be equal to its physical radius, (10.8) is useful as it provides guidance regarding the number of dipoles (N) that are necessary to accurately model the extinction from that particle. Similarly, for the special case of a nanorod, this expression reduces to

$$a_{\text{eff}} = \left(\frac{3}{4}\right)^{\frac{1}{3}} (r_r^2 L) \qquad (10.9)$$

Show in Fig. 10.2a is a nanorod target with a physical diameter 15 nm and length 40 nm, and which contains 6152 discrete dipoles. From (10.7)–(10.9), the lattice spacing of the target can be calculated as $d = 1.05$ nm. For targets composed of Au, this lattice spacing implies $|n|kd \approx 0.02 - 0.04$, which is well below the condition required by Draine and Flatau. To mimic structures that may occur within a

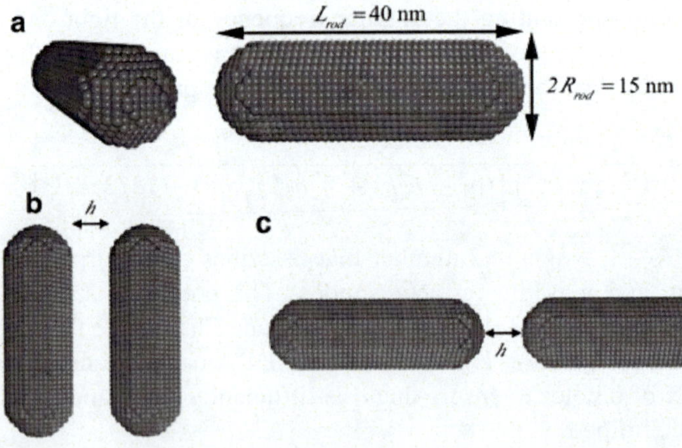

Fig. 10.2 DDA target structures. **a** A DDA target consisting of a single nanorod with $N = 6157$ dipoles (shown as spheres) representing a particle with a physical size of 15 × 40 nm. For this target, $a_{\text{eff}} = 25.6$ nm (10.7). **b** Two side-by-side oriented nanorods separated by a distance h, constructed from the target in (**a**). **c** Two end-to-end oriented nanorods, separated by a distance h

nanocomposite, such as aggregated or oriented nanoparticles, targets can be created using the nanorod target as a template, as demonstrated in Fig. 10.2b, c.

To calculate the scattering and absorption cross sections, the incident electric field at each lattice point and the polarization of each discrete dipole must be determined. Each dipole experiences an electric field

$$\mathbf{E}_{\text{inc}} = \mathbf{E}_0 \exp\left(\mathrm{i}\mathbf{k} \cdot \mathbf{r}_j - \mathrm{i}\omega t\right) \tag{10.10}$$

which represents the incoming, external electromagnetic radiation with frequency ω. Each discrete dipole interacts with this incoming electromagnetic field, and in turn, with all dipoles contained within the target. Therefore, the net electric field for dipole j is the difference of the two contributions, i.e.,

$$\mathbf{E}_j = \mathbf{E}_{\text{inc}} - \sum_{k \neq j} \mathbf{A}_{jk} \mathbf{P}_k \tag{10.11}$$

where $\mathbf{P}_k = \alpha_k \mathbf{E}_k$ is the polarization of dipole k, and

$$\mathbf{A}_{jk} = \frac{\exp(\mathrm{i}kr_{jk})}{r_{jk}} \times \left[k^2(\hat{r}_{jk}\hat{r}_{jk} - \mathbb{I}_3) + \frac{\mathrm{i}kr_{jk} - 1}{r_{jk}^2}(3\hat{r}_{jk}\hat{r}_{jk} - \mathbb{I}_3) \right] \tag{10.12}$$

In (10.12), $\mathbb{I}_3$ is the 3×3 identity matrix, $r_{jk} = |\mathbf{r}_j - \mathbf{r}_k|$, and $\hat{r}_{jk} = (\mathbf{r}_j - \mathbf{r}_k)/r_{jk}$. The diagonal terms $A_{jj} = \alpha_j^{-1}$, reducing (10.11) to

$$\sum_k \mathbf{A}_{jk} \mathbf{P}_k = \mathbf{E}_{\text{inc}} \tag{10.13}$$

This linear system is then solved using standard linear algebra methods to obtain the polarization for each dipole j, i.e., $\mathbf{P}_j$.

The absorption and scattering cross sections, C_{abs} and C_{sca}, respectively, are obtained from the incident electromagnetic field and the individual dipole polarizations. The extinction cross section is the sum of the scattering and absorption terms, $C_{\text{ext}} = C_{\text{sca}} + C_{\text{abs}}$. For a target composed of N_{DDA} dipoles, the absorption and scattering cross sections read:

$$C_{\text{abs}} = \frac{4\pi k}{|\mathbf{E}_0|^2} \sum_{j=1}^{N_{\text{DDA}}} \left\{ \mathrm{Im}\left[\mathbf{P}_j \cdot \left(\alpha_j^{-1}\right)^* \mathbf{P}_j^* \right] - \frac{2}{3} k^3 |\mathbf{P}_j|^2 \right\} \tag{10.14}$$

$$C_{\text{sca}} = \frac{4\pi k}{|\mathbf{E}_0|^2} \sum_{j=1}^{N_{\text{DDA}}} \mathrm{Im}\left(\mathbf{E}_j^* \cdot \mathbf{P}_j\right) - C_{\text{abs}} \tag{10.15}$$

where $\mathbf{E}_j^*$, $\mathbf{P}_j^*$, and α_j^* denote the complex conjugates of those quantities. The scattering and absorption efficiencies, Q_{sca} and Q_{abs}, are obtained by normalizing the respective cross sections by πa_{eff}^2.

Because it captures the effects of both scattering and absorption, $Q_{\text{ext}} = Q_{\text{sca}} + Q_{\text{abs}}$ is a convenient quantity to calculate to characterize the optical response of nanocomposites that contain nanoparticles with a plasmonic response. However, for anisotropic nanoparticles, such as gold nanorods, the extinction coefficient is polarization dependent. In particular, surface plasmons are excited only by *p*-polarized light (i.e., the electric field is parallel to the plane of incidence). For this reason, to correctly model the optical response of a rod-like object, Q_{ext} needs only to be calculated for two orientations of the rods: parallel and perpendicular with respect to the polarization of the incident light. For more anisotropic objects, a larger number of orientations of the object will need to be considered, while for a sphere, only one orientation is necessary to correctly model the optical response. Surface plasmon resonances (SPRs) are indicated by large enhancements in the scattering/absorption cross sections at particular wavelengths.

The behavior of the extinction cross section with respect to wavelength is primarily dependent on the incident electromagnetic field, the interactions between individual dipoles, and the relative refractive indices of the particles and matrix material. For this reason, DDA is an effective method not only for calculating the optical response of isolated nanoparticles within a polymer matrix, but also the optical response of self-assembled nanoparticles within nanocomposites. For instance, nanoparticles in close proximity to one another will exhibit surface plasmon coupling, which may result in a red or blueshift of the position of the surface plasmon resonance relative to the position in isolated particles. However, because the total number of dipoles scales with the number of nanoparticles in the DDA target, assemblies consisting of large numbers of nanoparticles are computationally expensive and in many instances, computationally intractable. In these cases, other techniques such as FDTD, described later in this chapter, may be more applicable.

10.2.2.2 Correction of the Dielectric Function

The previous discussion of the theoretical background of DDA did not explicitly consider the effect of material type on the extinction characteristics of nanoparticles. The effect of material type (e.g., gold or silver) on the optical properties determined by a DDA calculation can be seen in (10.5) and (10.6). The values of the material dielectric function directly modify the polarizibilities of the discrete dipoles. From the free electron model, the dielectric function of a material can be expressed as:

$$\varepsilon(\omega) = 1 - \frac{\omega_p^2}{\omega^2} \tag{10.16}$$

where ω_p is the plasma frequency of the material. This expression neglects several important contributions. First, it does not consider the effect of collisions of the electrons with the positively charged, stationary atomic cores. To include this effect, a relaxation time τ for the free electron gas is introduced into the model, leading to a corrected dielectric function:

$$\varepsilon_b(\omega) = 1 - \frac{\omega_p^2}{\omega(\omega + \mathrm{i}/\tau)} \tag{10.17}$$

where the subscript b designates that this is the dielectric function of the bulk material.

However, in the case that DDA is used to calculate the optical properties of polymer nanocomposites, it may be necessary to correct for the presence of the nanoparticle surface, from which electrons may collide. The result of such corrections is an additional surface relaxation term $\tau_a \approx a_{\mathrm{eff}}/v_F$, where v_F is the Fermi velocity. Physically, τ_a represents the timescale of collisions of electrons with the nanoparticle surface. The resulting dielectric function is

$$\varepsilon_s(\omega) = 1 - \frac{\omega_p^2}{\omega(\omega + \mathrm{i}/\tau + \mathrm{i}/\tau_a)} \tag{10.18}$$

where the subscript s designates that the dielectric function was corrected for surface damping.

Typically, only bulk dielectric data is available in the scientific literature. Hence, to obtain a dielectric function suitable for calculating the extinction spectra of nanoparticles, the bulk dielectric data must be corrected as:

$$\varepsilon_s(\omega) = \varepsilon_b(\omega) + \left[\frac{\omega_p^2}{\omega(\omega + \mathrm{i}/\tau)} - \frac{\omega_p^2}{\omega(\omega + \mathrm{i}/\tau + \mathrm{i}/\tau_a)}\right] \tag{10.19}$$

In practice, the dielectric function must be corrected each time the value of a_{eff} changes for a given DDA system. For gold nanoparticles, $\hbar\omega_p = 8.55$ eV, $\hbar/\tau = 0.108$ eV, and $v_F = 1.41 \times 10^6$ m/s.

10.2.2.3 Examples from the Literature

DDA has been used in several studies aiming to predict and interpret the optical response of polymer nanocomposites and similar materials. It is particularly useful in that targets can be constructed from the outputs of other modeling techniques, such as Monte Carlo or molecular dynamics simulations, and then examined with an appropriate DDA code. In addition, DDA has been applied to numerous experimental studies by constructing targets that are based on electron micrographs, for example. Here, we highlight a few such studies to illustrate the application of DDA to predict the optical properties of polymer nanocomposites.

DDA has seen a lot of applications in the characterization of surface plasmons in Au nanoparticles. Work from El-Sayed and coworkers demonstrated a "plasmon ruler" effect in which the wavelength of the surface plasmon shifted exponentially with the distance between pairs Au nanodiscs [14]. The relative shift scaled as $\Delta\lambda/\lambda_0 \sim \exp(-x/D\tau)$, where D is the nanodisc diameter, x is their separation distance, and τ is a unitless decay constant. Experiments found the decay constant $\tau = 0.18 \pm 0.02$ whereas DDA found $\tau = 0.23 \pm 0.03$, implying that plasmon coupling took place within distances equal to about 20% of the nanodisc size. For anisotropic nanoparticles, such as gold nanorods, the details of plasmonic coupling depend on both the separation of the nanorods as well as their relative orientations [15].

While the studies highlighted above considered nanoparticles in a vacuum or deposited on a surface in air, it is trivial to extend the application of DDA to polymer nanocomposites by taking into account the refractive index of the nanocomposite. Autophobic dewetting is a well-known phenomenon that can result in the aggregation of polymer-grafted nanoparticles within a chemically-identical polymer matrix [16, 17]. As a general guideline, nanospheres have been observed to aggregate when the degree of polymerization of the brush N is around three times smaller than that of the matrix P, i.e., $P > 3N$. For nanorods, it has been observed that aggregation occurs for $P > 2N$. A significant number of computational studies have found more detailed expressions for these transitions that depend on grafting density and other factors [1, 18]. A result of aggregation due to autophobic dewetting is that nanoparticles which exhibit surface plasmon resonances come into close contact with one another, and the surface plasmons in each nanoparticle begin to couple to those in others. Monte Carlo has been used to predict the equilibrium structures that appear due to this process. For instance, in the case of spherical nanoparticles, Monte Carlo has predicted that sheets, clusters, and other assemblies of nanoparticles are predominant, depending on various factors. For the case of nanorods, the particles generally form side-by-side aligned structures upon aggregation in polymer thin films, with an interparticle separation that can be modulated by varying the molecular weight of the grafted polymers. DDA is useful in these instances since the absorption and scattering of light can be calculated for representative structures that are either observed experimentally, or predicted from Monte Carlo simulations, for example. Hore et al. used a combination of Monte Carlo simulations, DDA calculations, and experiments to investigate the coupling between autophobic dewetting and the optical properties of polystyrene (PS)-functionalized Au nanorods within PS thin films. As seen in the optical spectra in Fig. 10.3, the theoretical results quantitatively capture many aspects of the experimental measurements, showing an increasing blue shift in absorption as nanorods become increasingly aggregated. However, because of the simplicity of the DDA calculations, the effects of polydispersity in the sample—both in terms of nanoparticle size and aggregate structure—are not effectively captured.

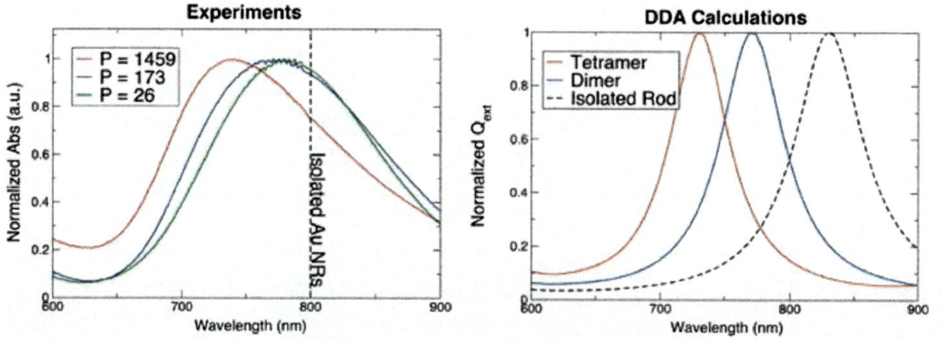

Fig. 10.3 (Left) UV-Vis absorption data for Au nanorod/polystyrene nanocomposites as a function of matrix degree of polymerization (*P*). As *P* increases, nanorods increasingly aggregate, leading to a blueshift in the absorption wavelength due to plasmonic coupling between nanoparticles. (Right) DDA calculations of the extinction efficiency of isolated rods, pairs of rods, and tetramers of rods to mimic increasing aggregation with molecular weight. The trends are in reasonable quantitative agreement with experiments. Reproduced with permission from [17], Copyright ©2014 American Chemical Society

10.2.3 Finite-Difference Time-Domain (FDTD) Method

The finite-difference time-domain method was introduced in the mid-1960s by Yee [19] FDTD can be applied to obtain time-dependent solutions of the electric and magnetic fields for arbitrary systems. A major characteristic of FDTD is that the fields are staggered spatially and temporally. FDTD is best suited to modeling systems that do not exhibit strong resonances—a situation that would require a large number of iterations to fully describe the evolution of the fields. To illustrate the fundamentals of the FDTD approach, we briefly review it here, but for a full description, the interested reader should consult more complete texts, such as those by Sullivan, Schneider, and others [20, 21]. In addition, several commercial and open source software packages exist that implement FDTD, including Lumerical, Angora, MEEP, and many others.

10.2.3.1 Theoretical Background

As its name suggests, FDTD is a finite difference technique that is solved numerically on a grid. Because of this, it is can be trivially parallelized for multi-core CPUs and GPUs using MPI/OpenMP, or CUDA/OpenCL, respectively. Implementation of FDTD requires the use of a Yee cell, which represents the manner in which the electric (**E**) and magnetic fields (**B**) are related to one another spatially. In the case of polymer nanocomposites, FDTD may be used to calculate the local electric and magnetic fields for analyzing materials for surface-enhanced Raman spectroscopy (SERS) applications, the reflectance/transmittance of a nanocomposite, or features relating to plasmonics such as the extinction cross section.

The Yee cell describes a particular manner in which to co-locate the **E** and **B** fields onto a cubic lattice. For each cubic cell, which is labeled by three indices *lmn*,

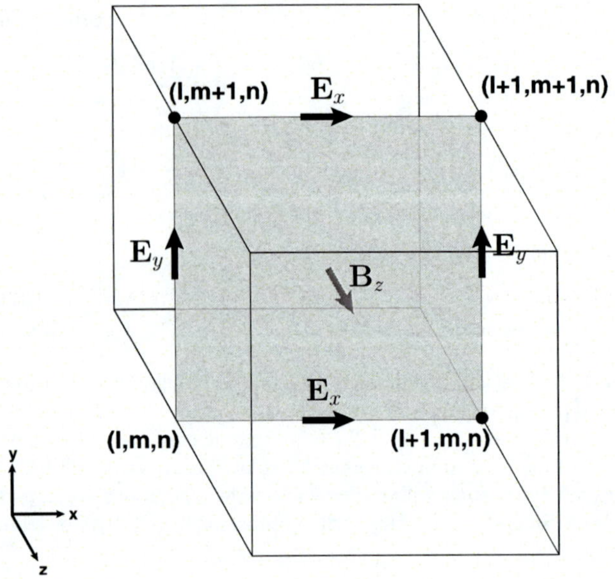

Fig. 10.4 Schematic of the Yee cell describing the positions of the electric and magnetic fields in an FDTD calculation. The electric and magnetic fields are staggered in space

the components of **E** are placed at the midpoint of each edge of the cube, while the components of **B** are placed in the centers of each face of the cube. This is shown schematically in Fig. 10.4. The Yee cell staggers **E** and **B** in space, and guarantees that each point of **E** is surrounded by components of **B**, and vice versa. The Yee cell satisfies the two divergence conditions in the macroscopic definition of Maxwell's equations:

$$\nabla \cdot \mathbf{H} = 0 \tag{10.20}$$

$$\nabla \cdot \mathbf{E} = \frac{\rho}{\varepsilon_0} \tag{10.21}$$

To completely describe the system of interest, knowledge of the conductivities and dielectric responses of each of the material components is necessary. The remaining Maxwell equations (i.e., the "curl equations") yield the time evolution of the **E** and **H** fields:

$$\frac{\partial \mathbf{E}}{\partial t} = \frac{1}{\varepsilon_0} \nabla \times \mathbf{H} \tag{10.22}$$

$$\frac{\partial \mathbf{H}}{\partial t} = -\frac{1}{\mu_0} \nabla \times \mathbf{E} \tag{10.23}$$

Note that (10.22) neglects the presence of any current density **J** that may appear in the system, and that $\mu_0 \mathbf{H} = \mathbf{B}$. μ_0 is the vacuum permeability. In addition, to perform the calculation for a real material, the dielectric constants ε of the material must also be accounted for in (10.22).

As a brief example of the implementation of FDTD using the concept of a Yee cell, consider applying FDTD in one dimension, for simplicity. In one dimension, (10.22) and (10.23) reduce to:

$$\frac{\partial E_x}{\partial t} = -\frac{1}{\varepsilon_0}\frac{\partial H_y}{\partial z} \tag{10.24}$$

$$\frac{\partial H_y}{\partial t} = -\frac{1}{\mu_0}\frac{\partial E_x}{\partial z} \tag{10.25}$$

To solve (10.24) and (10.25) numerically, both the spatial and temporal derivatives are replaced by finite differences, resulting in:

$$\frac{E_x^{\tau+1/2}(n) - E_x^{\tau-1/2}(n)}{\Delta t} = -\frac{1}{\varepsilon_0}\frac{H_y^\tau\left(n+\frac{1}{2}\right) - H_y^\tau\left(n-\frac{1}{2}\right)}{\Delta x} \tag{10.26}$$

$$\frac{H_y^{\tau+1}\left(n+\frac{1}{2}\right) - H_y^\tau\left(n+\frac{1}{2}\right)}{\Delta t} = -\frac{1}{\mu_0}\frac{E_x^{\tau+1/2}(n+1) - E_x^{\tau+1/2}(n)}{\Delta x} \tag{10.27}$$

where n is an index on the Yee cell in the z-direction (cf., Fig. 10.4), τ is the current time step, Δt is the size of a time step, and Δx is the lattice spacing. These equations can be cast in another form which provides update equations for the **E** and **H** fields:

$$E_x^{\tau+1/2}(n) = E_x^{\tau-1/2}(n) - \frac{\Delta t}{\varepsilon_0 \Delta x}\left[H_y^\tau\left(n+\frac{1}{2}\right) - H_y^\tau\left(n-\frac{1}{2}\right)\right] \tag{10.28}$$

$$H_y^{\tau+1}\left(n+\frac{1}{2}\right) = H_y^\tau\left(n+\frac{1}{2}\right) - \frac{\Delta t}{\mu_0 \Delta x}\left[E_x^{\tau+1/2}(n+1) - E_x^{\tau+1/2}(n)\right] \tag{10.29}$$

To perform the FDTD calculation, (10.28) and (10.29) are iterated in time τ for each lattice site n to trace the evolution of the fields within the system. Note that the staggered configuration of the Yee cell results in a staggering of the solutions in both time in space. For instance, the electric field **E** at site n is determined at time $(\tau + 1/2)$ using the previous value of **E** at $(\tau - 1/2)$ along with the magnetic field **H** at staggered lattice sites $(n + 1/2)$ and $(n - 1/2)$. The new value of **H** at time $(\tau + 1)$ is then determined from its value at time τ and the values of the electric field at time $(\tau + 1/2)$. The calculation proceeds in this staggered fashion until the desired temporal evolution of the fields is captured in the calculation. For this reason, strongly resonating systems will typically require a large number of iterations of (10.28) and (10.29) to fully capture the temporal evolution of the system, which may be prohibitively expensive from a computational standpoint. The extension from one dimension to three dimensions follows a similar procedure as shown above.

Although the process outlined above is relatively straightforward, in practice there are several additional considerations that must be taken into account to accurately model a polymer nanocomposite. First, the choices of Δx and Δt must be chosen to ensure that the calculation is stable and accurate captures the propagation of any electromagnetic waves. In particular, to account for the finite speed of light, Δt is chosen to be $\Delta t = \Delta x/\sqrt{n}c$, where c is the speed of light in a vacuum and n is the dimensionality of the system. In addition to these values, the boundaries of the calculation must be treated to, for example, absorb outgoing radiation. A commonly used boundary is the perfectly matched layer (PML), which absorbs radiation at the boundary to prevent unphysical reflections back into the system of interest. To analyze scattering of incident radiation by objects, it is common to use the total field/scattered field (TF/SF) source to separate outgoing electromagnetic radiation from an incoming plane wave. The scattering and absorption cross sections are calculated from the absorbed power (P_{abs}) of the total field and the power of the scattered field (P_{sca}) as

$$C_{sca} = \frac{P_{sca}}{P_0} \tag{10.30}$$

$$C_{abs} = \frac{P_{abs}}{P_0} \tag{10.31}$$

where P_0 is the power of the incident plane wave [22]. The total extinction cross section is the sum of the two contributions, $C_{ext} = C_{sca} + C_{abs}$. In practice, to analyze the scattering of electromagnetic radiation by a polymer nanocomposite, one would interrogate the material by subjecting it to pulses of different wavelengths and polarizations, and recording the average scattering and absorption at each wavelength.

10.2.3.2 Examples from the Literature

As with DDA, FDTD calculations have been used throughout the scientific literature to predict the optical properties of polymer nanocomposites. As one example that combines multiple theoretical techniques to understand a polymer nanocomposite, Wang et al. used FDTD to predict the plasmonic properties of Au nanorod/polystyrene composites as a function of nanorod separation and nanorod size [23]. The authors of this work combined experiments with self-consistent field theory (SCFT) calculations to demonstrate that the particles were more likely to aggregate as the polymer matrix molecular weight increased and/or the nanorod length increased. By analyzing the interparticle spacing from scanning electron microscopy (SEM) images, the authors found that nanoparticles exhibited a very uniform spacing of ca. 35 nm upon aggregation. Using this value for the interparticle separation, FDTD calculations were able to quantitatively reproduce the extinction characteristics of the nanocomposites by considering the scattering and absorption of incident radiation by a pair of Au nanorods. In this case, though, FDTD did not provide any particular advantage over DDA calculations, and the results of the two techniques were quite similar [17, 23].

In more complex situations, FDTD provides significant advantages over DDA in that it can accurately reproduce many aspects of the electromagnetic field. In a related example, Glor *et al.* used a combination of ellipsometry and FDTD calculations to probe the out-of-plane alignment and reorientation dynamics of Au nanorods that were embedded in a PMMA thin film [24]. FDTD calculations allowed the authors to determine the polarizability of a Au nanorod from the electric near-field distribution, and compare to ellipsometry measurements to determine the average angle between the nanorod orientation and that of the film/substrate. Another study by Wu et al. investigated the optical properties of cylinder-forming polystyrene-b-poly(4-vinyl pyridine) (PS-b-P4VP) films with Au or Ag nanoparticles within the cylindrical P4VP phase, with the objective of using the material as a meta-lens [25]. The distribution of nanoparticles within the cylindrical phases caused negative refraction of incident light, and FDTD calculations demonstrated the material could focus the light and split the beam.

10.3 Calculation of Electrical Properties

Beyond predicting the interaction of a nanocomposite with light, predicting the electrical properties, such as conductivity/resistance, of a polymer nanocomposite has garnered considerable attention. The desire to mass produce flexible electronics, for instance, has motivated the design and development of transparent conductor technologies.

Calculation of electrical conductivity in polymer nanocomposites is many times related to the concept of percolation of nanoparticles throughout a material. In the case of anisotropic nanoparticles such as carbon nanotubes, nanowires, or nanorods the percolation threshold may be low—especially as the aspect ratio $\nu = L/D$ increases. At and after percolation, conductive nanoparticles form one or more continuous, conductive paths through the material—leading to enhanced electrical conductivity.

10.3.1 Resistor Network Models

10.3.1.1 Theoretical Background

A common approach to modeling the electrical conductivity of a polymer nanocomposite is to co-locate structural information onto a simple cubic (3D) or square (2D) mesh of spacing Δr, and apply finite-difference techniques to determine the current through the material [26]. This "resistor network model" approach assumes that at each grid point $\mathbf{r}$ within the system, the material is either conducting or non-conducting, and described by a conductivity $\sigma(\mathbf{r})$ at each point. In this case, structural

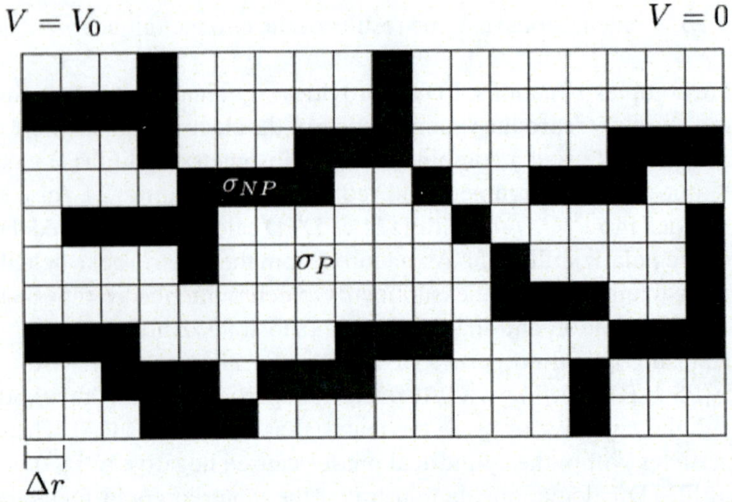

Fig. 10.5 Example of a 2D mesh, with mesh size Δr, containing a percolating network of nanowires of length $L = 3\Delta r$ in 2D (shown in black). The regions of the mesh containing nanowires have electrical conductivities σ_{NP}, while the surrounding polymer matrix (white regions) have conductivities σ_P. A potential difference of V_0 is applied across the material

information used to construct the arrangement of $\sigma(\mathbf{r})$ may come from Monte Carlo, coarse-grained particle dynamics models, or other techniques. This representation is illustrated in Fig. 10.5.

Given the spatial distribution of σ, the current $\mathbf{J}(\mathbf{r})$ can be calculated at each point through the system according to

$$\mathbf{J}(\mathbf{r}) = \sigma(\mathbf{r})\nabla V(\mathbf{r}) \tag{10.32}$$

subject to the condition that at each point $\nabla \cdot \mathbf{J}(\mathbf{r}) = 0$. This condition is equivalent to Kirchhoff's Law, stating that at each grid point i surrounded by points j,

$$\sum_j G_{ij}(V_i - V_j) = 0 \tag{10.33}$$

where the conductance

$$G_{ij} = \frac{\Delta r}{2}\left[\sigma(r_i) + \sigma(r_j)\right] \tag{10.34}$$

The resulting physical system, and its interpretation in terms of a resistor network, can be seen in Fig. 10.6.

Once the spatial dependence of $\sigma(\mathbf{r})$ is known, the current is calculated by applying the appropriate boundary conditions. A common approach is to set the voltage at a constant value of $V = V_0$ on one end of the system and $V = 0$ on the opposite end. After application of the boundary conditions, (10.34) can be evaluated at each grid

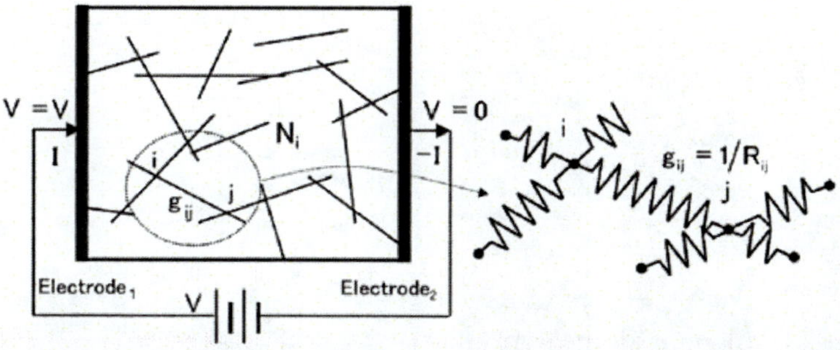

Fig. 10.6 Illustration of the translation of a carbon nanotube nanocomposite into a resistor network. Reproduced with permission from [27], Copyright ©2008 Elsevier

point to produce the conductance throughout the system, followed by the evaluation of (10.33) to obtain the voltage V_i at each point. A convenient, equivalent form of (10.33) is given by

$$V_i = \frac{\sum_j G_{ij} V_j}{\sum_j G_{ij}} \tag{10.35}$$

which implies a system of linear equations that can be solved with standard linear algebra techniques and software libraries. Finally, the current through the material can be calculated by evaluating

$$I = \sum_j G_{ij}(V_i - V_j) \tag{10.36}$$

for all points i at the boundary of the system [28].

10.3.1.2 Examples from the Literature

Resistor network models have been successful in describing the electrical behavior of several classes of polymer nanocomposites, including Ag nanowire nanocomposites, carbon nanotube (CNT) nanocomposites, graphene nanocomposites, and others. Hu et al. used a resistor network to model piezoresistance in CNT/epoxy nanocomposites, demonstrating that near the percolation threshold, CNT nanocomposites were more sensitive than more traditional strain gauges [27]. Dalmas et al. showed that this approach is also valid for curved, conductive fibrous networks, while others have shown its applicability to understand the effects of shear flow on electrical conductivity [29, 30].

Work from Winey et al. has used this model to quantify the conductivity of Ag nanowire composites, connecting both simulations and experimental measurements [31–33]. In a study by White et al., nanowire morphologies were generated randomly, and the conductivity through the nanocomposite was evaluated using a resistor net-

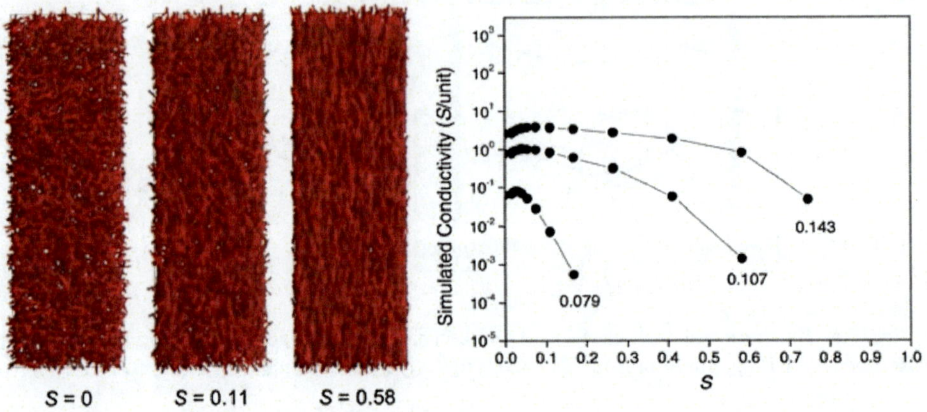

Fig. 10.7 (Left) Simulated morphologies of nanowires within a polymer nanocomposite as a function of their orientational order parameter S. (Right) Simulated nanocomposite conductivities from a resistor network model for nanowires as a function of S. The numbers to the right of each curve indicate the volume fraction of nanowires in each system. Reproduced with permission from [31], Copyright ©2009 American Physical Society

work model. As shown in Fig. 10.7, as the volume fraction of nanowires increased, their orientational order S increased, resulting in an increase in electrical conductivity. Amjadi and coworkers also employed a resistor network model to predict the performance of Ag nanowire/elastomer composites when used as strain sensors [34]. More recent work from Kwon et al. combined coarse-grained molecular dynamics simulations with a resistor network model to investigate the effect of nanoparticle shape and flexibility on the conductivity of polymer nanocomposites [35].

10.3.2 The Poisson-Nernst-Planck (PNP) Formalism

Although its use in the polymer nanocomposite field is not as widespread as in others, the Poisson-Nernst-Planck (PNP) formalism has shown promise as a reliable method to model ionic transport phenomena in a number of systems, such as with polymer membranes or solid-state nanochannels. In addition, it has been widely employed to interpret experimental investigations of polymer translocation. Looking to the future, this technique may be a viable approach for predicting and understanding macroscopic ionic currents from structural simulations of ionomer melts, for example. In taking such an approach, it is key to assume that the timescales of ionic motions are much faster than those of the surrounding polymers and nanoparticles. In other words, one must assume the polymers and nanoparticles are essentially stationary. In the solution state, as is the case when modeling polymer translocation, this assumption is typically true. However, in the melt state, this assumption may or may not apply, and so future researchers must take this into account.

10.3.2.1 Theoretical Background

At its core, PNP is a mean-field theory that couples the Poisson equation for the spatial distribution of electrostatic potential (ϕ) to the Nernst-Planck equation for predicting ion transport in response to that potential. Characteristic to the PNP approach is self-consistently solving the Poisson and Nernst-Planck equations on a grid until the desired level of error is achieved.

In a typical implementation, the electrostatic potential is first determined by solving the Poisson or linearized Poisson-Boltzman equation subject to the system's boundary conditions. For the case using the Poisson equation as an initial guess,

$$\nabla \cdot \varepsilon(\mathbf{r})\nabla\phi = -\frac{\rho}{\varepsilon_0} \tag{10.37}$$

where $\varepsilon(\mathbf{r})$ captures the spatial dependence of the dielectric constant throughout the system, and ρ is the charge density in the system. Note that ρ may contain both mobile charge density due to ions in the system, and fixed charges due to charged interfaces, etc. In aqueous systems, a commonly made approximation is that $\varepsilon(\mathbf{r}) = 80$ for regions containing solvent, and $\varepsilon(\mathbf{r}) = 2$ for insulating regions such as those composed of silicon nitrides, polymers, proteins, and the like. Lenart and coworkers recently used PNP theory to model ion transport through dendritic nanoparticles, and characterized the permeability of the nanoparticles by a hardness parameter h_{NP}. Under this approximation, the dielectric constant of regions containing the nanoparticles was calculated as $\varepsilon(\mathbf{r}) = h_{NP} \times 2 + (1 - h_{NP}) \times 80$, where the hardness parameter $0 \leq h_{NP} \leq 1$ [36]. At the boundaries of the system, a typical strategy is to apply a combination of periodic and/or Dirichlet boundary conditions to fix the value of the electrostatic potential ϕ.

Starting from a uniform value of $\phi = 0$, the Poisson equation can be solved to arbitrary precision through a successive over-relaxation (SOR) algorithm, where each new iteration of the potential at grid point i, ϕ_i^{n+1}, can be expressed in terms of the current value of the potential ϕ_i^n as:

$$\phi_i^{n+1} = \tilde{\phi}_i + (1 - \lambda)\phi_i^n \tag{10.38}$$

where the value

$$\tilde{\phi}_i = \left(\sum_j \phi_j^n \frac{2\varepsilon_i\varepsilon_j}{\varepsilon_i + \varepsilon_j}\right) \Big/ \left(\sum_j \frac{2\varepsilon_i\varepsilon_j}{\varepsilon_i + \varepsilon_j}\right) \tag{10.39}$$

with elements j being all adjacent points in the mesh [37]. Typically this process is iterated until the relative difference between ϕ^{n+1} and ϕ^n is less than 10^{-8}.

The initial value of the electrostatic potential is then used to solve the Nernst-Planck equation,

$$\nabla \cdot \left[D_i\nabla c_i(\mathbf{r}) + \frac{D_iZ_i}{k_BT}c_i(\mathbf{r})\nabla\phi(\mathbf{r})\right] = 0 \tag{10.40}$$

where D_i and c_i are the diffusion coefficients and concentrations, respectively, of ion type i, and Z_i is the charge carried by that ion. To more easily solve the Nernst-Planck equation, it can be converted into a form that resembles a Laplace-like equation through a Slotboom transformation: [38]

$$\psi_i(\mathbf{r}) = C_i \exp\left(-\frac{Z_i\phi(\mathbf{r})}{k_B T}\right) \tag{10.41}$$

leading to the equation

$$\nabla \cdot \varepsilon_{eff,i} \nabla \psi_i(\mathbf{r}) = 0 \tag{10.42}$$

The new dielectric-like function $\varepsilon_{eff,i}$ contains the diffusion coefficients

$$\varepsilon_{eff,i}(\mathbf{r}) = D_i \exp\left(-\frac{Z_i\phi(\mathbf{r})}{k_B T}\right) \tag{10.43}$$

The Nernst-Planck equation is solved to desired accuracy using the same approach outlined above for the Poisson equation. Typically, it is solved iteratively using an SOR approach until the error is on the order of 10^{-7}, after which the Poisson equation is solved using the newly determined concentration profiles. This self-consistent solution is repeated until the error in both equations is satisfactory. After obtaining a solution for $\phi(\mathbf{r})$ and $c_i(\mathbf{r})$, the ionic currents and ion fluxes can be calculated from the Nernst-Planck solution, i.e.,

$$j_i(\mathbf{r}) = -D_i\left[\nabla c_i(\mathbf{r}) + \frac{c_i(\mathbf{r})Z_i}{k_B T}\nabla\phi(\mathbf{r})\right] \tag{10.44}$$

10.3.2.2 Examples from the Literature

Examples applying the PNP formalism to understand polymer nanocomposites are rare in the literature—if they exist at all. However, several relevant studies do exist showing the applicability of this technique to understanding such materials. Meng et al. used PNP calculations to predict Li^+ conductivity in confined environments, such as in thin films of solid electrolyte [39]. Chan and coworkers also used PNP calculations to investigate the formation of ionic layers between graphene sheets, with the goal of understanding the flow of Li ions in the system [40].

10.4 Summary and Outlook

Looking to the future, polymer nanocomposites are likely to continue to enable new technologies and materials due to a large palette of particle types/functionalities, polymer materials, and self-assembly strategies. At this juncture, a mature suite of computational tools exists for predicting structural information in polymer nanocomposites—including molecular dynamics (MD), Monte Carlo (MC), dissipative particle dynamics (DPD), polymer field theory, and many others—that have been experimentally-validated under a number of conditions. In many instances, these techniques are suitable for coupling to others, like those described in this chapter, for calculating optical, electrical, and ionic transport characteristics in polymer nanocomposites. In fact, several examples exist in the literature that do exactly this, and it is likely that this approach will continue to be successful going forward.

In addition, the advancement of computer hardware, most notably the graphical processing unit (GPU), has significantly increased the speed with which these calculations can be performed. Many of the techniques discussed in this chapter are, at their core, finite-element techniques that can be trivially parallelized on GPU hardware. This in itself may enable more efficient computing of both nanocomposite structure and optical/electrical properties when those techniques are coupled to one another, and is sure to significantly advance our understanding of polymer nanocomposites and their physical properties.

Acknowledgements This work was supported in part by a National Science Foundation CAREER Award (DMR-1651002). The identification of software packages does not imply endorsement, nor that these are necessarily the best for a given purpose.

References

1. S.K. Kumar, N. Jouault, B. Benicewicz, T. Neely, Macromolecules **46**, 3199–3124 (2013)
2. M. Moniruzzaman, K.I. Winey, Macromolecules **39**, 5194–5205 (2006)
3. M.J.A. Hore, R.J. Composto, Macromolecules **47**, 875 (2013)
4. S.A. Maier, *Plasmonics: Fundamentals and Applications*, 1st edn. (Springer, New York, 2007), pp. 65–87
5. C.F. Bohren, D.R. Huffman, *Absorption and Scattering of Light by Small Particles, 1st end* (John Wiley & Sons, New York, 1998)
6. R.J. Martin, J. Modern Optics **40**, 2467 (1993)
7. B. Hourahine, K. Holms, F. Papoff, J. Phys.: Conf. Ser. **367**, 012010 (2012)
8. A. Dang, S. Ojha, C.M. Hui, C. Mahoney, K. Matyjaszewski, M.R. Bockstaller, Langmuir **30**, 14434 (2014)
9. A. Incel, T. Güner, O. Parlak, M.M. Demir, A.C.S. Appl, Mater. Interf. **7**, 27539–27546 (2015)
10. V.A. Markel, J. Optical Soc. Am. A **33**, 1244 (2016)
11. D.B. Stojanović, L. Brajović, D. Dramlić, V. Radmilović, P. Uskovoć, R. Aleksić, Prog. Org. Coatings **76**, 626 (2013)
12. W.R. Lenart, M.J.A. Hore, Nano-Structures & Nano-Objects **16**, 428 (2018)
13. E.M. Purcell, C.R. Pennypacker, Astrophys. J. **186**, 705 (1973)
14. P.K. Jain, W. Huang, M.A. El-Sayed, Nano Lett. **7**, 2080 (2007)
15. A.M. Funston, C. Novo, T.J. Davis, P. Mulvaney, Nano Lett. **9**, 1651 (2009)
16. P. Akcora, H. Liu, S.K. Kumar, J. Moll, Y. Li, B.C. Benicewicz, L.S. Shadler, D. Acehan, A.Z. Panagiotopoulos, V.A. Pryamitsyn, V. Ganesan, J. Ilavsky, P. Thiyagarajan, R.H. Colby, J.F. Douglas, Nature Mater. **8**, 354 (2009)
17. M.J.A. Hore, A.L. Frischknecht, R.J. Composto, ACS Macro Lett. **1**, 115 (2012)
18. A.L. Frischknecht, M.J.A. Hore, J. Ford, R.J. Composto, Macromolecules **46**, 2856 (2013)
19. K. Yee, IEEE Trans. Antennas Propag. **14**, 302 (1966)

20. D.M. Sullivan, *Electromagnetic Simulation using the FDTD Method*, 2nd edn. (Wiley, New Jersey, 2013)
21. J. B. Schneider, *Understanding the Finite-Difference Time-Domain Method*. 2017. http://www.eecs.wsu.edu/~schneidj/ufdtd. Date Accessed: Jan 2020
22. C. Tira, D. Tira, T. Simon, S. Astilean, J. Mol. Struc. **1072**, 137–143 (2014)
23. D. Wang, M.J.A. Hore, X. Ye, C. Zheng, C.B. Murray, R.J. Composto, Soft Matter **10**, 3404–3413 (2014)
24. E.C. Glor, R.C. Ferrier, C. Li, R.J. Composto, Z. Fakhraai, Soft Matter **13**, 2207 (2017)
25. L.Y.L. Wu, B. Leng, W. He, A. Bisht, C.C. Wong, *2013 IEEE 5th International Nanoelectronics Conference (INEC)* (Singapore, 2013), pp. 393–396
26. S. Kirkpatrick, Rev. Mod. Phys. **45**, 574 (1973)
27. N. Hu, Y. Karube, C. Yan, Z. Masuda, H. Fukunaga, Acta Mater. **56**, 2929–2936 (2008)
28. N. Hu, Z. Masuda, C. Yan, G. Yamamoto, H. Fukunaga, T. Hasida, Nanotechnology **19**, 215701 (2008)
29. F. Dalmas, R. Dendievel, L. Chazeau, J.-Y. Cavaillé, C. Gauthier, Acta Mater. **54**, 2923–2931 (2006)
30. A.E. Eken, E.J. Tozzi, D.J. Klingenberg, W. Bauhofer, J. Appl. Phys. **109**, 084342 (2011)
31. S.I. White, B.A. DiDonna, M. Mu, T.C. Lubensky, K.I. Winey, Phys. Rev. B **79**, 024301 (2009)
32. S.I. White, R.M. Mutiso, P.M. Vora, D. Jahnke, S. Hsu, J.M. Kikkawa, J.E. Fischer, K.I. Winey, Adv. Func. Mater. **20**, 2709–2716 (2010)
33. R.M. Mutiso, M.C. Sherrott, K.I. Winey, Phys. Rev. B **86**, 214306 (2012)
34. M. Amjadi, A. Pichitpajongkit, S. Lee, S. Ryu, I. Park, ACS Nano **8**, 5154–5163 (2014)
35. S. Kwon, H.W. Cho, G. Gwon, K. Kim, B.J. Sung, Phys. Rev. E **93**, 032501 (2016)
36. W.R. Lenart, W. Kong, W.C. Oltjen, M.J.A. Hore, J. Mater. Chem. B **7**, 6428–6437
37. D.S. Bolintineanu, A. Sayyed-Ahmad, H.T. Davis, Y.N. Kaznessis, PLoS Comput. Biol. **5**, e100027 (2009)
38. J.W. Slotboom, Electron Lett. **5**, 677–678 (1969)
39. D. Meng, B. Zheng, G. Lin, M.L. Sushko, Commun. Comput. Phys. **16**, 1298–1322 (2014)
40. Y. Chan, M. Saeed, S.-L. Lee, J.J. Wylie, Sci. Rep. **9**, 11712 (2019)

Chapter 11
Data-Driven Multiscale Science for Tire Compounding: Methods and Future Directions

Hongyi Xu, Richard J. Sheridan, L. Catherine Brinson, Wei Chen, Bing Jiang, George Papakonstantopoulos, Patrycja Polinska, and Craig Burkhart

Abstract Modern tire compound design is confronted with the simultaneous optimization of multiple performance properties, most of which have tradeoffs between the properties. In order to uncover new design principles to overcome these historical tradeoffs, multiscale compound experiment, physics, and simulation are being developed and integrated into next-generation design platforms across the tire industry. This chapter describes the efforts in our laboratories to quantify compound structures and properties at multiple scales—from nanometers to microns—and their application in compound simulations. This integration of experiment and simulation has been

H. Xu
Department of Mechanical Engineering, The University of Connecticut, Storrs, CT 06269, USA
e-mail: hongyi.3.xu@uconn.edu

R. J. Sheridan · L. Catherine Brinson
Department of Materials Science and Engineering, Duke University, Durham, NC 27708, USA
e-mail: richard.sheridan@duke.edu

L. Catherine Brinson
e-mail: cate.brinson@duke.edu

W. Chen
Department of Mechanical Engineering, Northwestern University, Evanston, IL 60208, USA
e-mail: weichen@northwestern.edu

B. Jiang · G. Papakonstantopoulos · C. Burkhart (✉)
Global Materials Science, and Tire and Performance Science Divisions, The Goodyear Tire and Rubber Company, Akron, OH 44306, USA
e-mail: craig.burkhart@goodyear.com

B. Jiang
e-mail: bing_jiang@goodyear.com

G. Papakonstantopoulos
e-mail: george_papakonstantopo@goodyear.com

P. Polinska
Global Materials Science, and Tire and Performance Science Divisions, The Goodyear Tire and Rubber Company, 7750 Colmar-Berg, Luxembourg
e-mail: patrycja_polinska@goodyear.com

V. V. Ginzburg and L. M. Hall (eds.), *Theory and Modeling of Polymer Nanocomposites*, Springer Series in Materials Science 310,
https://doi.org/10.1007/978-3-030-60443-1_11

found to be critical to highlighting the levers in data-driven multiscale compound design. We also provide a glimpse into the next set of capabilities, particularly from data science, which will impact future compound design.

11.1 Introduction

In the face of global climate change, governments, research institutes and the automobile industry are pursuing strategies to reduce CO_2 emissions from passenger and light-commercial vehicles. This effort requires complementary measures for car components which have a high impact on fuel consumption. Tires, which are the only vehicle component in direct contact with the ground, are responsible for 5–10% of the fuel consumption of a vehicle, mainly due to rolling resistance.

The European Commission (EC) plans to reduce the total vehicle energy consumption by 30% by 2030 [1] compared to 2021 by targeting several actions, including tire labeling. The tire labeling regulation, shown in Fig. 11.1, comprises three performance areas: fuel efficiency, wet grip, and noise. Lowering a tire's rolling resistance index will reduce fuel consumption, but on the other hand may have a negative influence on wet grip or other performance measures such as noise emissions. Optimizing these three areas simultaneously will require breaking through current tradeoffs into new material design frontiers.

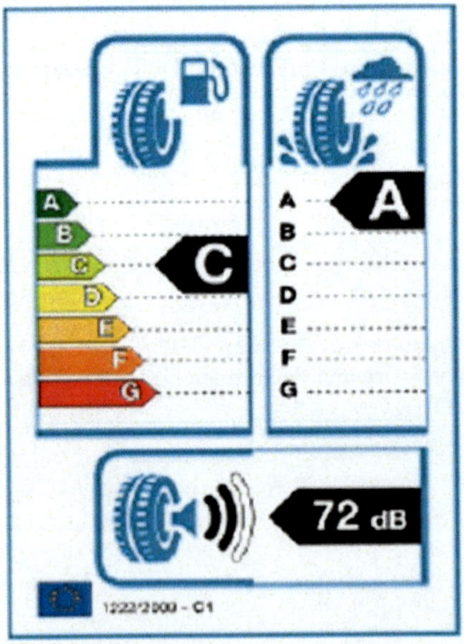

Fig. 11.1 Tire labeling showing specifications for fuel efficiency, wet grip and noise. Reproduced from [1]

While tire performance requires a complex interplay of design, compound science and manufacturing capability, the focus of this Chapter resides in compound science. From a materials standpoint, rubber compounds are complex materials, with chemical and structural features operating on multiple time and size scales. Compared to other materials-based industries, one of the differentiating aspects of tire technology is the time-tested sulfur vulcanization process which occurs during the manufacturing phase. This chemical process transforms the material properties from a flowing rubber formulation into a three-dimensional viscoelastic solid. In this step various additives (e.g., sulfur and accelerators) are dry-mixed and heated to affect crosslinking into an elastomeric network. The filler component achieves a reinforcement effect due to both physical interactions and additional chemical links with the elastomeric network [2–5], providing a large degree of strengthening of the rubber compound. Carbon blacks and amorphous silicas are the most widely used reinforcing nanofillers in the rubber industry. Without reinforcing fillers, tires would need to be replaced considerably more often, resulting in an increased burden on end-of-life-tire management, an increase in tire manufacturing energy requirements, as well as an increase in greenhouse gas emissions per vehicle unit-distance driven [6].

The mechanical properties of rubber compounds are dominated by a complex combination of multiscale chemistry and structure. The main structural entities include the vulcanized elastomer network (on the order of nanometers per crosslink), the filler-filler network (tens to hundreds of nanometers) and the filler-rubber interphase (tens of nanometers). The combination of the latter two structures creates the mechanical percolation network in a rubber compound. The most difficult structure to quantify is the filler-rubber interphase, which is formed by the mutual influences of the filler and rubber chemistries. This region of rubber surrounds the filler particles and has a mechanical response which may be strongly altered by its physical and/or chemical proximity to the filler surface. This structure has been investigated for over 50 years, sometimes being described in the compounding literature as "occluded rubber" [7]. Despite decades of indirect evidence and speculation, there exists limited direct mechanical evidence or visualization of such interphase properties. Experimental studies have recently indicated this interphase can extend from a few nanometers (nm) to over 50 nm [8, 9]. Understanding how these structures and scales interact with one another is critical to expanding current compound design frontiers.

11.1.1 Early Laboratory Compound Design Concepts

The historical practice of compounding was, and still is, a somewhat empirical occupation. Nevertheless, this empirical grounding led to realizing the connection between laboratory dynamic mechanical properties and performance. In one of the seminal tread compounding design papers of the mid-1980s, Nordsiek [10], promoted the concept of viscoelastic properties as critical to tread (and tire) performance.

Figure 11.2 demonstrates the design principles: a viscoelastic spectrum, in this case the normalized energy loss, tan δ, is swept as a function of temperature. The

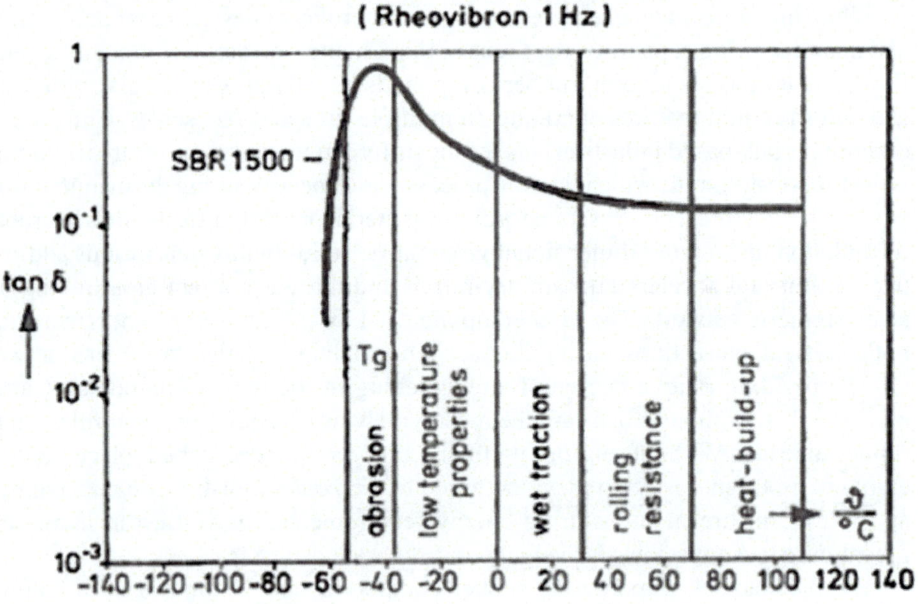

Fig. 11.2 Integral rubber concept. Tread performance is determined by temperature/frequency windows in tan δ spectra. Reproduced with permission from [10]. Copyright (1985) Huthig GmbH

breakthrough concept involved the realization of different temperature windows could be mapped onto various tread performance properties. In principle, one could tune each performance behavior—rolling resistance, wet traction, cold weather traction, by optimizing the tan δ spectrum in each window. Variants of this design principle are still used today.

Despite the obvious utility of this design idea, two major problems require resolution: (1) Next-neighbor performance regions, such as rolling resistance and wet traction, are difficult to decouple without chemical and structural intuition; (2) Given the very large recipe design space, it is not clear how compound recipe ingredients have to be selected and formulated to optimize each property. These problems have been traditionally solved by expert compounders, who by their experience and training map formulations onto properties. In reality, reducing this design knowledge to a quantitative physicochemical basis requires many domain knowledge sets: rubber chemistry and structure; filler chemistry, its structure and surface area; coupling agent chemistries for fillers with functional surfaces; the resultant filler-rubber interactions; sulfur vulcanization chemistry and its crosslinking networks; and finally, a computational platform which integrates such chemical and physical structures and properties into predictions. These domains must be mustered into a cohesive knowledge ecosystem in order to reveal deeper compound design principles.

A promising direction for merging these design needs is offered by the development of multiscale science, which incorporates high fidelity experiment, modeling and simulation as core knowledge resources in a data science framework. This field has emerged as one of the focal research areas in tire technology. Multiscale methods integrate, guide and optimize internal structure, and therefore provide design levers to understand and control material performance at multiple scales. Such qualities make this design approach highly attractive for next-generation rubber compounds. Driven by some of the advances outlined in this Chapter, recent developments of nanoscale experimental characterization methods as well as computational infrastructure are opening the way for the detailed elucidation of the fundamental compound structures and properties responsible for primary tire performance.

11.1.2 Chapter Scope

In Sect. 11.2, we present an overview of the data-driven experimental, mathematical, and computational tools which have been developed in our laboratories to understand nanoparticle-filled rubber composites. The linear viscoelastic properties of carbon-black-filled styrene-butadiene rubber composites are measured using dynamic mechanical analyses, and the resulting viscoelastic spectra are compared to simple models which ignore nanoscale structure such as particle distribution and interphase. Techniques to reconstruct compound filler microstructure are described, based on N-point correlation functions and particle descriptors. A computational analysis of the generation of representative volume elements (RVE) is presented, and an optimal statistical reconstruction is demonstrated. The interphase structure is then inferred based on the reconstruction of the viscoelastic measurement by optimization.

Finally, in Sect. 11.3, we give a glimpse into near-term future directions. Going forward, the integration of experimental, physics-based simulation and machine learning design methods are seen as critical for future compound design practices. Data science is central in the integration of multidisciplinary tire compounding environment. We see the development of these capabilities as critical to modern adaptive design approaches.

11.2 Methods Framework

We present a framework for the computational design of viscoelastic nanoparticle composites in order to achieve optimal material properties which cannot readily be found by traditional trial-and-error approaches. This framework integrates knowledge from experimental tests into well-defined numerical microstructure models to predict the properties of the nanocomposite. Specifically, a microstructure model is established to capture the complex microstructure characteristics of the composites, the properties of each phase in the microstructure is obtained from experimental tests, and simulations for overall response prediction demonstrate agreement with validation tests.

Dynamic mechanical analysis (DMA) is an established macro-scale technique to interrogate the linear viscoelastic response of materials through a gentle application of oscillatory strain while monitoring the force transmitted through a specimen in real time [11]. Using the principle of time–temperature superposition (TTS), the small-deformation mechanical response in terms of the complex modulus E^* or the loss tangent tan δ can be evaluated over many decades of frequency. Here, we use the tan δ peak as a characteristic marker of the glass transition, whether in the temperature or frequency domain. This provides a geometry- and material-agnostic point in the viscoelastic spectrum that can be tracked with high precision and accuracy.

Here we focus on a carbon-black-filled styrene-butadiene rubber (CB-SBR) composite [8]. Experimental DMA data from this system at various composite loadings is displayed in Fig. 11.3a, b, plotted in the temperature domain and frequency domain, respectively. The viscoelastic spectrum is a strong function of particle loading, and the tan δ peak is a particularly sensitive measure of changes in viscoelastic behavior. Note that a shift of the tan δ in temperature space moves opposite to the corresponding shift of tan δ in the frequency space. The change of tan δ with filler has been observed in a variety of systems by many independent groups—some of that change is due to the inherent elasticity of the filler itself, but the complete response is more complex [12–21]. Interactions between the nanofillers and

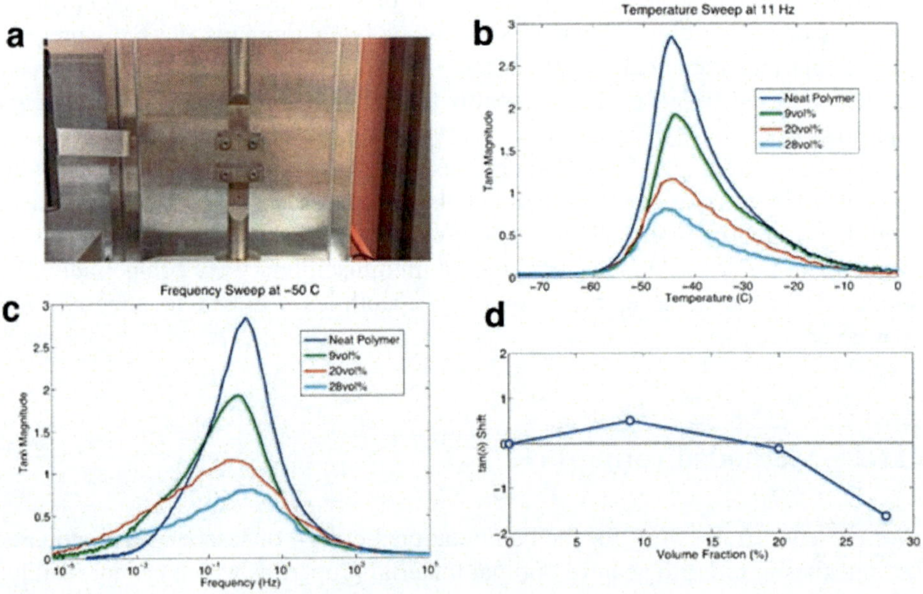

Fig. 11.3 Dynamic mechanical analysis (DMA) is a common experimental technique for determining viscoelastic properties of polymer-based materials. Experimental set-up (**a**). The tan δ spectra represented in the temperature (**b**) and frequency domains (**c**); each curve represents one sample and is not averaged. Corresponding glass transition temperature (T_g) shifts are plotted to show a shift towards higher temperatures at ~9 vol% followed by a shift towards lower temperatures (**d**) [8]. Adapted with permission from [8, 23]. Copyright (2012, 2016) Elsevier BV

the surrounding polymer chains alter the mobility of these polymer chains, resulting in a regime of "interphase" polymer, in which the material properties differ from the bulk matrix. Due to the high surface-to-volume ratio of nanofillers, the effects of this special region play an important role in the overall properties of nanocomposites, especially viscoelastic responses. Experimental observations have shown that the storage/loss modulus curve may be broadened or shifted in the time/temperature domain [22]. Since the properties of fillers are not time dependent, much of the improvement in the composite properties is ascribed to the interphase region. We will follow the study of Wood et al. [23] as an example and demonstration of the method to model tan δ shifts as a mechanical phenomenon due to the presence of filler and interphase.

11.2.1 Simple Representations of Filled Soft Composites

11.2.1.1 1D Scalar Models

The extent to which this CB-SBR compound can be modeled with very simple approaches should first be established. Wood [23] showed that a one-dimensional model of the system comprised of a perfectly elastic material (representing the carbon filler) in series with a Prony series material (representing the polymer matrix) is sufficient to recapitulate a tan δ shift to lower temperatures at high volume fractions (consistent with the shift to higher frequencies illustrated in Fig. 11.4). This shift in the tan δ peak occurs in the model when the particle modulus is substantially higher than the matrix modulus and represents the natural influence of the elastic particle on the relaxation behavior of the composite compared to the neat polymer.

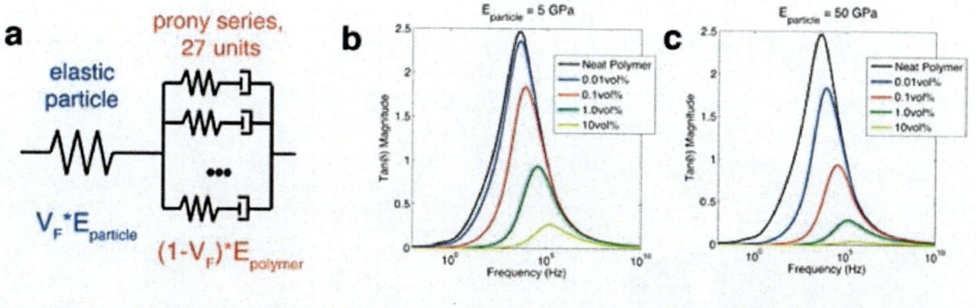

Fig. 11.4 A one-dimensional model is constructed that models an elastomeric composite with increasing volume fraction using an elastic spring in series with a 27 unit Prony series (**a**). When the particles are introduced (>0 vol%), the tan δ peaks shifts immediately to higher frequencies and reduces in magnitude. The magnitude of this shift and the drop in the maximum of tan δ is tied to the stiffness of the particle phase, 5 GPa (**b**) and 50 GPa (**c**). The shifting with increasing particle volume fraction is consistently towards increasing 'relaxation frequency' and thus decreasing relaxation time. Adapted with permission from [23]. Copyright (2016) Elsevier BV

No interphase is considered or necessary to obtain this shift in relaxation spectra toward shorter relaxation times. See Fig. 11.4.

11.2.1.2 3D Uniform Models

Although a one-dimensional model recovers important aspects of the experimental data and illustrates the essential effect of the elastic filler alone, several aspects must still be modeled, including the sensitivity of the tan δ peak position to volume fraction, microstructure and the behavior of the overall viscoelastic spectrum beyond the peak position. To increase the realism of the system, Wood created a three-dimensional finite element (FE) model that was otherwise comparable to the 1D model. A representative volume element (RVE) was generated by dispersing particle elements in a cube of matrix elements in a uniform, random spatial distribution. For several RVEs constructed this way with varying filler fractions, the viscoelastic spectrum was calculated. The location of the tan δ peak in temperature space for these models also decreases, but with a less strong dependence on filler volume fraction than the one-dimensional models, and more in-line with the experimental result. See Fig. 11.5.

In these models of elastic filler in a viscoelastic matrix, the temperature dependence of the tan δ peak monotonically decreased with increasing filler content, which fails to recover the initial increase at lower filler levels illustrated in Fig. 11.3 and other experimental data. In order to bridge this gap, additional realism was injected into the models by using a representative, reconstructed microstructure to account for particle distribution, following Deng et al. [8], and adding contributions from interphase [24, 25] polymer in the vicinity of the filler particles with altered relaxation times.

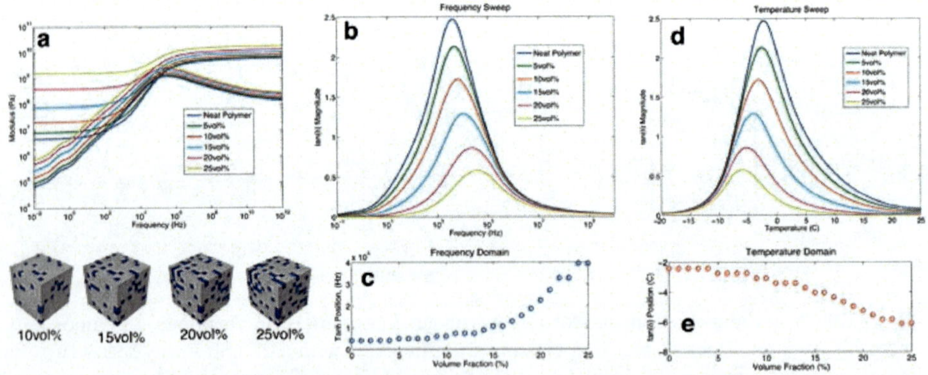

Fig. 11.5 A three-dimensional finite element model was constructed to represent a realistic tan δ behavior as particles concentration increases. The results—storage/loss modulus in (**a**) and tan δ in (**b**)—show a progressive shifting mechanism that both reduces tan δ magnitude and shifts the peak to higher frequencies (**c**). The tan δ shift to higher frequencies corresponds to a shift to lower temperatures (**d, e**). Adapted with permission from [23]. Copyright (2016) Elsevier BV

11.2.2 *Filler Microstructure Statistical Characterization and Stochastic Reconstruction*

The information of microstructure obtained in experimental observation is usually in the form of 2D image that cannot be used directly in material analysis and design due to the necessity of using 3D digital microstructure in material property analysis in order to obtain high accuracy in simulation results. The techniques of microstructure characterization are able to abstract quantitative descriptors from the image. Such descriptors are considered as the potential design variables due to their capability of capturing the key statistical features of the microstructure. Taken the prespecified descriptors as inputs, the reconstruction techniques generate multiple statistically equivalent random microstructure samples, which are passed into the material property analysis to obtain the microstructure-property relation.

There have been many other modeling frameworks for nanocomposites, operating at various length scales and with various assumptions [26–37]. However, none of these attempts to model the spatial distribution of particles, have shown a strong effect on the final properties of a composite [22, 38–40]. The understanding of these microstructure-property relationships appears to be necessary for composite material design.

To create an accurate representation of microstructure, a method is needed to generate FE geometries from experimental data taken from the samples of interest. Among nanoscale imaging techniques that provide sufficient spatial resolution, scanning and transmission electron microscopy (SEM and TEM, respectively) also have contrast in the CB-SBR system suitable for classifying pixels as either filler or matrix using simple binarization algorithms. It is possible to create FE geometries the directly from the binarized sample images, however a more rigorous and robust strategy is to "learn" the underlying particle distribution properties and generate new statistically representative samples for simulation, rather than testing a large number of specimens directly to gather statistics. Finally, with the geometry generated, it is possible to fit an interphase (size, shape, properties) to match DMA results from real composites. There are several published works using this strategy [41, 42] but we will follow Deng et al. [8] as an example here to introduce the framework, outlined in Fig. 11.6.

11.2.2.1 Microstructure Characterization and Reconstruction Based on Correlation Functions

The most widely used microstructure reconstruction techniques rely on the 2-point correlation function, which is defined as (in a two-phase random media):

$$S_2^i(x_1, x_2) \equiv \langle I^{(i)}(x_1) I^{(i)}(x_2) \rangle \tag{11.1}$$

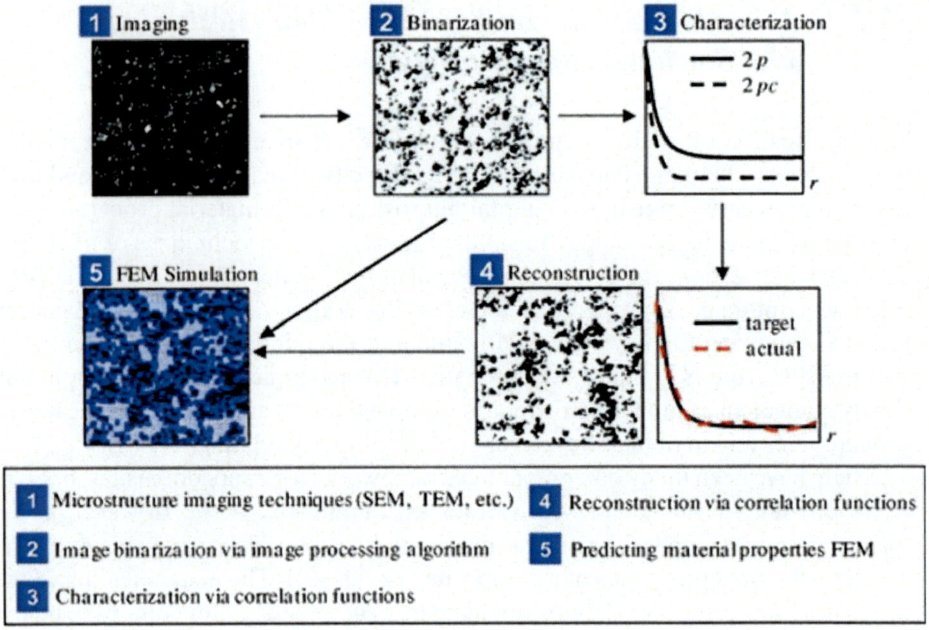

Fig. 11.6 The basic framework of data-driven modeling methods ("target" refers to the correlations of the digitized medium, "actual" are the correlations of reconstructed material). Adapted with permission from [8]. Copyright (2012) Elsevier BV

where I is an indicator function:

$$I^{(i)}(x) = \begin{cases} 1, & x \in \text{phase } 1 \\ 0, & x \in \text{phase } 0 \end{cases} \tag{11.2}$$

The physical meaning of the two-point correlation function is the probability of finding two points with a given distance r in the same phase of the random media. Therefore, two-point correlation is a function of distance r, and it can be denoted as $S_2^i(r)$. r can be any value from 0 to infinity, so the 2-point correlation function is usually truncated to a finite length by the microstructure image window size.

Based upon the definition, a two-point correlation function should follow several boundary conditions [43]:

- If $r = 0$, $S_2^i(r)$ equals to the volume fraction of composite;
- If $r \to \infty$, $S_2^i(r)$ approaches to the square of volume fraction;
- The derivative of $S_2^i(r)$ at $r = 0$ is equal to the surface area per unit volume.

Even though high accuracy can be obtained using 2-point correlation functions in microstructure characterization and reconstruction, the following challenges need further research efforts:

- The computational cost of microstructure reconstruction using 2-point correlation function is usually very high due to the large number of optimization iterations which adjust the image pixel by pixel in order to match the target 2-point correlation function. The high cost makes it impractical to apply this method on high-resolution image with a large window size, or reconstruction in dimension spaces higher than 2D.
- The vague physical meaning of the 2-point correlation function makes it hard to be used as design variables for microstructure. It is also difficult to relate the 2-point correlation function to the physics-based processing-microstructure model, the output of which is usually in the form of microstructure descriptors such as volume fraction, cluster number, radius, etc. Besides, a large amount of material analysis/design works have been done based on 2-point correlation function [43–45] or microstructure descriptors [46, 47]. A unified microstructure characterization and reconstruction technique is needed to relate these works with each other such that researchers can utilize existing data more efficiently.

Similarly, we can define other types of correlation functions (Fig. 11.7) [48–50]. The cluster correlation C_2^i is defined as the probability that finding two points with a distance r within one cluster of the filler phase. The surface correlation is defined as the probability that finding two points with a distance r located on the boundary of the filler phase. Surface correlations capture the morphology of interphase. The lineal path correlation is the probability that an entire line of length r lies in the filler phase. Different correlation function emphasizes on different aspects of microstructure features.

In this work, the digital microstructure image is represented by a 0–1 binary matrix. "1" represents filler, and "0" represents matrix. The purpose of characterization is to quantify the distribution status of "1" pixels in the space. There are two methods of evaluating correlation functions:

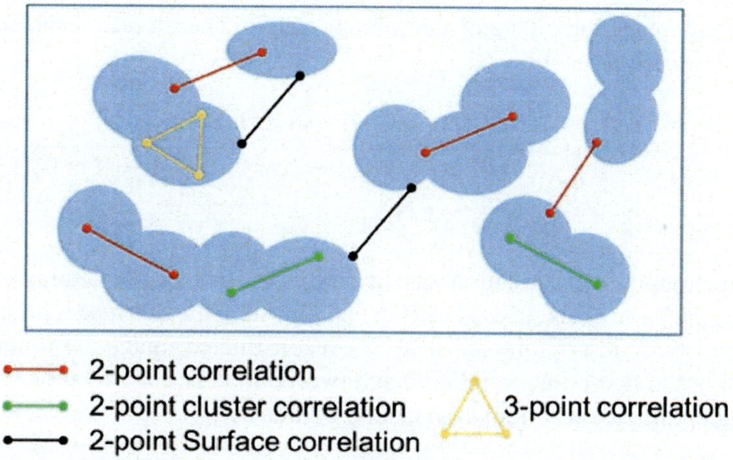

Fig. 11.7 Definition of different types of correlation functions

(1) The method using pixel-pixel distance histogram. This method follows the definition that 2-point correlation function is the normalized probability of finding two "1" pixels of distance r. Firstly, we calculate the distances between any two "1" pixels in the image. Based on the "1" pixel-pixel distance list, we obtain a frequency vector (histogram) of the pixel-pixel distance counts (how many pixel-pixel distances of 1, 2, 3 ... are observed from phase "1" in the image), which is noted as:

$$H\left(r = n | I^{(i)}(x) = 1, I^{(j)}(x) = 1\right), \quad i, j = 1, 2, 3, \ldots, n = 0, 1, 2, \ldots \tag{11.3}$$

where n is the distance between two pixels. The non-integer distances (e.g., the distance between two pixels on the diagonal direction is $\sqrt{2}n$) are rounded into the closest integer. i, j represents different "1" pixels.

(2) We calculate the distances between any two pixels in the image (no matter "1" and "0"). Based on the pixel-pixel distance list, we obtain a frequency vector (histogram) of the pixel-pixel distance counts (how many pixel-pixel distances of 1, 2, 3 are observed from the entire image), which is noted as:

$$H\left(r = n | I^{(i)}(x) = 1 \text{ or } 0, I^{(j)}(x) = 1 \text{ or } 0\right), \quad i, j = 1, 2, 3, \ldots, n = 0, 1, 2, \ldots \tag{11.4}$$

where n is the distance between two pixels. The non-integer distances (e.g., the distance between two pixels on the diagonal direction is $\sqrt{2}n$) are rounded into the closest integer. i, j represents different pixels (regardless of its value). $H\left(r = n | I^{(i)}(x) = 1, I^{(j)}(x) = 1\right)$ represents the number of pixel-pixel distance $r = n$ we can observe in "1" phase; $H\left(r = n | I^{(i)}(x) = 1 \text{ or } 0, I^{(j)}(x) = 1 \text{ or } 0\right)$ represents the highest possible number of pixel-pixel distance $r = n$ we can observe in the entire image regardless of the value of pixels. Therefore, we can calculate the 2-point correlation function as:

$$S_2^i(r = n) = \frac{H\left(r = n | I^{(i)}(x) = 1, I^{(j)}(x) = 1\right)}{H\left(r = n | I^{(i)}(x) = 1 \text{ or } 0, I^{(j)}(x) = 1 \text{ or } 0\right)}, \quad i, j = 1, 2, 3, \ldots, n = 0, 1, 2, \ldots \tag{11.5}$$

Microstructure reconstruction is an inverse problem. It can be naturally formulated as an optimization problem as shown in (11.6), where the discrepancies between the target statistical descriptors and that of a reconstructed image are minimized. For example, if the two-point correlation and two-point cluster correlation are used, the resulting optimization formulation can be written as:

$$E = \sum_r \left[\left(S_2^i(r) - \widehat{S}_2^i(r)\right)^2 + \left(C_2^i(r) - \widehat{C}_2^i(r)\right)^2\right] \tag{11.6}$$

where $\widehat{S}_2^i$ and $\widehat{C}_2^i$ are the resultant correlations of the reconstructed digitized medium at each of the discrete distance values r used in computationally determining the correlation functions. S_2^i and C_2^i are thus the target correlations to be realized. E is a fictitious "energy", in this case a multicomponent cost function, and is defined as a sum of squared differences between target correlations and those calculated from a generated microstructure. Therefore, the smaller the fictitious energy E, the closer the reconstructed microstructure is to the target. An illustrative example of the general microstructure reconstruction process is shown in Fig. 11.8. Traditional gradient-based optimization algorithms are unable to seek the global optimal solution due to the high nonlinearity and complexity of the resulting problem. Stochastic optimization methods such as simulated annealing [48, 50] and genetic algorithms [51] have been introduced to solve the resulting optimization problem. Here the simulated annealing algorithm is employed to reconstruct the microstructure of the two-phase material. The general formulation is reviewed as follows:

(1) Generating an initial binary microstructure. Based on the correlation functions, an initial microstructure with a given volume fraction of black pixels (secondary phase) can be randomly generated with uniform distribution. The black pixels will be randomly scattered within the image matrix while preserving the given volume fraction.
(2) Generating new binary microstructures. A sequence of trial realizations of microstructures is generated by randomly moving the black pixels within the image matrix.

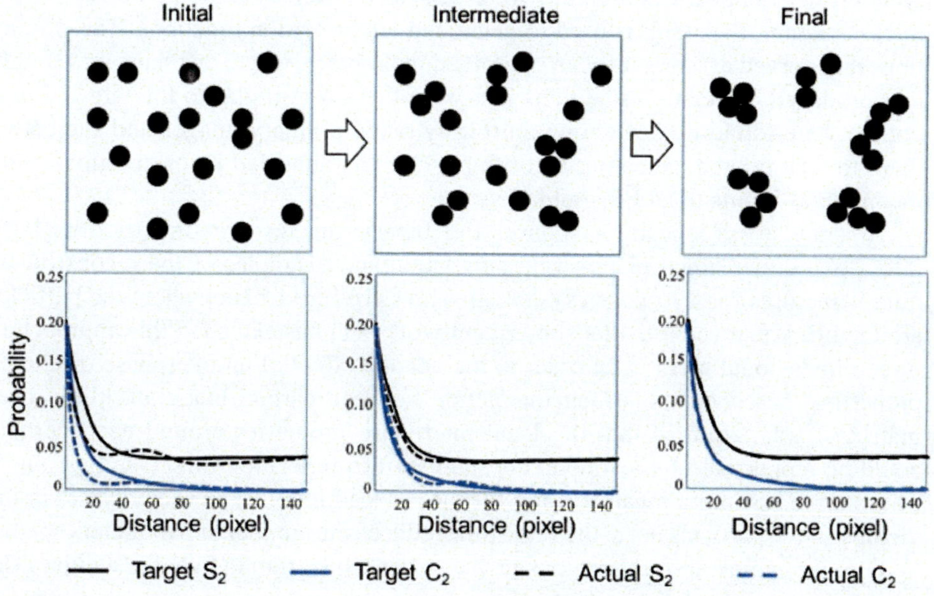

Fig. 11.8 Demonstration of the microstructure reconstruction process. Adapted with permission from [52]. Copyright (2013) Elsevier BV

(3) Accepting or rejecting the new generated microstructures. A realization of microstructure will be accepted with the probability:

$$p(\Delta E) = \begin{cases} 1, & \Delta E \leq 0 \\ \exp(-\frac{\Delta E}{kT}), & \Delta E > 0 \end{cases} \tag{11.7}$$

where $\Delta E = E' - E$ is the energy difference between the new and old realizations of microstructure, and T is a fictitious temperature in simulated annealing algorithm known as the cooling schedule.

(4) Iterative process termination. The iterative optimization process terminates once the temperature T is equal to zero or the energy E is lower than a predetermined value, say 1×10^{-5}. Otherwise, go to Step 2. The optimization-based stochastic reconstruction process is computationally expensive when it is applied to reconstruct high-resolution 3D microstructures. When other efficient MCR approaches (e.g., physical descriptor-based approach, random field-based approach, machine learning-based approach, etc. [53]) is employed, characterization of the reconstructed microstructures using multiple correlation functions is often needed for the purpose of validation.

11.2.2.2 2D Viscoelastic Reconstruction

In our realization of a data-driven framework (Fig. 11.8) [8], the reconstruction is based on correlation functions, specifically two-point radial correlation and two-point cluster radial correlation in two dimensions. After binarizing an image of the microstructure, the result is used to generate the correlation functions. These functions then served as a target for an optimization problem, where an image was sought that produced correlation functions that were the most similar to the target. Local optima were found by generating uniformly random binary images and subjecting them to a simulated annealing algorithm [52]. The annealed reconstructions were then used as inputs to an FE simulation.

To simulate composite response, the binary images are directly translated into a two-dimensional plane-strain simulation and examples of the reconstructed microstructures for the chosen system are shown in Fig. 11.10c, where the particles are the black pixels. Note that subsequent work [41] has extended this approach to three-dimensional models. In order to include the effect of an interphase of altered properties, based on the strong interaction between carbon black and the rubber matrix, it was assumed that the local interphase properties around each particle could be represented by a simple, constant shift in relaxation time (see Fig. 11.9). This simple assumption can account for the reduced mobility of the polymer in the vicinity of the particles but at the same time reduces the number of parameters needed to optimize against the real composite viscoelastic spectrum to two: the width of the interphase domain and magnitude of the shift factor.

Figure 11.10 shows the result of the composite predictions as the simulated interphase thickness was varied while the viscoelastic shift was maintained at one decade

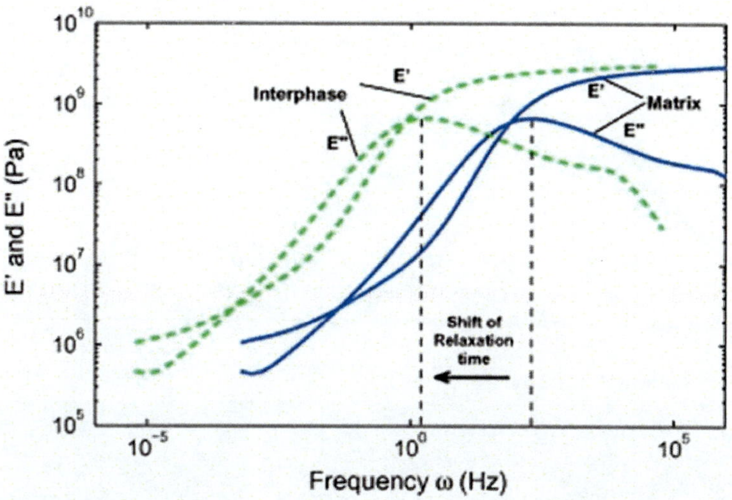

Fig. 11.9 Relation of material properties between interphase and bulk matrix by a simple shift in frequency/time space representing increase of relaxation times in the interphase domain relative to the matrix polymer. Adapted with permission from [8]. Copyright (2012) Elsevier BV

and filler fraction was held at 20%. The filled rubber literature reports a range of estimates for the interphase length, from 2 to 30 nm [2, 9, 54]. The tan δ shift at larger thicknesses (representing 50 and 75 nm interphase) is quite extreme compared to the experimental data, indicating that a single-layer (25 nm) thickness of interphase is the most appropriate. The role of the assumed shift factor for the interphase properties relative to the matrix (Ds) was also examined, as is illustrated in Fig. 11.11. Here, relaxation shifts from 0.5 to 2.0 decades are shown with the thickness held at one layer (25 nm). These shifts are in the range of possibility established experimentally for thermoplastics [55–57] and rubbers [58]. Based on these simulations, interphase polymer with a shift of 1.5 decades slower than that of the matrix material is the most appropriate to fit the experimental data. Overall, these results are reasonable, and similar results are also obtained with different approaches, such as the inverse modeling method [59].

11.2.2.3 Microstructure Characterization and Reconstruction Based on Physical Descriptors

A physical descriptor-based characterization and reconstruction approach [41, 60] is also proposed to enable statistical mapping of processing parameters to the physically meaningful statistical microstructure descriptors. Particle fillers of the nanocomposites are described using three categories of microstructure features: *composition*, *dispersion* and *geometry*. These three categories provide a complete representation of the microstructure.

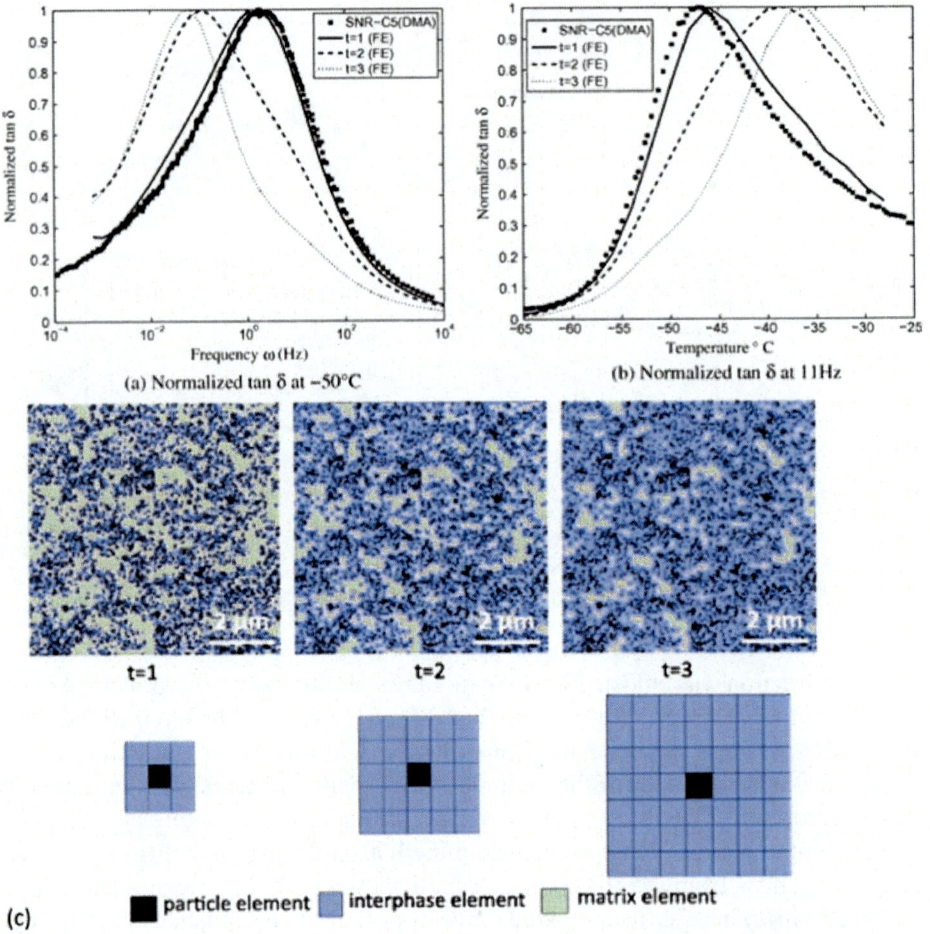

Fig. 11.10 **a, b** Comparison of predicted tan δ curves with experimental data: here the interphase properties are fixed while the interphase thickness is varied. **c** Visualization of the definition and effect of thickness for this simulation. Thickness t is in units of geometry mesh pixels, one pixel corresponds to approximately 25 nm. Adapted with permission from [8]. Copyright (2012) Elsevier BV

At the lowest fidelity level, *composition* indicates the percentage of each material constituent (e.g., volume fraction of nanoparticles). Composition is the lowest order of microstructure information as it only captures the homogenized response of an entire material system. One level down, *dispersion* represents the spatial distribution patterns of each phase. Local feature-induced material properties are reflected by the dispersion status of different phases in the material. For example, the nearest distance between carbon nanoparticles in polymer composites determines both the damping property and electrical conductivity. At the lowest level, *geometry* provides a more detailed description of the shape information of each phase (e.g., fillers in polymer composites). Geometry descriptors also have a strong impact on certain

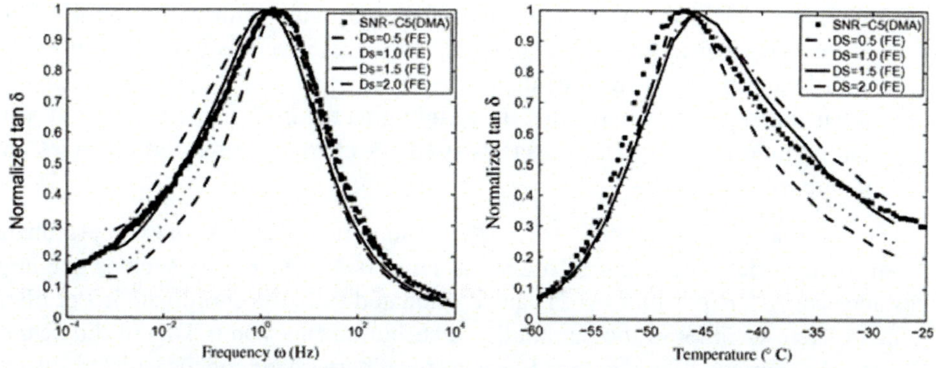

(a) Normalized tan δ at −50°C (b) Normalized tan δ at 11Hz

Fig. 11.11 Comparison of predicted tan δ curves against experimental data: here the interphase thickness is fixed while the interphase properties, decades of shift (Ds), varied. Increasing Ds indicates increasingly longer relaxation times for the interphase. Reproduced with permission from [8]. Copyright (2012) Elsevier BV

material properties [61]. For example, the pore size of porous material determines the yield point and critical load in microbuckling. It should be noted that due to the heterogeneity, descriptors used at each level are statistical, including mean and higher order moments such as variance, skewness etc.

The microstructure characterization process is illustrated by the flowchart in Fig. 11.12. The following information are obtained from image processing for characterization:

(1) A binary image that identifies different phases (black-polymer matrix, white-fillers);

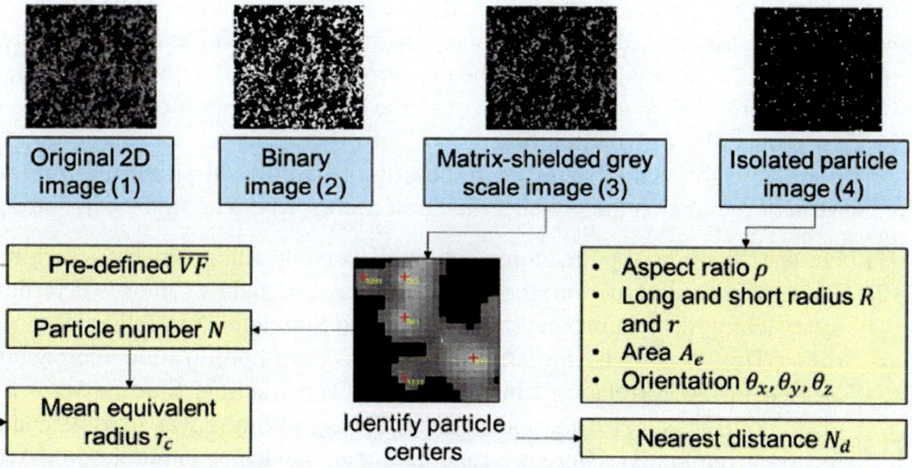

Fig. 11.12 2D microstructure characterization flowchart. Adapted with permission from [60]. Copyright (2014), Elsevier BV

(2) A matrix-shielded image that retains the grayscale information of the filler phase. The matrix phase is marked by "0". This image is used to identify the coordinates of the particle centers;
(3) An isolated particle image that only highlights the filler particles isolated from the particle clusters. This image is used for characterizing particle sizes and geometries.

The composition descriptor, $\overline{\text{VF}}$ of the nanoparticle filler, is known beforehand from the material processing information. Dispersion descriptors such as neighbor distances or Morisita's index [62] can be evaluated from the coordinate list of filler' centers. Here we choose the cumulative distribution function (CDF) of the nearest distance (n_d) among filler' centers. Based on the binary image and the isolated particle image, two geometry descriptors are evaluated: each particle's area A_e and the aspect ratio ρ. The geometry of filler is approximated with ellipse of major radius R and minor radius r. Aspect ratio ρ is defined as:

$$\rho = \frac{R_{2D}}{r} \tag{11.8}$$

Particle area A_e (mean $\overline{A_e}$) can be related to the equivalent radius of the particle r_c, and the total number of particles $\overline{N}$:

$$r_{c\_2D} = \sqrt{R_{2D} \times r} = \sqrt{\frac{A_e}{\pi}} \tag{11.9}$$

$$\overline{A_e} = \frac{L^2 \times \overline{\text{VF}}}{\overline{N_{2D}}} \tag{11.10}$$

where L is the side length of the microstructure image. Symbols with a bar overhead are deterministic variables; symbols without an overhead bar are random variables, which are described by distribution functions instead of a single value. Orientation ($\theta_x, \theta_y, \theta_z$) of the filler is assumed to be uniformly distributed with respect to the 3D Cartesian coordinate system. Statistics of all the geometry descriptors are represented by their probability distribution functions (PDF).

Figure 11.13 shows the process of 3D descriptor-based multi-phase microstructure reconstruction to match the pre-specified descriptors. There are four major steps:

(1) Pre-specification of target composition, dispersion, and geometry descriptors;
(2) Dispersion reconstruction: the location of each particle's center is determined by matching dispersion descriptors using the Simulated Annealing (SA) algorithm. This strategy is similar to the pixel-moving optimization approach that is used in the correlation function-based reconstruction. Starting from a set of randomly dispersed particle centers, the algorithm moves particle centers' locations randomly. After each random move, the dispersion descriptors (CDF of n_d) is re-evaluated and compared with the target values. Both the "good" moves that lead to a better matching in target dispersion descriptors, and the

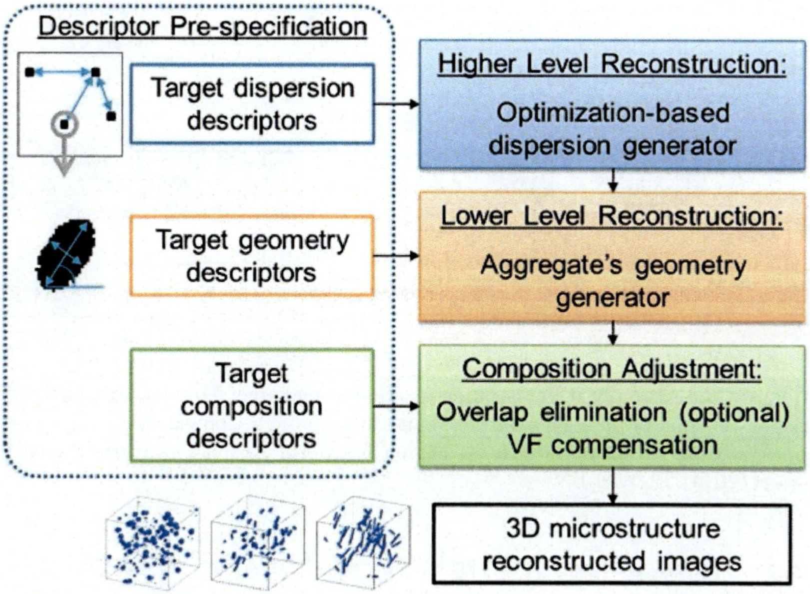

Fig. 11.13 Flowchart of multi-phase descriptor-based reconstruction. Adapted with permission from [60]. Copyright (2014) Elsevier BV

"bad" moves that lead to a worse matching, could be accepted, with probabilities determined by the Simulated Annealing (SA) schedule. This step generates an image where only the locations of filler' centers are indicated, referred to as "*center distribution map*".

(3) Geometry reconstruction: at the locations of particle centers, the algorithm reconstructs each particle based on the geometry descriptors. Values of each particle's geometrical descriptors are generated by sampling the statistical distribution functions obtained in characterization. The reconstructed geometry is placed onto a randomly selected particle center in the center distribution map.

(4) Composition adjustment: this is the final step of reconstruction when the digital image is fine tuned to match the composition descriptors (volume fraction of the filler phase). The volume fraction of the reconstructed image is adjusted by adding/subtracting pixels on phase boundaries.

The quality of the 3D reconstruction (Fig. 11.14) is assessed quantitatively by comparing the correlation functions of the 2D cross-sections from the 3D reconstruction with the original 2D image. The target 2D image and several 2D cut samples from the 3D reconstruction are shown in Fig. 11.7b. Figure 11.7a also compares the mean 2-point correlation functions of the 2D cut samples with the 2-point correlation function of 2D target image. Figure 11.7b compares the surface correlation functions.

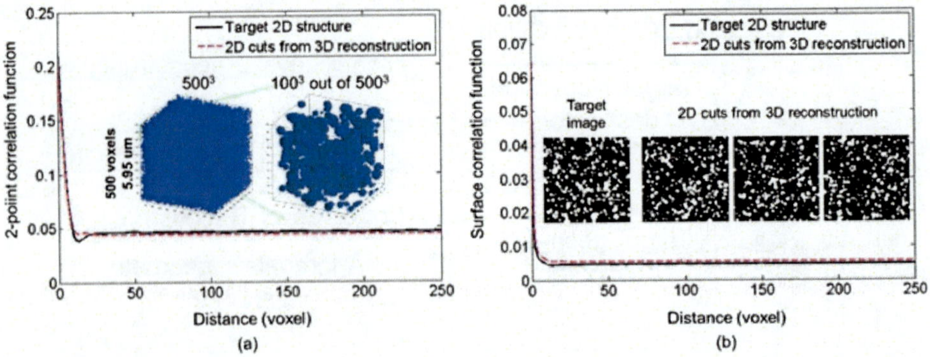

Fig. 11.14 Comparison of the correlation function of the target 2D microstructure and the mean correlation function of the 2D cuts from 3D reconstruction. **a** Comparison of 2-point correlation functions; **b** Comparison of surface correlation functions. Adapted with permission from [60]. Copyright (2014), Elsevier BV

11.2.3 Nanocomposite Properties and Statistical Study of the Representative Volume Element

In real applications, it is often computationally expensive to use a single, strictly qualified RVE size with high precision. Especially when dealing with three-dimensional FE simulations, the number of elements as well as the computation time increases cubically as the length of a side increases [58]. Statistical methods such as taking the average properties of multiple realizations make it possible to achieve higher precision at a smaller RVE size [these RVEs are sometimes referred to as statistical volume elements (SVEs)] [63–65]. Therefore, it is especially appealing to find optimal conditions avoiding one large RVE and instead using a smaller RVE with multiple realizations that reduces the computational cost but still satisfies the desired precision.

Here we introduce a microstructure-based modeling framework [8] for viscoelastic polymer nanocomposites applied to 3D microstructures characterized and reconstructed by a descriptor-based method. The determination of the smallest RVE size and the minimum number of realizations for a given volume size is demonstrated on a three-phase RVE containing 10% particles by volume. The influences of volume fraction (VF), composition (interphase layer) and dispersion of the particles on the RVE study results are also discussed in this section.

11.2.3.1 Comparison of Two Mesh Strategies: Voxel Mesh Versus Conformal Mesh

The selection of meshing strategy has a strong impact on material property predictions. In simulating composites, a trade-off often must be made between ease of creation of the mesh, accuracy of the structural representation, and computational efficiency of the models. Here we consider two mesh strategies, voxel mesh and conformal mesh, to understand their impacts on the prediction of material properties. The advantages of a voxelated mesh is that it is very easy to generate and apply to microstructures, while the drawback is that to obtain a good quality representation of a complex microstructure, very small voxels may be required which can result in large computational cost. The potential advantages of a conformal mesh is the ability to track complex microstructures more closely with the cost of more difficult meshing strategies and complications of appropriate mesh gradients which again may lead to larger computational costs. To explore these two options, a very simple model structure is established as a comparative study with only two cubic inclusions, surrounded by interphase and matrix material.

For the voxel mesh, the simulated properties using different mesh sizes are compared in Fig. 11.15. It is found that the impact of mesh size on the simulation result is negligible, so we can use coarser mesh to reduce the computational costs without reducing the prediction accuracy. On the other hand, a large discrepancy can be observed in the low frequency domain when the conformal mesh is used in simulations (Fig. 11.16). We also observe many distorted elements in the conformal mesh

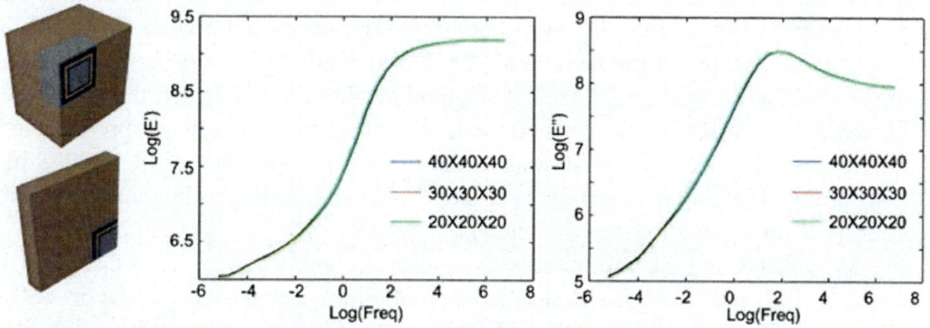

Fig. 11.15 Voxel mesh: simulation results using voxel meshes of different mesh sizes but the same VF. Adapted with permission from [58]. Copyright (2018) Elsevier BV

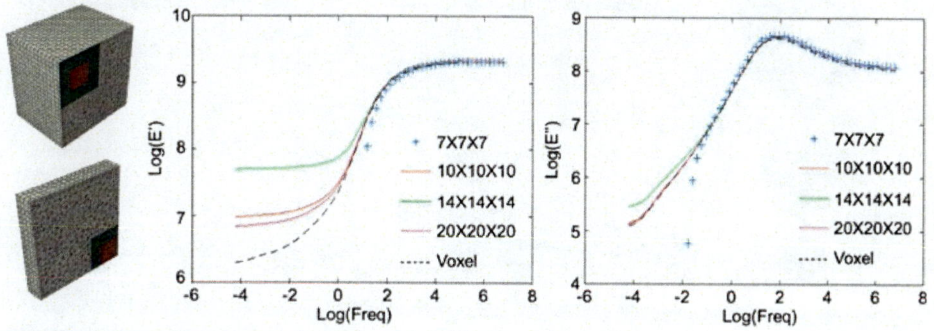

Fig. 11.16 Conformal mesh: simulation results with different mesh sizes but the same VF. Adapted with permission from [58]. Copyright (2018) Elsevier BV

or even mesh errors when applying the conformal mesh on more complex structures (results not shown here). Therefore, voxelated mesh is recommended in this study.

11.2.3.2 Relationship Between RVE Volume Size and the Variations in Simulated Material Properties

If unlimited computational resources were available, we can simply predict material properties with a large RVE model of a high resolution. However, in realistic cases, this strategy of unlimited computational resources and time is rarely the best solution. In order to achieve a balance between prediction accuracy and computational efficiency, microstructure models of different sizes are tested to understand the influence of RVE/SVE volume size on the predicted material properties [58]. At the outset, a sufficiently large microstructure RVE model is established to provide a reference property and computational cost. For each smaller volume size cut from this large benchmark model, a convergence study is conducted to guarantee a sufficient number of samples at that size. A trade-off between a single very large RVE and multiple smaller representative SVEs will be achieved with respect to computational cost and simulation accuracy. In order to quantitatively measure the mean and variance from the simulations, we selected the frequency point where tan δ reaches its peak and compared the mean and standard deviation of the peak value for each of the volume sizes in Fig. 11.17. The standard deviation between the simulations decreases as the RVE volume increases. However, even though we pre-selected a seemly large number of realizations in the FE simulations, the effective tan δ peak still shows a bias in the mean for very small volumes. This bias becomes negligible after the volume size reaches 10^3 voxels.

Also considering the computational costs presented in Fig. 11.17, we suggest 10^3 voxels as the optimal RVE size as a balance of computation time and the processing time corresponding to the number of realizations. One simulation with a volume of 60^3 voxels costs 2680 min, while 76 simulations with a volume of 10^3 voxels only

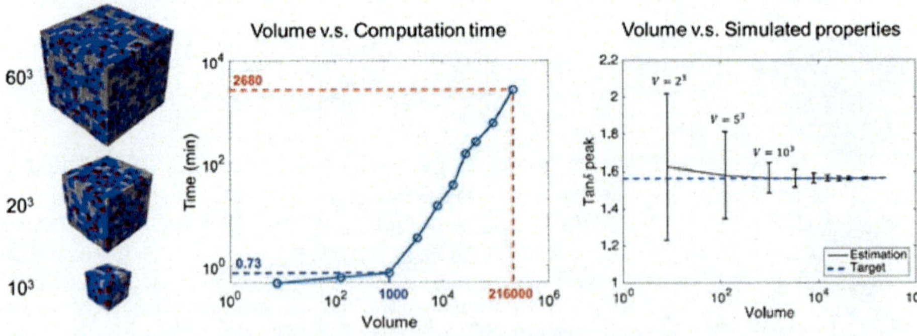

Fig. 11.17 Statistical study of the relation between RVE volume size and the predicted properties. Adapted with permission from [58]. Copyright (2018), Elsevier BV

cost 56 min while introducing less than 2% error. The use of sufficient realizations of $V = 10^3$ voxels instead of the full RVE represents a reduction in computational time of two orders of magnitude while still achieving the desired precision [58]. The ability to reduce computation time by several orders of magnitude will allow us to rapidly fill design of experiments (DOEs) for data mining and to consider more complex microstructures where large RVEs would be prohibitive in future studies.

11.2.3.3 Computational Analysis of Nanocomposites with FE Modeling

Finally, we establish realistic FE models based on the stochastically reconstructed 3D digital microstructures. Influences on the filled elastomers' properties mainly come from three aspects: the filler, the polymer matrix, and the interphase. The fillers are considered as elastic with Young's modulus of 73 (GPa), Poisson's ratio 0.19, and density 2.2 g/cm^3. The matrix is assigned linear viscoelastic material properties whose frequency response is given by the property curves in Fig. 11.8 obtained from physical experiments.

Here, as earlier (Fig. 11.9), the property of the interphase is assumed to be related to the property of the matrix by shifting and broadening in the frequency domain [11, 66]. The assumption for the shifting and broadening behaviors are based on our observation of the comparison between the master curves of the filled and unfilled systems in experiments [8]. In Fig. 11.18, the storage and loss modulus of the matrix (solid curves) are obtained from experiments and the interphase modulus (dashed curves) are then obtained by shifting those of the matrix to a lower frequency and applying a broadening factor.

We use ABAQUS to solve the RVEs given the material properties for each phase stated above. The dynamic moduli are calculated in the frequency-domain

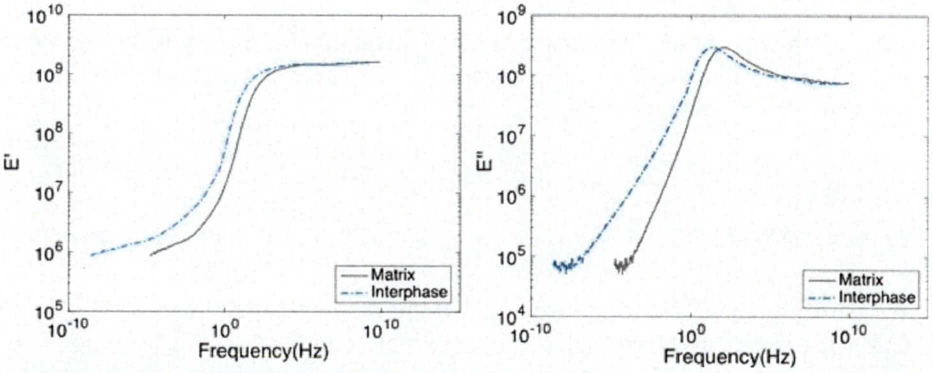

Fig. 11.18 Three-phase FE model and master curves including (left) storage modulus, (right) loss modulus. In these plots, the experimental data of the real matrix material is labeled "Matrix" and the "Interphase" curve is generated by artificially shifting the matrix curve to lower frequency and applying a broadening factor. Reproduced with permission from [58]. Copyright (2018) Elsevier BV

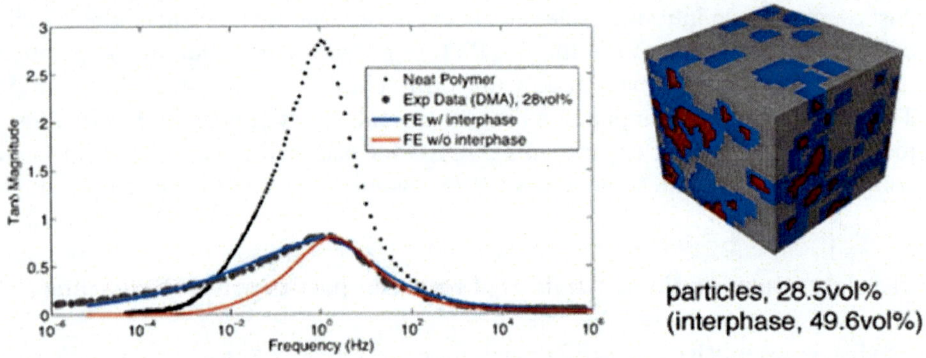

Fig. 11.19 Finite element models (60 × 60 × 60) with representative microstructures and interphase gradients match experimental results at high filler concentrations, even when tan δ appears to shift to higher frequencies, or lower temperatures [8]. We observe that the peak position, at this high filler loading, is nearly the same with or without interphase, therefore controlled by the presence of the stiff phase. Adapted with permission from [23]. Copyright (2016) Elsevier BV

viscoelastic module for small strain oscillations. In Fig. 11.19, we demonstrate the necessity of considering the interphase. Two RVE models, one with interphase (interphase VF = 49.6%) and one without interphase, are established for nanocomposites with a filler VF of 28.5%. The predicted material properties are validated with the experimental test results by DMA. Significant prediction error in the low frequency domain is observed in the model without interphase, while the model with interphase successfully captures the tan δ values in the full frequency domain.

Wood [23] used a two-layer strategy to generate the interphase gradient required to observe positive tan δ peak temperature shift [56]. Following the reconstruction strategy of Deng et al. and Xu et al. [8, 67], a reconstruction using SEM images of a CB-SBR was produced and transformed into a three-dimensional FE mesh. In Fig. 11.19, the effect of the optimal interphase gradient is contrasted with the same mesh without any interphase effect. The interphase mainly increases the loss of the material at lower frequencies, while the tan δ peak position and magnitude are dominated by the pure particle effect.

11.2.4 Summary

We have presented and demonstrated a framework of capabilities for the computational assessment of viscoelastic nanoparticle composites. We have done so to establish the important structural features needed to both replicate and extend our understanding of complex rubber compounds. From the use of dynamic mechanical analyses (DMA) and scanning electron microscopy (SEM), we provide structure–property relationships down to the nanometer scale. This combination of macro-scale experimentation, and nanoscale characterization, is necessary to link mesoscale-level structure to observed viscoelastic behavior.

This endeavor has also shown the importance of three structural entities to compound viscoelastic behavior: filler morphology, the far-field rubber matrix, and the filler-rubber interphase. Although it is considered a purely elastic material, and mechanically the simplest, the filler morphology itself is one of the most complex features of a rubber compound. Standard dry-mixing of a compound formulation are known widely to affect the overall filler morphology [68] and the subsequent mechanical properties. We have delineated a number of micrograph image analysis and reconstruction methods for filler morphology which are useful in the building of realistic compound models.

The second structural entity which is relevant for this style of mesoscale modeling and simulation is the far-field rubber matrix. It is perhaps the simplest morphological entity at the mesoscale, compared to the filler and interphase morphologies. In this light, unfilled rubber samples are constructed and analyzed solely by DMA methods. The construction of master curves, via generalized Maxwell models (i.e., Prony series constructions) are the main item contributed to the overall mesoscale continuum models.

Of the structural elements, the filler-rubber interphase, is the most difficult to ascertain. Its viscoelastic properties and geometrical extent in this work are determined by a set of methods of increasing complexity. The simplest methods allow for a shifting of the far-field rubber master curves to account for the general filler-rubber interaction effects, as well as an iterative estimation of the interphase size. This mechanical model has been demonstrated in the literature [56, 69–71] as a gradient in glass transition temperature, T_g, as a function of distance from the filler surface. This simplest interphase model, that of a single mechanical behavior, can only estimate the effective viscoelastic and geometric behavior. More complex interphase models, in which these mechanical gradients are captured by incorporating both master curve shifting and spectral broadening, have been shown to be quantitative. From a design perspective, the utilization of shifting and spectral broadening bring into play the ability to perform 'what-if' scenarios on this elusive structural element.

The ability to perform these 'what-if' scenarios with good accuracy is also dependent upon the size extent of the mesoscale models in play. The size scale, also known as the representative volume element (RVE) size, is somewhat dependent on the correlation lengths of the underlying filler morphologies. The number of finite-elements in the continuum model also define the accuracy of the simulation. There exists a tradeoff between number of elements and computational convergence. We have also shown the requirements for obtaining very good accuracy from the number of elements perspective, using sub-sized cells which we call statistical volume elements (SVEs). The reader may find the use of SVEs to be an attractive computational alternative to full-size, high-resolution RVEs.

Combining filler morphological reconstruction with interphase analysis and cell size allows a more complete design tool set for unraveling the impact of design intricacies of rubber compounds. This process allows one to vary the design scenario at hand. One can use these capabilities to obtain quantitative correspondence with known compound viscoelastic behavior. However, one can also use the same analyses in a discovery context, by varying the developed design space in such a way which allows a more extensive evaluation compared to experiment. Finally, there also exists the ability to examine features which are outside the realm of current experimental facility, which enable one to potentially search for limiting viscoelastic behaviors and/or demonstrating the value of a previously unconsidered design space.

11.3 Future Directions

In this final section, we illustrate the trends which are escalating in data-driven multiscale sciences, especially as they concern tire compounding. So far, we have emphasized mesoscale-level models which are physics-based. We see two additional major trends in compound design: (1) the increasing research into material and cheminformatics [72]; and, hand-in-hand (2) the explosion in the use of machine learning tools for chemical and material problems. These areas are complementary to the more traditional physics-based tools, and are part of an overall adaptive materials design ecosystem [73] for compounding.

11.3.1 Material Informatics and Cheminformatics

Data-centric informatics, which we consider to be a form of feature engineering, is being leveraged in compound materials science by the advent or accumulation of large-scale databases from external [74] and internal R&D sources, respectively. There are many elements which can be considered for this style of data environment, shown in Fig. 11.20.

The underlying principle of these environments is the centrality of design, leveraging the knowledge sets from physics-based simulations, data science, material characterization and chemistry, and theory. These wide-ranging environments require a level of data integration which historically has not been a priority. The implementation of 'data lakes' will be a key unifying element of such systems.

On another front, we also see the use of chemically relevant feature engineering, also known as cheminformatics, being leveraged for data-centric compound design solutions [74]. In Fig. 11.21, the incorporation of rubber structure and other structural elements are used in compound design. One of the major reasons for defining chemical features is the ability to enable large-scale data compression from typical corporate formulation databases. In machine learning applications without such feature engineering, it is not unusual to have input vectors in the many hundreds to describe compound formulations. Most industrially relevant compounds are comprised of

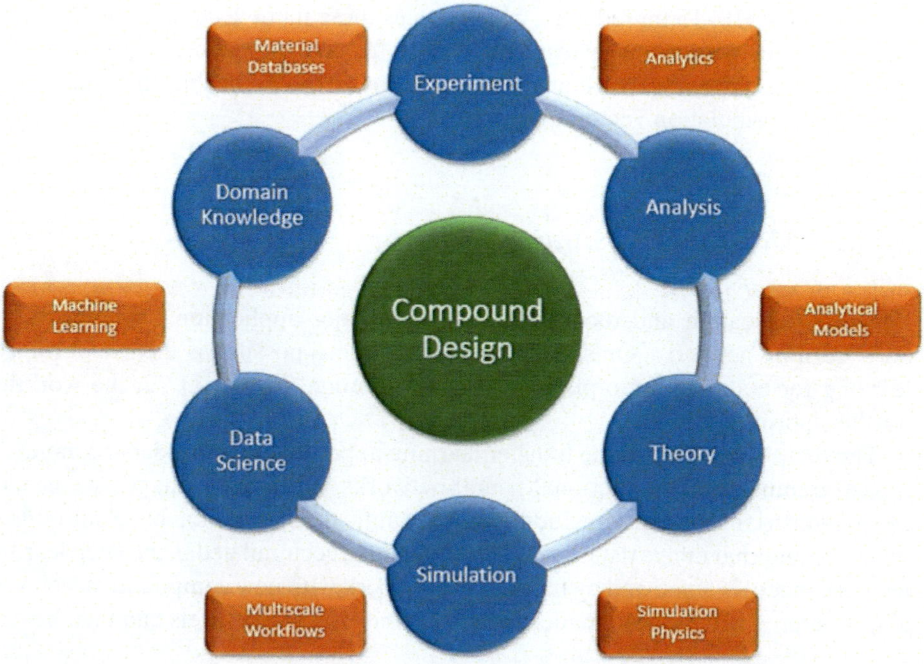

Fig. 11.20 Elements of a materials informatics environment for compound design. Such environments merge knowledge from many technical disciplines, tying the elements into a design framework at the core

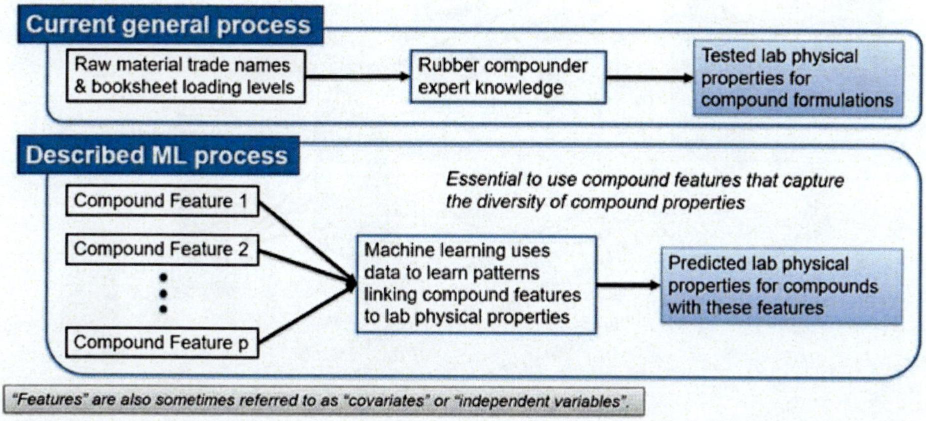

Fig. 11.21 General schemes for standard laboratory-based compound design versus cheminformatics enabled design. Note that the relevant chemical features (cheminformatics) are described and input into the machine learning process. This process requires existing data and is complementary to standard practices. Adapted from [72]. Copyright (2019) The Tire Society

approximately 10–15 ingredients, which means these input vectors, for each formulation, are very sparse. The resultant cheminformatics machine learning models also have the advantage of being potentially more generalizable than the sparse, large-scale formulation vector approach.

11.3.2 Machine Learning

Image classification and reconstruction are a major application of convolutional deep learning networks. One of the many adaptations possible is the use of transfer learning for compound morphological reconstruction. In Fig. 11.22, the workflow for this approach is shown.

The main advantage of the transfer learning approach is that of data economy. In typical trainings of convolutional neural networks, millions of images are needed. Most industrial concerns, including tire manufacturers, do not have millions of images in internal image databases. The knowledge accumulated in the deep learning network model is leveraged by training the network with new compound morphological structures from much smaller image databases. These models can then be used

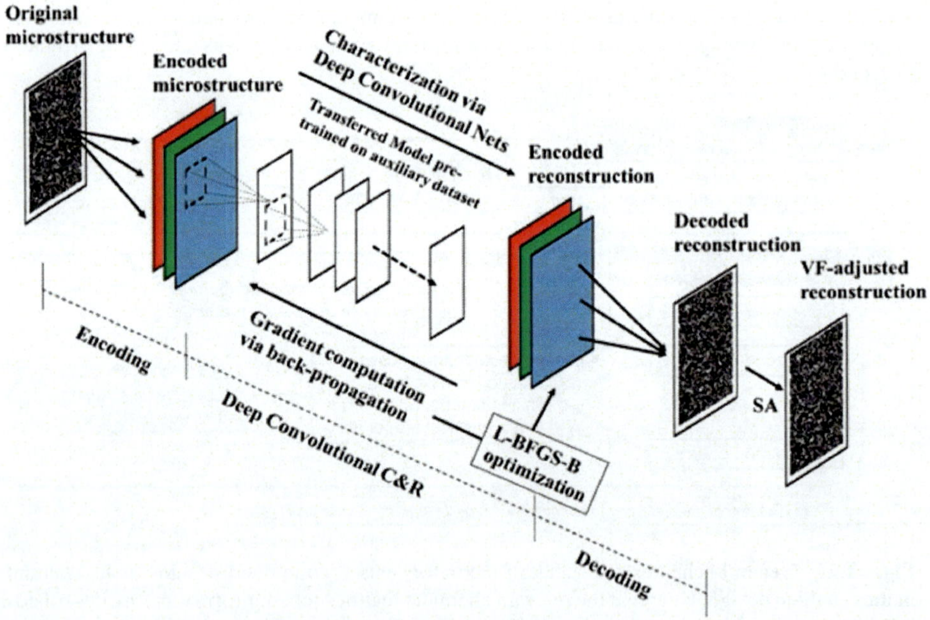

Fig. 11.22 Rubber compound microstructural reconstruction workflow. This method utilizes large image databases for the original VGG-19 deep learning networks. This large-scale deep learning network was transferred to the new application, in this case, rubber compound reconstruction. Reproduced with permission from [75]. Copyright (2018) Springer Nature Ltd

for image reconstruction and model building for simulations or for other machine learning-based structure–property relationships [75].

11.3.3 Summary

Moving outside of the strictly physics-based approach of this chapter, we have illustrated emerging trends in tire compound design which are complementary to our data-driven multiscale approach. Many materials design areas, outside of tire compounding, have leveraged extensive external and internal corporate databases, computational materials science at many levels, analytical characterization techniques and machine learning methodologies. We view these methods and data sources as complementary and of significant utility in general tire compound design.

The next 5–10 years should see the further integration of these environments into mainstream tire compound design, with each company utilizing these resources depending on their cultures and resources. The combination and integration into a cohesive compound design environment will greatly accelerate the development of new, optimal and robust compounds for tire technology. We look forward to seeing the materials and tire design worlds seamlessly integrated into fully functional virtual design ecosystems.

References

1. European Commission, Proposal on the labelling of tyres with respect to fuel efficiency and other essential parameters, Document 52018PC0296, 17 May 2018. https://eur-lex.europa.eu/legal-content/EN/TXT/?uri=CELEX:52018PC0296. Accessed on 12/09/2019; Commission takes action to reinforce EU's global leadership in clean vehicles, Press Release, 8 Nov 2017. https://europa.eu/rapid/press-release_IP-17-4242_en.htm. Accessed on 12/09/2019
2. L. Leblanc, Rubber–filler interactions and rheological properties in filled compounds. Prog. Polym. Sci. **27**, 627–687 (2002)
3. J. Fröhlich, W. Niedermeier, H.D. Luginsland, Compos. Part A Appl. Sci. Manuf. **36**, 449–460 (2005)
4. I. Morozov, B. Lauke, G. Heinrich, Comput. Mater. Sci. **47**, 817–825 (2010)
5. A. Mostafa, A. Abouel-Kasem, M.R. Bayoumi, M.G. El-Sebaie, J. Test. Eval. **38**, 347–359 (2010)
6. Reinforcing Fillers in the Rubber Industry, Fact Sheet 2012. European Tyre and Rubber Manufacturers' Association
7. A. Medalia, Effective degree of immobilization or rubber occluded within carbon black aggregates. Rubber Chem. Technol. **45**, 1171–1194 (1972)
8. H. Deng et al., Utilizing real and statistically reconstructed microstructures for the viscoelastic modeling of polymer nanocomposites. Compos. Sci. Technol. **72**, 1725–1732 (2012)
9. M. Qu et al., Nanoscale visualization and multiscale mechanical implications of bound rubber interphases in rubber-carbon black nanocomposites. Soft Matter **7**, 1066–1077 (2011)
10. K.H. Nordsiek, The "integral rubber" concept: an approach to an ideal tire tread rubber. Kautschuk Gummi Kunstoffe **38**, 178–185 (1985)

11. H.F. Brinson, L.C. Brinson, *Polymer Engineering Science and Viscoelasticity* (Springer, Berlin, 2008). https://doi.org/10.1007/978-0-387-73861-1
12. V. Arrighi, I.J. McEwen, H. Qian, M.B. Serrano Prieto, The glass transition and interfacial layer in styrene-butadiene rubber containing silica nanofiller. Polymer (Guildf) **44**, 6259–6266 (2003)
13. E. Donth et al., Fine structure of the main transition in amorphous polymers: entanglement spacing and characteristic length of the glass transition. Discussion of examples. Macromolecules **29**, 6589–6600 (1996)
14. J. Fröhlich, R. Thomann, R. Mülhaupt, Toughened epoxy hybrid nanocomposites containing both an organophilic layered silicate filler and a compatibilized liquid rubber. Macromolecules **36**, 7205–7211 (2003)
15. M. Jacob, B. Francis, S. Thomas, K.T. Varughese, Dynamical mechanical analysis of sisal/oil palm hybrid fiber-reinforced natural rubber composites. Polym. Compos. **27**, 671–680 (2006)
16. M.A. López-Manchado, J. Biagiotti, L. Valentini, J.M. Kenny, Dynamic mechanical and Raman spectroscopy studies on interaction between single-walled carbon nanotubes and natural rubber. J. Appl. Polym. Sci. **92**, 3394–3400 (2004)
17. M.A.L. Manchado, L. Valentini, J. Biagiotti, J.M. Kenny, Thermal and mechanical properties of single-walled carbon nanotubes–polypropylene composites prepared by melt processing. Carbon **43**, 1499–1505 (2005)
18. J.K. Mishra, I. Kim, C.-S. Ha, New millable polyurethane/organoclay nanocomposite: preparation, characterization and properties. Macromol. Rapid Commun. **24**, 671–675 (2003)
19. P. Pötschke, M. Abdel-Goad, I. Alig, S. Dudkin, D. Lellinger, Rheological and dielectrical characterization of melt mixed polycarbonate-multiwalled carbon nanotube composites. Polymer (Guildf) **45**, 8863–8870 (2004)
20. G. Tsagaropoulos, A. Eisenburg, Direct observation of two glass transitions in silica-filled polymers. Implications to the morphology of random ionomers. Macromolecules **28**, 396–398 (1995)
21. Y. Sun, Z. Zhang, K.-S. Moon, C.P. Wong, Glass transition and relaxation behavior of epoxy nanocomposites. J. Polym. Sci. Part B Polym. Phys. **42**, 3849–3858 (2004)
22. T. Ramanathan et al., Functionalized graphene sheets for polymer nanocomposites. Nat. Nanotechnol. **3**, 327 (2008)
23. C.D. Wood, A. Ajdari, C.W. Burkhart, K.W. Putz, L.C. Brinson, Understanding competing mechanisms for glass transition changes in filled elastomers. Compos. Sci. Technol. **127**, 88–94 (2016)
24. C.P. Tsui, C.Y. Tang, T.C. Lee, Finite element analysis of polymer composites filled by interphase coated particles. J. Mater. Process. Technol. **117**, 105–110 (2001)
25. F.T. Fisher, L.C. Brinson, Viscoelastic interphases in polymer–matrix composites: theoretical models and finite-element analysis. Compos. Sci. Technol. **61**, 731–748 (2001)
26. G.M. Odegard, T.S. Gates, K.E. Wise, C. Park, E.J. Siochi, Constitutive modeling of nanotube–reinforced polymer composites. Compos. Sci. Technol. **63**, 1671–1687 (2003)
27. G.D. Smith, D. Bedrov, L. Li, O. Byutner, A molecular dynamics simulation study of the viscoelastic properties of polymer nanocomposites. J. Chem. Phys. **117**, 9478–9489 (2002)
28. J.S. Smith, D. Bedrov, G.D. Smith, A molecular dynamics simulation study of nanoparticle interactions in a model polymer-nanoparticle composite. Compos. Sci. Technol. **63**, 1599–1605 (2003)
29. F.W. Starr, T.B. Schrøder, S.C. Glotzer, Molecular dynamics simulation of a polymer melt with a nanoscopic particle. Macromolecules **35**, 4481–4492 (2002)
30. D. Brown, V. Marcadon, P. Mélé, N.D. Albérola, Effect of filler particle size on the properties of model nanocomposites. Macromolecules **41**, 1499–1511 (2008)
31. N. Sheng et al., Multiscale micromechanical modeling of polymer/clay nanocomposites and the effective clay particle. Polymer (Guildf) **45**, 487–506 (2004)
32. G.D. Seidel, D.C. Lagoudas, Micromechanical analysis of the effective elastic properties of carbon nanotube reinforced composites. Mech. Mater. **38**, 884–907 (2006)

33. X.-L. Gao, K. Li, A shear-lag model for carbon nanotube-reinforced polymer composites. Int. J. Solids Struct. **42**, 1649–1667 (2005)
34. N.D. Alberola, P. Mele, Interface and mechanical coupling effects in model particulate composites. Polym. Eng. Sci. **37**, 1712–1721 (1997)
35. Z. Hashin, P.J.M. Monteiro, An inverse method to determine the elastic properties of the interphase between the aggregate and the cement paste. Cem. Concr. Res. **32**, 1291–1300 (2002)
36. R.A. Dickie, Heterogeneous polymer–polymer composites. I. Theory of viscoelastic properties and equivalent mechanical models. J. Appl. Polym. Sci. **17**, 45–63 (1973)
37. T. Mori, K. Tanaka, Average stress in matrix and average elastic energy of materials with misfitting inclusions. Acta Metall. **21**, 571–574 (1973)
38. R. Qiao, H. Deng, K.W. Putz, L.C. Brinson, Effect of particle agglomeration and interphase on the glass transition temperature of polymer nanocomposites. J. Polym. Sci. Part B Polym. Phys. **49**, 740–748 (2011)
39. J. Cho, I.M. Daniel, Reinforcement of carbon/epoxy composites with multi-wall carbon nanotubes and dispersion enhancing block copolymers. Scr. Mater. **58**, 533–536 (2008)
40. R.K. Duncan et al., Viscoelastic behavior of nanotube-filled polycarbonate: effect of aspect ratio and interface chemistry. Int. J. Smart Nano Mater. **1**, 53–68 (2010)
41. H. Xu, Y. Li, C. Brinson, W. Chen, A descriptor-based design methodology for developing heterogeneous microstructural materials system. J. Mech. Des. **136**, 051007 (2014)
42. Y. Wang et al., Identifying interphase properties in polymer nanocomposites using adaptive optimization. Compos. Sci. Technol. **162**, 146–155 (2018)
43. Y. Jiao, F.H. Stillinger, S. Torquato, Modeling heterogeneous materials via two-point correlation functions: basic principles. Phys. Rev. E **76**, 31110 (2007)
44. S. Torquato, *Random Heterogeneous Materials: Microstructure and Macroscopic Properties* (Springer, New York, 2013)
45. C.L.Y. Yeong, S. Torquato, Reconstructing random media. II. Three-dimensional media from two-dimensional cuts. Phys. Rev. E **58**, 224–233 (1998)
46. I.A. Morozov, B. Lauke, G. Heinrich, A novel method of quantitative characterization of filled rubber structures by AFM. KGK Kaut. Gummi Kunstst. **64**, 24–27 (2011)
47. L. Karásek, M. Sumita, Characterization of dispersion state of filler and polymer-filler interactions in rubber-carbon black composites. J. Mater. Sci. **31**, 281–289 (1996)
48. C.L.Y. Yeong, S. Torquato, Reconstructing random media. Phys. Rev. E **57**, 495–506 (1998)
49. S. Torquato, Optimal design of heterogeneous materials. Annu. Rev. Mater. Res. **40**, 101–129 (2010)
50. H. Xu, R. Liu, A. Choudhary, W. Chen, A machine learning-based design representation method for designing heterogeneous microstructures. J. Mech. Des. **137** (2015)
51. B.C. Collins, K. Matous, D. Rypl, Three-dimensional reconstruction of statistically optimal unit cells of multimodal particulate composites. Int. J. Multiscale Comput. Eng. **8**, 489–507 (2010)
52. Y. Liu, M.S. Greene, W. Chen, D.A. Dikin, W.K. Liu, Microstructure reconstruction for stochastic multiscale material design, in *Proceedings of the ASME 2011 International Design Engineering Technical Conferences and Computers and Information in Engineering Conference*, vol. 5. *37th Design Automation Conference, Parts A and B* (ASME, 2011), pp. 497–507. https://doi.org/10.1115/DETC2011-48323
53. R. Bostanabad et al., Computational microstructure characterization and reconstruction: review of the state-of-the-art techniques. Prog. Mater. Sci. **95**, 1–41 (2018)
54. P.F. Brune et al., Direct measurement of rubber interphase stiffness. Macromolecules **49**, 4909–4922 (2016)
55. J.A. Forrest, K. Dalnoki-Veress, The glass transition in thin polymer films. Adv. Colloid Interface Sci. **94**, 167–195 (2001)
56. C.J. Ellison, J.M. Torkelson, The distribution of glass-transition temperatures in nanoscopically confined glass formers. Nat. Mater. **2**, 695–700 (2003)

57. Y. Li, M. Kröger, W.K. Liu, Nanoparticle effect on the dynamics of polymer chains and their entanglement network. Phys. Rev. Lett. **109**, 118001 (2012)
58. A. Hu et al., Computational analysis of particle reinforced viscoelastic polymer nanocomposites—statistical study of representative volume element. J. Mech. Phys. Solids **114**, 55–74 (2018)
59. J. Gao et al., An inverse modeling approach for predicting filled rubber performance. Comput. Methods Appl. Mech. Eng. **357**, 112567 (2019)
60. H. Xu, D.A. Dikin, C. Burkhart, W. Chen, Descriptor-based methodology for statistical characterization and 3D reconstruction of microstructural materials. Comput. Mater. Sci. **85**, 206–216 (2014)
61. R. Tian et al., A multiresolution continuum simulation of the ductile fracture process. J. Mech. Phys. Solids **58**, 1681–1700 (2010)
62. M. Morisita, Iσ-index, a measure of dispersion of individuals. Res. Popul. Ecol. (Kyoto) **4**, 1–7 (1962)
63. A.S. Dalaq, D.W. Abueidda, R.K. Abu Al-Rub, I.M. Jasiuk, Finite element prediction of effective elastic properties of interpenetrating phase composites with architectured 3D sheet reinforcements. Int. J. Solids Struct. **83**, 169–182 (2016)
64. X. Yin, W. Chen, A. To, C. McVeigh, W.K. Liu, Statistical volume element method for predicting microstructure–constitutive property relations. Comput. Methods Appl. Mech. Eng. **197**, 3516–3529 (2008)
65. J. Zhao, O.Y. Kontsevoi, W. Xiong, J. Smith, Simulation-aided constitutive law development—assessment of low triaxiality void nucleation models via extended finite element method. J. Mech. Phys. Solids **102**, 30–45 (2017)
66. K.W. Putz, C.A. Mitchell, R. Krishnamoorti, P.F. Green, Elastic modulus of single-walled carbon nanotube/poly(methyl methacrylate) nanocomposites. J. Polym. Sci. Part B Polym. Phys. **42**, 2286–2293 (2004)
67. H. Xu et al., Stochastic reassembly strategy for managing information complexity in heterogeneous materials analysis and design. J. Mech. Des. **135** (2013)
68. A. Jean, D. Jeulin, S. Forest, S. Cantournet, F. N'Guyen, A multiscale microstructural model of carbon black distribution in rubber. J. Microscopy **241**, 243–260 (2011)
69. P. Rittigstein, R.D. Priestley, L.J. Broadbelt, J.M. Torkelson, Model polymer nanocomposites provide an understanding of confinement effects in real nanocomposites. Nat. Mater. **6**, 278–282 (2007)
70. M. Zhang, S. Askar, J.M. Torkelson, L.C. Brinson, Stiffness gradients in glassy polymer model composites: comparisons of quantitative characterization by fluorescence spectroscopy and atomic force microscopy. Macromolecules **50**, 5447–5458 (2017)
71. M. Tauban, J.-V. Delannoy, P. Sotta, D.R. Long, Effect of filler morphology and distribution state on the linear and nonlinear mechanical behavior of nanofilled elastomers. Macromolecules **50**, 6369–6384 (2017)
72. B. Kelly, B. Crookston, N. Isitman, C. Burkhart, *Machine Learning Guided Discovery of New Compounds*, Paper 6.2 (Tire Society Conference, Akron, OH, 2019)
73. D. Xue, P.V. Balachandran, J. Hogden, J. Theiler, D. Xue, T. Lookman, Accelerated search form materials with targeted properties by adaptive design. Nat. Commun. **7**, 11241 (2016)
74. Materials Genome Initiative, Center for Hierarchical Materials Design (CHiMaD). https://chimad.northwestern.edu/. Accessed on 12/9/2019
75. X. Li, Y. Zhang, H. Zhao, C. Burkhart, L.C. Brinson, W. Chen, A transfer learning approach for microstructure reconstruction and structure-property predictions. Sci. Rep. **8**, 13461 (2018)

Index

A
Absorption, 265
Adam and Gibbs theory, 127
Ag nanowire, 275
α relaxation, 85, 86, 89, 92
Anisotropic assembly, 38
Aspect ratio, 298
Associating polymer, 16
Atomic closures, 17

B
Bare nanoparticles, 103
Boron nitride, 244
Boundary curvature, 103
Bound layer, 83, 84, 92, 119
Bound polymer, 86–88, 94
Bound polymer layer, 81, 85, 91, 94
Bound rubber layer, 82
Broadband Dielectric Spectroscopy (BDS), 85, 90
Bruggeman model, 244, 245, 247

C
Carbon nanotube, 236, 240, 248, 249, 252, 275
Cellulose Nanocrystal (CNC), 185–187, 192, 197
Chain conformation, 3
Cheminformatics, 306
Clausius-Mossotti relationship, 263
"Cloaking" effect, 126
Cluster correlation, 291
CNC-PMMA, 187
Coarse-grained, 180, 181, 197
Coarse-grained polymer, 104
Coarse-graining, 181
Colloidal limit, 24, 27
Columnar, 32
Complex grafted nanoparticles, 61
Composition, 296
Concentrated Polymer Brush (CPB), 191, 192
Concentration tensor, 208–210
Conductance, 274
Conformal mesh, 300
Correction of the dielectric function, 266
Critical chain length, 189, 192

D
Data lakes, 306
Dead layer, 90, 94
Debye Waller Factor, 182
Density Functional Theory (DFT), 23, 25–28, 30, 33–35, 38, 39, 138, 156, 157
Density gradient, 114
Depletion attraction, 27
Descriptors, 289
Diblock copolymers, 34
Differential Scanning Calorimetry (DSC), 90
Diffusion coefficient, 87, 92, 93
Discrete Dipole Approximation (DDA), 260
Dispersed/aggregated, 3
Dispersion, 296
Dissipative Particle Dynamics (DPD), 141, 161, 163, 167
1D scalar models, 287
3D uniform models, 288

V. V. Ginzburg and L. M. Hall (eds.), *Theory and Modeling of Polymer Nanocomposites*, Springer Series in Materials Science 310,
https://doi.org/10.1007/978-3-030-60443-1

Dynamic mechanical analyses, 286, 304
Dynamic T_g, 123

E

Effective Medium Approximation (EMA), 236, 241, 245, 255
Electrical properties, 259, 273
Energy renormalization, 181–185, 197
Entropy-enthalpy compensation effect, 182
Equivalent continuum, 220, 224, 225–227
Eshelby, 206–208, 210, 227
Extinction, 261
Extinction cross section, 265

F

Failure, 214, 215, 228
Feature engineering, 306
Field-theoretic simulations, 46
Field theory, 47
Finite-Difference Time-Domain (FDTD), 260, 269
Finite Element Analysis (FEA), 215–219, 228–230, 246, 248–252, 255
Flory-Huggins lattice model, 34
Flory-Huggins χ parameter, 3
Flow battery, 133–135, 141, 168
Free interface, 106
Fuel cell, 133, 134, 139–141, 149, 157, 158, 167, 168
Fundamental Measure Theory (FMT), 25, 27, 39

G

Geometry, 296
Glass transition temperature, 118
Glassy, 90
"Glassy" (immobilized) layer, 83
Glassy layer, 83–85, 90, 94
Grafted nanoparticles, 58
Grafting density, 4

H

Hairy Nanoparticle (HNP), 27, 38
Homogenization, 208, 210–212, 226, 228
Hybrid particle-field theory, 47, 51

I

Ideal dispersion, 103
Imidazole, 140, 158, 159
Implementation of FDTD, 271
Interfacial density gradient scale, 125
Interfacial density scale, 109
Interfacial density variance scale, 116
Interfacial dynamics, 82, 86
Interfacial interaction strength, 103
Interfacial layers, 81, 85, 86, 91, 102
Interfacial mobility scale, 106
Interfacial thermal resistance, 251, 252, 254
Interphase, 226, 227, 283, 288
Interphase gradient, 304
Intra-molecular pair correlation function, 9
Inverse Boltzmann method, 181, 183–185
Ionic liquid, 134, 140, 157, 159
Ionic transport, 276
Ion transport, 260
Isotropic, 32

K

Kapitza radius, 242
Kapitza resistance, 242, 251, 255
Kirchhoff's Law, 274

L

Laminate, 212–214
Large-scale Atomic/Molecular Massively Parallel Simulator (LAMMPS), 187
Lineal path correlation, 291
Local Density Approximation (LDA), 25
Local interphase properties, 294
Local packing frustration, 115

M

Machine learning, 308
Material informatics, 306
Matrix relaxation time, 121
Maxwell-Garnett effective medium model, 262
Maxwell-Garnett (MG) model, 241
Maxwell's equations, 270
Mean square displacement, 183, 184, 188
Mesoscale DDFT (MDDFT), 161, 163, 169
Mesoscale model, 193, 197
Mesoscopic simulation, 134, 160, 161
Metamodel, 179, 188
MG model, 241, 243, 245
Micromechanical model, 240, 245, 251, 255
Micro-mechanics, 206, 226, 227
Microstructure characterization, 289
Mie theory, 260
Mobile layer, 94

Mobility, 84
Mobility gradient, 106
Molecular closures, 17
Molecular dynamics, 134, 138, 142, 150, 168, 181, 186, 188, 197
Monte Carlo field-theoretic simulations, 74
Monte Carlo simulations, 38
Mori-Tanaka, 209, 211, 226, 228

N
Nafion, 133, 136–157, 162, 165, 168, 169
Nanoparticle limit, 24
Nanoparticle-Organic Hybrid Materials (NOHM), 27
Nanoplatelets, 24
Nanoplates, 29
Nanorods, 24, 28
Nanospheres, 24, 34
Nearest distance, 298
Nernst-Planck equation, 277
Nuclear Magnetic Resonance (NMR), 82–84, 83, 90

O
Occluded rubber, 283
Optical absorption, 260
Optical properties, 259

P
Particle's area, 298
Percolation, 136, 139, 164–166, 244, 249, 250, 273
Perfectly Matched Layer (PML), 272
Physical descriptor-based characterization and reconstruction, 295
Plasmonic coupling, 260
Plastic solid, 32
2-point correlation function, 289
Poisson and Nernst-Planck equations, 277
Poisson-Boltzman equation, 277
Poisson-Nernst-Planck formalism, 260
Polarization, 266
Polybutadiene (PB), 186, 192, 194, 195
Polycarbonate (PC), 183, 184, 186, 192, 194, 195
Polyelectrolyte Membrane (PEM), 133–135, 137–141, 147, 161, 167–169
Polymer field theory, 46
Polymer-Grafted Nanoparticle (PGNP), 38
Polymer mobility, 82, 85, 86, 94
Polymer-nanoparticle interaction strength, 104
Polymer Reference Interaction Site Model (PRISM) theory, 1
Poly(methyl methacrylate) (PMMA), 186, 192, 194–196
Polystyrene (PS), 185, 186, 192, 194
Potential of Mean Force (PMF), 24, 194–197
Proton transport, 133, 135–137, 141, 143, 145, 149, 157–159, 168
pyPRISM, 1

R
Reconstructed microstructure, 288
Reconstruction, 289
Relaxation, 85
Relaxation profile, 113
Relaxation time, 84, 86, 87, 89, 91–93, 106
Representative Volume Element (RVE), 217, 220–225, 285, 305
Resistor network models, 260, 273
Resonance Enhanced Dynamic Light Scattering (REDLS), 89
RVE volume size, 302

S
Scanning Electron Microscopy (SEM), 289, 304
Scattering, 260
Scattering and absorption efficiencies, 266
Scattering cross sections, 261, 265
Second order Legandre parameter, 184
Segmental dynamics, 81, 83, 87, 90, 91, 94
Segmental mobility, 90
Segmental relaxation time, 106
Self-Consistent Field/Density Functional Theory (SCF-DFT), 23, 27, 33, 36, 38, 39
Self-Consistent Field Theory (SCFT), 25, 30
Self-consistent PRISM theory—molecular simulation, 11
Self-intermediate scattering function, 108
Semi-Dilute Polymer Brush (SDPB), 191, 192
Sheets, 38
Simulated annealing, 293, 298
Single-particle density, 25
Slotboom transformation, 278
Small Angle Neutron Scattering (SANS), 83, 94
Small Angle X-Ray Scattering (SAXS), 90
SPDB, 192

Stochastic optimization methods, 293
Storage and loss modulus, 303
Strings, 38
Strong-segregation theory, 38
Structure-property, 203, 230
Styrene-butadiene rubber, 286
Successive Over-Relaxation (SOR) algorithm, 277
Sulfonated poly(ether-ether ketone) (SPEEK), 137, 147, 160–167, 169
Surface correlation, 291
Surface plasmon resonance, 260
Surface plasmons, 260

T
Tan δ, 283
Temperature dependence, 288
T_g, 85, 87, 88, 91, 94
Theoretically-informed Langevin dynamics, 69
Thermal conductivity, 235, 236–252, 254, 255
Thermal Interface Materials (TIM), 236
Thermodynamic T_g, 123
Thin polymer film, 103
Three-dimensional Finite Element (FE) model, 288
Time of Flight (TOF), 84
Total Field/Scattered Field (TF/SF) source, 272
Transfer learning, 308
Transmission Electron Microscopy (TEM), 289
Two-dimensional plane-strain simulation, 294
Two-layer strategy, 304
Two-point correlation, 290

V
Viscoelastic spectrum, 288
Volume fraction, 296
Voxel mesh, 300

W
Weighted Density Approximation (WDA), 25, 35, 39
Wertheim perturbation scheme, 26
Wetting, 3

Y
Yee cell, 269